Fifth International Multi-Conference on Engineering and Technology (IMETI 2016)

MATEC Web of Conferences Volume 119 (2017)

Taichung, Taiwan
28 October - 1 November 2016

Editors:

Wen-Hsiang Hsieh

ISBN: 978-1-5108-4639-5

Printed from e-media with permission by:

Curran Associates, Inc.
57 Morehouse Lane
Red Hook, NY 12571

Some format issues inherent in the e-media version may also appear in this print version.

Printed by Curran Associates, Inc. (2017)

For additional information, please contact EDP Sciences – Web of Conferences at the address below.

EDP Sciences – Web of Conferences
17, Avenue du Hoggar
Parc d'Activité de Courtabœuf
BP 112
F-91944 Les Ulis Cedex A
France

Phone: +33 (0) 1 69 18 75 75
Fax: +33 (0) 1 69 28 84 91

contact-edps@webofconferences.org

Additional copies of this publication are available from:

Curran Associates, Inc.
57 Morehouse Lane
Red Hook, NY 12571 USA
Phone: 845-758-0400
Fax: 845-758-2633
Email: curran@proceedings.com
Web: www.proceedings.com

TABLE OF CONTENTS

Effect of sintering condition on the microstructure and electrical properties of lead-free $(Na_{0.5}K_{0.5})NbO_3$ - $Bi_{0.5}(Na_{0.83}K_{0.17})_{0.5}TiO_3$ ceramics

Chun Huy Wang[1,2,a]

[1]*Department of Electronic Engineering, Nan-Jeon University, Tainan City 737, Taiwan*
[2]*Graduate School of Engineering Science and Technology, Nan-Jeon University, Tainan City 737, Taiwan*

Abstract. In this work, the piezoelectric ceramic system of the $(1\text{-}x)(Na_{0.5}K_{0.5})NbO_3$ - $xBi(Na_{0.83}K_{0.17})TiO_3$ with composition close to the morphotropic phase boundary is studied. $(Na_{0.5}K_{0.5})NbO_3$ with 0~5 mole% $Bi(Na_{0.83}K_{0.17})TiO_3$ has been prepared following the conventional mixed oxide process. The effect of sintering time on the properties of $0.97(Na_{0.5}K_{0.5})NbO_3$-$0.03Bi(Na_{0.83}K_{0.17})TiO_3$ ceramics is discussed. For $0.97(Na_{0.5}K_{0.5})NbO_3$-$0.03Bi(Na_{0.83}K_{0.17})$TiO3 ceramics, the electromechanical coupling coefficients of the thickness mode kt and the piezoelectric constant d_{33} reach 0.46 and 155 p/CN, respectively, at the sintering of 1100 oC for 3 h. Dielectric and piezoelectric properties have maximum values at the sintering temperature of 1100 °C for 5 h. For $0.97(Na_{0.5}K_{0.5})NbO_3$ -$0.03Bi(Na_{0.83}K_{0.17})TiO_3$ ceramics, the electromechanical coupling coefficients of the thickness mode k_t and the piezoelectric constant d_{33} reach 0.56 and 190, respectively, at the sintering of 1100 °C for 5 h. The effect of prolonging the sintering time to the $0.97(Na_{0.5}K_{0.5})NbO_3$-$0.03Bi(Na_{0.83}K_{0.17})TiO_3$ system is a helpful method on ceramic processing to improve densification and properties.

1 Introduction

Lead-free piezoelectric materials, such as langasite single crystal, ferroelectric ceramics with perovskite structure, and Bi-layered structure oxides, have been reported [1-2]. Among these materials, $(Na_{0.5}K_{0.5})NbO_3$ [abbreviated as NKN] ceramics are considered good candidates for lead-free piezoelectric ceramics because of their strong piezoelectricity and ferroelectricity. The hot-pressed NKN ceramics (~99% of theoretical density) have been reported to possess a large piezoelectric longitudinal response (d_{33}~160 pC/N), a high planar coupling coefficient (k_p~45%), and a high phase transition temperature (T_c=420 °C) [1]. NKN ceramics sintered by ordinary sintering show relatively lower values (k_p=25%) due to difficulty in the processing of dense ceramics by ordinary sintering [2]. This issue has received significant attention and has been thoroughly investigated [3-5].

Nevertheless, dense NKN ceramics are difficult to obtain since their phase stability is limited to 1140 °C, which is close to the melting point [6]. Many researchers used hot pressing or spark plasma sintering [abbreviated as SPS] techniques to yield better quality ceramics [7]. The addition of some perovskite compounds to form solid solutions with NKN or synthesis by spark plasma sintering (SPS) has been carried out to obtain lead-free materials suitable for industrial applications. Perovskite

[a] Corresponding author : wangch@mail.nju.edu.tw

structure ($(Bi_{0.5}Na_{0.5})TiO_3$ [abbreviated as BNT] ceramics are considered as one of the candidates for lead-free piezoelectric ceramics because of its high ferroelectric properties and high Curie temperature (T_c= 320 °C). Some modifications of BNT ceramics have proved to be helpful by forming solid solution with other perovskite oxides. $Bi_{0.5}(Na_{0.83}K_{0.17})_{0.5}TiO_3$ is well-known lead-free piezoelectric materials, which has a tetragonal symmetry at room temperature and show better piezoelectric response. However, there have been few studies on the properties of NKN ceramics modified with $Bi_{0.5}(Na_{0.83}K_{0.17})_{0.5}TiO_3$ [abbreviated as BNKT]. Therefore, it is desirable to expect that the solid mixture of NKN (orthorhomic) and $Bi_{0.5}(Na_{0.83}K_{0.17})_{0.5}TiO_3$ (tetragonal) have a morphotropic phase boundary (abbreviated as MPB) and better piezoelectric property. However, there have been few studies on the properties of NKN ceramics modified with $Bi_{0.5}(Na_{0.83}K_{0.17})_{0.5}TiO_3$ [abbreviated as BNKT]. The aim of this study is not only to investigate the physical and electrical properties of the $(Na_{0.5}K_{0.5})NbO_3$- $Bi_{0.5}(Na_{0.83}K_{0.17})_{0.5}TiO_3$ [NKN-BNKT] system by the conventional mixed oxide method, but also to discuss the effect of sintering condition on the microstructure and electrical properties of NKN-BNKT ceramics

2 Methods and procedures

The $(1\text{-}x)(Na_{0.5}K_{0.5})NbO_3\text{-}xBi_{0.5}(Na_{0.83}K_{0.17})_{0.5}TiO_3$ [(1-x)NKN-xBNKT] (x=0.01, 0.02, 0.03, 0.04, and 0.05) ceramics were prepared by conventional mixed oxide method. The powders were calcined at 950 °C for 4 h. The disc samples were sintered in alumina crucible at 1100 °C for 3~7 h in air atmosphere. To measure relevant piezoelectric properties, the prepared ceramic samples were polarized in silicone oil at 160 °C under 4 kV/mm for 30 min. An X-ray diffractometer (Seimens D5000) using Cu Kα radiation was used to evaluate the crystal structure of the sintered ceramics. The room temperature dielectric constant at 1 kHz was measured by LCR meter. The piezoelectric properties were measured by a resonance-antiresonance method based on IEEE standards using an impedance/gain-phase analyzer (Angilent 4194). Samples for observations of the microstructure were polished and thermally etched. The microstructures were observed by a scanning electron microscope (SEM). The mean grain size was calculated by the line intercept method. The density was measured by Archimedes method.

3 Results and discussion

In this system, none of the samples showed deliquescence when exposed under water for a long period, indicating no existence of the unstable second phase. Figure 1 shows the X-ray diffraction [abbreviated as XRD] patterns of (1-x)NKN-xBNKT ceramics for x= 0.01, 0.02, 0.03, 0.04, and 0.05. All the compositions were sintered in air at 1100 °C for 3 h. The main identified phase matches the orthorhombic NKN with space group *Amm*2 at room temperature, and the results reveal that the orthorhombic structure is preserved. It can be concluded that the NKN-BNKT ceramics have orthorhombic structures in the case $x \leq 0.03$.With increasing of BNKT content, however, the structure changes from orthorhombic to tetragonal. A distinct tetragonal splitting can be seen when x=0.03. Therefore, it can be suggested that the specimens of x=0.03 correspond to the compositions located in the orthorhombic and tetragonal sides of the MPB, respectively, while the two phases coexist when x=0.03.The measured density and relative density of (1-x)NKN-xBNKT ceramics are shown as a function of x composition in Fig. 2. The measured density and relative density gradually increase with increasing x and reach a maximum value when x was added up to 0.03. Further increasing x, the measured density and relative density will decrease. The measured density of the sintered samples is 90–97% of the theoretical density. Figure 3 shows the microstructure of the (1-x)NKN-xBNKT ceramics with different BNKT contents. It can be seen in Fig. 6(a) that the 0.99NKN-0.01BNKT ceramics consist of small grains with a loose structure and a high porosity. It can be seen that (1-x)NKN-xBNKT ceramics with x=0.02~0.03 consist of larger grains and denser structure in Fig. 3(b)~3(c). The grain size of (1-x)NKN-xBNKT decreases with increasing the BNKT contents (x=0.04 to 0.05). Increasing the BNKT contents may cause lattice distortion and a resultant variation in composition, which leads to an inhomogeneous microstructure in Fig. 3(d) ~3(e). These conditions may increase the porosity and decrease the grain size. With increasing porosity and decreasing the grain size of (1-x)NKN-xBNKT ceramics, the piezoelectric and dielectric properties decrease gradually.

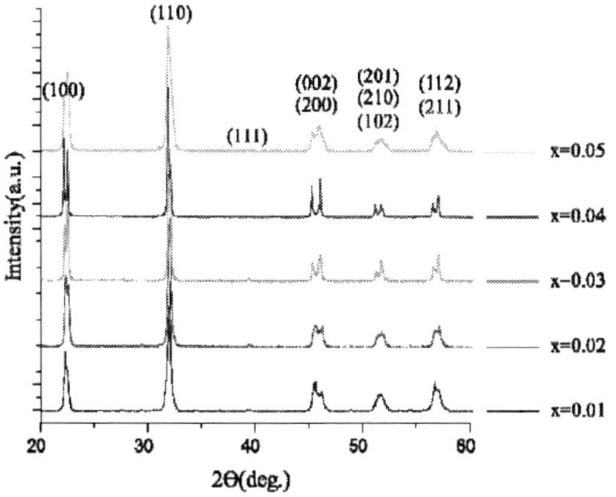

Figure 1. XRD patterns of (1-x)NKN-xBNKT ceramics system with different BNKT contents.

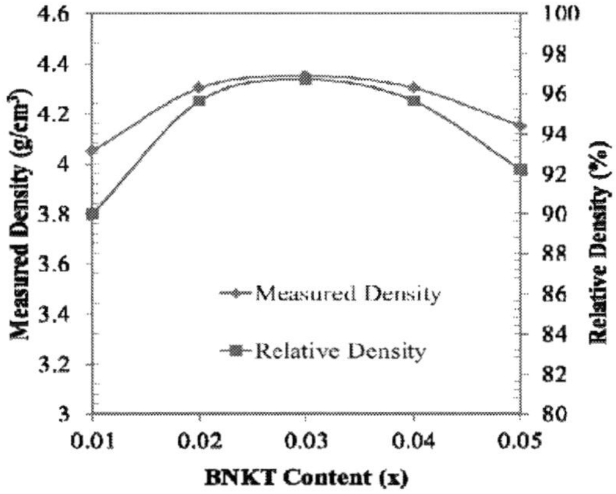

Figure 2. Measured density and relative density BNKT contents.

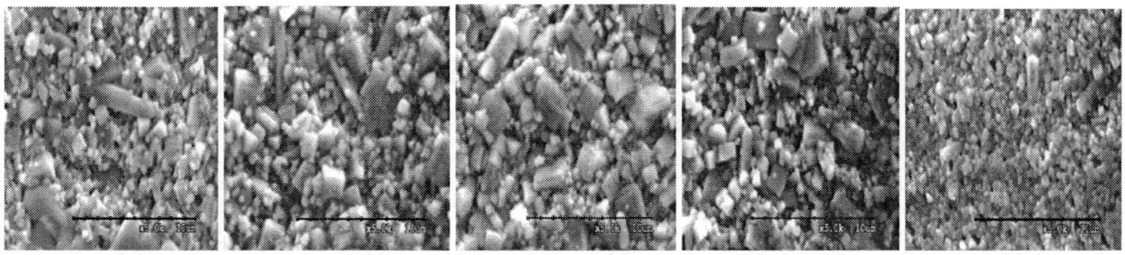

(a) x=0.01 (b) x=0.02 (c) x=0.03 (d) x=0.04 (e) x=0.05

Figure 3. The microstructure of (16-x)NKN-xBNKT ceramics with different BNKT contents: (a) x=0.01, (b) x=0.02, (c) x=0.03, (d) x=0.04, and (e) x=0.05. Bar= 10 μm.

The electromechanical coupling factor is related to different BNKT addition. The planar coupling factor (k_p) has been used extensively as a measure of the piezoelectric response of PZT type ceramics. It was found that k_p depended on the material parameters such as grain size, porosity, and chemical composition. Piezoelectric activity reaches a maximum value, when ceramic compositions are chosen near those of the morphotropic phase boundary. Usually, the piezoelectric activity is appreciated by the value of the electromechanically coupling factor in radial mode k_p. Therefore, k_p is the figure of merit of the piezoelectric activity and the square of that gives the efficiency of the conversion of electrical-mechanical energy. The planar coupling factor (k_p) and thickness coupling factor (k_t) of (1-x)NKN-xBNKT ceramics are shown as a function of x composition in Fig. 4. Coupling factor k_p and k_t gradually increase along with increasing x and reach maximum values when x was added up to 0.03. Further increasing x, k_p and k_t will decrease. In the experiment, the samples possess the optimal k_p and k_t of 29% of 46%, respectively. Figure 5 shows the dielectric constant and dielectric loss tangent (tan δ) of (1-x)NKN-xBNKT ceramics with different BNKT contents. Dielectric constant K^T_{33} of the samples increases with increasing x initially, reaches a maximum value (K^T_{33}=950) at x=0.03. Going on increasing x, K^T_{33} shows a obvious decrease. The variation trend of the dielectric loss tangent (tan δ) with BNKT contents is inverse to that of the measured density. The minimum value of the dielectric loss tangent (tan δ= 0.04) is found at the phase boundary (x=0.03).

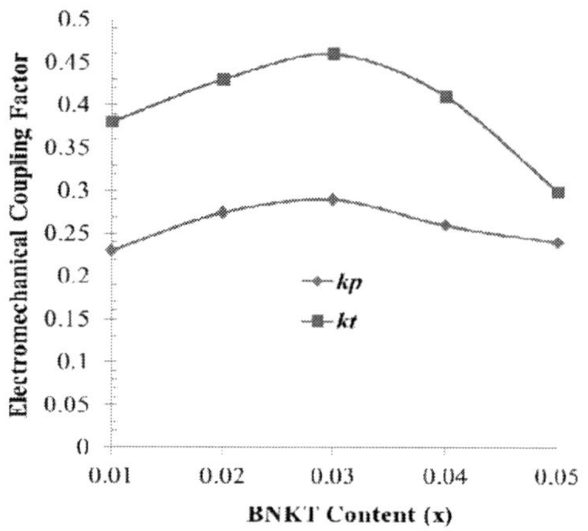

Figure 4. The electromechanical quality factor (kp and kt) of (1-x) NKN - xBNKT ceramics as a function of BNKT contents.

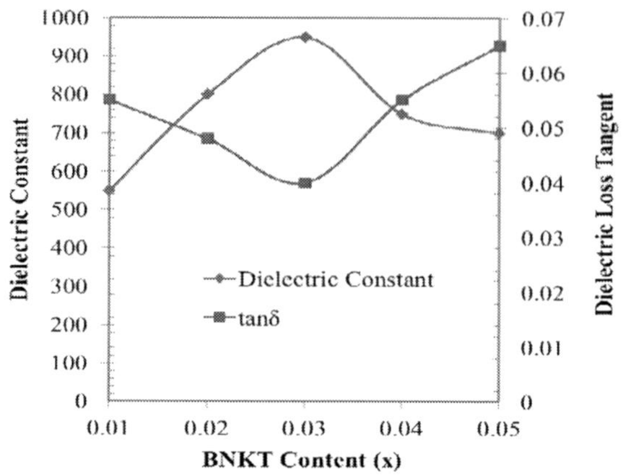

Figure 5. The dielectric constant and dielectric loss tangent (tanδ) of (1-x) NKN-xBNKT ceramics as a function of BNKT contents.

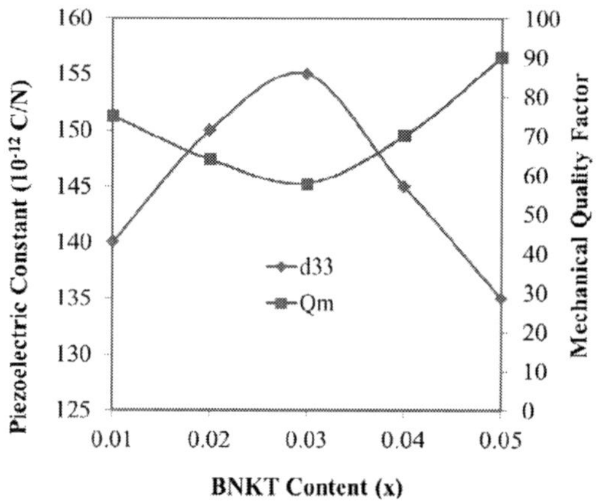

Figure 6. The piezoelectric constant and mechanical quality factor of (1-x)NKN-xBNKT ceramics as a function of BNKT contents.

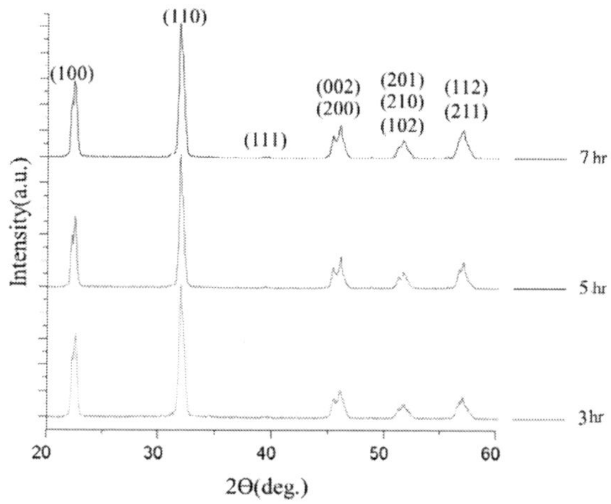

Figure 7. XRD patterns of 0.97NKN-0.03BNKT ceramics sintered at 1100°C for different time (a) 3 h(b) 5 h (c) 7 h.

Figure 6 shows the piezoelectric constant (d_{33}) and the mechanical quality factor (Q_m) of (1-x)NKN-xBNKT ceramics with different BNKT contents. The piezoelectric constant (d_{33}) gradually increase along with increasing x and reach a maximum value (d_{33}=155 pC/N) when x was added up to 0.03. Further increasing x, the d_{33} will decrease. The mechanical quality factor increases with an increase of BNKT contents. The variation trend of the mechanical quality factor (Q_m) with BNKT contents is inverse to that of the planar coupling factor (k_p). The content-depended effect of BNKT on the properties of NKN-BNKT ceramics can be attributed to two factors : the chemical modification and the crystalline structure influence of the material. When BNKT-content doped is relatively low, chemical modification caused by dissolution of BNKT into the NKN lattice plays a major role on the properties of NKN-BNKT ceramics. It is supposed that NKN is ABO_3 perovskite structure materials. When BNKT as an additive was added into ceramics, it is the most likely that the Bi^{3+} (1.4 Å ionic radius) could substitute for K^+ (1.64 Å) and Na^+ (1.39 Å) in the A site, while Ti^{4+} (0.61 Å) could substitute for Nb^{5+} (0.64 Å) in the B sites of ABO_3 perovskite structure. The substitute of one Bi^{3+} ion could creates two vacancy at the A site, which can relax the strain caused by reorinentation of ferroelectric domains.Therefore, domains in NKN-BNKT ceramics can be sufficiently reoriented during the polarization process. On the other hand, due to the difference of ionic radius between Ti^{4+} and Nb^{5+} ions, this substitution will result in an aberrance of crystalline structure, which benefits the reorientation of domains. That may be the explanation for the increase of piezoelectric properties. However, further increasing BNKT contents (>3 mole % BNKT), chemical modification play a major role on the properties of NKN-BNKT ceramics again. Increasing the BNKT contents, due to the difference of ionic radius among K^{+}, Na^{+} and Bi^{+3} ions in the A site, may cause lattice distortion and a resultant variation in composition, which leads to an inhomogeneous microstructure. These conditions may increase the porosity and decrease the grain size, as shown in Fig. 6(d)~6(e). With increasing porosity and decreasing the grain size of (1-x)NKN-xBNKT ceramics, the piezoelectric and dielectric properties decrease gradually. Figure 7 shows the XRD patterns of $0.97(Na_{0.5}K_{0.5})NbO_3$-$0.03Bi(Na_{0.83}K_{0.17})TiO_3$ [abbreviated 0.97NKN-0.03BNKT] with different sintering time (3, 5 and 7 h) at 1100 °C. The orthorhombic and tetragonal phase of 0.97NKN-0.03BNKT ceramics at room temperature is characterized on the XRD patterns in the 2θ ranges of 44~48°. The XRD analysis of sintered samples shows that 0.97NKN -0.03BNKT is sure of a perovskite structure and forms a solid solution. The BNKT appears to have diffused into the NKN lattice to form a solid solution, in which Ba^{+2} ion occupies (Na^{+},K^{+}) lattice and Ti^{+4} ion enters Nb^{+5} sites of NKN. As increasing the sintering time, the patterns of the 0.97NKN-0.03BNKT ceramics do not change in evidence.

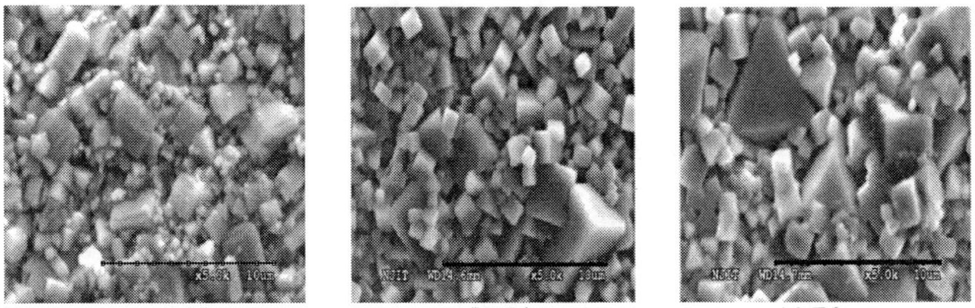

(a) 3 h. (b) 5 h. (c) 7 h.

Figure 8. SEM images of 0.97NKN-0.03BNKT ceramics sintered at 1100°C for different time (a) 3 h. (b) 5 h (c) 7 h. Bar= 10 μm

Figure 8 shows the microstructure of 0.97NKN-0.03BNKT ceramics sintered at 1100 °C for different time (a) 3 h (b) 5 h and (c) 7 h. It can be seen that the 0.97NKN-0.03BNKT ceramics consist of very small grains, with a relatively loose structure, and a high porosity in Fig. 8(a). However, the 0.97NKN-0.03BNKT ceramic at the sintering time Of 5 h are denser, with low porosity, and exhibit a larger and grain size of ~2.5 μm in Fig. 8(b). In Fig. 8(c), the irregular shape is found. According to "COBLE" models, "discontinuous" grain growth is the nomenclature selected here for the occurrence various called discontinuous or exaggerated, or second recrystallization. This may increase the porosity, attributed to a longer sintering time. Figure 9 to 12 show the physical and electrical properties of 0.97NKN-0.03BNKT sintered at 1100 °C for different time. The measured density and relative density of 0.97NKN-0.03BNKT sintered 1100 °C at different time were shown in Fig. 9. The measured density and relative density increase with increasing the sintering time until it reaches 5 h, then decreases above 5 h. The real density and relative density of the sintered samples is 95–98% of the theoretical density. The electromechanical coupling factor (k_p and k_t) of 0.97NKN-0.03BNKT sintered 1100 °C at different time were shown in Fig. 10. Coupling factor k_p and k_t gradually increase along with the sintering time and reach maximum values (k_p =0.32, k_t=0.56) when x was added up to 5 h. Further increasing the sintering time, k_p and k_t will decrease. The dielectric constant and dielectric loss tangent (tan δ) of 0.97NKN-0.03BNKT sintered 1100 °C at different time were shown in Fig 11. The dielectric constant increases with increasing the sintering time until it reaches 5 h, then decreases above 5 h. The variation trend of the dielectric loss tangent (tan δ) with BNKT contents is inverse to that of the measured density.The mechanical quality factor (Q_m) and piezoelectric constant (d_{33}) of 0.97NKN-0.03BNKT sintered 1100 °C at different time were shown in Fig.12. The mechanical quality factor (Q_m) and piezoelectric constant (d_{33}) reaches maximum values (Q_m =50, d_{33}=190 pC/N) at the sintering time of 5 h.

 DOI: 10.1051/matecconf/201711901001

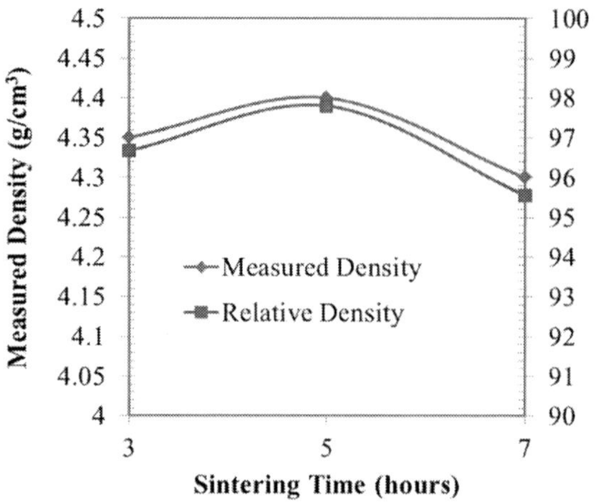

Figure 9. Measured density and relative density of 0.97NKN-0.03BNKT ceramics as a function of sintering time at 1100°C.

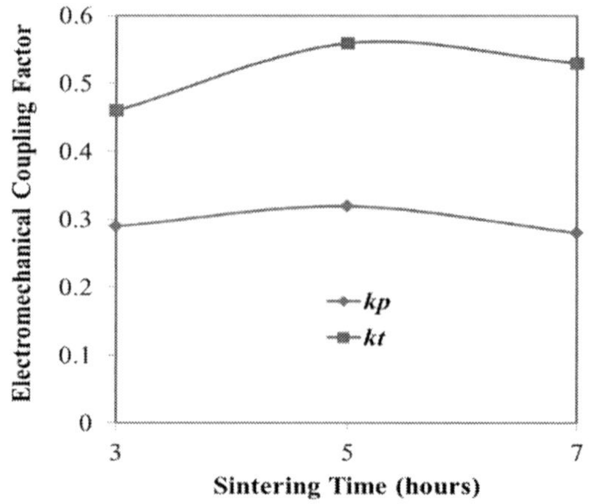

Figure 10. The electromechanical quality factor (k_p and k_t) of 0.97NKN-0.03BNKT ceramics as a function of sintering time at 1100 °C.

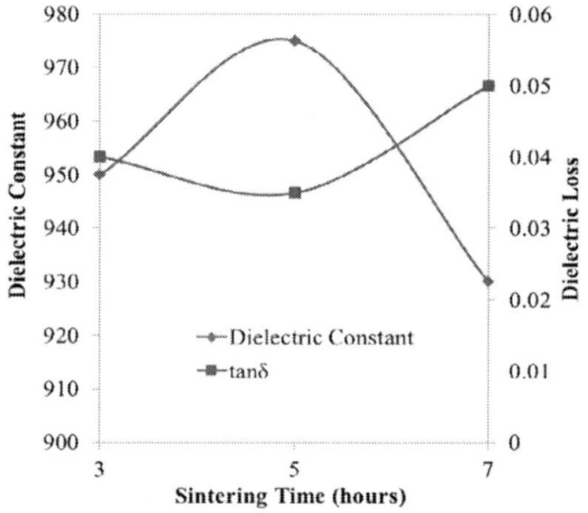

Figure 11. The dielectric constant and dielectric loss tangent (tanδ) of 0.97NKN-0.03BNKT ceramics as a function of sintering time at 1100 °C.

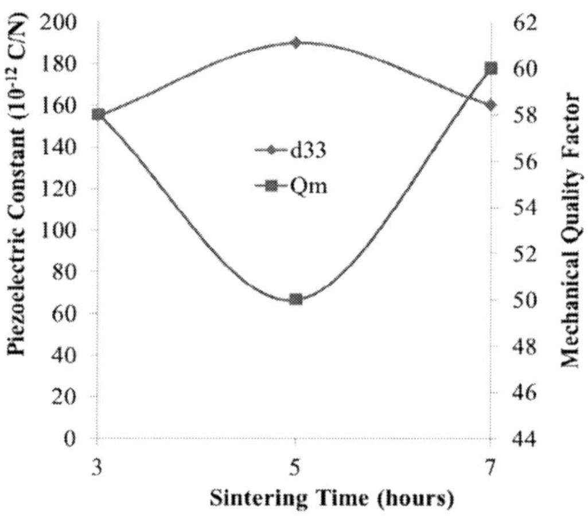

Figure 12. The piezoelectric constant (d_{33}) and Q_m of 0.97NKN-0.03 BNKT ceramics as a function of sintering time at 1100 °C.

4 Summary

The (1-x)NKN-xBNKT (x= 0.01, 0.02, 0.03, 0.04 and 0.05) binary system had been prepared by a conventional ceramic sintering technique. From the XRD patterns, a MPB between orthorhombic and tetragonal phase appears at x=0.03. For 0.97NKN-0.03BNKT ceramics, the electromechanical coupling coefficients of the planar mode k_p, the thickness mode k_t and the piezoelectric constant d_{33} reach 0.29, 0.46 and 155 pC/N, respectively, at the sintering of 1100 °C for 3 h.

The effect of sintering time on the properties of 0.97NKN-0.03BNKT ceramics was discussed. For 0.97NKN-0.03BNKT ceramics, the planar electromechanical coupling coefficient k_p, the thickness electromechanical coupling coefficient k_t and the piezoelectric constant d_{33} reach 0.32, 0.56 and 190 pC/N, respectively at the sintering of 1100 °C for 5 h. Prolonging the sintering time to the 0.97NKN-0.03BNKT system is a helpful method on ceramic processing to improve densification and properties. Our results show that 0.97NKN-0.03BNKT solid solution ceramics are promising lead-free ceramics for a wide range of electromechanical transducer applications.

References

1. G. Shirane, R. Newnham, and R. Pepinsky, Phys. Rev., **96**, 581 (1954)
2. M. D. Maeder, D. Damjanovic, and N. Setter, J. Electroceram., **13**, 385 (2004)
3. V. Lingwal, B. S. Semwal, and N. S. Panwar, Bull. Mater. Sci., **26**, 619 (2003)
4. H. Birol, D. Damjanovic, and N. Setter, J. Eur. Ceram. Soc., **26**, 861 (2006)
5. T. Takenaka, H. Nagata, and Y. Hiruma, Jpn. J. Appl. Phys., **47,** 3787 (2008)
6. S. Y. Chu, W. Water, Y. D. Juang, J. T. Liaw, and S. B. Dai, Ferroelectrics, **297**, 11 (2003)
7. M. Matsubara,T.Yamaguchi, K. Kikuta, and S. Hirano, Jpn. J. Appl. Phys., **44**, 6136 (2005)

Measurement of power and energy consumption of a competition-mobile-robot

Tsing-Tshih Tsung[1,a], Yun-Ju Chuang[2,b], Ho Chang[2], Chun-Lun Shen[3] and Yin-Tung Sun[2]

[1]*Department of Mechanical Engineering, Chinese Culture University, Taipei, Taiwan.*
[2]*Graduate Institute of Manufacturing Technology, National Taipei University of Technology, Taipei, Taiwan.*
[3]*Institute of Mechatronic Engineering, National Taipei University of Technology, Taipei, Taiwan.*

Abstract. Power and energy consumption are the two most important factors for successful operation; they also play important roles in performance identification.The measurement of power and energy consumption is a common test in the development process of a competition mobile robot. If the power of a competition mobile robot is not sufficient, the running time in the competition will be too long and winning the competition will not be possible. Thus, the power and energy consumption are basic and important measurement parameters for a competition mobile robot. In this paper, five types of hand-made competition mobile robots are successfully developed and their performances are measured. From the measurements, their powers and energy consumptions are evaluated and analyzed, respectively. The test results show large differences in the powers and energy consumptions of the five models, even though the same motors were used. The design and construction of the competition mobile robot are the key parameters that cause these huge differences. It is possible to develop the measurement techniques for power and energy consumption, quickly and precisely, to determine how to modify a competition mobile robot rapidly and efficiently to a condition optimal for a mobile robot competition.

1 Introduction

Robots can perform a large number of tasks for humans, using the different kinds of energy available today; therefore, there are many different types of robots, and many more will be developed in the future.

For the training and education of the mobile robots, many countries hold Robotics Competitions such as World Skills - Mobile Robotics, IEEE International Robot Hands on Competition and Symposium, and RoboCup as shown in figure.1

For a competition mobile robot, power and energy consumption are always the key factors in the competition. The power and energy consumption of a competition mobile robot play important roles in deciding the results of performance identification in a mobile robot competition.

In a robot competition, energy and time are key factors that decide the performance identification. For a mobile competition robot, many kinds of energies can be chosen, and many different types of

[a] Corresponding author : tttsung@gmail.com
[b] Corresponding author : andy2551468@gmail.com

DOI: 10.1051/matecconf/201711901002

power units can be used. The weight of the power unit that the robot can carry is limited in the competitions; moreover, the competition time is also limited. Therefore, in order to win the competition, the robot must be simple, fast, and must have low energy consumption.

Measurement of power is a common performance identification technique in a mobile robot competition. If the power of a competition mobile robot is not enough, the competition time will be too long and it will not be possible to win the competition. Thus, the measurements of power and energy consumption are the basic measurements for a competition mobile robot, and their results will be presented in this paper.

In the present study, five types of electricity as energy sources for the robots were produced. measured energy five robots working in individual consumption, energy and time analysis of different robots in each work.

The The measurement results of the final energy consumption and the amount of time, can be used in the robot competition for creating a robot arrangement conducive to the competition tasks, in order to benefit the ultimate task, which is winning the race.

Figure 1. Robotics Competitions

2 Types of Competition Mobile Robots

Five types of the competition robots were developed in this research. The power and energy consumption of each were measured and analyzed. The energy consumptions of each working process in the competition robots, and the operating times for the various durations of the various processes, were measured and analyzed in detail. Furthermore, their performances were identified. According to the performance identification result, the energy distribution mode will be determined and the control unit will be built.

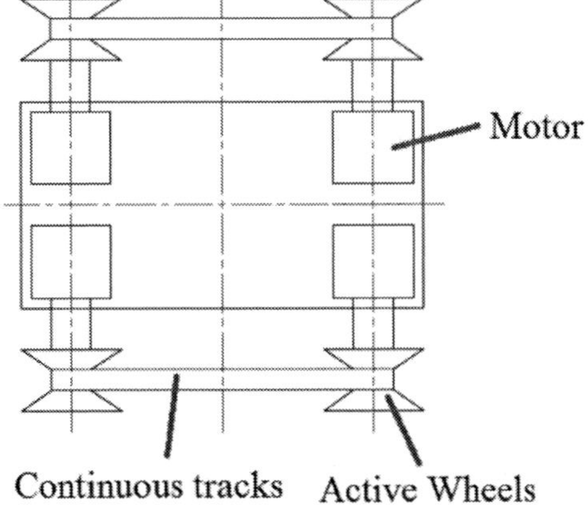

Figure 2. Sketch of Type-1 robots

2.1 Type-1 robot

Figure 2 shown the Type-1 robot has two continuous tracks, four active wheels, four motors, and a weight of 1.6 kg.The continuous track design can increase the friction while climbing the ramp.

2.2 Type-2 robot

Figure 3 shown the Type-2 robot has two continuous tracks, four active wheels, two idler wheels, four motors, and a weight of 1.5 kg.The idler-wheel design can adjust the elastic in the continuous tracks so that they work more smoothly.

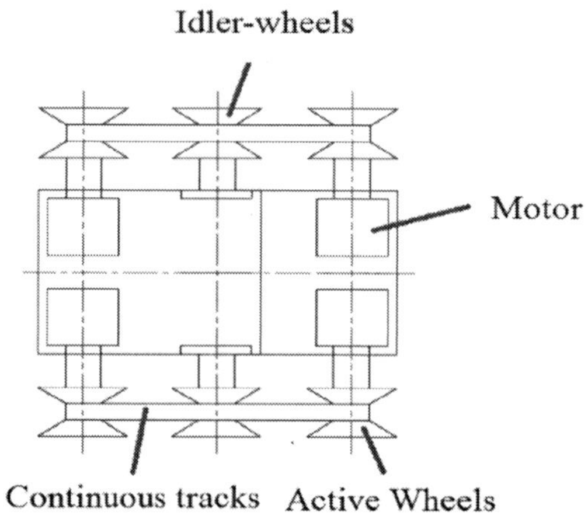

Figure 3. Sketch of Type-2 robots

2.3 Type-3 robot

Figure 4 shown the Type-3 robot has four active wheels, four motors, and a weight of 1.15 kg. The non-continuous track design can improve the speed.

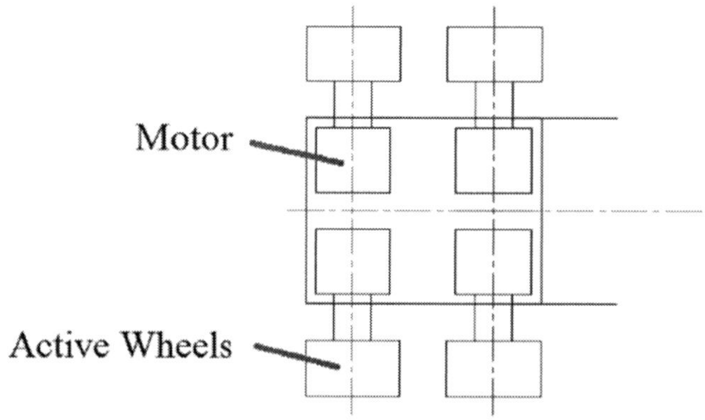

Figure 4. Sketch of Type-3 robots

2.4 Type-4 robot

Figure 5 shown the Type-4 robot has two active wheels, two passive wheels, two motors, and a weight of 3.7 kg. The cutting two-motor design will change the center of rotation, and the distance in between will affect the rotation.

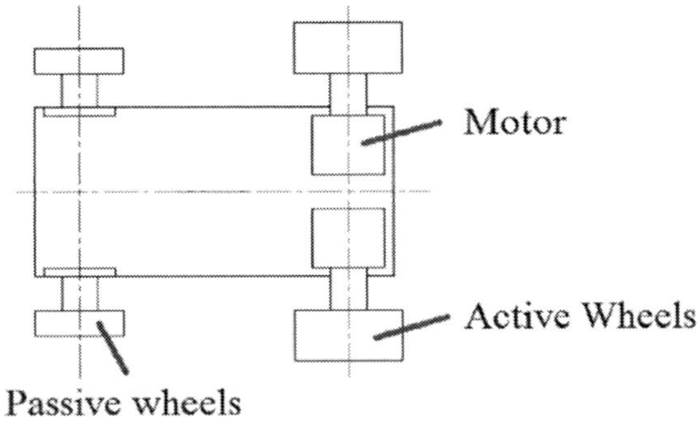

Figure 5. Sketch of Type-4 robots

2.5 Type-5 robot

Figure 6 shown the Type-5 robot has two active wheels, one passive wheel, two motors, and a weight of 4.2 kg. The single passive wheel enables a smoother rotation.

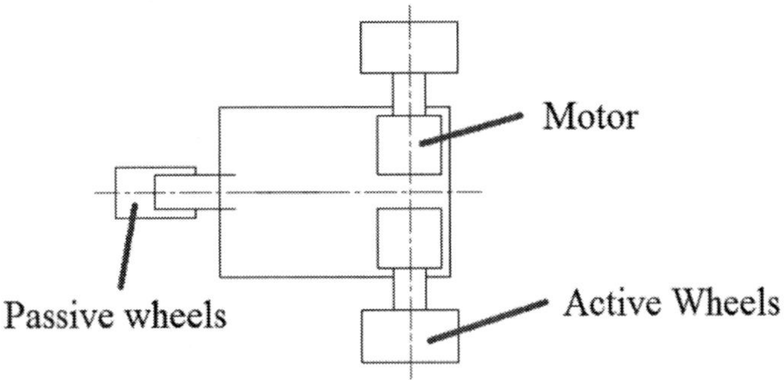

Figure 6. Sketch of Type-5 robots

3 Battery performance characteristics

Table 1 and Figure 7 show the performance of the battery. This battery used is of 12-V, 2.3-Ah capacity.

Table 1. Battery performance (YUASA, 2006)

Nominal Capacity (Ah)	NP2.3-12
20hr to 1.75vpc 30°C	2.3
Voltage	12
Energy Density (Wh.L.20hr)	76
Specific Energy (Wh.kg.20hr)	29
Int. Resistance (m.Ohms)	65
Maximum discharge (A)	23
Short Circuit current (A)	69

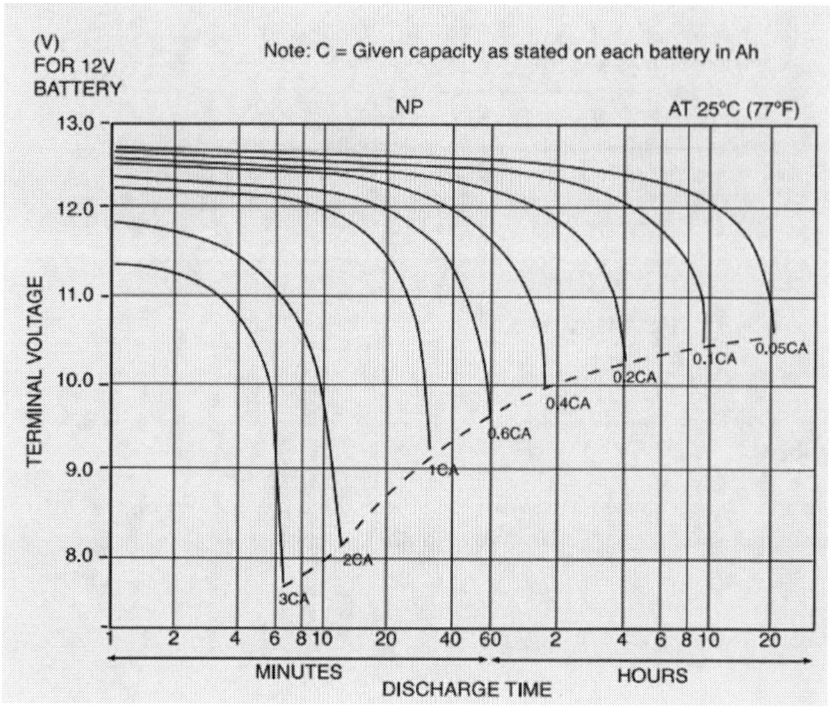

Figure 7. Battery performance (YUASA, 2006)

4 Test setup for competition mobile robot

The testing setup is based on the setup used in the World Skills Manufacturing Team Challenge, 2013 as shown in figure 8. It involves the task of climbing a ramp.

Figure 8. World Skills Manufacturing Team Challenge, 2013

This research measures the power and energy consumption of a competition mobile robot climbing a slope.

The testing setup is shown in Figure 9.

The research follows these steps for testing:

1. Measure the voltage at START.
2. Measure the voltage at END.

This test will measure the power and energy consumption of a competition mobile robot.

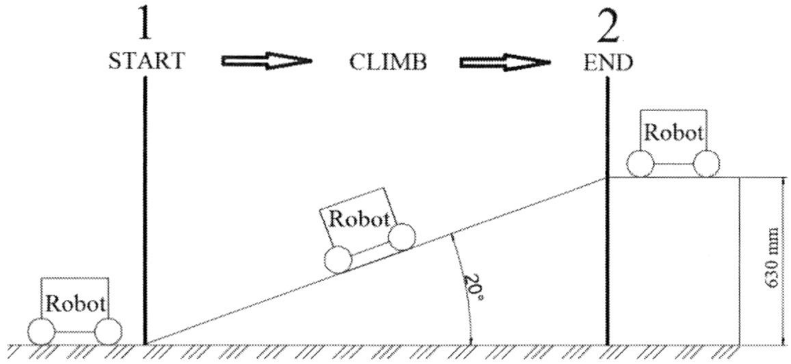

Figure 9. Ramp testing setup

5 Results and discussion

5.1 Type-1 robot

The results from the Type-1 robot are shown in Figures 10 to 12.

Figure 10 shown the numbers 1 to 10 energy consumption in the 0.05 to 0.1 V.

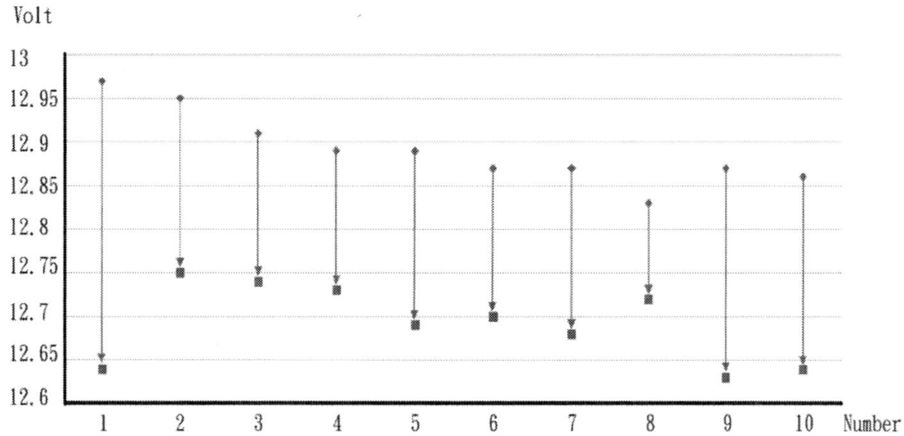

Figure 10. Energy consumption of Type-1 robot

Figure 11 shown the numbers 1 to 10 battery's voltage recovery in 0 to 0.08 V.

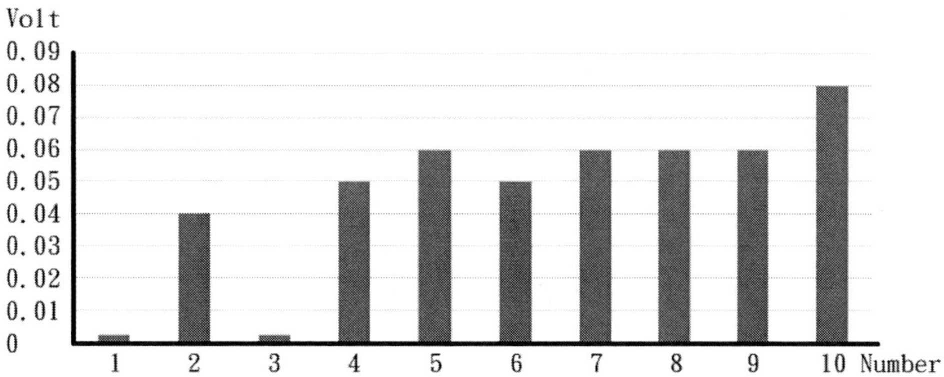

Figure 11. Battery's Volt recover of Type-1 robot

DOI: 10.1051/matecconf/201711901002

Figure 12 shown the numbers 1 to 10 power in 0.6 to 1.3 W.

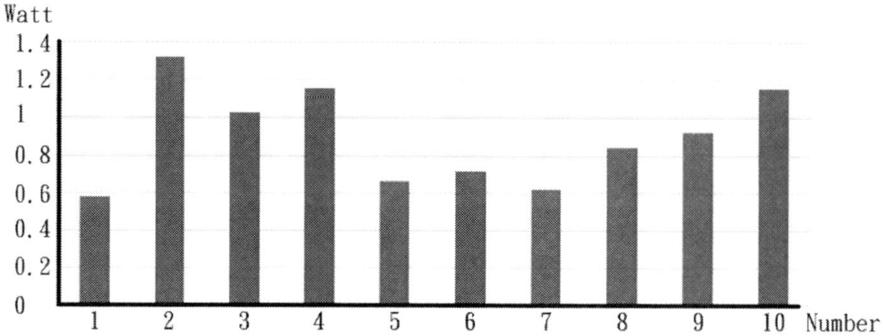

Figure 12. Power of Type-1 robots

5.2 Type-2 robot

The results from the Type-2 robot are shown in Figures 13 to 15.

Figure 13 shown the numbers 1 to 10 energy consumption in the 0.11 to 0.33 V.

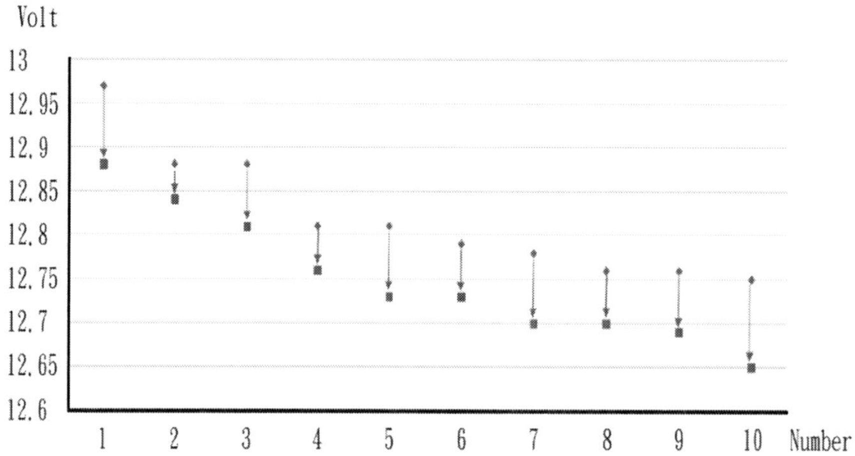

Figure 13. Energy consumption of Type-2 robot

Figure 14 shown the numbers 1 to 10 Battery's Volt recover in 0.15 to 0.31 V.

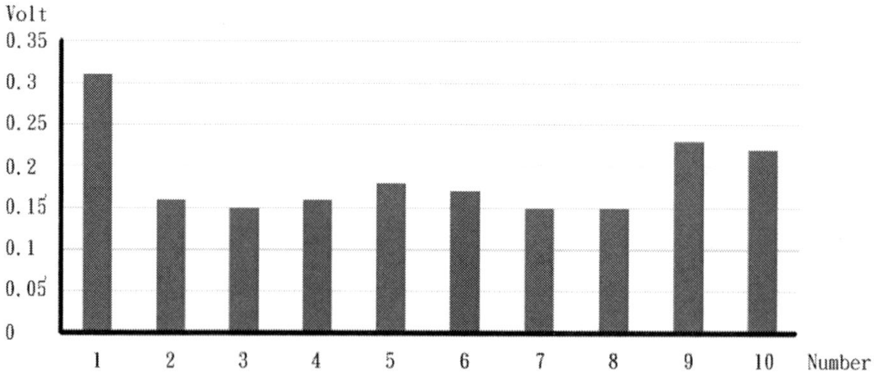

Figure 14. Battery's Volt recover of Type-2 robot

Figure 15 shown the numbers 1 to 10 Power in 0.28 to 0.47 W.

 DOI: 10.1051/matecconf/201711901002

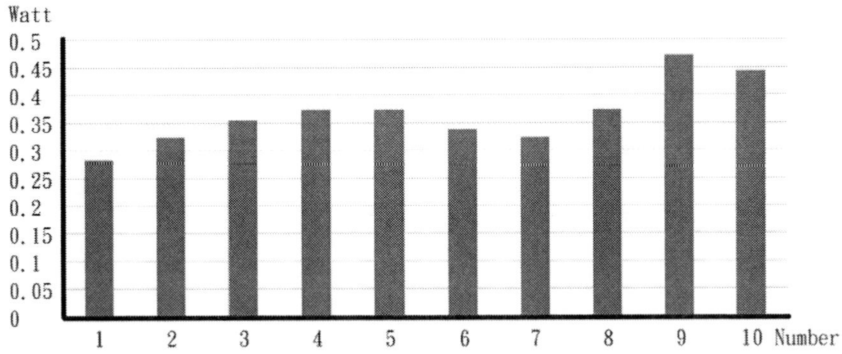

Figure 15. Power of Type-2 robots

5.3 Type-3 robot

The results from the Type-3 robot are shown in Figures 16 to 18.

Figure 16 shown the numbers 1 to 10 energy consumption in the 0.12 to 0.16 V.

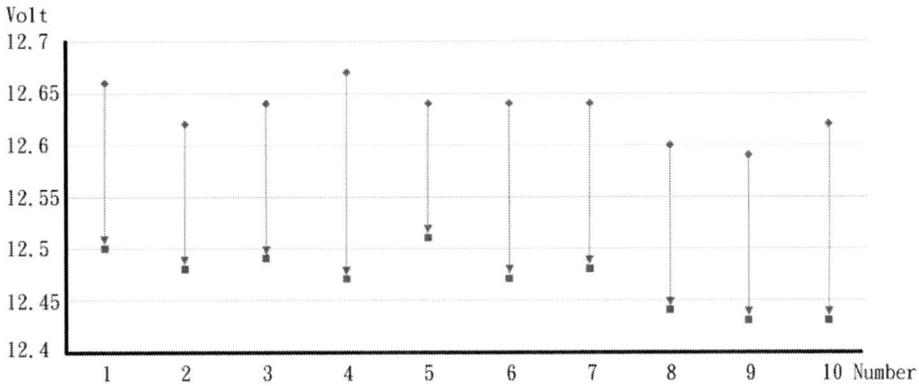

Figure 16. Energy consumption of Type-3 robot

Figure 17 shown the numbers 1 to 10 Battery's Volt recover in 0.12 to 0.2 V.

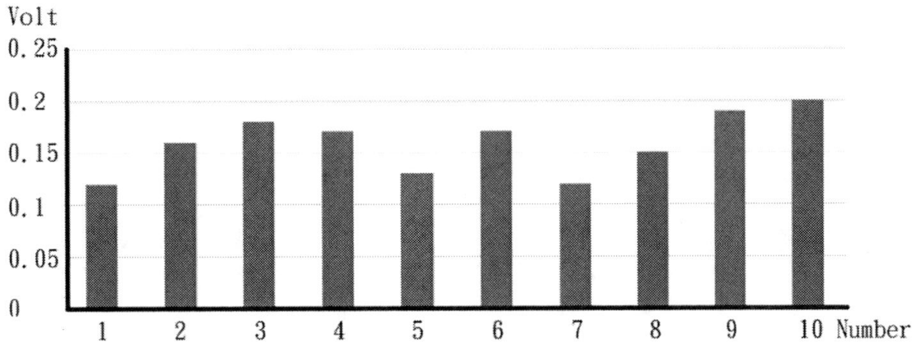

Figure 17. Battery's Volt recover of Type-3 robot

Figure 18 shown the numbers 1 to 10 Power in 0.4 to 0.69 W.

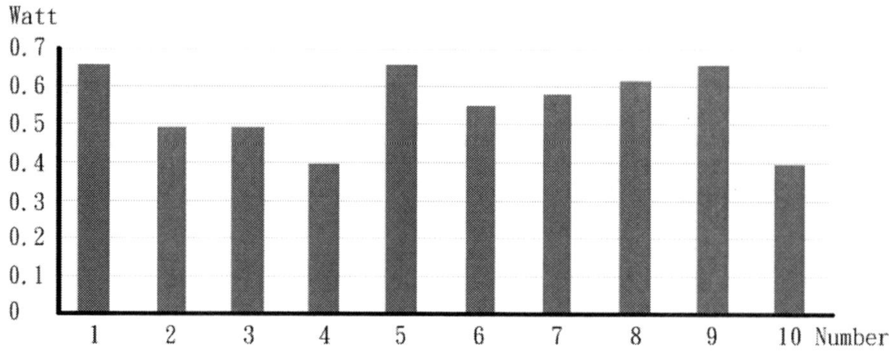

Figure 18. Power of Type-3 robots

5.4 Type-4 robot

The results from the Type-4 robot are shown in Figures 19 to 21.

Figure 19 shown the numbers 1 to 10 energy consumption in the 0.08 to 0.17 V.

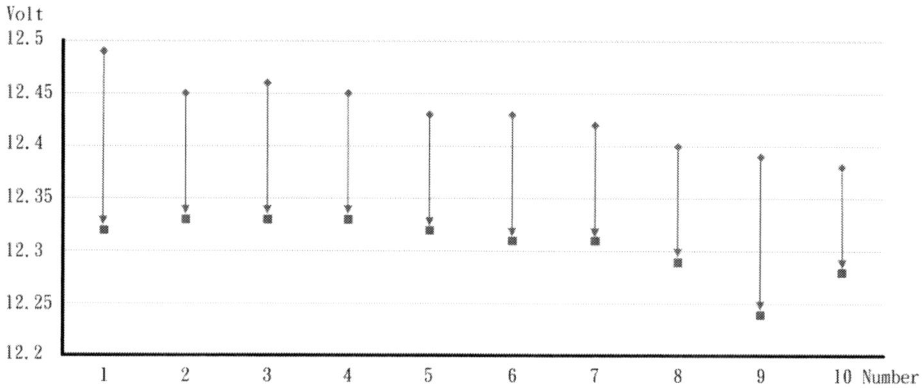

Figure 19. Energy consumption of Type-4 robot

Figure 20 shown the numbers 1 to 10 Battery's Volt recover in 0.09 to 0.26 V.

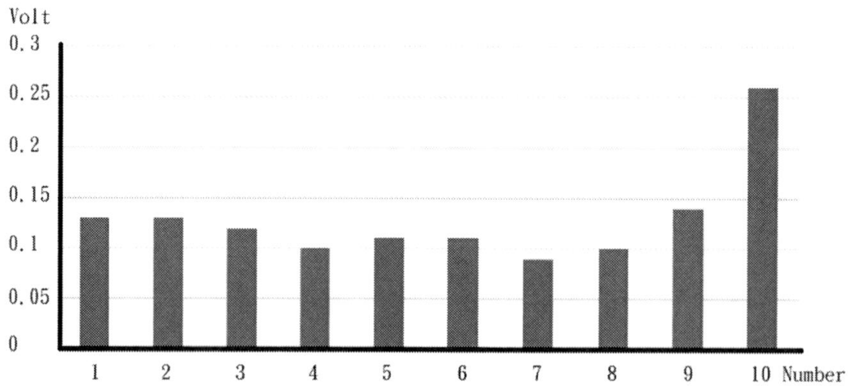

Figure 20. Battery's Volt recover of Type-4 robot

Figure 21 shown the numbers 1 to 10 Power in 1.14 to 1.76 W.

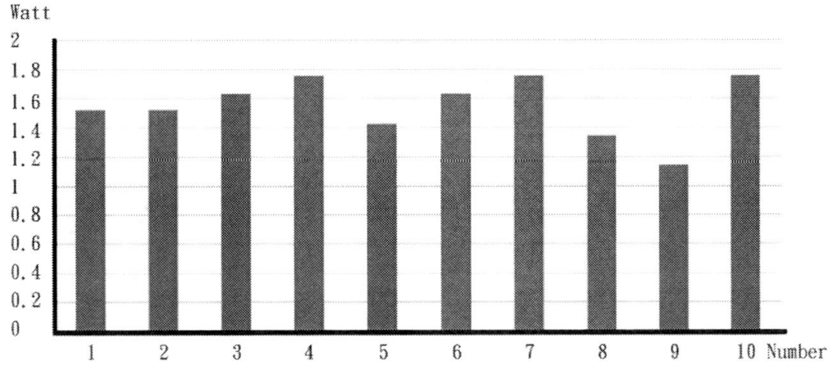

Figure 21. Power of Type-4 robots

5.5 Type-5 robot

The results from the Type-5 robot are shown in Figures 22 to 24.

Figure 22 shown the numbers 1 to 10 energy consumption in the 0.06 to 0.14 V.

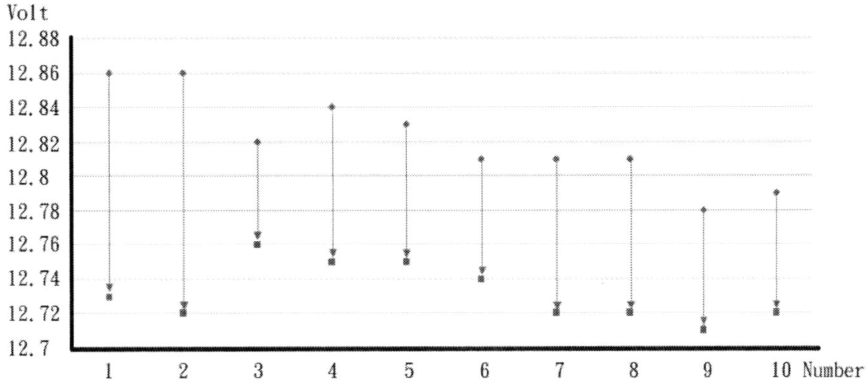

Figure 22. Energy consumption of Type-5 robot

Figure 23 shown the numbers 1 to 10 Battery's Volt recover in 0.06 to 0.13 V.

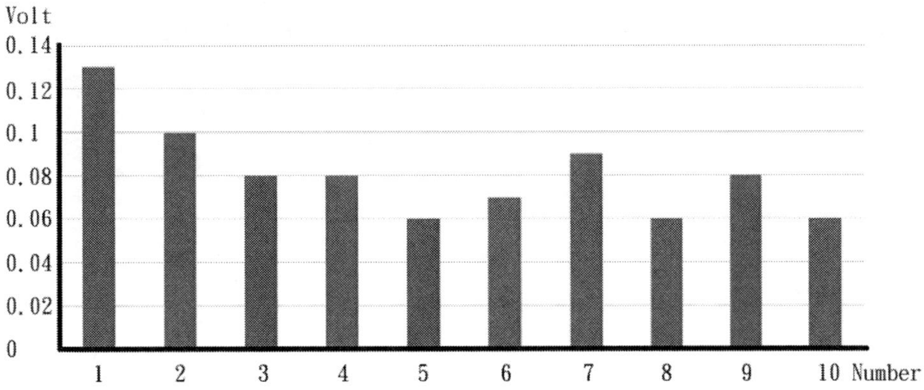

Figure 23. Battery's Volt recover of Type-5 robot

Figure 24 shown the numbers 1 to 10 Power in 1.73 to 3.7 W.

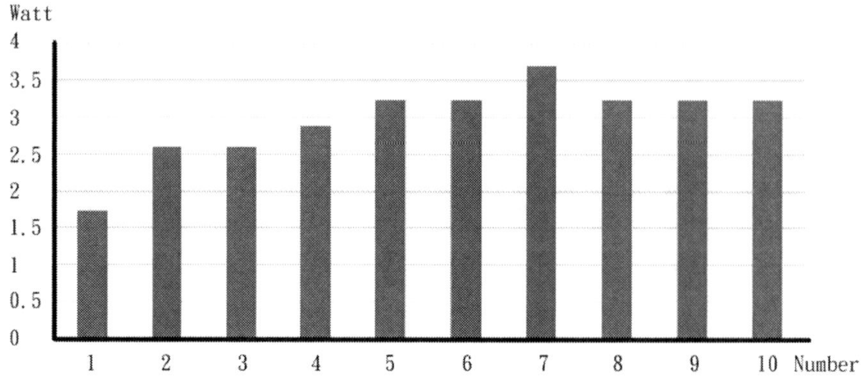

Figure 24. Power of Type-5 robots

6 Conclusions

In short, the measurement results can be used to demonstrate the performance of a robot as shown in table 2 :

1. The energy consumption of the Type-1 robot is the maximum, and that of the Type-2 robot is the minimum.
2. The Type-2 battery's recover is the maximum, and that of Type-1 is the minimum.
3. The power of the Type-5 robot is the maximum, and that of the Type-2 robot is the minimum.
4. The Type-3 robot is the lightest, and the Type-5 robot is the heaviest.
5. A higher energy does not guarantee a higher power.
6. Weight is not an important factor in energy consumption; the operating mechanism is a more important factor.
7. Energy consumption of Volt like Battery's recover, so Battery's recover is affected by energy consumption.
8. We can use more battery to intermittent discharge as the battery voltage can be recovered.

Table 2. Type-1 to Type-5 results

Robot	Energy Consumption(Volt)	Battery's Volt recove(Volt)	Power(Watt)	Weight(kg)
Type-1	0.05~0.1	0~0.08	0.6~1.3	1.6
Type-2	0.11~0.33	0.15~0.31	0.28~0.47	1.5
Type-3	0.12~0.16	0.12~0.2	0.4~0.69	1.15
Type-4	0.08~0.17	0.09~0.26	1.14~1.76	3.7
Type-5	0.06~0.14	0.06~0.13	1.73~3.7	4.2

References

1. Guardabrazo, T.A. and Gonzalez de Santos P, *Industrial Robot: An International Journal*, Building an energetic model to evaluate and optimize power consumption in walking robots, **31**(2), 201-208 (2004)
2. Lin, S.-J, On discharging modeling and capacity estimation of Li-Ion battery, Master Program of Electrical and Control Engineering (2005)
3. YUASA, Valve regulated lead-acid batteries short form brochure (2006)
4. Wong, Y.-T, Real-Time terrain adaptation control for quadrupled robot (2012)
5. Bruzzone, L. and Fanghella, P. Industrial Robot: An International Journal, Mantis: hybrid leg-wheel ground mobile robot, **41** (1), 26-36 (2014)

Analysis and optimization of a novel linear permanent Vernier motor

Yi Shao[a], Degang Lv and Ying Xie

School of Electrical and Electronic Engineering, Harbin University of Science and Technology, China

Abstract. In this paper, a novel linear permanent magnet vernier (LPMV) motor is proposed. The advantages of the proposed motor are high force density, high efficiency, simple structure, and low cost. Firstly, the structure and the operation principle of the proposed motor which adopts 12-slots-11-pair-poles structure are descried. Based on the air-gap permeance function, the components of air-gap flux density is discussed. Secondly, the validation of LPMV is analyzed by finite element method (FEM), and the basic electromagnetic performance such as air-gap flux density, back-EMF, detent force and thrust waveform are analyzed in detail. Finally, the relationship between the thrust force characteristics and the design parameters are analyzed to provide useful information to the designers of LPMV.

1 Introduction

Compared with conventional permanent magnet motors, the direct-drive permanent magnet motors have a lot of advantages. With the absence of a reduction gear, the problems, such as backlash, friction loss, and maintenance can be eliminated. In addition, to some extent, the complexities and failure possibility of the system can be lowered, thus, the reliability and efficiency are enhanced.
A coaxial magnetic gear has been proposed as a type of direct drive technology, which offers a high thrust density. However, the integrated machines employ two or more moving parts with multiple air-gaps; thus, they suffer from manufacture difficulty. To circumvent this problem, a new class of linear machines based on the field modulation principle of the magnetic gear, known as vernier machines, was introduced.

PMV motor was firstly proposed in 1995 by A. Ishizak. It made clear that the PMV motor can obtain high thrust at low speed and the steady thrust equation was derived in [1].Then, Toba and Lipo presented comprehensive design considerations of surface PMV motor for pursuing thrust maximizing [2]. After that various new PMV topologies have been proposed to further improve thrust density. In [3, 4], both the thrust density and power factor of the PMV motors has been improved by employing spoke magnets. In recent years, Vernier PM motor is attracting more and more attention due to their simple mechanical structure, high torque density, sinusoidal back electromotive force (EMF) waveform, and low pulsation ripple [5-11].

Contrasted with the traditional transmission way, the linear motor is omitted mechanical transmission part. And linear motor has simple structure, and there is no direct contact between mover and stator. So it has a low friction, low noise and long working life. The characteristic of high

[a] Corresponding author : 18645140103@163.com

precision can be easily achieved by the corresponding control method. So the linear motor is widely used in industrial processes, transportation, etc. For a variety of linear drive applications, high force density is essential.

As the rotation Vernier permanent magnet motors are investigated widely around the world, the research of VPM which is employed in linear machine field is few. By extending the idea to linear topology, a novel LPMV machine, which can offer simple structure and high force density, is proposed and analyzed in this paper. In order to analyze the electromagnetic performance, such as the thrust density, the thrust ripple, the detent force, etc., the influence of and some design parameters, such as the stator/rotor pole number, the slot opening ratio, the magnet thickness, the excitation current, etc. The performance will be fully investigated by finite-element analysis (FEA) in detail [12-17].

2 Structure and working principle of linear permanent magnet Vernier motor

2.1. Structure of linear permanent magnet Vernier motor

Fig.1 shows the topology of proposed LPMV motor. The long stator is made of simple iron core and permanent magnets (PM),and the PM is surface mounted on the stator. The mover is consisted of excitation winding and iron core.

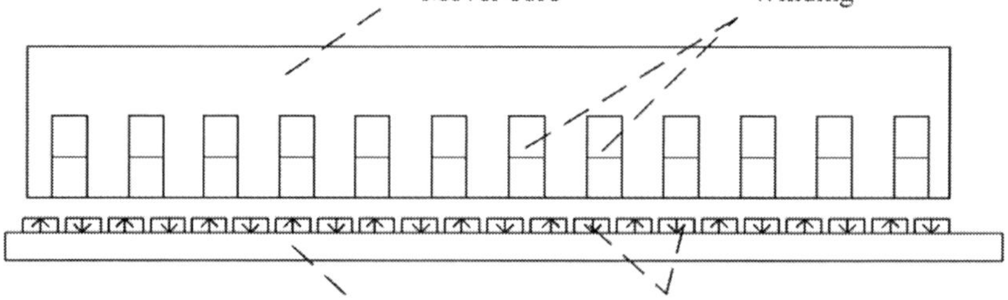

Figure 1. The novel linear permanent magnet vernier motor.

Different with regular PM machines, the slots and tooth of the stator for VPM machine are not only for embedding windings or serving as a part of the magnetic path, but also working as flux modulators, which has a significant influence on thrust performance of VPM machines. In order to maintain the same modulation effect along radial direction of a VPM machine, trapezoidal slot is desirable since the ratio of slot opening width to slot pitch remains constant. However, due to manufacturing difficulties, a radially proportional structure for an axial flux machine may not be a cost-effective option when silicon steel sheet are used to fabricate iron cores. Therefore, rectangular instead of trapezoidal slots are chosen for the proposed machine. On the rotor side, trapezoidal shaped magnets are preferred to keep the pole are unchanged and thus remain a balanced magnetic loading along radial direction.

2.2 Principle of linear permanent magnet Vernier motor

Stator coils with concentrated windings are excited by a three-phase balanced sinusoidal current and generate a travelling magnetic field. The magnetomotive force (MMF) of the A-phase can be expressed in the following equations:

$$F_{ca}=\frac{4\sqrt{2}NI}{\pi p}\sum_{m}\left\{\frac{1}{m}g_{u}(\mathrm{m})\sin(mp\theta)\sin(\omega \mathrm{t})\right\} \tag{1}$$

$$g_{u}(m)=\sin\frac{m}{2}\pi\cdot\sin\frac{m}{6}\pi \tag{2}$$

Where N is the number of coil turns in each phase, I is the of sinusoidal coil current (effective value),p is the pole pairs of the travelling magnetic field, θ is the mechanical angle, and m is a positive odd number. The angular velocity ω is equal to $2\pi f$,where f is the frequency of the sinusoidal current.

$$F_{c}=F_{0}\sum_{l}\left[\frac{1}{6l+1}g(l)\cos\{\omega t-(6l+1)p\theta\}\right] \tag{3}$$

$$F_{0}=\frac{6\sqrt{2}NI}{\pi p} \tag{4}$$

$$g(l)=\sin\frac{6l+1}{2}\pi\cdot\sin\frac{6l+1}{2}\pi \tag{5}$$

where l is an integer $(l=0,\pm1,\pm2,\pm3)$

The permeance of the stator teeth can be expressed in the following equation:

$$P_{sm}=P_{s0}+\sum_{m}P_{sm}\cos(mZ_{1}\theta) \tag{6}$$

Where P_{sm} is the amplitude of the permeance of the mth harmonic,Z_1 is the number of stator teeth. However,(6) can be simplified using its major components, and can be expressed as follows:

$$P_{s1}=P_{s0}+P_{s1}\cos(Z_{1}\theta) \tag{7}$$

Consequently the multiplication of (3) and (7) yields the flux density distribution

$$B_{sl}=F_{c}P_{s1}=F_{0}\sum_{l}\left\{\frac{1}{6l+1}g(l)\cos(\omega t-(6l+1)p\theta)\right\}\times\{P_{0}+P_{1}\cos(Z_{1}\theta)\} \tag{8}$$

The major components $(l=0,\pm1,\pm2,\pm3)$ of the flux density distribution are expanded and shown below:

$$B_{s(0)}=F_{0}P_{0}g(0)\cos(\omega t-p\theta)+\frac{F_{0}P_{1}g(0)}{2}\left[\cos\{\omega t+\theta(Z_{1}-p)\}+\cos\{\omega t-\theta(Z_{1}+p)\}\right] \tag{9}$$

$$B_{s(-1)}=\frac{F_{0}P_{0}g(-1)}{5}\cos(\omega t+5p\theta)+\frac{F_{0}P_{1}g(-1)}{5}\left[\cos\{\omega t+\theta(Z_{1}+5p)\}+\cos\{\omega t-\theta(Z_{1}-5p)\}\right] \tag{10}$$

$$B_{s(+1)}=\frac{F_{0}P_{0}g(1)}{7}\cos(\omega t-7p\theta)+\frac{F_{0}P_{1}g(1)}{7}\left[\cos\{\omega t+\theta(Z_{1}-7p)\}+\cos\{\omega t-\theta(Z_{1}+7p)\}\right] \tag{11}$$

$$B_{s(-2)} = \frac{F_0 P_0 g(-2)}{11}\cos(\omega t + 13p\theta) + \frac{F_0 P_1 g(-2)}{11}\left[\cos\{\omega t + \theta(Z_1 + 11p)\} + \cos\{\omega t - \theta(Z_1 - 11p)\}\right] \quad (12)$$

$$B_{s(+2)} = \frac{F_0 P_0 g(2)}{13}\cos(\omega t - 13p\theta) + \frac{F_0 P_1 g(2)}{13}\left[\cos\{\omega t + \theta(Z_1 - 13p)\} + \cos\{\omega t - \theta(Z_1 + 13p)\}\right] \quad (13)$$

In this paper, $Z_1 = 12$ and $p = 1$ are chosen. Therefore the second term (9) and the first term (12) generate the 11th flux harmonic, which travels in the opposite direction of the fundamental rotating magnetic field at 1/11 times the speed. At the same time, the third term(13) and the first term (9) generate the 13th flux harmonic, which travels in the same direction of the fundamental rotating magnetic field at 1/13 times the speed. The mover will synchronize with the 11th or 13th flux harmonic if the number of permanent magnet pole pairs chosen is 11 or 13.
As mentioned in the previous section, the mover of the vernier motor synchronizes with the flux harmonics of the magnetic field and rotates with a reduction ratio with respect to the angular velocity of the magnetic field.
Therefore in order to operate correctly, the following structural relationship shown in(14) between Z_1, p, and Z_2 has to be satisfied. The rotor will then rotate at ω_m , which is an angular velocity with a reduction ratio of p/Z_2 of the synchronous velocity ω /p,and is expressed in (15).

$$Z_2 = Z_1 \pm p \quad (14)$$

$$\omega_m = \frac{\omega}{p} \times \frac{p}{Z_2} = \frac{\omega}{Z_2} \quad (15)$$

3 Model validation

In order to verify the forgoing analysis about proposed LPMV, a finite element model is built and investigated by Maxwell_2D, the main design parameters of proposed motor are tabulated in table I and the structure of proposed LPMV is shown in Fig 2.

Table 1. Setting Word's margins.

Symbol	Parameter
Mover width	132mm
Mover pole pitch h_c	11mm
Mover Slot Width h_c	4.8mm
Mover Slot Depth g_c	6mm
PM thickness g_m	2.3mm
PM width h_m	4.8mm
Number of Slot/Pitch	12/22
Magnet material	NdFe35
Current Density	6A/mm^2
Thrust	235N
Mover Speed	1m/s
Each slot number of conductor	25
Air-gap length	1mm

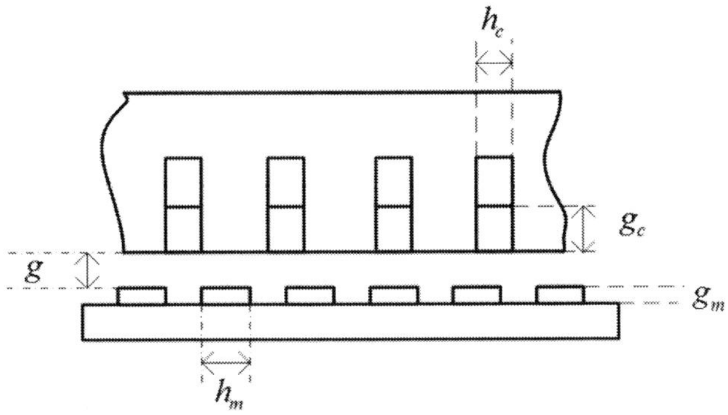

Figure 2. The simplified structure of proposed motor.

3.1 Magnetic field distribution

Fig 3. shows the magnet flux plot of a VPML machine at a different angle position, and it can be seen that the small rotor position displacement introduces a large flux variation. VPML machine has a same operation principle with magnetic gear motor which is called as "magnetic gear effects" .

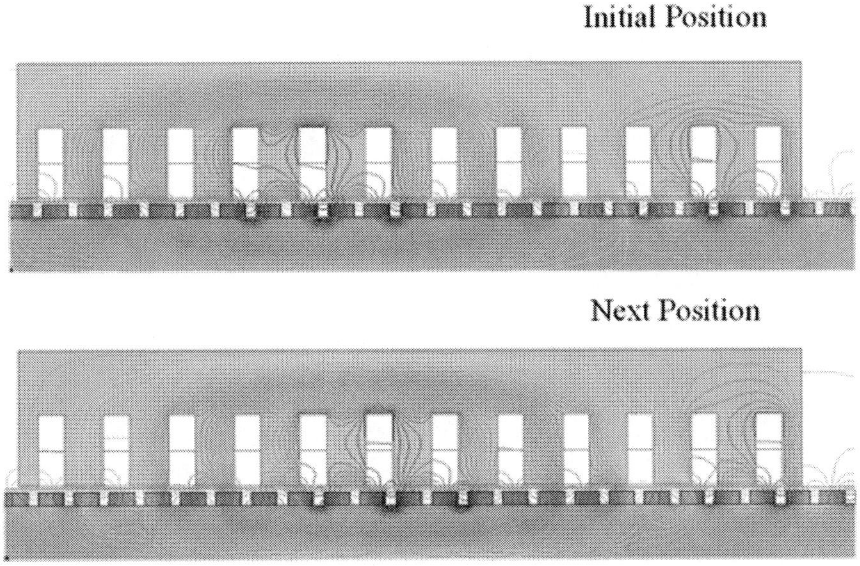

Figure 3. The magnetic field distributions of proposed motor

3.2 Air-gap flux density

The air-gap flux density waveform of the proposed VPML machine only due to PM is shown in Fig 4. It can be seen the maximum flux density of the proposed motor is 0.55T, and the harmonic analysis is done by Fast Fourier Transformation (FFT) as shown in Fig 5. the largest component of air-gap flux density is 0.29T of 11th harmonic. Respectively,Fig.5 shows the corresponding space harmonic spectrum. Obviously,the 11th harmonic makes the maximum.

 DOI: 10.1051/matecconf/201711901003

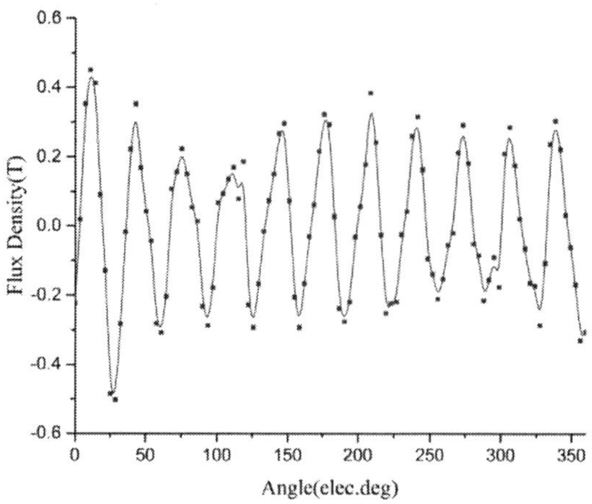

Figure 4. The Air-gap flux density distribution of proposed motor.

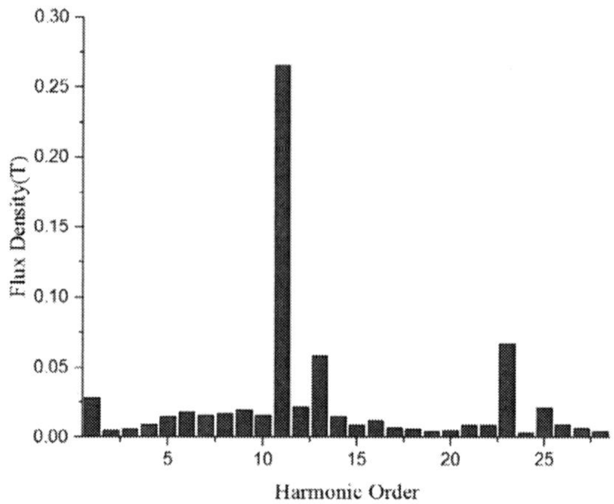

Figure 5. The corresponding space harmonic spectrum of proposed motor.

3.3 No-load Back-EMF and force characteristic

The no-load back-EMF waveform of the proposed motors at nominal speed of 1m/s is shown in Fig 6. the phase angle between a, b, and c is 120o, however, the amplitude of a is 76.7V, and it is smaller than 75.2Vof b and c, which is the result of end effect in the linear machine.

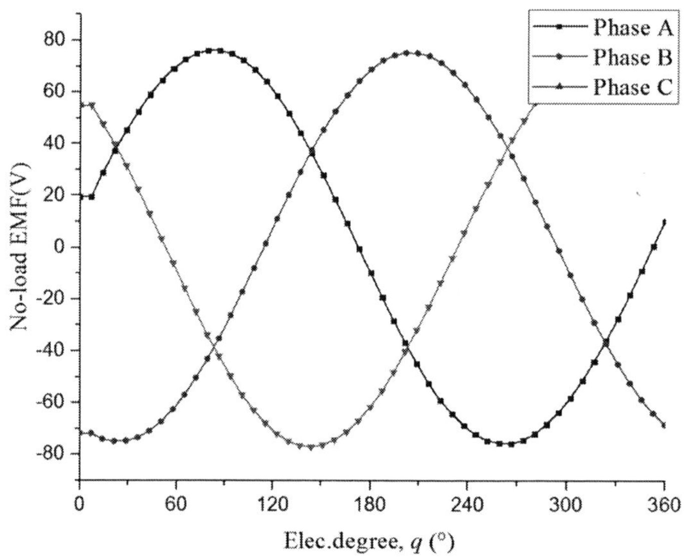

Figure 6. The no-load back-EMF waveform of the proposed motor.

Fig 7. Shows the characteristic of the thrust force and cogging force, and the average thrust is 235N, the thrust ripple is 7%. The thrust ripple includes detent force, cogging force and friction, and the amplitude of cogging force is 5.1N.

 DOI: 10.1051/matecconf/201711901003

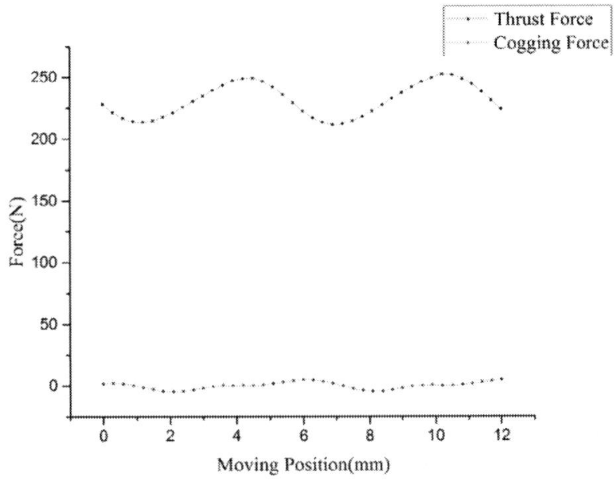

Figure 7. The force characteristic of the proposed motor.

4 Parameter analysis and its determination

In this section, the influence of several geometrical and physical parameters on the thrust force is investigated to provide useful information for the designers of permanent magnet vernier linear motor.

4.1 Influence of the air gap length

Fig 8 shows the thrust force characteristic with different air gap lengths. It can be seen that the thrust decreases gradually as the air gap increases, and this is a result of lower magnetic density under larger air gap length. When the air-gap length is 2mm, the amplitude of 11th harmonic of air-gap flux density is 1.38T, that is approximately 1/2 of the one under 1mm air-gap; the average thrust is 112 N, that is 1/2 of the one under 1mm air-gap.

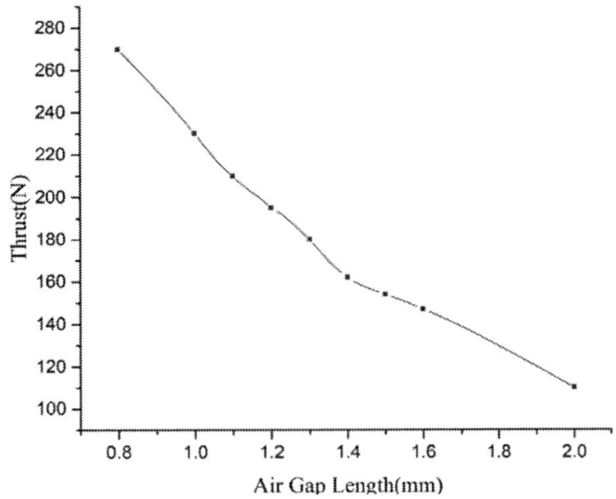

Figure 8. The thrust force characteristic with different air gap length.

4.2 Influence of the magnet dimension

The effect of the two magnet dimensions: the width of the permanent magnets h_m and the height of the permanent magnets b_m, are analyzed. The study is performed under the consideration of constant magnet volume. The pole-arc coefficient is calculated by:

$$\alpha = \frac{h_m}{\tau} \quad (16)$$

where τ is the pole pitch.

The pole-arc coefficient has an important influence on the air gap flux density waveform by changing the percentage of 11th harmonic which can significantly influence the average thrust. Meanwhile, the pole-arc coefficient also influences the thrust ripple. The analysis result is shown in Fig 9, the thrust is increased with the decrease of pole-arc coefficient, from 14 N under α =0.65 to 20N under α=0.9; the thrust ripple reaches the minimum value of 7% when α =0.8.

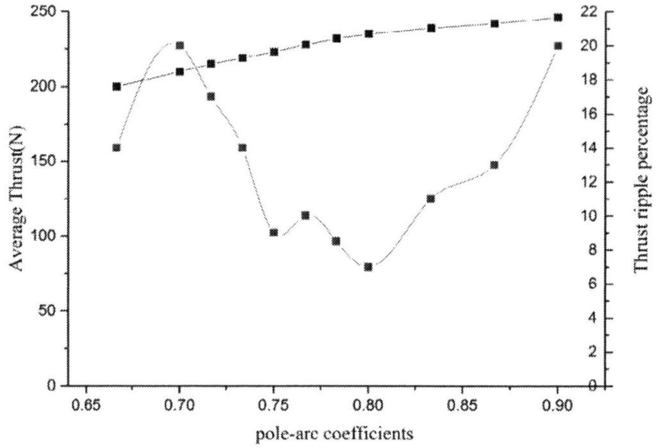

Figure 9. The thrust force characteristic with different pole-arc coefficients.

4.3 Influenced of the slot-opening factor

The slot-opening factor influences the distribution function of air-gap permeance, then influences the detent force and thrust ripple of motor. The analysis of slot-opening factor is done under the condition of same slot-area and slot-filling factor. From Fig 10, the thrust ripple is increased with the increasing of slot-opening factor when the current density is small as 6A/mm^2 and 12A/mm^2; then the thrust ripple is decreased with the increasing of slot-opening factor when the current density is 18A/mm^2, that is the result of saturation effect in the teeth part.

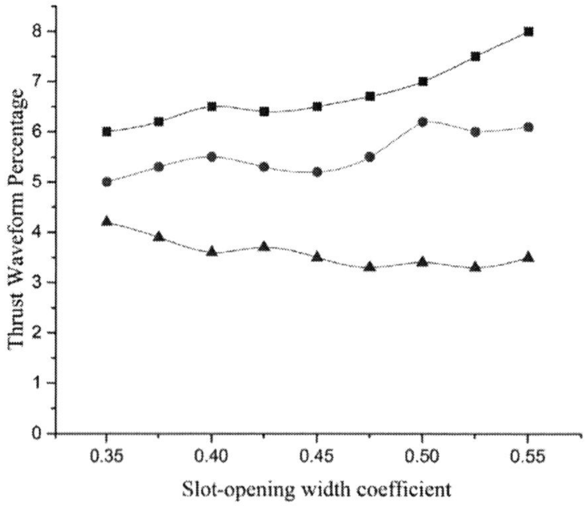

Figure 10. The thrust ripple with different slot-opening width coefficient.

4.4 Influenced of the excitation current

The curve of thrust versus current density is shown in Fig 11, it can be seen that the thrust increases linear with the increasing of current density when the current density is under 16A/mm2, then the thrust increases slowly as the effect of core saturation.

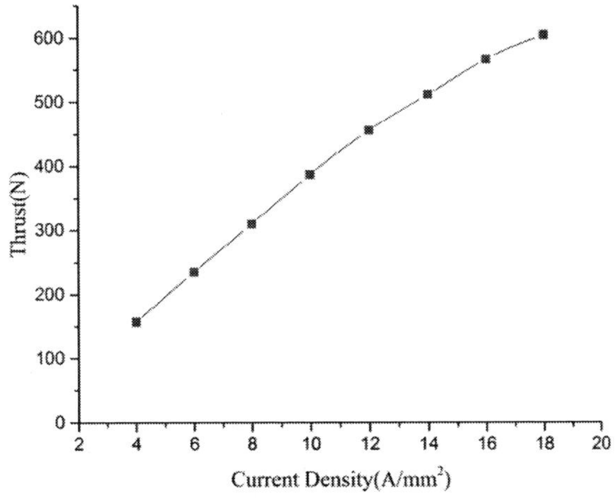

Figure 11. The thrust force characteristic with different current density.

5 Conclusion

A LVPM motor has been presented and described in this paper. FEA is employed to analyze the influence of the mover and stator pole number combination on the electromagnetic performance, such as the back EMF, the flux density, and the thrust ripple. The optimal stator/rotor pole pair for the maximum thrust is 1/11. Furthermore, the influence of design parameters including the slot opening ratio, the pole arc, and the magnet thickness on the torque capability has been investigated.It is found that the optimal pole arc for the maximum torque is around 0.81; and the optimal slot opening ratio is around 0.65. At last, it is worth to mention that these findings in this paper also can be easily extended and employed for the conventional rotate PMV motor designs.

References

1. A. Ishizaki, T. Tanaka, K. Takasaki, et al., Theory and optimum design of PM Vernier motor, International Conference on Electrical Machines and Drives, IET, 208-212 (1995)
2. A. Toba and T.A. Lipo, Generic torque-maximizing design methodology of surface permanent-magnet vernier machine, IEEE Transactions on Industry Applications, **36** (6), 1539-1546 (2000)
3. X. Li, K.T. Chau, and M. Cheng, Comparative analysis and experimental verification of an effective permanent-magnet Vernier machine, IEEE Transactions on Magnetics, **51** (7), 1-9 (2015)
4. D. Li, R. Qu, and T.A. Lipo, High power factor vernier permanent magnet machines, IEEE Transactions on Industry Applications, **50** (6), 1534-1540 (2013)
5. D.G. Dorrell, A.M. Knight, L. Evans, et al., Analysis and design techniques applied to hybrid vehicle drive machines-assessment of alternative IPM and induction motor topologies, IEEE Transactions on Industrial Electronics, **59** (10), 3690-3699 (2012)
6. I. Boldea, L.N. Tutelea, L. Parsa, et al., Automotive electric propulsion systems with reduced or no permanent magnets: an overview, IEEE Transactions on Industrial Electronics, **61** (10), 5696-5711 (2014)

7. W. Zhao, J. Zheng, J. Wang, et al., Design and analysis of a linear permanent- magnet vernier machine with improved force density, IEEE Transactions on Industrial Electronics, **63** (4), 1-1 (2015)
8. D. Li, R. Qu, J. Li, et al., Consequent pole, toroidal winding, outer rotor vernier permanent magnet machines, IEEE Transactions on Industry Applications, 1-1 (2015)
9. F. Magnussen and H. Lendenmann, Parasitic effects in PM machines with concentrated windings, IEEE Transactions on Industry Applications, **43** (5), 1223-1232 (2007)
10. S. Niu, K.T. Chau, J.Z. Jiang, et al., Design and control of a new double-stator cup-rotor permanent-magnet machine for wind power generation, IEEE Transactions on Magnetics, **43** (6), 2501-2503 (2007)
11. R. Qu, D. Li, and J. Wang, Relationship between magnetic gears and vernier machines, International Conference on Electrical Machines and Systems, IEEE, 1-6 (2011)
12. B. Kou, Y. Jin, L. Zhang, et al., Characteristic analysis and control of a hybrid excitation linear eddy current brake, Energies, **8** (7), 7441-7464 (2015)
13. K. Okada, N. Niguchi, and K. Hirata, Analysis of a vernier motor with concentrated windings, IEEE Transactions on Magnetics, **49** (49), 2241-2244 (2013)
14. S.L. Ho, S. Niu, and W.N. Fu, Design and comparison of vernier permanent magnet machines, IEEE Transactions on Magnetics, **47** (10), 3280-3283 (2011)
15. E. Barbisio, F. Fiorillo, and C. Ragusa, Predicting loss in magnetic steels under arbitrary induction waveform and with minor hysteresis loops, IEEE Transactions on Magnetics, **40** (4), 1810-1819 (2004)
16. M. Cheng, K.T. Chau, and C.C. Chan, Design and analysis of a new doubly salient permanent magnet motor, IEEE Transactions on Magnetics, **37** (4), 3012-3020 (2001)
17. Y. Du, M. Cheng, K.T. Chau, et al., Comparison of linear primary permanent magnet vernier machine and linear vernier hybrid machine, IEEE Transactions on Magnetics, **50** (11), 1-4 (2014)

MATEC Web of Conferences **119**, 01004 (2017) DOI: 10.1051/matecconf/201711901004
IMETI 2016

Life cycle guideline of petrochemical plant underground piping system

Jeng-Ywan Shih[1,a], Shen-Ying Hsiao[2] and Ta-Peng Chang[2]

[1]*Department of Chemical Engineering, Ming Chi University of Technology, New Taipei City 24301, Taiwan*
[2]*Department of Civil and Construction Engineering, National Taiwan University of Science and Technology, Taipei City 10607, Taiwan*

Abstract. According to statistics of petrochemical plant disaster, the type of underground pipeline leakage is the highest proportion, for example, Kaohsiung gas explosion in 2014 is a typical case. Therefore, improvement strategy of petrochemical plant underground piping system from both engineering and management becomes an important issue. Through reviewing regulations as well as surveying questionnaire, including kinds of piping materials, 3D drawing files, operation procedures, information sharing, etc., the findings show lack contact of integrated management with engineering executive and insufficient technical requirements are major defects. Overviewing current problems of domestic petrochemical plant underground piping system management, and comparing to international criteria and specifications, this research focuses on the of piping design, construction, operations, maintenance, and inspection. Then management procedures and engineering technical feasibility strategies are suggested. In addition, the proposed life cycle guideline in order to reduce the disaster incidence of petrochemical plant underground pipelines.

1 Introduction

The petrochemical industry deriving oil refining industry, producing basic petrochemical raw materials including olefins, aromatic hydrocarbons, alkanes and so on. Furthermore, serial petrochemical products manufactured [1]. In Taiwan all petrochemical industry output value of about 3.4 trillion, up 12000 vendors, employment of about 28.5 million people [2]. The petrochemical industry is an important pillar of economy; however industrial disaster events often take place, major types mainly in the highest fire, followed by contact with harmful substances and explosion [3]. In 2012 continuously fire and explosion incidents of Mailiao industrial park, resulting in refineries, alkylation plants, and other related factories in full shutdown, although not causing deaths, but has caused a high degree of international media concern [4]. The survey found that the pipeline system problems account for a large proportion in the safety management of the petrochemical plant, the pipeline system has often been ignored, mainly due to the pipeline system through various parts of the whole plant, the management is more complex, and it is more likely to be negligent maintenance of disasters arising from leakage. July 31, 2014 in Kaohsiung gas explosion events have resulted in 32 deaths and 321 people were injured [5], and the loss of property is more difficult to estimate, even more clearly in petrochemical underground pipeline safety is already urgent. In response to the impact

[a] Corresponding author: jengywan@gmail.com

of Kaohsiung gas explosion incident, the central and local governments to proceed with regulatory amendments, For example, Ministry of Economic Affairs in December, 2014 promulgated the "Underground pipelines Industrial Safety Management Reference Guide", July, 2015 Kaohsiung City Government set "Kaohsiung pipeline management and maintenance of existing industrial approach", etc. [6].

Petrochemical industry production of large-scale, operating more hazardous or harmful substances, and close production relationships between the various factories, in order to facilitate the storage and transportation of raw materials, mostly concentrated in the same area, with the process engineering control measures and the implementation of safety management procedures education and training, disaster prevention, prevention of accidents occurred outside, and industrial safety audit is focused on the monitoring and inspection process itself [7]. According to Chang's study, the proportion of petrochemical plant equipment damage to the pipeline system of the type accounted for 33% of the highest item, which is the main reason for corrosion of the pipeline [8]. For the petrochemical plant capable of safe operation, establishing proper pipeline system management system is a top priority should be the design, procurement, construction and operation of the petrochemical plant underground pipeline management mechanism and the corrosion protection and other engineering technology as key elements. From the various stages of the life cycle of underground pipelines petrochemical plant, selecting the appropriate underground pipeline management system and engineering technology, avoiding catastrophic event occurred in the past, making the total petrochemical plants underground piping system for safe use.

2 Operation guideline and expert interview

This study investigates the feasibility of establishing life cycle model of underground pipeline management system of technology and engineering, with reference to foreign petrochemical plant with safety criteria and domestic relevant regulations, drawing up underground pipelines petrochemical plant planning and design, construction, maintenance and inspection guideline, then conducting interviews with experts in industry, government and science, according to the draft amendment after recommendation and comparing Kaohsiung gas explosion incident crux, explaining the problem and proposing improvement strategy.

2.1 Outline of guideline

Outline of guideline is for petrochemical plant underground pipelines life cycle operation including plan, design, construction, maintenance, and inspection, the flowchart is shown in figure 1.

2.1.1 Plan and design

(1) Plan and design using Smart Plant (SP) software contains SP 3D (Three Dimension), SP PID (Piping and Instrument Diagram), SP I (Instrumentation) and SP M (Material) [9]. Smart Plant 3D graphics in 30%, 60%, 90% need to review SP 3D and SP PID correctness with the owners, and also in conjunction with the construction personnel audit constructability and maintainability. In addition, using geographic information system (GIS) to verify stratification and positioning of underground pipelines.
(2) Process flow diagram (PFD) is provided by licensor, then completing SP PID and SP I system in sequence, after considering safety, economy, operation, maintenance, constructability, environmental protection, and other factors, finishing the whole plant layout.
(3) While engineers perform design, calculation and graphics, complying with contract and specifications.
(4) Design calculations and analytical work are in accordance with the relevant design standards or comply with regulatory requirements. The use of software for design calculations should be tested

and verified, using the qualified and legitimate software, and confirming that the adequate scope, content and limitations. Pipeline design calculations must be made stress calculations for review.

(5) Matching list of the pipeline and pipe fittings must be in accordance with the material, shape, size, connection, and other characteristics, especially welding, thread and flange engaging.

(6) According to API 580 risk-based inspection (RBI) [10], establishing a risk-based inspection program, selecting the location and the amount of inspection, in order to facilitate pipeline thickness measurement.

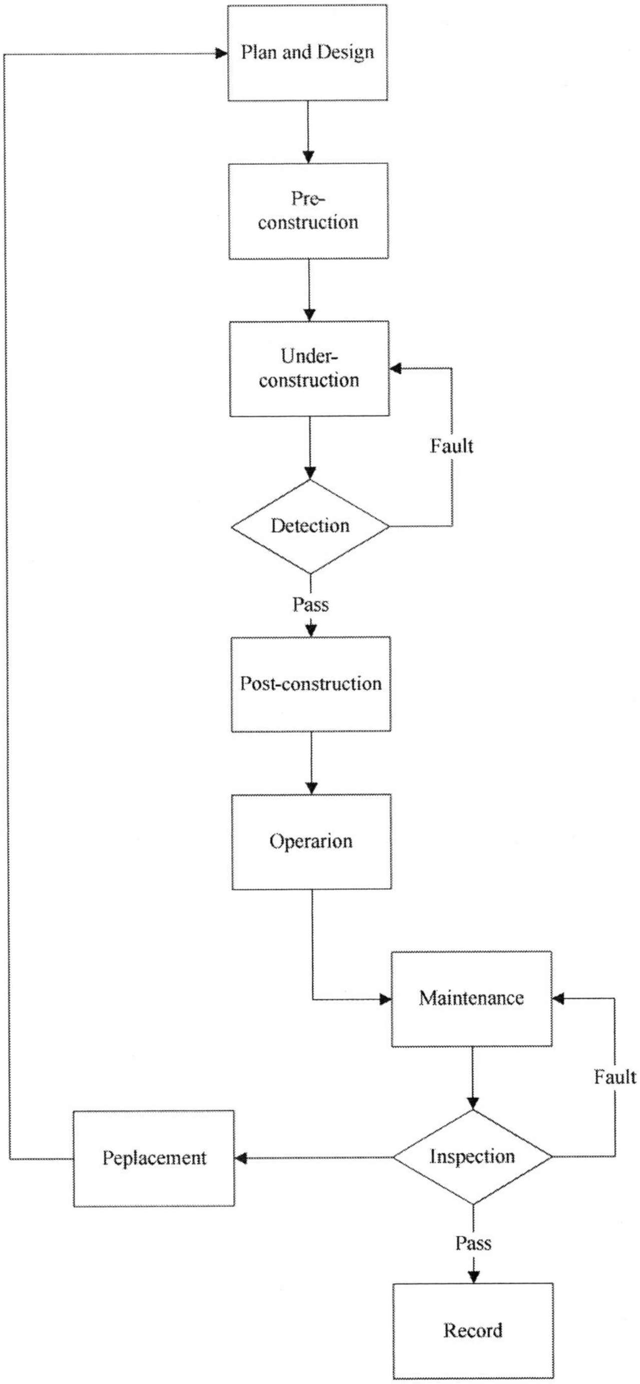

Figure 1. Flowchart of petrochemical plant underground pipelines life cycle operation process [6, 7]

2.1.2 Construction

(1) Applying flare and operation permissions before construction, according to constructability-oriented planning, arranging series schedule, reviewing the status of the pipeline material as well as related document.
(2) Underground Pipeline Construction jobs contain pipe and pipe fittings blasting, painting, coating, safety measures, trench excavation, prefabricated pipeline, pipeline lifting and welding, pipeline pressure test, establishing thickness database, cathodic protection, backfill and compaction.
(3) Underground pipeline construction subsequent job includes site clearance, built drawings, checklist, not only the implementation of self-inspections, but also in conjunction with the owners to examine and record full data, as a basis for acceptance.

2.1.3 Maintenance

Maintenance plan contains definitions exam type, such as an inner tube, an outer tube, operation and inspection; defined for each type of inspection and the execution time period; describing the inspection methods used for each type of inspection and non-destructive inspection techniques; description of test the location and status of non-destructive testing point condition; for the past inspection records describing those positions in the past there have been chipping record; taking once described the type of harm, to predict where may be damage in the future; describing the pipeline system possible type of defect, damage and injury rate acceptable standard; for a variety of types of defect, an appropriate description of non-destructive testing methods and testing capabilities; in accordance with relevant regulations and standards defined the maximum permissible during inspection; extending inspection requirements; according to the type of fluid, environmental damage and potential damage inspection adjustment period; description of requirement section of the pipe to check entry requirements and special tools.
(1) Operation requires Pay constantly attention to flow pressure of input and output ends output end, building a cloud platform for monitoring real-time information, if the abnormal situation happens, then taking emergency shutdown to avoid the damage.
(2) In order to obtain precise position and inspection of underground pipelines, so it had better use SP 3D and GIS for maintenance
(3) According to API 570 [11] of the pipeline system, performing thickness measurement and visual inspection, and the inspection cycle as shown in Table 1. Since climate of Taiwan is humid and rainy, considering environmental factor coefficient (0.5-0.9) to adjust the inspection cycle, i.e. the least duration established by multiplying the environmental factor coefficients, reducing maintenance frequency, and increasing safety.

Table 1. Inspection cycle [11]

Pipeline System	Thickness measurement	Visual inspection	Environmental factor coefficient
Type 1	5 years	5 years	0.5-0.9
Type 2	10 years	5 years	0.5-0.9
Type 3	10 years	10 years	0.5-0.9
Type 4	In accordance with the actual needs	In accordance with the actual needs	-

Notes:
Type 1: A significant impact on the safety and environmental leakage, such as C2, C3 and C4 line.
Type 2: Other categories not included in the pipeline, such as hydrogen, fuel and gas pipelines.
Type 3: It is not located in densely populated areas, such as pipeline of tank storage area.
Type 4: Non-flammable and non-toxic pipeline, such as steam, nitrogen, lubricants and water lines.

2.1.4 Inspection

Inspection contains pre-operation, inspection operation, inspection mode, etc. Items of cautions includes the relatively inaccessible underground pipeline set up radio frequency identification (RFID), confirming indeed visit of inspectors; using cameras equipped with GPS positioning at each checkpoint photos as evidence; for the same underground pipeline systems, each time different inspectors to avoid being deceived; every maintenance record must include the maintenance of underground pipeline photos; supervisors in charge of maintenance records shall sign off, if necessary checking part of the pipeline; the underground pipeline maintenance and replacement comply with contractual requirements.

2.2 Issues of expert interview

Table 2. Issues of expert interview for life cycle of petrochemical plant underground pipelines [9, 11]

Life cycle stages	Issues
Plan and design	Feasibility of underground pipelines unified design software using Smart Plant with SP 3D.
	SP 3D graphics in 30%, 60%, 90% period required in conjunction with the owners, reviewing the correctness, implementing constructability, checking the adequacy and maintainability.
	Domestic regulations and standards are suitable for petrochemical plant underground pipelines.
	For anti-corrosion coating type hot-melt, heat-resistant and cold-inclusive, which is the most suitable for Taiwan environment?
	Are there deficiencies for matching list contents of pipelines and pipe fittings?
Construction	Applying government permit before construction.
	The constructability oriented planning as well as scheduled construction sequence within the plan stage shortens the construction duration and obtain better quality.
	For pipes and pipe fittings material before installation must first blasting, using double layers of epoxy primer painting, each a minimum of 35μm.
	Avoiding trench because rainwater from entering the tube section, the use of impermeable plastic cover and tape to seal the nozzle orifice or at both ends of the use of temporary closure welding.
	After completion of the construction site, the real size of the pipeline immediately records in finished drawings, and updating SP 3D graphics and GIS system.
	Underground pipeline construction checklist is appropriate.
Maintenance	Notice on operational flow pressures of input and output ends, with cloud platform both sides can monitor real-time information each other.
	Considering Environmental factor coefficient from 0.5 to 0.9 for inspection cycle.
	Forcing owners and users of petrochemical pipelines to register.
Inspection	Detection methods are adequate for petrochemical plant underground pipelines.
	Close interval potential detection executes proceeds every five years.
	Mandatory provisions of API 570-check / change / repair / deprecated standards for petrochemical industry.
	Setting up RFID for the relatively inaccessible underground pipelines to confirm indeed visit of inspectors, using cameras with GPS to position each checkpoint, and introducing remote monitoring system to collect and control all information.

In accordance with petrochemical plants underground pipeline plan and design, construction, maintenance, and inspection, issues of expert interviews to develop the subject as shown in Table 2. Expert interviews via various petrochemical industry, related engineering companies, academia, government authorities and professional petrochemical plants underground pipeline inspection companies, namely 17 recommendations after finishing analysis, including plan and design 6, construction 2, maintenance 4 and inspection 5, amending life cycle guideline according to the proposed.

3 Crux of the problem and improvement strategy

Table 3. Crux of the problem and improvement strategy for life cycle of petrochemical plant underground pipelines operation process [9, 10]

Life cycle stages	Crux of the problem	Kaohsiung gas explosion incident	Improvement strategy
Plan and design	Domestic regulations for underground pipeline plan and design, providing 3D graphics and GIS system are not necessary. So, inconsistencies arising out of map data happen.	Serious problems of graphic management and lack of 3D information of underground pipeline.	Smart Plant as the plan and design software to facilitate unified graphical format and easy to integrate related industry. Also the use of GIS system with Smart Plant software, obtaining accurate position of underground pipeline. Under the policy of specific petrochemical zone, building new underground pipeline system as well as migrating existing underground pipeline to a common trench.
Construction	Domestic regulations for underground pipeline construction without a unified standard. Besides introducing international criteria, it had better to establish domestic standards.	At gas explosion site, the 4-inch propylene pipeline was enveloped inside the drainage culvert, not only the patch was found, but also any signs of corrosion prevention layers.	Enforcing structure and material inspection. Reviewing the construction and quality report. While underground pipeline installation, using SP 3D graphics, GIS system, and full supervision.
Maintenance	Maintenance of underground pipelines inside plant is easier than those outside plant. Due to the extensive area, it is difficult to audit. Furthermore, it rarely excavating to maintain, because applying permission.	After completion of the pipeline, implementing close interval potential detection twice, first in 1991, second in 1997. Propylene line belongs to type 1 pipeline, measuring the thickness and making visual inspection every five years. Furthermore, lack of input and output data base.	Establishing a cloud platform of operation, showing real-time information instantly. Preparation of maintenance plans, and according to maintenance records, checking with RFID in order to avoid missing. The use of ultrasonic gauge, measuring the thickness, and then calculating the remaining life of the pipeline.
Inspection	Kaohsiung road excavation management regulations and petroleum management law are no express provision.	The completion of the pipeline in 1991, executing close interval potential detection twice. If only local inspection, it is difficult to detect.	Firstly, applying SP 3D and GIS system to verify the correct position, and using smart pig to check the inner and outer thickness. Finally, performing specific point thickness inspection in accordance with API 580.

Petrochemical plant operation problems are presented in life cycle stages of plan, design, construction; maintenance and inspection, with corresponding to Kaohsiung gas explosion incident, finally proposed improvement strategy are shown in Table 3.

4 Conclusions

(1) Life cycle guideline is practical for eliminating the disaster incidence of petrochemical plant underground pipelines.
(2) At the period 30%, 60%, and 90% is the appropriate timing for the overall project, together with owners to review the correctness of 3D graphics.
(3) The constructability oriented planning as well as construction sequence schedule during plan stage not only shorten the construction period, but also obtaining better quality.
(4) All time pay attention to the flow pressure at input and output end on the operation, establishing a cloud platform for data base.
(5) For relatively inaccessible underground pipeline can be set up RFID, using cameras equipped with GPS positioning at each checkpoint photos as evidence to confirm the pipeline inspection, and introducing remote monitoring system for collecting and controlling all information.
(6) Close interval potential detection and cathodic protection belong to a wide range of protection; proceed every 5 years for low-risk pipeline, and 3 years for high-risk pipeline.

References

1. C.F. Lai and T.H. Chang, Win situation of environmental protection and economy, Science development, **455**, 28-35 (2010)
2. Commerce, Industry and Services Census Report, Directorate General of Budget, Accounting and Statistics, Executive Yuan, ROC, 22-50 (2013)
3. C.W. Cheng and H.C. Yao, A study of occupational disaster survey and prevention in petrochemical industry of Taiwan, Industrial safety and health, **293**, 43-63 (2013)
4. Z.S. Kao, C.Z. Chen, and C.M. Hsu, A study of hazard prevention for contractual operations in the petrochemical industry with one industrial park as case study, Journal of occupational safety and health, **19** (2), 319-329 (2011)
5. Disaster Management White Paper, Executive Yuan ROC, 40-45 (2015)
6. M.C. Dan, C.S. Li, and W.S. Li, The current status of underground pipeline safety management in Taiwan, Civil and Hydraulic Engineering, **43**, 39-46 (2016)
7. C.M. Chang and C.T. Chen, The study of the grading and inspection system for safety management in petrochemical industry, Institute of Labor, Occupational Safety and Health, 5-8 (2015)
8. R.R. Chang, Study for remaining life estimation of localized and high temperature corrosion and inspection strategy for piping of refinery plant, Dissertation, National Yunlin University of Science and Technology (2004)
9. http://www.intergraph.com/products/ppm/smartplant/
10. API 580, Risk-Based Inspection, Second Edition, American Petroleum Institute (2009)
11. API 570, Piping Inspection Code In-service Inspection, Rating, Repair, and Alteration of Piping System, Third Edition, American Petroleum Institute (2009)

 DOI: 10.1051/matecconf/201711901005

Memory allocation and computations for Laplace's equation of 3-D arbitrary boundary problems

Tswn-Syau Tsay[a]

Associate Professor, Asia University, Taichung 41354, Taiwan

Abstract. Computation iteration schemes and memory allocation technique for finite difference method were presented in this paper. The transformed form of a groundwater flow problem in the generalized curvilinear coordinates was taken to be the illustrating example and a 3-dimensional second order accurate 19-point scheme was presented. Traditional element-by-element methods (e.g. SOR) are preferred since it is simple and memory efficient but time consuming in computation. For efficient memory allocation, an index method was presented to store the sparse non-symmetric matrix of the problem. For computations, conjugate-gradient-like methods were reported to be computationally efficient. Among them, using incomplete Choleski decomposition as preconditioner was reported to be good method for iteration convergence. In general, the developed index method in this paper has the following advantages: (1) adaptable to various governing and boundary conditions, (2) flexible for higher order approximation, (3) independence of problem dimension, (4) efficient for complex problems when global matrix is not symmetric, (5) convenience for general sparse matrices, (6) computationally efficient in the most time consuming procedure of matrix multiplication, and (7) applicable to any developed matrix solver.

1 Introduction

Computation iteration schemes of the transformed form of a groundwater flow problem in generalized curvilinear coordinates are studied in this paper. The groundwater flow governing equation [1] der ived from Darcy's Law and equation of continuity is

$$\nabla\cdot(\underline{\underline{K}}\cdot\nabla\phi)=0 \tag{1}$$

where ϕ = piezometric head and $\underline{\underline{K}}$ = hydraulic conductivity tensor. As the free surface of the unconfined aquifer is especially interested in this research, the governing is transformed in the computational domain for a more accurate solution. Hence, the governing equation becomes [2]

[a] Corresponding author : tswnsyau@gmail.com

$$
\begin{aligned}
&-\sum_k\sum_k\sum_m\sum_n\sum_l\sum_j\sum_i K_{ij}\frac{\partial x_j}{\partial \xi^l}\frac{\partial x_i}{\partial \xi^n}g^{lm}g^{kk}\Gamma^n_{kk}\phi_{\xi^m} \\
&+\sum_k\sum_m\sum_n\sum_l\sum_j\sum_i K_{ij}g^{lm}g^{kn}\left(\frac{\partial x_j}{\partial \xi^l}\frac{\partial x_i}{\partial \xi^n}\right)_{\xi^k}\phi_{\xi^m} \\
&+\sum_k\sum_m\sum_n\sum_l\sum_j\sum_i K_{ij}\frac{\partial x_j}{\partial \xi^l}\frac{\partial x_i}{\partial \xi^n}g^{kn}\left(g^{lm}\right)_{\xi^k}\phi_{\xi^m} \\
&+\sum_k\sum_m\sum_n\sum_l\sum_j\sum_i K_{ij}\frac{\partial x_j}{\partial \xi^l}\frac{\partial x_i}{\partial \xi^n}g^{kn}g^{lm}\phi_{\xi^m\xi^k}=0.
\end{aligned}
\tag{2}
$$

where and are axes of the generalized curvilinear coordinates, Γ^k_{ij} is the Christoffel tensor, and g^{ij} is contravariant metric tensor. In addition to the complex of the governing equation in the generalized curvilinear coordinate, the free surface is boundary condition is highly non-linear which can not be simplified except small free gradient[1].

The numerical solution of this governing partial differential equation, subject to the initial and boundary conditions, can be solved by dividing the domain into a collection of points or elemental cells, discretizing the equation to a set of algebraic equations on this collection, and solving the resulting linear system of equations. Traditional element-by-element methods (e.g. SOR) are preferred since it is simple and memory efficient but time consuming in computation. In recent years, conjugate-gradient-like methods were reported to be computationally efficient [3, 4]. Among them, using incomplete Choleski decomposition as preconditioner was reported to be a much better method for iteration convergence [5]. Conjugate gradient like methods were applicable to many fields[6-9]. In this paper, an index method about memory allocation was developed and applied to a 3-dimensional arbitrary boundary problem. Some numerical experiments had been accomplished for ground water mounding problem using conjugate-gradient-like methods in the generalized curvilinear coordinates.

2 Methods and materials

2.1 The governing equation and boundary condition

The experimental law of Darcy can be expressed as

$$\underline{q}=-\underline{\underline{K}}\cdot\nabla\phi \tag{3}$$

where $\underline{q}$ is the specific flux vector. For impressible fluid

$$\nabla\cdot\underline{q}=0 \tag{4}$$

The governing equation (Eq. (1)) results from substituting Darcy's law into the equation of continuity.The free surface boundary condition of groundwater includes (1) zero pressure condition and (2) prescribed flux condition. It is assumed that the free surface separates a 100% saturation from a region of zero saturation. Applying the zero pressure condition on the free surface to the definition of the piezometric head gives

$$\phi=z \tag{5}$$

which shows that the piezometric head on the free surface is equal to the elevation head. The unsteady free surface boundary condition is

$$\frac{\partial F}{\partial t}+\left(\frac{q-N}{n_e}\right)\cdot\nabla F=0 \tag{6}$$

where $\underline{N}$ is the vertical recharge rate, n_e is the effective porosity and F is the free surface function.

2.2 The generalized curvilinear coordinates

Transforming the problem and solving it in a regular domain is a technique that has been widely used to solve problems with complex shapes of boundaries such as ducts, engine intakes, and complete aircraft or automobiles. The procedure involves generating a grid in the physical domain, then transforming the physical domain (Cartesian coordinate) into the computational domain (generalized curvilinear coordinate), and finally solving the problem in the computational domain. Grid generation for the physical domain is a key step to obtaining accurate, reasonable solutions. In the development of aerospace engineering, conformal mapping has played an important role. The basic idea of conformal mapping is to transform an arbitrary contour into a circle using analytical functions so that problems of potential flow can be solved; this method utilizes the invariance of solutions to Laplace's equation mapped by analytic functions. This technique is suitable and efficient; yet applications to fluid dynamics were rare before 1970's because of the computational difficulties. Conformal mapping was also applied in the field of ground water mechanics. Many closed-form solutions for simplified ground water flow problems were developed before computers were widely used [10]. The field of computational fluid dynamics has rapidly developed with the use of computers. Since the computational difficulties can be overcome, the equations of motion can be solved numerically considering pressure, viscous shear and compressibility. To undertake accurate calculations, a well-organized computational grid is necessary. In the early 1970's, the theory of numerically-generated, boundary-conforming coordinate systems was developed in order to meet the need for a well-organized grids of arbitrarily-shaped geometry.

For the finite difference method, the imposition of boundary conditions on a domain of complex geometry requires a complicated interpolation of the data on local grid lines [11] and typically, a local loss of accuracy in the computational solution. Such difficulties motivated the introduction of a mapping or transformation from the physical domain (x, y, z) (rectangular coordinate domain) to a computational domain (ξ, η, ζ) (generalized coordinate domain). A distorted region in the physical domain is mapped into a rectangular region in generalized coordinate space (Figure 1).

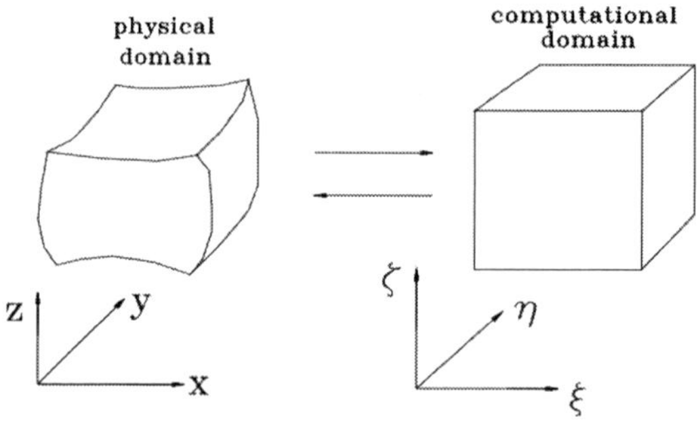

Figure 1. Correspondence of the physical and generalized coordinate domain[12].

It is assumed that there is a unique, single-valued relation between the physical domain and the generalized coordinate domain; for a three-dimensional problem, these relations are 3

$$\xi=\xi(x,y,z),\ \eta=\eta(x,y,z) \text{ and } \zeta=\zeta(x,y,z) \tag{7}$$

with the inverse relations

$$x = x(\xi,\eta,\zeta),\ y = y(\xi,\eta,\zeta) \text{ and } z = z(\xi,\eta,\zeta) \tag{8}$$

The relation between the physical domain and the computational domain can be expressed by the well-known Jacobian matrix, $\underline{\underline{J}}$, and its inverse, $\underline{\underline{J}}^{-1}$, defined as

$$\underline{\underline{J}} = \begin{bmatrix} \frac{\partial\xi}{\partial x} & \frac{\partial\xi}{\partial y} & \frac{\partial\xi}{\partial z} \\ \frac{\partial\eta}{\partial x} & \frac{\partial\eta}{\partial y} & \frac{\partial\eta}{\partial z} \\ \frac{\partial\zeta}{\partial x} & \frac{\partial\zeta}{\partial y} & \frac{\partial\zeta}{\partial z} \end{bmatrix} \tag{9}$$

and

$$\underline{\underline{J}}^{-1} = \begin{bmatrix} \frac{\partial x}{\partial\xi} & \frac{\partial x}{\partial\eta} & \frac{\partial x}{\partial\zeta} \\ \frac{\partial y}{\partial\xi} & \frac{\partial y}{\partial\eta} & \frac{\partial y}{\partial\zeta} \\ \frac{\partial z}{\partial\xi} & \frac{\partial z}{\partial\eta} & \frac{\partial z}{\partial\zeta} \end{bmatrix} \tag{10}$$

The governing equation transform is performed by the relations between physical domain and the generalized curvilinear domain [12-14]. In tensor notation, a point in Cartesian coordinates is given by $x_i(x_1, x_2, x_3)$ and in generalized curvilinear coordinates is given by $\xi^i(\xi^1, \xi^2, \xi^3)$.

The directed line segment $\underline{dr}$, joining two points an incremental distance apart (Figure 2), is written

$$\underline{dr} = \sum_i \underline{g}_i d\xi^i \tag{11}$$

which defines the base vector , $\underline{g}_i$, in curvilinear coordinates.

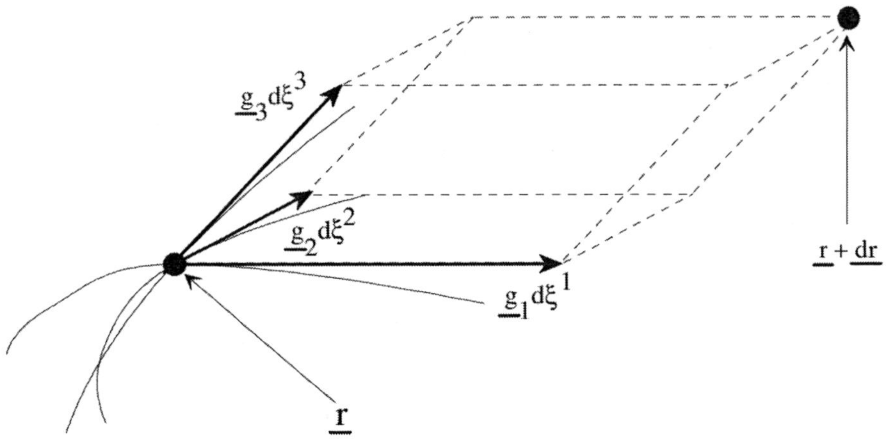

Figure 2. Base vectors, $\underline{g}_i$ in generalized curvilinear coordinates[12].

Thus

$$\underline{g}_i = \frac{\partial}{\partial \xi^i}\underline{r} = \sum_j \frac{\partial x_j}{\partial \xi^i}\frac{\partial}{\partial x_j}\underline{r} = \sum_j \frac{\partial x_j}{\partial \xi^i}\underline{\delta}_j \tag{12}$$

where $\underline{g}_i$ and $\underline{\delta}_j$ are covariant base vectors in curvilinear and Cartesian coordinates, respectively. $\underline{g}_i$ can be written as a linear combination of $\underline{\delta}_j$, are tangent to the coordinate curves but are not of unit length. As the covariant base vectors are not necessarily orthogonal in generalized curvilinear coordinates, contravariant base vectors, $\underline{g}^i$, are introduced by

$$\underline{g}^i = \nabla \xi^i = \sum_j \frac{\partial \xi^i}{\partial x_j}\underline{\delta}_j \tag{13}$$

The contravariant vectors, $\underline{g}^i$, are perpendicular to the surface ξ^i = constant. From Eqs. (12) and (13),

$$\underline{g}^i \cdot \underline{g}_j = \delta^i_j \tag{14a}$$

$$\underline{g}_i \cdot \underline{g}^j = \delta_i^j \tag{14b}$$

where δ_i^j is the Kronecker delta, which is 0 if $i \neq j$ and 1 if i = j. The covariant metric tensor is defined by

$$\underline{g}_i \cdot \underline{g}_j = g_{ij} = \sum_{k=1}^{3} \frac{\partial x_k}{\partial \xi^i}\frac{\partial x_k}{\partial \xi^j} \tag{15a}$$

and the contravariant metric tensor is given by

$$\underline{g}^i \cdot \underline{g}^j = g^{ij} \tag{15b}$$

The determinant of g_{ij} is called g and $\sqrt{g} = \underline{g}_1 \cdot (\underline{g}_2 \times \underline{g}_3) = \|\underline{\underline{J}}^{-1}\|$. The cross (vector) product of any two base vectors is

$$\underline{g}_i \times \underline{g}_j = \sum_k \sqrt{g}\,\varepsilon_{ijk}\,\underline{g}^k \tag{16a}$$

$$\underline{g}^i \times \underline{g}^j = \sum_k \frac{1}{\sqrt{g}}\varepsilon_{ijk}\,\underline{g}_k \tag{16b}$$

where $\varepsilon_{ijk} = 1$, if ijk=123,231, 312; $\varepsilon_{ijk} = -1$, if ijk=321,132, 213; and $\varepsilon_{ijk} = 0$, if any two indices are the same. The contravariant metric tensor can be evaluated from the covariant metric tensor as

$$g^{il} = \underline{g}^i \cdot \underline{g}^l = \frac{1}{g}(\underline{g}_j \times \underline{g}_k)\cdot(\underline{g}_m \times \underline{g}_n) = \frac{1}{g}(g_{jm}g_{kn} - g_{jn}g_{km}) \tag{17}$$

DOI: 10.1051/matecconf/201711901005

Applying the relations between Cartesian coordinates and the generalized curvilinear coordinates, the governing equation can be derived as Eq. (2).

The steady state free surface boundary condition can be transformed from Eq. (6) to be

$$-\sum_{m}\sum_{l}\sum_{i}\sum_{j}\sum_{k} K_{ij} F_{x_m} \frac{\partial x_m}{\partial \xi^k} \frac{\partial \xi^l}{\partial x_j} \frac{\partial \xi^k}{\partial x_i} \phi_{\xi^l} - \sum_{i}\sum_{k}\sum_{m} N_i F_{x_m} \frac{\partial x_m}{\partial \xi^k} \frac{\partial \xi^k}{\partial x_i} = 0 \tag{18}$$

This numerical model was verified by Hele-Shaw model and showed good agreements [2].

2.3 Numerical implementation and the sparse matrix

The governing equation for ground water mounding domain is a Laplacian type equation. Assuming that every block of the aquifer is homogeneous, the governing equation in the generalized curvilinear domain is Eq. (2). In two-dimensional problem, the discretization is carried out by a second-order accurate, nine-point scheme (Figure 3a). Information at point (I, J) is approximated by its surrounding eight points (i.e. points (I-1, J-1), (I-1, J), ... , and (I+1, J+1)). For a three- dimensional problem, a second-order accurate scheme is used as well. Approximate information at (I, J) is from its surrounding 18 points (Figure 3b).

According to the governing equation and boundary condition of the numerical model stated above, a set of linear system of equations can be found

$$Ax = b \tag{19}$$

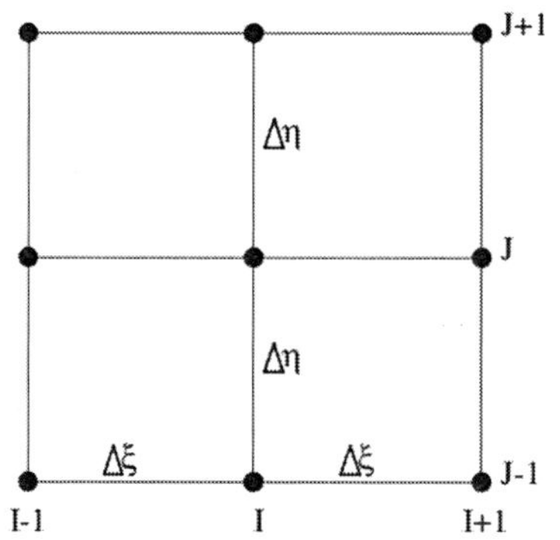

Figure 3a. Two-dimensional: second order accurate nine-point scheme.

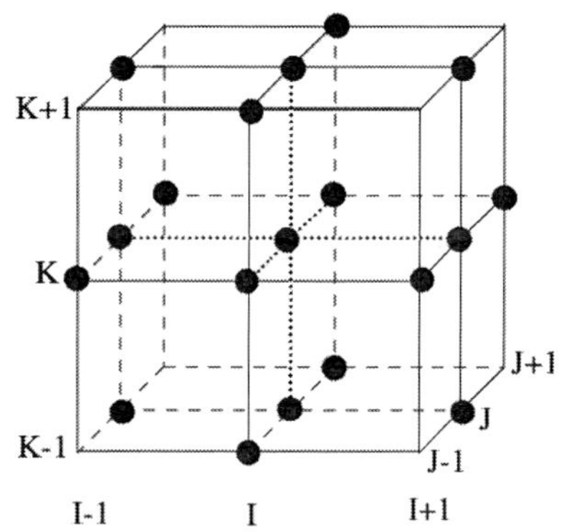

Figure 3b. Three-dimensional: second order accurate 19-point scheme.

For a 5 by 4 grid problem after transformed in the generalized curvilinear coordinates (Figure 4a), matrix A (Figure 4b) is apparently to be sparse and non-symmetric if the problem domain contains two types of soil property. The soil properties are different at either side of node 3, 8, 13, and 18. In the following matrix A, c denotes value from constant head boundary condition, g denotes value from governing equation, b denotes value from no flow or free surface boundary condition, and h denotes value from flow through heterogeneity condition. The numerical implementation flow chart is described in Figure 5. Systems of linear equations of "guessed free surface" can be solved by conjugate-gradient-like methods and its extensions, which were proposed in this research.

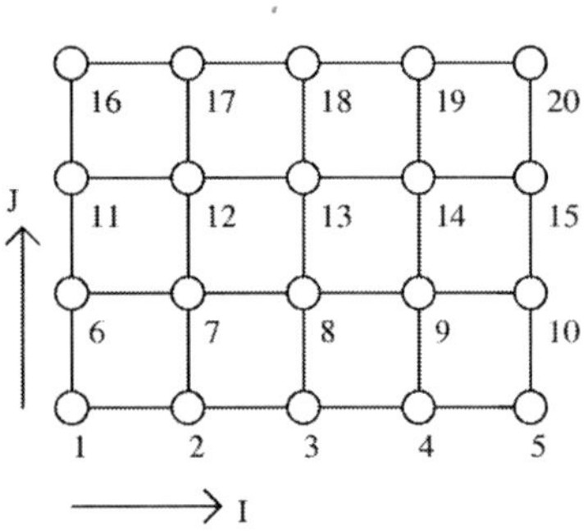

Figure 4a. A uniform 5 by 4 grid domain in generalized curvilinear coordinates.

$$
\begin{bmatrix}
c&0&0&0&0&0&0&0&0&0&0&0&0&0&0&0&0&0&0&0\\
b&b&b& & & &b& & & & &b& & & & & & & &0\\
 &b&b&b& & & &b& & & & &b& & & & & & &0\\
 & &b&b&b& & & &b& & & & &b& & & & & &0\\
 & & & &c& & & & & & & & & & & & & & &0\\
 & & & & &c& & & & & & & & & & & & & &0\\
g&g&g& & &g&g&g& & & &g&g&g& & & & & &0\\
 & &h& & &h&h&h&h&h& & &h& & & & & & &0\\
 & &g&g&g& & &g&g&g& & &g&g&g& & & & &0\\
 & & & & & & & & &c& & & & & & & & &0&0\\
 & & & & & & & & & &c& & & & & & & &0&0\\
 & & & & &g&g&g& & &g&g&g& & &g&g&g&0&0\\
 & & & & & & &h& & &h&h&h&h&h& & &h&0&0\\
 & & & & & & &g&g&g& & &g&g&g& & &g&g&g\\
 & & & & & & & & & & & & & &c& & & & & \\
 & & & & & & & & & & & & & & &c& & & & \\
 & & & & & &b& & & & &b& & & &b&b&b& & \\
 & & & & & & &b& & & & &b& & & &b&b&b& \\
 & & & & & & & &b& & & & &b& & & &b&b&b\\
 & & & & & & & & & & & & & & & & & & &c
\end{bmatrix}
$$

Figure 4b. Matrix A of a 5 by 4 grid domain in generalized curvilinear coordinates.

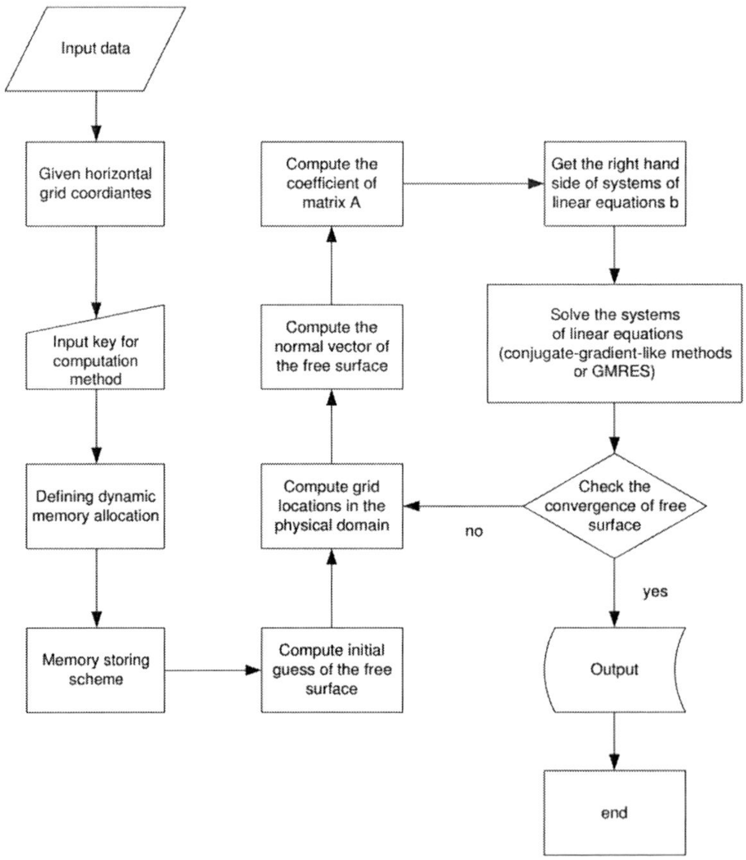

Figure 5. Computational flow chart for groundwater free surface modelling.

2.4 The conjugate-gradient-like methods

The conjugate-gradient-like methods, which are based upon the method of descent, did not converge satisfactorily even though the computations were accelerated. Therefore, preconditioning was used. Preconditioning solves the system of linear equations, transformed from Eq. (19),

$$\overline{A}\overline{x} = \overline{b} \tag{20}$$

in which the eigenvalues of $\overline{\mathbf{A}}$ are more clustered than those of A [15]. Two preconditioners were used: diagonal scaling and incomplete Choleski decomposition. Diagonal scaling is very simple and easy to program since the matrices in the preconditioning algorithm are divided by the diagonal matrix of A. In the incomplete Choleski decomposition, a modification by lumping the neglected values of the off-diagonal elements to the diagonal elements is also used [16]. The idea of preconditioning is illustrated in the following description. From Eqs. (19) and (20),

$$\overline{A} = B^{-T}B^{-1}A,\ \overline{x} = x, and\ \overline{b} = B^{-T}B^{-1}b \tag{21}$$

Thus

$$BB^{T}\overline{A} = A \tag{22}$$

Let $M = BB^T$, then M^{-1} is the preconditioner. B is usually evaluated from A. For example, $A=BB^T$ for a symmetric matrix. This matrix is useful because $M^{-1}A$ is closer to an identity matrix than A (Strikwerda, 1989). Algorithms for conjugate-gradient-like methods can be found in many places [5, 17, 18].

3 Results

The conjugate-gradient-like methods were used as a pilot study of solving Eqs. (2) and (18). Figure 6 is a sketch of the two-dimensional flow field and boundary conditions for steady recharge and flow through a two-layered aquifer with a horizontal, impervious bottom and finite, constant head lateral boundaries.

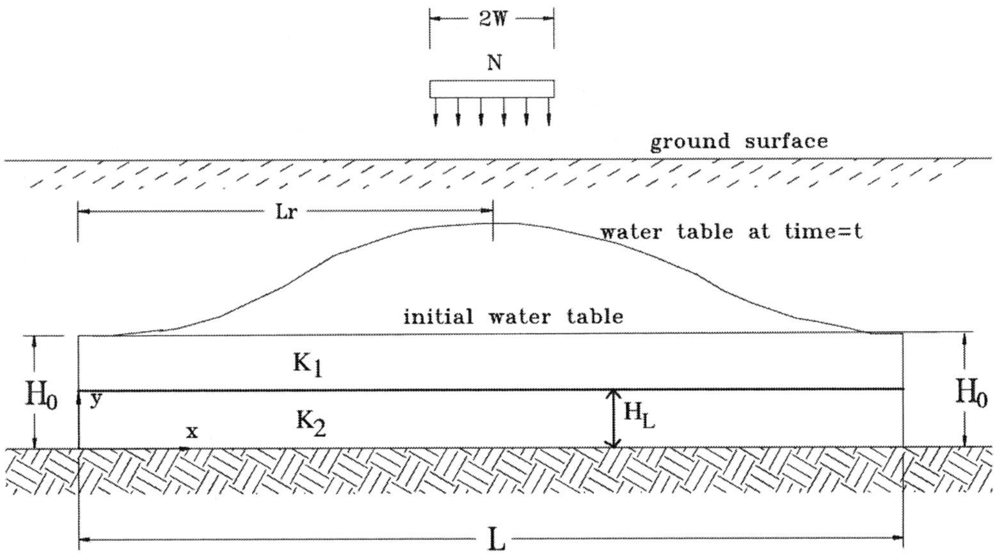

Figure 6. Configuration of ground water mounding model.

The governing equation is $\nabla \cdot (\underline{\underline{K}} \cdot \nabla \phi) = 0$ as was stated previously.

The boundary conditions are:
(1)Dirichlet boundary condition on x = 0 and x = L (constant head);
(2)Neumann boundary condition on the impervious bottom (no flow);
(3)the free surface boundary conditions include
(a) the pressure is zero, and

(b) $$\frac{\partial F}{\partial t} + \frac{q \cdot \nabla F}{\theta} = \begin{cases} \dfrac{\underline{N} \cdot \nabla F}{\theta} & \textit{for area receiving recharge} \\ 0 & \textit{for area not receiving recharge} \end{cases}$$

Assuming the thickness of the lower layer is H_L, the internal boundary conditions on the vertical location of H_L satisfies equivalent pressure head on either aquifer and equivalent flux through this "internal" boundary between the lower and the upper aquifer. The problem was actually solved in the generalized curvilinear domain and the governing equation and boundary should be transformed to be Eqs. (2) and (18)

Computations of the steady state ground water mound location by different methods were compared by the number of iterations for various schemes to solve the system of linear equations to achieve an error less than 10^{-6}. The procedure requires an initial guess for the water table elevation, computation of ϕ in the flow domain, and an iteration scheme.

This example had a recharge width of 10 m with its center located 100 m from the left boundary and steadily supplying water at a rate (N) of 0.001 cm/sec to an aquifer with 40 m initial saturated thickness (H_0) and 200 m long (L). The hydraulic conductivity of the upper aquifer (K_1) was 0.002 cm/sec and lower aquifer (K_2) is 0.003 cm/sec, respectively. The computation grid for the problem is 57 by 57. All the computation is set to be double precision and a convergence criterion of 10^{-6} is used. Four iterative schemes and the sky-line method were compared for the above example:
(1) ICG: preconditioned conjugate gradient by incomplete Choleski decomposition
(2) IGCR(n): preconditioned generalized conjugate residual by incomplete Choleski decomposition with inner loop = n
(3) DGCR(n): preconditioned generalized conjugate residual by diagonal scaling with inner loop = n
(4) SOR: point successive-over-relaxation

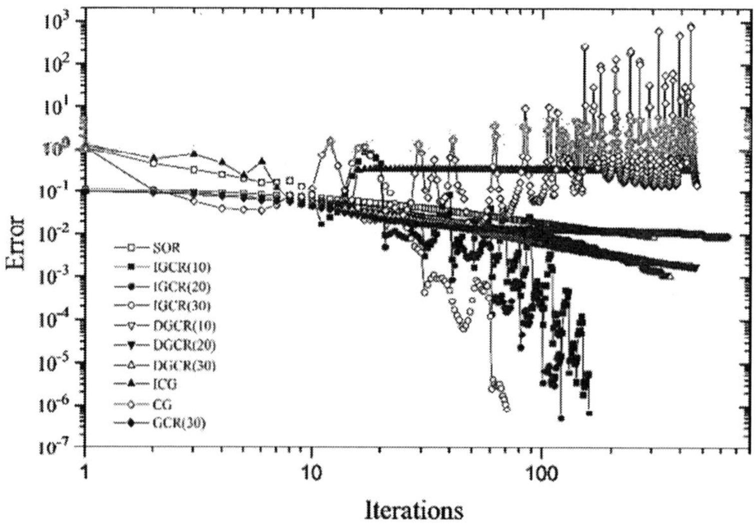

Figure 7. Error vs. iterations for various iterative schemes applied on a 57 by 57 grid

Figure 7 shows number of iterations vs. their error by various iterative schemes for a 57 by 57 grid problem. It is indicated that IGCR(30) is the best method from error convergence. Figure 7 also shown

DOI: 10.1051/matecconf/201711901005

that IGCR(10) and IGCR(20) were able to solve this non-symmetric matrix while other methods were not suitable (i.e., SOR, GCR(30), DGCR(10), DGCR(20) and DRCR(30)) or were divergent (i.e., ICG and CG) as CG and ICG only works for solving systems of linear equations with symmetric coefficient matrices. It is evident that the application of preconditioner reduces the iterations (comparing GCR(n), DGCR(n) and IGCR(n)). In this case, using incomplete Choleski decomposition showed a great improvement.

The conjugate-gradient-like and Sky-Line (SL) methods were used to solve a three-dimensional ground water mounding model. This example used a rate of 0.05 cm/sec recharge on an area $1m^2$ for a homogeneous aquifer (K=0.1 cm/sec) with initial thickness 15 m. The aquifer has horizontal dimensions of 28 m by 28 m (i.e., 784 m^2 area). Three methods are tested by comparing their iterations vs. error. ICG is not valid as shown in Figure 8. For error convergence, IGCR(n) is the best (most efficient) method. Since the three-dimensional global matrix has a very wide band width, there are a large number of zero terms stored when SL is used. LU decomposition of SL is no longer efficient because there are many more operations than for IGCR(n). IGCR(n) displays advantages in computational efficiency and memory efficiency over SL for the three-dimensional finite difference model.

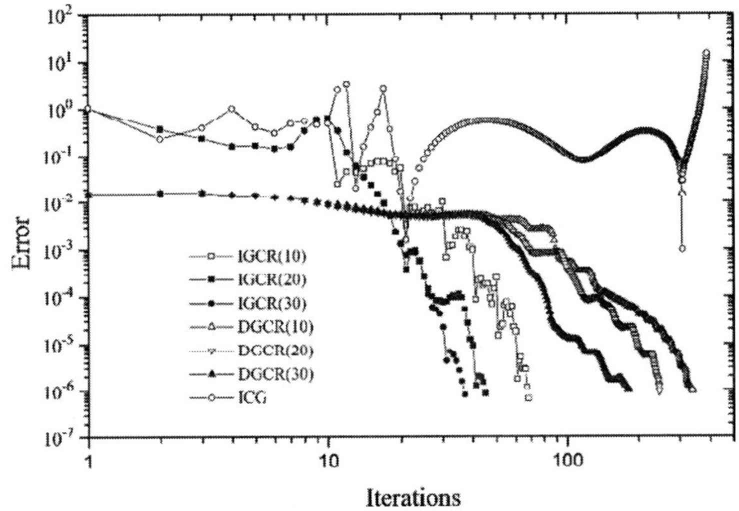

Figure 8. Error vs. iterations for various iterative schemes applied on a 22 by 22 by 22 grid

4 Discussion and conclusion

The 3-dimensional, finite difference model showed that IGCR(n) is the most efficient method comparing to other methods in this paper. Besides, The Index Method is an efficient way to store a sparse symmetric or non-symmetric matrix. Its application to IGCR(n) shows a significant reduction in computations compared to other methods. In general, the index method has the following advantages:

(1) adaptable to various governing and boundary conditions;
(2) flexible for higher order approximation;
(3) independence of problem dimension;
(4) efficient for complex problems when global matrix is not symmetric;
(5) convenience for general sparse matrices;
(6) computationally efficient in the most time consuming procedure of matrix multiplication (since only non-zero terms are stored); and
(7) applicable to any developed matrix solver.

Acknowledgement

The author wishes to express his sincere appreciation to Professor Shen-Haw Ju in National Cheng Kung University, Taiwan for his help on developing the memory allocation method in this research.

References

1. J. Bear, *Dynamics of Fluids in Porous Media, Dover Books on Physics and Chemistry*, New York: Dover, **17**, 764 (1988)
2. T.S. Tsay and J.A. Hoopes, , Numerical simulation of ground water mounding and its verification by Hele-Shaw model, Computers & Geosciences, **24** (10), 979-990 (1998)
3. M. Bercobier and A. Rosenthal, Using the conjugate gradient method with preconditioning for solving FEM approximations of elasticity problems, Engineering Computation, **3**, 77-80 (1986)
4. E. Suetomi and H. Sekimoto, Conjugate gradient like methods and their application to fixed source neutron diffusion problem, Journal of Nuclear Science and Technology, **26** (10), 899-912 (1989)
5. U.R.B. Obeysekare, M.B. Allen, R.E. Ewing, and J.H. George, Application of conjugate-gradient-like methods to a hyperbolic problem in porous-media flow, International Journal for Numerical Methods in Fluids, **7** (6), 551-566 (1987)
6. J. Golebiowski and J. Forenc, Parallel computations of the step response of a floor heater with the use of a graphics processing unit. Part 1: models and algorithms, Bulletin of the Polish Academy of Sciences-Technical Sciences, **61** (4), 943-948 (2013)
7. H.L. Lee, W.L. Chen, W.J. Chang, and Y.C. Yang, Estimation of surface heat flux and temperature distributions in a multilayer tissue based on the hyperbolic model of heat conduction, Computer Methods in Biomechanics and Biomedical Engineering, **18** (14), 1525-1534 (2015)
8. X. Wang, F. Liu, L.C. Jiao, J. Wu, and J. Chen, Incomplete variables truncated conjugate gradient method for signal reconstruction in compressed sensing, Information Sciences, **288**, 387-411, (2014)
9. G. Yuan and M. Zhang, A three-terms Polak–Ribière–Polyak conjugate gradient algorithm for large-scale nonlinear equations, Journal of Computational and Applied Mathematics, **286**, 186-195, (2015)
10. M.E. Haar, *Ground Water and Seepage*, New York: McGraw-Hill Book Co. Inc., 315 (1962)
11. R. Singh, Prediction of mound geometry under recharge basins, Water Resources Research, **12** (4), 775-780 (1976)
12. R.B. Bird, R.C. Armstrong, and O. Hassager, *Dynamics of Polymeric Liquids*, **1**, 2nd Ed.: Fluid Mechanics, New York: John Wiley & Sons, 649 (1987)
13. R. Aris, *Vectors, Tensors, and the Basic Equations of Fluid Mechanics*, Englewood Cliffs, N.J.,: Prentice-Hall, 286 (1962)
14. J.F. Thompson, Z.U.A. Warsi, and C.W. Mastin, *Numerical Grid Generation : Foundations and Applications*, New York: North-Holland : Elsevier Science Pub. Co. [distributor], **15**, 483 (1985)
15. J. Strikwerda, *Finite Difference Schemes and Partial Differential Equations*, Second Edition, Other Titles in Applied Mathematics, Society for Industrial and Applied Mathematics, 439, (2004)
16. T.A. Manteuffel, The shifted incomplete Choleski factorization, Sandia Laboratories, SAND, Livermore Lab., 78-82 (1978)
17. Y. Saad and M.H. Schultz, A generalized residual algorithm for solving non-symmetric linear systems, SIAM Journal on Scientific and Statistical Computing, **7**, 856-869 (1986)
18. F.S.B.F. Oliveira and K. Anastasiou, An efficient computational model for water wave propagation in coastal regions, Applied Ocean Research, **20**, 263-271 (1998)

 DOI: 10.1051/matecconf/201711901006

Customized products and cloud service information system development research

Chien-Wen Hung[a]

Department of Information Management, Chia-Nan University of Pharmacy & Science, Tainan, Taiwan

Abstract. This study presents a cloud service customized product information system to enable businesses to provide customized product marketing on the Internet to meet consumer demand for customized products. The cloud service of the information system development strategic framework proposed in this study contains three elements: (1) e-commerce services, (2) promotion type modules, and (3) cloud services customized promotional products. In this study, a mining cloud information system to detect customer behavior is proposed. The association rules from relational database design are utilized to mine consumer behavior to generate cross-selling proposals for customer products and marketing for a retailing mall in Taiwan. The study is composed of several parts, as follows. A market segment and application of association rules in data exploration techniques (Association Rule Mining) and sequence-like exploration (Sequential Pattern Mining), efficient analysis of customers, consumer behavior, identification of candidates for promotional products, and using cloud service delivery and evaluation of targets to evaluate candidates for promotional products for production. However, in addition to cloud service customized promotional products, the quantity of promotional products sales varies for different customers. We strive to achieve increased customer loyalty and profits through the use of active cloud service customized promotional products.

1 Introduction

In recent years, cloud computing has always been a hot topic in the field of information. Gartner research pointed out that the world's cloud computing in 2015 the output value of 162.1 billion US dollars, compared with 1136 billion US dollars in 2014 increased by 29.9%. Cloud computing is expected in 2016 the number will reach 193.2 billion US dollars. Therefore, in the information industry trends, the cloud computing is the only way to open up the market.

For enterprises, in addition to easy access to data, information technology maintenance costs, such as office software, operating systems, hardware, can be greatly reduced. Therefore, it is possible to reduce the running cost and improve the industrial competitiveness. Cloud computing represents the advent of a new business era characterized by increased productivity and flexibility in business activities. At the same time, cloud computing analytics tools to meet customer needs are a commercially competitive IT tool [16].

When the Internet environment becomes more mature, commercial activities of traditional industries experienced significant change. Many industries compete to invest in the e-commerce

[a] Corresponding author : cwhong@mail.cnu.edu.tw

market by making direct contact with the customer and the enterprise. E-commerce (Electronic Commerce, EC), which is a year-round Internet market without geographical borders restrictions is formed and created a substantial market opportunity. Doing business over the Internet is a new commercial activity, where customers just move the mouse to go online and shop. Given the rapid development of the technology in Internet business transactions, using the current business model, customer transaction behavior, parity, purchase, payment, and others on the Internet can be analyzed faster, more convenient, more efficient, and more economical. According to the Institute for Information Industry survey, internet sales of goods gradually increase owing to lower internet transaction prices compared with those of the physical channel. There is a lot of money-saving information that can be shared, which has resulted in the continued growth of internet trading "e-commerce" type of site visitors. The most popular "e-commerce" type websites include music, travel, shopping, computer 3C, and fashion and beauty websites [5].

To enable customers and enterprises to feel the benefits of network marketing, one solution is to provide customized customer service and efficient means of information technology to collect, analyze, produce, and meet customer needs. Therefore, this study proposes a cloud-based customized customer service information technology to help companies to analyze quickly and accurately the customer's consumption behavior. The proposed system provides results of the analysis, which help generate active promotional products and provide customers with a lower promotional price. The system saves time for customers when buying goods partly because the individual needs are met, increasing the probability of turnover for customers and companies, which creates a win–win situation [14].

In this study, a mining Cloud computing system to detect customer behavior is proposed. The association rules from relational database design are utilized to mine consumer behavior in order to generate cross-selling proposals for a customized products design and marketing for a retailing mall in Taiwan. The other sections of the study are arranged as follows. In Section 2, we describe the Literature review and background of the case firm and summarize the goal of the research project of the case firm. Section 3 introduces the proposed Cloud Information data mining system, which contains a system framework and relational database design. Section 4 introduces the data mining process, result analysis, and customized products design. Discussions and conclusions are presented in Section 5.

2 Literature review

Cloud computing provides the following features [24]:

(1) Cost：Because information infrastructure is usually provided by third-party vendors, there is no need to purchase information devices for one-time or non-intensive information needs.

(2) Virtual information environment：Cloud computing environment used in the basic information technology can share the information server and storage function equipment to improve utilization.

(3) a shared information environment：that allows a large number of user groups to share resources and costs, allowing centralized information infrastructures to increase peak load times and information system utilization

(4) Scalability and flexibility：to provide cloud service resource flexibility, the user can load the peak load time engineering design content

(5) Information devices and storage location independence：support users through a Web browser from anywhere to access cloud services, regardless of their use of equipment or their access to the service network location.

Cloud computing architecture framework is shown in Figure 1.

This architectural framework focuses on providing the concepts and functions that are used in cloud computing. This cloud computing architecture is available from the perspective of IT and Information Security Professionals. The cloud architecture framework includes unique cloud features such as virtualization and cloud services, as well as multiple users deploying the cloud services model. [27]

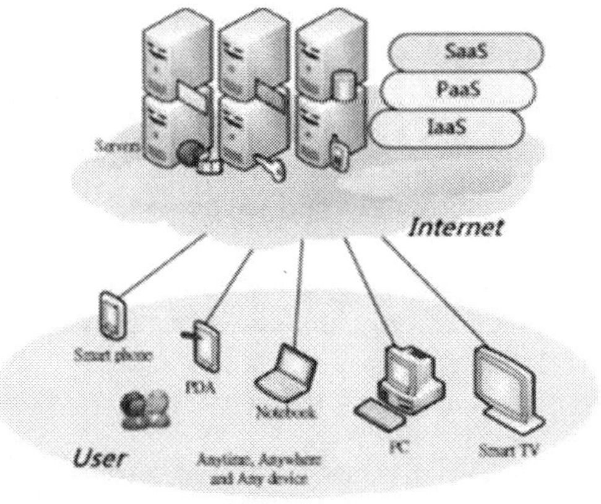

Figure 1. Cloud computing architecture framework [27].

Enterprises in cloud service environment providing customized customer service have become popular. However, quick and proper understanding of customer needs to provide customized services should be a priority of companies. Therefore, the proposed customer service for customized applications and the use of data mining techniques to collect information from a large number of promotional products to meet customer needs will result in proactive customized information to the customer to save time searching for products. The system can enhance the competitiveness of enterprises and increase corporate profits [17].

SaaS is a model that provides software through the Internet. Instead of purchasing software, users can use the Web-based software to manage their business activities without having to maintain the software. The service provider manages and maintains the software, For many small businesses, SaaS is the best way to adopt advanced technology, which eliminates the need for companies to buy, build and maintain infrastructure and applications. In recent years, the rise of SaaS has given traditional packaged software vendors real pressure.

PaaS is a platform as a service (Platform as a Service). Platform as a service is a cloud computing service that provides a computing platform and a solution stack as a service. At the typical level of cloud computing, the platform as a service layer is between software as a service and infrastructure as a service.

Platform-as-a-Service provides the ability for users to deploy and create cloud infrastructure to clients, or to use programming languages, libraries, and services. Users do not need to manage and control the cloud infrastructure, including the network, servers, operating systems or storage, but need to control the upper application deployment and application hosting environment.

PaaS takes the software development platform as a service and delivers it to the user as a software-as-a-service (SaaS) model. Therefore, PaaS is also an application of the SaaS model. However, the emergence of PaaS can accelerate the development of SaaS, especially to accelerate the development of SaaS application speed.

Infrastructure as a Service (IaaS) is the software that consumers use to process, store, network and various basic computing resources, deploy and execute operating systems or applications, and so on. Clients can deploy and run processing, storage, networking, and other basic computing resources at will, without the need to purchase network devices such as servers and software. They cannot control or control the underlying infrastructure, but can control operating systems, storage devices, Deployed applications

This paper is to use PaaS architecture to collect customer consumption data and to analyze the interaction of purchased goods. This study aims to construct a cloud service customized product selection information system to provide a reference to the industry.

This study is composed of the following segments:

(1) Using relevant literature examples, the collection and collation of cloud computing services for customized selection of merchandise subject to a business or organization are discussed to create an application of cloud computing services.
(2) Using relevant literature examples, the selected cloud computing services customized commodity index is collected and collated. Using questionnaire results form experts after the screening, the effect of cloud services customized commodity indexes and the construction of a cloud computing services customized selection of merchandise information system architecture are summarized using factor analysis with association rules and the neural network approach.
(3) Cloud services customized product selection system development and empirical analysis are used to strengthen the plan of the proposed cloud computing services customized merchandise selection of models and analyze the practicality of the system to provide cloud computing services model that allows corporate information.

The cloud services customized catalog is better than the traditional catalog and has two basic benefits [18]:

(1) The cloud services customized catalog offers a wide range of information and possessed more depth content. The amount of information it can provide is far more than the traditional catalog and does not have traditional catalog layout space limitations.
(2) The cloud services customized catalog has better timeliness. The content of the e-catalog can be immediate or frequently updated, providing consumers with the latest information. In contrast, the traditional catalog update takes a year or a few months below the content cycle.

3 Methodology

The customized products framework of the system is shown in Figure. 2,

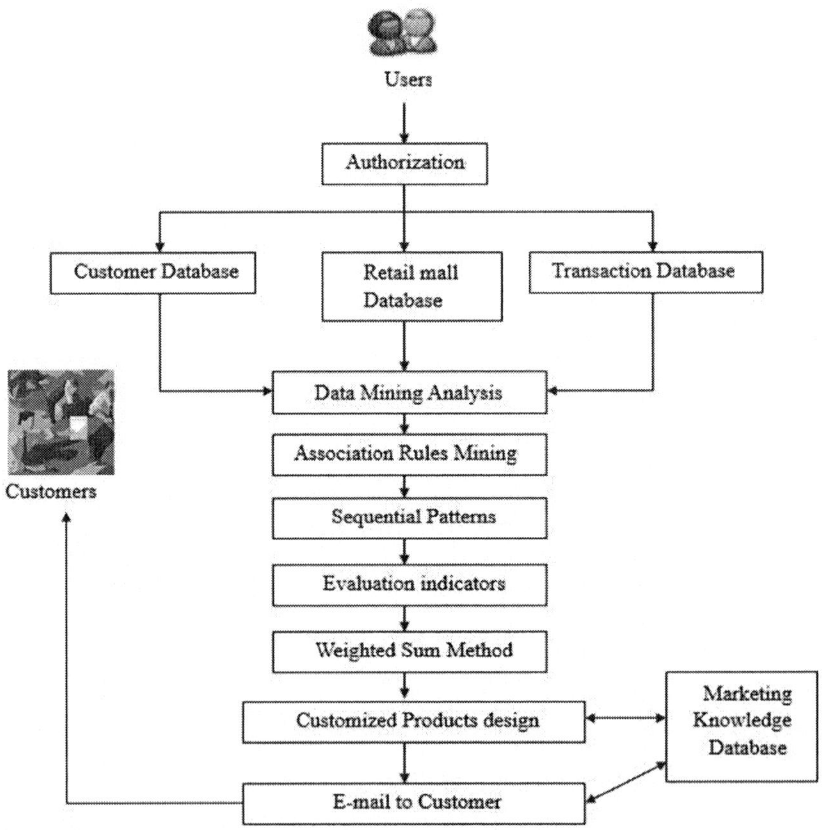

Figure 2. Customized products cloud Information System.

The relational database management system is used to conduct the data mining which consists of four steps:

Step1: Input the system code by certification and open the databases, which consist of the customer database, the retail mall database, and the transaction database.

Step 2: Analyze the data according to the database.

Step 3: Design the customized products according to the analysis result.

Step 4: E-mail the customized products electronic catalog to the relevant branches and to the customers.

3.1 Association rule mining

The association rules are proposed by Agrawal and Srikant (1995). Association rule mining is widely used for analyzing the product items purchased by consumers. It is also used to support sales promotion and marketing segmentation [7-8]. The association rule is represented by X➔Y where X and Y are a set of items. This rule means that the transaction records in a business database that contain X tend to contain Y. A large number of valid algorithms for mining association rules have been proposed [2-4].

In this study, a mining system to detect customer behavior is proposed. The association rules from relational database design are utilized to mine consumer behavior.

In the consumer purchase of computers and memory, for example, the association rules are as follows: Computer ➔ Memory [support = 20%, confidence = 80%]

The formula shows that 20% of the entire transaction database will buy a computer and memory, while 80% of the total customers who purchase computers will buy them together with memory. The following steps are used to determine the association rules:

(1) First, find a collection of high-frequency items (Large Item set). This collection of support must be greater than the user customized minimum support (Minimum Support).

(2) Second, use a collection of items produced by high-frequency generating association rules.

Currently, many algorithms can identify the high-frequency items associated with a collection of rules, such as Apriori [1,18]. Apriori is the most commonly used and best-known algorithm; therefore, its data analysis will be used in this study.

3.2 Sequential patterns

Sequential patterns are proposed by Agrawal and Srikant (1995). The algorithm of the association rules has similar concept, mainly using items to identify the order of occurrence of the information hidden in the data. The same customer at different points in time has many transaction records, which form a sequence. Each sequence set for the project formed an ordered set (Order Set). The sequential patterns are expressed as: A ➔ B; where A and B are arbitrary database collection items, and $A \cap B = \varphi$. The conditions for the establishment of sequential patterns as association rules comprise two kinds of parameter values, namely, support and confidence. Support represents some entries in the database that appear to scale. Reliability is the extent of the credibility of this sequence. Therefore, meaningful sequence patterns are greater than the degree of support and confidence threshold values set by the user.

Consumers have to buy notebooks and color laser printers, for example. The sequential pattern can be expressed as follows: Notebook ➔ color laser printer [support = 15%, confidence = 60%]. The formula shows that 15% of the entire transaction database of customers that buy the notebook will buy a color laser printer after a period of time; the 60% denotes the customers that will buy the notebook and then buy a color laser printer.

The identification of sequential patterns with the same association rules comprises the following two steps:

(1) First, find a collection of high-frequency items. This collection of support must be greater than the user-customized minimum support.

(2) Second, using step (1), the resulting collection of high-frequency items generated sequential patterns.

Currently, many algorithms can identify high-frequency sequential patterns collection of items, such as Apriori [1,12,29], and other algorithms. Apriori is the best known and the most commonly used algorithm.

This study will use its data analysis. The concept of association rules with similar processing steps are as follows:

(1) First, find the database to the customer number (Customer_id) primary key and the transaction time for the second sort key.
(2) Identify all projects that meet the minimum support set in the known collection of high-frequency items. This item set corresponds to a set of consecutive integers for subsequent comparison.
(3) The use of high-frequency items by Apriori algorithm collections. The minimum support to find a sequence, i.e. Large Sequence, must be met.
(4) High-frequency items from the collections are used to find the longest sequence.

3.3 Evaluation indicators

To assess the candidate promotional products, we propose the following three evaluation indicators: profit targets, customer satisfaction index, and turnover rate indicators. All index values must be planned after being dropped to a lesser extent.

(1) Profit targets: According to the sales of goods reached by the proportion of the target content, through the administrator to enter the appropriate average as a target value.
(2) Customer satisfaction indicators: In addition to the considerations outside business perspective, customer relationship management must also be given importance. If only corporate interests are considered, it is likely to cause bias; therefore, the customer's point of view, such as the customer satisfaction index, must also be considered. The customer can have the right not to be satisfied, which is in the target range of −1.0 to 1.0.
(3) Turnover rate indicator: This indicator is used to assess the extent of acceptance of the promotional products. The turnover rate is defined as follows. In the range 0.0 to 1.0, Turnover rate = number of promotional products accepted / Promotion Views.

3.4 Weighted sum method

We use multiple criteria decision analysis methods in the basic weighted sum method to calculate the final assessment score of the promotional products. Findings that show poor sorting and reporting were excluded. Suppose there are m candidate promotional products and n evaluation indicators, each candidate Promotional Products weighted total score is calculated using the following formula.

$$A_{WSM-score} = \max_i \sum_{j=i}^{n} a_{ij} w_j \text{ , } for\ i = 1, 2, \ldots, m, \tag{1}$$

Where a_{ij} is the i-th candidate promotional products, j is the index value of the actual indicators, and w_j is the j-th index value of importance [6].

4 Customized Products Analysis

Profile of the case：The retail mall was established in 2000. The retail mall specializes in Daily life products. It has been a leading provider of personalized daily life products and related products for consumers. In the past decade, the firm has been proud of its products quality and the value expected by customers. The retail mall main customers are female consumers.

From our interviews with the company marketing managers, the customized products design is the fundamental promotional tool for the firm. The customized products is designed by the marketing office based on decisions from department meetings. According to the promotion schedule, approximately three month is required for the marketing office to design and produce customized products. Furthermore, the customized products is redesigned seasonally. For example, more customized products is dedicated to facial care products and decorations products. Hence, at the beginning of the customized products would contain a hundred kinds of facial care products and 60 types of the decorations products. Upon the completion of the customized products catalog, copies of the customized products catalog are delivered to each branch five days before it is released to the public.

The customized products catalog is designed and produced by the marketing office based on aggregated information from head office, but branch office are without authority to make decisions on customized products catalog content. Hence the customized products catalog contents cannot provide a customer-oriented segmentation service for the different branch office needs. Furthermore, the cross selling customized products promotion is still neglected in current promotion policy. In addition to the traditional electronic catalog, they have another promotion mode: making advertisements for electronic magazines and electronic newspapers for the Branch Anniversary.

The data was collected from June 2015 to May 2016. The database of the system consists of three major parts, namely, customer data, product data, and transaction data.

In this study, we propose a cloud service information system prototype using data mining techniques to help enterprises find suitable promotional products for each customer in the cloud customer database. The customized products cloud service information system architecture diagram is shown in Figure 3.

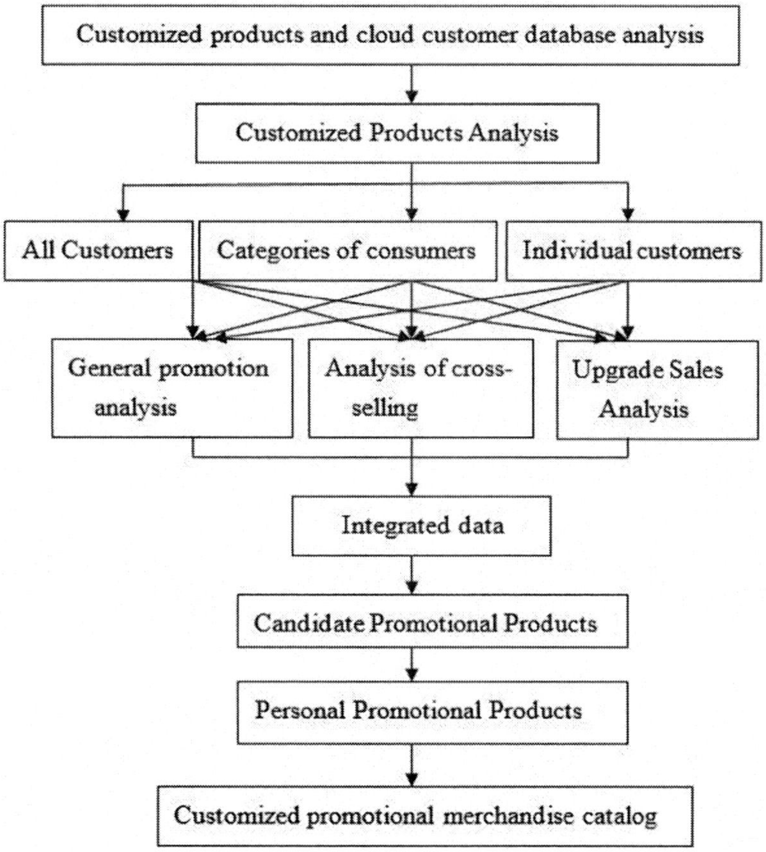

Figure 3. Customized products Cloud Service Information System Architecture Diagram.

Each step is detailed as follows.

Step 1: Create a basic cloud customer database: This step is the main characteristic attribute in the collection of user data. It records all possible influence factors of the customer buying behavior attribute data. In this study, the cloud customer database is composed of two parts, namely, basic information and preference category. C1 to Ci-1 are the basic properties of the cloud customer information, such as gender, education, salary, and others. Ci records the user's preferences for product categories, such as facial cleanser, Shampoo, and so on.

Step 2: Cloud pre-processing of customer databases and program code conversion: This step is mainly for the removal of any customer purchase records in the cloud and the retrieval of the required fields from the cloud customer database. Then, the cloud customer attribute data are matched with the users to do the coding to attribute on the clustering, which is represented as follows:

$$\text{Ci-1} = \{A_1 A_2 \ldots A_K\}, A_k \in \{0,1\},\ 1 \le k \le K$$

$$\text{If } A_h = 1 \text{ then } A_k = 1, \forall k \le h \text{ and } A_k = 0, \forall k > h \circ$$

Step 3: Association rules: In this study, the most commonly used method of association rules is Apriori. This algorithm searches from all cloud customer transaction database to identify products that are purchased together. Meanwhile, this approach is also applied to each customer groups and individual customer transaction data to explore the various customer groups and each individual customer consumption behavior. The transaction data table is shown in Table 1.

Table 1. Transaction data table.

Project ID	Product project collection	Customer ID	Category
1	{P8,P15,P18}	A105001	A
2	{P2,P5,P15,P18}	A105001	A
3	{P3,P8,P9}	A105002	A
4	{Q8,Q9,Q15}	A105002	B
5	{Q3,Q5,Q12}	A105003	B
6	{Q12}	A105005	B
7	{R6,R18}	A105005	C
8	{R2,R6,R10}	A105007	C
9	{Q9,Q17}	A105007	B
10	{Q5,Q10,Q12,Q20}	A105008	B
11	{R8,R20}	A105004	C
12	{R6,R20,R23}	A105004	C
13	{R23}	A105004	C
14	{R3,R6,R15}	A105006	C
15	{R5,R17,R18,R20}	A105006	C

All the customer correlation patterns for the product {P3, P8} and {P15, P18} are obtained, including all customers who tend to purchase product P3 and product P8 P18 P15 and products purchased together. In addition, Table 2 shows that Customer ID belongs customer group A. With the association of ethnic A rule mining, we found that this population of customers often purchase products P15 and P18. Similarly, the application of association rules to individual customers (e.g. Customer ID example) identifies which customers would like to buy products P2 and P5.

Thus, in this example, a pattern of all associated products for three different types of customers (all customers, customer groups, and individual customers) is discovered and shown in Table 2.

Table 2. Three kinds of customer type pattern-associated products.

Customer Type	Pattern associated products
All customers	{P3,P8}{P15,P18}
A customer groups	{P15,P18}
Individual customers (Customer 1)	{P15,P18}

Step 4: Sequential patterns

Sequential patterns can be found after customers buy certain products to purchase certain other products. In this study, the use of sequential patterns from all of the customer's transaction database to identify what customers often order or purchase. Similarly, this method can also be applied to each customer groups and individual customers to explore the various customer groups and the sequence of individual customer buying patterns.

All the product pattern sequences are shown in Table 3. We acquired all of our customer's sequential pattern of the product {P3, P14} and {P8, P14}. As stated previously, Customer A105001 belongs to customer group A. Using a sequence of ethnic pattern mining, we found that this group of customers often purchases products P5 P8.

Table 3. Three kinds of customer type sequential pattern products.

Customer Type	Pattern associated products
All customers	{P3,P14}{P8,P14}
A customer groups	{P8,P5}
Individual customers (Customer 1)	–

Step 5: Evaluation Index of promotional products

All products are calculated with a weighted total score. Product promotion can be determined by fractional order. In addition, warning threshold values (ρ) can also be set to remove a low fraction of products. Table 4 shows the candidate promotional products. For example, assume that the threshold value is 0.3; then, the sub-standard promotional products will be excluded. Finally, the price of product promotion order of {P16}> {P3, P9}>{P4, P8} > {P3, P8}; In addition, the product promotion sequence number is {P5}> {P9}> {P3}. In summary, when the quantity of the purchase reaches the threshold set by the policy, it becomes preferential.

Table 4. Various candidate promotional products index score and weighted total score.

Candidate promotional products		Customer satisfaction index	Turnover rate index	Weighted total score
Price of Product	{P3,P8}	0.48	0.58	0.52
	{P15,P18}	0.46	0.52	0.49
	{P4}	0.43	0.28	0.48
	{P9}	0.18	0.53	0.46
	{P4,P8}	0.52	0.61	0.54
	{P3,P9}	0.63	0.65	0.56
	{P4,P9}	0.32	0.46	0.47
	{P16}	0.83	0.66	0.69
	{P16,P18}	0.45	0.35	0.47
Quantity of Product	{P2}	0.36	0.36	0.46
	{P9}	0.65	0.75	0.67
	{P3}	0.48	0.68	0.63
	{P5}	0.83	0.86	0.68
	{P8}	0.46	0.67	0.48
	{P10}	0.42	0.27	0.46

Step 6: Customized promotional merchandise products

If all of the candidate product promotions are sent to customers, it may cause excessive opposite effect; therefore, not all candidate promotional products are sent to customers. In this system, we use the weighted sum of the three assessment methods and indicators to sort the promotional products, which can be divided into the price of products and the number of products. Figure 4,and Figure 5 shows that policymakers, according to market experience or preference, were given three kinds of indicators with different weights to calculate the sum of the weighted product candidate scores.

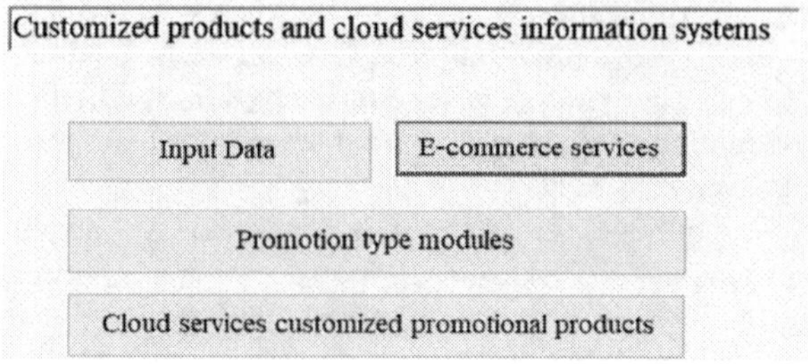

Figure 4. Customized product and Cloud services information systems

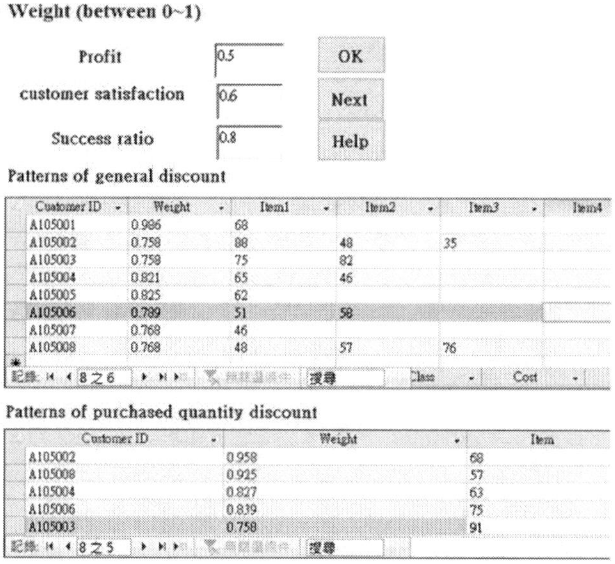

Customer ID	Weight	Item1	Item2	Item3	Item4
A105001	0.986	68			
A105002	0.758	88	48	35	
A105003	0.758	75	82		
A105004	0.821	65	46		
A105005	0.825	62			
A105006	0.789	51	58		
A105007	0.768	46			
A105008	0.768	48	57	76	

Customer ID	Weight	Item
A105002	0.958	68
A105008	0.925	57
A105004	0.827	63
A105006	0.839	75
A105003	0.758	91

Figure 5. Customized commodity product promotions.

Therefore, when these three indicators by policy makers are found, the needs of individual customer's promotional products shall be considered in the current business situation. The main objective of the promotion is to determine the appropriate weight of each index. The system helps enterprises to efficiently deliver the customized commodity product promotion, as well as really meet the business situation and needs.

5 Discussion and Conclusion

The Cloud customer database developed in this research serves as the information system. After importing the benefits that can be found in a number of products imported, we found that the number is high, which is due to the adoption of a strategic customized services. Product customers must first purchase a lot or a combination. Therefore, after the import volume greatly improved, the total sales and gross profit of the product increased, as shown in Table 5, Figure 6, Figure 7, Figure 8, Figure 9.

The detailed results are shown in Figure 6, the use of cloud services information system customized products, the total sales of its products have increased. Figure 7 shows the use of cloud

DOI: 10.1051/matecconf/201711901006

services information system customized products, the gross profit of its products have increased. Figure 8 shows the use of cloud services information system customized products, the sales improvement rate have increased. Figure 9 shows the use of cloud services information system customized products, the sales improvement increase in gross profit rate have increased.

The results of mining enabled the sales ranking of product as main promotional products and the designing of the customized products promotional catalog based on products combined. Because some customers had bought two products at the same time, in order to encourage customers to purchase a third one, which belongs to the customer product preference list.

Table 5. Cloud services information system and improvement rate of the total sales and Gross Profit.

Customized product portfolio ratio		P2 Product	P9 Product	P10 Product	P3 Product	P4 Product
Before using cloud services information system	Total sales(quantity)	21836	32652	38741	42578	36843
	Gross Profit(Price)	125,635	142,789	227,345	176,285	257,425
After using cloud services information system	Total sales(quantity)	29586	35821	43258	48762	42517
	Sales improvement rate	*26.19%*	*8.85%*	*10.44%*	*12.68%*	*13.35%*
	Gross Profit(Price)	168,523	182,560	281,405	237,851	316,580
	Improvement in gross profit rate	*25.45%*	*21.79%*	*19.21%*	*25.88%*	*18.69%*

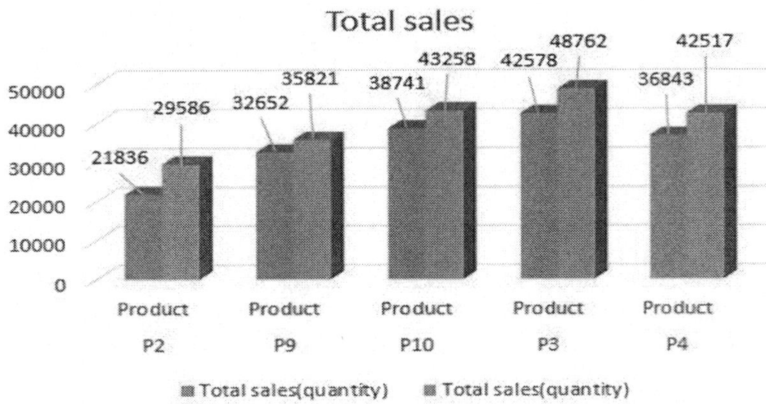

Figure 6. Comparison of total sales after the cloud services information system.

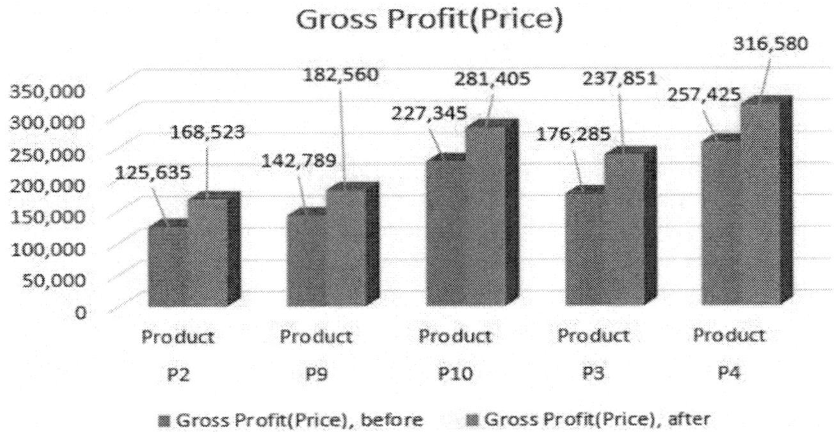

Figure 7. Comparison of Gross Profit after the cloud services information system.

 DOI: 10.1051/matecconf/201711901006

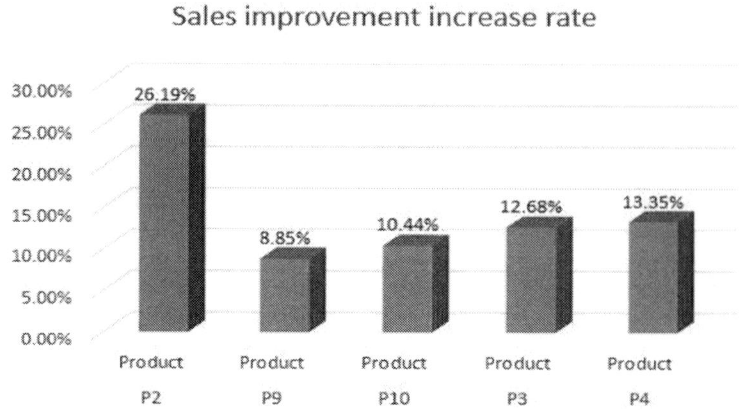

Figure 8. Comparison of sales improvement increase rate after the cloud services information system

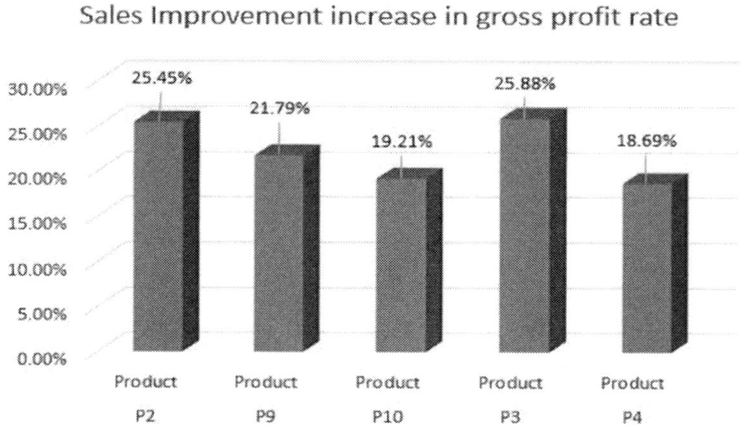

Figure 9. Comparison of sales improvement increase in gross profit rate after the cloud services information system

This research provides a cloud services information systems to enhance the sales of the case firm products by changing the original promotion method. It helps supervisors and employees make better decisions about promotional activities by providing them with useful knowledge from the new system proposed here. After applying customized products promotional catalogs in marketing, it is helpful to the company to catch on to the market acceptance of products and to know product sales conditions through the data mining system so as to mend marketing strategy and achieve the goal of product sales.

Besides, the data mining system also can correct the design of customized products catalogs according to market sales data and design customized products promotional catalogs of different periods. Thus, applying the data mining system effectively cannot only allow understanding of customers' purchasing behavior, but also provide knowledge of the cognition related to customers' purchasing for the policymaker to make the best policy.

In this paper, we developed a cloud services information system that can detect changes in customer behavior based on customer transaction data and sales products data. The advent of data mining has enhanced the accuracy of predicting customer purchase behavior. Mining for changes in customer purchase behavior is useful for satisfying customer needs in a dynamic internet business environment.

In this study, the association rules are used to identify links between customer profiles and products purchased. It provides marketing managers with a useful tool to rapidly search for valuable information based on customer transaction cloud information. Thus, it enables marketing managers to rapidly establish marketing strategies to enhance sales and profit.

For future research direction, such as sales promotion strategy and pricing strategy improvement, customers must be anticipated in different ways for cluster analysis. Considering the quick changes in customer preferences, companies must be able to grasp immediately the change in the customers' needs for customer retention. Therefore, dynamic data mining is essential, that is, the contents of the cloud customer database will have a transaction when the cloud service information system must correctly reflect the new database in the hidden data and reanalyze the customer's data.

Future study focus includes (1) pricing strategy and (2) dynamic data mining in both directions for further extensions and discussion to meet the business requirements. In the "pricing strategy", the competitor's pricing strategy may be taken into account. As for the "Dynamic Data Mining" in future studies, it is to allow enterprises to more securely grasp the customer by constant analysis of consumer behavior.

References

1. D.A. Aaker, Measuring brand equity across products and markets, California Management Review, **38** (3), 103-120 (1996)
2. R. Agrawal and R. Srikant, Fast algorithms for mining association rules, Proc. of the 20th International Conference on Very Large Databases, Santiago, Chile ,USA (1994)
3. R. Agrawal and R. Srikant, mining sequential pattern, Proc. of the 11th International Conference on Data Engineering, **12**, 3-14 (1995)
4. S.S. Anand, A.R. Patrick, J.G. Hughes, and D.A. Bell, A data mining methodology for cross-sales, Knowledge-Based Systems, **10**, 449-461 (1998)
5. A. Berson, S.J. Smith, and K. Thearling, Building Data Mining Applications for CRM, McGraw-Hill, USA (2015)
6. G.A. Carpenter and S. Grossberg, The ART of adaptive pattern recognition by self-organizing neural network, IEEE Computer, **21** (3), 77-88 (1988)
7. S.W. Changchien, C.F. Lee, and Y.J. Hsu, On-line personalized sales promotion in electronic commerce, Expert Systems with Applications, **27** (1), 35-52 (2004)
8. S.W. Changchien and T.C. Lu, Mining association rules procedures to support on-line recommendation by customers and products fragmentation, Expert Systems with Applications, **20** (4), 325-335 (2001)
9. Y.H. Cho, J.K. Kim, and S.H. Kim, A personalized recommender system based on web usage mining and decision tree introduction, Expert Systems with Applications, **23**, 329-342 (2002)
10. R. David and V. François, Enterprise information systems state of the art: Past, present and future trends, Computers in Industry, **79**, 3-13 (2016)
11. M. Gibbert, M. Leibold, and P. Gilbert, Five styles of customer knowledge management, and how smart companies use them to create value, European Management Journal, **5**, 459-469, (2002)
12. J. Han, J. Pei, B. Mortazavi-Asl, Q. Chen, U. Dayal, and M.C. Hsu, Free span: frequent pattern-projected sequential pattern mining, Proc. of the International Conference on Knowledge Discovery and Data Mining, Boston, MA, USA (2015)
13. D. Hoffman and T.P. Novak, A new marketing paradigm for 102 electronic commerce, The Information Society, **13**, 43-54 (2010)
14. G.R, Iyer, A.D. Miyazaki, D. Grewal, and M. Giordano, Linking web-based segmentation to pricing tactics, Journal of Product & Brand Management, **11** (5), 288-302 (2012)
15. A.N. Keller and M.R. Genesereth, Using information to create a housewares virtual Catalog, International Journal of Electronic Commerce and Business Media, **7** (4), 41-44 (2007)
16. W. Kim, Cloud computing: today and tomorrow, Journal of Object Technology, **8** (1), 65-72, (2009)
17. R.D. Lawrence, G.S. Almasi, V. Kotlyear, M.S. Viveros, and S.S. Duri, Personalization of supermarket product recommendations, Data Mining and Knowledge Discovery, **8**, 11-32, (2015)
18. H. Leonard, L.R. Eduardo, and V. Stefan, A cloud brokerage approach for solving the resource management problem in multi-cloud environments, Computers & Industrial Engineering, **95**, 16-26 (2016)

19. C. Lin and C. Hong, Using customer knowledge in designing electronic catalog, Expert Systems with Applications, **34**, 119-127 (2008)
20. D. Nash and A. Stema-Karwat, An application of DEA to measure branch cross selling efficiency, Computer Operations Research, **5**, 385-392 (2013)
21. L.P. Shan and L. Jae-Nam, Using E-CRM for a unified view of the customer, CACM, **12**, 95-99 (2003)
22. M.J. Shaw, C. Subramaniam, G.W. Tan, and M.E. Welge, Knowledge management and data mining for marketing, Decision Support Systems, **31**, 127-137 (2011)
23. K. Stanoevska-Slabeva and B. Schmid, Internet electronic product catalogs: an approach beyond simple keyword and multimedia, Computer Networks, **32**, 701-715 (2015)
24. S. Subashini and V. Kavitha, A survey on security issues in service delivery models of cloud computing, Journal of Network and Computer Applications, **34** (1), 1-11 (2011)
25. C.F. Surprenant and M.R. Solomon, Predictability and personalization in the service encounter, Journal of Marketing, **51**, 86-89 (2007)
26. L. Terveen, W. Hill, B. Armento, D. McDonald, and J. Creter, PHOAKS: a system for sharing recommendations, Communications of the ACM, **40** (3), 59-62 (2015)
27. J.M. Tsai and S.W. Hung, A novel model of technology diffusion: System dynamics perspective for cloud computing, Journal of Engineering and Technology Management, **33**, 47-62 (2014)
28. E. Triantaphyllou, Multi-criteria decision making methods: a comparative study, Kluwer Academic Publishers, USA (2015)
29. T.R. Xiang, J.F. Kong, L. Hao, and Q. George, Cloud-enabled real-time platform for adaptive planning and control in auction logistics center, Computers & Industrial Engineering, **84**, 79-90 (2015)

Clustering with fuzzy supervised algorithm

Fong-Jhu Yih[1], Yuan-Horng Lin[2] and Jeng-Ming Yih[3,a]

[1]*Department of Information Engineering, National Chung Cheng University, Chiayi 621, Taiwan*
[2]*Department of Mathematics Education National Tai-Chung University of Education, Taichung City 403, Taiwan*
[3]*Department of General Education, Min-Hwei Junior College of Health Care Management, Tainan 736, Taiwan*

Abstract. In GK-algorithm, modified Mahalanobis distance with preserved volume was used. However, the added fuzzy covariance matrices in their distance measure were not directly derived from the objective function. A Fuzzy C-Means algorithm based on Mahalanobis distance (FCM-M) was proposed to improve those limitations of GG and GK algorithms, but it is not stable enough when some of its covariance matrices are not equal. In this paper, an improved Supervised Clustering Algorithm Based on FCM by taking a new threshold value and a new convergent process is proposed. The experimental results of real data sets show that our proposed new algorithm has the best performance. Not only replacing the common covariance matrix with the correlation matrix in the objective function in the Supervised Clustering Algorithm.

1 Introduction

To overcome the drawback due to Euclidean distance, we could try to extend the distance measure to Mahalanobis distance (MD). However, Krishnapuram and Kim (1999)pointed out that the Mahalanobis distance can not be used directly in clustering algorithm. Gustafson-Kessel (GK) clustering algorithm and Gath-Geva (GG) clustering algorithm were developed to detect non-spherical structural clusters. In GK-algorithm, a modified Mahalanobis distance with preserved volume was used. However, the added fuzzy covariance matrices in their distance measure were not directly derived from the objective function. In GG algorithm, the Gaussian distance can only be used for the data with multivariate normal distribution. To add a regulating factor of each covariance matrix to each class in the objective function, and deleted the constraint of the determinants of covariance matrices in the GK algorithm, the Fuzzy C-Means algorithm based on Mahalanobis distance (FCM-M).Fuzzy partition clustering is a branch in cluster analysis and it is widely used in pattern recognition. Among many well- known fuzzy partition clustering algorithms, Bezdek's Fuzzy C-Means (FCM) (1981), Pal, Pal and Bezdek's Possibility C-Means (PCM) (1993), and Pal, Pal and Bezdek's Fuzzy Possibility C-Means (FPCM) (1997) are all based on Euclidean distance measure for clustering. Hence, those fuzzy partition clustering algorithms can only be used for the data set with the same super spherical shape for each class. Instead of using Euclidean distance measure, Gustafson and Kessel (1979) proposed the G-K algorithm which employs the Mahalanobis distance. It is a fuzzy partition clustering algorithm which can be used for the classes with different geometrical shapes in the data set. However, without the prior information of the shape volume for each class, the G-K

[a] Corresponding author : yih@mail.ntcu.edu.tw

algorithm can only be utilized for the classes with the same volume. In other words, if any dimension of a class is greater than the number of samples in the class, the estimated covariance matrix of the class may not be fully ranked. Hence, the algorithm will induce the singular problem for the inverse covariance matrix. This is an important issue need be addressed when we use the G-K algorithm for clustering. To overcome the issues, a new solution is proposed. A regulating factor of the covariance matrix is added to each class in the objective function, and the constraint of the determinant of the covariance matrices defined in the G-K algorithm is removed. Furthermore, the FCM-AM algorithms included two algorithms, FCM-M and FCM-CM, proposed by our previous works (Hsiang-Chuan Liu, Jeng-Ming Yih, Shin-Wu Liu, 2007, 2008).

2 Fuzzy Partitions Clustering Algorithms Based on Euclidean Distance

The popular fuzzy c-means algorithm based on Euclidean distance function converges to a local minimum of the objective function, which can only be used to detect spherical structural clusters. Gustafson-Kessel clustering algorithm and Gath-Geva clustering algorithm were developed to detect non-spherical structural clusters. However, Gustafson-Kessel clustering algorithm needs added constraint of fuzzy covariance matrix, Gath-Geva clustering algorithm can only be used for the data with multivariate Gaussian distribution. The objective of a fuzzy clustering algorithm is to partition the data into clusters so that the similarity of data objects within each cluster is maximized and the similarity of data objects among clusters is minimized. In the objective function based methods, the objective function is a function of data matrix, membership matrix and prototypes of clusters. It measures the overall dissimilarity of data objects within each cluster. Hence, by minimizing the objective function, we can obtain the best partition of the data set.

2.1 Fuzzy C-Means Algorithm

To overcome the drawback due to Euclidean distance, we could try to extend the distance measure to Mahalanobis distance (MD). However, Krishnapuram and Kim (1999) pointed out that the Mahalanobis distance can not be used directly in clustering algorithm. Fuzzy C-Means Algorithm (FCM) which objective function of FCM is given.

2.2 Fuzzy Clustering

Clustering technique plays an important role in data analysis and interpretation. Fuzzy clustering is a branch in clustering analysis and it is widely used in the pattern recognition field. Fuzzy clustering algorithms can only be used to detect the data classes with the same super spherical shapes. To overcome the drawback due to Euclidean distance, we could try to extend the distance measure to Mahalanobis distance (MD). However, Krishnapuram and Kim (1999) pointed out that the Mahalanobis distance can not be used directly in clustering algorithm. Gustafson-Kessel (GK) clustering algorithm and Gath-Geva (GG) clustering algorithm were developed to detect non-spherical structural clusters. In GK-algorithm, a modified Mahalanobis distance with preserved volume was used. However, the added fuzzy covariance matrices in their distance measure were not directly derived from the objective function. In GG algorithm, the Gaussian distance can only be used for the data with multivariate normal distribution. To add a regulating factor of each covariance matrix to each class in the objective function, and deleted the constraint of the determinants of covariance matrices in the GK algorithm, the Fuzzy C-Means algorithm based on Mahalanobis distance (FCM-M) was proposed, and then For improving the stability of the FCM-M clustering results, Replace all of the covariance matrices with the same common covariance matrix in the objective function in the FCM-M algorithm.

2.3 The FCM-AM Algorithm and Its Special Cases

Using the Liu-algorithm, we can obtain the objective function of the Fuzzy C-Means algorithm based on Alternative Mahalanobis distances (FCM-AM) as following.

2.4 FCM-NM Algorithm

In this study, a fuzzy clustering algorithm called Fuzzy C-Means algorithm based on normalized Mahalanobis distance (FCM-NM) is used, which was improved by normalizing for each feature in the objective function and also replacing the threshold in the FCM-CM algorithm. We can obtain the objective function of FCM-NM as following:

3 Experiment Real Data

In this study, Linear algebra test for university students is designed by author. The instrument consists of 19 dichotomous items which measure 6 concepts. The data set used in the experimental study is an educational data from university students in Taiwan. There are 231 university students from Taiwan in this test. The tool consist of six concepts, its contents are shown in Table 1.

Table 1. The content of Concepts

Classes	Concepts
1	Operation of matrix
2	System of linear equations
3	Determinants
4	Vector space and the property of R^n
5	Eigen-value and eigenvector
6	Geometry of linear algebra

Applying fuzzy clustering algorithms as we mentioned above, the clustering performances of each algorithm are calculated with same fuzzier m = 2 and the clustering accuracies are compared and shown in Table 2.

Table 2. The content of Concepts

Algorithm	Accuracies (%)
FCM	0.738
FCM-AM	0.823
FCM-NM	0.894

The Mean clustering Accuracies of 100 different initial value sets of FCM, FCM-AM and FCM-NM for the Dataset was shown in TABLE 2. From this table, we can find that the performance of FCM algorithm always worse than FCM-AM for above dataset. Although the performance of FCM-NM algorithm is better than which of FCM-AM algorithm in the dataset. In other words, our proposed two algorithms, FCM-NM and FCM-AM are better than FCM algorithm. Hence, the new algorithm, FCM-NM, has the best performance.

4 Conclusion

Clustering technique plays an important role in data analysis and interpretation. It groups data into clusters so that the data objects within a cluster have high similarity in comparison to one another, but are very dissimilar to those data objects in other clusters. The well-known FCM is based on Euclidean distance function, which can only be used to detect spherical structural clusters. GK algorithm and GG algorithm were developed to detect non-spherical structural clusters. However, the former needs added constraint of fuzzy covariance matrix, the later can only be used for the data with multivariate Gaussian distribution. Three improved Fuzzy C-Means algorithm based on different Mahalanobis distance, called FCM-M, FCM-CM, and FCM-NM were proposed by our previous works. In this paper, a further improved Fuzzy C-Means algorithm based on a normalized Mahalanobis distance (FCM-NM) by taking a new convergent process is proposed. The experimental result of the real data set shows that our proposed new algorithm has the best performance.

References

1. J.C. Bezdek, *Pattern Recognition with Fuzzy Objective Function Algorithms*, Springer Science & Business Media, 65-70 (1981)
2. H.C. Liu, J.M. Yih, D.B. Wu, and S.W. Liu, Fuzzy possibility c-mean clustering algorithms based on complete mahalanobis distances, International Conference on Wavelet Analysis and Pattern Recognition, **1**, 50-55 (2008)
3. R. Krishnapuram and J. Kim, A note on the Gustafson-Kessel and adaptive fuzzy clustering algorithm, IEEE Transactions on Fuzzy Systems, **7** (4), 453-461 (1999)
4. D.E. Gustafson and W.C. Kessel, Fuzzy clustering with a fuzzy covariance matrix, *Proc. IEEE Conf. Decision Contr. San Diego*, CA, 761-766 (1979)
5. I. Gath and A.B. Geva, Unsupervised optimal fuzzy clustering, IEEE Trans. Pattern Anal. Machine Intell, **11** (7), 773-781 (1989)
6. J.C. Dunn, A fuzzy relative of the isolated data process and its use in detecting compact, well-separated clusters, J. Cybern, **3** (3), 32-57 (1973)
7. F. Höppner, F. Klawonn, R. Kruse, and T. Runkler, *Fuzzy Cluster Analysis*, John Wiley and Sons (1999)
8. H.C. Liu, J.M. Yih, and S.W. Liu, Fuzzy C-means algorithm based on mahalanobis distances and better initial values, *Proceedings of the 10th Joint Conference& 12th International Conference on Fuzzy Theory & Technology*, **1**, 1398-1404 (2007)
9. H.C. Liu, J.M. Yih, D.B. Wu, and S.W. Liu, Fuzzy C-means algorithm based on "complete" Mahalanobis cistances, *Proceedings of International conference on Machine Learning and Cybernetics*, **7** (6), 3569-3574 (2008)
10. H.C. Liu, J.M. Yih, W.C. Lin, and T.S. Liu, Fuzzy C-means algorithm based on PSO and Mahalanobis distances, International Journal of Innovative Computing, Information and Control , **5** (12B), 5033-500 (2009)
11. H.C. Liu, J.M. Yih, W.C. Lin, and D.B. Wu, Journal of Multiple Valued Logic & Soft Computing, **15**, 581-595 (2009)
12. B. Balasko, J. Abonyi, and B. Feil, Fuzzy clustering and data analysis Toolbox for use with Matlab from http://www.mathworks.com/ matlabcentral/ fileexchange/7473 (2013)
13. H.C. Liu, B.C. Jeng, J.M. Yih, and Y.K. Yu, Fuzzy C-means algorithm based on standard mahalanobis distances, *Proceedings of the 2009 International Symposium on Information Processing* (*ISIP'09*), 422-427 (2009)
14. C. Ding, T. Li, and W. Ping, On the equivalence between non-negative matrix factorization and probabilistic latent semantic indexing, Computational Statistics and Data Analysis, **52** (8), 3913-3927 (2008)

A hierarchical network model for network topology design using genetic algorithm

Chunlin Wang[1], Ning Huang[1,a], Shuo Zhang[2], Yue Zhang[1] and Weiqiang Wu[1]

[1]*School of Reliability and Systems Engineering, Beihang University, Beijing, China*
[2]*China Ship Development and Design Center, Wuhan, China*

Abstract. Network topology design has directly impact on network construction costs and network performance. Majority of current network topology design take the network physical topology parameters into consideration, such as the reliability and cost constraints, ignoring the actual traffic information on the logical network. Moreover, the network traffic exhibits self-similar feature over large time scales, which is complicated and is difficult to predict. In this paper, firstly, a hierarchical network model is proposed that consists of the upper logical topology and the lower physical topology. The logical topology describes the self-similar traffic information on the network and the ON/OFF model is adopted to model the self-similar traffic. The lower physical topology represents the connection relationship between the all kinds of network devices and links. Then, taking advantage of the hierarchical network model, a novel network topology design method based on the genetic algorithm is proposed, which aimed at obtaining the network with minimum delay under certain reliability and cost constraints. Finally, a practical example is presented to verify the effectiveness and the accuracy of our network topology design method. Results show that our method obtains better results than the other methods.

1 Introduction

With the development of network communication technique, the types of network equipment are increasing and the network size is becoming larger. More importantly, the increasing complexity of network traffic appears on the communication network. All these factors have increased the difficulty of network topology design. However, the topology of the communication network is decisive; once improper, the reliability will decrease and the network performance will also decrease [1]. For the communication network, the real-time performance, reliability and construction cost are three decisive indicators which need to be considered in the network design phase [2, 3].

There have been many related researches on the communication network topology design. And these works can be generally divided into the following two categories. The first category focuses on the physical topology design. They often consider the topology design for minimizing the total cost subject to a constraint on the network reliability or for maximizing the network reliability while a constraint on the total cost [4]. For example, Shao [5] proposed a forest research algorithm to optimization of a computer network expansion with a reliability constraint. Altiparmak [6] and Pierre

[a]Corresponding author : hn@buaa.edu.cn

[7] introduced a genetic algorithm-based scheme to find an optimal solution when maximized the network reliability subject to certain cost constraint. The above related researches often focus on the physical network, ignoring the network traffic transmission on the network. The second category considers the co-design of both physical topology and logical topology. Such as, Fencl [8] tried to optimize the physical and logical topology simultaneously, and real-time performance and fault-tolerance level are considered in the LAN topology design. Tian [1] took advantage of the bi-level programing theory [9] to optimize the reliability and the real-time performance simultaneously under cost constraint. Usually, such simultaneous design is treated as a multi-objective process [10], which is hard to ensure the ultimate optimal solution.

From the related research, the second category begins to consider the network traffic in the topology design. However, the traffic they described is too simple and the network size is small. In [2, 8, 10] the traffic load as a known constant value in advance and then the traffic matrix is applied to the network. But the traffic on the communication networks is complicated and is difficult to predict [11]. Moreover, the self-similar feature of the traffic has been proved to be the most important feature of network traffic [12, 13], which is ignored in the network topology design.

In this paper, a hierarchical network model is proposed to solve the communication network topology design problem. The lower level is physical topology network. The upper level is logical topology which is the result of self-similar traffic transmission. The self-similar traffic transmission determines the network real-time performance, and the physical topology network is subject to the total cost and reliability constraints. Then, a network topology design method using the genetic algorithm is presented, which aimed at obtaining the network with minimum delay under certain reliability and cost constraints. Finally, an illustrative example is presented to verify the effectiveness of our communication network topology design method.

2 Hierarchical network model

Networks tend to have a two-layer or multi-layer structure in the real-world. For the communication network, the two-layer structure is commonly used in current research [14, 15]. Here, a hierarchical network model is proposed to solve the network topology design problem. The upper layer is the logical topology and the lower layer is the physical topology in our model. The lower physical topology is the actual network, including the physical devices and links. The upper logical topology represents the different traffic flows, which will determine the network performance. The hierarchical network model can be seen in Figure 1.

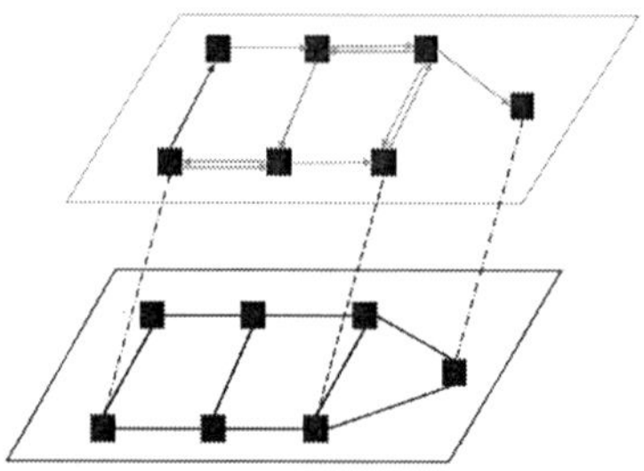

Figure 1. Hierarchical network model.

2.1 Physical Network topology

The lower physical topology describes the connection relationship between all kinds of network devices and links. The communication network can be modeled as an undirected graph *G*, with *N* nodes and *E* edges. Mathematically, it can be depicted by an $N \times N$ adjacency matrix $\{w_{ij}\}_{N\times N}$. If there is an edge between node *i* and node *j*, the $w_{ij} = 1$; otherwise $w_{ij} = 0$.For the actual network design, the network resources are often limited due to certain cost constraints. Moreover, the network must meet some basic reliability requirements, such as several specified nodes can be connected. In this paper, the cost constraint and reliability requirement are both taken into consideration in our network topology design. For the cost constraint, it is reflected in the constraint on the total number of links in the network. That is because the number and the price of the nodes often have been identified before the topology design. Essentially, the network topology design is to determine the connection relationship between each node. Then the total number of links is an important parameter to reflect the cost constraint. For the reliability requirement, all nodes in the network need to be able to communicate with each other, primarily because the connectivity is the basic premise for network operation.

Then, the cost constraint and reliability requirement are two main considerations in our network topology design, which limit the total number of links and require the network connectivity respectively.

2.2 Logical Network topology

Logical network describes the different traffic flows on the physical network and the self-similarity of network traffic has been proved to be the most important feature of network traffic. Different self-similar traffic models have been proposed in related research, such as the heavy-tailed ON/OFF model, wavelets model and α -stable traffic model. Among these models, the heavy-tailed ON/OFF model has been widely used to generate self-similar traffic due to its mathematical simplicity and well explanation of the self-similar feature.

For the ON/OFF model, it assumes that the source alternates between an ON-period and an OFF-period [16]. During the ON-period, packets are generated and sent from the source to the destination at a constant rate v, while during the OFF-period, no packets are transmitted. This process can be called as alternating renewal process. Generally, each source is considered as identically independently distributed for the ON-period and OFF-period. Then, it is only need to specify the distribution of durations of the ON-period and OFF-period for each source.

Pareto distribution is a typical heavy-tailed distribution. Let $f(x)$, $F(x)$ denote the probability density function and the cumulative distribution function for the durations of ON-period and OFF-period. The probability density function (*pdf*) of Pareto distribution is:

$$f(x) = \alpha k^{\alpha} x^{-\alpha-1}, 0 < k \le x \tag{1}$$

The cumulative distribution function can be obtained as:

$$F(x) = P[X \le x] = 1-(k/x)^{\alpha} \tag{2}$$

Where k represents the minimum value of the random variables, α determines the mean and variance of the random variable.

The self-similar traffic transmission determines the network real-time performance. Here, an average packet delay [17] is introduced to evaluate network real-time performance. It can be calculated as follows:

$$T(x) = \frac{1}{\gamma}\sum_{i=1}^{m}\frac{f_i}{C_i - f_i} \tag{3}$$

Where γ is the total arrival rate into the network in bits per second; m is the number of links; f_i and C_i are the assigned traffic load and capacity of link i respectively.

2.3 Problem Formulation

According to our hierarchical network model above, network topology design can be regard as a network topology optimization design problem with certain constraints. The objective function is the average packet delay which is decided in the logical network. The constraints include the cost constraint and reliability requirement in physical network. That is to say, the network topology should be designed that the objective function is optimal under certain cost and reliability constraints. This problem can be described as:

$$\begin{aligned} &Min \quad T(x) \\ &s.t. \, Cost(G(N,E)) \le C_{max} \\ &\quad Rel(G(N,E)) \ge R_{min} \end{aligned} \tag{4}$$

Where $Cost(G(N,E))$ is the actual construction cost of links in network, which is reflected by the number of links in later examples; C_{max} is the maximum value of cost; $Rel(G(N,E))$ is all-terminal connective reliability in this paper; R_{min} is the minimum value of network reliability.

This network design problem has been shown to be NP-hard [2]. Then, several heuristic algorithms have been proposed to deal with the specific topology design problems, including Genetic Algorithm (GA), Neural Network, Tabu Search, and Ant Colony Optimization. Especially, genetic algorithm has a great advantage in efficiency and implementation of solving the optimization problems with constraints and multi-objective. In this paper, we employ a genetic algorithm for our network topology design.

3 Network topology design

A genetic algorithm is based on the theory of natural evolution and has the following steps: encoding and decoding, initial population generation, evaluation, selection, crossover, mutation. First of all, a set of the initial individuals is generated, forming the initial population. Then, a series of genetic operators modifies this population repeatedly, including the selection, crossover and mutation. The modified population is the offspring for the current population. At last, find out the optimal solution in the iterative process, thanks to the fitness function.

3.1 Encoding and Decoding

The chromosome that describes network topology is a vector i with l elements.

$$i = \{i_j : i_j \in \{0,1\}\}, j = 1,...,l$$

Where N is the number of nodes in the network. $l = N(N-1)/2$. Vector i characterize the upper triangular part of the adjacency matrix P because the adjacency matrix is symmetrical. The specific encoding and decoding process can be briefly described from the following example:

The chromosome is the upper triangular part of the adjacency matrix P:

Then, the vector $\{1\,1\,0\,0\,0\,1\,1\,1\,1\,0\}$ is the chromosome which can represent the network topology in Figure 2. For the decoding process, the chromosome is converted into a matrix through a reverse process.

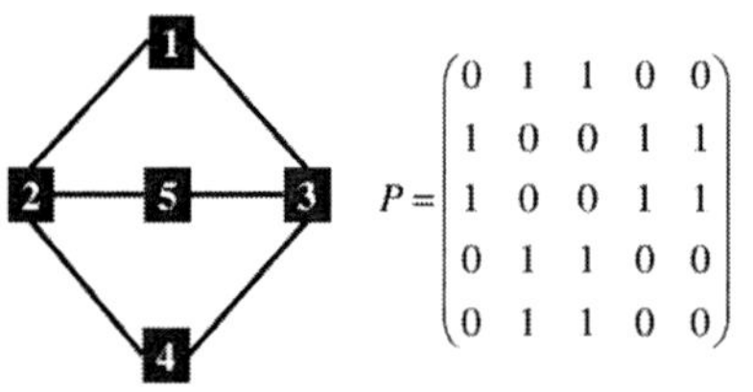

Figure 2. An example topology and adjacency matrix.

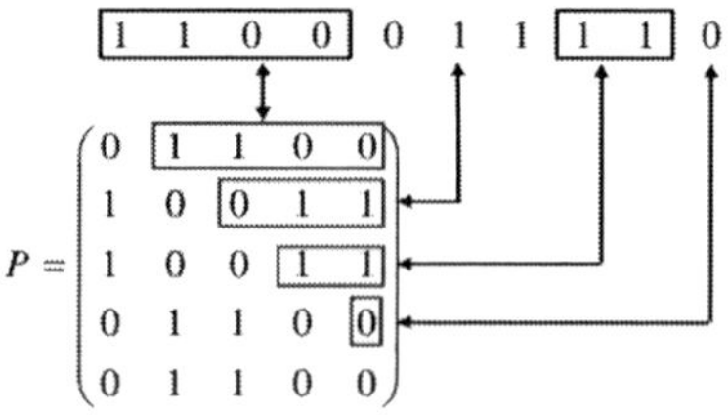

Figure 3. The chromosome encoding.

3.2 Initial Population

Generally, the initial population is generated by completely random topologies. However, the random topologies can't guarantee all-terminal are connected for our topology design. Reference [18] proposed a feasible problem solution for the initial population generation. Firstly, design a ring topology connecting all nodes of the network, thus guaranteeing that all initial solutions are feasible. The ring topology is randomly generated for each individual of the initial population. Afterwards, t links are added to the ring topology, connecting t pairs of randomly selected nodes.

3.3 Fitness Function

We revise the objective function (average packet delay $T(x)$) as a fitness function of the GA, denoted by $Fit(x)$. The fitness function can be expressed as (5) because the lower the objective value, the higher the fitness will be:

$$Fit(x) = \begin{cases} C - T(x), & if \quad T(x) < C \\ 0 \quad , & if \quad T(x) \geq C \end{cases} \tag{5}$$

Where C is a relatively large constant value, and the selection of this value can refer to the allowed maximum value of objective function.

3.4 Selection, Crossover and Mutation

In the selection phase, pairs of individuals are chosen for crossover. Usually, individuals are selected based on their fitness value. In this paper, the selection process is based on a roulette-wheel selection scheme, meaning that chromosome with larger fitness value has greater probability of being selected into the mating population.

In the crossover operation, pairs of individuals previously selected are combined, giving rise to another pair of new individuals (offspring). Here, the single point crossover method is selected for the crossover operation. Firstly, a crossover place is randomly selected for the two selected individuals. Then, the two left sides of individuals are copied to offspring 1 and offspring 2, respectively. The right sides of each code shall be exchanged for the offspring1 and offspring 2. An example is displayed in Figure 4.

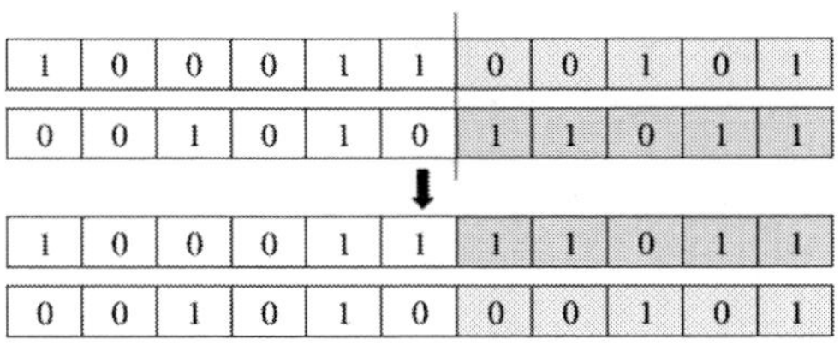

Figure 4. Crossover operation.

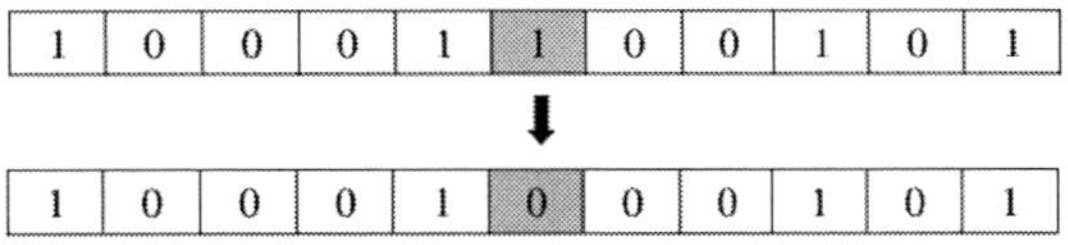

Figure 5. Mutation operation.

In the mutation operation, a mutated chromosome is chosen at random according to the mutation probability. Then, a simple exchange of 0's to 1's, or vice versa, at random locations of the genetic code. This mutation operation has the goal of increasing the diversity of the population. An example is shown in Figure 5.

After these series of operations (selection, crossover and mutation), the next generation is created. Taking the cost constraint of the physical topology into consideration, some newborn individuals may not satisfy this constraint. So, it is needed to analyse these newborn individuals and discard those that do not meet the constraints. At last, an optimal network topology can be obtained through certain number of iterations.

4 Numerical results and analysis

4.1 Illustrative Example

The test of our topology design is made for a medium sized network with the number of nodes *N*=16. The related network traffic assumptions as follows: Each node can be regarded as an ON/OFF source which sends certain packets to all other nodes in the network. The packet transmits along the shortest path from the source to destination. And the related parameters for ON/OFF traffic model are: $\alpha = 1.1$, $k = 1$, $v = 1$.

For the test of network topology design, our objective function is the average packet delay $T(x)$ and the physical layer has a number of constraints, including the number of edges in the network $M < 30$, which reflects the network cost constraint, and edge capacity $C_i = 60$.

Lastly, the experiment was done with the following GA parameters: population size $pop\_size = 50$; crossover probability $P_c = 0.4$; mutation probability $P_m = 0.05$; the number of iterations is 200. The constant value $C = 5$ in the fitness function.

The result of genetic algorithm for our topology design is shown in Figure 6.

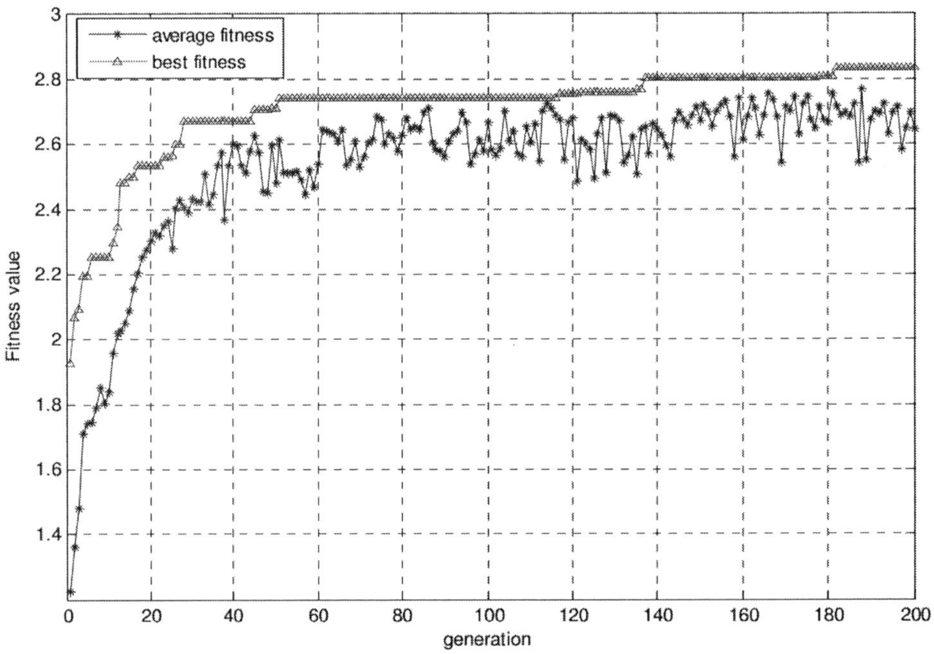

Figure 6. The results of example.

In Figure 6, the red curve stands for the best fitness value in each generation and the blue curve represents the average fitness value for each generation. It can be seen that the best fitness value reaches the maximum value 2.83 and tends to converge after about 180 generations. Then the optimal chromosome in the iterative process is:

100000010000001101000000100000000010000000100010010000100000

000001001001001110000000100000001100101100000100101000100101

The optimal network topology also can be obtained after decoding the optimal chromosome. The adjacency matrix of the optimal network is as follows:

$$\begin{bmatrix}
0 & 1 & 0 & 0 & 0 & 0 & 0 & 0 & 1 & 0 & 0 & 0 & 0 & 0 & 0 & 1 \\
1 & 0 & 1 & 0 & 1 & 0 & 0 & 0 & 0 & 0 & 0 & 1 & 0 & 0 & 0 & 0 \\
0 & 1 & 0 & 0 & 0 & 0 & 0 & 0 & 1 & 0 & 0 & 0 & 0 & 0 & 0 & 0 \\
0 & 0 & 0 & 0 & 1 & 0 & 0 & 0 & 1 & 0 & 0 & 1 & 0 & 0 & 0 & 0 \\
0 & 1 & 0 & 1 & 0 & 1 & 0 & 0 & 0 & 0 & 0 & 0 & 0 & 0 & 0 & 0 \\
0 & 0 & 0 & 0 & 1 & 0 & 1 & 0 & 0 & 1 & 0 & 0 & 1 & 0 & 0 & 1 \\
0 & 0 & 0 & 0 & 0 & 1 & 0 & 1 & 1 & 0 & 0 & 0 & 0 & 0 & 0 & 0 \\
0 & 0 & 0 & 0 & 0 & 0 & 1 & 0 & 1 & 0 & 0 & 0 & 0 & 0 & 0 & 0 \\
1 & 0 & 1 & 1 & 0 & 0 & 1 & 1 & 0 & 1 & 1 & 0 & 0 & 1 & 0 & 1 \\
0 & 0 & 0 & 0 & 0 & 1 & 0 & 0 & 1 & 0 & 1 & 0 & 0 & 0 & 0 & 0 \\
0 & 0 & 0 & 0 & 0 & 0 & 0 & 0 & 1 & 1 & 0 & 1 & 0 & 0 & 1 & 0 \\
0 & 1 & 0 & 1 & 0 & 0 & 0 & 0 & 0 & 0 & 1 & 0 & 1 & 0 & 0 & 0 \\
0 & 0 & 0 & 0 & 0 & 1 & 0 & 0 & 0 & 0 & 0 & 1 & 0 & 1 & 0 & 0 \\
0 & 0 & 0 & 0 & 0 & 0 & 0 & 0 & 1 & 0 & 0 & 0 & 1 & 0 & 1 & 0 \\
0 & 0 & 0 & 0 & 0 & 0 & 0 & 0 & 0 & 0 & 1 & 0 & 0 & 1 & 0 & 1 \\
1 & 0 & 0 & 0 & 0 & 1 & 0 & 0 & 1 & 0 & 0 & 0 & 0 & 0 & 1 & 0
\end{bmatrix}$$

4.2 Accuracy verification

To illustrate the accuracy of our approach, numerical experiments of the network topology design under different cost constraints are also provided. The number of edges in the network is limited as $M < 30$, $M < 32$, $M < 34$ respectively. And the result is as follows:

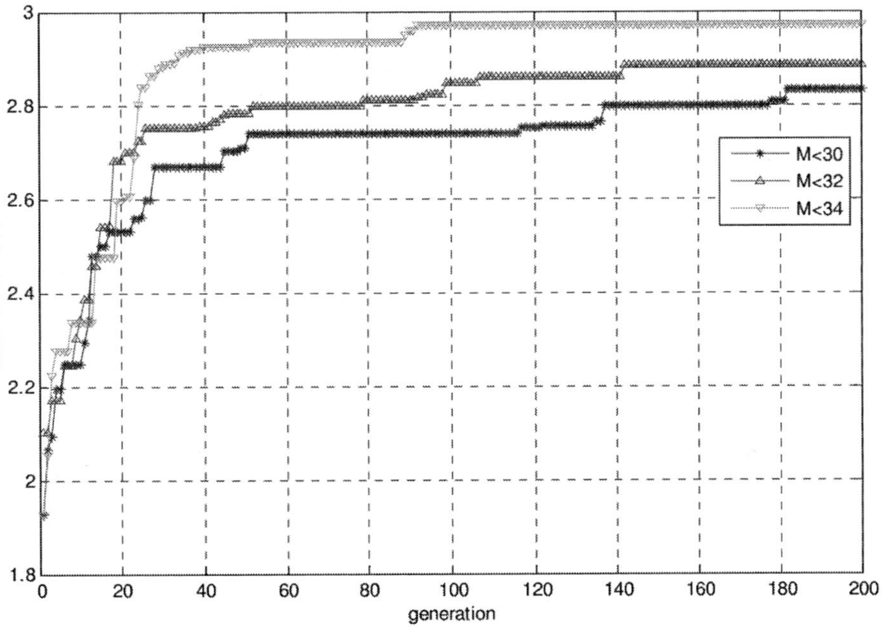

Figure 7. The results for different M.

With the increase of the number of edges *M*, the maximum of the best fitness value is increasing. The optimal solutions under different constraints can be obtained by using our topology design method. And the results of average packet delay *T*(*x*) for these three optimal topologies are shown in Table 1.

Table 1. The average packet delay

	M<30	*M*<32	*M*<34
average packet delay *T*(*x*)	*T*=0.0342	*T*=0.0333	*T*=0.0319

As shown in Table 1, with more edges are allowed in the network design (more cost is allowed), the generated network topology has better performance. This conclusion is consistent with the actual situation.

Moreover, several method comparisons also have been provided to demonstrate the accuracy of our topology design approach. Firstly, the classic square lattice network model is introduced to analyse comparatively (seen as Figure 8). Then, the second is the topology generation method proposed in Reference [18], which is adopted for our initial population generation. We guarantee the same network constraints for different approaches and then compare the network performance of different methods. The result is shown in Table 2.

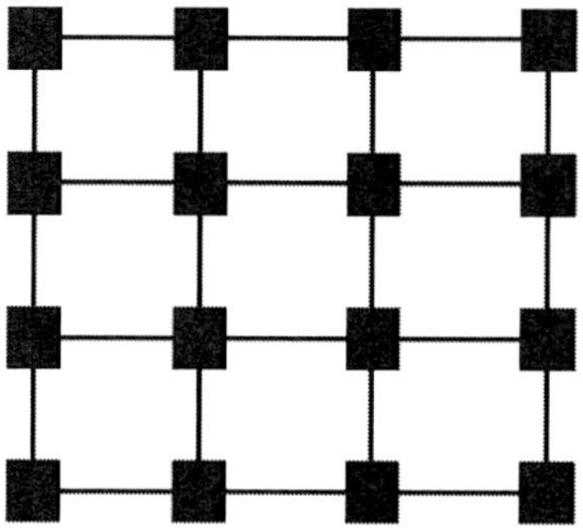

Figure 8. Lattice network model.

Table 2. The comparison results

	Our method	**Lattice network**	**Reference [18]**
average packet delay *T(x)*	*T*=0.0412	*T*=0.0507	*T*=0.04677

It can be seen that the network average packet delay of our design method is the smallest under same design constraints. That is to say, the topology design method proposed in this paper has better effectiveness and practicality compared to the other methods.

5 Conclusion

In this paper, a general hierarchical network model is proposed to describe the problem of communication network topology design. The lower level of the hierarchical model describes the cost and reliability constraints. And the upper level determines the network performance when self-similar traffic is loaded on the network. Based on the genetic algorithm, the topology design solution is presented in the paper, which is aimed at designing the optimal network topology under certain reliability and cost constraints. Finally, a practical example is presented to illustrate our topology design method. What's more, numerical comparative analyses also have been carried out to verify the accuracy of our proposed topology design method. The result justifies that our method has certain guiding significance to communication network topology design. Our future research will focus on some more realistic physical layer constraints, such as considering the distance between nodes and the heterogeneity of nodes.

References

1. Y. Tian, H. Dong, L. Jia, Y. Qin, and S. Li, Math. Probl. Eng, **2014** (2014)
2. L. Zhang, M. Lampe, and Z. Wang, Frequenz, **66**,159 (2012)

3. Y. Jiang, R. Li, R. Kang, and N. Huang, International Conference on Quality, Reliability, Risk, Maintenance, and Safety Engineering, **245** (2012)
4. F.M. Shao, X.S. Shen, and P.H. Ho, IEEE Trans. Reliab., **54**, 421 (2005)
5. F.M. Shao and L.C. Zhao, Comput. Math. Appl., **35**, 17(1998)
6. F. Altiparmak, B. Dengiz, and A.E. Smith, IEEE International Conference on Systems, Man, and Cybernetics, **5**, 4676 (1998)
7. S. Pierre and G. Legault, IEEE T. Cybern., **28**, 249 (1998)
8. T. Fencl, P. Burget, and J. Bilek, Control Eng. Practice., **19**, 1287(2011)
9. Z. Gao, J. Wu, and H. Sun,Transp. Res. Pt. B-Methodol., **39**, 479 (2005)
10. C.S. Wang and C.T. Chang, IEEE-ACM Trans. Netw., **16**, 680 (2008)
11. Y. Wang, C. Tian, S. Wang, and W. Liu, Internet Conference of China., **80** (2014)
12. W.E. Lel, M.S. Taqqu, W. Willinger, and D.V. Wilson, IEEE-ACM Trans. Netw., **2**, 1 (1994)
13. V. Paxson and S. Floyd, IEEE-ACM Trans. Netw., **3**, 226 (1995)
14. M. Kurant and P. Thiran, Phys. Rev. Lett., **96**, 138701 (2006)
15. Y. Zhuo, Y. Peng, C. Liu, Y. Liu, and K. Long, Physica A., **390**, 2401 (2011)
16. W. Willinger, M. S. Taqqu, and R. Sherman, IEEE-ACM Trans. Netw., **5**, 71 (1997)
17. E. Szlachcic, IEEE International Conference on Dependability of Computer Systems., 150 (2006)
18. J.F. Labourdette, E. Bouillet, and R. Ramamurthy, IEEE-ACM Trans. Netw., **13**, 906 (2005)

Green options for anti-sulfate of slag cement concrete containing pozzolans

Chien-Jou Yang, Yuan-Yuan Chen[a]

Department of Architecture, Hwa Hsia University of Technology, New Taipei City 235, Taiwan

Abstract. This study mainly adopted densified mixture design algorithm (DMDA) with pozzolans (fly ash and slag), different fineness slag cement (1:1; MF40 and HF40) and Type I cement (E40) to construct the mixtures for w/cm=0.40, and applied ACI 211.1R and Type II cement as control group (CII40, w/c=0.40). Life cycle inventory of LEED suggested by PCA for cementitious materials (kg/m^3) contained cement use, CO_2 emission, raw materials, energy consumption, compressive strength, and immersed in different concentration Na_2SO_4 solution. Results showed cement content, CO_2 emission, raw materials and energy consumption for E40, MF40 and HF40, with respect to CII40, were 14% ~ 26%, 14% ~ 26%, 13% ~ 26% and 17%~28%. At 28 days, compressive strength(all mixtures) were greater than 41MPa. Repeatedly 25 cycles, specimens immersed in 5000ppm Na_2SO_4 solution and oven-dried at 105°C, the exterior had no damage, and weight loss (*n*) and pulse velocity change (*nv*) were less than -1% and -5%. But in saturated Na_2SO_4 solution, the *n* and *nv* were ranged from -0.91% (E40) to -2.62% (MF40) and -6.7% (E40) to -10.9% (MF40). The exterior had been obviously scaling (chalking) or spalling at the second (CII40), the fifth (MF40 and HF40) and the ninth cycle (E40). The comprehensive evaluation of green options for anti-sulfate indicated that the merits of all mixtures were respectively E40 > HF40 > MF40 > CII40.

1 Sulfate attack and pozzolans

Concrete structure exposed in different concentration sulfate solution environment(submersion or immersion) were due to the chemical reaction of sulfate-attack. The reaction of hydration product of ettringite due to extreme expansion at the hardened stage is a very severe destructive ambient worldwide [1-3]. Especially, the infrastructure constructed under the ground water level were easy to suffer the aggressive ion attack. The sulfate ions concentration (in ground water) and the $Ca(OH)_2$ content in the paste of concrete played important roles, and the permeability of concrete was the same [2, 4-7]. According to ACI 318-11 [8], the suggestion of chapter 4 indicated exposure to various concentration sulfate ion, the mixture must be simultaneously considered the compressive strength, water-to-cementitious ratio, cement type used and pozzolans [9-15], etc. The effect of global greenhouse gas emission had a serious issue. After the 2015 Paris Global Climate Summit Meeting [16], internationally the carbon dioxide reduction and sustainability is in full swing taking actual action.

[a] Corresponding author : cyy@cc.hwh.edu.tw

The U.S. Green Building Council (USGBC) is transforming the way which design, build, maintain and operate our buildings, homes and communities [17]. LEED, or Leadership in Energy and Environmental Design, is changing the way we think about how buildings and communities are planned, constructed, maintained and operated. Leaders around the world have made LEED the most widely used third-party verification for green buildings, with around 1.85 million square feet being certified daily. LEED-certified buildings are resource efficient. They use less water and energy and reduce greenhouse gas emissions. The newest version of LEED is designed to be more flexible and improve the overall user experience including materials (Focuses on materials to get a better understanding of what's in them and the effect those components have on human health and the environment), performance-based, smart grid and water efficiency [18-20]. In Taiwan, the Architecture and Building Research Institute promoted EEWH green building evaluation system included ecology, energy saving, waste reduction (CO_2 reduction indicator and construction waste reduction indicator, non-metallic recycled material) and health [21]. LEED for Green options using life cycle inventory was according to Portland Cement Association(PCA) suggestion including the carbon dioxide emission for cement and others cementitious materials of the mix, raw materials used for manufacture process, energy consumption during manufacturing, properly engineering properties including compressive strength development and sustainable durability behaviors to analyze and assess, and applied the most proper mixture from green option for the life cycle inventory [22-24].
The structural concrete with cementitious materials included cement, fly ash, slag, silica fume and other pozzolans used widely in buildings design as specification. Cement industry has been always among the largest CO_2 emission sources. Second only to the water, concrete is the most used material in the world. Almost 5% of global CO_2 emissions are caused by cement plants, while 900 kg CO_2 is emitted to the atmosphere for producing one ton of cement (now 800 kg) [25, 26]. The basis of the carbon dioxide emission of cement includes two parts, one is from consuming the electricity, oil and others fuels to heat until about 1500 ^{o}C, and the other is from the raw materials (lime + clay) decomposing due to high temperature at 900 ^{o}C ($CaCO_3 \rightarrow CaO + CO_2$). The CO_2 emission of two parts are about the same [1,2]. Pozzolans can substitute to cement with high pollution in manufacturing process. The carbon footprint of pozzolans from post-industry production can be regarded as zero. The carbon dioxide emission all have be classified to their industrial products. Pozzolans from post-industry after high temperature treatment was included fly ash, ground-granulated blast furnace slag (Blast-furnace slag is water quenching and grinding to powder. According to PCA suggested the slag CO_2 emitted about 2% of cement [26]), silica fume, etc, above a billion metric tons in 2015 [27, 28]. Pozzolans are an important option for carbon dioxide reduction and sustainable development [29-31]. This paper focuses on an infrastructure for underground transport structure located at the Kaohsiung city in the southern area of Taiwan. The structures were exposed to groundwater containing 3000 ppm for sulfate ion concentration [32]. According to Table 4.2.1(Exposure categories and classes) of ACI 318-11(Building code requirements for structure concrete), the mix proportion applied DMDA [33] method incorporating slag cement (slag : cement=1:1; slag including medium and high fineness) and pozzolans at a w/cm of 0.40 when exposed in sulfate environments. The ACI 211.1 [34] design method was used as a control group for w/c=0.40. The life cycle inventory of cementitious materials in concrete was suggested by PCA as including cement content, CO_2 emission, raw materials, energy consumption and the properties with compressive strength and sulfate resistance [25].

2 Materials, mix proportion and experiments

2.1 Materials

The properties for cementitious materials were according to ASTM C150 as shown in Table 1. The ground granulated blast-furnace slag (produced by the China Hi-Ment Corp.) and Type F fly ash(acquired from Tai-Power Hsin-Ta Thermal Power Plant) conformed to ASTM C989 and ASTM C618 as shown in Table 1. Coarse and fine aggregates met ASTM C33 requirements were as shown in

Table 2. Water was potable water conformed to ASTM C94. The high range water-reducing agents was produced by Sika corporation and used as a high flow agent and conformed to ASTM C494 Type G.

Table 1. Properties of cementitious materials.

Item	Type I	Type II	SC^{MF}	SC^{HF}	Fly ash	$Slag^{MF}$	$Slag^{HF}$
SiO_2	21.5	22.4	-	26.8	51.2	34.0	-
Al_2O_3	4.8	4.6	-	9.2	24.3	14.7	-
Fe_2O_3	3.1	4.9	-	1.9	6.1	0.3	-
CaO	62.4	60.7	-	37.8	6.3	42.0	-
MgO	2.9	3.1	-	5.4	1.6	6.3	-
SO_3	2.1	2.0	-	0.8	0.6	0.4	-
Na_2O	0.22	0.10	-	0.44	0.15	-	-
K_2O	0.70	0.50	-	0.41	0.27	-	-
L.O.I(%)	-	-	-	-	4.9	0.3	-
Fineness(cm_2/g)	3,622	3,420	3,686	4,602	3,110	4,302	5651
G_s	3.14	3.13	3.06	3.05	2.32	2.87	2.88

MF: Medium Fineness Slag Cement (Slag : Cement(I) = 1:1);
HF: High Fineness Slag Cement (Slag : Cement(I) = 1:1).

Table 2. Properties of aggregates.

Properties	Corse Aggregate*	Fine Aggregate**
$G_s(SSD)$	2.64	2.63
$G_s(OD)$	2.63	2.62
Absorption(%)	1.1	2.2
$D_{max.}(mm)$	12.5	-
F.M.(%)	6.08	2.82
$UW_{dry}(kg/m^3)$	1,536	-

*: Crushed stone
**: River sand

2.2 Mix proportion

Table 3. Mixture proportion (kg/m^3)

Materials, kg/m^3	E40[1]	MF40[2]	HF40[3]	CII40[4]
Cement	141	71	71	548
Slag	94	165	165	-
Fly ash	140	140	140	-
Coarse Aggregate	957	957	957	845
Fine Aggregate	936	936	936	749
Water(Included HRWRA)	150	150	150	219
$UW_{concrete}$	2418	2419	2419	2361

1.Used ordinary Portland cement Type I;
2.Used medium Fineness Slag cement (Slag : Cement = 1 : 1);
3.Used high Fineness Slag cement (Slag : Cement = 1 : 1);
4.Used ordinary Portland cement Type II.

This paper applied Densified mixture design algorithm(DMDA) [35] method and a w/c(m)=0.40 to acquire the cement, slag, fly ash, sand, stone and water (with SP) amount as per kg/m^3 (E40). Slag cement (1:1) of different fineness (medium and high fineness shown in Table 1) was used to substitute the cement content of E40 for acting as MF40 and HF40. Finally, using ACI 21.1-12 [36] method with Type II cement designed as a control group(CII40). The mixture proportions were shown in Table 3.

2.3 Experiment

Experiments including compressive strength according to ASTM C31 with 10cmφx20cmH which were curing in saturated lime water and 5000ppm Na_2SO_4 solution, and performed the tests at 7, 28, 56 and 182 days. Anti-sulfate was with 10cmφx20cmH specimens cured in saturated lime water 28 days, then, respectively immersed in 5000ppm and saturated sulfate solution for 24 hours, and oven-dry at 105°C for 24 hours (ASTM C88-13), that was one cycle and repeatedly 25 cycles to measure the weight loss (*n*, ASTM C127), longitudinal pulse velocity variance(*nv*, ASTM C597) and exterior changes. The calculations were as shown in Formula 1-3, and the deterioration of exterior surface due to scaling or spalling was observed.

$$n = \frac{W_2 - W_{12}}{W_1} \times 100\% \quad (1)$$

Where, W_1: Weight for oven-dry 24 hours at 105°C;
W_2: Weight for immersing in a solution for 24 hours, after which the specimen was taken out and dried the surface.

$$v = \frac{L}{\mu_s}(m/s) \quad (2)$$

Here, L: Averaged length of specimen(*m*)
μ_s : Transferring time from both ends($\bullet 10^{-6}$ *second*).

$$nv = \frac{v_{lw} - v_{ppm}}{v_{lw}}(\%) \quad (3)$$

Here, v_{lw}: Velocity for curing in saturated lime water;
V_{ppm}: Velocity for immersed in *x* ppm Na_2SO_4 solution, and took out and dried the surface.

3 Results and discussions

3.1 Cement use

With respect to the CII40 group being 100%, the cement content of E40, MF40, and HF40 were as shown in Table 4. The reduction rate for E40, MF40, and HF40 were 26%, 14%, and 14%, respectively. The cement content of mixtures designed with the DMDA method were 74%~86% less than that of CII40. Incorporation of pozzolans and slag cement in the mixtures allowed a dramatic reduction in the amount of cement. However, it still must be compared with other performances. Pozzolans regarded as post-industrial recycled products are very useful mineral admixtures and ecological materials in concrete. According to suggestions by the PCA, the life cycle inventory for various mixtures of cementitious materials including cement, slag cement, and other pozzolans were compared to CO_2 emissions, raw materials, energy consumption, and engineering design indexes containing factors such as compressive strength and anti-sulfate property as shown in Tables 5-7.

Table 4. The amount of various cementitious materials in concrete mixtures per kg/m^3.

Materials, kg/m^3	E40	MF40	HF40	CII40
Cement	141	71	71	548
Slag	94	164	164	-
Fly ash	140	140	140	-
Rate(%)	26	14	14	100

3.2 CO_2 emission

PCA had performed that CO_2 emission of cement manufacture were 927 kg/per 1000 kg, GGBFS 2% emission of cement, and fly ash and silica fume (both carbon footprint as 0%, because of the CO_2 emission were owing to their front-products of electricity and steel). Calculation results indicated that E40, MF40, and HF40, with respect to CII40 equal to 100%, were 26%, 14%, and 14%, respectively. The mixtures that added pozzolans and slag cement significantly were 74% ~ 86% less than that of CII40. It was apparent that the reduction in the amount of cement was also accompanied by a reduction in CO_2 emissions. Concrete used DMDA mix method incorporating a higher pozzolans content (E40, MF40, and HF40) could be more carbon reduction.

Table 5. CO_2 emission for cementitious materials in concrete mixtures(kg/m^3).

Materials, kg/m^3	E40	MF40	HF40	CII40
Cement	131	66	66	508
Slag	1.7	3.0	3.0	0
Fly ash	0	0	0	0
Total	132.7	69.0	69.0	508.0
Relative Value(%)	**26**	**14**	**14**	**100**

CO_2 Emission of cement: 927 kg/1000 kg of cement;
CO_2 Emission of slag: 2% of 927 kg/1000 kg of cement;
CO_2 Emission of fly ash: None.

3.3 Raw materials

Cement is an industry for higher natural resource requirements. The main raw materials including limestone and clay (weight, 3:1) for the manufacturing process consume a large amount of natural resources. With reference to PCA reports, the raw materials required to produce 1,000 kg of cement consume more than 1,613 kg of natural materials. For the nature was an intrusive destruction, air pollution and consumption of ecological resource were very serious problems. The consumption of raw materials (limestone and clay) for cement in kg/m^3 were shown in Table 6. With respect to CII40 being 100%, E40, MF40, and HF40 were 26%, 13%, and 13%, respectively. Mixtures designed with the DMDA method showed that the raw material consumption of cementitious materials significantly decreased by 74% ~ 87%. For the cement industry, both pozzolans and slag cement play an effect in ecology.

Table 6. The raw materials used for cementitious materials in concrete(kg/m^3).

Materials, kg/m^3	E40	MF40	HF40	CII40
Cement	227	115	115	884
Slag	-	-	-	-
Fly ash	-	-	-	-
Total	227	115	115	884
Relative Value(%)	**26**	**13**	**13**	**100**

Raw materials: 1613 kg/1000 kg of cement

3.4 Energy consumption

The PCA literature states that the energy consumption for cement manufacturing is 4.8 GJ/1,000 kg of cement, and grinding process consumed for granulated blast-furnace slag was 14% of cement per 1,000kg. Since the energy consumption was due to their front-products, electricity and steel, fly ash, and silica fume were regarded as being zero. According to Table 7, the energy consumption of cementitious materials in the mixtures, with respect to CII40 being 100%, were 28% (E40), 17% (MF40), and 17% (HF40). E40, HF40, and HF40 used the DMDA method, and compared to CII40, consumed 72 ~ 83% less energy for the cementitious compounds. The expected effect on energy saving of CII40 which adopted the ACI 211.1 method was much less.

Table 7. Energy consumption for cementitious materials in concrete mixture(kg/m^3).

Materials, GJ/t(1000kg)	E40	MF40	HF40	CII40
Cement	0.677	0.341	0.341	2.631
Slag	0.063	0.110	0.110	-
Fly ash	0	0	0	-
Total	0.740	0.441	0.441	2.631
Relative Value(%)	**28**	**17**	**17**	**100**

Cement Energy Consumption: 4.8 GJ/1000 kg Cement;
Slag Energy Consumption: 14% of 4.8 GJ/per 1000 kg cement.

3.5 Compressive strength effect

The life cycle inventory (front four items) was used to assess the effect of design and production for cementitious materials in the mixtures. The hardened properties and anti-sulfate durability were used to evaluate and predict service performance. This study measured the compressive strength by curing in saturated lime water and immersing in a 5,000 ppm sulfate solution.

The compressive strength development (7d, 28d, 56d, and 182 days) for curing in saturated lime water and 5,000 ppm sulfate solution were as shown in Table 8. At 28 days, E40, HF40, and HF40 showed compressive strengths of 47 MPa and 48MPa, 43 MPa and 45 MPa, and 48 MPa and 47 MPa, respectively. Meanwhile, those of CII40 were 47 MPa and 47 MPa. The variance of two curing ambient were less than 5%. All the mixtures were suitable in environments exposed to high concentrations of sulfate solution (ACI 318-13). Meanwhile, at 182 days, the compressive strengths of E40, MF40, HF40, and CII40 were 71 MPa and 72 MPa, 62 MPa, and 63 MPa, 67 MPa and 66 MPa, 62 MPa and 62 MPa, respectively. The amount of cement for CII40 was too high to increase the compressive strength. Pozzolans and slag cement for anti-sulfate concrete had a very important effect in lowering the heat of hydration and refining the microstructure to be denser to improve long-term properties. The compressive strength ratios at 182 days, with respect to CII40 being 100%, were 115% and 113% for E40, 100% and 102% for MF40, 108% and 105% for HF40 as shown in Table 9. E40 with pozzolans increased the compressive strength. MF40 and HF40 incorporating pozzolans and slag cement indicated an increased fineness and a better strength enhancement. The finer blast-furnace slag grinding following water quenching was a useful application for ecology and carbon reduction.

Table 8. Compressive strength development (curing in saturated lime water/5000ppm Na_2SO_4 solution).

Compressive Strength(MPa)	E40	MF40	HF40	CII40
7 days	34/36	29/33	31/35	40/37
28 days	47/48	43/45	48/47	47/47
56 days	54/55	55/55	56/56	54/54
182 days	71/70	62/63	67/66	62/62

Table 9. Strength ratio with respect to CII40 as 100%(curing in saturated lime water/5000ppm Na_2SO_4 solution).

Compressive Strength(%)	E40	MF40	HF40	CII40
7 days	85/97	73/89	78/95	100/100
28 days	100/102	92/96	100/100	100/100
56 days	100/102	102/102	104/104	100/100
182 days	115/113	100/102	108/105	100/100

3.6 Sulfate resistance effect

In order to ensure that the effect for mixtures immersed in different concentration sulfate solution, this study was according to ASTM C88-10 repeated exposure to the wet-dry extreme environment to observe the exterior and measure the difference for weight loss and longitudinal pulse velocity. Specimens were immersed in a solution for 24 hours and oven-dried for 24 hours at 105°C, and this cyclic process was repeated for 25 cycles. The exterior was observed to gage whether scaling, spalling, or nothing occurred. Upon oven-drying, the weight and longitudinal pulse velocity of the specimens were measured. Test results were as shown in Table 10.

3.6.1 Exterior changes

The specimens were immersed in a 5,000 ppm Na_2SO_4 solution and oven-dried, with this process being repeated for 25 cycles. Upon careful examination of the external surface, the exteriors of all the mixtures were observed to be neither damaged nor had defects. The ASTM C88-10 method showed that the force of expansion developed from the sulfates during multiple immersions was somewhat similar to the expansion of water during freezing. Oven-drying the specimens at 105 °C was similar to freezing and immersing them in the sulfate solution was similar to a sulfate attack at an ambient temperature. The method was constructed at extreme exposure environments to accelerate the speed of destruction.

However, when the specimens were immersed in a saturated solution of Na_2SO_4, the exteriors displayed differences among the surfaces. Mix CII40 underwent scaling from the second cycle; MF40 and HF40 began spalling significantly after the fifth cycle; and E40 commenced slight spalling (just like tiny pitting) during the ninth cycle. From the exterior observation, the mixtures showed that E40 had the best anti-sulfate resistance performance due to the compressive test and immersion in saturated sulfate solution, while MF40 and HF40 were superior to CII40 [35].

3.6.2 Weight loss

The specimens during the extreme test process, especially during oven-drying at 105°C and the chemical reaction of the sulfate attack to form ettringite induced scaling, spalling, and weight loss as shown in Table 10. Following immersion in a 5,000 ppm solution of Na_2O_4, the weight loss rates (n) of the mixtures were less than -1% after the 25 cycles. At -1%, CII40 had the highest weight loss rate and no exterior abnormalities. Other mixtures also had the same behaviors, and E40 had the smallest n value as shown in Figure 1. Following immersion in the saturated solution of Na_2SO_4, n_{E40}, n_{MF40}, n_{HF40}, and n_{CII40} were 0.91% (24 cycles), -2.62% (20 cycles), -2.12% (24 cycles), and -2.05% (24 cycles), respectively. The test results showed that n_{E40} was less than -1%, and the others were all greater than -2%. A value of -2.62% was recorded for n_{MF40} as shown in Figure 1. E40 had the smallest n value due to weight loss.

Table 10. The weight loss and pulse velocity change for specimen with various mixtures.

Mix	5000ppm Na_2O_4 solution					Saturated Na_2O_4 solution				
	n %	N cycles	V m/s	n	*nv* %	*n* , %	N (cycles)	V m/s	n	*nv* %
E40	-0.55	19	4500*	12	-1.7	-0.91	24	4280	22	-6.5
MF40	-0.62	7	4400*	25	-2.7	-2.62	20	4030	11	-10.9
HF40	-0.71	9	4480*	25	-4.4	-2.12	24	4250	11	-9.3
CII40	-1.00	24	4300*	15	-2.6	-2.05	24	3945	5	-10.7

*: The ultrasonic pulse velocity of specimens cured in lime water, at 28days, were respectively 4576, 4524, 4685 and 4474m/s.

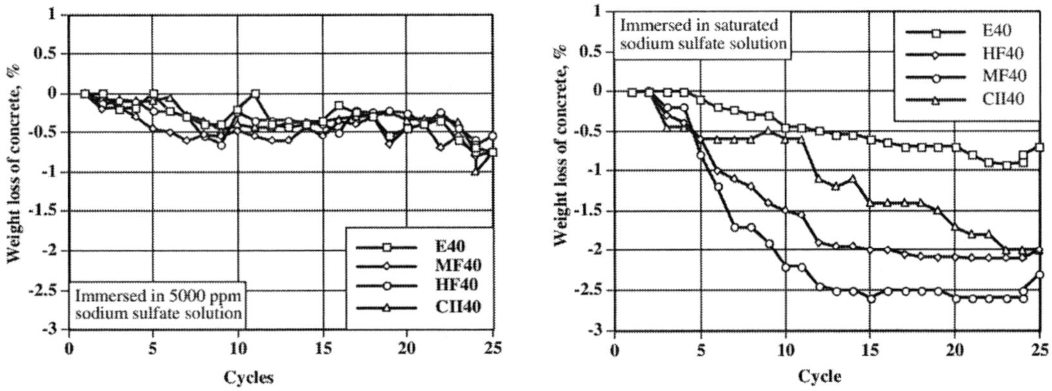

Figure 1. Weight loss development due to immersion in a 5,000 ppm saturated solution of Na_2SO_4 and oven-drying repeatedly for various mixtures.

3.6.3 Ultrasonic pulse velocity(UPV) variance

The ultrasonic pulse velocity (UPV) transferred through the longitudinal direction of a specimen to detect interior defects such as (micro)cracks, voids, and honeycombs, can indirectly measure the quality of concrete. Moreover, according to Whitehurst [37] in his nondestructive testing concrete review, the pulse velocity for good and excellent concrete quality reported the UPV ranges from 3,660 m/s to 4,570 m/s. In Table 11 and Figure 2, specimens immersed in the 5,000 ppm Na_2SO_4 solution showed that the lowest UPV (see Figure 2, left) for E40, MF40, HF40, and CII40 were 4,500 m/s, 4,400 m/s, 4,480 m/s, and 4300 m/s, respectively; meanwhile, in the saturated Na_2SO_4 solution, the UPV dropped down to 4,280 m/s, 4,030 m/s, 4,250 m/s, and 3,945 m/s, respectively. The specimens cured in lime water at 28 days were 4,576 m/s (E40), 4,524 m/s (MF40), 4,685 m/s (HF40), and 4,474m/s (CII40). Repeatedly immersed in 5,000 ppm Na2SO4 solution and oven-dry; and repeatedly immersed in saturated and oven-dried ambient. The variance was -1.7%, -2.7%, -4.4%, -3.9% and -6.5%, -10.9%, -9.3%, -11.8%. The *nv* values were less than 5% for specimens repeatedly immersed in the 5,000 ppm Na_2SO_4 solution and oven-dried. Mixtures designed with pozzolans, slag cement, Type II cement, w/cm=0.40, and compressive strength, could have an improved anti-sulfate property.

But the result of immersion for saturated sulfate solution ambient had significantly different, either high concentration accelerating sulfate attack or due to oven-dried at 105 °C. The deterioration, with respect to curing in saturated lime water, showed all *nv* values to be above 5%. Meanwhile, MF40 and CII40 had values above 10%: nv_{MF40} (-10.9%) and nv_{CII40} (-11.08%) as shown in Table 10 and Figure 2 (right). Repeatedly, the extremely exposed ambient with chemical attack of Na_2SO_4 and oven-dried influenced the exterior integrity and interior compactness. These factors were due to incongruous force of wet-dry and expansion-attraction between the paste and aggregates. From the results, nv_{E40} of -6.5% was the lowest among the mixtures. The variables in mixtures were C_3S and C_3A content. In Table 11, with respect to CII40 as 100%, E40, HF40 and MF40 which displayed C_3S and C_3A amount were much less than CII40, and hydration products of CH were much fewer. Pozzolanic reaction of these mixtures (excluded CII40) with oven-dry process accelerated the compactness of interior structure. The cement paste for CII4 mix was even higher than others. The objects of suffered attack were more widely, and the water amount was higher enough to loosen the unit weight (see Table 3). The UPV for before and later experimental results extended the minimum values, but was completely greater 3,660 m/s. The mixtures in this study displayed that interior quality were better. E40 had a better anti-sulfate attack performance. The order of sulfate resistance in terms of decreasing performance was E40, HF40, MF40, and CII40. In the study, the mixtures with DMDA method incorporating pozzolans, slag cement and lower water-to-cementitious ratio might be a viable approach.

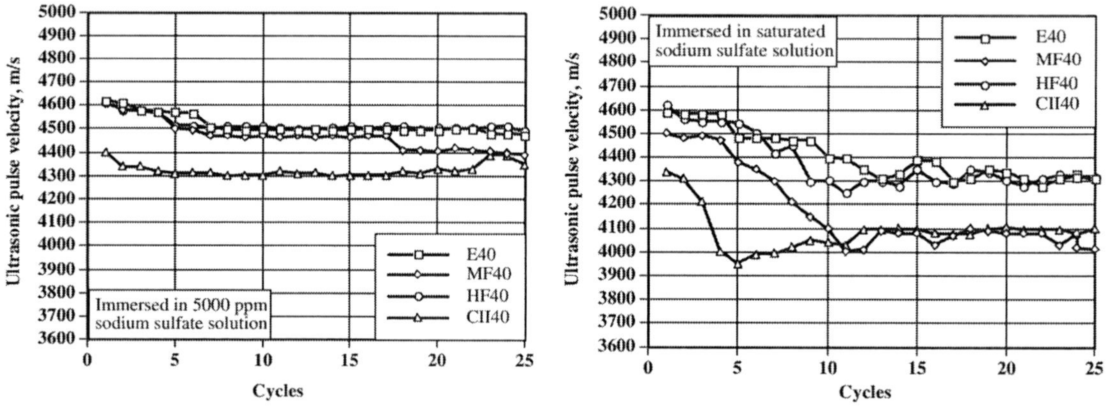

Figure 2. Ultrasonic pulse velocity development of immersed in 5000ppm and saturated solution of Na_2SO_4 and oven-dry repeatedly for various mixtures.

Table 11. The weight and ratio for clinker composition in the mixtures.

Mix	E40	MF40	HF40	CII40
Cement(kg/m^3)	141	71	71	548
$C_3S^\#$	30.7	15.4	15.4	205.5
$C_2S^\#$	78.0	39.0	39.0	215.9
$C_3A^\#$	10.7	5.4	5.4	21.4
$C_3S(\%)^*$	15.0	7.5	7.5	100.0
$C_2S(\%)^*$	36.1	18.1	18.1	100.0
$C_3A(\%)^*$	50.1	25.1	25.1	100.0

#: The component ratio of C_3S, C_2S and C_3A are referenced in Table 1.
*: With respect to CII40 as 100%.

3.7 Assessment of green index

Results of life cycle inventory for concrete mixtures with various pozzolans are given in Table 12. The MF40 and HF40 mixtures displayed a higher carbon reduction than E40. From the view point of cement content, CII40 had a large amount of cement used in the mixture, Type II cement was more expensive than Type I, and the properties were not better as expected. The results of CO_2 emissions, raw materials, and energy consumption were similar to cement used. The differences for compressive strength cured in saturated lime water and immersed in a 5,000 ppm solution of Na_2SO_4 were all below 5%. The mixtures in this study with various design methods could be applied in extreme sulfate attack environments. The effect of repeatedly immersing in the 5,000 ppm solution and oven-drying for 25 cycles had similar results. For repeatedly immersing in the saturated Na_2SO_4 solution and oven-drying for 25 cycles, E40 showed the best performance, others were HF40, MF40, and CII40. These mixtures met the requirements for environments with serious sulfate exposure. Integrating carbon reduction, hardened properties, and durability, E40 and H40 with the DMDA method containing a large amount of pozzolans enhanced the long-term properties and anti-sulfate performance.

Table 12. The comprehensive evaluation of green options for different mixtures(%).

Parameter(%)		E40	MF40	HF40	CII40
Cement use		26	14	14	100
CO_2 Emission		26	14	14	100
Raw Materials		26	13	13	100
Energy Consumption		28	17	17	100
Compressive Strength	28 days(N^*)	100	92	100	100
	28 days(a^*)	102	96	100	100
	56 days(N^*)	100	102	104	100
	56 days(a^*)	102	102	104	100
	182 days(N^*)	115	100	108	100
	182 days(a^*)	113	102	105	100
Sulfate resistance (cycles, deterioration)	28 days(1^{**})	➢ 25	➢ 25	➢ 25	➢ 25
	Exterior damage	None	None	None	None
	n (1^{**})	-0.55	-0.62	-0.71	-1.00
	nv (1^{**})	-1.7	-2.7	-4.4	-2.6
	28 days(2^{***})	9	5	5	2
	Exterior damage	Initial spalling	Initial spalling	Initial spalling	Initial chalking
	n (2^{***})	-0.91	-2.62	-2.12	-2.05
	nv (2^{***})	-6.50	-10.90	-9.30	-10.70

N*: Cured in 5000ppm solution of Na_2SO_4.
a*: Cured in saturated lime water.
1**: Repeatedly immersed in 5000ppm solution of sodium sulfate and oven-dried.
2***: Repeatedly immersed in saturated solution of sodium sulfate and oven-dried.

4 Conclusions

In this study, mixture proportions adopted a densified mixture design algorithm (DMDA) method, a water-to-cement (w/cm) of 0.40, various amounts of pozzolans, and slag cement of different fineness values (E40, MF40, and H40). The ACI design method acted as a control group (CII40). Experiments conducted included compressive strength and repeated immersion in 5,000 ppm and saturated Na_2SO_4 solutions. Green options for anti-corrosion of the life cycle inventory of Leadership in Energy and Environmental Design (LEED) suggested by the Portland Cement Association (PCA) included CO_2 emissions, raw materials consumption, energy consumption, compressive strength, and immersion effect of the Na_2SO_4 solution (variance of weight loss and pulse velocity). Analysis and test results were as follows:

1. The cement content for E40, MF40, and HF40 were 26%, 14%, and 14%, respectively, with respect to CII40 being 100%.
2. CO_2 emissions for E40, MF40, HF40, and CII40 were 133 kg/m^3, 69 kg/m^3, 69 kg/m^3, 508 kg/m^3, respectively. A mix proportion incorporating the correct amount of pozzolans and slag cement can greatly decrease cement use and reduce CO_2 emissions.
3. Raw materials for cement manufacturing for E40, MF40, HF40, and CII40 were 26% 13%, 13%, and 100%, respectively. The effect of pozzolans used in the mixture can decrease the ecological impact.

4. Energy consumption of cement manufacturing for E40, MF40, and HF40 were 28%, 17%, and 17%, with respect to CII40 being 100%. The mixtures incorporating pozzolans showed energy saving benefits.
5. After 28 days, curing in saturated lime water and the 5,000 ppm sulfate solution showed no visible change in the compressive strength and all mixtures were suitable in extreme sulfate solution environments.
6. Specimens were repeatedly immersed in 5,000 ppm and saturated Na_2SO_4 solutions, and were oven-driing at an ambient temperature of 105°C. Results showed that the exterior had no change in the 5,000 ppm Na_2SO_4 solution. All the mixtures were properly designed. In the saturated Na_2SO_4solution, exterior damage appeared in the second cycle (CII40, initial scaling), the fifth cycle (MF40 and HF40, initial spalling), and the ninth cycle (E40, initial spalling).
7. After repeated immersion in the 5,000 ppm Na_2SO_4 solution, all weight loss variance (n) values were less than 1%, and the ultrasonic pulse velocity variance (nv) values were less 5%. After repeated immersion in the saturated Na_2SO_4 solution, n values ranged from -0.91% (E40, the smallest) to -2.62% (MF40, the largest); meanwhile, nv values ranged from -6.7% (E40, the smallest) to -10.9% (MF40, the largest).
8. The comprehensive evaluation of green options for anti-sulfate performance of the mixtures in serious sulfate exposure environments indicated the following: E40 > HF40 > MF40 > CII40.

References

1. P.K. Mehta and P.J. Monteiro, *Concrete: microstructure, properties, and materials, fourth edition* (McGraw-Hill Education, New York, 2014)
2. S. Mindess, J.F. Young, and D. Darwin, *Concrete, second edition* (Pearson, London, 2002)
3. A.M. Neville, *Properties of concrete, fifth edition* (Prentice Hall, San Francisco, USA, 2012)
4. A. M. Hossack and M.D.A. Thomas, Cem. Concr. Res., **73**, 136 (2015)
5. M.L. Nehdi, A.R. Suleiman, and A.M. Soliman, Cem. Concr. Res., **64**, 42 (2014)
6. C. Yu, W. Sun, and K. Scrivener, Cem. Concr. Res., **43**, 105 (2013)
7. Z. Makhloufi, S. Aggoun, B. Benabed, E. H. Kadri, and M. Bederina, Cem. Concr. Compos., **65**, 186 (2016)
8. ACI Committee 318, *Building code requirements for structural concrete (ACI 318-11) and commentary* (ACI, Farmington Hills, MI, 2011)
9. F. Mittermayr, M. Rezvani, A. Baldermann, S. Hainer, P. Breitenbücher, J. Juhart, C.A. Graubner, and T. Proske, Cem. Concr. Compos., **55**, 364 (2015)
10. A. Skaropoulou, K. Sotiriadis, G. Kakali, and S. Tsivilis, Cem. Concr. Compos., **37**, 267 (2013)
11. L. Jiang and D. Niu, Concr. Build. Mater., **117**, 88 (2016)
12. S. Mirvalad and M. Nokken, Constr. Build. Mater., **84**, 19 (2015)
13. N. Ghafoori, M. Najimi, H. Diawara, and M.S. Islam, Constr. Build. Mater., **78**, 85 (2015)
14. I. Sirisawat, W. Saengsoy, L. Baingam, P. Krammart, and S. Tangtermsirikul, Constr. Build. Mater., **64**, 39 (2014)
15. A. Merida and F. Kharchi, Procedia Eng., **114**, 832 (2015)
16. United Nations Framework Convention on Climate Change, Paris Climate Change Conference, Nov. (2015) http://unfccc.int/meetings/paris_nov_2015/meeting/8926.php
17. A.P. Gurgun, R. Komurlu, and D. Arditi, Procedia Eng., **118**, 1145 (2015)
18. A.M. Rashad, Int. J. Sustain. Built Environ., **4**, 278 (2015)
19. H. Dekkiche and A. Taileb, Procedia Eng., **145**, 844 (2016)
20. C.L. Champagne and C.B. Aktas, Procedia Eng., **145**, 380 (2016)
21. F. Asdrubali, G. Baldinelli, F. Bianchi, and S. Sambuco, Build. Environ., **86**, 98 (2015)
22. Architecture and Building Research Institute, Ministry of the Interior, *Green building evaluation manual-basic version* (Architecture and Building Research Institute, Taipei, Taiwan, 2015)
23. M.L. Marceau, M.A. Nisbet, and M.G. VanGeem, *Life cycle inventory of Portland cement manufacture, PCA R&D Serial No. 2095b* (Portland Cement Association, Skokie, IL, 2005)

24. M.A. Nisbet, M.L. Marceau, and M.G. VanGeem, *Environmental life cycle inventory of Portland cement concrete, PCA R&D Serial No. 2137a* (Portland Cement Association, Skokie, IL, 2002)
25. M.L. Marceau, M.A. Nisbet, and M.G. VanGeem, *Life cycle inventory of Portland cement concrete, PCA R&D Serial No. 3007* (Portland Cement Association, Skokie, IL, 2007)
26. L. Acquaye, A. Chini, and M. Danso-Amoako, *Green options for high temperature cured concrete through the use of fly ash and blast furnace slag* (33rd Conference on Our World in Concrete & Structures, Singapore, August 25-27, 2008)
27. K. Scrivener, Chem. World. **March**, 62 (2008)
28. M.E. Haque, Int. J. Waste Resour., **3**, 22 (2013)
29. Y. Wei, Q. Wang, G. Ying, Z. Luo, and Y. Ninomiya, *Characteristics and utilization of fly ash from coal-fired power plant in China* (2015 World of Coal Ash Conference, Nasvhille, TN, May 5-7, 2015)
30. C. Meyer, Cem. Concr. Compos., **31**, 601(2009)
31. M. Nili and A.M. Salehi, Constr. Build. Mater., **24**, 2108 (2010)
32. N. Cobîrzan, A.A. Balog, and E. Moşonyi, Procedia Technol., **19**, 506 (2015)
33. K.W. Cheng, *The study on the engineering properties of densify concrete with high volume slag content* (Master Thesis, Department of Construction Engineering, National Taiwan University of Science and Technology, Taipei Taiwan, ROC, 2002)
34. Y.Y. Chen, B.L.A. Tuan, and C.L. Hwang, Constr. Build. Mater., **47**, 340 (2013)
35. Y.Y. Chen, C.C. Wu, and K.W. Cheng, J. Archit., **91**, 21 (2015)
36. ACI Committee 211, *Standard practice for selecting proportions for normal, heavyweight, and mass concrete (Reapproved 2009)* (ACI, Farmington Hills, MI, 1991)
37. E.A. Whitehurst, *Soniscope tests concrete structures* (Portland Cement Association, Chicago, 1951)

Green options for anti-corrosion of high strength concrete incorporating ternary pozzolan materials

Yuan-Yuan Chen[a], Chien-Jou Yang

Department of Architecture, Hwa Hsia University of Technology, New Taipei City 235, Taiwan

Abstract. This paper applied the densified mixture design algorithm(DMDA) method by incorporating ternary pozzolans (fly ash, slag and silica fume; mix I and mix II) to design high strength concrete (HSC) mixtures with w/cm ratios from 0.24 to 0.30. Concrete without pozzolans was used as a control group (mix III, w/c from 0.24 to 0.30), and silica fume (5%) was added as a substitute for part of the cement and set as mix IV. Experiments performed compressive strength, four-point resistance meter to measure the conductivity, and rapid chloride ion penetrability tests (ASTM C1202) were assessed the anti-corrosion. The life cycle inventory of LEED suggested by the PCA indicated the green options for cementitious materials. Results showed that mix I and II indicated cement used, CO_2 reduction, raw materials and energy consumption all decreased more 50% than mix III, and mix IV was 5% less. The compressive strength and anti-corrosion levels showed that mix I and II were better than mix III and IV, and with ternary pozzolans could enhance the long-term durability (92 days) due to a resistivity greater 20 KΩ-cm and a charge passed lower than 2000 Coulombs. HSC with an appropriate design could reduce the carbon footprint and improve the durability.

1 High strength concrete and corrosion

The definition of high-strength concrete (HSC) is the specified strength greater 41 MPa (6000 psi) [1-3]. HSC mixtures have two key characteristics: a low water-to-cement (cementitious) ratio and a high heat of hydration [1]. These factors lead to the thermal stress between the hydration products of the paste and aggregates, and cause inconsistent deformation of the interface, in turn forming many micro and thermal cracks [4-7]. That is a severe durability problem for HSC, especially when exposed in or adjacent marine environment. Corrosion is the most important and critical issue for concrete structures in Taiwan (because the country was surrounded by the sea). Many factors influenced the heat of hydration of HSC. The most commonly used method for creating HSC was incorporating pozzolans (post-industry products- such as fly ash, blast-furnace slag and silica fume; natural such as calcined clay, calcined shale, and meta-kaolin) in the mixture to decrease its cement content and enhance the fresh, hardened properties and long-term durability [4, 7]. Water, oxygen and chloride ions play important roles in the corrosion of embedded steel and the cracking of concrete. It is obvious that permeability of concrete is the key to control the various processes involved in these phenomena. Concrete mixture properties to ensure low permeability including low water-to-cementitious (w/cm), adequate cement content, control of aggregate size and grading, and use of mineral admixtures [4].

[a] Corresponding author : cyy@cc.hwh.edu.tw

The ACI 318-11 structural concrete building code is based on the exposure classes presented in Table 4.2.1(P: Requiring low permeability; C: Corrosion protection of reinforcement). It states that concrete mixtures should comply with the most restrictive requirements according to Table 4.3.1 (P1: In contact with water where low permeability is required; C2: Concrete exposed to moisture and an external resource of chlorides from deicing chemicals, salt, brackish water, seawater, or spray from these resources). For permeability requirements, the P1 class should have a w/cm less than 0.50 and a specified strength (f'_c) was greater than 4000 psi and the C2 class should have a w/cm below 0.40 and f'_c was above 5000 psi [8]. Proper consolidation(compactness) and curing of concrete are equally essential [4]. One standard method that provides a performance-based indicator of low permeability of concrete is ASTM C1202 for anti-corrosion experiments on permeability and corrosion resistance, which has been found to be reliable in laboratory evaluation than for acceptance in the field. The use of ASTM C1202 to test concrete mixtures proposed for use provides additional information on the performance of these mixtures [9]. For the Confederation Bridge linking Prince Edward Island with New Brunswick in mainland Canada (1997), the anti-corrosion performance of the concrete exterior exposed to a seawater environment showed that it was free crack, and that the higher electrical resistance is much more important than a lower permeability [10, 11].

2 Pozzolans and LEED

The main ingredients of concrete are coarse aggregates, fine aggregates, cementitious materials (cement and pozzolans), and water (included liquid admixture) [4-6]. The carbon footprint for concrete mainly emitted constitute was cement, contributes 5 per cent of annual anthropogenic global CO_2 production [12, 13]. According to a 2015 statistic of the worldwide cement production by major producing countries from 2010 to 2015 [14], China, India, and the United States were among the global top three. The production exceeded more than 40 billion metric tons in 2015. For every 1000 kg of cement manufactured, CO_2 emissions were reduced from 927 kg(before) to 800 kg(now) [12, 15]. The cement industry is a high-intensity manufacturing process. Every 1000 kg of cement produced consumes 1613 kg of raw materials (containing 72% of limestone, 13% of cement rock, 3% of shale, 4% of clay, and 2% of sand). The temperature at the lower end of the kiln is 1870 °C. In the hottest part of the kiln, the raw materials reach a temperature of 1480°C and melt. The molten material from the kiln, known as clinker, is ground into cement. A small amount of gypsum is added to the cement to control its setting time. The weighted average energy consumption, including fuel and electricity, during cement production is 4.8 GJ/metric ton of cement [16-20]. To process slag, it is from a blast-furnace during the production of pig iron, can be used a supplementary cementitious material as a partial replacement for Portland cement in concrete, It quenched with water and ground. One metric ton of slag uses about 14% of energy to produce one metric ton of cement. Additionally, one metric ton of slag produces about 2% of CO_2 emissions to produce one metric ton of cement [16, 21]. The carbon footprint for fly ash and silica fume were regarded as being zero because of recycling and the prevention of air pollution. CO_2 emissions were produced by their main products, electricity and siliceous-steel.

The supplementary cementitious materials (fly ash, ground granulated blast-furnace slag, silica fume, and others pozzolans) used for concrete could contribute to green design as follows: (1) As industrial by-products, their use as a partial replacement for Portland cement does not contribute to the energy and CO_2 impacts of cement in concrete; (2)Raw material usage is reduced in the manufacture of concrete; (3) Landfill disposal is reduced and there is an increased use of recovered industrial materials; and (4) SCMs improve concrete service life through greater concrete durability [21-24]. The commonly used materials include fly ash (C and F classes), ground granulated blast-furnace slag, and silica fume, especially fly ash and slag. Worldwide production exceeded over a billion metric tons. These post-industry products were recycled, reused, and reduced to avoid air pollution and to waste a large amount of landfill disposal fields. In many countries, pozzolans have been a recycled resource incorporated in concrete to enhance their properties. Pozzolans were beneficial for improving the durability of anti-corrosion, sulfate resistance, alkali-silica reaction, and regarded as the fifth required material in concrete.

The Leadership in Energy and Environmental Design(LEED) Green Building Rating System was developed by the U.S. Green Building Council (USGBC) in 1998. The LEED rating system is a voluntary, consensus-based, mark driven rating system based on existing proven technology [25-28]. It evaluates environmental performance from a whole building perspective over a building's life cycle, providing a definitive standard for what constitutes a "green building." The rating system is organized into five environmental categories: Sustainable Sites, Water Efficiency, Energy & Atmosphere, Materials & Resources, and Indoor Environmental Quality. An additional category, namely Innovation & Design Process, addresses sustainable building expertise as well as design measures not covered under the five environmental categories [25]. LEED is a performance-based system where credits are earned for satisfying criteria designed to address specific environmental impacts inherent in the design, construction, and operation and maintenance of buildings. For the LEED rating system for New Construction and Major Renovations (LEED-NC2.2) different levels of green building certification are awarded based on the total credits earned rating including Certified, Silver, Gold, and Platinum. The LEED system is designed to be comprehensive in scope yet simple in operation. The use of fly ash and slag can earn about 6 possible points (Sustainable Sites- 1 point, Material & Resources - 4 points, and Innovation in Design - 1 point) [29, 30].

The green options for anti-corrosion of high strength concrete with quaternary cementitious materials (cement, fly ash, slag, and silica fume) applied a densified mixture design algorithm(DMDA) and ACI 211.1R methods to examine the mixture feasibility. The life cycle inventory for the concrete mixture suggested by the Portland Cement Association (PCA) [31-35] and properties comprised of cement used, CO_2 emission, raw materials, energy consumption, compressive strength, surface resistivity, and charge passed.

3 Materials, Mix design and Experiments

3.1 Materials

The component and properties of cementitious materials are respectively matched ASTM C150 (cement), ASTM C618 (fly ash), ASTM C989 (slag) and ASTM C1240(silica fume), as shown in Table 1. Coarse and fine aggregates met ASTM C33 requirements, as shown in Table 2. Water was potable water conformed to ASTM C94. The high range water- reducing agents was produced by Sica Chem. Corp. used as a high flow agent and conformed to ASTM C494 Type G.

Table 1. Properties of cementitious materials.

Item	Cement	Fly ash	Slag	Silica fume
SiO_2	22.0	51.2	34.9	97.7
Al_2O_3	5.6	24.3	13.5	0.7
Fe_2O_3	3.4	6.1	0.3	0.1
CaO	62.8	6.3	41.8	0.4
MgO	2.6	1.6	7.2	0.4
SO_3	2.1	0.6	1.7	0.3
f-CaO	1.1	-	-	-
TiO_2	0.5	-	-	-
Na_2O	0.44	-	-	-
K_2O	0.78	-	-	-
V_2O_5	0.05	-	-	-

Table 2. Properties of aggregates.

Properties	Corse Aggregate*	Fine Aggregate**
$G_s(SSD)$	2.64	2.63
$G_s(OD)$	2.63	2.62
Absorption(%)	1.1	2.2
$D_{max.}(mm)$	12.5	-
F.M.(%)	6.08	2.82
$UW_{dry}(kg/m^3)$	1,536	-

*: Crushed stone
**: River sand

3.2 Mix proportion

The DMDA [36] and ACI 211.1-09 [37] method applied to design the high strength concrete mixtures with water-to-cementitious material(w/c(m)) ratios from 0.24 to 0.30 and different water contents. Ternary pozzolans were incorporated into mix I (5% slag) and mix II (FA: Slag = 7:3) for a higher slag content. ACI 211.1 without pozzolans was set as the control group (mix III), and 5% of cement was substituted using silica fume in mix IV. The results of the mix design were as shown in Tables 3-5.

Table 3. Mixtures proportion for I groups(kg/m^3).

Materials kg/m^3	I-246	I-247	I-248	I-276	I-277	I-278	I-306	I-307	I-308
Cement	284	376	466	217	299	379	197	238	310
Slag	16	21	26	12	17	21	11	13	17
Fly Ash	266	238	220	264	248	231	264	255	240
Silica Fume	15	20	21	12	16	20	10	13	16
Coarse Aggregate	753	700	648	777	728	680	776	750	706
Fine Aggregate	957	890	824	988	926	865	986	954	897
Water (Incl. HRWR)	137	157	177	136	156	176	144	156	175
Rate(%)	**29**	**39**	**48**	**25**	**35**	**44**	**26**	**31**	**40**

*With respect to mix III × 100%.

Table 4. Mixtures proportion for II groups(kg/m^3).

Materials kg/m^3	II-246	II-247	II-28	II-276	II-277	II-278	II-306	II-307	II-308
Cement	242	338	431	212	259	342	155	197	271
Slag	87	87	87	85	85	85	84	84	84
Fly Ash	220	204	189	221	213	199	227	219	206
Silica Fume	22	26	31	20	22	26	18	19	23
Coarse Aggregate	831	772	715	833	804	751	856	828	779
Fine Aggregate	882	820	760	885	853	797	909	879	827
Water (Incl. HRWR)	138	157	177	146	156	176	144	156	175
Rate(%)	**25**	**35**	**45**	**25**	**30**	**40**	**20**	**26**	**35**

*With respect to mix III × 100%.

Table 5. Mixtures proportion for III and IV groups(kg/m^3).

Material kg/m^3	III-2411	III-2711	III-3011	IV-2411	IV-2711	IV-3011
Cement	963	856	770	915	813	731
Slag	-	-	-	-	-	-
Fly ash	-	-	-	-	-	-
Silica fume	-	-	-	48	43	39
Coarse Aggregate	633	633	633	633	633	633
Fine Aggregate	471	560	632	413	509	585
Water (Non-HRWRA)	231	231	231	231	231	231
Rate(%)*	**100**	**100**	**100**	**95**	**95**	**95**

*With respect to mix III × 100%.

3.3 Experiments

The life cycle inventory for anti-corrosion characteristics by LEED according to the PCA included cement content, CO_2 emissions, raw materials, energy consumption, compressive strength (ASTM C39 at 7, 28, 56 and 92 days), anti-corrosion properties containing electrical resistivity(four-point resistivity meter made by Swiss Proceed at 7, 28, 56, and 92 days) upon contact with different concrete sections, and a rapid chloride ion penetrability test(ASTM C1202, at 92 day) [9] as shown in Table 6.

Table 6. Criteria of RCPT by ASTM C1202.

Measured Value, Coulombs	Chloride Ion Penetrability
> 4000	High
4000 - 2001	Moderate
2000 - 1001	Low
1000-101	Very Low
< 100	Negligible

4 Results and discussions

4.1 Cement use

With respect to mix III as being 100%, the cement content of mix I, mix II, and mix IV were as shown in the bottom of Tables 3-5. It indicated that mix IV contained 5% less cement. The decreasing rates for mix I and mix II were 25% ~ 48% and 20% ~ 45%, respectively. The amount of cement for mix I and mix II designed by the DMDA method were 50% less than mix III. Pozzolans used in the mixtures could dramatically reduce the cement content. However, it must still be compared with each property. According to suggestions from the PCA, the life cycle inventory for various mixtures of cementitious materials including cement and others pozzolans were compared to CO_2 emissions, raw materials, and energy consumption as shown in Tables 7-15.

4.2 CO_2 emission

Tables 7-9 show that the CO_2 emissions of cementitious material were 927 kg/1000 kg of cement, GGBFS 2% emission of cement, and the carbon dioxide footprint from fly ash and silica fume regarded as 0%. Calculation results showed that CO_2 emissions for mix I and mix II, with respect to mix III as being 100%, were 20% ~ 45% and 20% ~ 40%, respectively. The mixtures significantly decreased 50% more than mix III. The CO_2 reduction rate for mix IV was only 5% similar to the cement reduction. A reduction in the cement content caused a reduction in CO_2 emissions. High strength concrete mixtures with multi-pozzolans (mix I and mix II) may allow more carbon reduction.

Table 7. Life cycle inventory for LEED - CO_2 emission of binders for mix I.

Life cycle inventory- CO_2 emission of binders for concrete mix(kg/m^3)									
Material kg/m^3	I-246	I-247	I-248	I-276	I-277	I-278	I-306	I-307	I-308
Cement	263	349	432	201	277	351	183	221	287
Slag	0	0	0	0	0	0	0	0	0
Fly ash	0	0	0	0	0	0	0	0	0
Silica fume	0	0	0	0	0	0	0	0	0
Total	263	349	432	201	277	351	183	221	287
Rate(%)*	**29**	**39**	**48**	**25**	**35**	**44**	**26**	**31**	**40**

Cement CO_2 Emission: 927 kg/1000kg Cement;
Slag CO_2 Emission: 2% of 927 kg/1000kg Cement;
Fly Ash and Silica Fume CO_2 Emission: None.
*With respect to mix III × 100%.

Table 8. Life cycle inventory for LEED - CO_2 emission of binders for mix II.

Life cycle inventory- CO_2 emission of binders for concrete mix(kg/m^3)									
Material kg/m^3	II-246	II -247	II I-248	II -276	II -277	II -278	II -306	II -307	II -308
Cement	224	313	400	197	240	317	144	183	251
Slag	2	2	2	2	2	2	2	2	2
Fly ash	0	0	0	0	0	0	0	0	0
Silica fume	0	0	0	0	0	0	0	0	0
Total	226	315	402	199	242	319	146	185	253
Rate(%)*	**25**	**35**	**45**	**25**	**30**	**40**	**20**	**26**	**35**

Cement CO_2 Emission: 927 kg/1000kg Cement;
Slag CO_2 Emission: 2% of 927 kg/1000kg Cement;
Fly Ash and Silica Fume CO_2 Emission: None.
*With respect to mix III × 100%.

Table 9. Life cycle inventory for LEED -CO_2 emission of binders for mix III&IV.

Life cycle inventory- CO_2 emission of binders for concrete mix(kg/m^3)						
Material, kg/m^3	III-2411	III-2711	III-3011	IV-2411	IV-2711	IV-3011
Cement	893	794	714	848	754	678
Slag	0	0	0	0	0	0
Fly ash	0	0	0	0	0	0
Silica fume	0	0	0	0	0	0
Total	893	794	714	848	754	678
Rate(%)*	**100**	**100**	**100**	**95**	**95**	**95**

Cement CO_2 Emission: 927 kg/1000kg Cement;
Slag CO_2 Emission: 2% of 927 kg/1000kg Cement;
Fly Ash and Silica Fume CO_2 Emission: None.
*With respect to mix III × 100%.

4.3 Raw materials

The pozzolans used in the concrete were post-industry waste recycled resource materials. Raw materials mainly including limestone and clay (3:1) for the manufacturing process of cement consumed a large amount natural resources. According to the PCA, raw materials required to produce 1000 kg of cement had to consume more than 1613 kg of natural resources. The intrusive destruction of nature due to the territorial integrity were destroyed and air pollution were serious problems. The consumption of raw materials for cementitious materials in kg/m^3 were as shown in Tables 10-12. With respect to mix III as being 100%, the amount of raw materials consumed by mix I, mix II, and mix IV were 25% ~ 48%, 20% ~ 45% and 95%, respectively. Mixtures designed with the DMDA method showed that the raw materials of cementitious materials consumed significantly decreased more than 50%; meanwhile, the reduction of mix IV was very limited.

Table 10. Life cycle inventory for LEED – Raw materials of mix I for binders.

Life cycle inventory- Raw materials consumption of binders for concrete mix I									
Material kg/m^3	I-246	I-247	I-248	I-276	I-277	I-278	I-306	I-307	I-308
Cement	458	606	752	350	482	611	318	384	500
Slag	-	-	-	-	-	-	-	-	-
Fly ash	-	-	-	-	-	-	-	-	-
Silica fume	-	-	-	-	-	-	-	-	-
Total	458	606	752	350	482	611	318	384	500
Rate(%)*	**29**	**39**	**48**	**25**	**35**	**44**	**26**	**31**	**40**

Raw Materials: 1613kg/1000kg Cement.
*With respect to mix III x 100%.

Table 11. Life cycle inventory for LEED – Raw materials of mix I for binders.

Life cycle inventory- Raw materials of binders for concrete mix II									
Material, kg/m^3	II-246	II-247	II-248	II-276	II-277	II-278	II-306	II-307	II -308
Cement	390	545	695	342	418	552	250	318	437
Slag	-	-	-	-	-	-	-	-	-
Fly ash	-	-	-	-	-	-	-	-	-
Silica fume	-	-	-	-	-	-	-	-	-
Total	390	545	695	342	418	552	250	318	437
Rate(%)*	**25**	**35**	**45**	**25**	**30**	**40**	**20**	**26**	**35**

Raw Material : 1613 kg/1000 kg Cement.
*With respect to mix III x 100%.

Table 12. Life cycle inventory for LEED – Raw materials of mix I for binders.

Life cycle inventory- Raw materials of binders for concrete mix III & IV						
Material kg/m^3	III-2411	III-2711	III-3011	IV-2411	IV-2711	IV-3011
Cement	1553	1381	1242	1476	1311	1179
Slag	-	-	-	-	-	-
Fly ash	-	-	-	-	-	-
Silica fume	-	-	-	-	-	-
Total	1553	1381	1242	1476	1311	1179
Rate(%)*	**100**	**100**	**100**	**95**	**95**	**95**

Raw Material : 1613kg/1000kg Cement.
*With respect to mix III x 100%.

4.4 Energy consumption

By PCA reports, the energy consumption for cement manufacturing was 4.8 GJ/1000 kg of cement and ground granulated blast-furnace slag was 14% per 4.8 GJ/1000 kg of cement for slag grinding. Since the energy consumption was owing to their front of the main products, electricity and siliceous-steel, fly ash and silica fume were regarded as having zero energy consumption. According to Tables 13-15, the energy consumption of cementitious materials in the mixtures with respect to mix III being=100%, were 26% ~ 49% (mix I), 22% ~ 41% (mix II), and 95% (mix IV), respectively. Mix I and mix II used the DMDA method and in comparison to the cementitious materials with mix III, they consumed 50% less energy.

Table 13. Life cycle inventory for LEED - Energy consumption of binders for mix I.

Life cycle inventory- Energy consumption of binders for concrete I									
Material kg/m^3	I-246	I-247	I-248	I-276	I-277	I-278	I-306	I-307	I-308
Cement	1.36	1.80	2.24	1.04	1.44	1.82	0.95	1.14	1.49
Slag	0.01	0.01	0.02	0.01	0.01	0.01	0.01	0.01	0.01
Fly ash	-	-	-	-	-	-	-	-	-
Silica fume	-	-	-	-	-	-	-	-	-
Total	1.37	1.81	2.26	1.05	1.45	1.83	0.96	1.15	1.50
Rate(%)*	**30**	**39**	**49**	**26**	**35**	**45**	**26**	**31**	**41**

Cement Energy Consumption: 4.8GJ/1000kg Cement;
Slag Energy Consumption: 14% of 4.8GJ/1000kg Cement.
*With respect to mix III × 100.

Table 14. Life cycle inventory for LEED - Energy consumption of binders for mix II.

Life cycle inventory- Energy consumption of binders for concrete II									
Material kg/m^3	II-246	II-247	II-248	II-276	II-277	II-278	II-306	II-307	II-308
Cement	1.16	1.62	2.07	1.02	1.24	1.64	0.74	0.95	1.30
Slag	0.06	0.06	0.06	0.06	0.06	0.06	0.06	0.06	0.06
Fly ash	-	-	-	-	-	-	-	-	-
Silica fume	-	-	-	-	-	-	-	-	-
Total	1.22	1.68	2.13	1.08	1.30	1.70	0.80	1.01	1.36
Rate(%)*	**26**	**36**	**46**	**26**	**32**	**41**	**22**	**27**	**37**

Cement Energy Consumption: 4.8GJ/1000kg Cement;
Slag Energy Consumption: 14% of 4.8GJ/1000kg Cement.
*With respect to mix III × 100%.

Table 15. Life cycle inventory for LEED - Energy consumption of binders for mix III & IV.

Life cycle inventory- Energy consumption of binders for concrete III&IV						
Material kg/m^3	III-2411	III-2711	III-3011	IV-2411	IV-2711	IV-3011
Cement	4.62	4.11	3.70	4.39	3.90	3.51
Slag	-	-	-	-	-	-
Fly ash	-	-	-	-	-	-
Silica fume	-	-	-	-	-	-
Total	4.62	4.11	3.70	4.39	3.90	3.51
Rate(%)*	**100**	**100**	**100**	**95**	**95**	**95**

Cement Energy Consumption: 4.8GJ/1000kg Cement;
Slag Energy Consumption: 14% of 4.8GJ/1000kg Cement.
*With respect to mix III × 100%.

4.5 Compressive strength effect

The front four-items for life cycle inventory were used to assess the effect of design and production for cementitious materials in the mixtures. Next, tests were executed the performance of hardened properties and durability. This paper performed compressive strength development, surface resistivity, and rapid chloride ion penetrability tests to evaluate anti-corrosion.

The compressive strength development (7, 28, 56, and 92 days) were shown in Tables 16-18. Mix I and mix II with the DMDA method at 28 days showed that the strength ranged from 47 MPa to 73.6 MPa and 52.4 MPa to 72.9 MPa, respectively. Mix III and mix IV with the ACI 211.1 method indicated strengths from 50.7 MPa to 54.9 MPa and 53.9 MPa to 54.1 MPa, respectively. The cement content of mix III and IV were as twice as that of mix I and II, and the compressive strength increased slowly. At 92 days, mix I, II, III, and IV had strengths of 75.9 MPa ~ 98.4 MPa, 65.9 MPa ~ 99.5 MPa, 62 MPa ~ 65.7 MPa, and 65 MPa ~ 69.7 MPa, respectively. The cement amount for mix III was too high to rarely enhance the compressive strength. Mix IV had the same result. Pozzolans in high strength concrete had a very important effect in lowering the heat of hydration and refining the microstructure to be denser to improve the long-term properties. For the compressive strength ratio as shown in Table 19-21, with respect to mix III as being 100%, the strength ratios were 122% ~ 150%(mix I), 106% ~ 151%(mix II), and 105%~106%(mix IV) at 92 days. Mix I and mix II with quaternary cementitious materials increased the compressive strength. Meanwhile, that of mix IV only increased a little.

Table 16. Compressive strength development for mix I(MPa).

Ages	I-246	I-247	I-248	I-276	I-277	I-278	I-306	I-307	I-308
7 days	55.6	60.2	62.7	39.1	43.9	51.0	29.6	36.5	43.0
28 days	67.8	80.2	73.6	57.5	60.4	65.0	47.0	49.6	59.1
56 days	86.3	95.1	92.2	71.9	72.4	84.0	61.5	66.9	70.4
92 days	96.5	98.4	96.9	83.6	83.6	89.1	75.9	76.3	83.1

Table 17. Compressive strength development for mix II(MPa).

Ages	II-246	II-247	II-248	II-276	II-277	II-278	II-306	II-307	II-308
7 days	40.3	50.2	63.6	35.0	43.9	48.1	30.0	39.4	35.9
28 days	79.9	70.3	72.8	63.3	61.0	64.4	58.7	60.7	52.4
56 days	93.3	92.0	84.8	82.5	82.2	77.6	65.9	67.5	65.7
92 days	99.1	99.5	98.5	93.0	85.2	83.8	65.9	79.8	76.6

Table 18. Compressive strength development for mix III&IV(MPa).

Ages	III-2411	III-2711	III-3011	IV-2411	IV-2711	IV-3011
7 days	55.9	52.3	48.8	53.0	52.7	51.9
28 days	59.5	54.9	50.7	53.9	54.0	54.1
56 days	62.7	61.2	60.7	67.7	66.0	62.0
92 days	65.7	64.4	62.0	69.7	68.5	65.0

Table 19. Relative percent of compressive strength for mix I(as mix III=100 %).

Ages	I-246	I-247	I-248	I-276	I-277	I-278	I-306	I-307	I-308
7 days	99	108	112	75	84	98	61	75	88
28 days	114	135	124	105	110	118	93	92	117
56 days	138	152	147	117	118	137	101	110	116
92 days	147	150	150	130	130	138	122	123	134

Table 20. Relative percent of compressive strength for mix II(as mix III=100 %).

Ages	II-246	II-247	II-248	II-276	II-277	II-278	II-306	II-307	II-308
7 days	72	90	114	67	90	92	62	81	74
28 days	134	118	122	115	111	117	116	120	103
56 days	149	147	135	135	134	127	109	111	108
92 days	151	151	150	144	132	130	106	129	124

Table 21. Relative percent of compressive strength for mix III & IV(as mix III=100 %).

Ages	III-2411	III-2711	III-3011	IV-2411	IV-2711	IV-3011
7 days	100	100	100	95	101	106
28 days	100	100	100	91	98	107
56 days	100	100	100	108	108	102
92 days	100	100	100	106	106	105

4.6 Electrical resistivity

If the concrete structure compactness is high, there are fewer pores; thus, the connection is more difficult. As a result, the conducting channels are decreased or lengthened, and electrical resistivity is enhanced. The conductivity of concrete is mainly influenced by the moisture content. Concrete specimens were cured in saturated lime water, extracted from water, and then dried (interior saturation condition) to measure the surface resistivity at different points of various mixtures. Tables 22-24 show the resistivity development of different mixes. During the early ages (7 days), surface resistivity had a slow development. Only II-248 and IV-2711 groups had more than 20 KΩ-cm low corrosion occurrence threshold values. After 28 days, the resistivity for mix I, II, and IV groups incorporating various pozzolans significantly increased more than 20 KΩ-cm. Adding pozzolans could make the concrete interior microstructure denser to decrease the connection channels. At 92 days, the resistivity of a low w/cm ratio of 0.24 for mix I and mix II displayed poorer conductivity properties (all greater than 100 KΩ-cm). Mix II groups with a higher pozzolans-binder ratio and adjusted fly ash-slag ratio might increase the compressive strength and resistivity as shown in Tables 17 and Tables 23. According to ACI 318-11, the design criteria for anti-corrosion of concrete should have a lower the water-to-cementitious ratio and proper content of pozzolans added. In this study, mixtures with multi-pozzolans and lower w/cm ratios were identified the design results and performance similar to ACI 318-11.

All mixes, with respect to 20 KΩ-cm, indicated that the resistivity ratio in comparison to 20 KΩ-cm as shown in Table 25-27. At 92 days, the ratios expressed that mix I, mix II, mix III, and mix IV groups were respectively 400% ~ 705%, 350% ~ 705%, 85% ~ 105% and 230% ~ 290%. Mixtures without pozzolans might have difficulty in avoiding corrosion.

Table 22. Surface resistivity of mix I (KΩ-cm).

Ages	I-246	I-247	I-248	I-276	I-277	I-278	I-306	I-307	I-308
7 days	5	17	17	7	11	10	6	7	8
28 days	52	74	80	38	51	53	25	36	42
56 days	94	95	97	69	74	81	44	52	65
92 days	141	137	126	91	94	91	80	82	82

Table 23. Surface resistivity of mix II (KΩ-cm).

Ages	II-246	II-247	II-248	II-276	II-277	II-278	II-306	II-307	II-308
7 days	14	15	21	11	8	13	6	9	12
28 days	73	75	82	64	61	75	43	42	56
56 days	98	94	100	97	90	89	72	75	72
92 days	141	134	129	138	124	113	99	94	70

Table 24. Surface resistivity of mix III & IV (KΩ-cm)

Ages	III-2411	III-2711	III-3011	IV-2411	IV-2711	IV-3011
7 days	11	10	8	17	20	14
28 days	16	15	13	41	40	30
56 days	20	18	16	50	47	39
92 days	21	19	17	58	56	46

Table 25. Rate of surface resistivity for mix I (as 20 KΩ-cm = 100%).

Ages	I-246	I-247	I-248	I-276	I-277	I-278	I-306	I-307	I-308
7 days	55	85	85	35	55	50	30	35	40
28 days	260	370	400	190	255	265	125	180	210
56 days	470	475	485	345	370	405	220	260	325
92 days	705	685	630	455	470	455	400	410	410

Table 26. Rate of surface resistivity for mix II (as 20 KΩ-cm = 100%).

Ages	II-246	II-247	II-248	II-276	II-277	II-278	II-306	II-307	II-308
7 days	70	75	105	55	40	65	30	45	60
28 days	365	375	410	320	305	375	215	210	280
56 days	490	470	500	485	450	445	360	375	360
92 days	705	670	645	690	620	565	495	470	350

Table 27. Rate of surface resistivity for mix II (as 20 KΩ-cm = 100%).

Ages	III-2411	III-2711	III-3011	IV-2411	IV-2711	IV-3011
7 days	55	50	40	85	100	70
28 days	80	75	65	205	200	150
56 days	100	90	80	250	235	195
92 days	105	95	85	290	280	230

4.7 Charge passed

The final options of the life cycle inventory for high strength concrete in this study performed a rapid chloride penetrability test (RCPT) according to ASTM C1202 to evaluate the durability of anti-corrosion after 92 days. The test results of the RCPT were as shown in Table 28. Mix I and Mix II were 292 C ~ 828 C and 251 C ~ 555 C, respectively. From the anti-corrosion on chloride ion penetrability in Table 5, these mixtures showed the rating of chloride ion migration as being at a very low level. Mix III and mix IV were 1003 C ~ 2091 C and were attributed to a rating of Moderate to Low. Mixtures incorporating pozzolans and applying the DMDA method could enhance the compactness and refine the pore structure of the conducting channels causing a decrease or lengthening [11].

Table 28. Test results of rapid chloride ion penetrability(Coulombs).

Mix I	I-246	I-247	I-248	I-276	I-277	I-278	I-306	I-307	I-308
92 days	292	324	515	414	448	540	517	653	828
Mix II	II-246	II-247	II-248	II-276	II-277	II-278	II-306	II-307	II-308
92 days	251	311	352	308	362	402	423	504	555
Mix III & IV	III-2411	III-2711	III-3011	IV-2411	IV-2711	IV-3011	-	-	-
92 days	1977	2003	2045	1090	1596	1841	-	-	-

Test results of the mixtures used 2000 C (Low level permeability) as the denominator to calculate the corresponding percentage in Table 29. Mix I, mix II, mix III, and mix IV were 15% ~ 41%, 13% ~ 27%, 99% ~ 102%, and 55% ~ 92%, respectively. The effects of anti-corrosion with pozzolan mixtures were much better than when no pozzolans were used. Mix II (with a higher amount of slag) was superior to mix I. Further, mix IV with ACI 211-11 and adding silica fume were beneficial in anti-corrosion, but the traditional mix method for mix III without any pozzolans was not suitable for the exposed marine environment. Green options for anti-corrosion mixtures of high strength concrete included the following seven items containing cement used, CO_2 emissions, raw materials, energy consumption, compressive strength, the surface resistivity, and charge passed. The first four items were used to evaluate the carbon reduction of cementitious material mixtures, and the last three items were applied to assess the sustainable development of high strength concrete for anti-corrosion. The carbon emissions and resource consumption for mix I and mix II decreased by more than 50%, and mix IV had a reduction only 5%, all with respect to mix III. Sustainable development of anti-corrosion properties in this study was regarded as the resistivity and charge passed. The former is a non-destructive test method adopted in the laboratory and field having a simple, light, and fast technique to obtain a lot of useful data; meanwhile, the latter is a widely used destructive test performed in the lab to evaluate the conductivity of concrete.

Table 29. Percent of rapid chloride ion penetrability for mixes (as 2000 C = 100%).

Mix I	I-246	I-247	I-248	I-276	I-277	I-278	I-306	I-307	I-308
92 days	15	16	26	21	22	27	26	33	41
Mix II	II-246	II-247	II-248	II-276	II-277	II-278	II-306	II-307	II-308
92 days	13	16	18	15	18	20	21	25	28
Mix III & IV	III-2411	III-2711	III-3011	IV-2411	IV-2711	IV-3011	-	-	-
92 days	99	100	102	55	80	92	-	-	-

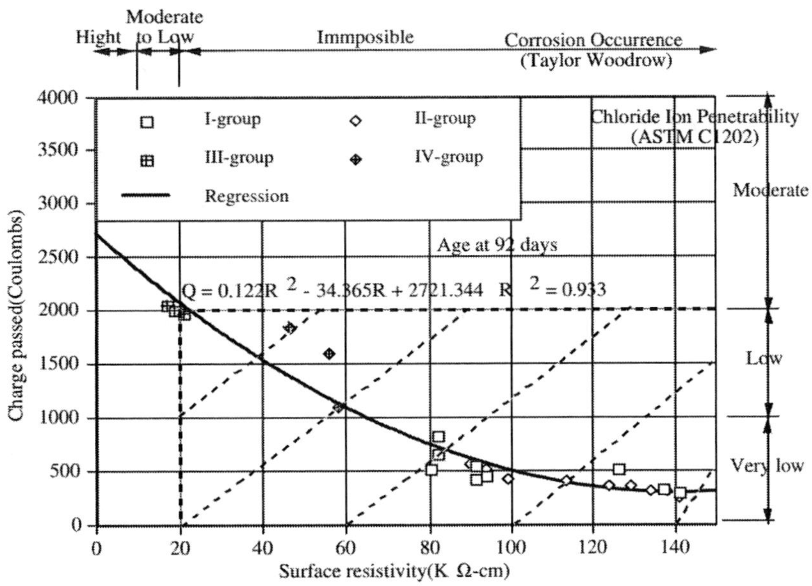

Figure 1. Probability of sustainable potential of green options for various mixtures according to Taylor Woodrow (resistivity, R) and ASTM C1202 (RCPT, Q).

According to the occurrence of corrosion suggested by Woodrow Tayler [38] and the criteria for chloride ion penetrability with ASTM C1202, green options for anti-corrosion performance of high strength concrete are listed in Table 24-26 and Table28. For all mixtures in the paper, the relationship between the resistivity and charge passed were as shown in Figure 1 and Formula 1. The correlation r^2 was equal to 0.933, which belonged to a highly relevant level.

$$Q = 0.122R^2 - 34.365R + 2721.344 \quad r^2 = 0.933 \tag{1}$$

Sustainable criteria for anti-corrosion at 92 days were that the resistivity should be greater than 20 KΩ-cm and the charge passed should be less than 2000 C. Mixtures for mix I, mix II, mix IV, and III-2411 matched the requirements; however, III-2711 and III-3011 mixtures did not. High strength concrete with a lower water-to-cement ratio (III-2411) could meet the criteria for anti-corrosion, but the cement content was too high to maintain the performance demands. The water content of mix I and mix II (See Table 3-4 and Table 29), w/cm was fixed, the higher the water amount, the lower the charge passed could be displayed the trend. That's water used in mixtures dominated the durability issue. CO_2 emissions and resource consumption for the cementitious materials in the mix displayed neither carbon reduction nor sustainability. Multi-pozzolans incorporated in the mixtures (mix I and mix II groups) designed with the DMDA method could enhance the long-term performance and sustainable development during the service life. In order to obtain a carbon dioxide reduction and sustainability for concrete mixture, a proper mix design and an addition of the correct amount of pozzolans will present an optimal option.

5 Conclusions

Mixture proportions adopted the densified mixture design algorithm (DMDA) method and water-to-cementitious material (w/cm) ratios from 0.24~0.30 with various amounts of pozzolans (mix I & II). ACI design method acted as control groups (mix III) and added silica fume (mix IV groups). Green options for anti-corrosion of the life cycle inventory of the Leadership in Energy and Environmental Design (LEED) suggested by the Portland Cement Association (PCA) included cement used, CO_2 emissions, raw materials, and energy consumption, the compressive strength and surface resistivity development and a rapid chloride ion penetrability test (RCPT), respectively. Test results are summarized as follows:

1. The DMDA applied packing between aggregates by fly ash and slag to enhance the density of aggregate structures for mix I and II. The cement content used was 25% ~ 48% and 20% ~ 45% than that of mix III. Meanwhile, mix IV was 95%.
2. CO_2 emissions for mix I and mix II groups, with respect to mix III, were 25% ~ 48% and 20% ~ 45%. Meanwhile, mix IV was 95%. CO_2 emissions were similar to the ratio of cement content used.
3. With respect to mix III, raw materials of cement production for mix I, II, and IV were 25% ~ 48%, 20% ~ 45%, and 95%, respectively.
4. With respect to mix III, the energy consumption of cement manufacturing for mix I, II, and IV was 26% ~ 49%, 22% ~ 46%, and 95%, respectively.
5. At 28 days, the compressive strength of mix I and mix II were all greater than 100% with respect to mix III. However, that of mix IV was 91% ~ 107%. After 56 days, all the mixtures were above 100%. For II-24 groups (w/cm=0.24~0.30), the compressive strength ratio exceeded 150% at 92 days. This clearly indicated the effect of incorporating pozzolans.
6. At 28 days, the surface resistivity for mix I, II and IV groups, were all greater than 20 KΩ-cm. This was a high potential of anti-corrosion. For mix III after 56 days, only III-2411 was larger than 20 KΩ-cm, others had a high corrosion risk in the field of application.
7. RCPT results showed that all mixtures, except III-2711 and III-3011, were below 2000 C, had a lower chloride ion penetrability, and higher anti-corrosion levels. Mix I and II indicated that the charge passed were all less than 1000 C to highlight the benefits of pozzolans.
8. This paper attempted to indicate the options of sustainable potential for anti-corrosion of high strength concrete mixtures. Let $Q_s \leqq 2000$ C and $R_s \geqq 20$ KΩ-cm at 92 days. All the mixtures, except III-2711 and III-3011, could match the above requirements. In order to achieve carbon dioxide reduction and sustainability for a concrete mixture, a proper mix design and addition of the correct amount of pozzolans will present an optimal option.

References

1. ACI Committee 363, *State-of-the-art report on high-strength concrete* (ACI, Farmington Hills, MI, 1997)
2. P. Paultre and D. Mitchell, ACI Concr. Int., **25**, 76 (2003)
3. L. Andrew, C. Wonchang, M. Amir, R. Sami, and Z. Paul, ACI Mat. J., **106**, 1 (2009)
4. P.K. Mehta, *Concrete: microstructure, properties, and materials, fourth edition* (McGraw-Hill Education, New York, 2014)
5. A.M. Neville, *Properties of concrete, fifth edition* (Prentice Hall, San Francisco, USA, 2012)
6. S. Mindess, J.F. Young, and D. Darwin, *Concrete, second edition* (Pearson, London, 2002)
7. J. Gajda and M. VanGeem, ACI Concr. Int., **24**, 59 (2002)
8. ACI Committee 318, *Building code requirements for structural concrete (ACI 318-11) and commentary* (ACI, Farmington Hills, MI, 2011)
9. American Society for Testing and Materials, *C1202-10: Standard test method for electrical indication of concrete's ability to resist chloride ion penetration* (ASTM International, West Conshohocken, PA, 2010)
10. D. Laszlo, Concr. Int., **20**, 66 (1998)
11. C.L. Hwang, *High performance concrete theory and practice* (James Publisher, Taipei, 2014)
12. K. Scrivener, Chem. World., **March**, 62 (2008)
13. K. Scrivener, Ind. Concr. J., **88**, 11 (2014)
14. World cement production, 2010-2015, The Statistics Portal (2016) http://www.statista.com/statistics/219343/cement-production-worldwide/
15. A.P. Gurgun, R. Komurlu, and D. Arditi, Procedia Eng., **118**, 1145 (2015)
16. C. Meyer, Cem. Concr. Compos., **31**, 601 (2009)
17. A. Petek Gursel, E. Masanet, A. Horvath, and A. Stadel, Cem. Concr. Compos., **51**, 38 (2014)
18. E. Aprianti, P. Shafigh, S. Bahri, and J. N. Farahani, Constr. Build. Mater., **74**, 176(2015)
19. M. Nili and A.M. Salehi, Constr. Build. Mater., **24**, 2108 (2010)
20. N. Cobîrzan, A.A. Balog, and E. Moşonyi, Procedia Technol., **19**, 506 (2015)
21. M.M. Hossain, M.R. Karim, M. Hasan, M.K. Hossain, and M.F.M. Zain, Constr. Build. Mater., **116**, 128 (2016)
22. R.E. Rodríguez-Camacho and R. Uribe-Afif, Cem. Concr. Res., **32**, 1851 (2002)
23. K. Celik, C. Meral, M. Mancio, P. Kumar Mehta, and P.J.M. Monteiro, Constr. Build. Mater., **67**, 14 (2014)
24. B.B. Sabir, S. Wild, and J. Bai, Cem. Concr. Compos., **23**, 441 (2001)
25. H. Dekkiche and A. Taileb, Procedia Eng., **145**, 844 (2016)
26. C.L. Champagne and C.B. Aktas, Procedia Eng., **145**, 380 (2016)
27. F. Asdrubali, G. Baldinelli, F. Bianchi, and S. Sambuco, Build. Environ., **86**, 98 (2015)
28. A.P. Gurgun, R. Komurlu, and D. Arditi, Procedia Eng., **118**, 1145 (2015)
29. M.L. Marceau, M.A. Nisbet, and M.G. VanGeem, *Life cycle inventory of Portland cement manufacture, PCA R&D Serial No. 2095b* (Portland Cement Association, Skokie, IL, 2005)
30. L. Acquaye, A. Chini, and M. Danso-Amoako, *Green options for high temperature cured concrete through the use of fly ash and blast furnace slag* (33rd Conference on Our World in Concrete & Structures, Singapore, August, 25-27, 2008)
31. M.A. Nisbet, M.L. Marceau, and M.G. VanGeem, *Environmental life cycle inventory of Portland cement concrete, PCA R&D Serial No. 2137a* (Portland Cement Association, Skokie, IL, 2002)
32. M. Marceau and M.G. VanGeem, *Life cycle inventory of slag cement concrete, CTL Project No. 312012* (Construction Technology Laboratories, Skokie, IL, 2003)
33. ECCO EV27, *LEED Green Building rating system and concrete* (Environment Council of Concrete Organizations, Skokie, IL, 2002)
34. M.L. Marceau, M.A. Nisbet, and M.G. VanGeem, *Life cycle inventory of Portland cement concrete, PCA R&D Serial No. 3007* (Portland Cement Association, Skokie, IL, 2007)
35. American Concrete Institute, *Slag cement in concrete and mortar, ACI 233R-03* (ACI, Farmington Hills, MI, 2003)

36. Y.Y. Chen, B.L.A. Tuan, and C.L. Hwang, Constr. Build. Mater., **47**, 340 (2013)
37. ACI Committee 211, *Standard practice for selecting proportions for normal, heavyweight, and mass concrete (Reapproved 2009)* (ACI, Farmington Hills, MI, 1991)
38. T. Woodrow, *Marine durability survey of the Tongue Sands Tower. Concrete in the oceans Technical Report No. 5* (Taylor Woodrow Research Laboratories, Canada, 1980)

Study on the factors that influence the intention of college students to participate in the green building

ChienJen Hung [1,a], Chenchen Lai[1] and Meiyan Chen[1,2]

[1]*Department of Civil Engineering, Fuzhou College of Foreign Studies and Trade, Fuzhou*
[2]*PhD program, Department of Business Administration, National Chengchi University, Taipei City 11605, Taiwan*

Abstract. The concept of green building has been extended for a period of time on both sides of the Taiwan Strait. However, there still is curiosity and skepticism in Fujian on green building. While in Taiwan, villages are filled with green-evolved houses and modern environmental farmhouses. With Theory of Planned Behavior as the framework, this study investigates influences of college students' attitude toward green building, subjective norms of green building and perceived behavioral control of green building on the intention to participate in the green building. Findings show that college students' attitude toward green building and the perceived behavioral control of green building significantly influence the intention to participate in the green building, but "subjective norms of green building" doesn't have significant influence. The reason is probably that college students are at the rebellious stage and have high self consciousness and independent viewpoints.

1 Introduction

Green building means that, on the basis of green culture, try best to choose materials of the nature and design energy-saving and comfortable living space in the process of construction behavior in order to achieve environmental protection and retard environmental deterioration. Therefore, a building should meet not only basic functional requirements but also fundamental requirements of green architecture, i.e. focusing on human physical and mental needs as well as providing people living in the building with convenience, comfort and physical and mental health. In this paper the intention of college students to participate in the green building mainly refers to their intention to promote and change houses into green buildings. The concept of green building has been popularized for many years on both sides of the Taiwan Straits. In Taiwan, many representative green buildings have been erected. In Fujian province of Mainland China, green building has been supported by the government and its promotion is still in progress. With regard to the government's policy and support of environmental protection, the State Council issued Comments on Accelerating the Development of Energy-Saving and Environment-Friendly Industries in August 2013, which clearly defines four key missions for enhancing development of energy-saving and environment-friendly industries.This is extremely significant for Chinese government's responsibilities in the environment. Evaluation standard for green building (GB/T50378-2006)[1], published and implemented in 2006, clearly describes definition, evaluation indicators and evaluations methods of green building. Since enforcement, this standard has

[a] Corresponding author : kukusau8@foxmail.com

been used to effectively guide practical works of green building in China. In Implementation Scheme for Green Building Action of Fujian Province, it is clearly indicated that the major goal of this plan is to make 40% of newly-constructed urban and town buildings meet requirements of green building standard in 2020. "Accelerating green building technology innovation", "striving to develop emerging industries like green construction materials", "launching publicity and education" and other similar works are also mentioned.

Thanks to promotion by government and media in these years, common people have become familiar with green building. However, before cost is reduced to an acceptable level, there is a big gap from people's actual buying green building or their intention to renovate original house into green building. After completing studying, college students will become the core force of the society. Their behaviors exert great influence on the society. With students from a private university in Fujian province as the objects, this study is aimed to (1) investigate college students' intention to participate in green building; (2) investigate influences of different factors such as college students' attitude toward green building and subjective norms of green building on the intention to participate in the green building; (3) discuss and formulate management implications.

2 Literature review

Definitions of green building are available in many papers. For example, as described by W. Zhang & B. Li. (1998) [2], green building system is a new architecture system established based on principle of virtuous circle of ecosystem, with green economy as the foundation, green society as the connotation, green technology as the support and green environment as the mark. According to Evaluation standard for green building (2006) [1], within its total life cycle, green building should maximally save resources (energy, land, water and materials conservation), protect environment, reduce pollution, provide people healthy, practical and high-efficiency use space, and achieve harmony with the nature. "Green building" is a term encompassing strategies, techniques, and construction products that are less resource-intensive or pollution-producing than "regular" construction. (A. J. Hoffman & R. Henn, 2008). [3]

Many researches on green building have been published in recent years. Through investigating rainwater collecting facilities of green buildings in housing estate of Changsha City, W. Zhang, Y. Huang, W. Xiao, L. Yang & Z. Xiong (2015) [4] suggested that roofing rainwater collecting system features relative small investment and easy management and maintenance, and thus can be a part of energy-saving building. X. Chen & Y. Li (2015) [5] studied application of green building theory into design of ruins museum. They pointed out, to enhance our abilities of ecological architecture design, it's necessary to use the experience of other countries for reference, stick to human-centered principle of design, and construct ecological-type ruins museum adapt to China. X. Qin & L. Jing (2013) [6] conducted descriptive statistical analysis and inferential statistical analysis using SPSS, in order to investigate evaluation of risk factors of green building during the entire life cycle. The present study chooses college students as the objects, and discusses factors that influence the intention to participate in green building.

In theory of reasoned action, behavioral intention is an antecedent factor of behavior and influenced by Attitude and Subjective Norm. Theory of planned behavior adds Perceived Behavioral Control into theory of reasoned action and includes act of the will in the discussed model, and thus can better predict behavior (J. Zhang & Q. Zheng, 2012) [7]. Researches with regard to application of theory of planned behavior include study by Lao Kefu and Wu Jia (2013) [8], which used theory of planned behavior to discuss influencing mechanism of green consumer behavior. They found subjective norm and perceived behavioral control of green consumption significantly influenced consumer's intention to participate in green consumption and such intention also has significant influence on green consumption behavior; however, attitude toward green consumption doesn't exert direct and significant influence on the intention to participate in green consumption. H. Zhang, C. Bai & C. Li. (2011) [9] utilized theory of reasoned action and theory of planned behavior to investigate consumers' intention to go shopping online; findings showed attitude and perceived behavioral control significantly influenced online shopping intention but influence of subject norm was not significant. L. Lin & X. Bai. (2014) [10] used theory of planned behavior to investigate university students' delay behavior. They pointed out, more active attitude toward specified task results in stronger perceived behavior control, stricter subjective norm of accomplishing tasks on schedule, stronger behavioral intention and reduced delay behaviors. Theory of planned behavior is a good integrated theoretical framework for delay behavior researches. F. Boudreau&G. Godin (2009) [11] used theory of planned behavior to investigate patient's physical activity intention.

3 Research hypotheses

Based on theory of reasoned action, the present study attempts to investigate influences of college students' green building attitude (GBA), green building subjective norm (GBSN) and green building perceived behavioral control (GBPC) on green building behavioral intention (GBBI). Following hypotheses are proposed on the basis of literature data construction.

Hypothesis 1 (H1): there is positive interrelationship between college students' GBA and GBSN. According to theory of planned behavior put forward by Ajzen (1991) [12], positive interaction effects exist between human's attitude toward certain behaviors and their subjective norm. For college students, green building is a rational behavior constricted by subjective conditions. Their GBA and GBSN influence each other. Once more active in green building, college students will be more likely to feel a greater influence by social norms. Meanwhile, in case of more active attitude toward environmental protection, college students will perceive stronger influence from the society or relatives and friends. Consequently, interrelationship between GBA and GBSN is positive.

Hypothesis 2 (H2): there is positive interrelationship between college students' GBSN and GBPC. According to theory of planned behavior put forward by Ajzen (1991) [12], positive interaction effects exist between human's subjective norm of certain behaviors and their perceive behavioral control. With stronger GBSN, college students will make more preparations for green building and form stronger GBPC. Also, in case of more sufficient conditions for green building, college students will more strongly hope that more people can recognize and participate in green building in order to generate a better ecological environment. Therefore, GBSN and GBPC affect each other and such mutual influence is positive.

Hypothesis 3 (H3): there is positive interrelationship between college students' GBA and GBPC. According to theory of planned behavior put forward by Ajzen (1991) [12], positive interaction effects exist between human's attitude toward certain behaviors and their perceive behavioral control. College students who hold active attitude toward green building behavior are willing to make more preparations and have stronger perceived behavioral control. College students who have stronger GBPC will form stronger expectation of good ecological environment and become more active in green building. Consequently, interrelationship between GBA and GBPC is positive.

Hypothesis 4 (H4): college students' GBA positively influences GBBI.

It is describe in Ajzen's (1991) [12] theory of planned behavior that people's attitude toward certain behaviors will influence their behavioral intention. Occurrence of GBBI firstly depends on GBA. More active attitude is associated with consumer's stronger desire for green building. Therefore, college students' GBA has positive influence on their green building intention.

Hypothesis 5 (H5): college students' GBSN positively influences GBBI.

As mentioned in Ajzen's (1991) [12] theory of planned behavior, people's subjective norm of certain behaviors will influence their behavioral intention. Mostly, green building behavior is a kind of reasoned action. Consumers who perceive greater pressure from surrounding people will generate a stronger intention of green building. College students' GBSN also exerts positive influence on their intention to participate in green building.

Hypothesis 6 (H6): college students' GBPC positively influences GBBI.

According to Ajzen's (1991) [12] theory of planned behavior, people's perceived behavioral control of certain behaviors will positively influence their behavioral intention. This is an important expression of influence of personal behavior on objective condition constrictions. Green building behavior is a kind of action constricted by certain objective conditions. If consumers feel they have more active capabilities and conditions for conducting green building behavior, they will form a stronger intention to participate in green building. Therefore, GBPC positively influences GBBI.

4 Questionnaire design and survey result

Table 1. Setting of questionnaire items (variables).

GBA	Q11	I agree that green building can reduce urban heat island effect.
	Q12	I agree that green building can insulate heat, lower down temperature and make the space inside house refreshing and comfortable.
	Q13	I agree that green building can beautify environment.
	Q14	I agree that green building can protect environment, reduce pollution and provide people healthy and nature-friendly building.
GBSN	Q21	I think teacher's behaviors and attitudes about green building influence my behavior and attitude.
	Q22	I think parents or relatives' behaviors and attitudes about green building influence my behavior and attitude.
	Q23	I think friends' behaviors and attitudes about green building influence my behavior and attitude.
	Q24	I think the government's behaviors and attitudes about green building influence my behavior and attitude.
	Q25	I think TV media's behaviors and attitudes about green building influence my behavior and attitude.
GBPC	Q31	1. When I can't find trash can outside, I will temporarily wait and throw the rubbish when trash can is available.
	Q32	2. When friend conducts behavior of wasting earth resources, I will persuade him/her not to do so.
	Q33	3. I persist in CO2 emission reduction and energy saving whenever and wherever.
	Q34	4. I often use mass transportation for saving natural resources.
GBBI	Q01	I want to change my house into a green building that can insulate heat, lower down temperature and provide a refreshing and comfortable home.
	Q02	I want to change my house into a green building for beautifying environment.
	Q03	I want to participate in promotion of green building for protecting environment of the earth.
	Q04	I'm willing to change my house into a green building for becoming a pioneer in environmental protection.
	Q05	I want to own a green building.

This questionnaire is mainly for inquiring college students' green building behaviors such as attitude, subjective norm, perceived behavioral control and intention. Questions are set according to Ajzen's (1991) [13] theory of planned behavior and related literatures (L. Lin & X. Bai. ， 2014 ; H. Zhang, C. Bai & C. Li. ， 2011) [9, 10] (see table 1). There are four questions (variables) about college students 'GBA- Q11, Q12, Q13 and Q14, five questions (variables) about GBSN-Q21, Q22, Q23, Q24 and Q25, four questions (variables) about GBPC-Q31, Q32 、 Q33 and Q34, and five questions (variables) about GBBI- Q01, Q02, Q03, Q04 and Q05. In addition, interviewees' basic data are asked, including gender, age, family income, major, department and parents occupation. Five-point Likert scale is used. Option and corresponding score of each question (variable) is as the following: quite agree (5 points), agree (4 points), no opinion (3 points), disagree (2 points) and quite disagree (1 point).

Survey of this study was done at a private university of Fujian Province from April 8 to April 15, 2016. Six teachers sent 497 questionnaires to 11 departments. 443 valid questionnaires were recycled, accounting for 89.13%. Survey was performed by means of randomly choosing students and asking them to complete questionnaire on site.

In valid questionnaires, male and female students made up 33.7% and 66.1% respectively. In terms of age, students born in 1997, 1996, 1995, 1994, 1993 and other years accounted for 10.4%, 15.8%, 30%, 34.1%, 8.4 % and 1.1% respectively. With regard to father's occupation, industry, commerce, agriculture, public office, teacher, freelance work, unemployment, others and no statement accounted for 14%, 27.4%, 27.8%, 6.2%, 1.8%, 15.9%, 2.1%, 4.7% and 1.8% respectively. With respect to place of household registration, Fujian, other province and no statement accounted for 91.2%, 5.6% and 3.2% respectively. In terms of family income, "below 4000", "4000-6000", "6000-10000", "10000-15000", "above 15000" and "no statement" accounted for 21.4%, 24.4%, 20.5%, 12.4%, 11.7% and 9.5% respectively. Regarding department, English, Asia-Europe language, economy & trade, logistics, information, business & administration, financing & accounting, engineering, television & film, arts & design, and statement accounted for 9.5%, 9.9%, 8.4%, 10.6%, 8.8%, 9.0%, 8.8%, 5.6%, 9.3%, 9.3%, 9.5% and 1.4% respectively.

In analysis of the interviewee data variable "intention to participate in green building" there was significant different between different genders (P=0.021，*) and female is superior to male student. Interviewees showed obvious differences in monthly family income (P=0.003，**). Analysis by scheffe method demonstrated, students from family with monthly income below 4000 (RMB) were superior to those from family with monthly income of 10000-15000 (RMB) in terms of the intention to participate in green building. There were also significant differences between interviewees studying in different departments (P=0.012，*). Engineering department presented the highest score, followed by arts & design department. Aisa-Europe Language Department had the lowest score. However, analysis by scheffe method showed no significant difference. Interviewees' age and their parents' occupation didn't lead to significant difference.

5 Modeling fitting and hypothesis testing

5.1 Date reliability and validity testing

Table 2. Test indicators and fitting results of measurement model.

Latent	Observation	Cronbach	Estimate	Std.	CR	P
GBA	Q11	0.855	1.000			
	Q12		1.086	.070	15.426	***
	Q13		1.077	.073	14.772	***
	Q14		1.132	.071	16.051	***
GBSN	Q21	0.899	1.000			
	Q22		1.015	.045	22.709	***
	Q23		1.012	.044	23.066	***
	Q24		.810	.049	16.466	***
	Q25		.760	.046	16.407	***
GBPC	Q31	0.823	1.000			
	Q32		1.101	.061	18.063	***
	Q33		.819	.060	13.603	***
	Q34		.970	.070	13.849	***
GBBI	Q01	0.856	1.000			
	Q02		1.244	.089	14.013	***
	Q03		1.241	.087	14.285	***
	Q04		1.379	.095	14.469	***
	Q05		1.000	.088	11.390	***

This study used SPSS20.0 to carry out reliability testing of all of the 18 observation variables. The obtained reliability testing indicator- Cronbach α value was 0.899. Cronbach α values of four variables- GBA (Q11, Q12, Q13 and Q14), GBSN (Q21, Q22, Q23, Q24 and Q25), GBPC (Q31, Q32, Q33 and Q34) and GBBI (Q01, Q02, Q03, Q04 and Q05) were 0.855, 0.899, 0.823 and 0.856 respectively (as listed in table 2), indicating good reliability of data. This study used AMOS 20.0 to analyze confirmatory factors. Analysis results showed that standardized estimates (Std. Estimate) of observation variables corresponding to GBA, GBSN, GBPC and GBBI all exceeding the benchmark value-0.5 and all estimates were of significant level (P <0.001, ***) (as listed in table 2), indicating latent variables of the present study had good convergent validity. Comparison of square root of Average Variance Extracted (AVE) of each latent variable and correlation coefficient of each latent variable calculated based on results of confirmatory factor analysis demonstrated, GBA, GBSN, GBPC and GBBI were highly correlated with each other. This conforms to viewpoint of theory of planned behavior (Ajzen, 1991). Furthermore, correlation coefficient of latent variable is not larger than square root of AVE (as listed in table 3). According to above testing results of convergent validity and discriminant validity, survey data of this study passes validity test.

Table 3. Discriminant validity testing of latent variables.

	GBA	**GBSN**	**GBPC**	**GBBI**
GBA	**0.77**	←	$\sqrt{AVE}$	
GBSN	0.33	**0.80**		
GBPC	0.43	0.42	**0.75**	
GBBI	0.51	0.43	0.75	**0.75**

5.2 Fitting results

This study used AMOS20.0 to perform structural equation model fitting for the data of the obtained 443 questionnaires and the theoretical model of green building behavioral intention. Table 4 shows fitness test indicators of the fitting. Freedom degree of chi-square value of this study is 3.081, near the ideal boundary value-3. RMSEA value is 0.069. RMSEA value between 0.05 and 0.08 is good. Besides, other important fitness test indicators of this study, including NFI, GFI, CFI, RFI, IFI, TLI and AGFI, are all greater than or close to 0.9, suggesting very good fitness evaluation results. Theoretical model and survey data of this study can be subject to good fitting (as listed in table 4).

Table 4. Fitness test indicators

Power of statistical test	**Indicator**	**Standard value**	**Green building intention model**	**Reach or not**	**Remarks**
Absolute fitness	X^2		397.451	√	
	X^2/df	1~5	3.081	√	
	GFI	>0.9	0.910	√	
	AGFI	>0.9	0.880		Close to
	RMR	<0.08	0.038	√	
	RMSEA	<0.08	0.069	√	
Incremental fitness	NFI	>0.9	0.912	√	
	NNFI(TLI)	>0.9	0.927	√	
	CFI	>0.9	0.938	√	
	RFI	>0.9	0.895		Close to
	IFI	>0.9	0.939	√	
Simplified fitness	PNFI	>0.5	0.769	√	
	PGFI	>0.5	0.686	√	

5.3 Analysis of test results

Fitting results of hypothesis model and survey data of this study also demonstrate the three variables, GBA, GBSN and GBPC, significantly influence each other (p<0.001, ***). Hypothesis 1 (H 1), hypothesis 2 (H2) and hypothesis 3 (H3) are all supported. In addition, indicators of hypothesis 4 (H4) and hypothesis 6 (H6) also present significant influences (p<0.001, ***) and these two hypotheses are supported. However, with respect to hypothesis 5 (H5), college students' GBSN has no significant influence on GBBI. As per Ajzen's (1991) theory of planned behavior, people's subjective norm of certain behaviors has direct and important influence on their behaviors. However, findings of the present study demonstrate college students' GBSN doesn't exert significant influence on their intention to participate in green building (as shown in table 5) (see figure 1).

Table 5. Interrelations between each variable.

			Estimate	S.E.	C.R.	P
GBBI	<---	GBPC	.573	.063	9.154	***
	<---	GBA	.216	.047	4.571	***
	<---	GBSN	-.017	.027	-.641	.521
GBSN	<-->	GBA	.164	.029	5.688	***
GBA	<-->	GBPC	.143	.022	6.402	***
GBSN	<-->	GBPC	.224	.034	6.656	***

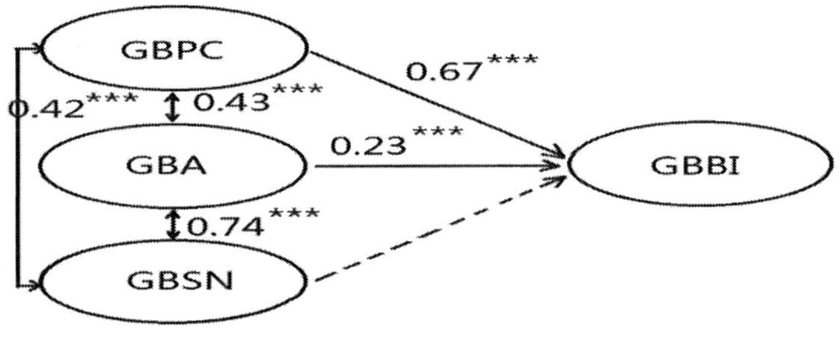

Figure 1. Structural equation of the intention to participate in green building.

6 Conclusions

Based on the theory of planned behavior proposed by Ajzen, this study investigates whether Fujian college students' "green building attitude", "green building subjective norm" and "green building perceived behavioral control" influence the intention to participate in the green building. A total of 443 valid questionnaires are obtained. Interviewees with different genders, family monthly incomes and majors show obvious differences in the intention to participate in green building. Analysis of structural equation model (SEM) shows GBA and GBPC have significant positive influence on GBBI but GBSN doesn't present such significant influence. The reason might be that college students are at the rebellious adolescence and unwilling to follow the direction of social norms, and have high self consciousness and own independent opinions. Therefore, GBSN cannot exert significant influence on the intention to participate in green building. L. Lin and X. Bai. (2014) [10] made a research of university students' school work delay behavior and found these students' subjective norms don't have significant influence on behavioral intention. For analyzing consumers' intention to go shopping online, H. Zhang, C. Bai & C. Li. (2011) [9] investigated young and highly-educated netizen populations and found influence of subjective norm on online shopping intention was not significant. All these results support that young people's high self consciousness lead to insignificant influence of subjective norm on behavioral intention.

From results of this study we can see, students' more positive GBA and GBPC are associated with more positive intention to participate in the green building. Consequently, organizations such as governmental agencies and schools are suggested to promote green building in life and studying so that students will perceive close relationship between green building and current life. Furthermore, benefits brought by green building to the environment can also be emphasized so that students will realize practicing of green building can not only maintain environmental ecology but also enhance self image. Only in this way, students will have improvement in GBA, GBPC and GBBI.

The present study has some limitations. It doesn't conduct large-scale sampling. Objects of survey only include students of a university. There is no analysis of common people of the whole society. In the future, researchers should conduct survey among students of universities and colleges throughout the country so as to achieve more diverse and extensive samples and discussion space. In addition, more detailed influencing factors are not investigated. Future related researches are suggested to conduct further study of other factors.

References

1. Y. Wang, JCST, **7**, 14-17 (2006)
2. W. Zhang and B. Li., JSTPP, **5**, 16-17 (1998)
3. A. J. Hoffman and R. Henn, Org. & Env., **21** (4), 390-419 (2008)
4. W. Zhang, Y. Huang, W. Xiao, L. Yang, and Z.J. Xiong, JHAS, **12**, 75-77 (2015)
5. X. Chen and Y. Li., J. Urb. and Arch., **33**, 10-11 (2015)
6. X. Qin and L. Jing, CCEJ, **8**, 123-135 (2013)
7. J. Zhang and Q. Zheng, JCE, **18** (1), 77-81 (2012)
8. K. Lao and J. Wu, JFS, **2**, 91-100 (2013)
9. H. Zhang, C. Bai, and C. Li, JSS, **25** (9), 130-135 (2011)
10. L. Lin and X. Bai, CJCP, **22** (5), 855-859 (2014)
11. F. Boudreau and G. Godin, IJBNPA, **6** (1), 35-35 (2009)
12. I. Ajzen, OBHDP, **50** (2), 179-211 (1991)

MATEC Web of Conferences **119**, 01012 (2017) DOI: 10.1051/matecconf/201711901012
IMETI 2016

Analysis and simulation on two types of thrust reversers in an aircraft engine

Feng Tian, Jingwu He, Yuexi Xiong and Yanxiong Zhao[a]

School of Aeronautical Science and Engineering, Beijing University of Aeronautics and Astronautics, Beijing, China

Abstract. With rapid development of new composite material and manufacturing , innovative engineering solutions are supplied to the advanced nacelle, such as integrated propulsion system(IPS) , carbon-fiber composite inner skin by single-piece molding process , which offers a reduction in fuel burn and less noise produced by engines. The advanced nacelle has an O-duct thrust reverser demonstrator whose composite structure is in the form of an "O" as opposed to the traditional "D-duct". A comparative study is to be conducted to investigate the differences between the latest O-duct and conventional D-duct in numerical approaches. To focus on the quantitative analysis of thrust reverser's operation, this paper mainly uses CATIA/Digital Mock Up(DMU) to simulate under deployment and stowed conditions of two different thrust reverser. After comparing the structural weight, the design models of blocker door are built for kinematic analysis of relevant mechanism and simulation. The results show that simplified design and elimination of multiple interfaces generates weight saving, O-duct improves airflows within the engine, meanwhile D-duct has excellent cost effective and maintainability.

1 Introduction

1.1 Thrust reverser

Modern large passenger aircraft have great promotion on both the maximum takeoff gross weight and takeoff speed, more runway length are required to takeoff and landing. To face with the ever growing airport flow and varieties of emergency conditions, we need reliable deceleration system to reduce distance of landing or rejected takeoff effectively. Till now on, there are many ways for airliner to slow down, such as wheel brake, spoiler for aerodynamic braking, thrust reverser system, drag parachute and so on [2]. Though most of modern airliners are equipped with thrust reversers, the design of thrust reversers has influence on engine nacelle's weight and wing performance, as well as cruise speed. Those parameters decide overall operation and maintenance cost to a certain extent. GE's research suggests that the weight of thrust reverser accounts for thirty percent of the whole nacelle, especially for turbofan engines whose fan diameter are over 2.5 meters, and the air leakage and pressure loss caused by the thrust reverser leads to 0.5~1.0% increase in specific fuel

[a] Corresponding author : hejingwu@buaa.edu.cn

consumption during the flight. According to the economic evaluation of BOEING, using thrust reverser on B767 aircraft need extra 125000 dollars per year besides the normal running expenses.

But thrust reverser are still essential, because of many advantages over other braking systems. Not only can thrust reverser keeps efficient speed deceleration until the aircraft stops, but also they're not affected by wet weather and icy runway conditions, that's what make it unique. The ability to work at such wide range of condition enhance the safety factor of aircraft [2]. It supplies credible retarding force when the airplane is made to force land, rejected takeoff or landing at bad environment. To cooperate with the thrust reverser reduces losses of other devices, although it can be used repeatedly.

Table 1. The effects of thrust reverser.

Condition / Reduce (%)	Dry	Wet	Icy
Landing	56	35	18
Rejected takeoff	66	45	33

The thrust reverser is located in the fan or exhaust section of the engine for eliminating forward thrust by deflecting fan air or exhaust gasses in the forward direction can be used for stopping during landing and a RTO(rejected takeoff).

1.2 D-duct & O-duct

Traditional thrust reverser involved target type, clam-shell type and cold stream type. Cascade or cold stream reverser has better structural integrity, versatility and also the steady air exhaust up to the 60~70% of maximum engine thrust [1]. Apparently it's only appropriate for turbofan engine. The blocker door mechanism as a significant part of thrust reverser directly impact the reverse thrust and system's efficiency [7]. According to the Design Manual of Airplane and related information, equation for thrust efficiency and reverse thrust is as follow.

$$\eta_r = \frac{axial\,force\,caused\,by\,reverse\,flow}{total\,thrust} \times 100\% \quad (1)$$

η_r——thrust efficiency

The motion of blocker door, time required to fully open and the last thrust degree all depends on how blocker door mechanism designed [3]. So the main point of this paper is to develop integrated design methodologies for the thrust reverser with simple models for analysis & simulation.

The D-duct system uses left and right T/R halves that attach to the engine strut. Tension latches and inner V-blades hold the haves together [4]. Thrust reverser is a translating sleeve and cascade design. Each thrust reverser has a wall with a translating sleeve of the left and right half of the ducted fan. The two sleeves on each engine work simultaneously. However, the two sleeves are independent of each other. The T/R sleeve is composite assembly with an outer sleeve and an inner sleeve (inner wall). The outer sleeve makes the exterior contour of the T/R. The inner sleeve makes the outer wall of the fan duct. Also, the inner sleeve holds the T/R blocker doors. The inner and outer sleeves are fastened together at the aft end.

An evident feature of the O-duct is its 330-dge single-piece carbon composite inner skin, which is produced by an innovative molding process. Not like D-duct which divide into two parts. This kind of architecture simplify the translating cowl which is composed of outer and inner sleeve. Without the latches it loses some weight. Another change is that O-duct cancel the rod exposed in the bypass, replacing with complicated block door mechanism setting up under cascades. While drives the translating cowl is moving backwards that driven by electric actuating cylinder, it also bring the blocker doors to work position with certain degrees, consequently turn the direction of air flow.

2 Methodology & procedure

The modeling and simulation is based on the structural model of CFM series engine's thrust reverser.3D models of two kinds of thrust reversers are built in CATIA. In order to ensure the integrality of simulation and analysis dependability, the whole nacelle is modelled to improve the quality of T/R models. As far as we know, thrust reverser has three typical structure, and thrust reverser's research value is that large transport airplane need it. By the mean time, the blocker door mechanism is not always same. So the discussion is between common D-duct and latest O-duct. Focus on the simulation model and kinematics analysis. The research method could be used at other similar thrust reverser system. According the model in CATIA, kinematics simulation and dynamical analysis is conducted in DMU and ABAQUS. The kinematics analysis contains the law of joints, movement parameter measurement and analysis. After which the model is imported into ABAQUS for dynamic analysis, the dynamic simulation displays the load spectrum of blocker door mechanism.

3 Structural model

Some parameters of nacelle have been developed (shown in Table 2). The parameters detailed explanation with reference to Fig.1. So as the thrust reverser. Try to ensure the two thrust reversers' subordinate structure in the same size, appropriate simplification is also conductive to the weight comparing.

Table 2. Computation of geometrical parameters.

Symbol	Dimension	Estimating relationship
DIH	Intake highlight diameter	$DIH=0.037W_a+32.2$
M_H	Maximum height of main cowl	$M_H=1.21D_F$
LC	Main cowl length	$LC=[2.36D_F-0.01(D_FMMO)^2]$
DFO	Main cowl diameter at fan exit	$DFO=(0.00036\mu W_a+5.84)^2$
DMG	Gas generator cowl diameter at fan exit	$DMG=(0.000475\ \mu W_a+4.5)^2$

The result is calculated and shown in the Table 3.

Table 3. Calculation result.

Symbol	Result
M_H	2395mm
LC	4098.8mm
DFO	118.18lb/sec
DMG	124.06lb/sec
DIH	2012mm

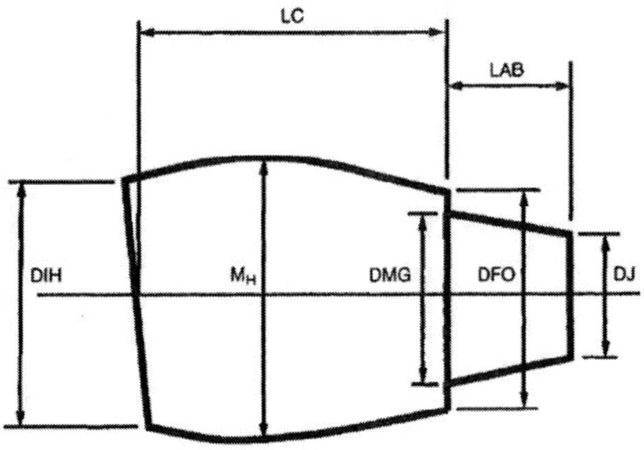

Figure 1. Typical of the nacelle.

3.1 D-duct

The thrust reverser sleeve assemblies are located on the aft part of the engine. They are stowed in a forward position during flight. When the thrust reverser is actuated, the sleeves move to the aft. This moves the blocker doors to block the flow of air between the fan duct assembly and the thrust reverser sleeve assemblies [5], then exposes the cascade vanes. The air then flows out through the cascade vanes which direct the air forward to create reverse thrust [6].

On the basis of D-duct's structure, main components of thrust reverser are fixed, on which the model is accomplished in CATIA(see Figs.2-3).

main components of each thrust reverser half :

- Translating Sleeve
- Inner Wall
- 3 Hydraulic actuators and 2 sync shafts
- 6 Cascades
- 5 Blocker doors
- 5 Blocker door drag links
- 1 T/R opening actuator on pylon
- Torque box
- Fan duct inner wall
- 3 Access doors
- Upper and lower sliders and tracks

As the model shown, some shape parameters of thrust reverser(Table 4) and the stroke of translating sleeve under different working condition are measured in CATIA.

Table 4. Leading particulars of T/R.

Height(above ground)	730mm
Length	1553mm
Maximum diameter	1240mm

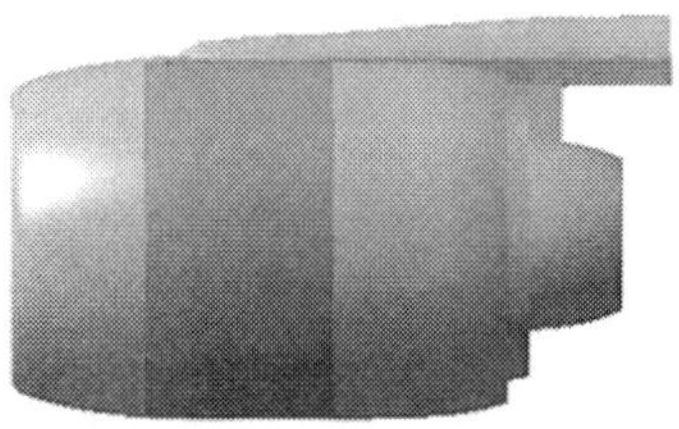

Figure 2. Stowed.

Figure 3. Deployed

3.2 O-duct

The O-duct is similar to the "D" type thrust reverser on some level. It's also consist of translating cowl, inner panel, actuator, blocker door mechanism, cascades, torque box, access doors and sliders. But with the structure changing of translating cowl from two parts into single one, more blocker doors are set up, which means more cascades and no more tension latches. Same as the blocker door mechanism, it cancels blocker door drag links, which contributes to decreasing aerodynamic drag. Above all, on the one hand structural simplification and different blocker door mechanism let every part of nacelle integrated better. On the other side, new technology is usually not perfect. The use of composite material and improved thrust reverser raise the T/R efficiency, but also left an uncertainty in terms of maintenance and economy.

All illustrated in Figs.4-5, from the comparison we see that O-duct has larger and more complete are of airflow than D-duct, which imply the air flow more smoothly through the bypass with less bifurcation. Due to it's compact structure.

DOI: 10.1051/matecconf/201711901012

Figure 4. D-duct

Figure 5. O-duct

4 Simulation & result

4.1 DMU simulation

Digital Mock Up or DMU is a concept that allows the description of a product in 3D. The goal of doing DMU is to reduce time-to-market by identifying potential issues during the design, save product development costs by minimizing the number of physical prototypes needed to build, finally increase product quality. Figure 6 represent the work flow and key steps of DMU simulation.

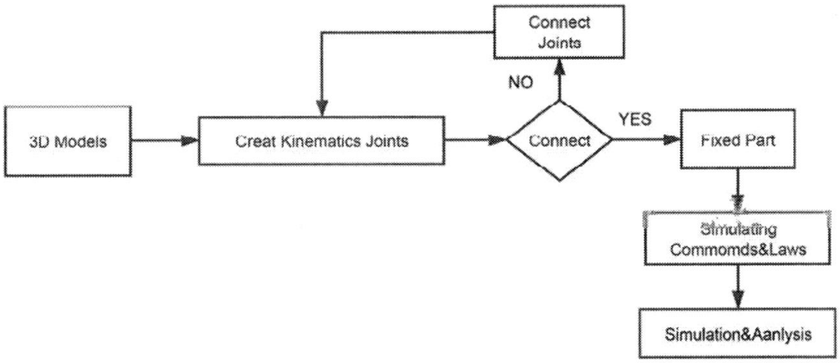

Figure 6 . The DMU simulation flow chart.

4.2 Result

By simplifying the O-duct block door mechanism [4], establishing kinematic pairs, setting up the relationship with time, the DMU simulation is conducted in CATIA. In the simulation, both the unfolding time of two types thrust reverser is set for 1.5 second, and it holds for about 10 seconds at operating position. From the result of simulation, it's clear to understand how block door works. For instance, each block door mechanism of O-duct is mainly consist of five parts except the block door itself (Figure 7). To unlock the sensor during the DMU simulation, CATIA outputs corresponding curve of motion parameter based on the position of measurement point and reference point. Figure 8 has it's measurement point on the telescopic rod for the curve of liner speed.

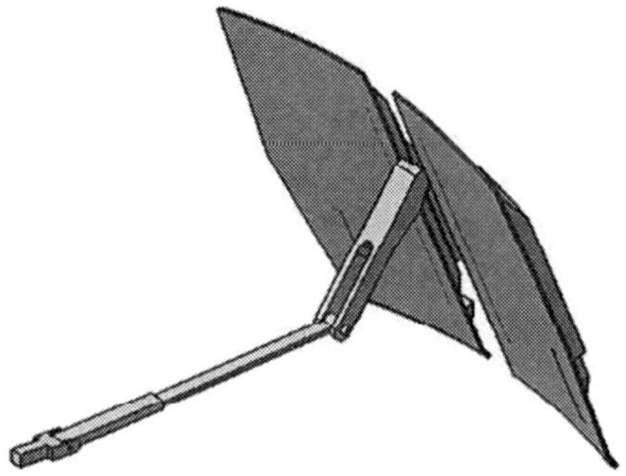

Figure 7. O-duct block door mechanism

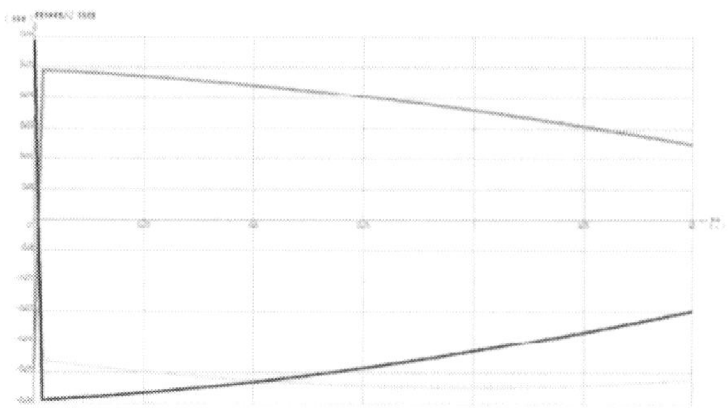

Figure 8. Liner speed of reference point

5 Dynamical analysis

5.1 Reverse thrust

After getting the exit flow of cascade and bypass, the thrust of bypass can be calculated. The exit angle α is one of the key parameters, which determined by the geometry of cascade.

$$R_{ft} = \frac{G_{zt}}{g}(C_5 - V) - \frac{G_{ft}}{g}\left(C_5 + C_{ft}\cos\alpha\right) \tag{2}$$

R_{ft}- reverse thrust
G_{ft}- when thrust reverser works, the gas flow through T/R
C_5- speed of exit gas flow
V- landing speed
A- exit angle
G_{zt}- the gas flow through engine
g- gravity acceleration

It's defined that the reverser thrust counts about 30 percent of maximum thrust, the pressure on block door is 0.0035Mpa.

5.2 Dynamical simulation

To have a convenient and accurate simulating method, simplified the model of block door mechanism is needed. Holding the necessary features, then transfer the CATIA data into ABAQUS for meshing.

Figure 9. O-duct's block door.

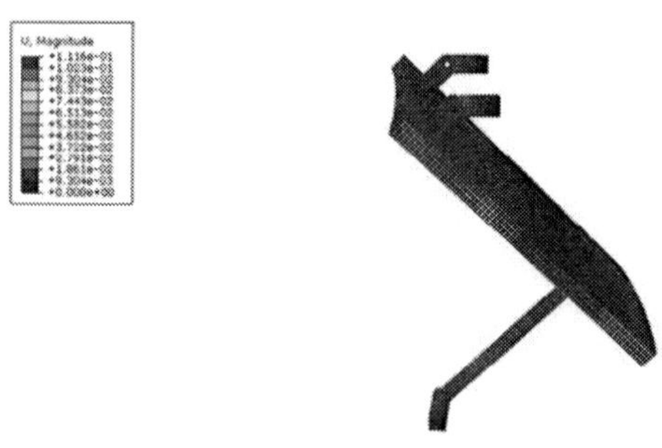

Figure 10. D -duct's block door.

The material properties for structural analysis are aluminium alloy with a Young's modulus of 71Gpa and a density of 2800kg/m2. The pressure is applied to the block door using the calculation results. Assuming that two kinds of block door have same opening time and reverser angle, and in the similar condition of airflow, ignore the influence of cascades [8]. Figure 9-10 shows a contour plot representing the deformations. The maximum displacement both on the block door.

Figure 11-12 show a contour plot representing the stress, the maximum stress occurs on joints which tie the block door with translating sleeve. The maximum stress value of O-duct is 91Mpa, which almost twice the value of D-duct,57.1 Mpa.

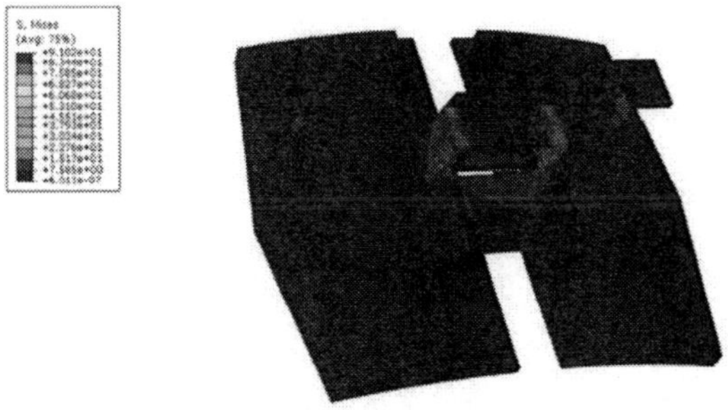

Figure 11. Stress contours of O-duct's block door.

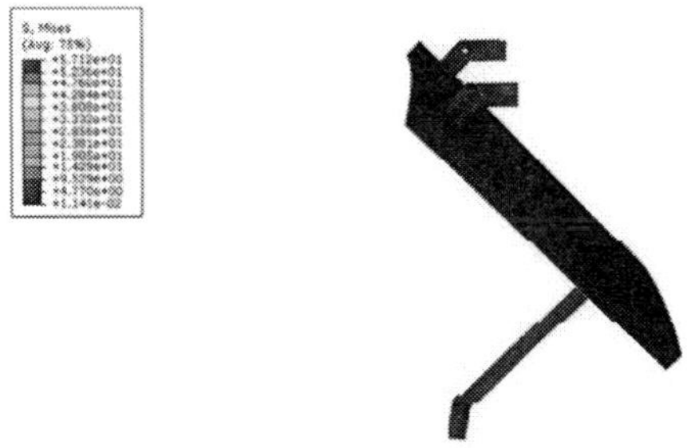

Figure 12. Stress contours of O-duct's block door.

6 Conclusions

This paper carried out a contrastive study of two types of thrust reverser, through the kinematic and dynamic simulation of models which built in CATIA. Based on which, the structure, movement and stress analysis is conducted. The finally consequence imply that if leaves out the pylon, for the structure varies from each other, D-duct is heavier with sorts of tension latches and connectors. In the process of DMU simulation, it's evident that O-duct requires more time to work at full open position, and some trouble may happen with it's complicated block door mechanism. As the result of dynamic simulation, maximum displacement point is located in the edge of block door, while the stress concentrate on the links between translating sleeve and the block door mechanism.

References

1. Q. Xuhao, W. Zhanxue, L. Zengwen, Z. Xiaobo, and S. Jingwei, Machinery Design & Manufacture. J., **11** (2013)
2. J.A. Yetter, Why Do Airlines Want and Use Thrust Reversers[R], NASA.TM-10 (1995)
3. S. Béchu, N. Gascon, S. Roche, M. Prioul, L. Albarede, P. Lasgorceix, and M. Dudeck, AIAA, AIAA-2000-3524 (2000)
4. Z. Shaojun, W. Hanping, H. Jiangjun, S. Libing, and J.Sichuan. Ordnance, **3** (2015)
5. D.J. Strash, J.M. Summa, J.H. Frank, and R. Standish, J. Propul Power, **16** (2000)
6. Z. Yunhao, Eriqitai, International Coference on Industrial Control and Electronics Engineering (2012)
7. E.G. Rulz, T. Mahmod, V. Sethi, and P. Pilidis, *Proceedings of ASME Turbo Expo*, GT2012-68810 (2012)
8. Z. Guodong and W. Qiang, J. Aerosp Power, **27** (2012)

Research on typical topologies of a tubular horizontal-gap passive magnetic levitation vibration isolator

Yiheng Zhou[a], Baoquan Kou, Xiaobao Yang, Jun Luo and He Zhang

School of Electrical Engineering and Automation, Harbin Institute of Technology, China

Abstract. Magnetic levitation vibration isolators have attracted more and more attention in the field of high-precision measuring and machining equipment. In this paper, we describe a tubular horizontal-gap passive magnetic levitation vibration isolator. Four typical topologies of the tubular horizontal-gap passive magnetic levitation vibration isolator are proposed. The analytical expression of magnetic force is derived. The relationship between levitation force, force density, force ripple and major structural parameters are analysed by finite element method, which is conductive to the design and optimization of the tubular horizontal-gap passive magnetic levitation vibration isolator. The force characteristics of different topologies of the tubular horizontal-gap passive magnetic levitation vibration isolator are compared and evaluated from the aspect of force density, force ripple and manufacturability. In comparison with conventional passive magnetic levitation vibration isolators, the proposed tubular horizontal-gap passive magnetic levitation vibration isolator shows advantage in higher force density.

1 Introduction

High-precision measuring and machining technology is one of the most important research topics in modern industry. Besides the motion control strategies [1], vibration control is also one of the significant factors influencing the high-precision measuring and machining quality [2]. Owing to the advantages of wide vibration isolation bandwidth and low vibration transmissibility, active vibration isolation systems, which integrates actuators with passive gravity compensation devices, has great potential in high-precision application fields like microscope and lithography. The commonly used passive gravity compensation device includes mechanical springs, air bearings, and passive magnetic levitation vibration isolators. Among these methods, the mechanical spring is subject to mechanical coupling, which limits the positioning accuracy of the machining equipment. The air bearing is a good choice for gravity compensation because of its stable characteristics and mature technologies. However, the system structure will be very complex if air bearings are used in some measuring or machining equipment requiring a high vacuum condition [3], such as extreme ultraviolet lithography. In contrast, owing to the advantages of no mechanical contact, vacuum compatibility (no need of liquid or gas auxiliary equipment) and low natural frequency, passive magnetic levitation vibration isolators have drew increased interest in recent years [4-15].

[a] Corresponding author : sethvoler@163.com

In this paper, a tubular horizontal-gap passive magnetic levitation vibration isolator (THPMLVI) is proposed, and a variety of typical topologies of the THPMLVI are studied to obtain higher high force density, which will reduce the volume of permanent magnet and the cost of THPMLVI. The paper is organized as follows. In Sec. 2, the typical topologies and operation principle of THPMLVI are introduced. In Sec. 3, the analytical expression of magnetic force is derived. In Sec. 4, the relationship between levitation force, force density, force ripple and major structural parameters are analysed by finite element method, which is conductive to the design and optimization of the THPMLV. The characteristics of different topologies of the THPMLV are compared and evaluated from the aspect of force density, force ripple and manufacturability.

2 Typical topologies and operating principle

The proposed tubular horizontal-gap passive magnetic levitation vibration isolator consists of three components, i.e., outer stator, mover, and inner stator, as shown in Fig. 1 (a). The mover consists of an aluminium plate and a magnet array. Both the outer and inner stators consist of an aluminium plate and a magnet array. There are four typical topologies of THPMLVI, i.e., NS-NS topology, Halbach-NS topology, NS-Halbach topology, and Halbach-Halbach topology, as shown in Fig. 1. For the NS-NS topology of THPMLVI, its mover consists of 3 radially magnetized magnets, and each of its stator consists of 4 radially magnetized magnets. For the Halbach-NS topology of THPMLVI, its mover consists of 3 radially magnetized magnets, and each of its stator consists of 4 radially magnetized magnets and 5 axially magnetized magnets. For the NS-Halbach topology of THPMLVI, its mover consists of 3 radially magnetized magnets and 8 axially magnetized magnets, and each of its stator consists of 4 radially magnetized magnets. For the Halbach-Halbach topology of THPMLVI, its mover consists of 3 radially magnetized magnets and 8 axially magnetized magnets, and each of its stator consists of 4 radially magnetized magnets and 5 axially magnetized magnets. The magnetization direction of the magnets are shown as arrows.

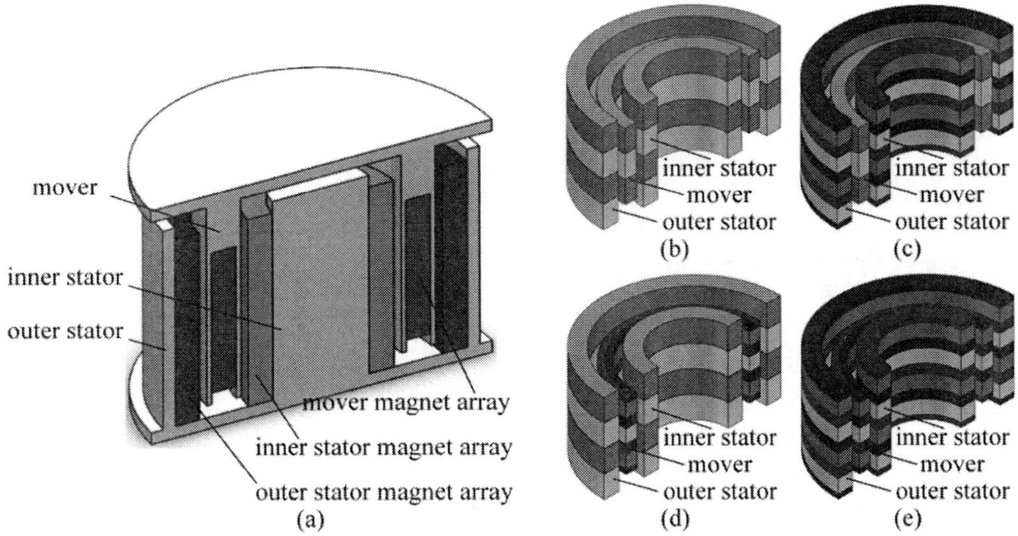

Figure 1. Typical topologies of the proposed THPMLVI. (a) basic structure, (b) NS-NS topology, (c) Halbach-NS topology, (d) NS-Halbach topology, (e) Halbach-Halbach topology.

The operating principle of THPMLVI is based on the attraction and repulsion between permanent magnets. Take the NS-NS topology as example, the mover of THPMLVI is suspended above the stators by magnetic force between mover magnet array and stator magnet arrays. When the mover moves relative to the stators along the axial direction, the magnets with the same magnetization direction will generate attractive force with positive stiffness, and the magnets with opposite magnetization direction will generate repulsive force with negative stiffness, thus the total levitation force is near-constant. When the mover moves relative to the stators along the radial direction, due to the symmetry, the total levitation force is also near-constant. Through adopting the alternating magnetization direction and dual-stator structure, the f THPMLVI can generate near-constant levitation force and near-zero stiffness in six degree of freedom stroke, which is useful for improving the vibration isolation performance of active vibration isolation system.

In order to study the force characteristics of THPMLVI, the magnetic force between two mover magnet array and stator magnet arrays should be firstly calculated. According to the equivalent current model [16], the magnetic field generated by stator magnet arrays can be calculated as (1). Thus the magnetic force acting on the mover magnet array can be calculated as (2) [17]. The levitation force F, stiffness k, natural frequency f of the THPMLVI can be expressed as (3). As shown in (3), compared to stiffness k, the force ripple dF/F is a better indicator which can reflect its vibration isolation performance, because the natural frequency f is directly related to vibration isolation ability. Therefore, force ripple is analysed instead of stiffness in the next section.

$$\boldsymbol{B} = \nabla\times \boldsymbol{A}, \quad \boldsymbol{A} = \frac{\mu_0}{4\pi}\oint_S \frac{\boldsymbol{j}_M}{|r-r'|}da + \frac{\mu_0}{4\pi}\oint_V \frac{\boldsymbol{J}_M}{|r-r'|}d^3x \tag{1}$$

$$\boldsymbol{F} = \int_S (\boldsymbol{j}_m \times \boldsymbol{B})ds + \int_V (\boldsymbol{J}_m \times \boldsymbol{B})dv \tag{2}$$

$$k = \frac{dF}{ds}, \quad f = \frac{1}{2\pi}\sqrt{\frac{k}{m}} = \frac{\sqrt{g}}{2\pi}\sqrt{\frac{dF/F}{ds}} \tag{3}$$

3 Force characteristics optimization and comparison

The relationship between levitation force, force density, force ripple and major structural parameters are analysed by finite element method in this section, which is conductive to the design and optimization of the tubular horizontal-gap passive magnetic levitation vibration isolator. The major structural parameters of four topologies of THPMLVI are shown in Fig. 2. The force characteristics like levitation force, force density, and force ripple are mainly influenced by major structural parameters like the width of stator magnets d_s, d_{s1}, d_{s2}, d_{s3}, width of mover magnets d_m, d_{m1}, d_{m2}, d_{m3}, thickness of stator magnets h_s, and thickness of mover magnets h_m. In addition, width of stator magnets d_{s1}, d_{s2}, d_{s3} are expressed by stator magnet coefficient λ_s, d_{s1}=0.5(1-λ_s)d_s, d_{s2}=$\lambda_s d_s$, d_{s3}=(1-λ_s)d_s, and width of mover magnets d_{m1}, d_{m2}, d_{m3} are expressed by mover magnet coefficient λ_m, d_{m1}=0.5(1-λ_m)d_m, d_{m2}=$\lambda_m d_m$, d_{m3}=(1-λ_m)d_m. The aforementioned design parameters are optimized with some structural parameters fixed, i.e., the radius of mover magnets is 40 mm, considering the stroke and manufacturability, the air gap length is chosen to be 2.5 mm, and in order to ensure low natural frequency d_s= d_m.

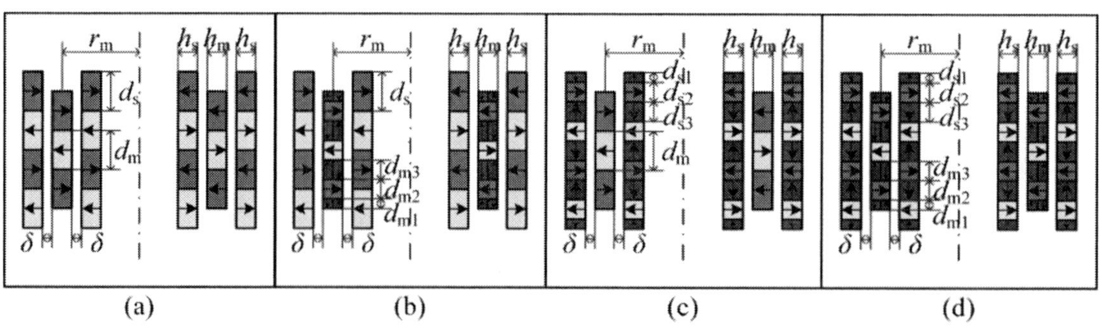

Figure 2. The major structural parameters of typical topologies of THPMLVI.

For the NS-NS topology of THPMLVI shown in Fig. 2 (b), through changing the thickness of mover magnets h_m and width of stator magnets d_s, the levitation force and force density are calculated as shown in Fig. 3. For a certain width of stator magnets d_s, with the increase in thickness of mover magnets h_m, the levitation force increases, the force density first increases and then decreases. The maximum fore density 11.72 N/cm3 occurs when h_m and d_s are about 8mm and 10mm, respectively. With the optimal width of stator magnets d_s of 10 mm, the levitation force and force density with respect to the thickness of stator magnets h_s and thickness of mover magnets h_m are calculated, as shown in Fig. 3. For a certain thickness of mover magnets h_m, with the increase in thickness of stator magnets h_s, the levitation force increases, the force density first increases and then decreases. The maximum fore density 14.00 N/cm3 occurs when h_m and h_m are about 4mm and 6mm, respectively. Through changing the thickness of mover magnets h_m, thickness of stator magnets h_s and width of stator magnets d_s, the force ripple are calculated as shown in Fig. 3. With the increase in thickness of mover magnets h_m, thickness of stator magnets h_s, the force ripple is almost unchanged. With the increase in width of stator magnets d_s, the force ripple decreases significantly.

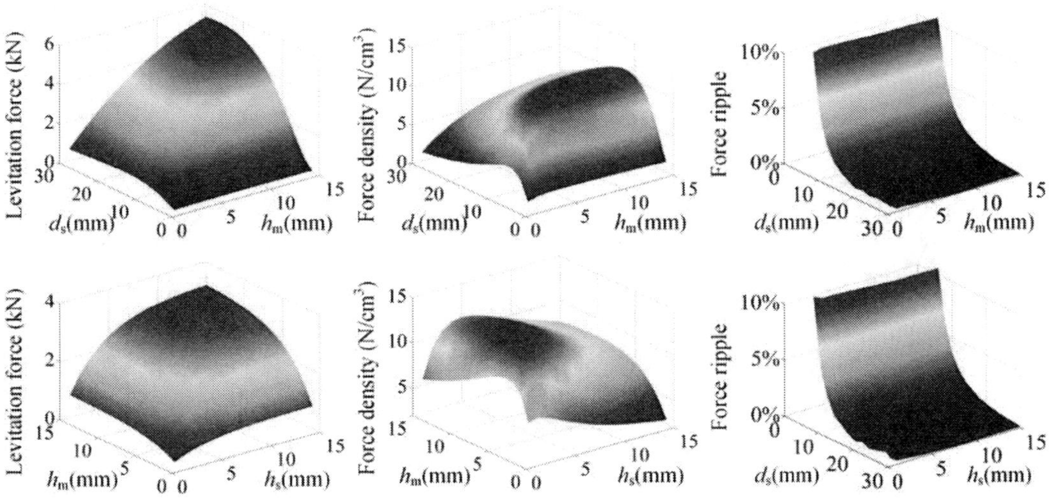

Figure 3. Force characteristics of NS-NS topology vs. d_s, h_m and h_s.

For the Halbach-NS topology of THPMLVI shown in Fig. 2 (c), through changing the thickness of mover magnets h_m and width of stator magnets d_s, the levitation force and force density are calculated as shown in Fig. 4. For a certain width of stator magnets d_s, with the increase in thickness of mover magnets h_m, the levitation force increases, the force density first increases and then decreases. The maximum fore density 16.11 N/cm^3 occurs when h_m and d_s are about 12mm and 9mm, respectively. With the optimal width of stator magnets h_m of 9 mm, the levitation force and force density with respect to the stator magnet coefficient λ_s and width of stator magnets d_s are calculated, as shown in Fig. 4. For a width of stator magnets d_s, with the increase in stator magnet coefficient λ_s, both the levitation force and force density first increase and then decrease. The maximum fore density 16.58 N/cm^3 occurs when d_s and λ_s are about 12mm and 0.4, respectively. With the optimal d_s and λ_s of 12mm and 0.4, the levitation force and force density with respect to the thickness of stator magnets h_s and thickness of mover magnets h_m are calculated, as shown in Fig. 4. For a certain thickness of mover magnets h_m, with the increase in thickness of stator magnets h_s, the levitation force increases, the force density first increases and then decreases. The maximum fore density 19.46 N/cm^3 occurs when h_s and h_m are about 4mm and 7mm, respectively. Through changing the stator magnet coefficient λ_s, thickness of mover magnets h_m, thickness of stator magnets h_s and width of stator magnets d_s, the force ripple are calculated as shown in Fig. 11. For a certain width of stator magnets d_s, with the increase in stator magnet coefficient λ_s, the force ripple decreases. With the increase in thickness of mover magnets h_m, thickness of stator magnets h_s, the force ripple is almost unchanged. With the increase in width of stator magnets d_s, the force ripple decreases significantly.

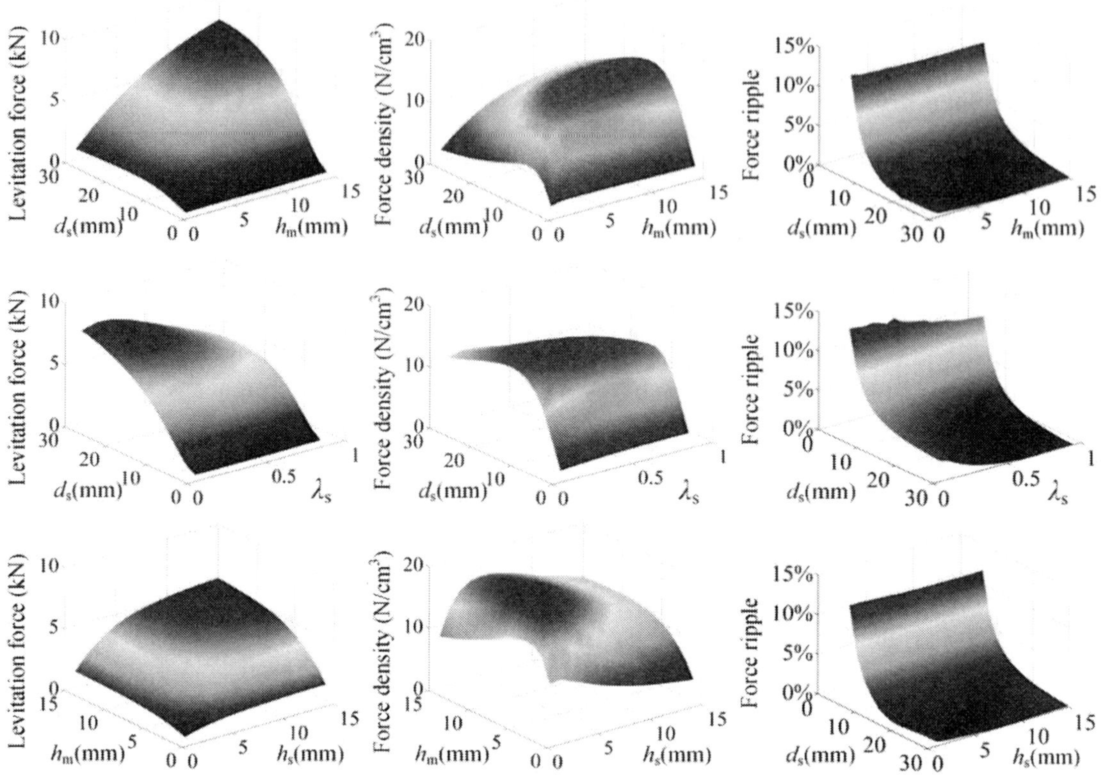

Figure 4. Force characteristics of Halbach-NS topology vs. d_s, h_m, h_s and λ_s.

For the NS-Halbach topology of THPMLVI shown in Fig. 2 (d), through changing the thickness of mover magnets h_m and width of stator magnets d_s, the levitation force and force density are calculated, as shown in Fig. 5. For a certain width of stator magnets d_s, with the increase in thickness of mover magnets h_m, the levitation force increases, the force density first increases and then de decreases. The maximum fore density 13.27 N/cm^3 occurs when d_s and h_m are about 10mm and 11mm, respectively. With the optimal width of stator magnets h_m of 11 mm, the levitation force and force density with respect to the mover magnet coefficient λ_m and width of stator magnets d_s are calculated, as shown in Fig. 5. For a width of stator magnets d_s, with the increase in mover magnet coefficient λ_m, both the levitation force and force density first increase and then decrease. The maximum fore density 13.26 N/cm^3 occurs when d_s and λ_m are about 10mm and 0.6, respectively. With the optimal d_s and λ_m of 10mm and 0.6, the levitation force and force density with respect to the thickness of stator magnets h_s and thickness of mover magnets h_m are calculated, as shown in Fig. 15. For a certain thickness of mover magnets h_m, with the increase in thickness of stator magnets h_s, the levitation force increases, the force density first increases and then decreases. The maximum fore density 15.29 N/cm^3 occurs when h_s and h_m are about 4mm and 9mm, respectively. Through changing the mover magnet coefficient λ_m, thickness of mover magnets h_m, thickness of stator magnets h_s and width of stator magnets d_s, the force ripple are calculated as shown in Fig. 5. For a certain width of stator magnets d_s, with the increase in mover magnet coefficient λ_m, the force ripple decreases. With the increase in thickness of mover magnets h_m, thickness of stator magnets h_s, the force ripple is almost unchanged. With the increase in width of stator magnets d_s, the force ripple decreases significantly.

For the Halbach-Halbach topology of THPMLVI shown in Fig. 2 (e), through changing the thickness of mover magnets h_m and width of stator magnets d_s, the levitation force and force density are calculated, as shown in Fig. 6. For a certain width of stator magnets d_s, with the increase in thickness of mover magnets h_m, the levitation force increases, the force density first increases and then de decreases. The maximum fore density 18.76 N/cm^3 occurs when d_s and h_m are about 12mm and 14mm, respectively. With the optimal width of stator magnets h_m of 14 mm, the levitation force and force density with respect to the stator magnet coefficient λ_s and width of stator magnets d_s are calculated, as shown in Fig. 19. For a width of stator magnets d_s, with the increase in stator magnet coefficient λ_s, both the levitation force and force density first increase and then decrease. The maximum force density 18.76 N/cm^3 occurs when d_s and λ_s are about 12mm and 0.5, respectively. With the optimal width of stator magnets h_m and stator magnet coefficient λ_s of 14 mm and 0.5, the levitation force and force density with respect to the mover magnet coefficient λ_m and width of stator magnets d_s are calculated, as shown in Fig. 6. For a width of stator magnets d_s, with the increase in mover magnet coefficient λ_m, both the levitation force and force density first increase and then decrease. The maximum fore density 18.99 N/cm^3 occurs when d_s and λ_m are about 12mm and 0.6, respectively. With the optimal d_s, λ_s, λ_m of 12mm, 0.5, 0.6, the levitation force and force density with respect to the thickness of stator magnets h_s and thickness of mover magnets h_m are calculated, as shown in Fig. 6. For a certain thickness of mover magnets h_m, with the increase in thickness of stator magnets h_s, the levitation force increases, the force density first increases and then decreases. The maximum fore density 21.28 N/cm^3 occurs when h_s and h_m are about 4mm and 9mm, respectively. Through changing the stator magnet coefficient λ_s, mover magnet coefficient λ_m, thickness of mover magnets h_m, thickness of stator magnets h_s and width of stator magnets d_s, the force ripple are calculated as shown in Fig. 6. With the increase in stator magnet coefficient λ_s, mover magnet coefficient λ_m, thickness of mover magnets h_m, thickness of stator magnets h_s, the force ripple is almost unchanged. With the increase in width of stator magnets d_s, the force ripple decreases significantly.

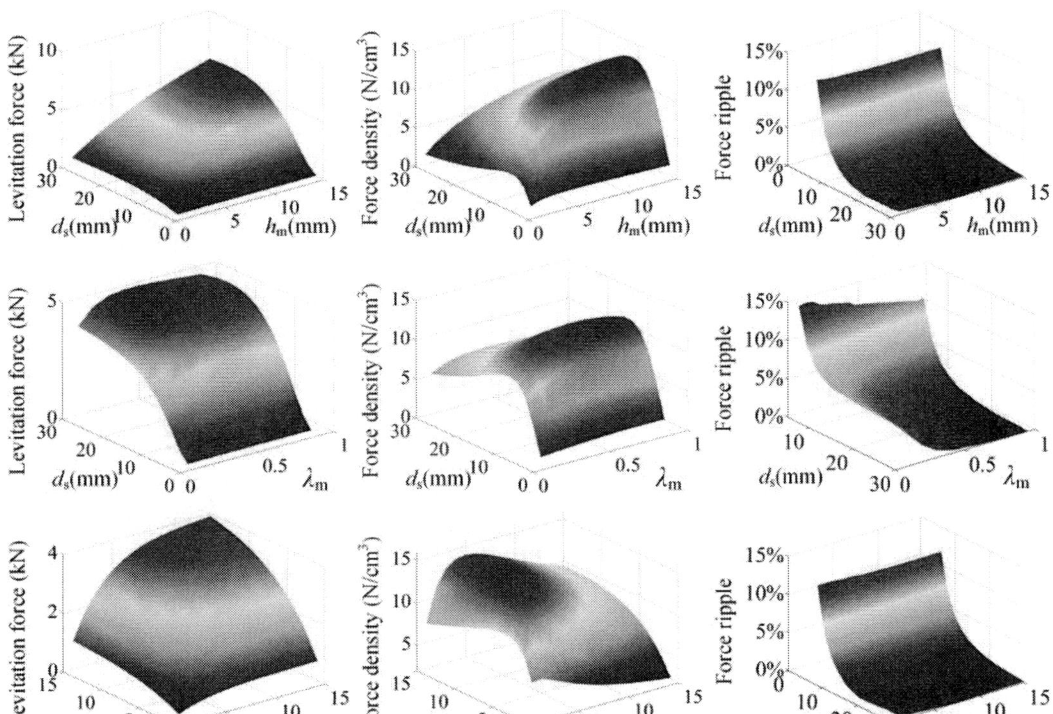

Figure 5. Force characteristics of NS-Halbach topology vs. d_s, h_m, h_s and λ_m.

The maximum force density of NS-NS topology, Halbach-NS topology, NS-Halbach topology and Halbach-Halbach topology are 14.00 N/cm^3, 19.46 N/cm^3, 15.29 N/cm^3 and 21.28 N/cm^3, separately. It can be concluded that the Halbach-Halbach topology and Halbach-NS topology possesses much larger force density than NS-Halbach topology and NS-NS topology. This shows that the Halbach permanent magnet array on the stator can effectively increase the force density, but the Halbach permanent magnet array on the mover hardly increase the force density. This is because the Halbach permanent magnet array on the stator can effectively enhance the magnetic field, but the Halbach permanent magnet array on the mover suffers serious magnetic flux leakage. The minimum force ripple of NS-NS topology, Halbach-NS topology, NS-Halbach topology and Halbach-Halbach topology are nearly the same. The problem is that high force density and low force ripple can't be reached at the same time. This is because low force ripple requires large width of stator magnets d_s more than 30mm, but high force density requires width of stator magnets d_s about 10mm.

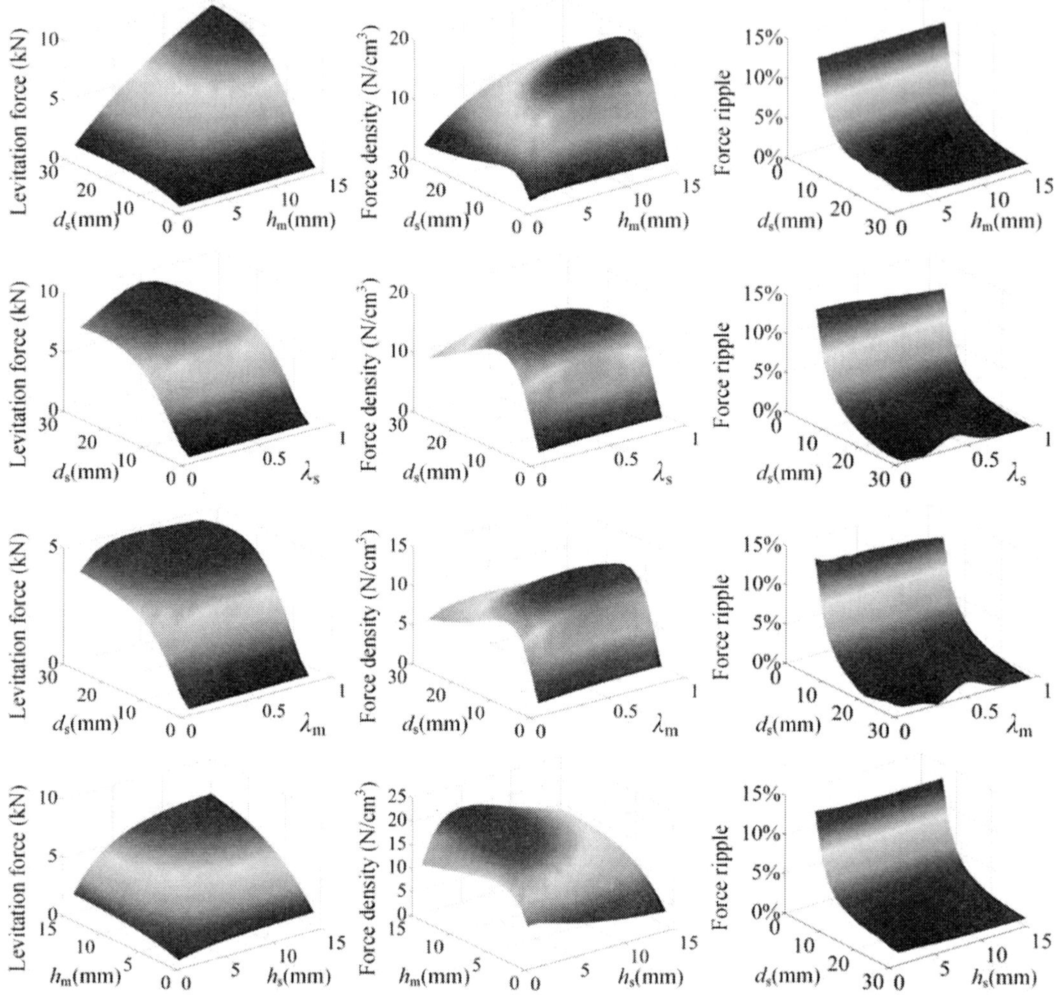

Figure 6. Force characteristics of Halbach -Halbach topology vs. d_s, h_m, h_s, λ_s and λ_m.

4 Conclusions

In this paper, a novel tubular horizontal-gap passive magnetic levitation vibration isolator (THPMLVI) has been investigated. The typical topologies and operation principle of THPMLVI are introduced. The analytical expression of magnetic force is derived. Further researches of the THPMLVI are taken under a finite element model. The relationship between levitation force, force density, force ripple and major structural parameters of the THPMLVI are analysed, which is conductive to the design and optimization of the THPMLVI. The force characteristics of four topologies of the THPMLV are compared and evaluated from the aspect of force density, force ripple and manufacturability, their major structural parameters are optimized to obtain higher high force density, which will reduce the volume of permanent magnet and the cost of THPMLVI. The following conclusions can be drawn:

(1) For the topologies of the THPMLVI, the relationship of force density is as follows: Halbach-Halbach topology > Halbach-NS topology > NS-Halbach topology > NS-NS topology. The maximum force density of NS-NS topology, Halbach-NS topology, NS-Halbach topology and Halbach-Halbach topology are 14.00 N/cm3, 19.46 N/cm3, 15.29 N/cm3 and 21.28 N/cm3, separately. The Halbach permanent magnet array on the stator can effectively increase the force density.

(2) The minimum force ripple of NS-NS topology, Halbach-NS topology, NS-Halbach topology and Halbach-Halbach topology are nearly the same. High force density and low force ripple are contradictory.

(3) For the topologies of the THPMLVI, the order of good manufacturability is as follows: NS-NS topology > Halbach-NS topology > NS-Halbach topology > Halbach-Halbach topology.

(4) The topology and major structural parameters of the THPMLVI should be determined according to the actual technical requirement, and the force density, force ripple and manufacturability should be taking into account at the same time.

Acknowledgments

This work was supported in part by the National Natural Science Foundation of China under Project 51507034.

References

1. C. Hu, B. Yao, and Q.Wang, IEEE Trans. Ind. Electron., **58**, 5195-5205 (2011)
2. L. Yu, C. Rao, and L. Du, *Proc. ISECS 2nd Int. Conf. Comput.Contr. Manage.*, 368-371 (2009)
3. E. Puppin and V. Fratello, Rev. Sci. Instrum., **73**, 4034-4036 (2002)
4. Y. Zhu, Q. Li, D. Xu, and M. Zhang, IEEE Tran. Magn., **48**, 2228-2238 (2012)
5. W. Robertson, B. Cazzolato, and A. Zander, IEEE Tran. Magn., **41**, 3826-3838 (2005)
6. W. Robertson, M. Kidner, B. Cazzolato, and A. Zander, J. Sound Vib., **326**, 88-103 (2009)
7. W. Robertson, B. Cazzolato, and A. Zander, J. Sound Vib., **331**, 1331-1341 (2012)
8. J. Janssen, J. Paulides, E. Lomonova, and B. Delinchant, J. Yonnet. Mechatronics, **23**, 197-203 (2013)
9. D. van Casteren, J. Paulides, J. Janssen, and E. Lomonova, IEEE Trans. Ind. Appl., **51**, 204-210 (2015)
10. D. van Casteren, J. Paulides, and E. Lomonova, Archives of Electrical Engineering, **64**, 593-604 (2015)
11. Y. Zhu, Q. Li, D. Xu, C. Hu, and M. Zhang, Rev. Sci. Instrum., **83**, 098108 (2012)
12. D. Xu, Q. Yu, J. Zhou, and S. Bishop, J. Sound Vib., **332**, 3377-3389 (2013)
13. W. Wu, X. Chen, and Y. Shan, J. Sound Vib., **333**, 2958-2970 (2014)
14. K. Shin. International Journal of Precision Engineering and Manufacturing, **15**, 439-445 (2014)
15. Y. Zheng, X. Zhang, Y. Luo, B. Yan, and C. Ma, J. Sound Vib., **360**, 31-52 (2016)
16. E. Fuilani, S. Reznik, and A. Kroll, IEEE Trans. Magn., **31**, 844-851 (1995)
17. E. Fuilani, IEEE Trans. Magn., **29**, 2295-2301 (1993)

Dynamic analysis of a geared rotor-bearing system with translational motion due to shaft deformation under residual shaft bow effect

Ying-Chung Chen[1, a] and Chun-Ku Kuo[2]

[1]*Department of Aeronautical and Mechanical Engineering, Air Force Academy, Kaohsiung 820, Taiwan*
[2]*Department of Mechanical Engineering, Air Force Institute of Technology, Kaohsiung 820, Taiwan*

Abstract. In this paper, the dynamic response of a spur geared rotor-bearing system has been studied when the gear set has translational motion due to shaft deformation and residual shaft bow effect. A new dynamic model for the geared rotor-bearing system, considering translational motion and residual shaft bow effect, is proposed, in which the distance between the centers of two gears varies with time. Therefore, the proposed model regards gear pressure angle and contact ratio as time-varying variables, while the previous model regards them as constant. A finite element model of the geared rotor-bearing system developed, the equations of motion are obtained by applying Lagrange's equation. After deriving nonlinear equations of motion for the geared rotor-bearing system, the dynamic responses are computed by applying the fourth-order Runge-Kutta numerical method. The dynamic response including lateral response of two gears, gear pressure angle and contact ratio with different phase angle between two residual shafts are investigated. The results show that the magnitude of the residual shaft bow, the phase angle between two residual shafts will significantly affect gear pressure angle and contact ratio. This new model produces more accurate dynamic responses in comparison to those of the previous model.

1 Introduction

The geared rotor-bearing system is one of the main mechanisms for modern power transmission. It is often coupled with bearing systems, aviation generator, gas turbines, turbo-generators, power generation and power extraction. On account of the increasing demand for high speed and accurate transportation, the research in the field of geared rotor dynamics is very important. In practice, the shafts would deform after a certain period of service time. The dynamic analysis of the geared rotor-bearing system with translational motion due to shaft deformation under residual shaft bow effect is investigated in this paper.

Lund [1] described a method for calculating the coupled torsional-lateral vibrations in a geared system of rotors. Nelson [2] studied effects of shear deformation and rotary inertia on rotor-bearing systems by utilizing the Timoshenko beam theory and finite element method. Kahraman et al. [3] developed a finite element model of a geared rotor system supported by flexible bearings. Shiau et al. [4] analyzed the coupled bending and torsional vibration of geared rotors as well as the effect of axial

[a] Corresponding author : ycchen80@gmail.com

 DOI: 10.1051/matecconf/201711901014

torque on bending vibrations. Lee et al. [5] developed the coupling of lateral and torsional vibration for the geared rotor-bearing system. The effect of residual shaft bow on the unbalance response of a single mass flexible rotor was investigated by Nicholas et al. [6]. Shiau and Lee [7] examined the dynamic response of a simply supported rotor under the effects of disk skew, unbalanced mass and residual shaft bow. The effect of the residual shaft bow on unbalance response of rotor-bearing system was investigated by Shiau et al. [8]. Kim et al. [9] analyzed the dynamic response of a spur gears by considering the pressure angle and the contact ratio as time-varying variables and a translational motion in the gear set due to bearing deformation. The effects of residual shaft bow and viscoelastic supports on dynamic response of a rotor-bearing system were investigated by Kang et al. [10]. Chen et al. [11] analyzed the interaction between gears, shafts and journal bearings in a geared rotor-bearing system. An optimization procedure for vibration reduction of a spur geared rotor-bearing system with residual shaft bow was investigated by Chen et al. [12].

2 Modeling of the system

The configuration of a geared rotor-bearing system is shown in Figure 1. Two uniform flexible shafts are of length L_1 and L_2. The residual bows of shaft 1 and 2 are δ_{r1} and δ_{r2}. The gear pair is mounted on the shafts and an external torque M exerts upon the driving gear. The contacting mesh force is represented by the gear mesh stiffness k_m and damping c_m along the pressure line. Four bearings are modeled as flexible elements with damping and stiffness denoted as c_j^b and k_j^b. A single shaft system with a rigid disk is shown in Figure 2, in which the fixed reference frame, $X-Y-Z$, is used to describe the system motion. Five degrees of freedom V, W, α, B, Γ are considered at each nodal point of the shaft, where V and W are lateral displacements along Y and Z directions, respectively, B and Γ are rotational displacements, α is the torsional displacement, and the axial translational vibration is neglected. The displacement due to the residual shaft bow, shown in Figure 1 is assumed as a second-order polynomial [11]:

$$\delta_j(s) = 4\delta_{rj}\frac{s}{L}\left(1-\frac{s}{L}\right), \quad j=1,2 \tag{1}$$

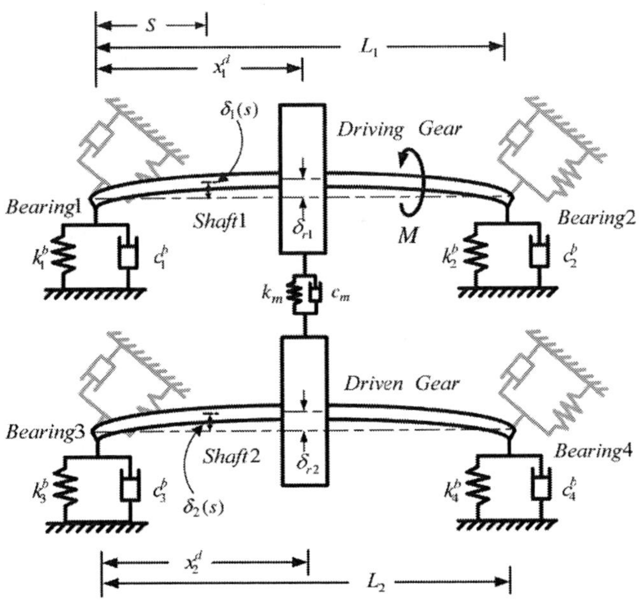

Figure 1. The configuration of a geared rotor-bearing system.

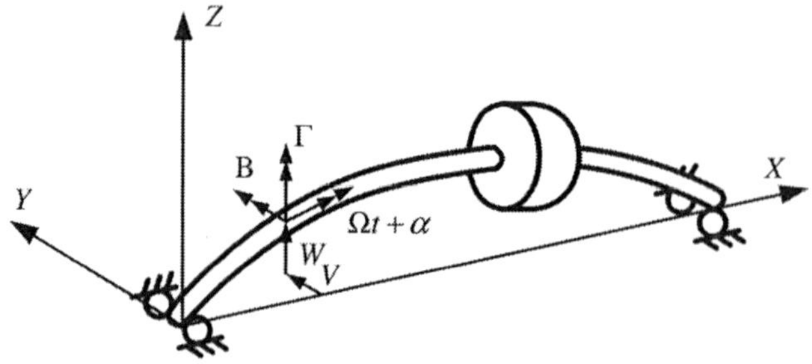

Figure 2. Typical rotor configuration and coordinates.

Figure 3 shows the configuration and generalized coordinates for the pair of gears [9], which are mounted on the shafts as shown in Figure1. The mass, transverse mass moment of inertia, polar mass moment of inertia of the driving gear and driven gear are, respectively, m_{d1} , I_{dD1} , I_{dp1} , and m_{d2} , I_{dD2} , I_{dp2} . The motion of the gear set can be defined with ten generalized coordinates, which are the same as the node coordinates of two shafts, where the gears are mounted. The dashed and solid circles represent the gear pair before and after motion, respectively. O_1 and O_2 are the initial centers of each shaft without residual bow. O_1' and O_2' are the centers under residual shaft bow effect. C_1 and C_2 are the axis centers after the motion. The mass centers of gears are denoted by G_1 and G_2 . The magnitudes of the eccentricity are denoted by e_1 and e_2 , and the eccentricity angles are denoted as Φ_{d1} and Φ_{d2} . The angular coordinates of the driving and driven gears are defined as

$$\theta_i(t) = \pm\Omega_i t + \alpha_{di}(t) + \Phi_{di} \tag{2}$$

where $i = 1$ and 2 mean the driving gear and driven gear, respectively. Ω_i is the spin speeds of shaft. The magnitudes of the residual shaft bow are denoted by δ_{r1} and δ_{r2} , and the residual angles are denoted as Φ_{r1} and Φ_{r2} . The angular coordinates of the driving and driven shafts are defined as

$$\theta_{ri}(t) = \pm\Omega_i t + \alpha_{di}(t) + \Phi_{ri} \tag{3}$$

where $i = 1$ and 2 mean the driving shaft and driven shaft, respectively. The displacement vectors for the mass centers of the gears are defined as

$$\hat{r}_i(t) = [V_{di}(t) + \delta_{ri}\cos\theta_{ri} + e_i\cos\theta_i(t)]\hat{j}_i + [W_{di}(t) + \delta_{ri}\sin\theta_{ri} + e_i\sin\theta_i(t)]\hat{k}_i \tag{4}$$

where $\hat{j}_i$ and $\hat{k}_i$ are the unit vectors in X_i and Y_i axes. The distance between centers, d , is changed to d' after the motion and residual shaft bow effect, and d' is defined as

$$d'(t) = \begin{Bmatrix} [V_{d2}(t) - V_{d1}(t) + d + \delta_{r2}\cos\theta_{r2}(t) - \delta_{r1}\cos\theta_{r1}(t)]^2 \\ +[W_{d2}(t) - W_{d1}(t) + \delta_{r2}\sin\theta_{r2}(t) - \delta_{r1}\sin\theta_{r1}(t)]^2 \end{Bmatrix}^{0.5} \tag{5}$$

In Figure 4, the gear pair is modeled as the equivalent stiffness k_m and damping c_m along the pressure line between the teeth. The equivalent stiffness is treated as time-varying coefficients in this paper. The pressure line is defined as the common tangent line of the base circles for the gear set, and the pressure angle ϕ_p is defined as

$$\phi_p(t) = \cos^{-1}\left(\frac{R_1 + R_2}{d'(t)}\right) \tag{6}$$

where R_1 and R_2 are the radii of base circles of gears.The angle of the driven gear relative to the driving gear is represented by

$$\psi(t) = \tan^{-1}\left(\frac{W_{d2} - W_{d1} + \delta_{r2}\sin\theta_{r2} - \delta_{r1}\sin\theta_{r1}}{V_{d2} - V_{d1} + d + \delta_{r2}\cos\theta_{r2} - \delta_{r1}\cos\theta_{r1}}\right) \tag{7}$$

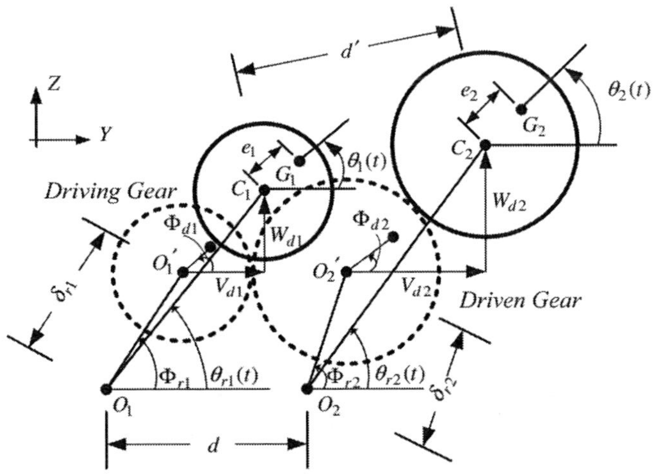

Figure 3. The configuration and generalized coordinates for gear pair.

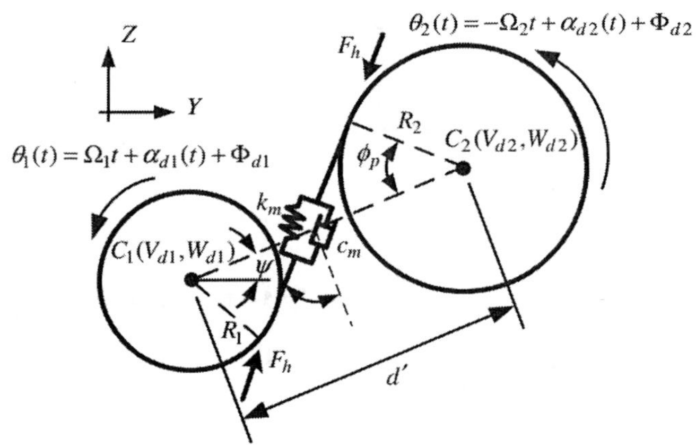

Figure 4. Gear mesh model of the gear pair.

The pressure angle and the gear position angle in this paper are affected by the translational motions of the gear pair and residual shaft bow effect while the previous studies neglected. According to the above assumption, the gear mesh is modeled as the equivalent stiffness and damping. The gear mesh deformation along the pressure line can be written as

$$\begin{aligned}\delta(t) = &(V_{d2} - V_{d1} + \delta_{r2}\cos\theta_{r2} - \delta_{r1}\cos\theta_{r1})\sin(\phi_p - \psi) \\ &+ (W_{d2} - W_{d1} + \delta_{r2}\sin\theta_{r2} - \delta_{r1}\sin\theta_{r1})\cos(\phi_p - \psi) - (R_1\alpha_{d1} + R_2\alpha_{d2})\end{aligned} \tag{8}$$

The gear mesh force along the pressure line can be expressed as

$$F_h(t) = k_m\delta + c_m\dot{\delta} \tag{9}$$

The mesh stiffness model in Kim et al. [9] is used in this paper, and is described as a periodic function with mesh period T_m in Figure 5. The succeeding pair of teeth should contact immediately when one pair of teeth lose the contact with each other. The contact ratio m_p is presented to measure this overlapping action. There are two pairs of contacting teeth when time t is in the range from $(n-1)T_m$ to $(m_p+n-2)T_m$, and k_u is the maximum mesh stiffness in this mesh process. There is only one pair of contacting teeth if t is in the range from $(m_p+n-2)T_m$ to nT_m, and k_l is the minimum mesh stiffness. The definition of T_m and m_p are defined as

$$T_m = \frac{2\pi}{N_1\Omega_1} = \frac{2\pi}{N_2\Omega_2} \tag{10}$$

$$m_p(t) = \frac{\sqrt{A_1^2 - R_1^2} + \sqrt{A_2^2 - R_2^2} - d\sin\phi_p(t)}{p_b} \tag{11}$$

where N_1 and N_2 are the numbers of teeth for gears, A_1 and A_2 are the radii of the addendum circles, and p_b is the base pitch.

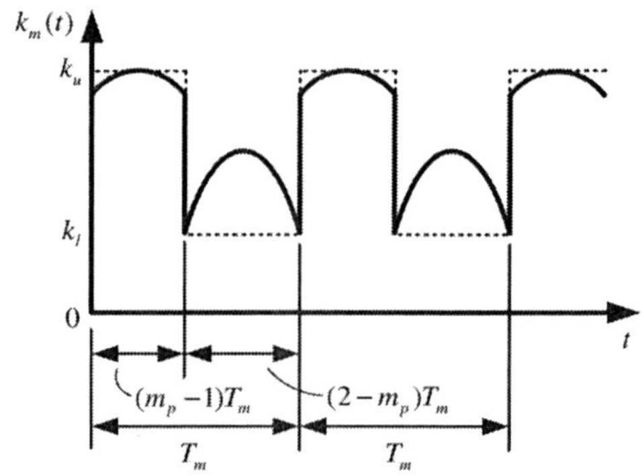

Figure 5. Gear mesh stiffness with the contact ratio m_p and the mesh period T_m.

3 Governing equation of the system

The components of this geared rotor-bearing system include the gear pair, bearing supports and rotor shafts. The equations of motion are derived for each component as follows:

3.1 Gear pair

The gear pair in this paper is assumed as two rigid disks, so the kinetic energies of driving gear and driven gear can be described as

$$T_i^g = \frac{m_{di}\dot{\hat{r}}_i^2 + I_{dp,i}(\pm\Omega_i + \dot{\alpha}_{di})^2 + I_{dD,i}\dot{\mathrm{B}}_{di}^2 + I_{dD,i}\dot{\Gamma}_{di}^2 + I_{dp,i}(\pm\Omega_i + \dot{\alpha}_{di})(\dot{\mathrm{B}}_{di}\Gamma_{di} - \dot{\Gamma}_{di}\mathrm{B}_{di})}{2} \tag{12}$$

The components of the gear mesh forces on the driving gear and the moment about the geometric center can be expressed

$$F_{hy1} = -F_h \sin(\phi_p - \psi) = -(k_m \delta + c_m \dot{\delta}) \sin(\phi_p - \psi) \tag{13}$$

$$F_{hz1} = -F_h \cos(\phi_p - \psi) = -(k_m \delta + c_m \dot{\delta}) \cos(\phi_p - \psi) \tag{14}$$

$$T_{h1} = -R_1 F_h = -R_1 (k_m \delta + c_m \dot{\delta}) + M \tag{15}$$

For the driven gears, similar expressions can be obtained. The corresponding equation of motion can be obtained as

$$\begin{bmatrix} [M_1^g] & 0 \\ 0 & [M_2^g] \end{bmatrix} \{\ddot{q}^g\} + \Omega_1 \begin{bmatrix} [G_1^g] & 0 \\ 0 & \frac{N_{t1}}{N_{t2}}[G_2^g] \end{bmatrix} \{\dot{q}^g\} + k_m [S_h]\{q^g\} = \{F^g\} \tag{16}$$

where $\{q^g\} = \{V_{g1} \quad W_{g1} \quad \mathrm{B}_{g1} \quad \Gamma_{g1} \quad \alpha_{g1} \quad V_{g2} \quad W_{g2} \quad \mathrm{B}_{g2} \quad \Gamma_{g2} \quad \alpha_{g2}\}^T$, $[M_1^g]$ and $[M_2^g]$ are mass matrices of driving gear and driven gear, $[G_1^g]$ and $[G_2^g]$ are gyroscopic matrices. The matrix $[S_h]$ is the stiffness term due to gear mesh effect. The vector $\{F^g\}$ is the force vector due to gear eccentricity and external torque.

3.2 Shaft

For shafts, two-noded element is used for the finite element formulation in this paper. As shown in Figure 6, five degrees of freedom are considered at each nodal point. The kinetic energy and potential energy of the shaft element can be expressed as

$$T^e = \frac{1}{2}\int_0^l \left\{\rho A(\dot{V}_s^2 + \dot{W}_s^2) + I_D(\dot{\mathrm{B}}_s^2 + \dot{\Gamma}_s^2) + I_P(\Omega_i + \dot{\alpha}_s)^2 + I_P(\Omega_i + \dot{\alpha}_s)(\dot{\mathrm{B}}_s \Gamma_s - \dot{\Gamma}_s \mathrm{B}_s)\right\} ds \tag{17}$$

$$U^e = \frac{1}{2}\int_0^l \left\{EI_D\left[(\mathrm{B}_s')^2 + (\Gamma_s')^2\right] + GI_P(\alpha_s')^2 + \kappa' GA\left[(V_s' - \Gamma_s)^2 + (W_s' + \mathrm{B}_s)^2\right]\right\} ds \tag{18}$$

where ρ, A, κ', I_D and I_p are the mass density, cross-section area, shear factor for the circular cross-section, transverse mass moment of inertia and polar mass moment of inertia for the shaft, respectively. E, G are the Young's modulus and shear modulus of the shaft. By using the Lagrangian approach, the motion equations of a shaft element can be obtained as:

$$\left([M^e] + [M^r]\right)\{\ddot{q}^e\} + \Omega\left([G^e] + [G^r]\right)\{\dot{q}^e\} + \left([K^e] + [K^r]\right)\{q^e\} = \{F^r\} \tag{19}$$

where $\left\{q^e\right\} = \{V_{e1} \quad W_{e1} \quad \mathrm{B}_{e1} \quad \Gamma_{e1} \quad \alpha_{e1} \quad V_{e2} \quad W_{e2} \quad \mathrm{B}_{e2} \quad \Gamma_{e2} \quad \alpha_{e2}\}^T$, $[M^e]$, $[G^e]$ and $[K^e]$ are mass, gyroscopic and stiffness matrices of the shaft element, respectively; $[M^r]$, $[G^r]$, $[K^r]$ and $\{F^r\}$ are mass, gyroscopic, stiffness matrices and forcing vector due to the residual shaft bow, respectively. The details are given in Shiau et al. [8].

3.3 Bearing

In this paper, four bearing supports are assumed as isotropic. The potential energy and virtual work of jth bearing support can be expressed as

 DOI: 10.1051/matecconf/201711901014

$$U_j^b = \frac{k_j^b V_{bj}^{\ 2} + k_j^b W_{bj}^{\ 2}}{2} \text{ and } \delta W_j^{nc} = -c_j^b \dot{V}_{bj} \delta V_{bj} - c_j^b \dot{W}_{bj} \delta W_{bj} \tag{20}$$

By using the Lagrangian approach, the motion equations of jth bearing support can be obtained as

$$\begin{bmatrix} c_j^b & 0 \\ 0 & c_j^b \end{bmatrix} \{\dot{q}_j^b\} + \begin{bmatrix} k_j^b & 0 \\ 0 & k_j^b \end{bmatrix} \{q_j^b\} = \begin{Bmatrix} 0 \\ 0 \end{Bmatrix} \tag{21}$$

where $\{q_j^b\} = \{V_{bj} \quad W_{bj}\}^T$ and $j = 1 \sim 4$.

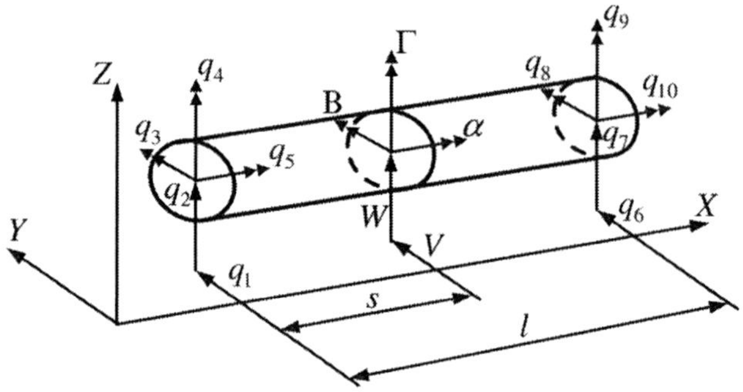

Figure 6. Shaft element and coordinate system.

3.4 System equation of motion

The system equation of motion is obtained by assembling equations of each component of the system, and can be expressed as:

$$[M^s]\{\ddot{q}^s\} + \left(\Omega[G^s] + [C^s]\right)\{\dot{q}^s\} + [K^s]\{q^s\} = \{F^s\} \tag{22}$$

where $\{q^s\}$ is the state vector, $[M^s]$, $[G^s]$, $[C^s]$ and $[K^s]$ are the mass, gyroscopic, damping and stiffness matrices of the system, respectively. $\{F^s\}$ is the forcing vector due to gear mesh, external torque , gear eccentricity and residual shaft bow effect.

Table 1. Parameters of the spur gear pair.

	Driving gear	Driven gear
Number of teeth	$N_1 = 20$	$N_2 = 30$
Base circle radius [mm]	$R_1 = 18.8$	$R_2 = 28.2$
Addendum circle radius [mm]	$A_1 = 22$	$A_2 = 32$
Mass [g]	$m_1 = 78.4$	$m_2 = 176.5$
Mass moment of inertia [kg-m^2]	1.39×10^{-5}	7.01×10^{-5}
Eccentricity angle[degree]	$\Phi_{d1} = 0$	$\Phi_{d2} = 0$
Gear mesh damping coefficient [N-sec/m]	$c_m = 1.8$	
Mesh stiffness during one-pair contact [N/m]	$k_l = 0.75 \times 10^8$	
Mesh stiffness during two-pair contact [N/m]	$k_u = 1.25 \times 10^8$	

DOI: 10.1051/matecconf/201711901014

4 Results and discussion

In this section numerical simulations are performed for a geared rotor-bearing system. The parameters of the spur gear pair and rotor-bearing system are shown in Table 1 and Table 2, which are referred to Kim et al. [9] and Chen et al. [12], respectively. The nonlinear gear mesh stiffness shows in Figure 5 are considered and Ω_1 is selected as 3000(rpm). The radius of two shaft are the same and equal to $r_{s1} = r_{s2} = 0.0185$ (m), the residual angles of two shaft are the same and equal to $\Phi_{r1} = \Phi_{r2} = 0$. The external torque M is selected as 45(N-m). Figure 7 shows the lateral responses of driving gear and driven gear with different magnitudes of residual shaft bow. It is seen that the lateral response increases as the magnitude of residual shaft bow increases.

Table 2. Parameters of the rotor-bearing system.

Shaft length [mm]	$L_1 = L_1 = 254$
Gear pair position [mm]	$x_1^d = x_2^d = L_1 / 2$
Shaft density [kg/m^3]	$\rho = 7800$
Young's modulus [N/m^2]	$E = 2.07 \times 10^{11}$
Shear modulus [N/m^2]	$G = 8.28 \times 10^{10}$
Poisson ratio	$\nu = 0.25$
Bearing damping coefficient [N-sec/m]	$c_{yy} = c_{zz} = 1 \times 10^3$
Bearing stiffness [N/m]	$k_{yy} = k_{zz} = 1 \times 10^8$

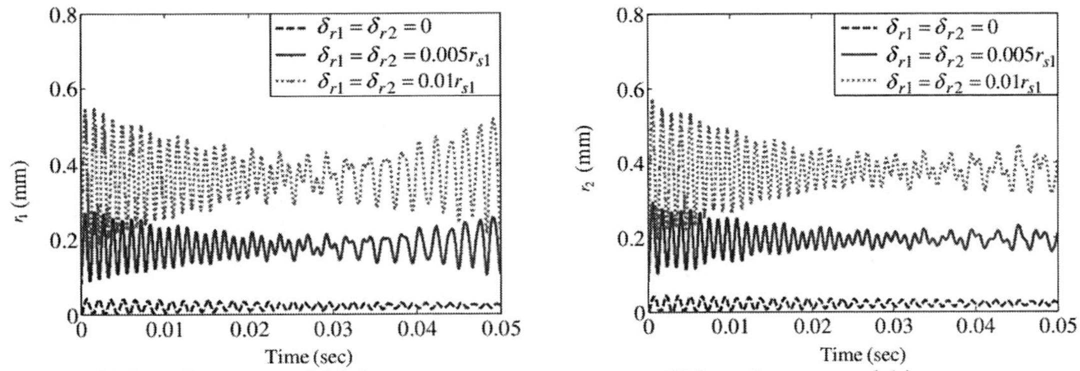

(a) lateral response of driving gear (b) lateral response of driven gear

Figure 7. Lateral response of driving gear and driven gear with different magnitudes of residual shaft bow.

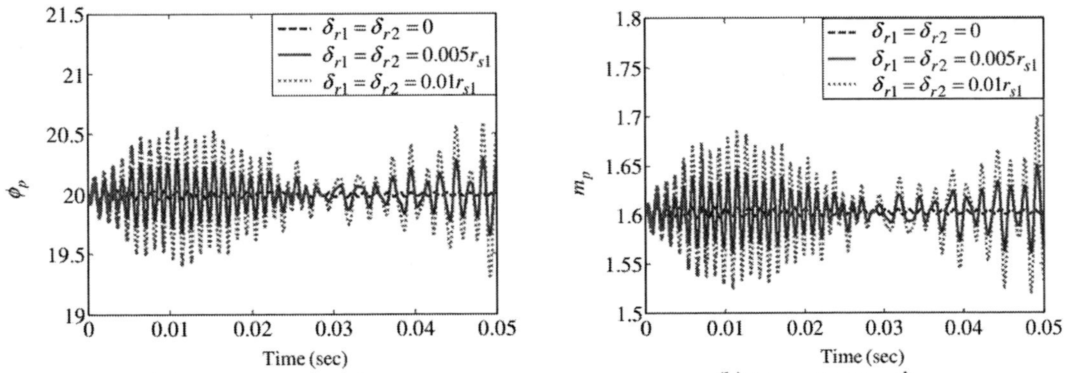

(a) gear pressure angle (b) gear contact ratio

Figure 8. Gear pressure angle and contact ratio with different magnitudes of residual shaft bow.

Gear pressure angle and contact ratio with different magnitudes of residual shaft bow are shown in Figure 8. With residual shaft bow increases, gear pressure angle and contact ratio enhance. The effect of residual shaft bow would be significant and should not be ignored while δ_{r1} and δ_{r2} were large enough.

Figure 9 and Figure 10 show gear pressure angle and contact ratio with different external forces and magnitudes of residual shaft bow. In this section the external torque M is selected as 10 and 45(N-m). When the external force is higher, the effect of residual shaft bow on gear pressure angle and contact ratio will be enlarged.

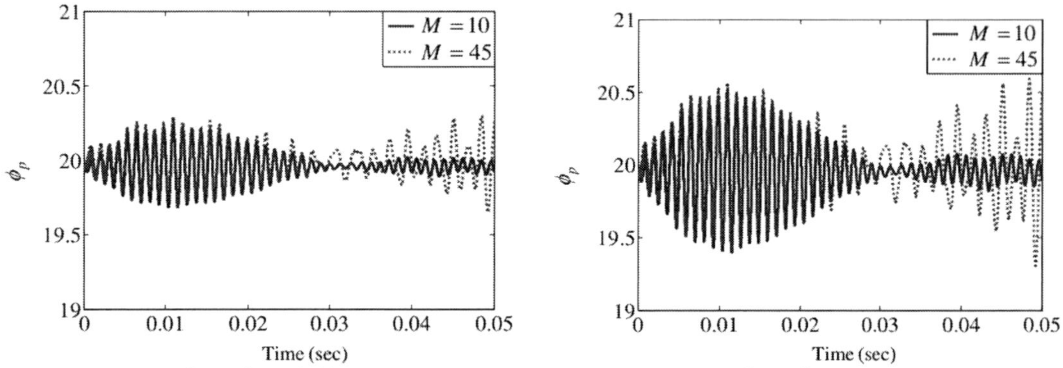

(a) $\delta_{r1} = \delta_{r2} = 0.005 r_{s1}$ (b) $\delta_{r1} = \delta_{r2} = 0.01 r_{s1}$

Figure 9. Gear pressure angle with different external force and magnitudes of residual shaft bow.

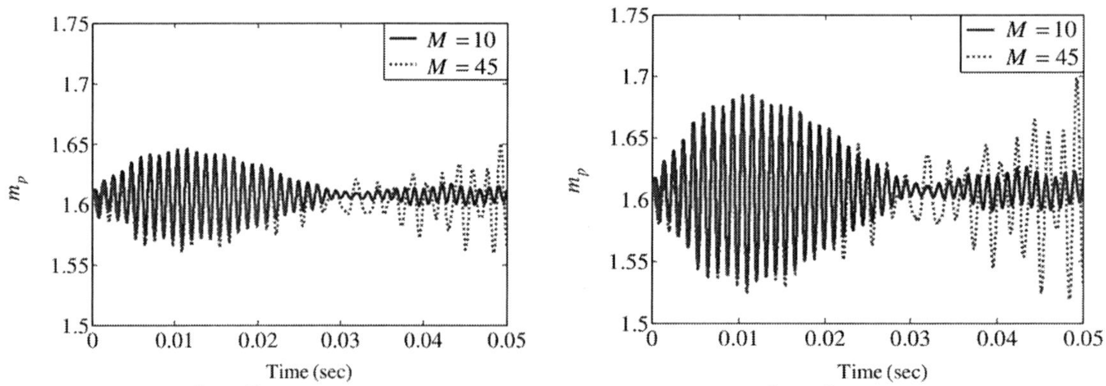

(a) $\delta_{r1} = \delta_{r2} = 0.005 r_{s1}$ (b) $\delta_{r1} = \delta_{r2} = 0.01 r_{s1}$

Figure 10. Gear contact ratio with different external force and magnitudes of residual shaft bow.

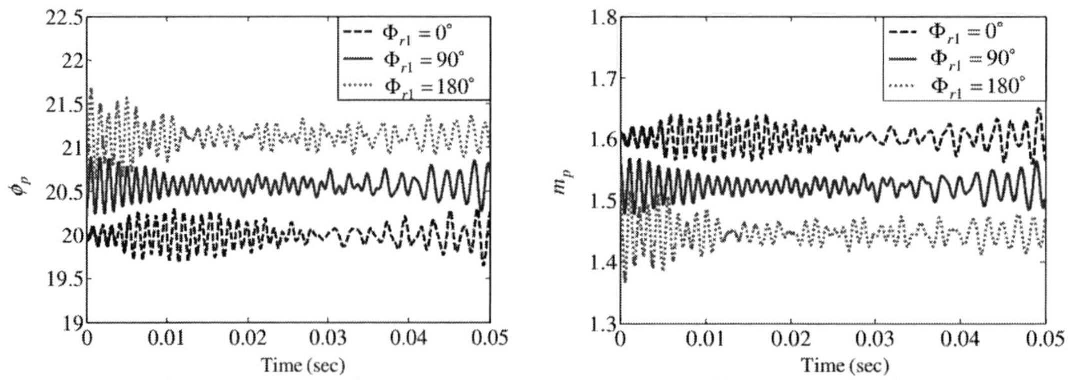

(a) gear pressure angle (b) gear contact ratio

Figure 11. Gear pressure angle and contact ratio with different residual angles of shaft 1.

In the followings, the residual shaft bow is considered for $\delta_{r1} = \delta_{r2} = 0.005 r_{s1}$, the residual angle of shaft 2 is still selected as $\Phi_{r2} = 0$. Gear pressure angle and contact ratio with different residual angles of shaft 1 are shown in Figure 11. Gear pressure angle is increased when the residual angle of shaft 1 is increased. Gear contact ratio is reduced when the residual angle of shaft 1 is increased. It means that residual angle significantly affects gear pressure angle and contact ratio.

5 Conclusions

This paper investigates a new dynamic model for spur geared rotor-bearing system which is considering translational motion and residual shaft bow effect. This new model which is deficient in the previous researches produces more accurate dynamic responses in comparison to those of the previous model. Based on the equations of motion, the dynamic responses were computed by applying the fourth-order Runge-Kutta numerical method. With residual shaft bow increases, lateral response, gear pressure angle and contact ratio enhance. With higher external force, the effect of residual shaft bow on gear pressure angle and contact ratio will be enlarged. Residual angle significantly affects the gear pressure angle and contact ratio. The effect of residual shaft bow would be significant and should not be ignored when the residual shaft bow was large enough.

References

1. J. W. Lund, J. Mech. Des., **100**, 535-538 (1978)
2. H. D. Nelson, J. Mech. Des. **102**, 793-803 (1980)
3. A. Kahraman, H. N. Ozguven, D. R. Houser, and J. J. Zakrajsek, J. Mech. Des., **114**, 507-514 (1992)
4. T. N. Shiau, S. T. Choi, and J. R. Chang, Mech. Mach. Theory, **33**, 761-783 (1998)
5. A. S. Lee, J. W. Ha,and D. H. Choi, J. Sound Vibr., **263**, 725-742 (2003)
6. J. C. Nicholas, E. J. Gunter, and P. E. Allaire, J. Eng. Power., **98**, 171-181 (1976)
7. T. N. Shiau, and E. K. Lee, J. Vib. Acoust., **111**, 170-178 (1989)
8. T. N. Shiau, E. K. Lee, and Y. C. Chen, *IGTI* GT2006-90435 (2006)
9. W. Kim, H. H. Yoo, and J. C. Hung, J. Sound Vibr., **329**, 4409-4421 (2010)
10. C. H. Kang, W. C. Hsu, E. K. Lee, and T. N. Shiau, Mech. Mach. Theory, **46**, 264-275 (2011)
11. Y. C. Chen, C. H. Kang, and S. T. Choi, J. Vibroeng., **14**, 1141-1150 (2012)
12. Y. C. Chen, C. H. Kang, and S. T. Choi, J. Chin. Soc. Mech. Eng., **36**, 471-480 (2015)

MATEC Web of Conferences **119**, 01015 (2017) DOI: 10.1051/matecconf/201711901015
IMETI 2016

Exploring the benefit of farming slideshow to agri-business marketing

Yung-Hsing Peng[a], Po-Cheng Huang, Ying-Hsu Chen, Chia-Cheng Lin and Ming-Hsiang Wang

Innovative DigiTech-Enabled Applications and Services Institute, Institute for Information Industry, Taipei 106,Taiwan

Abstract. In Taiwan agriculture, most of the production management is done by paperwork. This kind of approach is labor-costing, and unable to give real-time information, which makes it difficult to disclose sufficient information from production, and therefore results in the problem of food safety. To break the bottleneck, the Institute for Information Industry (III) develops the Smart Agri-management Platform (S.A.M.P.), providing cloud services for agricultural production and sale management. In this paper, we adopt the S.A.M.P. to collect farming records containing images, date, field location, and task name. Traditionally, farming records are used to generate a table of texts for traceability. However, it is difficult to arouse the interest from customers by merely using texts. Therefore, this paper proposes the mechanism of farming slideshow, a new presentation of traceability by dynamic arrangement to farming records on S.A.M.P. The experiment is performed in the Yongling organic farm in Kaohsiung, generating over 30,000 vegetable packages supporting farming slideshow. According to 42 valid questionnaires from customers, 83% of them express higher willing to buy products with farming slideshow than those with only texts. Therefore, food traceability with slideshow can offer benefit to agri-business marketing, which is a positive reference for future agriculture.

1 Introduction

In recent years, food safety has become an important social issue in Taiwan, which arouses more and more public attention. Since people beware of what they eat, this market trend definitely influences Taiwan agriculture: the origin of most Taiwanese food. Currently, there are two main certificates for agricultural products, which are the traceable agricultural products (TAP), and the CAS Taiwan organic label. These certificates are issued by certification bodies, such as the Tse-Xin Organic Certification Corporation (TOC), or the National Chung Hsing University. During the certification, the environment of the farmland will be carefully examined, and it is necessary for farmers to regularly provide their farming records, which are usually managed by paperwork. The motivation of such certificate is to show the safety of agricultural products to customers, which promotes the trading of safe products. Taking the TAP as an example, customers can trace the farming records by inputting the unique query code attached on packaged fruits or vegetables. However, to our knowledge, the presentation of such farming records is merely text that contains the date and the name of farming task.

[a] Corresponding author : pengyh@iii.org.tw

For general customers, the name of farming task can be a professional term not easy to comprehend. In addition, the recording habit of farmers deeply affects the reliability of farming data. When records are collected by paperwork, the date and the name of task can be improperly filled, and the recognition of hand-written text also results in some hardness for data process. Because of the above phenomenon, it is difficult to arouse the interest from customers even if the code of traceability is provided. In other words, the present solution for processing farming records can hardly benefit the sale of agri-products. Therefore, it is worth study to develop a service that fulfils both the need of certification and marketing.

Recently, the Institute for Information Industry (III) develops the Smart Agri-management Platform (S.A.M.P.) [0] that provides innovative cloud services for agricultural production and sale management. For agricultural data analysis, many interesting researches have also been investigated, such as the crop growth estimation [2-5], farmland climate tracking [6-8], and crop price prediction [9-11]. There are similar systems developed for agricultural data analysis, such as the Farmlogs[12] and KisanHub[13]. However, the utilization of agricultural data for marketing is scarcely discussed. In this paper, we propose the mechanism of farming slideshow, which is a new presentation of traceability achieved by dynamic arrangement to farming records on S.A.M.P. The organization of this paper is as follows. In Section 2, we describe the methodology for creating farming slideshow. In Section 3, we propose the experimental results in Yongling organic farm. Finally, we give some conclusions and future work in Section 4.

2 Methodology

In this section, we first introduce the architecture of production management using S.A.M.P. After that, we briefly describe the implementation for generating the farming slideshow.

2.1 Architecture

The architecture of the production management using S.A.M.P. is shown in Figure 1. The S.A.M.P. utilizes front-end apps and back-end web applications for data acquisition and production management, respectively. With S.A.M.P., managers can supervise each field (farmland) with ease, and assign proper tasks to farmers of concerns. Next, farmers can confirm the content of task and report the completion of task with their mobile phone. In this way, the farming data are recorded during the management, which eliminates the paperwork.

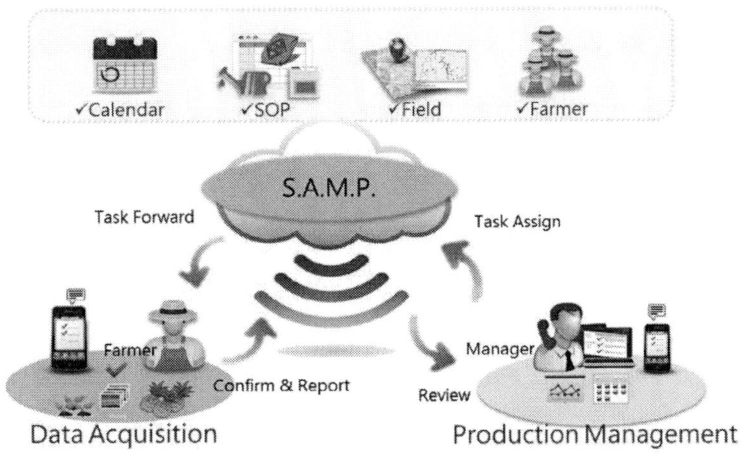

Figure 1. The architecture of the production management using S.A.M.P.

2.2 Implementation

Currently, the server of S.A.M.P. is maintained by the Institute for Information Industry, and the farming slideshow is implemented as an embedded service on S.A.M.P. The material used for generating the slideshow include (1)the name of the agri-enterprise, (2)the location of the production site (field), (3)the period of cultivation, (4)the names and the dates for finished tasks, (5)the name of the farmer who completed the task, and (6)the farming image of the task. The above information can be collected in different stages when managers and farmers are using S.A.M.P. First, the name of the agri-enterprise is retrieved from the user profile. The location of fields can be specified by drawing polygons with our web application, or by referring to the GPS signal returned from mobile phone. The period of cultivation is configured by production managers when they are making the planting schedule. The names of tasks are also defined by the managers when they are establishing the SOP for cultivation. Meanwhile, the dates of finished tasks depend on the time when farmers report their completion of tasks, and the names of farmers can be easily mapped by their login accounts. Finally, the farming images are from the reported pictures that farmers take with their mobile phones. On S.A.M.P., the collected information is first represented in table form. With proper re-organizing, the information can come in the form of slideshow with QR code. The screenshots of our implementation is given as Figure 2.

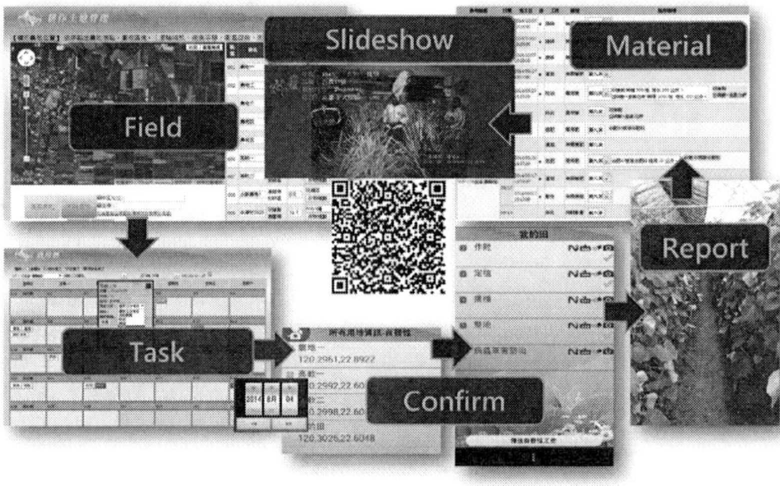

Figure 2. The implementation for the farming slideshow on S.A.M.P.

3 Experimental Results

To perform the experiment, we cooperate with the Yongling organic farm in Kaohsiung. The experiment began from October 1st, 2015 to December 31st, 2015. With the app of S.A.M.P., the staff in Yongling collected farming data via their mobile phones, reporting the dates, names, and images of tasks to the platform. By processing these data, the S.A.M.P. generates 57 farming slideshow for 16 different vegetables planted in 30 greenhouses. On average, each slideshow covers a 28-day cultivation period from seedling to harvest. During the experiment, the produced farming slideshows are further transformed into 30,000 QR codes attached on packed vegetables of weight 250g. To investigate the effect of farming slideshow for agri-product marketing, Yongling designed a questionnaire for customers who are interested to buy organic vegetables. In total, the questionnaire was given to 55 customers. According to 42 valid returns, farming slideshow offers better understanding of agricultural production to 95% customers, and 83% of customers express higher willing to buy products with farming slideshow than those with only text descriptions. The results of questionnaire are given in Table 1.

Table 1. The questionnaire for exploring the benefit of farming slideshow.

No.	Survey	Statistics
1	You used to purchase organic agri-products.	95.3%
2	You used to read some QR code with your mobile phone.	69.8%
3	You prefer to buy agri-products with TAP or CAS labels than those without any labels.	88.4%
4	You used to buy agri-products that support on-line traceability for farming records.	60.5%
5	You used to trace the farming records spontaneously. (If your answer is "No", please go to question No. 7.	14.0%
6	You used to trace farming records with images.	7.0%
7	Farming slideshows provide more messages than text descriptions do.	86.0%
8	Farming slideshow can effectively disclose the information of production process.	100.0%
9	The cost for supporting farming slideshow is higher than that for supporting text descriptions.	74.4%
10	Farming slideshow gives you better understanding for the production of agri-products than pure text does.	95.3%
11	You are happy to see that farming slideshow will someday replace the certification with only texts.	100.0%
12	Revealing the production process can increase your willing to buy the product.	90.7%
13	Farming slideshow increases your willing to buy agri-product.	83.7%
14	Farming slideshow helps you quickly review the agricultural production.	87.2%
15	Farming slideshow helps you clearly understand the agricultural production.	88.4%
16	How much extra money (in Taiwanese dollars) are you willing to pay for each packed vegetable, for obtaining its farming slideshow.	1.48

In Figure 3, we provide some sample labels in Yongling, and readers can watch farming slideshows by scanning the QR codes. Note that for the purpose of publication, the product names in Figure 3 are translated from the original Chinese labels.

Figure 3. The sample labels produced for the experiment in Yongling organic farm.

4 Conclusions

In this paper, we propose the farming slideshow as a service for agri-business marketing. This service is implemented on S.A.M.P., and its benefit is verified by an experiment on Yongling organic farm. From the experimental results, one can see that farming slideshow successfully draws the attention from most customers. In addition to the public attention for information disclosure, Table 1 expresses two interesting phenomenon. First, though farming slideshow seems to take additional efforts, there are only 75% customers believe that it will increase production cost. That is, about 25% customers notice that the convenience brought by S.A.M.P. can reduce laboring cost on doing paperwork. In addition, customers are willing to pay little extra money for purchasing products with farming slideshow, and all of them are happy to see that farming slideshow will someday replace the text description. Finally, the statistics for question 5 is only 14%, which implies that farming slideshow is better to be displayed actively on screen, in order to catch the eyes from customers. In summary, the proposed farming slideshow achieves a good initial result on agri-business marketing. In the future, we would like to extend the idea to food processing industry, to make a farm-to-table disclosure of food information.

Acknowledgment

This study is conducted under the "Online and Offline integrated Smart Commerce Platform (3/4)" of the Institute for Information Industry which is subsidized by the Ministry of Economy Affairs of Taiwan. The authors would like to express their appreciation to the staff of Yongling organic farm for their kind support on farming data collection and questionnaire design for customers.

References

1. The Smart Agri-management Platform (S.A.M.P., http://shennong.ideas.iii.org.tw/SAMP/)
2. A. Kross, H. McNairn, D. Lapen, M. Sunohara, and C. Champagne, Assessment of rapid eye vegetation indices for estimation of leaf area index and biomass in corn and soybean crops, International Journal of Applied Earth Observation and Geoinformation, **34**, 235-248 (2015)
3. Y.H. Peng, C.S. Hsu, and P.C. Huang, Harvest forecasting with environmental information for cucumbers cultivated in net houses, Journal of Advanced Agricultural Technologies, **2**, 83-87 (2015)
4. Y.H. Peng, C.S. Hsu, P.C. Huang, and Y.D. Wu, An effective wavelength utilization for spectroscopic analysis on orchid chlorophyll measurement, *Proc. of the 2014 IEEE International Conference on Automation Science and Engineering*, Taipei, Taiwan, 716-721 (2014)
5. P. Rischbeck, S. Elsayed, B. Mistele, G. Barmeier, K. Heil, and U. Schmidhalter, Data fusion of spectral, thermal and canopy height parameters for improved yield prediction of drought stressed spring barley, European Journal of Agronomy, **78**, 44-59 (2016)
6. C.S. Hsu, Y.H. Peng, and P.C. Huang, Climate tracking for farmland using open data from the central weather bureau in Taiwan, Journal of Advances in Information Technology, **6**, 227-232 (2015)
7. B. Lalic, M. Francia, J. Eitzinger, Z. Podraščanin, and Ilija Arsenić, Effectiveness of short-term numerical weather prediction in predicting growing degree days and meteorological conditions for apple scab appearance, Meteorological Applications, **23**, 50-56 (2016)
8. M. Knapp, Enhanced measurement of weather and climate in natural areas through integrated networks, Transactions of the Kansas Academy of Science, **119**, 64-68 (2016)
9. H. Liu and S. Shao, India's tea price analysis based on ARMA model, Modern Economy, **7**, 118-123 (2016)
10. Y.H. Peng, C.S. Hsu, and P.C. Huang, Developing crop price forecasting service using open data from Taiwan markets, *Proc. of the 2015 Conference on Technologies and Applications of Artificial Intelligence*, Tainan, Taiwan, 172-175 (2015)
11. Y.H. Peng, C.S. Hsu, P.C. Huang, An investigation of spacial approaches for crop price forecasting in different Taiwan markets, *Proc. of the 2015 Conference on Technologies and Applications of Artificial Intelligence*, Tainan, Taiwan, 176-179 (2015)
12. The Farmlogs system (https://farmlogs.com/)
13. The KisanHub system (https://www.kisanhub.com/)

Effects of auricular therapy to body composition in young asian women

Ching-Hsiu Hsieh[1,a], Tzu-Jung Tseng[2] , Mei-Yu Huang[3] and Pei-Ying Chuang[4]

[1]*Department of Nursing, Chang Gung University Of Science and Technology, Assistant Professor, Chiayi, Taiwan*
[2]*Department of Nursing, Chang Gung University Of Science and Technology, Doctor, Chiayi, Taiwan*
[3]*Department of Nursing, Yuanpei University of Medical Technology, Hsinchu, Taiwan*
[4]*School of Nursing, University of Maryland, Assistant Professor, USA*

Abstract. This study investigated the efficacy of auricular therapy using semen vaccariae (SV) via auricular point sticking on weight loss, changes to waist circumference, and waist-to-hip ratio.
The study used a randomized design with one control group and one experimental group consisting of young Asian women with waist circumferences $\geq$ 80 cm.
The total sample size was 50 young women who ranged in age from 18 to 20 years-old.
Each participant attended weekly ten-minute sessions during which auricular acupressure treatment was performed. During treatment, the control group had 0.5 cm^2 adhesive tape placed onto their auricular acupoints, while the experimental group had the SV stick applied to the auricular acupoints. Dependent variables were collected at baseline and follow-up for four weeks.
The control group showed significantly increased body weight (+0.08 kg, $p \leq 0.05$) and waist circumference (+0.20 cm , $p \leq 0.05$); however, the SV group showed significantly decreased body weight (-1.9120 kg, $p \leq 0.05$), waist circumference (-4.9200 cm, $p \leq 0.05$) and waist-to-hip ratio (-0.0297, $p \leq 0.05$) during four weeks of treatment.
Auricular acupressure with SV may be an efficacious option to include in weight loss programs for young women.

1 Introduction

Many studies have shown that excess weight and obesity, especially abdominal obesity, are associated with numerous chronic health problems, increased mortality rates of cardiovascular or cancer, and depressive symptoms or moderate-to-severe depressive symptoms [1-3].

In the United States and worldwide, obesity is a health issue and public health problem [1]. In 2004, the United States Department of Health and Human Services (DHHS) reported that overweight adolescents have a 70% chance of becoming overweight or obese adults (U.S. DHHS, 2004) [4]. If one or more parents are overweight or obese, the risk of becoming overweight or obese in adulthood increases to 80% [5]. Thus, to prevent adult obesity, strategies must target young adults [4]. Not only is auricular acupressure of traditional Chinese medicine remedies convenient, safe and effective, this method can be used to treat obesity-related symptoms such as swelling, constipation, insomnia, dysmenorrhea, headache and digestive illness. Due to the relative safety and broad applications of this

[a] Corresponding author : chinghsiuh@yahoo.com

technique, auricular acupressure has become a popular weight loss method; however, the extent of its efficacy for weight loss and obesity treatment has not been fully studied.

Waist circumference (WC) is an anthropometric measure of central obesity in population studies [6] because high WC values are also positively correlated with cardiovascular complications [7]. According to the classifications adopted by the American Heart Association (AHA) and the National Heart, Lung, and Blood Institute (NHLBI) of the United States, elevated WC in males is defined as equal to or greater than 102 cm and is defined in females as equal to or greater than 88 cm (AHA and NHLBI, 2009) [8]. For the purpose of this study, WC $\geq$ 80 cm in females was used, conforming to the Bureau of Health Promotion, Department of Health, R.O.C. (Taiwan) (2006) [9].

In traditional Chinese Medicine (TCM), "Qi" is thought to be the energy that circulates throughout the body. Traditional acupuncturists activate Qi by inserting needles into specific points (acupoints) on the body [10-12]. Practitioners of Chinese medicine believe that an "excess" or "deficiency" of Qi can be normalized by the specific stimulation of select acupoints [10-13]. Also, acupoint stimulation activates the metabolism of individuals [14]. One of the tenets of TCM is that acupuncture treatment may modify central nervous system neurotransmitter levels through the stimulation of acupoints [10-11].

Auricular acupuncture or acupressure therapy has been used as an adjunctive therapeutic modality in complementary medicine practice worldwide [10-16]. In TCM, acupoints of the ear are believed to connect to the internal organs of the human body [17, 23]. Auricular acupuncture stimulation utilizes needles, pellets (Semen Vaccariae), magnetic pearls and other materials. In many countries, auricular acupuncture therapy is the TCM method most frequently used for obesity treatment [18]. A large body of research shows that the most effective auricular acupuncture points for treating obesity are the Shenmen, Mouth, Stomach, Endocrine, and Small Intestine points. Specifically, stimulating the Shenmen point calms patients and induces a degree of sedation. Stimulating the auricular mouth acupoint reduces feelings of anger and has been used to treat oral ulcers. The stomach and endocrine auricular points diminish appetite and reduce the sensation of hunger, which causes feelings of satiety and fullness, promoting weight loss [1,17-23]. The small intestine point treats dyspepsia, which may ultimately contribute to weight loss [23].

The plant pellet Semen Vaccariae (SV), termed Wang-Bu-Liu-Xing by the Chinese and commonly known as vaccaria seeds, are the dried seeds of Vaccaria segetalis ripe fruit. More than thousands of years ago, Chinese medicine was used to treat amenorrhoea and breast infections [23]. One study reported the systematic analysis of SV metabolism and the distribution of its chemical constituents after oral administration [24]. Also, SV is comprised of small round seeds that are commonly used in auricular taping, and the average SV size is approximately 0.13 cm [23]. SV is easily purchased from every Chinese herb market. SV sticks apply pressure to the skin both via acupuncture and finger pressure treatment when placed onto the skin points. In general, auricular therapy using SV has been found to be cost-effective and hygienic, with few side effects; therefore, SV may be more appealing to clients than many other complementary therapies for weight reduction [23].

2 Methods

2.1 Demographics and study design

This study was designed to test the effects of auricular acupressure by SV on body composition factors such as weight loss, WC and waist-to-hip ratio (WHR) in young Asian females. The total intervention duration was four weeks. The sample consisted of 55 participants who ranged in age from 18 to 20 years-old and were recruited from universities in northern Taiwan. The participants were randomly assigned to one of two groups (control or experimental) by a computer-generated list, which was held confidential; therefore, no one except the principal investigator knew the treatment allocation of the patients. Ethical approval was obtained from their schools, and informed consent was obtained from each participant. All participants were free to withdraw from the study at any time. Potential benefits of the study included a reduction in participant weight and WC, which have the potential to improve health status and overall well-being.

2.2 Intervention

For the purpose of this study, a WC $\geq$ 80 cm in females was used as a selection criterion because this parameter conforms with the Bureau of Health Promotion, Department of Health, R.O.C. (Taiwan) (2006) criteria. The participants were randomly assigned to one of two groups. Both groups received health education describing an appropriate reduced calorie diet and lifestyle modifications such as methods to increase activity levels. During treatment, the control group had adhesive tape (a $0.5cm^2$ piece of tape) placed onto their auricular acupoints, while the experimental group received the SV stick applied onto the auricular acupoints. The auricular acupoints used were the Shenmen, Mouth, Stomach, Small Intestine and Endocrine points. Each week, the SV or adhesive tape was placed onto one ear, and the following week, the SV or tape only was applied to the other ear. No specific changes to diet or physical activity were prescribed. Each participant met with the study coordinators once a week for ten minutes for a total of four weeks. Body composition measurements, including body weight, WC, hip circumference and WHR, were recorded for each participant each week as shown in Figure 1, which depicts a flow chart of the study design.

3 Data Analysis

The data were analyzed using the SPSS software (SPSS 19.0. for Windows) the General Linear Mixed-Effects Model (GLMM). Weight loss, WC and WHR measurements were collected for each group at each weekly meeting and were also assessed for significance. Mean and standard deviations of BW, WC and WHR score were compared among subjects who were treated with each group. In the primary study outcome measures of differences between both groups were analyzed using linear mixed model analysis of variance with group and meeting time as fixed effects. for continuous outcomes. Normally distributed variables and generalized estimating equations for categorical variables. The results were considered to be significant at $p < 0.05$.

4 Results

Table 1. Mean difference in body weight (kg) from baseline to the completion of the intervention (N=50).

Week	Control Group	p	Experimental Group	p
	Body Weight Mean Difference		Body Weight Mean Difference	
Baseline - 1st Week	+0.1080	* $\leq$ 0.05	-0.4680	* $\leq$ 0.05
Baseline - 2nd Week	+0.0680	* $\leq$ 0.05	-1.1440	* $\leq$ 0.05
Baseline - 3rd Week	+0.1120	* $\leq$ 0.05	-1.6240	* $\leq$ 0.05
Baseline - 4th Week	+0.0800	* $\leq$ 0.05	-1.9120	* $\leq$ 0.05

* $p \leq 0.05$

After four weeks of auricular therapy, 5 participants had withdrawn from the study for a variety of reasons, and these 5 subjects were not included in the final analysis. The final analysis includes 25 subjects in the control group and 25 subjects in the treatment group.

In the control group, the average body weight remained the same during the first two weeks (BW=+0.0680 kg); however, the average body weight significantly increased (BW = +0.0800 kg) during the third and fourth observation weeks ($p \leq 0.05$). In contrast, in the experimental group, the average body weight was significantly reduced ($p \leq 0.05$) from the first week to the fourth week (BW = -0.4680 kg during the first week; -1.1440 kg during the second week; -1.6240 kg during the third week; and -1.9120 kg during the fourth week) as shown in Table 1. Furthermore, the participants who received auricular acupressure treatment with the SV were noted to have decreased body weight from week-to-week.

The WC was also measured weekly during the four-week program duration. After four weeks, the WC of the control group increased significantly by +0.2000 cm ($p \leq 0.05$); therefore, similar to the overall body weight, simply applying adhesive tape to auricular acupoints increased the WC after four weeks (Table 2). In contrast, the WC of the treatment group decreased significantly by -4.9200 cm ($p \leq 0.05$), and the participants who received auricular acupressure with SV were noted to have decreased WC from week-to-week.

Table 2. Mean difference of WC (cm) from baseline to the completion of the intervention (N=50)

Week	Control Group	p	Experimental Group	p
	WC Mean Difference		WC Mean Difference	
Baseline – 1st Week	+0.4400	.000	-1.0480	* ≤ 0.05
Baseline - 2nd Week	+0.3800	.000	-2.6000	* ≤ 0.05
Baseline - 3rd Week	+0.7000	.000	-3.6800	* ≤ 0.05
Baseline - 4th Week	+0.2000	.000	-4.6800	* ≤ 0.05

* $p \leq 0.05$

The WHR was also measured weekly. After four weeks, the WHR of the control group showed a non-significant increase by 0.0003 ($p = 0.728$), which is in contrast to the body weight and WC, which increased significantly in the control group (Table 3). The WHR of the SV treatment group decreased significantly by 0.0297 ($p \leq 0.05$) over the course of the study. The participants who received continuous auricular acupressure treatment with SV were noted to have decreased WHR from week-to-week (Table 3).

Table 3. Mean difference of WHR from baseline to the completion of the intervention (N = 50)

Week	Control Group	p	Experimental Group	p
	WHR Mean Difference		WHR Mean Difference	
Baseline - 1st Week	+0.0009	.496	-0.0070	* ≤ 0.05
Baseline - 2nd Week	-0.0007	.782	-0.0172	* ≤ 0.05
Baseline - 3rd Week	+0.0038	.951	-0.0199	* ≤ 0.05
Baseline - 4th Week	+0.0003	.728	-0.0297	* ≤ 0.05

* $p \leq 0.05$

5 Discussion

The current study demonstrates that after 4 weeks of auricular **acupressure** treatment using SV, the body weight of the participants significantly decreased. This result confirmed that repeated auricular acupressure decreases body weight, which has been supported by other research studies. Chien and colleagues (2011) used electrical stimulation of acupoints on 21 obese patients over treatment periods ranging from 12 weeks and reported weight loss ranging from 63.48 ± 4.9 kg to 62.40 ± 4.81 kg, decreased waist circumference from 84.25 ± 2.69 cm to 82.33 ± 2.27 cm, and decreased waist to hip ratio from 0.84 ± 0.03 to 0.83 ± 0.02 [17]. Additionally, Richards and Marley (1998) assessed the efficacy of ear acupuncture treatment on weight loss in a 4-week study, and their results showed an average weight loss of 4.0 ± 1.4 kg ($p < 0.05$) in the treatment group [20].

Beyond its effects on weight loss, this study suggests that acupressure may have additional benefits to anthropometry. Lee, Jacobs, Schreiner, Iribarren and Hankinson (2007) showed that WC and WHR predict coronary artery calcification (CAC). In the current study, 4 weeks of auricular acupressure treatment using SV significantly decreased participant WC and WHR, which may then translate to reduced risk of CAC [24].

6 Conclusion

Cho and colleagues (2009) reported that stimulation via acupuncture, acupressure and heating (moxibustion) can be used to treat obesity [26]. Reduction of body weight and decreasing the prevalence of obesity in society are critical health issues [25]. Research has shown that a relationship exists between abdominal obesity and subclinical atherosclerotic vascular disease in multiple race, sex and age groups. Abdominal obesity, measured by waist size or WHR, is related with early atherosclerosis as measured by the presence of CAC [24]. Thus, the current study is important because the findings show that abdominal obesity can be safely and effectively reduced by auricular acupressure.

Other auricular acupoints will be considered in the future study. The sample size of our study is similar to the size in many previous research studies that reached similar conclusions. However, the participants in the current study were not representative of both genders. This study was exclusively comprised of women participants, and future studies should specifically recruit male participants to better reflect the general population.

In summary, auricular acupressure using SV decreased body weight, WC and WHR after just four weeks of treatment. Therefore, auricular acupuncture offers an effective and economical alternative for the treatment of obesity in contrast to more conventional treatment methods such as exercise, diet control, medicine and surgery. However, the effects of long-term treatment and follow-up after treatment are necessary for future studies. Future studies should also consider the level of baseline body weight on the effectiveness of the treatment, and the extent to which body weight can be modulated by auricular acupressure with SV.

References

1. S. K. Kumanyika, E. Obarzanek, N. Stettler, R. Bell, A. E. Field, S. P. Fortmann, B. A. Franklin, M. W. Gillman, C. E. Lewis, W. C. Poston, J. Stevens, and Y. Hong, Population-based prevention of obesity: the need for comprehensive promotion of healthful eating, physical activity, and energy balance: a scientific statement from American Heart Association Council on Epidemiology and Prevention, Interdiscuplinary Committee for Prevention, Circulation **118** (4), 428-464 (2008)
2. G. Zhao, E. S. Ford, C. Y. Li, J. Tsai, S. Dhingra, and L. Balluz, Waist circumference, abdominal obesity, and depression among overweight and obese U.S. adults: national health and nutrition examination survey 2005-2006, BMC Psychiatry **11**, 130 (2011)
3. C. Zhang, K. M. Rexrode, R. M. VanDam, T. Y. Li, and F. B. Hu, Abdominal obesity and th risk of all-cause, cardiovascular, and cancar mortality: sixteen years of follow-up in US women, Circulation, **117** (13), 1658-1667 (2008)
4. National Center for Chronic Disease Prevention and Health Promotion, Defining overweight and obesity, http://www.cdc.gov/nccdphp/dnpa/obesity/defining.htm (2004)
5. National Health and Nutrition Examination Survey (NHANES) 1999-2000, Hyattsville, MD, US Department of Health and Human Services, Centers for Disease Control and Prevention, National Center for Health Statistics (2002)
6. R. O. Moreira, K. F. Marca, J. C. Appolinario, and W. F. Coutinho, Increased waist circumference is associated with an increased prevalence of moon disorders and depressive symptoms in obese women, Eating and weight Disorders, **12** (1), 35-40 (2007)

7. A. Onat, G. S. Avei, M. M. Barlan, H. Uyarel, B. Uzunlar, and V. Sanaoy, Measures of abdominal obesity assessed for visceral adiposity and relation to coronary risk, International Journal of Obesity and Related Metabolic Disorders, **28** (8), 1018-1025 (2004)
8. American Heart Association and the National Heart, Lung and Blood Institute, Metabolic Syndrome, http://www.americanheart.org/presenter.jhtml?identifier=4756 (2009)
9. Bureau of Health Promotion, Department of Health, R.O.C.(Taiwan), Metabolic Syndrome, http://www.bhp.doh.gov.tw/BHPnet/Portal/Them_Show.aspx?Subject=200712250023&Class=2 &No=200712250120 (2006)
10. T. Oleson, Differential application of auricular acupuncture for myofascial, autonomic, and naturopathic pain, Medical Acupuncture, **9**, 23-28 (1998).
11. T. Oleson, Auriculotherapy stimulation for neuro-rehabilitation, NeuroRehabilitation, **17** (1), 49-62 (2002)
12. A. Vickers and C. Zollman, ABC of complementary medicine: acupuncture, British Med. J., **319** (7215), 973-976 (1999)
13. E. R. Mitchell, *Fighting drug abuse with acupuncture*: *the treatment that works*, Berkeley : Pacific View Press (1995)
14. J. Li, Clinical experience in acupuncture treatment of obesity, Journal of Traditional Chinese Medicine, **19** (1), 48-51 (1999)
15. B. L. Frank and N. Soliman, Obesity treatment through auricular therapy and auricular medicine, Medical Acupuncture, **14** (1), 33-35 (2002)
16. C. H. Hsieh, The effects of auricular acupressure on weight loss and serum lipid levels in overweight adolescents, The American Journal of Chinese Medicine, **38** (4), 675-682 (2010)
17. L. W. Chien, M. H. Lin, H. Y. Chung, and C. F. Liu, Transcutanous electrical stimulation of acupoints changes body composition and heart rate variability in postmenopausal women with obesity, Evidence-Based Complementary and Alternative Medicine, **2011** (9), 862121 (2011)
18. G. Stux and B. Pomeranz, *Basics of acupuncture*, 4th ed. Springer-Verlag: Berlin (1998)
19. R. Mazzoni, E. Mannucci, S. M. Rizzllo, V. Ricca, and C. M. Rotella, Failure of acupuncture in the treatment of obesity: a pilot study, Eating Weight Disorders, **4** (4), 198-202 (1999)
20. D. Richards and J. Marley, Stimulation of auricular acupuncture points in weight loss, Australia Family Physician, **27** (2), S73-S77 (1998)
21. T. S. Shafshak, Electroacupuncture and exercise in body weight reduction and their application in rehabilitating patients with knee osteoarthritis, The American Journal of Chinese Medicine, **23** (1), 15-25 (1995)
22. R. P. Steiner, N. Kupper, and A. W. Davis, Obesity and appetite control: comparison of acupuncture therapies and behavior modification, *Proc. of International Forum on Family Medicine Education, Society of Teachers of Family Medicine*, Kansas City, 313-326 (1983)
23. S. N. Chang, *Chinese Acupuncture*, 3rd ed. Beijing: People Sanitation (1995)
24. S. Sang, A. Lao, Z. Chen, J. Uzawa, and Y. Fujimoto, Chemistry and bioactivity of the seeds of Vaccaria segetalis, in: Oriental Foods and Herbs, American Chemical Society, 279-291 (2003)
25. C. D. Lee, D. R. Jacobs Jr., P. J. Schreiner, C. Iribarren, and A. Hankinson, Abdominal obesity and coronary artery calcification in young adults: the coronary artery risk development in young adults (CARDIA) study, The American Journal of Clinical Nutrition, **86** (1), 48-54 (2007)
26. J. M. Ripp and S. Hess, The role of physical activity in the prevention and management of obesity, Journal of the American Dietetic Association, **98** (10), S31-S38 (1998)
27. S. H. Cho, J. S. Lee, L. Thabane, and J. Lee, Acupuncture for obesity: a systematic review and meta-analysis, International Journal of Obesity, **33** (2), 183-196 (2009)

A study of the pinch performance and subjective satisfaction with chopsticks of different lengths

Rain Chen[1,a], Min Wang[2] and Jo-Han Chang[3]

[1]*Department of Visual Communication Design, Southern Taiwan University of Science and Technology, Tainan City 710, Taiwan*
[2]*Graduate School of Digital Content and Applied Design, Southern Taiwan University of Science and Technology, Tainan City 710, Taiwan*
[3]*Department of Industrial Design, National Taipei University of Technology, Taipei City 106, Taiwan*

Abstract: Total 40 college students (16 males and 24 females), with an average age of 22, were surveyed in this study. The participants are highly familiar with the use of chopsticks and were tested with their handedness (95% of the participants are right-handers and 5% of the participants are left-handers). The control variable in this study is chopsticks 6 millimeters in diameter. Instruments used in this experiment were four pairs of chopsticks of different lengths:120 mm (Type A), 150 mm (Type B), 180 mm (Type C), 210 mm (Type D). Each participant was requested to randomly use these 4 pairs of chopsticks to pick up peanuts (10 millimeters in diameter). Least Significant Difference (LSD) was utilized for measuring quantity of pinched peanuts and participants' satisfaction at four different types of chopsticks.. The research results revealed that quantity of pinched peanuts showed significant difference (p=0.017) when the length of chopsticks changed from 120 mm to 180 mm. In terms of satisfaction, it presented significant difference (p=0.001) among four types, and participants showed higher satisfaction at longer chopsticks. Correlation Analysis will be further applied in future study, with a larger quantity and a random approach, and the future study can be used to find out the most appropriate chopsticks length for the reference of chopsticks manufacturers.

1 Introduction

At least 1.5 billion people worldwide use chopsticks every day as eating utensils. Chopsticks were developed in the Eastern World due to living conditions and environmental factors in ancient times while knives and forks were their counterparts in the Western World. Since chopsticks are not commonly used in Western countries, ergonomic studies of them are relatively scarce in comparison with those of forks and knives (Riley and Cochran, 1980; Cochran and Riley, 1986a, 1986b).

In terms of chopstick-related researches, the effect of chopstick lengths on the pinch performance was the first issue that had been systematically studied. Hsu and other researchers believed that the length of chopsticks has a significant influence on their performance in pinching food and suggested that the best chopstick lengths for adults and children are 240 mm and 180 mm respectively (Hsu and Wu, 1991). Chan's study suggested that the optimum chopstick diameter is around 6 mm (Chan, 1999). Wu, believed that the best diameter of chopstick tip is 4 mm, whereas he discovered the

[a] Corresponding author : ocean@stust.edu.tw

influence of chopstick tip angle on food-pinching performance and advised on a 2° tip angle as a complement to 4 mm diameter tip (Wu, 1995) .

Chopsticks on the market have a variety of cross-section shapes, but circular and square shapes are the most common types. Chen classified chopsticks according to their appearances such as a tip with carved grooves and cross-section shapes such as circular and square. Then he put forward six common combinations for experiment. It was discovered that chopsticks with circular cross-section shape have better food-pinching performance and chopsticks with carved grooves on tips have better performance of pinching food than those without grooves (Chen, 1998). Wu and other researchers also compared the performance of four cross-section shapes of chopstick handles, which were: circular, rectangular, hexagonal, and rounded square, and it was discovered that cross-section shapes which are circular and rounded square have the best performance (Wu and Tsai, 2012). A summary of the related studies is shown in Table 1.

Table 1. A summary of studies of chopsticks.

Researcher	Independent Variables	Control Variables	Criteria Measured	Outcome
Hsu and Wu (1991)	*Chopstick lengths for adults: 180 mm, 210 mm, 240 mm, 270 mm, 300 mm, 330 mm *Chopstick lengths for children: 150 mm, 180 mm, 210 mm, 240 mm, 270 mm, 300 mm	*Handle diameter: 6 mm *Tip length: 80 mm *Tip diameter: 3.5 mm *Tip slant angle: 1°	*Food-pinching performance *Performance of pulling food apart *Subjective evaluation *Subjective ranking	*Optimum length for children: 180 mm *Optimum length for adults: 240 mm *Optimum length for restaurants: 210 mm
Wu (1995)	*Handle diameters: 4mm, 6mm, 8mm *Tip slant angles: 0°, 2°, 4°, 6° ∘	*Total length: 210 mm *Handle length: 180 mm *Tip length: 30 mm (disposable chopsticks)	*Food-pinching performance *Performance of pulling food apart *Food-cutting performance *Food-piercing performance *Subjective evaluation *Subjective ranking	*Optimum handle diameter: 6 mm *Optimum tip diameter: 4 mm *Optimum slant angle of tip: 2°
Chan (1999)	*Handle diameters: 2mm, 4mm, 6mm, 8mm, 10mm *Target: red beans, peanut kernels	*Total length: 240 mm *Tip length: 80 mm *Tip slant angle: 1°	*Food-pinching performance *Subjective ranking	*Optimum handle diameter: 6 mm
Wu and Tsai (2012)	*The way of holding chopsticks: pincer-like, scissor-like *Cross-section shapes of handle: circular, rectangular, square, hexagonal, rounded square	*Total length: 230 mm *Tip diameter: 2 mm *Tip length: 115 mm	*Food-pinching performance *Performance of pulling food apart *Food-cutting performance *Subjective evaluation *Subjective ranking	*Pincer-like holding works best *Handles of circular and rounded square shapes work best
Lin (2002)	*The way of holding chopsticks: pincer-like, scissor-like	*Handle diameter: 6 mm *Tip diameter: 4 mm	*Food-pinching performance *Food-pulling force *Subjective	*Bamboo and wood work better than the other substances; tip

	*With or without carved grooves on the tips	*Tip length: 30 mm *Tip slant angle: 2° *Substance: bamboo, wood, stainless steel, plastic	evaluation *Subjective ranking	carved grooves are preferable *Pincer-like holding works better in terms of food-pinching performance and subjective evaluation
Chen (1998)	*Circular handle and circular tip with carved grooves (1 mm wide and 0.5 mm deep)	*Tip slant angle: 2°	*Performance of pulling food apart *Food-pinching performance *Food-moving performance *Subjective evaluation *Subjective ranking	*Circular tip with carved grooves works best *For chopsticks used at home, circular handle and circular tip with carved grooves work best *For chopsticks used at restaurants, circular handle and square tip work best

After the reviews of other studies of chopsticks and their food-serving performance, this research focuses on the influence of chopstick lengths on food-pinching and users' subjective. It is hoped that experiments of this research will help identify the most appropriate chopstick length, and which may serve as an important guidance for manufacturers in producing chopsticks.

2 Method

This research was an attempt to assess the influence of different chopstick lengths on food-serving performance through experiments. The independent variables were four pairs of chopsticks of different lengths, the dependent variables were food-pinching performance and subjective evaluations, and the control variables were chopstick diameters, shapes, and substances.

2.1 Subjects

A total of 40 college students, 40% male and 60% female, with an average age of 22, were invited to take part in the experiment. This research was designed to have volunteers participate in the experiment; therefore, the number of sexes was not brought under control. An additional reason was that according to some relevant studies, there are no significant differences between males and females when they use chopsticks. In the experiment of this research, subjects operated chopsticks in ways they normally do in their daily life with their habitual hands (95% of them are right-handers and 5% left-handers.) The experiment and its purpose were explained to subjects by the researcher in advance. After it was confirmed that all subjects understood and agreed to the experiment, they were asked to provide personal information and conduct the experiment.

2.2 Chopsticks

Four pairs of chopsticks of different lengths were used in the research, including 120 mm (Type A), 150 mm (Type B), 180 mm (Type C), and 210 mm (Type D). Based on results of relevant studies, it appeared that bamboo is the best substance for chopsticks as it provides more pinching accuracy regardless of people's holding methods (Lin, 2002). Therefore, bamboo was the choice of chopstick substance in the experiment. In addition to the material of bamboo, all four pairs of chopsticks had a 6 mm diameter handle, a 4 mm diameter circular tip with 2° slant angle and carved grooves (Chen, 1998).

2.3 Instruments and materials

Equipment used in the experiment included four pairs of chopsticks, a stopwatch, peanut kernels, a plate, and a cup. The chopsticks had a handle 6 mm in diameter, a circular tip 4 mm in diameter with shallow carved grooves (see Table 2).

Table 2. Experiment equipment and specifications.

Item	Specifications	Quantity
Bamboo chopsticks	210 mm, 180 mm, 150 mm, and 120 mm in length	4 pairs
Stopwatch	Standardized stopwatch	1
Peanut kernels	10 mm in diameter	Several
Plate	400 mm in diameter	1
cup	72 mm in diameter and 110 mm in height	1

2.4 Measurements and experiment procedure

The aim of this research was to understand the influence of chopstick lengths on the pinching performance and users' subjective evaluations, so the most appropriate chopstick length was to be judged from pinched quantity of peanut kernels and the subjective evaluation of subjects. What was measured in the experiment included (1) quantity of peanut kernels pinched by subjects given the same amount of time (larger quantity signified less time required for each pinch and easier operation of chopsticks), and (2) satisfaction level of subjects after the experiment by using a subjective preference questionnaire to understand their psychological needs. There were five satisfaction levels in the questionnaire: 5 stood for "most satisfied," and 1 stood for "least satisfied."

In preparation stage, to encourage subjects' cooperation and to smooth the process of the experiment, the researcher explained the content of the experiment and procedures in advance. Then subjects were asked to provide their basic information before starting the experiment. Before the formal experiment began, subjects were allowed to adjust their seats to the height they were comfortable with and choose a random pair of chopsticks for a one-minute practice. Once the experiment began, subjects were given chopsticks of different lengths on a random basis and were required to pinch peanut kernels from a 400 mm plate in front of them and put those peanut kernels in a cup (72 mm in diameter and 110 mm in height) placed below their mouth to simulate distance and height in real-world eating. The chopsticks were provided in random orders to eliminate the learning effect. Each test was to last for one minute, and no more pinching was allowed after the researcher gave a stop signal. Subjects had to rely solely on chopsticks with no other supplementary actions allowed during the test. To avoid fatigue, a two-minute break was arranged between consecutive tests. The experiment would count as completed only when subjects had finished the pinching test with all four sets of chopsticks of different lengths.

After the test, quantity of peanut kernels was counted by subjects, and subjects were then asked to immediately fill out a subjective evaluation questionnaire concerning their preferences (according to their level of satisfaction, a score of 1 to 5 was given, with 5 standing for "most satisfied" and 1 standing for "least satisfied").

3 Results

The statistical method of ANOVA was employed in the study, in which independent variables were chopsticks of four different lengths (120 mm as Type A, 150 mm as Type B, 180 mm as Type C, and 210 mm as Type D) and dependent variable was quantity pinched in one minute. The results showed a significant difference among four groups (p=0.001). A follow-up Least Significant Difference (LSD) analysis showed that when the length of chopsticks changed from 120 mm to 180 mm, a significant difference in quantity pinched was observed (p=0.017). The 180 mm group had the largest average quantity pinched (Mean=28.0), followed by the 210 mm group (Mean=26.5), the 150 mm group (Mean=25.6), and the 120 mm group (Mean=24.8). In terms of level of satisfaction, the 210 mm group had the highest score, and the average score decreased with length, which means that the longer chopsticks, the higher level of satisfaction found in subjects. Results are shown in Table 3 and 4.

Table 3. Pinching performance and subjective satisfaction score.

Type	Quantity		Satisfaction	
	Mean	Std. Deviation	Mean	Std. Deviation
Type A=120 mm	24.8500	5.45165	3.0000	1.08604
Type B=150 mm	25.6500	5.75816	3.2750	1.06187
Type C=180 mm	28.0250	5.46076	3.3750	1.25448
Type D=210 mm	26.5000	6.82379	4.0500	1.19722
Total	26.2563	5.96241	3.4250	1.20559

Table 4. Significance in the dependent variable.

Dependent Variable		Quantity			Satisfaction		
(I) Type	(J) Type	Mean Difference (I-J)	Std. Error	Sig.	Mean Difference (I-J)	Std. Error	Sig.
Type A	Type B	-0.80000	1.31938	0.545	-0.27500	0.25773	0.288
	Type C	-3.17500*	1.31938	0.017	-0.37500	0.25773	0.148
	Type D	-1.65000	1.31938	0.213	-1.05000*	0.25773	0.000
Type B	Type A	0.80000	1.31938	0.545	0.27500	0.25773	0.288
	Type C	-2.37500	1.31938	0.074	-0.10000	0.25773	0.699
	Type D	-0.85000	1.31938	0.520	-0.77500*	0.25773	0.003
Type C	Type A	3.17500*	1.31938	0.017	0.37500	0.25773	0.148
	Type B	2.37500	1.31938	0.074	0.10000	0.25773	0.699
	Type D	1.52500	1.31938	0.250	-0.67500*	0.25773	0.010
Type D	Type A	1.65000	1.31938	0.213	1.05000*	0.25773	0.000
	Type B	0.85000	1.31938	0.520	0.77500*	0.25773	0.003
	Type C	-1.52500	1.31938	0.250	0.67500*	0.25773	0.010

4 Discussion and Conclusion

The study proved that the length of chopsticks has a significant influence on the pinch performance. When the length of chopsticks changed from 120 mm to 180 mm, a significant difference in quantity pinched was observed (p=0.017). In terms of the total quantity pinched, however, Type C worked better than the others (Types A, B and D). When it came to the subjective evaluation by subjects, the subjective level of satisfaction increased with the length of chopsticks.

On the whole, a length between 18 and 21 centimeters is recommended for the design of chopstick. In the future, the study will be extended in a larger scale and in a more randomized manner with the implement of correlation analysis. It is expected to find the most appropriate chopstick length which may serve as important information for manufacturers in mass-production of chopsticks.

Chopsticks have been an important utensil for Chinese people and some Asian people for centuries, and there are billions of chopstick users every day. This research only focused on the length of chopsticks, but in the future, researchers can further study the influence of the shape of chopsticks on the pinch performance (Koreans use flat chopsticks, Japanese use polygonal chopsticks, and Chinese use round chopsticks) (Chiu, 2013). Moreover, the influence of material of chopsticks on the pinch performance can be also studied.

Acknowledgements

This work was partially supported by the National Science Council, Taiwan, under the Grant No. MOST 104-2221-E-218-015-MY2; Southern Taiwan University of Science and Technology, Taiwan, under the Grant No. R105118.

References

1. M.W. Riley and D.J. Cochran, Handles for sharp tools, *In Proceedings of the Symposium on Occupational Safety Research and Education*, 67-74 (1980)
2. D.J. Cochran and M.W. Riley, The effects of handle shape and size on exerted force, Human Factors, **28** (3), 253-265 (1986)
3. D.J. Cochran, M.W. Riley, An evaluation of knife handle guarding, Human Factors, **28** (3), 295-301 (1986)
4. S.H. Hsu and S.P. Wu, An investigation for determining the optimum length of chopsticks, Applied Ergonomics, **22** (6), 395-400 (1991)
5. T. Chan, A study for determining the optimum diameter of chopsticks, International Journal of Industrial Ergonomics, **23**, 101-105 (1999)
6. S.P. Wu, Effects of the handle diameter and tip angle of chopsticks on the food-serving performance of male subjects, Applied Ergonomics, **26** (6), 379-385 (1995)
7. Y.L. Chen, Effects of shape and operation of chopsticks on food-serving performance, Applied Ergonomics, **29** (4), 233-238 (1998)
8. S.P. Wu and W.H. Tsai, Handle cross dection and chopsticks grasping mode effects on food-serving performance, Journal of Work and Leisure, **3** (1), 1-10 (2012)
9. T.F. Lin, The study of the effects of pinching operation, material of chopsticks, and carved groove on food-serving performance, Theory of Huafan University (2002)
10. Y.J. Chiu, Study of Food Culture In Japan and Taiwan- Around The Chopsticks, Theory of I-Shou University (2013)

Effects of combined growth of biogenic and xenobiotic substrates on degradation of xenobiotic by activated sludge

Nguyen Phuc Thien[1,a], Hoang Thi Hoai[1], Doan Quang Tri[2,3] and Chen Yi Ching[1]

[1]*Department of Environmental Engineering, Da-Yeh University, Changhua 51591, Taiwan*
[2]*Lowland Technology International (LTI) Member, Saga University, Japan*
[3]*National Center for Hydrometeorological Forecasting (NCHMF), Vietnam*

Abstract. The purpose of this study was to research about supplementation of different concentrations of the substrate on the degradation rate of xenobiotic and to determine the optimal concentrations of the auxiliary substrates that are most beneficial of xenobiotic degradation rate. 2,4-dichlorophenol acid (2,4-D) was used representative xenobiotic organic compounds, while peptone and sugar used for auxiliary substrates. The activated sludge was completely break down 100 mg/l of 2,4-D for three consecutive times. The different concentrations between biogenic substracts of sucrose and peptone were fed separately or combined into the medium containing 200 mg/l of 2,4-D and 140 mg SS/l of activated sludge. The results showed that sugar and peptone could affect 2,4-D degradation rate to several different degree at different concentrations. In separate supplementation, 2,4-D degradation completed within 25 hours, 40 mg/l sugar and 150 mg/l peptone concentrations were found to be the optimal concentrations. In combined case, 2,4-D was consumed totally within 20 hours and the optimal concentration of the combined sugar and peptone concentrations were 40 and 150 mg/l, respectively.

1 Introduction

Xenobiotic organic compounds such as phenoxy acid herbicides are foreign to most indigenous microorganisms, including those in activated sludge. In addition to their hard-to-treat nature, xenobiotics can be uncoupled of the microorganisms' metabolism of biogenic substrates. The uncoupling, or the energy-spilling reactions, occurs when catabolism energy is dissipated or diverted away from anabolism through the metabolism of a substrate in bacterial cells [1]. Although xenobiotics containing wastewater can be suitably treated using biological methods [2,3,4,5], the xenobiotic nature of the pollutant requires the treatment plant microorganisms (typically activated sludge) to go through an acclimation phase before the microorganisms evolve the degradation capability for treating the influent xenobiotics. Thus, biodegradation method is used widely to treat xenobiotics in water treatment methods. One of the popular xenobiotics is 2,4-Dichlorophenoxyacetic acid (2,4-D), a common joint systemic herbicide used in the control of broadleaf weeds. Therefore, the target xenobiotic organism pollutant was 2,4-D and activated sludge was an agent treatment in this study. The activated sludge process reduced or eliminated wastewater toxicity from a variety of sources along a wastewater collection system [6,7]. Eckenfielder [8] suggested that aquatic toxicity

[a] Corresponding author : d0405601@cloud.dyu.edu.tw

data on specific organic chemicals should not be applied to industrial wastewater effluents because many of these compounds were removed during biological treatment. Several authors [9,10] suggested that the fed-batch technique could evaluate toxicity to activated sludge with substrate continuously added to a batch reactor and accumulation of substrate followed with time. Toxic or inhibitory substrates exhibited an upward slope in the substrate accumulation curve, and nonlinear curve fitting techniques were used to determine inhibition constants.

Activated sludge biomass grown on the feed of biogenic substrate must be in a healthy physiological condition [11]. However, inconsistent results are found in literature about the effects of biogenic organics on the degradation of man-made xenobiotics or hydrocarbons. There have been some study cases of both beneficial and adverse effects of biogenic organics on xenobiotic degradation. The beneficial good cases include: citrate on toluene [12]; natural amino acids on mono-substituted phenol [13]; natural organics such as manure on dichloro two chloros and a nitro-herbicides [14]; fatty acids on soil hydrocarbon [15]; pyruvate on naphthalene [16]. Conversely, adverse cases are also found: glucose or amino acids in on xylene and toluene [17]; ethanol in on benzene, toluene and xylene [18]; glycolic acid and glucose on p-cresol [19]; yeast extract and milk on 3-nitrobenzoate, 4-chlorobenzoate, 4-chlorophenol [20]. Diauxic growth is a phenomenon occurs when microorganism is grown on cultural medium with the presence of two substrates, one of which is easier for the microorganism to metabolize. Sugar is consumed first, and it leads to a rapid growth. Only after the easier more natural substrate has been exhausted, the cells switch to the second, resulting in two separate growth phases. There are some previous studies which the effects of diauxic growth on 2,4-D acclimation and degradation [21]. The results showed that: If we feed sugar with a proper concentration and at the proper time, it can make 2,4-D acclimation lag time shorten. On the contrary, it can make the acclimation lag time longer. This is explained that sugar is an easier substrate than 2,4-D for the microbial cell to utilize, and thus the biomass temporarily ignores or escapes from the xenobiotic acclimation stress. This pause in of the process lets the xenobiotic still be intact until sugar is consumed entirely completely, thereby causing the elongation of the acclimation lag time.

The purpose of this study, therefore, was to determine effects of auxiliary substrates on 2,4-D degradation and detail profiles of ATP that contained in activated sludge cells during the cells' degradation of a model xenobiotic compound 2,4- dichloro-phenoxyacetic acid (2,4-D). Sugar and peptone were supplemented with into cultural medium contained 2,4-D to determine the optimal concentrations of each the auxiliary substrates in separate or double. Generally, sugar and peptone are used as primary major carbon and nitrogen sources in a medium of microbial cultivation. However, 2,4-D plays a role as a carbon source in this study. Thus, sugar and peptone were used as the representation of the auxiliary substrates. In this study case, sugar sucrose and feed of 2,4-D, peptone and feed of 2,4-D with peptone are fed transferred into the cultural medium at the same time. Therefore, we can find out which concentration is the most effective on 2,4-D degradation. Results of this study can supplement for the knowledge about 2,4-D biodegradation, and apply to actual real water treatment system. The efficiency and applicability consist of economic efficiency when the methods are applying to actual xenobiotic treatment systems. However, this study can be used as references for studies of relative compounds and other xenobiotic organics.

2 Materials and methods

As mentioned above, 2,4-D was used to be target substrate, and activated sludge was used to be agent treatment. Peptone and sugar are considered as auxiliary substrates that could affect 2,4-D degradation by activated sludge.

2.1 Suspended solid (SS) measurements

Suspended solid is considered as a representation of microbe biomass in samples. It is measured one time per day during 2,4-D degradation in this study. The procedure is used, including the following steps: 1. The filter papers (pore size 0.45 μm) are washed by using distilled water and vacuum filtration; 2. After washing, the papers are dried about 4-5 hours at 105^0 C in a drying cabinet; 3. Scaling the dried the papers before using to filter samples, getting paper weight (M_0 - g); 4. Putting V (ml) sample on the dried filter paper and filtering by vacuum filtration to separate the solid from the liquid; 5. Drying again about 4- 5 hours at 105^0 C in the drying cabinet; 6. Scaling again the dried sample to get paper and SS weight (M_1 - g); 7. Calculating SS (mg/l) is showed in Equation 1:

$$SS = \frac{(M_1 - M_0)x100}{0.001\,xV} \tag{1}$$

where: SS is suspended solid weight (mg/l); M_0 is the weight of the initial dried paper (g); M_1 is a weight of the dried paper after filtering sample (g); V is a volume of the sample (ml).

2.2 Experimental method to obtain acclimated activated sludge

Activated sludge was fed in separate mediums corresponding to different periods as follows: The first period was to cultivate activated sludge in a sequence batch reactor with supplementing daily sucrose 100mg/l, peptone 18mg/l, $FeCl_3$ 1mg/l, NH_4Cl 30mg/l, K_2HPO_4 200mg/l, KH_2PO_4 156mg/l and $MgSO_4$ 31.26mg/l. The second period was to make the activated sludge acclimate to 2,4-D. After feeding in the continuous medium, 140 mg/l of the activated sludge was transferred to batch medium with 100 mg/l 2,4-D as sole carbon resource until the activated sludge degraded completely 2,4-D in three times. The original activated sludge spent about seven days degrading completely 100 mg/l 2,4-D in the first time. However, it only spent about two days in the second time, and one day for in the third time on this process. The obtained activated sludge was called A. A would be used major experiments with sugar and peptone supplementation.

2.3 Experimental methods with (sugar) sucrose and peptone

To determine effects of auxiliary substrates supplementation on 2,4-D degradation, the study conducted the experiments in separated and combined case. Experimental symbols: A stands for the reactor only includes A and 2,4-D; S_x stands for the reactor includes A, x mg/l Sugar and 2,4-D; P_y stands for the reactor includes A, y mg/l Peptone and 2,4-D; P_yS_x stands for the reactor includes A, y mg/l Peptone x mg/l Sugar and 2,4-D. All experiments were conducted with 200 mg/l 2,4-D and 140 mg/l A. In separated case: Each substrate was supplemented concurrently with 2,4-D into A medium with different concentrations. Supplemental sugar concentrations consisted of 20, 40, 60, 80, 100, 150 mg/l, while supplemental peptone concentrations were 20, 40, 100, 150, 200, 300 mg/l. In combined case: Both sugar and peptone are supplemented concurrently with 2,4-D into A medium with different couple concentrations. Supposing that: 0 stands for the best concentrations; "-"stands for the lower concentrations than the best concentrations; "+" stands for the higher concentrations than the best concentrations. From the supposition, the matrix of sugar and peptone concentrations is shown in Table 1.

Table 1. Matrix of sugar and peptone concentrations.

Sugar \ Peptone	-	0	+
-	- -	- 0	- +
0	0 -	0	0 +
+	+ -	+ 0	+ +

2,4-D concentration and SS were measured each 5 hours during the degradation process.

2.4 2,4-D degradation rate calculation

This study used Equation 2 to calculate 2,4-D degradation rate.

$$R = \frac{\Delta S}{\Delta t} = \frac{(S_i - S_{i+n})}{(t_{i+n} - t_i)} \tag{2}$$

where: R stands for 2,4-D degradation rate (mg 2,4-D/l.hr); S is 2,4-D concentration; t stands for time (hour); i stands for first time mark, the point was chosen at 0hr; i+n stands for the second time mark, the point was chosen at 20hrs or 15hrs for separate cases or combined case, respectively.

3 Results and discussion

3.1 Effects of different sugar concentrations on 2,4-D degradation rate

SS increased from 140 mg/l to 270 mg/l, depending on sugar concentration (figure 1a). Figure 1b showed that 2,4-D degradation completed within 25 hours. SS grew on higher sugar concentrations were always higher greater than the growth in on lower concentrations at the same time.

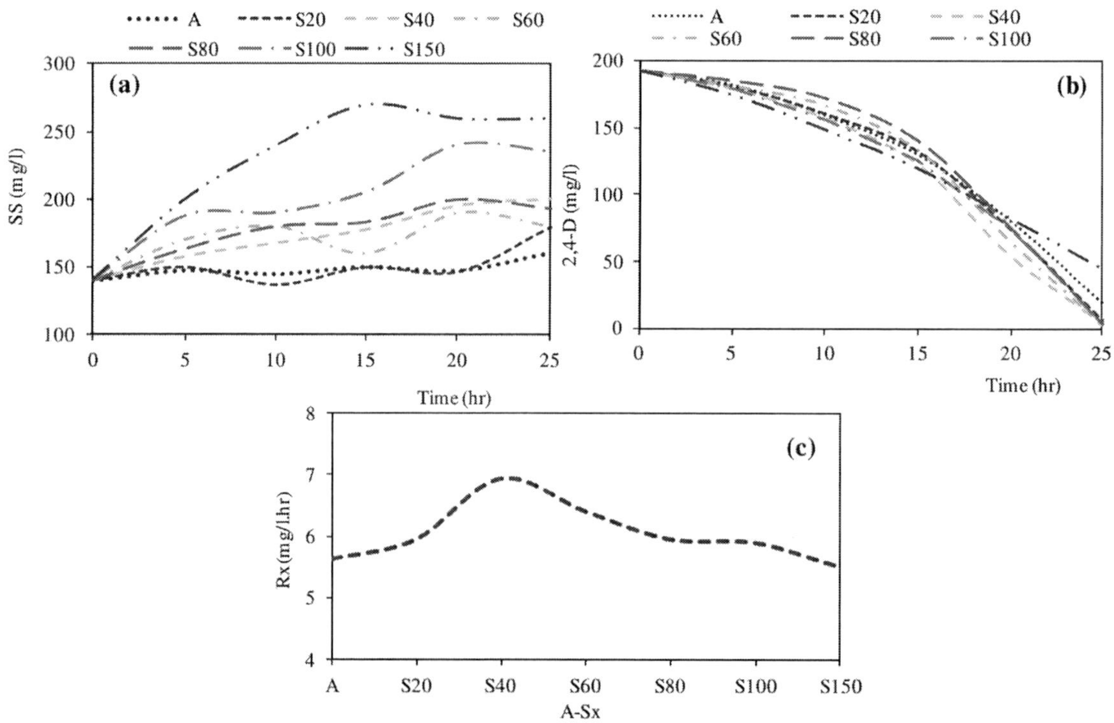

Figure 1. (a) 2,4-D degradation with sugar supplementation; (b) Activated sludge growth (SS) of sugar supplementation; (c) 2,4-D degradation rate of sugar supplementation.

The reason is that sugar is easily consumed by microorganisms and releases a higher amount of energy. The energy could help the sludge grow faster. 2,4-D degradation rates were calculated, and shown in figure 1c. Figure 1c showed that the reactor included 40 mg/l sugar, had the highest rate of 2,4-D degradation at 6.94 mg/l.hr. For sugar concentrations that were less than 40 mg/l, the rate increased when sugar concentration increased. Because the microorganisms can easily metabolize sugar, and releases a large of energy. The energy could help the sludge grow faster. But the trend would be not similar for very high sugar concentrations such as 150 mg/l. This could be explained by the diauxic growth process. Sugar is a better substrate than 2,4-D for microorganisms to utilize, and thus the biomass temporarily ignores or escapes from the xenobiotic degradation stress. This interference of the degradation lets the xenobiotic still being ignored until sugar is consumed completely. Thus, 40 mg/l sugar concentration is supposed the optimal sugar concentration to enhance 2,4-D degradation rate.

3.2 Effects of different peptone concentrations on 2,4-D degradation rate

Figure 2a showed that 2,4-D degradation completed within 25 hours. The growth of activated sludge also increased when peptone concentration increased. However, sugar made the activated sludge grow faster than peptone did at similar concentration. Therefore, sugar might release more energy than peptone did when microorganisms metabolized these substrates (figure 2b).

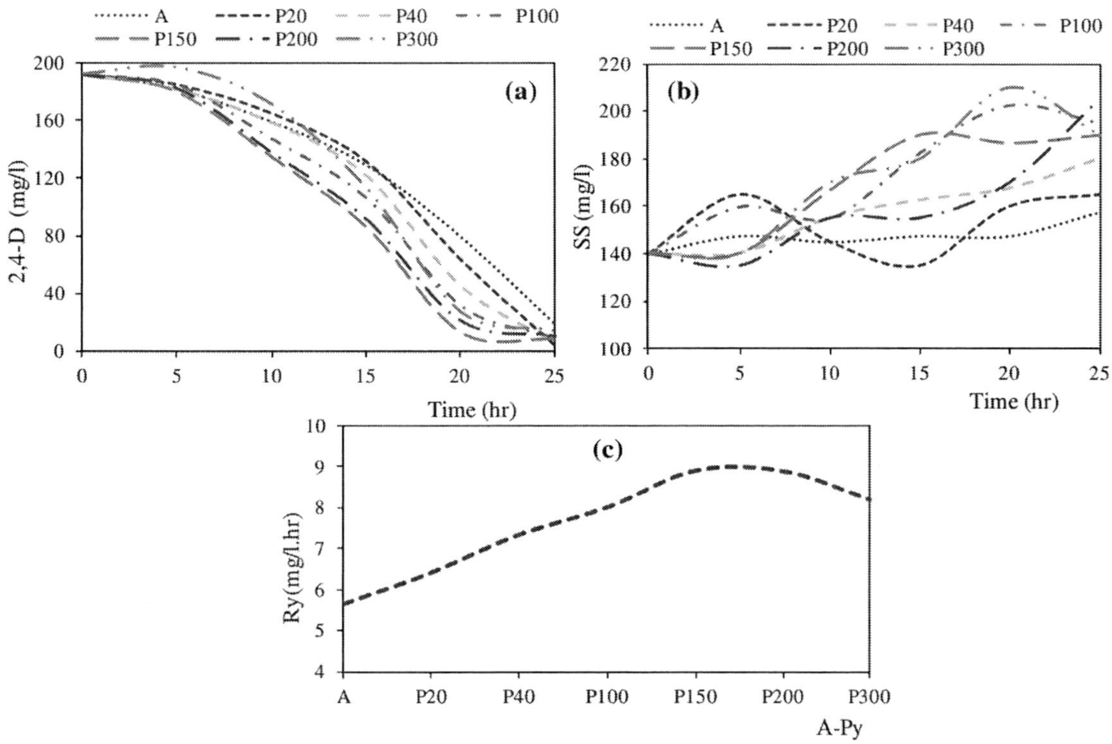

Figure 2. (a) 2,4-D degradation with peptone supplementation; (b) Activated sludge growth (SS) with peptone supplementation; (c) 2,4-D degradation rate of peptone supplementation.

Figure 2c showed that 2,4-D degradation rate increased when peptone concentrations were increased until peptone concentration reached higher 150 mg/l. This case was not like sugar because there was a range of peptone concentrations between 150 and 200 mg/l which have similar effect on 2,4-D degradation. There is been an assumption that peptone does not cause diauxic growth as sugar. Hence, the optimal peptone concentration for the best 2,4-D degradation rate was 150 mg/l.

3.3 Effects of combined auxiliary substrates on 2,4-D degradation rate

From the results obtained previously, 40 mg/l sugar concentration and 150 mg/l peptone concentration were supposed to be the optimal concentrations. This section presents the combined case which used both substrates: sugar and peptone were supplemented concurrently with different couple concentrations. These different coupled concentrations were designed as shown in Table 2.

 DOI: 10.1051/matecconf/201711901018

Table 2. Matrix of sugar and peptone concentrations.

Sugar (S) / Peptone (P)	S_1 (20 mg/l)	S_2 (40 mg/l)	S_3 (100 mg/l)
P_1 (100 mg/l)	P_1S_1	P_1S_2	P_1S_3
P_2 (150 mg/l)	P_2S_1	P_2S_2	P_2S_3
P_3 (200 mg/l)	P_3S_1	P_3S_2	P_3S_3

For combined sugar and peptone supplementation, 2,4-D degradation completed within about 20 hours (fugures 3a-3c); especially, the activated growth increases appreciably from 140 mg/l to 300 mg/l (figure 3d). It means reasonable effects of sugar and peptone on 2,4-D degradation and the activated sludge growth are significant. The growth was obviously resulted from the auxiliary substrates, shown in figure 3.

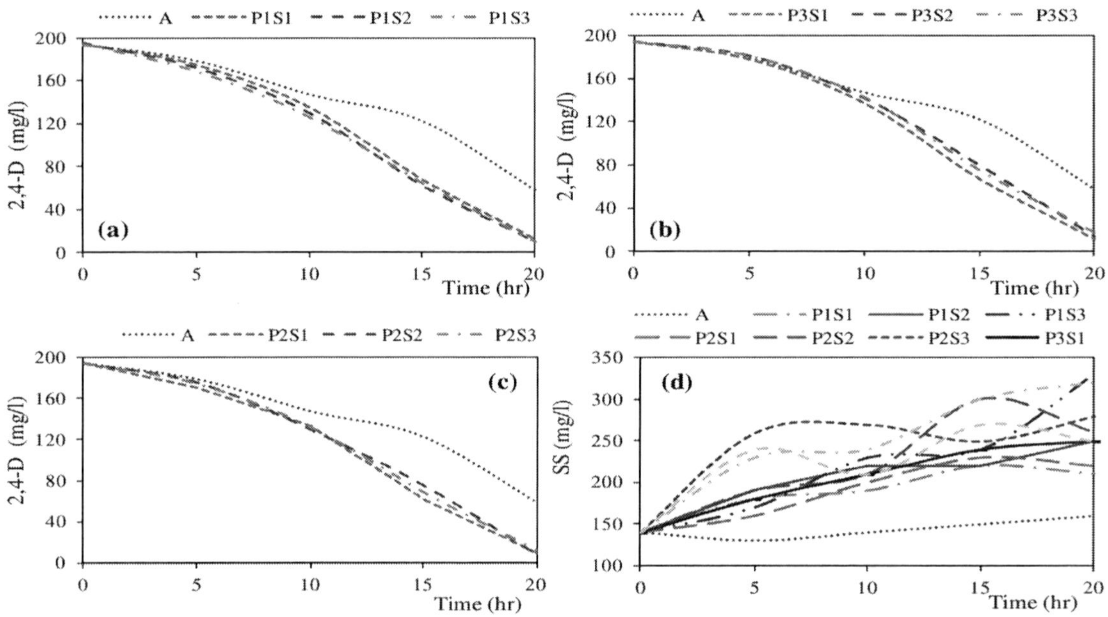

Figure 3. (a) 2,4-D degradation with sugar and peptone supplementation - P_1S_x; (b) 2,4-D degradation with sugar and peptone supplementation - P_3S_x; (c) 2,4-D degradation with sugar and peptone supplementation - P_2S_x; (d) Activated sludge growth (SS) with sugar and peptone supplementation.

Figure 4 showed that couple of 150 mg/l peptone and 40 mg/l sugar concentrations are supposed as the optimal concentrations with the rate is 8.99 mg/l.hr. In this case, it is very difficult to identify a certain trend of the rate because of the combination of two auxiliary substrates. Thus, this study only determines the optimal concentrations of combined auxiliary substrates.

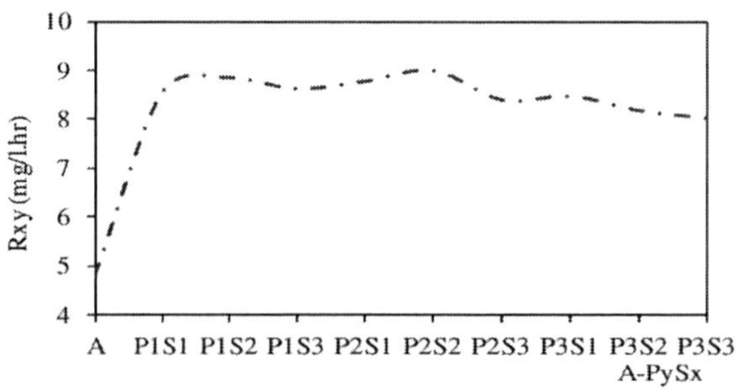

Figure 4. 2,4-D degradation rate with sugar and peptone supplementation.

4 Conclusions

1. For separate supplementation, both sugar and peptone could help 2,4-D degradation rate. 2,4-D degradation completed within 25 hours. The optimal concentrations of sugar and peptone were supposed 40 mg/l and 150 mg/l with 6.94 mg/l.hr and 150 mg/l.hr of 2,4-D degradation rate, respectively. These rates were higher than the rate of un-supplemented reactor (A).
2. For combined supplementation, 2,4-D degradation completed within 20 hours. The optimal concentration of the combined sugar and peptone supplementation is 40 mg/l and 150 mg/l for sugar and peptone, respectively. 2,4-D degradation rate at the optimal rate was 8.99 mg/l.

Acknowledgements

We are thankful to Prof.Chong, Nuyk-Ming for giving us this opportunity and facilities to carry out this study. The first author would like to give special thanks to Da-Yeh University, who provides financial support for this study.

References

1. J.B. Russell, The energy spilling reactions of bacteria and other organisms, Mol. Microbiol. Biotechnol., **13** (1-3), 1-11 (2007)
2. N.P. Hill, A.E. MacIntyre, R. Perry, and J.N. Lester, Water Res., **20**, 45-52 (1986)
3. M. Ettala, J. Koskela, and A. Kiesila, Water Res., **26**, 797-804 (1992)
4. S. Meric, G. Eremektar, F. Ciner, and O. Tünay, J. Hazard. Mater., **101** (2), 147-155 (2003)
5. H. Chin, P. Elefsiniotis, and N. Singhal, J. Environ. Eng. Sci., **4** (1), 57-63 (2005)
6. C.L. Logue, et al., J. Water Pollut. Control Fed., **61**, 632 (1989)
7. S. Joann, O.M. Richard, H.S. Joseph, A.S. Weber, and D.A. Michael, Activated Sludge, Res. J. Water Pollut. Control Fed., **62** (4), 398-406 (1990)
8. W.W. Jr. Eckenfelder, *Proceedings of The 43rd Ind. Waste Conf., Purdue Univ., West Lafayette, Ind.*, **1**, 73-78 (1989)
9. J. Patoczka, et al., *Proc. 43rd Ind. Waste Conf., Purdue Univ., West Lafayette, Ind.*, **51**, 256-265 (1989)
10. A.T. Watkin and W.W.Jr. Eckenfielder, Water Sci. Technol. (G.B.), **21**, 593 (1989)
11. N.M. Chong, M.L. Luong, and C.S. Hwu, Bioresour. Technol., **104**, 181-186 (2012)
12. E.M. Harrison and J.F. Barker, J. Contam. Hydrol., **1**, 349-373 (1987)
13. R.J. Shimp and F.K. Pfeander, Appl. Environ. Microbiol., **49**, 394-401 (1985)
14. T.B. Moorman, J.K. Cowan, E.L. Arthur, and J.R. Coats, Biol. Fertil. Soils, **33**, 541-545 (2001)
15. E.C. Nelson, M.V. Walter, I.D. Bossert, and D.G. Martin, Environ. Sci. Technol., **30**, 2406-2411 (1996)
16. K. Lee, J. Hazard. Mater., **105** (1-3), 157-167 (2003)
17. C.M. Swindoll, C.M. Aelion, and F.K. Pfaender, Appl. Environ. Microbiol., **54**, 212-217 (1988)
18. H.X. Corseuil, C.S. Hunt, R.D.S. Ferreira, and P.J.J. Alvarez, Water Res., **32**, 2065-2072 (1998)
19. D.L. Lewis, H.P. Kollig, and R.E. Hodson, Appl. Environ. Microbiol., **51**, 598-603 (1986)
20. Z. Hu, R.A. Ferrainab, J.F. Ericsonb, and B.F. Smetsa, Water Res., **39**, 3501-3510 (2005)
21. N.M. Chong, Bioresour. Technol., **100** (23), 5750-5756 (2009)

Removal of basic dyes from solution using coconut shell charcoal

Siriwan Srisorrachatr[1,2,a], Pariwat Kri-arb[1], Sirawit Sukyang[1] and Chutima Jumruen[1]

[1]*Department of Chemical Engineering, Faculty of Engineering, Srinakharinwirot University, Bangkok 10110, Thailand*
[2]*The Graduate School of Srinakharinwirot University, Bangkok 10110, Thailand*

Abstract. This research aims to use agricultural waste as an adsorbent for dye removal from textile wastewater. Coconut shell charcoal (CSC), an agricultural waste, was used as dye adsorbent whilst basic yellow13 (BY13) and basic red14 (BR14) were used as representative dye used in textile manufactory. The removal of BY13 and BR14 from aqueous solution by CSC was investigated in batch adsorption at room temperature. Two parameters investigated in this research were pH of dyes solution; pH 2, 7, and 11, and particle size of adsorbents; 510-700 μm and 1000-2000 μm. The adsorption model, Langmuir and Freundlich were also examined. It was found that by CSC, BY13 and BR14 had maximum removal percentage of 23.6 and 55.7 at pH 11. Size of adsorbent had also shown the effect on dye removal, i.e.; increasing size with decreasing removal capability. The dye removal experimental data were fitted to Langmuir adsorption model for both dyes with maximum adsorption capacity (q_m) for BY13 and BR14 of 19.76 and 22.93 mg/g, respectively. It can be implied that the adsorption is monolayer. It can conclude that CSC can be used as adsorbent for basic dye removal from aqueous solution.

1 Introduction

Textile and handicraft dyeing use large volumes of water in wet processing operation and thereby, generates substantial quantities of wastewater containing large amount of dissolved dyestuff and other products such as dispersing agent, dye-bath carriers, salts, emulsifiers, leveling agent and heavy metals. It is nowadays well known that there is the need of treating dye wastewater by not only concerning primary and secondary treatment of organic matter and suspended solid removal but also a tertiary treatment mostly for residual colour removal. The removal of remaining dye from dye wastewater is currently of great importance in recent years. Dyestuffs have complex chemical structures, and are hard to degrade biologically. Basic dyes are used by several industries, such as textile, paper and colouring products. Although many conventional methods such as precipitation, ion exchange, solvent extraction, electrochemical treatment, biosorption, filtration and adsorption are used for dye removal from aqueous solution, these methods are either expensive or incomplete for removal of high concentration of contaminants and high energy consumption. Adsorption is a common one of the most powerful and extensively used realistic methods for dyestuff removal from wastewater and

[a] Corresponding author : siriwans@g.swu.ac.th

discharges high water quality effluents without dyestuffs. Powdered activated carbon (PAC) has been the most widely used adsorbent for dyestuffs removal from wastewater. However, PAC is expensive and its use usually requires regeneration and reactivation procedure. Development and use of low cost adsorbents has been investigated by many researchers for effective removal of dyestuffs from wastewater. In Thailand, there are large amounts of coconut shell from coconut agro-industry waste. Although they can be used as organic fertilizer or solid fuel, for another benefit for using these materials, they can be used as the adsorbents in rural areas or small communities to conserve their environments. Exploring effective and low-cost adsorbent from agricultural waste may contribute to environmental sustainability and offer benefits for future commercial applications. The costs of activated carbon prepared from biomaterials are very low compared to the cost of commercial activated carbon. Many basic dyes, Basic Fuchsin dye [1], Basic Blue 3 [2] and Basic Red 9 [3] were studied with using agricultural wastes, in order to investigate the dye removal capacity. Another type of agricultural waste-adsorbent, coconut based adsorbent, are also interesting and many research papers were published. Coconut mesocarp was studied as an adsorbent for dye removal, such as Acid Blue158 (AB158) and Acid Yellow13 (AY13) [4] and textile dyes [5]. Low cost coconut shell based activated carbon (CSAC) was also examined for removal of Reactive Blue 19 (RB19) dye adsorption from aqueous solution [6]. Activated carbon from coconut shell was prepared via H_2SO_4 activation (CSAC) and its ability to remove textile dyes; maxilon blue GRL and direct yellow DY 12, from aqueous solutions were studied [7]. In this study, we attempt to investigate the removal ability of basic dyes from solution by coconut shell charcoal, which is agricultural waste in the community. The basic dyes used in this study are Basic Yellow 13 and Basic Red 14, which are those of dyestuff often used by the textile dyeing.

2 Materials and methods

2.1 Materials and chemicals

Basic Yellow 13 (BY13; $C_{20}H_{23}ClN_2O$) and Basic Red 14 (BR14; $C_{23}H_{26}ClN_3$), shown in Figure 1, were obtained from President Chemical Co., Ltd and were used without any purification. NaOH (VWR International Limited), HCl (from Fluka Chemica) were used for adjusting dye solution pH. All the chemicals used are analytical reagent grade and all the standards and reagent solutions were prepared with distilled water. Coconut shell charcoal (CSC) was purchased from BK Black Company, Bangkok, Thailand. The CSC with particle size of 510-700 µm was examined for Iodine number and Methylene Blue number. It was found that the values are 705.68 mg/g and 272.12 mg/g for Iodine Number and Methylene Blue Number, respectively.

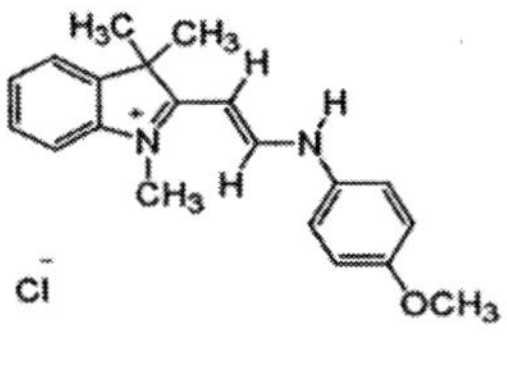

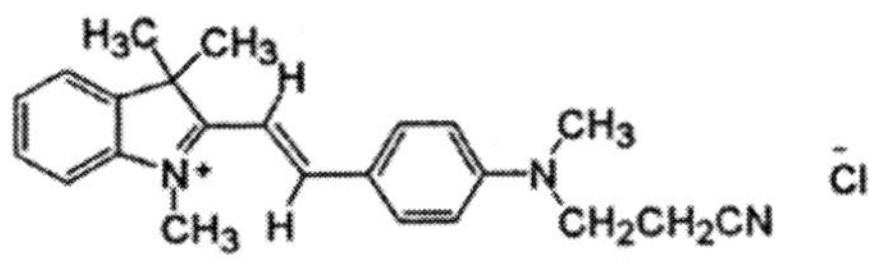

Figure 1. Chemical structure of (a) Basic Yellow 13 (BY13), [8] and (b) Basic Red 14 (BR 14), [9].

2.2 Batch adsorption experiments

All experiments in this study were examined as batch adsorption at room temperature. The parameters affecting on the removal capacity were investigated, such as pH of the solution and adsorbent size. A stock solution, 1000 mg/L Basic Yellow 13 (BY13) and Basic Red 14 (BR14), were prepared, and then diluted to 100 mg/L for initial concentration. The actual dye concentration was calculated from the calibation curve using spectrophotometrical method (SHIMADSU UV-visible spectrophotometer model UV-1610 (Japan) at 435 and 515 nanometers, respectively. The pH of the dye solution was adjusted to desired value by addition of dilute HCl or NaOH solutions. For batch study, 1.0 gram of coconut shell charcoal (CSC), particle size of 510-700 micrometer, was mixed with 200 mL of 100 mg/L BY13 in elremenyer flask. The mixture was shaked on orbital shaker (Gallenkamp orbital shaker, England) with constant speed of 150 rpm at room temperature. The dye solutions were then withdrawn at determined time interval periodically and seperated from the adsorbents by micro-syringe filtration. The resulting dye solution was determined by SHIMADSU UV-visible spectrophotomer. The effect of solution pH on dye removal was also studied at pH 2, 7 and 11. Then optimum pH was chosen for studying effect of CSC size on dye removal. The sizes of adsorbent used for study were 510-700 μm and 1000-2000 μm. The dye removal percentage and the amount of adsorbed dye (q_e), were calculated by equation (1) and (2), respectively.

$$\% \, Removal = \frac{C_i - C_f}{C_i} \times 100 \tag{1}$$

$$q_e = \frac{C_i - C_f}{W} \times V \tag{2}$$

where C_i and C_f are the initial and final concentrations (mg/L) of dye, W is the mass (g) of adsorbent and V is the volume of dye solution (L).

2.3 Adsorption isotherm

Adsorption isotherm experiment was carried out by mixing 1.0 gram CSC and 200 mL of dye solution of various concentration; 200, 250, 300, 400 and 500 mg/L at optimum pH and shaking at 150 rpm. Mixture of shaking dye and CSC was allowed to reach equilibrium about 2 hours, and then resulting solution was examined using UV-visible spectrophotometer for remaining dye concentration. For Langmuir adsorption model and Freundlich adsorption model, the calculated values of adsorbed dye (q_e) and remaining dye concentration (C_e) were plot according to equation (3) and (4), respectively.

Langmuir model:
$$q_e = \frac{q_m K_L C_e}{(1 + K_L C_e)}$$

and can be linearized to be

$$\frac{C_e}{q_e} = \frac{C_e}{q_m} + \frac{1}{K_L q_m} \tag{3}$$

Freundlich model:
$$q_e = K_F C_e^{\frac{1}{n}}$$

which can be written in the linear form as

$$\log q_e = \log K_F + \frac{1}{n} \log C_e \tag{4}$$

where C_e is equilibrium concentration (mg/L), q_e is adsorption capacity (mg/g), q_m is maximum adsorption capacity (mg/g), K_L is Langmuir adsorption constant and K_F is Freundlich adsorption constant. The plot of C_e/q_e versus C_e for Langmuir adsorption model will give the straight line with slope of $1/q_m$ and intercept of $1/(K_L q_m)$. For Freundlich adsorption model, plot of log q_e against log C_e also will be linear relationship. The relative coefficients of these models were calculated using linear least-squares fitting.

3 Results and discussion

3.1 Effect of pH on dye removal

The effect of pH on the removal of BY13 and BR14 by CSC was studied at pH 2, 7 and 11 wtih an initial dye concentration of 100 mg/L. The BR14 removal percentages and removal capacity at the various pH were plotted in Figure 2. The values of percent removal of BR14 by CSC are 10.50, 29.40 and 55.70 at pH 2, 7 and 11, respectively. Whereas the removal percentages of BY13 are 29.40 at pH 2, 45.87 at pH, and 56.28 at pH 11, respectively. It can see that the maximum of both dye removals occur at pH 11 and the removal percentage was decreased as pH lowered. The higher dye removal at higher pH may be due to the present negative charge of OH^- ions on adsorbent surface interacting with the positive dye ions at the adsorption site. At acidic pH, a significantly high electrostatic repulsion exists between the positively charged surface of the adsorbent and positively charged ionic dye.

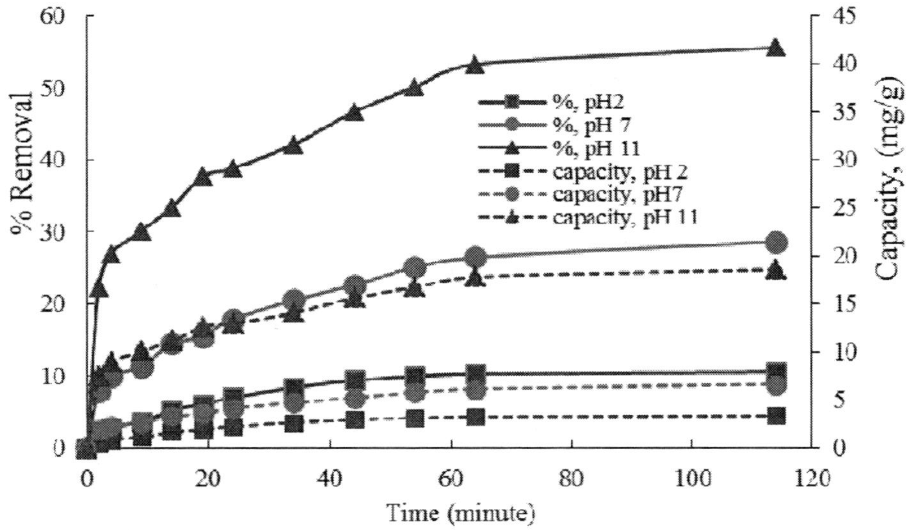

Figure 2. Effect of pH for BR14 removal by CSC of 500-710 μm at various times.

3.2 Effect of adsorbent size on dye removal

The size of adsorbent was varied in order to examine the effect of particle size on removal capacity. The experiment was carried out at pH 11, the optimum pH, with variation of 500-710 μm and 1000-2000 μm CSC for BY14 and BR14 dye removal. From the experiment at pH 11, it was found that the smaller size showed the higher removal capacity because of the larger specific surface of the smaller size. The experimental results were showed in Figure 3.

3.3 Adsorption isotherm

Adsorption isotherm describes the interaction between adsorbate and adsorbent materials. The experiment was carried out at pH 11 with 200 mL solution of adsorbate concentration of 200, 250, 300, 400 and 500 mg/L at pH 11. The experimental data were fitted to Langmuir adsorption model (equation 3) and Freundlich adsorption model (equation 4) as shown in Figure 4 for BY13 and Figure 5 for BR14, respectively.

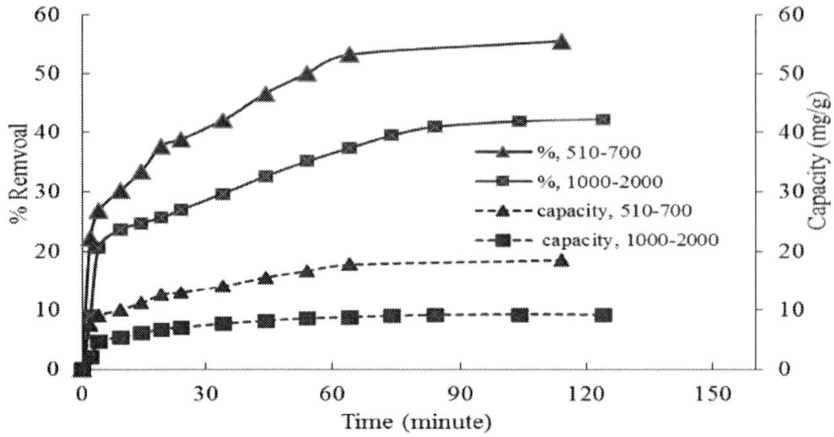

Figure 3. Effect of adsorbent sizes on BR14 removal by CSC, pH 11 at various times.

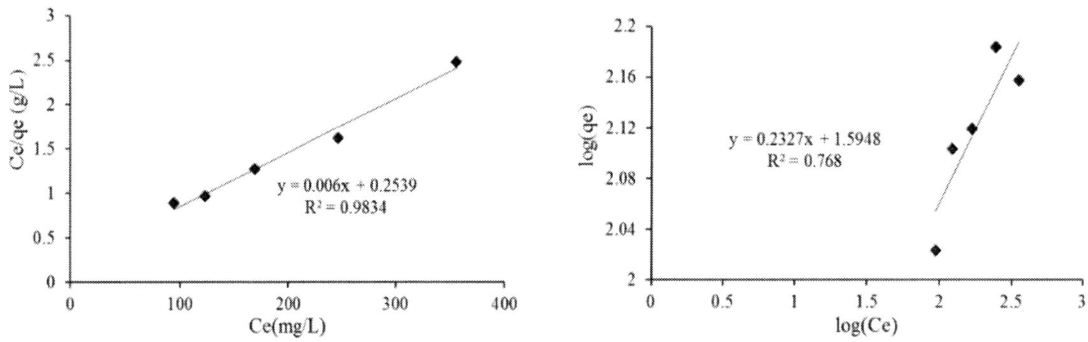

Figure 4. Plot of Langmuir and Freundlich adsorption model for BY13 with CSC, 510-700μm, pH 11.

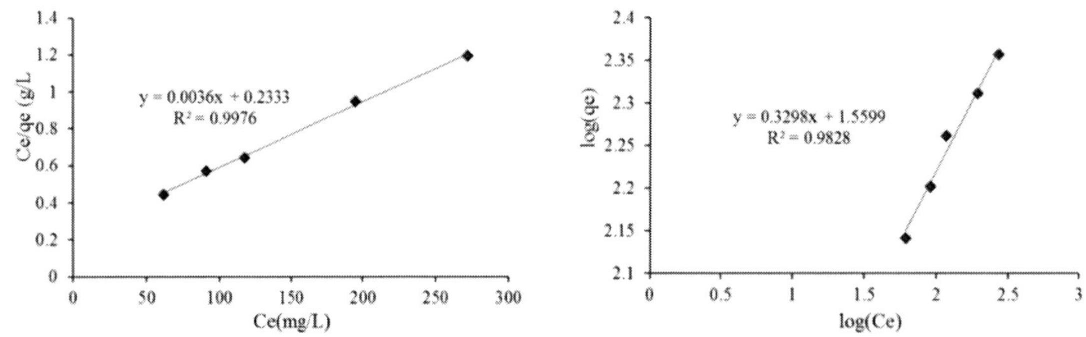

Figure 5. Plot of Langmuir and Freundlich adsorption model for BR14 with CSC, 510-700μm, pH 11.

From the figures, it can see that the experimental data fitted to Langmuir adsorption model for both BR14and BY13. The caculated values for BR14 and BY13 in both model are presented in Table 1. Maximum adsorption capacity, q_m, were 22.93 and 19.76 mg/g and Langmuir adsorption constant, K_L, were 1.87 x10^{-2} and 1.99 x10^{-2} L/ mg for BR14 and BY13, respectively. It can be interpreted that the interaction between CSC and dye was ionic interaction with monolayer adsorption. Regarding to properties of CSC, Iodine Number and Methylene Blue Number, it could be pronounced that the CSC could be employed as a promising bio-adsorbent for basic dye removal from wastewater.

Table 1. Calculated values from adsorption model of BR14 and BY13 by coconut shell charcoal at pH 11.

Calculated values from Adsorption model	BR14	BY13
Langmuir:		
Slope	0.004	0.006
Intercept	0.233	0.253
q_m	22.93	19.76
K_L x10^2	1.86	1.99
R^2	0.998	0.983
Freundlich:		
Slope	0.329	0.233
Intercept	1.559	1.595
n	3.08	4.29
K_F x10^2	36.29	39.33
R^2	0.983	0.768

4 Conclusions

The removal of Basic Yellow 13 and Basic Red 14 by CSC were investigated. The solution pH as well as adsorbent size can affect to the removal capacity. The removal at pH 11 is the most comparing among that of pH 2 ,7 and 11. Basic environment made the dye-CSC interaction more stronger than that of neutral or acidic solution. Larger size adsorbent shows lesser removal capapcity.The adsorption isotherm fitted to Langmuir model and maximum adsorption capacity, q_m, are found to be 19.76 mg/g and 22.93 mg/g and Langmuir adsorption constant, K_L, are 1.99 x 10^{-2} and 1.86.04x10^{-2} L/mg for Basic Yellow 13 and Basic Red 14, respectively. Thus these studies show that CSC, the disposed solid waste, can be effectively used as an alternative for commercial activated carbons for the removal of basic dye from aqueous solution.

Acknowledgements

Faculty of Engineering, Srinakharinwirot University is acknowledged for budget and equipment support. Assistant Professor Dr.Chanchira Jubsilp is also appreciated for managing the pictures.

References

1. M. E. Haddad, Journal of Taibah University for Science, **10**, 664-667 (2016)
2. S.L. Chan, Y.P. Tan, A.H. Abdullah, and S.T. Ong, J. Taiwan Inst Chem E., **31**, 306-315 (2016)
3. O. Duman, S. Tunç, and T.G. Polat, Micropor Mesopor Mat, **210**, 176-184 (2015)
4. S. Srisorrachatr, *Annual Conference on Engineering and Technology*, *Proc. Con.* (Nagoya, Japan, 2015)
5. S. Srisorrachatr, N. Pimpipat, and P. Senakham, *Pure and Applied Chemistry International Conference*, *Proc. Con. PACCON*, 663-666 (Chiang Mai, Thailand, 2012)
6. A.U. Isah, G. Abdulraheem, S. Bala, S. Muhammad, and M. Abdullahi, International Biodeterior. Biodegradation, **102**, 265-273 (2015)
7. A.S. Aljeboree, A.N. Alshirifi, and A.F. Alkaim, Arabian J. Chem. (2014) DOI: 10.1016/j.arabjc. 2014.01.020
8. http://www.worlddyevariety.com/basic-dyes/basic-yellow-13.html
9. http://www.worlddyevariety.com/basic-dyes/basic-red-14.html

A generalized DEMATEL theory with a shrinkage coefficient for an indirect relation matrix

Hsiang-Chuan Liu[1], Ben-Chang Shia[2], Yih-Chang Ou[3,a] and Huan-Wen Su[4]

[1]*Department of Biomedical Informatics, Asia University, Taichung 41354, Taiwan*
[2]*Big Data Research Center & School of Management School of Health Care Administration, Taipei Medical University, Taipei 11031, Taiwan*
[3]*Department of Finance, Ling Tung University, Taichung 40852, Taiwan*
[4]*Graduate School of Business Administration, Fu Jen Catholic University, Xinzhuang Dist., New Taipei City 24205, Taiwan*

Abstract. In this paper, a novel decision-making trial and evaluation laboratory (DEMATEL) theory with a shrinkage coefficient of indirect relation matrix is proposed, and a useful validity index, called Liu's validity index, is also proposed for evaluating the performance of any DEMATEL model. If the shrinkage coefficient of an indirect relation matrix is equal to 1, then this new theory is identical to the traditional theory; in other words, it is a generalization of the traditional theory. Furthermore, the indirect relation is always considerably greater than the direct one in traditional DEMATEL theory, which is unreasonable and unfair because it overemphasizes the influence of the indirect relation. We prove in this paper that if the shrinkage coefficient is equal to 0.5, then the indirect relation is less than its direct relation. Because the shrinkage coefficient belongs to [0.5, 1], according to Liu's validity index, we can find a more appropriate shrinkage coefficient to obtain a more efficient DEMATEL method. Some crucial properties of this new theory are discussed, and a simple example is provided to illustrate the advantages of the proposed theory.

1 Introduction

A decision-making trial and evaluation laboratory (DEMATEL) was developed by Gabus and Fontela [1] and Fontela and Gabus [2] for resolving complex and difficult problems. To date, DEMATEL has been widely used as one of the most effective tools for solving cause–effect relationships among evaluation factors [3-10], but the indirect relation of a DEMATEL model is always markedly greater than its direct relation, which is unreasonable and unfair because it overemphasizes the influence of the indirect relation. In this paper, to overcome this drawback, we propose an external shrinkage coefficient for constructing a reduced indirect relation matrix, such that the new indirect relation matrix is less than its direct relation matrix, and for evaluating the performance of any DEMATEL model. We also propose a useful validity index, called Liu's validity index, for evaluating the performance of any DEMATEL model. Subsequently, a generalized DEMATEL is derived and shown to be more useful and effective than previous DEMATEL models.

[a] Corresponding author : ycou@teamail.ltu.edu.tw

2 Traditional DEMATEL model

The procedure of the traditional DEMATEL method is briefly introduced as follows:
Step 1: Calculate the initial direct relation matrix Q

If n factors are asked by N experts to evaluate the degree of direct influence between two factors through pair-wise comparison, then the degree to which expert e perceives factor i as affecting factor j is expressed as

$$q_{ij}^{(e)}, e=1,2,..,N \,.\; q_{ij}^{(e)} \in \{0,1,2,3,4\}, \quad i,j=1,2,...,n \tag{1}$$

For each expert e, an individual direct relation matrix is constructed as

$$Q_e = \left[q_{ij}^{(e)}\right]_{n\times n}, e=1,2,..,N, \; q_{ii}^{(e)}=0, i=1,2,...,n \tag{2}$$

and we can obtain their average direct relation matrix, which is called the initial direct relation matrix Q;

$$Q=\left[q_{ij}\right]_{n\times n}=\frac{1}{N}\sum_{e=1}^{N}Q_e, \; q_{ij}=\frac{1}{N}\sum_{e=1}^{N}q_{ij}^{(e)}, \; i,j=1,2,...,n \tag{3}$$

Step 2: Calculate the direct relation matrix A

$$A=\left[a_{ij}\right]_{n\times n}=\lambda^{-1}Q, \; \lambda=\max_{1\le i,j\le n}\left\{\sum_{j=1}^{n}q_{ij}, \sum_{i=1}^{n}q_{ij}\right\} \tag{4}$$

where $\quad a_{ii}=0, i=1,2,...,n, \quad 0\le a_{ij}\le 1, i\ne j, i,j=1,2,...,n$

and

$$0\le\sum_{i=1}^{n}a_{ij}, \sum_{j=1}^{n}a_{ij}\le 1, \; i,j=1,2,...,n \tag{5}$$

Step 3: Calculate the indirect relation matrix B and the total relation matrix T
On the basis of Markov chain theory, we have

$$\lim_{k\to\infty} A^k = 0_{n\times n} \tag{6}$$

$$B=\left[b_{ij}\right]_{n\times n}=\lim_{k\to\infty}\left[A^2+A^3+...+A^k\right]=A^2\left(I-A\right)^{-1}, \tag{7}$$

$$T=\left[t_{ij}\right]_{n\times n}=A+B=\left[\left(a_{ij}+b_{ij}\right)\right]_{n\times n} \tag{8}$$

Step 4: Calculate the relation degree and prominence degree of each factor

$$r_i=\sum_{j=1}^{n}t_{ij}, \; c_i=\sum_{k=1}^{n}t_{ki}, \quad i=1,2,...,n \tag{9}$$

Where r_i indicates the total dispatch of both direct and indirect effects that factor i has on the other factors, and c_i indicates the total direct and indirect effects that factor i has on the other factors.

The relation degree of factor i is expressed as

$$x_i = r_i - c_i, \quad i = 1,2,\ldots,n \tag{10}$$

The prominence degree of factor i is expressed as

$$y_i = r_i + c_i, \ i = 1,2,\ldots,n \tag{11}$$

The relation–prominence matrix is expressed as

$$\left(x_i, y_i\right)_{i=1}^{n} \tag{12}$$

Step 5: Set the threshold value (α)
To eliminate some minor effect elements in matrix T to derive the impact–relations map, Ou Yang et al. [4] suggested the following threshold value:

$$\alpha_Y = \frac{1}{n^2}\sum_{i=1}^{n}\sum_{j=1}^{n} t_{ij} \tag{13}$$

Li and Tzeng [5] suggested a more informative threshold value, α_M, on the basis of their maximum mean de-entropy algorithm.
Step 6: Build a cause–effect relationship diagram
If $t_{ij} > \alpha_Y, or\ t_{ij} > \alpha_M$, then factor i is a net dispatch node of factor j, and factor j is a net receive node of factor i, which is expressed as

$$\left(x_i, y_i\right) \rightarrow \left(x_j, y_j\right), or \left(x_i, y_i\right) \leftarrow \left(x_j, y_j\right). \tag{14}$$

The graph of $\left(x_i, y_i\right)_{i=1}^{n}$, which includes the net direct edges, can present a cause–effect relationship diagram.

3 Critical properties of the generalized DEMATEL model

In this paper, we consider the critical properties of the traditional DEMATEL model to derive a more effective DEMATEL model, and then propose a generalized DEMATEL model. The critical properties of the generalized DEMATEL model are briefly described as follows.
Theorem 1. Critical properties of the generalized DEMATEL model
If

$$\text{(a)}\ A = \left[a_{ij}\right]_{n\times n} = \left[a_{IJ}^{(1)}\right]_{n\times n}, A^2 = \left[a_{IJ}^{(2)}\right]_{n\times n} = \left[\sum\nolimits_{k=1}^{n} a_{ik}a_{kj}\right]_{n\times n}, A^{m+1} = \left[a_{IJ}^{(m+1)}\right]_{n\times n} = \left[\sum\nolimits_{k=1}^{n} a_{ik}^{(m)} a_{kj}\right] \tag{15}$$

where

$$a_{ii} = 0, i = 1,2,\ldots,n, \ a_{ij} \geq 0, \ i,j = 1,2,\ldots,n, \sum\nolimits_{j=1}^{n} a_{ij} \leq 1, \sum\nolimits_{i=1}^{n} a_{ij} \leq 1 \tag{16}$$

$$\text{(b)}\ \delta^{(1)} = \max_{1\leq i,j\leq n}\left\{\sum_{j=1}^{n} a_{ij}, \sum_{i=1}^{n} a_{ij}\right\} = 1, \quad \delta^{(m)} = \max_{1\leq i,j\leq n}\left\{\sum_{j=1}^{n} a_{ij}^{(m)}, \sum_{i=1}^{n} a_{ij}^{(m)}\right\}, m \geq 2 \tag{17}$$

DOI: 10.1051/matecconf/201711901020

$$\text{(c)}\ \ B=\left[b_{ij}\right]_{n\times n}=A^2\left(I-A\right)^{-1},\ \mu=\max_{1\le i,j\le n}\left\{\sum_{j=1}^{n}b_{ij},\sum_{i=1}^{n}b_{ij}\right\} \tag{18}$$

$$\text{(d)}\ \ B_d=\left[b_{ij}^{(d)}\right]_{n\times n}=dA^2\left(I-dA\right)^{-1},d\in[0.5,1],\ \mu_d=\max_{1\le i,j\le n}\left\{\sum_{j=1}^{n}b_{ij}^{(d)},\sum_{i=1}^{n}b_{ij}^{(d)}\right\} \tag{19}$$

then

$$\text{(i)}\quad \delta^{(m+1)}\le\delta^{(m)}\le 1,\ m\ge 2 \tag{20}$$

$$\text{(ii)}\quad \delta^{(2)}<1\Rightarrow\delta^{(m+2)}<\delta^{(m)}\le 1,\ m\in N \tag{21}$$

$$\text{(iii)}\quad \mu=\lim_{m\to\infty}\sum_{s=2}^{m}\delta^{(m)},\ \mu_d=\lim_{m\to\infty}\sum_{s=2}^{m}d^{s-1}\delta^{s} \tag{22}$$

$$\text{(iv)}\quad d=0.5\Rightarrow 0<\mu_{0.5}\le 1 \tag{23}$$

$$\text{(v)}\quad d=\frac{1}{1+\delta^{(2)}}\Rightarrow\mu_{0.5}\le\mu_d\le 1 \tag{24}$$

Proof

(i) $\sum_{j=1}^{n}a_{ij}^{(2)}=\sum_{j=1}^{n}\left(\sum_{k=1}^{n}a_{ik}a_{kj}\right)=\sum_{k=1}^{n}\left(\sum_{j=1}^{n}a_{kj}\right)a_{ik}\le\sum_{k=1}^{n}a_{ik}=\sum_{j=1}^{n}a_{ij}\le 1$

$\Rightarrow\sum_{j=1}^{n}a_{ij}^{(2)}\le\sum_{j=1}^{n}a_{ij}\le 1,\ \textit{in general},\sum_{j=1}^{n}a_{ij}^{(m+1)}\le\sum_{j=1}^{n}a_{ij}^{(m)}\le 1$

Similarly, $\sum_{i=1}^{n}a_{ij}^{(m+1)}\le\sum_{i=1}^{n}a_{ij}^{(m)}\le 1,$ we can obtain $\delta^{(m+1)}\le\delta^{(m)}\le 1,\ m\ge 2.$

(ii) $\sum_{j=1}^{n}a_{ij}^{(3)}=\sum_{j=1}^{n}\left(\sum_{k=1}^{n}a_{ik}a_{kj}^{(2)}\right)=\sum_{k=1}^{n}\left(\sum_{j=1}^{n}a_{kj}^{(2)}\right)a_{ik}=\left(\sum_{j=1}^{n}a_{kj}^{(2)}\right)\sum_{k=1}^{n}a_{ik}>\sum_{j=1}^{n}a_{ij}\le 1$

$\sum_{j=1}^{n}a_{ij}^{(4)}=\sum_{j=1}^{n}\left(\sum_{k=1}^{n}a_{ik}^{(2)}a_{kj}^{(2)}\right)=\sum_{k=1}^{n}\left(\sum_{j=1}^{n}a_{kj}^{(2)}\right)a_{ik}^{(2)}=\left(\sum_{j=1}^{n}a_{kj}^{(2)}\right)\sum_{k=1}^{n}a_{ik}^{(2)}>\sum_{k=1}^{n}a_{ik}^{(2)}\le 1$

In general, $\sum_{j=1}^{n}a_{ij}^{(m+2)}<\sum_{j=1}^{n}a_{ij}^{(m)}\le 1,$ we can obtain $\delta^{(m+2)}<\delta^{(m)}\le 1,\ m\in N.$

(iii) $\mu=\max_{1\le i,j\le n}\left\{\sum_{j=1}^{n}b_{ij},\sum_{i=1}^{n}b_{ij}\right\}=\lim_{m\to\infty}\sum_{k=1}^{m}\max_{1\le i,j\le n}\left\{\sum_{j=1}^{n}a_{ij}^{(s)},\sum_{i=1}^{n}a_{ij}^{(s)}\right\}=\lim_{m\to\infty}\sum_{s=1}^{m}\delta^{(s)}$

Similarly, we have $\mu_d=\lim_{m\to\infty}\sum_{s=2}^{m}d^{s-1}\delta^{s}$.

(iv) We know that $\mu_d=\lim_{m\to\infty}\sum_{s=2}^{m}d^{s-1}\delta^{(s)}\le\delta^{(2)}\lim_{m\to\infty}\sum_{s=2}^{m}d^{s-1}=\delta^{(2)}\frac{d}{1-d}$.

Because $0<\delta^{(2)}\le 1,$ if $\frac{d}{1-d}\le 1,$ then $\mu_d=\lim_{m\to\infty}\sum_{s=2}^{m}d^{s-1}\delta^{(s)}\le\delta^{(2)}\frac{d}{1-d}\le 1;$

$$\text{that is, } d=\frac{1}{2}\Rightarrow 0<\mu_d=\mu_{0.5}\le 1. \tag{25}$$

(v) Because $\mu_d = \lim_{m\to\infty}\sum_{s=2}^{m} d^{s-1}\delta^{(s)}$, is an increasing function of d and $\delta^{(2)}$ and

$$0<\delta^{(2)}\le 1,\ \frac{d}{1-d}\le 1,\ \text{then}\ \mu_d = \lim_{m\to\infty}\sum_{s=2}^{m} d^{s-1}\delta^{(s)} \le \delta^{(2)}\frac{d}{1-d}\le 1,$$

$$\text{let}\ \delta^{(2)}\frac{d}{1-d}=1 \Rightarrow \frac{d}{1-d}=\frac{1}{\delta^{(2)}} \Rightarrow 0.5\le d=\frac{1}{1+\delta^{(2)}}\le 1,\ \text{then}\ \mu_{0.5}\le\mu_d\le 1. \tag{26}$$

4 Method of the generalized DEMATEL model

Because the indirect relation of a traditional DEMATEL model is always markedly greater than its direct relation, it overemphasizes the influence of the indirect relation. To overcome this drawback, given the aforementioned properties, an external shrinkage coefficient, *d*, is provided for constructing a reduced indirect relation matrix, and a useful validity index, called Liu's validity index, is also proposed for evaluating the performance of any DEMATEL model. Subsequently, a generalized DEMATEL model is derived and shown to be more useful and effective than previous DEMATEL models.

$$B_d=\left[b_{ij}^{(d)}\right]_{n\times n}=dA^2(I-dA)^{-1},\ d\in\left[\frac{1}{2},1\right] \tag{27}$$

$$T_d=\left[t_{ij}^{(d)}\right]_{n\times n}=A+B_d=\left[\left(a_{ij}+b_{ij}^{(d)}\right)\right],\ d\in\left[\frac{1}{2},1\right] \tag{28}$$

The relation–prominence of the DEMATEL (A,B_d) is defined as

$$R(A,B_d)=\left(x_i^{(d)},y_i^{(d)}\right)_{i=1}^{n}=\left(r_i^{(d)}-c_i^{(d)},r_i^{(d)}+c_i^{(d)}\right)_{i=1}^{n} \tag{29}$$

$$\text{where}\quad r_i^{(d)}=\sum_{j=1}^{n}t_{ij}^{(d)},\ c_i^{(d)}=\sum_{k=1}^{n}t_{ij}^{(d)},\ i=1,2,\ldots,n \tag{30}$$

$$\alpha_Y=\frac{1}{n^2}\sum_{i=1}^{n}\sum_{j=1}^{n}t_{ij}^{(d)} \tag{31}$$

If $t_{ij}^{(d)}>\alpha_Y, or\ t_{ij}^{(d)}>\alpha_M$, then factor i is a net dispatch node of factor *j*, and factor *j* is a net receive node of factor i, which can be expressed as

$$\left(x_i^{(d)},y_i^{(d)}\right)\to\left(x_j^{(d)},y_j^{(d)}\right) \tag{32}$$

The graph of $(x_i,y_i)_{i=1}^{n}$, which includes the net direct edges, can present a cause–effect relationship diagram.

If $d = 1$, then the new DEMATEL$(A,\ B_d)$ is simply the traditional DEMATEL$(A,\ B)$.

If $d = 0.5$, then $\max_{1 \le i, j \le n} \left\{ \sum_{j=1}^{n} b_{ij}^{(d)}, \sum_{i=1}^{n} b_{ij}^{(d)}, \right\} \le 1$, and the new DEMATEL $\left(A, B_d\right)$ is feasible because the influence of its indirect relation is no greater than the influence of its direct relation.
To evaluate the performance of any DEMATEL model, a useful validity index, called Liu's validity index, is proposed as follows:

$$V_L\left(A, B_d\right) = 1 - \frac{1}{1 + 5\sqrt{\sum_{i=1}^{n} \sqrt{\left(x_i^{(d)} - \bar{x}_d\right)^2 + \left(y_i^{(d)} - \bar{y}_d\right)^2}}} \tag{33}$$

where $\bar{x}_d = \frac{1}{n} \sum_{j=1}^{n} x_i^{(d)}$, $\bar{y}_d = \frac{1}{n} \sum_{j=1}^{n} y_i^{(d)}$

Note that Liu's validity index can also be used for comparing the performance of any two DEMATEL models.
Example 1.
If a direct relation matrix of a DEMATEL model is obtained as

$$A = \left[a_{ij}\right]_{4\times4} = \begin{bmatrix} 0 & 0.36 & 0.32 & 0.32 \\ 0.32 & 0 & 0.34 & 0.30 \\ 0.34 & 0.30 & 0 & 0.30 \\ 0.28 & 0.28 & 0.30 & 0 \end{bmatrix} \tag{34}$$

then we can obtain the following indirect relation matrix:

$$B = \left[b_{ij}\right]_{4\times4} = A^2\left(I - A\right)^{-1} = \begin{bmatrix} 3.9448 & 3.8483 & 3.9290 & 3.7995 \\ 3.7483 & 3.8237 & 3.7991 & 3.6907 \\ 3.6807 & 3.6963 & 3.8253 & 3.6331 \\ 3.4499 & 3.4478 & 3.4963 & 3.4508 \end{bmatrix} \tag{35}$$

where $\mu = \max_{1 \le i, j \le n} \left\{ \sum_{j=1}^{n} b_{ij}, \sum_{i=1}^{n} {}_{ij} \right\} = 15.5215 \gg \max_{1 \le i, j \le n} \left\{ \sum_{j=1}^{n} a_{ij}, \sum_{i=1}^{n} a_{ij}, \right\} = 1 \cdot$ (36)

This indicates that the DEMATEL model overemphasizes the influence of the indirect relation. Subsequently, we can obtain the total relation matrix, $T = \left[t_{ij}\right]_{4\times4}$, and the relation–prominence matrix, $R(A, B)$, respectively, as follows:

$$T = \left[t_{ij}\right]_{4\times4} = A + B = \left[\left(a_{ij} + b_{ij}\right)\right]_{4\times4} = \begin{bmatrix} 3.9448 & 4.2083 & 4.2490 & 4.1195 \\ 4.0683 & 3.8237 & 4.1391 & 3.9907 \\ 4.0207 & 3.9963 & 3.8253 & 3.9331 \\ 3.7299 & 3.7278 & 3.7963 & 3.4508 \end{bmatrix} \tag{37}$$

$$R(A, B) = \left(x_i, y_i\right)_{i=1}^{n} = \begin{pmatrix} 0.7579 & 32.2851 \\ 0.2658 & 31.7780 \\ -0.2344 & 31.7850 \\ -0.7893 & 30.1989 \end{pmatrix} \tag{38}$$

$$\alpha_Y = \frac{1}{4^2} \sum_{i=1}^{4} \sum_{j=1}^{4} t_{ij} = 3.9390 \tag{39}$$

$$T^* = \begin{bmatrix} 3.9448^* & 4.2083^* & 4.2490^* & 4.1195^* \\ 4.0683^* & 3.8237 & 4.1391^* & 3.9907^* \\ 4.0207^* & 3.9963^* & 3.8253 & 3.9331 \\ 3.7299 & 3.7278 & 3.7963 & 3.4508 \end{bmatrix} \tag{40}$$

$t_{ij} > \alpha_Y = 3.9390 \Rightarrow x_i \to x_j$ indicates that factor x_i is a net dispatch node of factor x_j.

The graph of $(x_i, y_i)_{i=1}^n$, which includes the net direct edges, can present a cause–effect relationship diagram.

$$V_L(A,B) = V_L(A,B_1) = 1 - \frac{1}{1+5\sqrt{\sum_{i=1}^{n}\sqrt{(x_i-\bar{x})^2+(y_i-\bar{y})^2}^2}} = 0.9015 \tag{41}$$

If we consider a new DEMATEL model with a shrinkage coefficient of 0.5, we can obtain the responding indirect relation matrix $B_{0.5}$ and its total relation matrix $T_{0.5}$, respectively, as follows:

$$B_{0.5} = 0.5A^2(I-0.5A)^{-1} = \begin{bmatrix} 0.2538 & 0.1993 & 0.2144 & 0.2047 \\ 0.2012 & 0.2450 & 0.2002 & 0.2012 \\ 0.1916 & 0.2020 & 0.2450 & 0.1971 \\ 0.1882 & 0.1879 & 0.1872 & 0.2182 \end{bmatrix}, \tag{42}$$

$$\text{then } \mu_{0.5} = 0.8722 < \max_{1\le i,j\le n}\left\{\sum_{j=1}^{n} a_{ij}, \sum_{i=1}^{n} a_{ij}\right\} = 1 \le \mu = 15.5215 \tag{43}$$

Formula (29) shows that the indirect relation is already less than its direct relation; we can then obtain its total relation matrix and relation–prominence matrix, respectively, as follows:

$$T_{0.5} = A + B_{0.5} = \begin{bmatrix} 0.2538 & 0.5593 & 0.5344 & 0.5247 \\ 0.5212 & 0.2450 & 0.5402 & 0.5012 \\ 0.5316 & 0.5020 & 0.2450 & 0.4971 \\ 0.4682 & 0.4679 & 0.4872 & 0.2182 \end{bmatrix} \tag{44}$$

$$\alpha_Y = \frac{1}{4^2}\sum_{i=1}^{4}\sum_{j=1}^{4} t_{ij}^{(0.5)} = 0.4735 \tag{45}$$

$$T_{0.5}^* = \begin{bmatrix} 0.2538 & 0.5593^* & 0.5344^* & 0.5247^* \\ 0.5212^* & 0.2450 & 0.5402^* & 0.5012^* \\ 0.5316^* & 0.5020^* & 0.2450 & 0.4971^* \\ 0.4682^* & 0.4679^* & 0.4872^* & 0.2182 \end{bmatrix} \tag{46}$$

$$R(A, B_{0.5}) = \left(x_i^{(0.5)}, y_i^{(0.5)}\right)_{i=1}^n = \begin{pmatrix} 0.0973 & 3.6469 \\ 0.0334 & 3.5718 \\ -0.0312 & 3.5824 \\ -0.0996 & 3.3526 \end{pmatrix}$$

DOI: 10.1051/matecconf/201711901020

If $t_{ij}^{(0.5)} > \alpha_Y = 0.4735$, *then* $x_i^{(0.5)} \rightarrow x_j^{(0.5)}$ indicates that factor i is a net dispatch node of factor j.

The graph of $\left(x_i^{(0.5)}, y_i^{(0.5)}\right)_{i=1}^{n}$, which includes the net direct edges, can present a cause–effect relationship diagram.

$$V_L\left(A, B_{0.5}\right) = 1 - \frac{1}{1 + 5\sqrt{\sum_{i=1}^{n}\sqrt{\left(x_i^{(0.5)} - \bar{x}_{0.5}\right)^2 + \left(y_i^{(0.5)} - \bar{y}_{0.5}\right)^2}}} = 0.7653 \quad (47)$$

Hence, DEMATEL$\left(A, B_{0.5}\right)$ is more effective than DEMATEL$\left(A, B\right)$. Furthermore, according to Liu's validity index, we can choose another shrinkage coefficient in [0.5, 1] to find the superior DEMATEL model.

5 Conclusion

In this paper, we show that the indirect relation is always considerably greater than the direct relation in traditional DEMATEL theory; consequently, it overemphasizes the influence of the indirect relation. Additionally, a novel DEMATEL theory with a shrinkage coefficient of an indirect relation matrix is proposed, and a useful validity index, called Liu's validity index, is also proposed for evaluating the performance of any DEMATEL model. We prove that for any DEMATEL model, if the shrinkage coefficient is equal to 0.5, then the indirect relation is less than its direct relation, and because the shrinkage coefficient belongs to [0.5, 1], according to Liu's validity index, we can determine a more appropriate shrinkage coefficient to obtain a more effective DEMATEL method. Some crucial properties of this new theory are discussed, and a simple example is provided in this paper to illustrate the advantages of the proposed theory.

References

1. A. Gabus and E. Fontela, *Perceptions of the World Problematique*: *Communication Procedure, Communicating with those Bearing Collective Responsibility* (Switzerland Geneva, 1973)
2. E. Fontela and A. Gabus, *The DEMATEL observer* (Switzerland Geneva, 1976)
3. G.H. Tzeng, C.H. Chiang, and C.W. Li, Expert. Syst. Appl., **32**, 4 (2007)
4. Y.P.O. Yang, H.M. Shieh, J.D. Leu, and G.H. Tzeng, Int. J. Oper. Res., **5**, 3 (2008)
5. C.W. Li and G.H. Tzeng, Expert. Syst. Appl., **36**, 6 (2009)
6. C.L. Lin and G.H. Tzeng, Expert. Syst. Appl., **36**, 6 (2009)
7. J.I. Shieh, H.H. Wu, and K.K. Huang, Knowl-based. Syst., **23**, 3 (2010)
8. D. Sumrit and P. Anuntavoranich, Int. T. J. Eng. Manage. App. Sci. Technol., **4**, 2 (2013)
9. C.W. Wang, Y.H. Lin, C.L. Chung, and M.T. Lee, Expert. Syst. Appl., **39**, 5 (2012)
10. E. Falatoonitoosi, Z. Leman, S. Sorooshian, and M. Salimi, Res. J. Appl. Sci. Eng. Technol., **5**, 13 (2013)

Do peak/slack seasons influence semiconductor machine outliers? A back propagation neural network analysis

Keng-Chieh Yang[1,a], Chia-Hui Huang[2] and Po-Hong Shih[3]

[1]*Department of Information Management, Hwa Hsia University of Technology, New Taipei City 235, Taiwan*
[2]*Department of Business Administration, National Taipei University of Business, Taipei City 100, Taiwan*
[3]*Institute of Information Management, National Chiao Tung University, Hsinchu 300, Taiwan*

Abstract. The semiconductor industry is often affected by the economy impact, which also influences the production schedule planning. The back propagation neural network model has the advantages of great precision and effectiveness. This research uses Novellus Vector Machine and its Remote Process Controller (RPC) function to collect the data. This study detects the gas transmission pressure of chamber. We uses fault detection and classification (FDC) to analyze the model. FDC can detect the deviations of the machine parameters when the parameters deviate from the original value and exceed the range of the specification. This study adopts back propagation neural network model and gray relational analysis as tools to analyze the data and detect the semiconductor machine outliers. The findings indicate that peak seasons have less outliers than slack seasons.

1 Introduction

Semiconductor industry often faces peak seasons and slack ones and this may impact the manufacturing and business operation. Generally the first quarter is the slack season, and the third quarter is the peak season [10]. The iSuppli research report also pointed out that the revenues of semiconductor industry in the first quarter is usually lower than that of the third or fourth quarter in the previous year. In the first quarter the number of orders is also less than that of the fourth quarter in the previous year. The utilization of foundry equipment is also put baptism of fire. Slack season will make the equipment and person idle [13,18]. The impact of idle equipment will influence the yield rate in the future. This research tries to study the yield rate of equipment in the peak seasons and slack ones, especially when equipment is shut down and restarted.

2 Methods and materials

Manufacturing process quality control is a set of procedures intended to ensure that a product or service adheres to a defined set of quality criteria or meets the requirements of the client [11]. In this section, we introduce quality control methods in the semiconductor wafer manufacturing process. Our study introduces the basic concepts of FDC and back propagation neural network in the Semiconductor Industry.

---

[a] Corresponding author : andesyoung.tw@gmail.com

2.1 Quality control in semiconductor manufacturing

The semiconductor manufacturing factory is usually using SPC to detect the quality of real-time processes. These information are often used to adjust the machine parameters after the event. SPC control charts are important tools to diagnose abnormalities of process [5,16]. Using these charts, engineers can understand the quality of the wafer. If the output of numerical data are between the upper bound (UCL) and lower bound (LCL), it means the quality is qualified. Otherwise, the quality is flaw.

Control charts are one of the most commonly used methods of Statistical Process Control (SPC), which monitors the stability of a process [17]. The main functions of a control chart include the data points, a centerline (mean value), and upper and lower limits (bounds to indicate where a process output is considered "out of control") [5, 8]. They visually display the fluctuations of a particular process variable, such as gas pressure, in a way that lets the engineer easily determines whether these variations fall within the specified process limits.

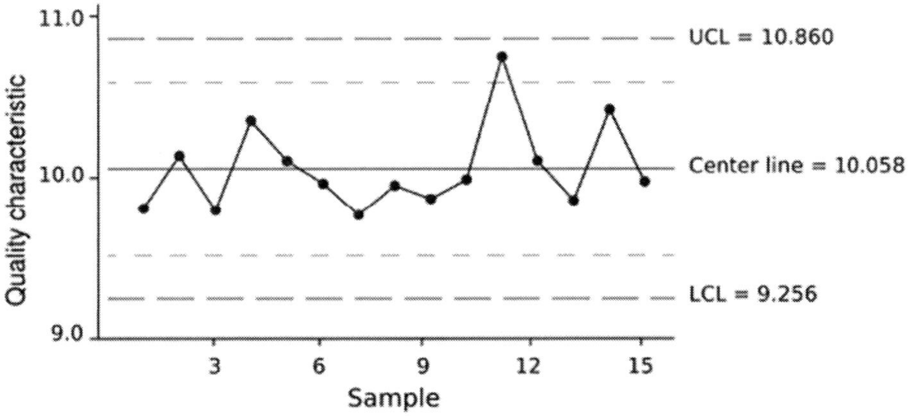

Figure 1. Sample of SPC control chart.

Fault Detection and Classification (FDC) contains two functions: fault detection (FD), and the fault classification (FC). Engineers are based on results of FD test to take some necessary actions. Different fault requires different corrective actions, while FC function is classified fault based on statistics Eigen-values so that engineers can quickly refer to the machine error code and restore the machine to normal state within the most minimum time [9].

In the semiconductor wafer process, when the machine produces a certain number of wafer, some parameters will drift from original ones. So, at this moment, FDC can detect deviations within a short time. When the parameters deviate from the original value, and may be beyond the range of the set interval, this needs to apply the Run-to-Run adjustments to modify the parameters directly and continuously collecting the running parameters of the machine and constant feedbacks [1]. Based on the previous activities of quality control, engineers can adjust the machine parameters to ensure that the production are within normal operations [15-16].

Engineers use FDC monitor to ensure the information of production status correct, including the manufacturing process, machine operating conditions, parameters, and use of the recipe. Engineers must check the machine status before the operation in issues; otherwise, when the production is finished, it would lead to the business loss. FDC monitor can avoid the waste of production capacity, reduce failures and ensure the producing yield increases [1, 19].

2.2 Back propagation neural network

The back propagation (BP) neural network algorithm is a multi-layer feed forward network trained according to error back propagation algorithm and is one of the most widely applied neural network models [2, 6]. BP network can be used to learn and store a great deal of mapping relations of input-output model, and no need to disclose in advance the mathematical equation that describes these mapping relations. Its learning rule is to adopt the steepest descent method in which the back propagation is used to regulate the weighted value and threshold value of the network to achieve the minimum error sum of square [12, 14]. This paper focuses on the analysis of the characteristics and mathematical theory of BP neural network and also points out the shortcomings of BP algorithm as well as several methods for improvement.

BP algorithm is a method to monitor learning. It utilizes the methods of mean square error and gradient descent to realize the modification to the connection weight of network [7]. The modification to the connection weight of network is aimed at achieving the minimum error sum of squares [4, 6]. In this algorithm, a little value is given to the connection value of network first, and then, a training sample is selected to calculate gradient of error relative to this sample.

The BP learning process can be described as follows [3]:

(1) Forward propagation of operating signal: the input signal is propagated from the input layer, via the hidden layer, to the output layer. During the forward propagation of operating signal, the weighted value and offset value of the network are maintained constant and the status of each layer of neuron will only exert an effect on that of next layer of neuron. In case that the expected output cannot be achieved in the output layer, it can be switched into the back propagation of error signal.

(2) Back propagation of error signal: the difference between the real output and expected output of the network is defined as the error signal; in the back propagation of error signal, the error signal is propagated from the output end to the input layer in a layer-by-layer manner. During the back propagation of error signal, the weighted value of network is regulated by the error feedback. The continuous modification of weighted value and offset value is applied to make the real output of network closer to the expected one.

3 Results and discussion

This study uses Artificial Neural Back-Propagation Network to predict the remote process controller (RPC) transferring pressure of the Novellus Vector equipment. The data is analysed from two different levels, including: the length of a Network training period and slack and peak season. The mean square error (MSE) was separately calculated for each category.

3.1 Back-propagation network parameters

The setup of network parameters in an Artificial Neural Network includes learning trials, learning rate, and the momentum correction coefficient. By setting up the learning rate at 0.1-0.3 for dynamic change, the momentum correction coefficient at 0.01-0.02, the specific setup values differ according to the network convergence status. When the MSE of the learning trail diminishes by 10-3/3000 epochs, this means the network learning process is completed. An average rate would be around 1,000-3,000 to achieve such standards.

Since the application of forward selection procedure will change the number of neurons in the hidden layer which leads to different groups of combinatorial optimization, this application should determine the number of neurons in the hidden layer prior to the selection process to prevent the process from being overly complicated.

To decide the number of neurons in the hidden layer, we should take the previously selected variable and substitute it into the network for training. Network training requires the exactly same setup, which includes learning trials, learning rate, and the momentum correction coefficient for a final decision. In this study, the number of neurons in the hidden layer is tested by setting the range at 1-2 times of the number of layers in the network. For example, if there were 10 input variables, then the number of layers would be 10. This research tests 1 to 20 number of neurons in the hidden layer by achieving the optimal result for the forward selection procedure that follows.

3.2 Network training in peak and slack seasons

Semiconductor industry has peak and slack seasons that change in cycle according to the time of the year. Right after Chinese New Year as often, the industry tends to be in the slack season since most of the clients have stocked up during the Christmas holidays already and do not need new supplies yet. Therefore, January and February are the slackest periods in the semiconductor industry. In contrast, when the clients use up their stock around March and April, the semiconductor industry starts to summit, and with May and June being the peak season. The months August to October are also another summit due to the client's need to stock up for the Christmas holidays.

(1) Slack season network training:

1,683 records of data were collected from the Transfer Pressure of the Novellus Vector from January 1st, 2010 to March 31st. The tested data retrieved from April 1st 2011 to April 5th consisted of 70 records. The learning times were set to 3,000, learning rate set to 0.1, and the momentum correction coefficient set to 0. Since the training results of the Artificial Neural Back-Propagation Network do not always gain the same results, this study conducts the same experiment 10 times to ensure the network stability.

The results of the tests show that each round of the training and tests are slightly different. However, all the results converge within the 800th cycle. As shown in figure 3, the overall performance is fairly well with the MSE= 0.0151 and RMSE=0. 6256. Table 1 shows that by comparing the Mable Train-R data and Target data and by depicting the network training prediction output figure. Figure 2 depicts the default mode and its outlier can be predicted.

Table 1. Network training and performance tests of slack season.

Hidden neuron	train - MSE	train - R
1	0.0142	0.6414
2	0.0146	0.6470
3	0.0152	0.6472
4	0.0146	0.6234
5	0.0149	0.6223
6	0.0157	0.6609
7	0.0158	0.5474
8	0.0153	0.6034
9	0.0148	0.6097
10	0.0155	0.6534
Average	0.0151	0.6256

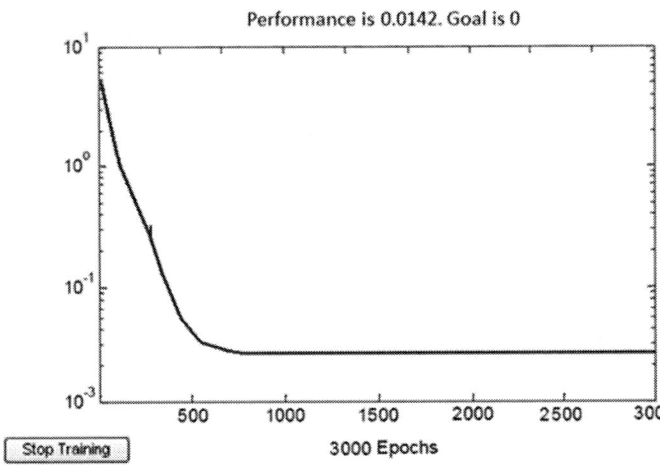

Figure 2. Network convergence figure of the slack seasons.

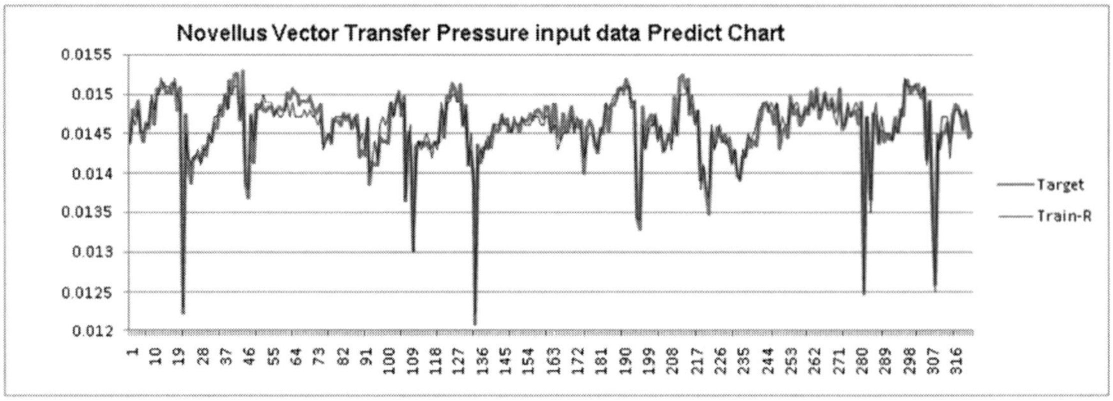

Figure 3. Network training prediction output figure of the slack seasons.

(2) Network training in the peak season

2,341 records of data were collected from the Transfer Pressure of the Novellus Vector from July 1st, 2010 to September 30th. The tested data retrieved from October 1st 2011 to October 5th consisted of 70 records. The learning times were set to 3,000, learning rate set to 0.1, and the momentum correction coefficient set to 0. Since the training results of the Artificial Neural Back-Propagation Network do not always gain the same results, this study conducts the same experiment 10 times to ensure the network stability.

Table 2. Network training and performance tests of peak seasons.

Hidden neuron	train - MSE	train - R
1	0.0119	0.6865
2	0.0123	0.6925
3	0.0129	0.6927
4	0.0123	0.6673
5	0.0126	0.6661
6	0.0134	0.7073
7	0.0135	0.5862
8	0.0130	0.6460
9	0.0125	0.6527
10	0.0132	0.6993
Average	0.0128	0.6696

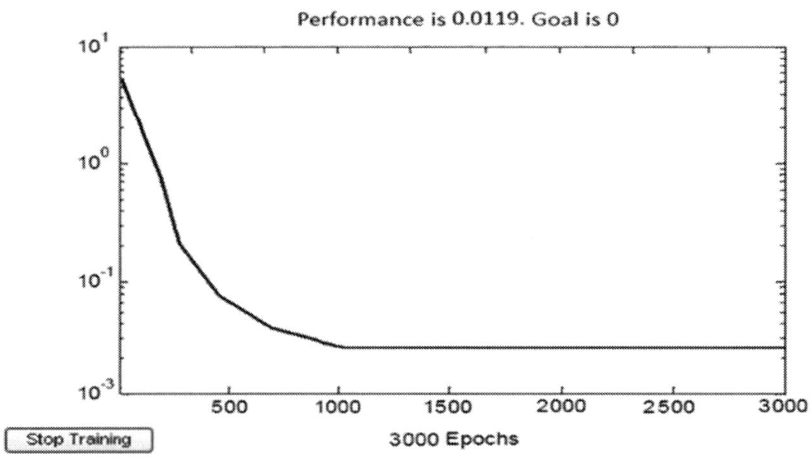

Figure 4. Network convergence figure of the peak seasons.

 DOI: 10.1051/matecconf/201711901021

The results of the tests show that each round of the training and tests are slightly different. However, all the results converge within the 1,050th cycle. As shown in figure 5, the overall performance is fairly well with the MSE= 0.0128 and RMSE=0. 6696. Table 2 shows that by comparing the Mable Train-R data and Target data and by depicting the network training prediction output figure. Figure 4 depicts the default mode and its outlier can be predicted.

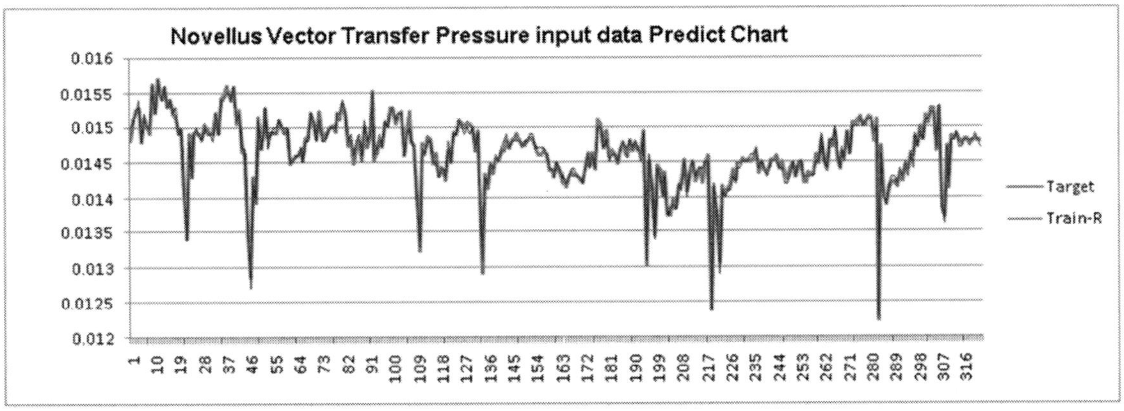

Figure 5. Network training prediction output figure of the peak seasons.

4 Conclusion

This study uses the mean square error (MSE) of the training output data to analyze if training results are acceptable and converged. The MSE equation can be seen in (1) shown below. The smaller MSE means the better network training results. In the network training process, the MSE might be unstable if the network is not converged. To improve such situation, the network learning rate and momentum correction coefficient should be readjusted to reach convergence.

This study analyzes the network training and prediction results by using the correlation coefficient (R) and MSE (equation (1) and (2) shown). The n in the equation signifies the total number of data inputs, μ notes the arithmetic mean, σ notes the standard deviation, the subscript i notes the serial of data, subscript t and a note the actual value and the network output value. This study analyzes the network output by calculating the correlation coefficient of the network trail output and the actual value and chooses the higher one as the optimal result. As shown in table 3, we take the trail network data and categorize it by network training time, low and high peak network training and calculate each the MSE of these data.

$$MSE = \frac{1}{N}\sum_{i=1}^{n}(t_i - a_i)^2 \tag{1}$$

$$R = \frac{(\sum t_i a_i) n \mu_t \mu_a}{(n-1)\sigma_t \sigma_a} \tag{2}$$

Table 3. The comparison of different types of training mode and performance tests.

	Peak season	Slack season
Train - MSE	0.01279	0.01510
Train - R	0.66965	0.62563

The results of the network training found that the MSE of the peak season was smaller than that of slack one. Hence, we can conclude that peak and slack seasons will influence semiconductor machine outliers.

This study uses the Artificial Neural Back-Propagation Network model to detect the outliers in semiconductor machines. Due to the complicity of the process technology and many types of machines in the semiconductor industry, we chose the machine, Novellus Vector, which had the hardest gas control and pressure control for the network training. This study focuses on semiconductor factories as our research subjects. The results are based upon these factories. We suggest that future studies choose a different industry to conduct their study and also compare the differences between industries.

Acknowledgement

This paper was partially support by Ministry of Science and Technology, Taiwan. Grant No. 102-2815-C-146-003-H and No. 104-2410-H-146-001 -

References

1. J.P. Card, M. Naimo, and W. Ziminsky, Run-to-run process control of a plasma etch process with neural network modelling, Qual Reliab Eng. Int., **14** (4), 247-260 (1998)
2. F.C. Chen, Back-propagation neural networks for nonlinear self-tuning adaptive control, IEEE Contr Syst Mag, **10** (3), 44-48 (1990)
3. W.C. Chen, A.H. Lee, W.J. Deng, and K.Y. Liu, The implementation of neural network for semiconductor PECVD process, Expert Syst. Appl., **32** (4), 1148-1153 (2007)
4. J.E. Dayhoff and J.M.D. Leo, Artificial neural networks, Cancer, **91** (S8), 1615-1635 (2001)
5. C.M. Fan, R.S. Guo, S.C. Chang, and C.S. Wei, SHEWMA: an end-of-line SPC scheme using wafer acceptance test data, IEEE T Semi Conduct M,**13** (3), 344-358 (2000)
6. A. Goh, Back-propagation neural networks for modelling complex systems, Artif Intell Med, **9** (3), 143-151 (1995)
7. H.C. Pu and Y.T. Hung, Use of artificial neural networks: Predicting trickling filter performance in a municipal wastewater treatment plant, Envion Manage, **6**, 16-27 (1995)
8. C.D. Himmel and G.S. May, Advantages of plasma etch modelling using neural networks over statistical techniques, IEEE T Semi Conduct M, **6** (2), 103-111 (1993)
9. S.J. Hong, W.Y. Lim, T. Cheong, and G.S. May, Fault detection and classification in plasma etch equipment for semiconductor manufacturing e-Diagnostics, IEEE T Semi Conduct M, **25** (1), 83-93 (2012)
10. SC-IQ: Semiconductor Intelligence, Semiconductors down 2.7% in '12, may grow 7.5% in '13., http://www.semiconductorintelligence.com/?p=756 (2013)
11. G.S. May and C.J. Spanos, *Fundamentals of Semiconductor Manufacturing and Process Control*, John Wiley & Sons (2006)
12. K. Mehrotra, C.K. Mohan, and S. Ranka, *Artificial Neural Networks*, the MIT Press (1997)
13. MIC. Output of Taiwan's Semiconductor Industry to Reach Approx. US$35 Billion in 2009, Says MIC. from http://web.iii.org.tw/News/news_more?id=11, Accessed 03 March 2015 (2009)
14. J. Li, J. Cheng, J. Shi, and F. Huang, Brief introduction of back propagation (BP) neural network algorithm and its improvement, Advances in Computer Science and Information Engineering, 553-558 (2012)
15. B.S. de Ugarte, A. Artiba, and R. Pellerin, Manufacturing execution system-a literature review, Prod Plan Control, **20** (6), 525-539 (2009)
16. Y.E. Shao, C.J. Lu and C.C. Chiu, A fault detection system for an autocorrelated process using SPC/EPC/ANN and SPC/EPC/SVM schemes, Int. J. Innov. Comput. Int., **7** (9), 5417-5428 (2011)
17. G. Smith, Statistical Process Control and Quality Improvement: Prentice Hall, **576** (1998)

18. J.Z. Wu, Inventory write-down prediction for semiconductor manufacturing considering inventory age, accounting principle and product structure with real settings, Computers & Industrial Engineering, **65** (1), 128-136 (2011)
19. H.H. Yue, S.J. Qin, R.J. Markle, C. Nauert, and M. Gatto, Fault detection of plasma etchers using optical emission spectra, IEEE T Semi Conduct M, **13** (3), 374 385 (2000)

Information system design in the advertising industry

Hsin Chi Ko[a]

Department of Mass Communication, Chang Jung Christian University, Tainan 71101, Taiwan

Abstract. In all fields of business that use information technology, an information management system can improve the efficiency of administration, creating systematic content records and storing data as the foundation for effective management and goal development. However, there is one field in which there is no research on information systems, namely the advertising industry. Thus, this study aims to identify the features of advertisements, and explore the concepts of an information system for the advertising field, adopting a qualitative research method to consider the various designs and menu content of an information system for copy writers.

1 Introduction

Advertising belongs to the field of mass communication. It is about delivering messages, with the aim of improving lives and increasing cultural competence, and is an applied social science that can benefit the life of the general public. The purpose of advertising is to improve quality of life and increase knowledge and facilitate cultural exchange.

Advertising is an indirect tool for convincing people. Its coverage is very broad, including fields such as business, public welfare, and promotion. It attracts people's attention through the elements of images and text. Messages delivered through advertising are full of cultural meaning, offering information related to life, culture, economics, etc., incorporating knowledge and economic values. Although people's living conditions have improved with the advancement of technology, the ideal of pursuing a wonderful life remains. Thus, advertising not only aims to introduce new products to society, but also to portray life in an active way. If people felt nothing after seeing an advertisement, it would be a crisis for advertising. A successful advertisement is one that can achieve the goal of moving people. The core of its design must be some element that can touch people's hearts. Furthermore, what the advertisement presents must be authentic. Only authenticity can lead to a sound reputation. Thus, increasing the general public's understanding of the essence of advertising can help to enhance its meaning.

Advertising is often combined with social movements and is closely related to society. In times of prosperity, advertising is presented with a lively and vigorous atmosphere in response to the demands of society. In times of economic decline, advertisements with unique features can still efficiently increase people's willingness to make a purchase. From the perspective of commercial advertisements, due to the strong competition in the market, the content of advertisements is designed to address people's needs. Advertising is also the bridge between social cultures and people. People need to

[a] Corresponding author : hsin@mail.cjcu.edu.tw

retrieve information through advertisements, which reveal the features of products. For the above reasons, advertising is a phenomenon of modern life.

Technological advances and people's living standards are associated with each other. Technology can aid in meeting society's demand for quality. In today's era of digital technology, advertisers have all started to develop related marketing strategies to follow the trend in technological development. The importance of designing an advertising information system is therefore clear.

An issue advertisers have been dealing with actively is to design advertising strategies according to modern conceptions while consolidating the social and cultural role of advertising. A review of the literature show that previous studies in Taiwan have researched numerous advertising-related issues, as follows: introspection in advertising [1], creative issues [2], government decree advertising [3], the persuasiveness of advertisements [4], image advertising [5], applications of rhetorical skills [6], images in advertisements [7], election advertising [8], advertising education [9], advertisement figures [10], public welfare advertising [11], advertisement humor [12], regional advertising [13], advertising effects [14], advertising communication culture [15], advertising and gender [16], advertising and photography [17], advertising sentiment [18], and advertising strategies [19], internet advertising [20]. However, there has as yet been no study on advertising information systems. Thus, in the age of the information economy, studying innovative concept of information systems is necessary, indeed essential.

2 Methods

This qualitative study employed content and concept analysis [21] to examine 200 advertisements collected from the print media in Taiwan published in 2013. The contents of the advertisements were analyzed through the following five steps: (i) reading the content of the advertisements; (ii) analyzing the aspects that construct the content, recording them, defining them, naming them based on the definitions, and listing the indicators for each construct; (iii) interpreting and diagnosing the overall content of the advertisement to explore its focus, the design method applied, and the style formed, naming the style based on its definition in accordance with the principle of choosing only one style to represent each advertisement, and listing the characteristics of the content that form this style; (iv) analyzing the copy writer's design of the advertisement, using indicators including literal meaning, lexical category, tone, tonal pattern, rhyme, pun, rhetorical skill, pronunciation, style, charm, and vigor, to determine the design type of the copy writer and name that design type in accordance with the principle of choosing only one style to represent each copy writer; (v) based on the results of the content analysis, design the information system menu and plan for related information.

The study thus comprised two stages employing the above research methods. The first stage was the research and analysis stage, and the second was the construction of the advertising information system. The first stage aimed to address the following research issues through analysis: (1) identify the main elements of the content of advertisements; (2) categorize the style of advertisements; (3) categorize the copy writer's design type. The second stage integrated the advertisement concepts based on the research results as the basis for the design of an information system. The issues dealt with in this stage include: (4) design of the advertising information system and display interface; (5) plot the main aspects of the advertisement content and plan the information system for the indicator menu; (6) plot the advertisement style and plan the information system for the characteristics menu; (7) plan the information system for copy writer design; (8) design an interactive model and process to address satisfaction with the completed advertising product from the customer side and the system side.

3 Results

The research results obtained are discussed according to the two stages. In the first stage, the three research issues are resolved. The results include: (i) five aspects related to advertisement content; (ii) eight advertisement styles; (iii) eight copy writer design types. The research results in the second stage are related to the five issues regarding the advertising information system display: (i) the main interface of the information system; (ii) the main menu interface of the advertising information system; (iii) the style menu interface of the advertising information system; (iv) the copy writer design type menu interface; (v) the interactive model regarding satisfaction from the customer side and the system side.

3.1 The five main aspects of advertisement content

The five main aspects of advertisement content identified are layout design, text arrangement, color application, pictures, and images. Of these, the first three are necessary for all advertisements, whereas the necessity of pictures and images depends on the style of the design. There are three indicators for each of the five aspects, as follows:

- layout: mainly text, mainly graphs, both text and graphs;
- text: title, slogan, and main text;
- color: main color, harmonious color, and depth; picture: figure, scene, and specific item;
- picture: figure, scene, and specific item;
- image: symbolic image, decorative image, and designated image.

The frequencies of the application of the five aspects is shown in Table 1.

Table 1. Frequencies of the application of the five aspects.

Aspect	Layout	Text	Color	Picture	Image
Frequency	200	200	200	186	138
Percentage	100%	100%	100%	93%	69%

3.2 The eight advertisement styles

A total of eight advertisement styles were identified: sense, function, affection, entertainment, intellectuality, literature & art, education, and value. Table 2 shows the frequencies of the styles.

Table 2. Frequencies of the eight advertising styles.

Style	Sense	Function	Affection	Entertainment
Frequency	200	200	200	186
Percentage	100%	100%	100%	93%
Style	Intellectuality	Literature & art	Education	Value
Frequency	16	6	26	9
Percentage	8%	3%	13%	4.5%

As can be seen, the most common style is sense. Each style has three characteristics. Sense includes the following: perception, i.e., fostering the feeling among audience members that they are personal recipients of the advertisement content; impression, which involves enhancing the fresh impression in the audience's mind; atmosphere, creating the effect of making the audience feel they are personally on the scene. Function incorporates the following: convenience, namely that the product or service being advertised can make life easier; efficacy, i.e., the effectiveness and benefits of the product or service; comparison, giving the audience a deeper impression through a comparison. Affection is related predominantly to messages of warm feelings, focusing on psychological factors:

happiness; love; aspiration, delivering the expectation of a dream coming true. Entertainment encompasses: humanity, delivering information on a humane event helping to relieve pressure; activity, delivering information on a tourist activity; theme, delivering information on a leisure activity of any kind of theme. Intellectuality includes: history, describing some meaningful history; culture, sharing knowledge regarding life and culture, such as knowledge about dietetic hygiene; specificity, describing knowledge from a specific professional field. Literature and art comprises: story, telling a touching story through exquisite wording; meaning, touching people using the art of language vividly; form, using different literary forms, such as classic poetry, modern poetry, and essays, to improve literary and artistic quality. Education includes: learning, aimed at teenagers as the target group to teach them something; goal, aiming to help the audience achieve a goal, such as obtaining a license or improving language proficiency; specialty, aiming to help the audience develop a professional skill. Finally, value encompasses: public welfare, through raising funds; behavior, for example, advising the audience not to drink and drive; promotion, describing an opinion or a claim from an individual or a unit in society.

3.3 The eight copy writer design types

There are eight types of copy writer design, which exert different influences: (1) vitality, featuring a fresh life force and dynamics, creating an appealing resonance; (2) moving power, using a moving plot as the main design skill to achieve the effects of tugging at the audience's heart strings and generating sympathetic responses; (3) humor, generating a knowing smile, meaning that the audience already knows the implied meaning before smiling, and thus the effect is achieved naturally; (4) amazement, impressing audience with a surprising and novel message that is hard to forget; (5) aesthetics, delighting the audience's visual sense; (6) reflection, triggering the audience to memorize the message delivered through association; (7) guidance, achieving a guiding effect by integrating an influential person, event, or item into the design goal; (8) change, increasing the quality of life, or fulfilling a hope, triggering action on the part of the audience. The frequency of application of the eight design types is summarized in Table 3.

Table 3. Frequency of application of the eight copy writer design types.

Type	Vitality	Moving power	Humor	Aesthetics
Frequency	66	36	16	30
Percentage	33%	18%	8%	15%
Type	Amazement	Reflection	Guidance	Change
Frequency	12	6	13	21
Percentage	6%	3%	6.5%	10.5%

3.4 The design and interface of the advertising information system

The advertising information system designed based on the research results above contains two domains, the customer side and the system side, and three constructs, which include the five aspects of advertisement content, the eight styles, and the eight copy writer design types. The mapping between the customer side and the system side is shown in Figure 1.

 DOI: 10.1051/matecconf/201711901022

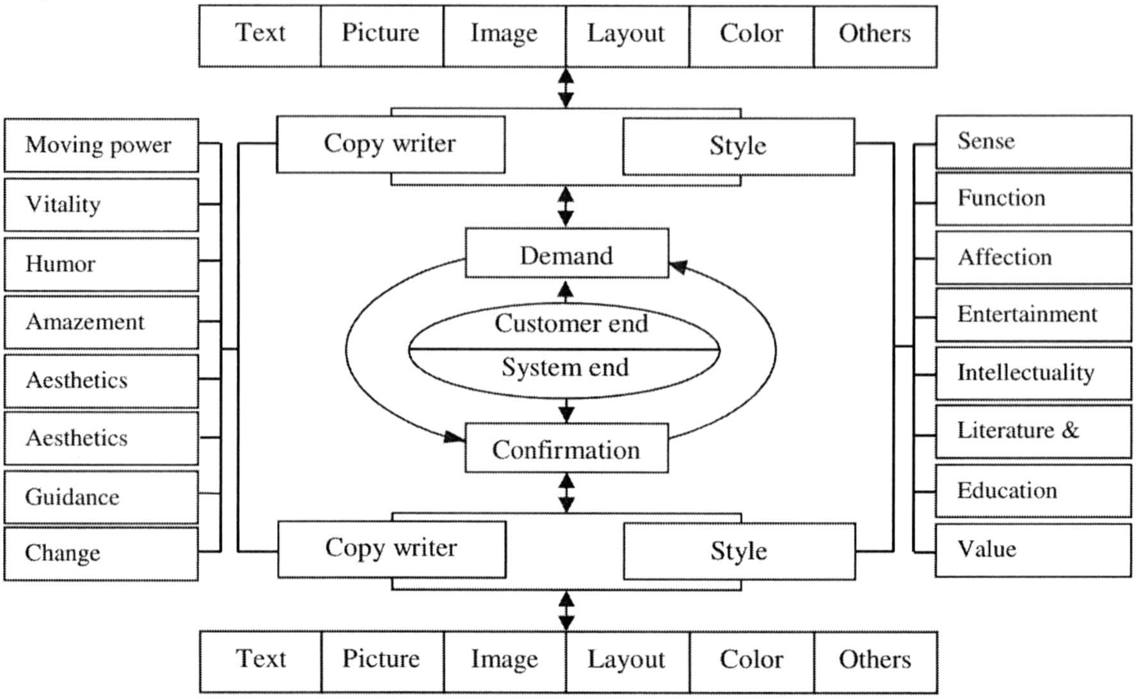

Figure 1. The interface of the advertising information system.

The menus in the information system are not designed as drop-down menus but as lists. The advantage is the visual effect, in that visitors from the customer side can see every item at first sight. Also, they can reflect on or compare the meanings of the items on the menu. For example, for each of the five aspects, there are three indicator menus, which are displayed instead of hidden, as shown in Figures 2 and 3. Moreover, to help visitors obtain information related to each menu, the design is that when clicking on a menu, the information related to the secondary menu will pop up, including example, frequency of use, date of use, percentage of use, number of satisfied use, number of modification, feedback, etc. The interface of the menu for the five main aspects of advertisement content and the interface of the menu for the eight advertising styles are shown in Figure 2 and Figure3.

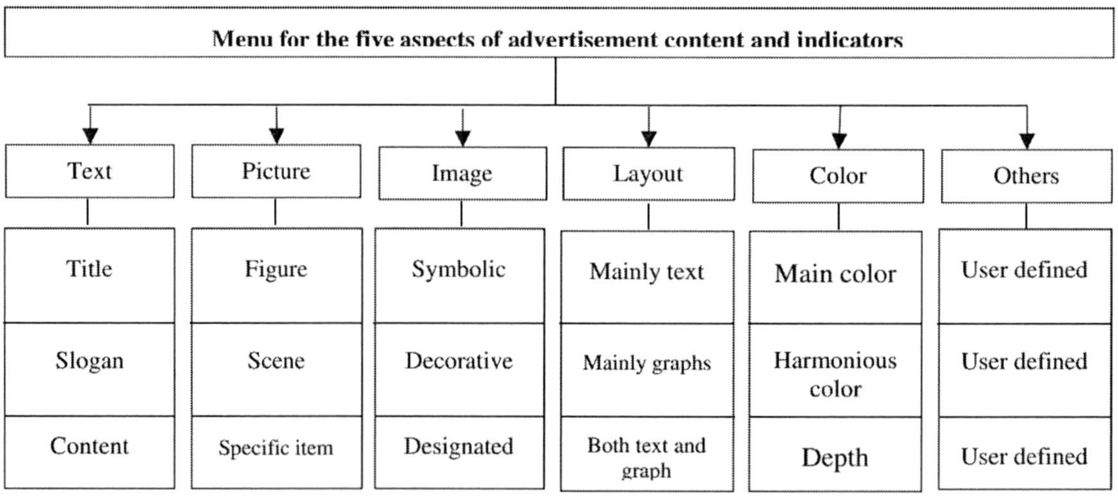

Figure 2. Interface of the menu for the five main aspects of advertisement content.

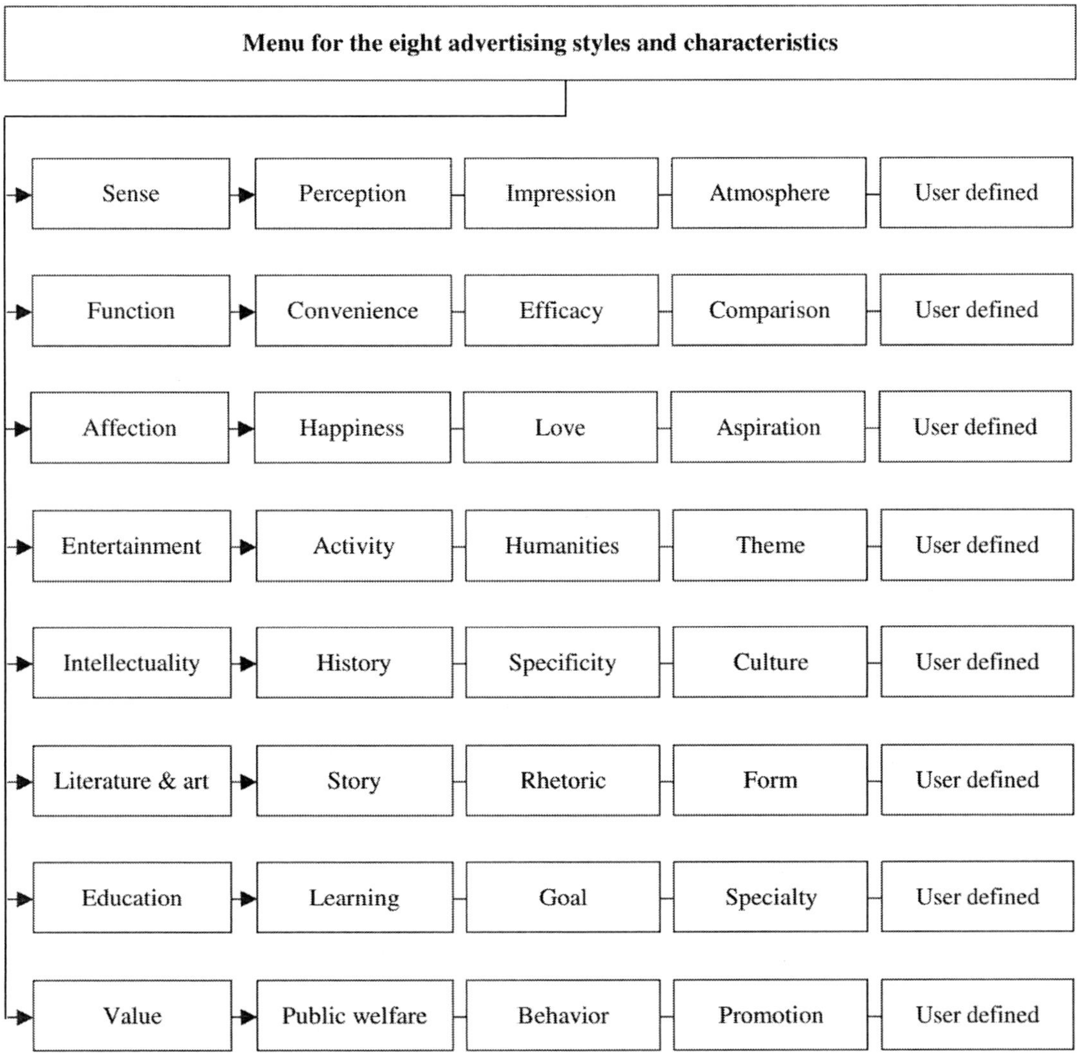

Figure 3. Interface of the menu for the eight advertising styles.

3.5 Interactive mode between the customer side and the system side

The main feature of the management design of the advertising information system is that the system provides the choices of five aspects of advertisement content, eight styles, and eight copy writer design, as well as user defined options through menus with examples. To avoid and reduce non-specific communications, the requirement form on the customer side offers detailed and specific descriptions for all the menus, so that users can decide which action to take after fully understanding the options. Should a user on the customer side not be satisfied with a completed product, he or she can review the requirement form to check the selected items at the system end. If the problem is at the system end, the company is responsible for modification. If the cause of the problem is a new requirement from the customer side, and there is no record in the form, the user can go through the requirement form process all over again and submit the new form to the system. The system automatically records every detail related to the menus and requirements for future inquiry and as a reference for structural data analysis. Figure 4 shows how the advertising information system works.

 DOI: 10.1051/matecconf/201711901022

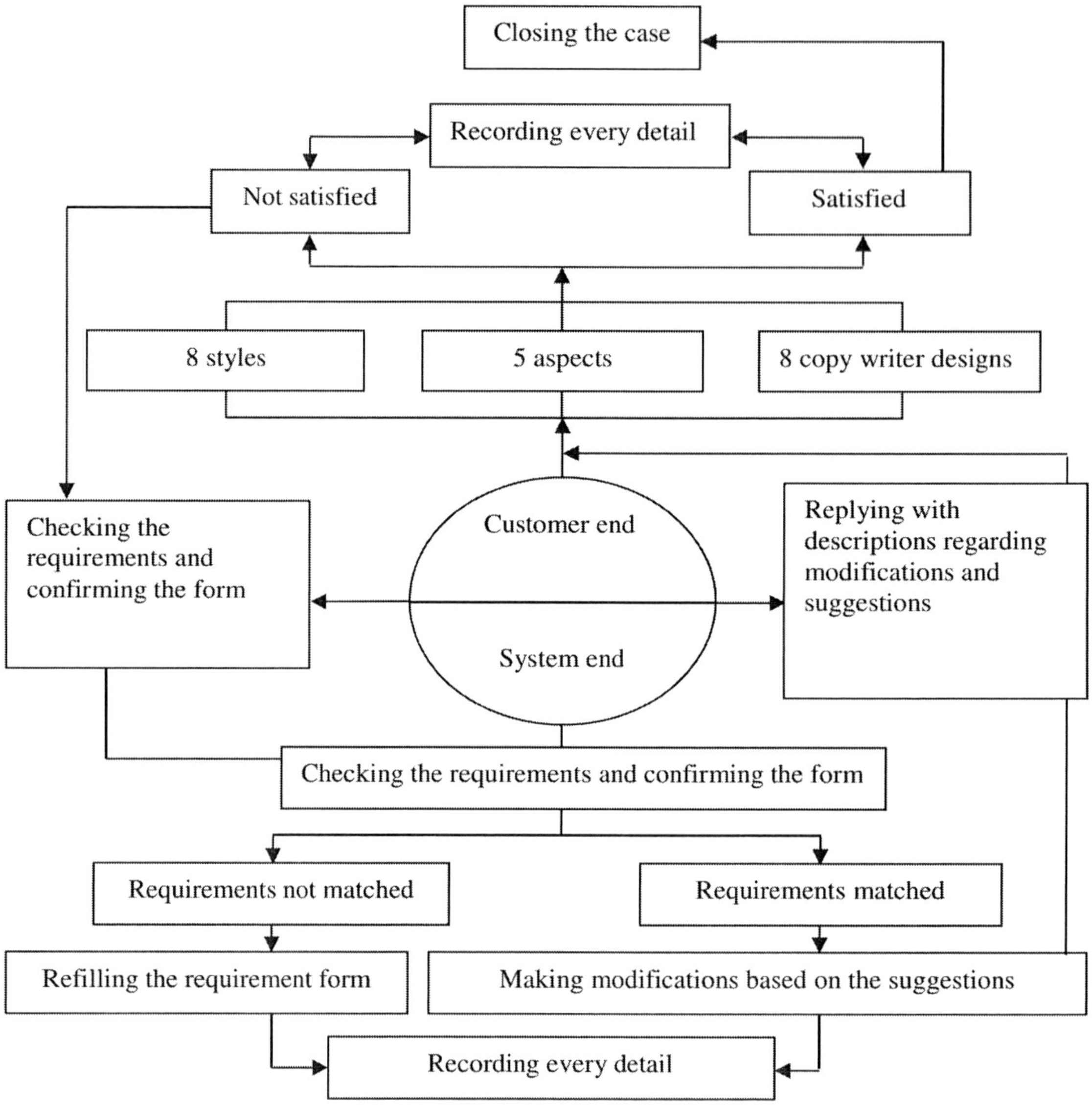

Figure 4. Interactive mode between the customer side and the system side.

4 Discussion

In the age of the knowledge economy, one of the important indicators of modern culture is the management and application of the combination of technology and information. As a result, the research and development of computer software, related applications, and management is very current, bringing about the age of the information economy. No studies in the field of advertising have thus far been related to the development of an advertising information system. This study aims to explore the design of an advertising information system, based on the concept of innovation, and integrating menus and various options for styles and designs. This advertising information system offers characteristics and aspects that can be applied to design in the hope of making a start in this field. The advertising information system is merely a platform linking the system side and the customer side, and enabling communication and interaction. To come up with a successful advertisement, a designer is still required to operate the advertising information system. The convenience and efficiency of the advertising information system designed by this study are improved through the concept of innovative management.

5 Conclusion

Advertising can do more than just deliver a message. It has the potential to create a trend. Thus, the style of an advertisement must be unique to gain the public's attention. Moreover, computers are now very popular and a part of people's lives, resulting in the age of the advertisement economy. The use of an advertising information system is a trend in the field of advertising innovation and management. Therefore, this study has designed an advertising information system comprising eight styles and eight copy writer designs based on the innovation, authenticity and accuracy. The application of the concept of innovative technology management helps to improve the efficacy of the advertising information system. With the design of various interfaces of functions and indicators, this study hopes to present the best descriptive effort for applications in practical advertisement design.

References

1. M.C. Chi, J. Adv., **8**, 171-186 (2003)
2. Y.G. Wu, J. Adv. Publ. Rela., **30**, 63-84 (2008)
3. C.T. Lai, J. Adv. Publ. Rela., **29**, 1-28 (2008)
4. S.C. Lin and C.H. Weng, J. Adv., **13**, 59-75 (2008)
5. Y.F. Chien and K.H. Wang, J. Adv. Publ. Rela., **26**, 27-59 (2006)
6. L.S. Chang, J. Adv., **11**, 83-94 (2006)
7. C.F. Chen, J. Adv. Publ. Rela., **35**, 75-112 (2011)
8. S.S. Chung and T.S. Niu, J. Adv., **11**, 107-131 (2006)
9. W.Y. Chang, J. Adv., **8**, 107-129 (2003)
10. C.C. Tsai, J. Adv. Publ. Rela., **25**, 107-129 (2006)
11. W.S. Chi and C.R. Bai, Cont. Comm., **6**, 66-68 (2005)
12. Y.N. Lin, B.Y. Chen, and C.C. Lin, Venu. Comm. J., **15**, 213-231 (2010)
13. C.W. Chang and S.K. Cho, J. Comm. Desi., **13**, 213-229 (2009)
14. I.P. Chiang and C.C. Chuang, Mana. Info. Comp., **3**, 222-234 (2014)
15. B.J. Zhang and J. Zhen, Univ. Lig. Indu., Social Science, **6**, 75-77 (2005)
16. T.F. Chen, J. H. C. Info. Comm., **8**, 79-109 (2011)
17. S.J. Chang and K.H. Chen., CCL Tech. J., **159**, 30-37 (2014)
18. C.L. Wu and S.F. Hsieh, J. CAGST, **2009**, 319-333 (2009)
19. K.K. Fan and C.J. Chen, J. Cult. Crea. Indu. Res., **2**, 211-222 (2012)
20. I.C. Wang and Y.F. Chien, J. Adv. Publ. Rela., **24**, 31-55 (2005)
21. W.K. Wang, E. Research Methods (Wunan, TPE, 2001)

MATEC Web of Conferences **119**, 01023 (2017) DOI: 10.1051/matecconf/201711901023
IMETI 2016

Study on the KSFs of TTQS introduction in international hotels in Taiwan

Ying-Chang Chen[a]

Graduate School of Health Industry Management, Ching Kuo Institute of Management and Health, Keelung City 20301, Taiwan

Abstract. This study explores the KSFs of TTQS, as introduced in international hotels in Taiwan. The method to carry out this study was using a survey AHP method. The findings of this study are, as follows: (1) Order of importance of the four dimensions: organizational management, training design, training execution, and result assessment, shows that when introducing TTQS, tourist hotels must first focus on reforming the organizational system. They should change the previous model to authorize training of the operational department. The training department should not play a passive role, and must establish a systematic training model with documents to dominate, as well as assist and audit training operations. (2) Among the 19 key factors of the sub-dimensions, the top 5 are high-ranking supervisors' high degree of support and participation, TTQS knowledge of persons in charge, connection between target members' and organizational needs, specific training goals before training, and regular training organization experience before introduction of TTQS. Based on the order, "high-ranking supervisors' support and participation" is the most important factor for organizations to introduce the TTQS system.

1 Research background and motives

On March 7, 2013, the World Economic Forum (WEF) announced the Travel & Tourism Competitiveness Report 2013. Taiwan was listed at the Top 33 among 140 countries It demonstrates the performance of Taiwan regarding policies of the tourism industry and promotion of tourism. Since 2009, the Executive Yuan has implemented economic policies, such as six major new industries and ten key service industries, which include a tourism excellence project. Ten key service industries refer to the internationalization of food and exhibition industries. In addition, according to the statistics of the Tourism Bureau, in 2012, the number of tourists in Taiwan grew significantly, and surpassed 7.3 million person-times. The foreign exchange of tourism reached 340 billion NTD and the result was impressive[1].

Based on ISO10015, Investors in People (IIP) of Britain, occupational training policies of Europe, active occupational training policies of Australia, and challenges encountered by Taiwan regarding a global knowledge economy society, in 2005, the Workforce Development Agency established the "Taiwan TrainQuali Scorecard", which was then renamed as the Taiwan TrainQuali System (TTQS),

[a] Corresponding author : proteus5520@gmail.com

in order to match the training management of enterprises, organizations, and training institutions in Taiwan, enhance the connection between training quality and the world, reinforce the professional competence of human resources training and development of the service industry, and strengthening labor force and national competitiveness. However, regarding the type of industry, in 2012, only 7 international hotels applied for the project. This demonstrates that few tourist hotels participated in and passed evaluation.

The research purposes are shown, as follows:

1. To recognize previous educational training and motivation to introduce TTQS in international hotels.
2. To analyze KSFs of TTQS introduced in international hotels.
3. To construct items of KSFs of TTQS introduced in international hotels.
4. To decide the relative weights of KSFs of TTQS introduced in international hotels.

2 Literature review

2.1 Educational training system of international hotels

In recent years, the development of international hotels is vigorous, thus, the labor force must increase in order to balance supply and demand in the labor market. International hotels in the past encountered various problems, such as, a lack of a labor force, a high turnover rate, and employees' poor performance, most of which was due to poor human resources development, meaning they should improve the practice of educational training[2]. Some scholars have suggested that currently, educational training of international hotels is mostly based on apprenticeships, where experienced employees instruct freshmen or offer courses designed for the immediate issues of business development. These courses are not regular and are not necessarily in demand by employees in different departments[3]. However, complete pre-service training and on-the-job training will enhance employees' work efficiency, facilitate increased competence to become medium and high-ranking supervisors, lower work stress, and help employees develop careers. Therefore, systematic educational training should be designed by such systems[4].

Educational training in enterprises must be based on the effectiveness of output, and evaluation of educational training is the most common method to recognize overall training results. There are two main purposes: one is to continuously improve training processes and effects, and the other is to serve as reference for training decision-making. Assessment of training performance is a circular system with dimensions and steps. The purpose is to thoroughly recognize and reference current affairs to adjust previous methods in order to respond to changes of future markets and continuously pursue maximum effectiveness of educational training in enterprises[5]. According to research in recent years, educational training in hotels is based on employees' organizational commitment, and it considers the effects of employees' emotions and affections. Hence, educational training in hotels can be regarded as a multi-dimensional framework, including training motivation, training acquisition, training benefit, and training support. The results demonstrated that all such measures positively influence organizational commitment[6].

2.2 TTQS (Talent Quality-management System)

Vocational training enhances the number of courses and strengthens training quality. In Britain, the policy of training quality assessment is based on the WTO ISO10015 system and Investment in People (IIP). Since 2005, in order to reinforce the quality and performance of occupational training, the Taiwanese government established the "Taiwan TrainQuali Scorecard" to enhance effective national human capital investment, and strengthen human resources competitiveness and the results of occupational training. In 2013, in order to enhance the overall organizational competence of occupational training units and increase talent development and investment, the original name was changed to TTQS (Talent Quality-management System). TTQS was established upon ISO 10015 and IIP. The purpose is to enhance national talent capital and promote business and professional occupational training institutions. Hence, three objectives are developed: (1) corresponding assessment; (2) differential assessment; (3) connection between training performance and organizational development goals[7]. According to the three objectives, the TTQS classifies the training processes of PDCA of ISO 10015 into five circular and associated dimensions (see Figure 1). These five dimensions are Plan, Design, Do, Review, and Outcome (PDDRO). According to the research of scholars and the Business Training Network, the meanings of PDDRO are as shown in Figure 1[8-10].

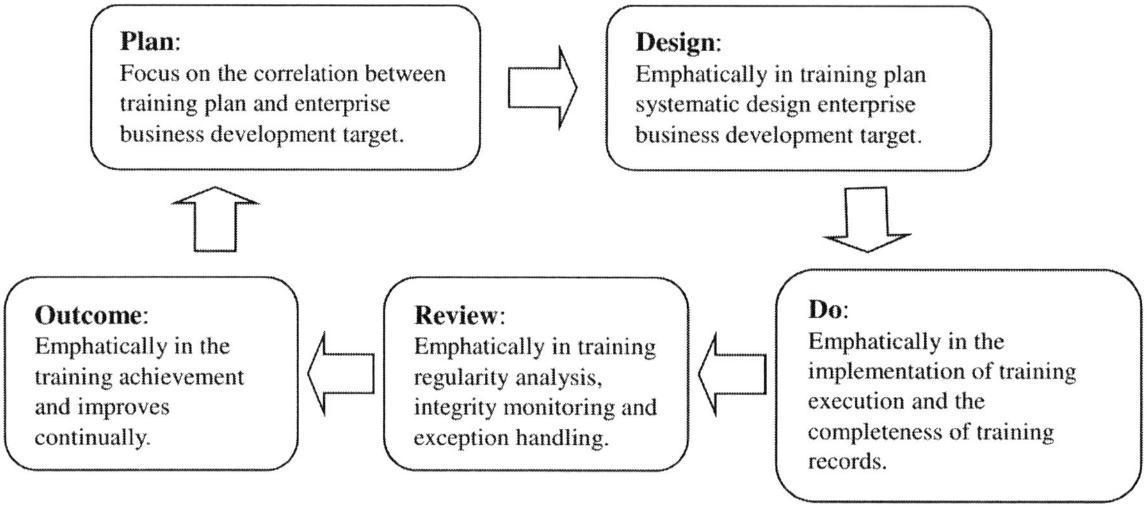

Figure 1. Five circular dimensions of TTQS.

2.3 KSFs (Key Success Factors)

KSFs appeared first in "Management Information Crisis", as published by Daniel in 1961. The success factors are defined, as follows: it is the work which should be excellent in order to be successful. Most industries have 3~6 success factors, and in order to be successful, enterprises must have excellent performance in these key factors. Hence, key success concepts become the important principles of the design of an information system.

KSFs are usually adopted when studying the relationship between industrial characteristics and organizational strategy. Aaker (1984) suggested that KSFs are the most important competitive advantages, assets, and techniques of enterprises, and they can be the base of competition[11]. The advantages of successful enterprises must be based on the KSFs of the industry. When an enterprise controls the KSFs of an industry, it will establish sustainable competitive advantages. Therefore, once organizations master certain fields and achieve prominent competence in comparison to others, they will have relative success factors in industrial environment. According to the research of Alazmi and Zairi[3], for enterprises, satisfying result in certain fields will guarantee successful performance. These activities are KSFs. Brotherton, Heinhuis, Miller, and Modema (2002) suggested that in order to completely accomplish an objective, enterprise should first establish KSFs[12]. According to Eindt, Jeffcoate, and Chappell (2002), in limited fields, variables of satisfying results that guarantee organizational success are KSFs[13]. Based on the views of previous scholars, KSFs are the key factors that enterprises must consider during industrial analysis. Successful enterprises usually are advantageous in the fields of KSFs, and when they seize competitive advantages in industry, operations will be sustainable.

Based on the definitions of scholars and related literature review, Chen (2011) suggested the following characteristics of KSFs: (1) KSFs change with time; (2) KSFs are different for different subjects of industry, goods, and markets; (3) KSFs will change with life cycle; (4) KSFs should be based on future trends[14]. Therefore, when studying KSFs, managers should consider the characteristics of the industry, as well as different times or life cycles of products, in order to obtain competitive advantages. Ho (1990) suggested that four methods are commonly adopted to confirm the number of KSFs, including regression analysis, factor analysis, Delphie, and AHP. AHP is the most common[15].

3 Research method

3.1 Research process

The research process is as shown in Figure 2. First, by collection, analysis, and review of primary literatures, the researcher generalized the analysis of KSFs in TTQS introduced in international hotels. Based on the theory, this study developed the dimensions and principles of a case interview outline. Secondly, it selected and invited the human resource supervisors of international hotels and TTQS assessment members. Based on the previous dimensions, it discussed and consulted experts regarding the questionnaire in order to validate the propriety of dimensions and adjust them to develop the final AHP questionnaire. Secondly, it selected the subjects. After retrieving the questionnaires, the researcher calculated the order of the factors, according to Expert Choice, to lead to the research results.

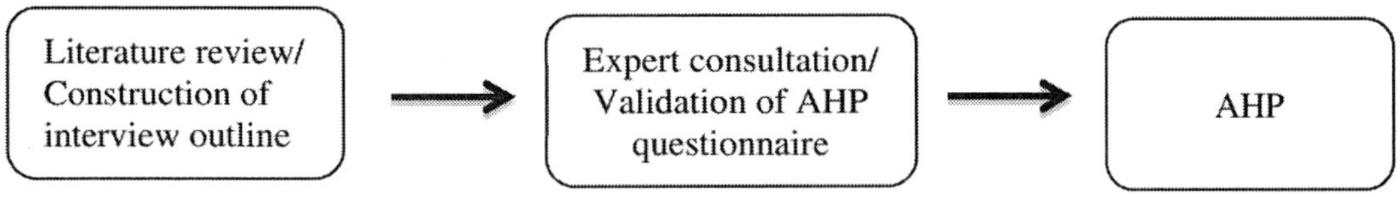

Figure 2. Research process

3.2 Subjects of experts

The interviews lasted from March 20 to June 10, 2014. There were ten subjects, including 3 human resource supervisors that have participated in TTQS assessment in international hotels, and 7 assessment members or consultants of international hotels. Among the assessment members, 2 have backgrounds in the food and hospitality industry.

3.3 AHP questionnaire distribution and effectiveness analysis

By literature review and hierarchical framework of interview results, this study established the AHP questionnaire. The purpose was to determine the weights and rankings of KSFs of TTQS introduced in international hotels. The questionnaires were distributed from July 1 to November 10, 2014. Regarding the number of questionnaires distributed, according to the suggestion of Saaty and Forman (1996), when dealing with decision-making by AHP, the number of experts should few, 5~15 is appropriate[16]. Therefore, in this study, questionnaires were distributed to 6 human resource supervisors that have participated in TTQS assessment in tourist hotels, and 9 assessment members or consultants of international hotels. A total of 15 questionnaires were distributed and return rate was 100%. The figures obtained from questionnaires were analyzed by Expert Choice. The valid return rate was 100%, and all questionnaires were tested by C.I. and C.R. Therefore, in this study, there are 15 valid questionnaires, and the screening principle is shown, as follows:

(1) C.I.: C.I.>0.1

(2) C.R.: C.R.>0.1

4. Data analysis and results

4.1 Literature review and questionnaire development

Table 1. KSFs of ISO10015 or TTQS introduced in enterprises.

Researchers	KSFs of ISO 10015 or TTQS introduced in enterprises	Research Method
Wang(2006)	1. HRD personnel should possess more knowledge and skills related to the process of training design 2. Support of management	Observation research
Lin(2006)	1. Stakeholders' active participation 2. Investigation of demands of business training 3. Construction standard of employees' competence 4. Planning and design competence of training course 5. Connection between training process and business performance	Observation research
Lin, Chin, Huang, Huang (2007)	1. Support of management 2.Employees' understanding and participation in the overall strategy of educational training of the company 3. Stakeholders' active participation 4. Continuous improvement 5. Systematic database	SPSS quantitative statistics
Peng(2007)	1.High-ranking supervisors' participation 2. Educational training plan should match operational strategy 3. Completeness of documents and records 4. Construction of standard business processes and training system 5. Person in charge must have complete competence and skill of competence analysis and design of training courses 6. Setting of KPI indicators 7. Training upon salary, examination, or promotions	Observation research
Chen (2008)	1. Support and determination of high-ranking supervisors 2. All employees' participation and identification 3. Educational training and employee quality 4. Validation and assistance of guidance unit 5. Customer orientation and communication 6. Usage of information system 7. Precise planning and thorough execution 8. Current system (such as ISO related system) and previous successful experience 9. Document control and writing 10. Continuous improvement 11. Confirmation of auditing and adjustment effect	Interview
Huang(2009)	1. Support and determination of high-ranking supervisors 2. Team members' execution ability of human capital 3. Connection between training demands and organizational performance 4. Training result enhances organizational performance 5. Project members have communication capacity 6. Employees' learning and action plan before training	Analytic hierarchy process (AHP)
Yeh and Tsai (2010)	1. High-ranking supervisors' high commitment and active involvement in IIP 2. One professional human resource employee in the organization as the key promoter of IIP 3. Successful and in-depth strategy communication in the organization 4. Members have ownership and sense of belonging to the institution, which effectively lowers employees' turnover rate 5. Organization can transform IIP into important management tool of	Observation research

	organizational reform and implement it 6. IIP system is further combined with organizational culture 7. Organization has continuous improvement	
Pai and Kung (2012)	1. Leaders participation, and implementation is from top to the bottom in order to enhance the determination of introduction. 2. Person in charge should have intention to learn knowledge related to TTQS. 3. Work Union should recognize market demand and communicate with lecturers regarding the design of courses. 4. Assistance from consultants of TTQS.	Analytic hierarchy process (AHP)
Huang et al.(2013)	1. Support and determination of high-ranking supervisors 2. Total employees' participation and identification 3. Educational training and employee quality 4. Validation and assistance of guidance unit 5. Customer orientation and communication 6. Usage of information system 7. Precise planning and thorough execution 8. Current system (such as ISO related system) and previous successful experience 9. Document control and writing 10. Continuous improvement 11. Confirmation of auditing and adjustment effect	

Table 2. Definitions and descriptions of key factors of TTQS introduced in tourist hotels.

Ultimate goal	Level 1	Code	Level 2
KSFs of TTQS introduced in international hotels	A. Organizational management	A1	High-ranking supervisors' high degree of support and participation
		A2	Active acquisition of TTQS related knowledge of persons in charge
		A3	Regular training organization experience before introduction of TTQS
		A4	High cognition of department supervisors regarding training responsibility
		A5	Complete documents related to training
		A6	Assistance from consultants of TTQS
	B. Training design	B1	Connection between competence demand of different levels and categories of posts of the course
		B2	Specific training goals before practice of training
		B3	Learners' recognition of course objectives before the course
		B4	Department supervisors' participation in course design
		B5	Lecturers' participation in course design
		B6	Connection between the demands of target learners and the organization
	C. Training execution	C1	Offering a positive environment for course learning by persons in charge
		C2	Employees' learning and action plan before training
		C3	Complete training record by systematic administrative management
		C4	Arrangement of assistants of courses in order to guarantee course execution
	D. Result assessment	D1	Complete review to continuously improve quality
		D2	Setting of KPI indicators
		D3	Training upon salary, examination, or promotions

Source: Compiled by this study

Results of data collection and literature of ISO10015 introduced in organizations, IIP of the U.K., and TTQS, are as shown in Table 1. Upon review of previous literature and key factors, regarding validity, before distributing questionnaires, this study authorized 3 human capital supervisors that participated in TTQS assessment, and 7 assessment members or consultants in international hotels, in order to examine the expert validity of the questionnaire. By interview or e-mail, this study deleted or added the descriptions of some items of the questionnaire. Finally four dimensions are developed: organizational management, training design, training execution, and result assessment (Level 1), including 19 sub-items (Level 2). The definitions and descriptions of KSFs are as shown in Table 2.

4.2 Relative weights of KSFs

By four dimensions, as obtained from previous interview results, including 16 KSFs, this study establishes the hierarchical structure of AHP and compares the relative weights in order to obtain the degrees of factors. The analytical results of the relative weights of the four dimensions, as well as the relative weights of the total KSFs, are as shown in Table 3. The results indicate 19 KSFs in four dimensions. Based on the suggestions of Saaty and Forman (1996), inconsistency of each level is C.R. (Consistence Ratio). When C.R. is lower than 0.1, consistency is acceptable. This study conducts analysis by Expert Choice, and obtains C.R.=0.017 of total dimensions. The value is lower than 0.1; hence, the consistency of matrix is high[16].

Table 3. Relative weights of the four levels and sub-factors of KSFs of TTQS introduced in tourist hotels.

Ultimate goal	Level 1	Weight	Level 2	Weight
KSFs of TTQS introduced in international hotels	A. organizational management	0.487	A1 High-ranking supervisors' high degree of support and participation	0.371
			A2 Active acquisition of TTQS related knowledge of persons in charge	0.225
			A3 Regular training organization experience before introduction of TTQS	0.127
			A4 High cognition of department supervisors regarding training responsibility	0.108
			A5 Complete documents related to training	0.046
			A6 Assistance from consultants of TTQS	0.123
	B. training design	0.286	B1 connection between competence demand of different levels and categories of posts of the course	0.122
			B2 Specific training goals before practice of training	0.257
			B3 Learners' recognition of course objectives before the course	0.111
			B4 Department supervisors' participation in course design	0.063
			B5 Lecturers' participation in course design	0.081
			B6 Connection between the demands of target learners and the organization	0.366
	C. training execution	0.127	C1 Offering a positive environment for course learning by persons in charge	0.243
			C2 Employees' learning and action plan before training	0.288
			C3 Complete training record by systematic administrative management	0.202
			C4 Arrangement of assistants of courses in order to guarantee course execution	0.267
	D. result assessment	0.109	D1 Complete review to continuously improve quality	0.235
			D2 Setting of KPI indicators	0.376
			D3 Training upon salary, examination, or promotions	0.389

Source: Compiled by this study

5 Conclusion and suggestions

5.1 Conclusions

In conclusion, the researcher will first generalize the findings and results discussion. Based on the rankings of the relative weights of the sub-factors of the dimensions, this study develops a weight framework for KSFs of TTQS introduced in tourist hotels (Table 4), as follows:

Table 4. Ranking of weights of KSFs of TTQS introduced in tourist hotels.

Ultimate goal	Level	Weight	Level 2	Weight	Ranking
KSFs of TTQS introduced in international hotels	A. organizational management	0.478	A1 High-ranking supervisors' high degree of support and participation	0.371	1
			A2 Active acquisition of TTQS related knowledge of persons in charge	0.225	2
			A3 Regular training organization experience before introduction of TTQS	0.127	3
			A6 Assistance from consultants of TTQS	0.123	4
			A4 High cognition of department supervisors regarding training responsibility	0.108	5
			A5 Complete documents related to training	0.046	6
	B. training design	0.286	B6 Connection between the demands of target learners and the organization	0.366	1
			B2 Specific training goals before practice of training	0.257	2
			B1 Connection between competence demand of different levels and posts of the course	0.122	3
			B3 Learners' recognition of course objectives before the course	0.111	4
			B5 Lecturers' participation in course design	0.081	5
			B4 Department supervisors' participation in course design	0.063	6
	C. training execution	0.127	C2 Employees' learning and action plan before training	0.288	1
			C4 Arrangement of assistants of courses in order to guarantee course execution	0.267	2
			C1 Offering a positive environment for course learning by persons in charge	0.243	3
			C3 Complete training record by systematic administrative management	0.202	4
	D. result assessment	0.109	D1 Training upon salary, examination, or promotions	0.389	1
			D2 Setting of KPI indicators	0.376	2
			D3 Complete review to continuously improve quality	0.235	3

Source: Compiled by this study

5.1.1 Importance rankings of the four dimensions

According to the analytical results of the dimensions at the first level, the rankings of importance are, as follows: organizational management (0.478), training design (0.286), training execution (0.127), and results assessment (0.109). The previous four dimensions have importance ranking of KSFs of

TTQS introduced in tourist hotels. The results show that according to all experts, organizational management is the priority of KSFs of TTQS introduced in tourist hotels. The result matches the findings of several scholars[9, 17-20]. Based on the previous result, when introducing TTQS, tourist hotels must first focus on reform of the organizational system, and change the previous model to authorize training for the business department. The training organization department should not be passive, and must establish a systematic training model with documents in order to dominate, assist, and audit training.

5.1.2 Analytical results of hierarchical weights of sub-factors

The second level includes 19 sub-dimensions. When analyzing weights, the researcher divides them into hierarchical weights and total weights. Hierarchical weights mean the relative importance of plans in the next level, as based upon the standard of the previous level, which shows the importance of plans, as based upon the criteria, meaning the prior vector upon the criteria of the previous level. Therefore, the sum of the hierarchical values upon the same criterion must be 1.0.

Regarding organizational management, the weight of high-ranking supervisors' high support and participation (0.371) is the highest; followed by: active acquisition of TTQS related knowledge of persons in charge (0.225), regular training organization experience before introduction of TTQS (0.127), assistance from consultants of TTQS (0.123), high cognition of department supervisors regarding training responsibility (0.108), and complete documents related to training (0.046). Based on research, TTQS is systematic educational training, and the introduction of TTQS means to introduce new ISO. Reform is required in organizational behavior. All organizational reform requires high-ranking supervisors' significant support and participation, and the results of this study indicate the same. In addition, understanding of TTQS knowledge of persons in charge is second. When introducing TTQS, they should not lack agreement or understanding. Therefore, regarding communication with high rank or introduction to employees, persons in charge should understand more about TTQS. Therefore, TTQS can be integrated with an original training model in order to avoid the obstacles of the introduction of a new system. In addition, when introducing the TTQS system, the assistance of a TTQS consultant and previous training of organization experience will accelerate the introduction.

Regarding training design, the weight of the connection between the demands of target learners and the organization (0.366) is the highest, followed by: specific training objectives before training (0.257), connection between competence demand of different levels and posts of the course (0.122), learners' recognition of course objectives before class (0.111), lecturers' participation in course design (0.081), and department supervisors' participation in course design (0.063). Based on the results, organizational demand is the priority of training for organizations, meaning the right learners and the right courses are the most important. Therefore, connection between the demands of target learners and the organization, specific training objectives before training, and connection between competence demand of different levels and posts of the course are the top 3 of the training design. In addition, in order to meet the competence gap of the business department by training, we suggest that lecturers and department supervisors can participate in course design.

Regarding training execution, the weight of the establishment of employees' learning and action plans before training (0.288) is the highest, followed by: Arrangement of assistants of courses in order to guarantee course execution (0.267), offering a positive environment for course learning by persons in charge (0.243), and complete training record by systematic administrative management (0.202). Based on the results, the establishment of employees' learning and action plan before training is the most important. This is consistent with the connection with organizational demand in training design. Secondly, arrangement of assistants in the course, as well as offering good training locations, will result in successful training.

Regarding the results assessment, the weight of training upon salary, examination, or promotions (0.389) is the highest, followed by: setting of KPI indicators (0.376) and complete review to continuously improve quality (0.235). According to the results, training upon salary, examination, or promotion will enhance the learners' training motivation. In addition, by setting KPI indicators, training and business departments will have a reference, and learn how to arrange a proper labor force and time. However, the results demonstrate that the score of continuous quality improvement by complete review is the lowest, which shows that enterprises pay attention to short-term effectiveness of training, and is the gap with TTQS that focuses on continuous improvement.

5.1.3 Analytical results of total weights of sub-factors

Table 5. Overall relative weights and rankings of KSFs of TTQS introduced in international hotels.

KSFs of TTQS introduced in international hotels	Overall weights	Ranking
A1 High-ranking supervisors' high degree of support and participation	0.177	1
A2 Active acquisition of TTQS related knowledge of persons in charge	0.108	2
B6 Connection between the demands of target learners and the organization	0.105	3
B2 Specific training goals before practice of training	0.074	4
A3 Regular training organization experience before introduction of TTQS	0.061	5
A6 Assistance from consultants of TTQS	0.059	6
A4 High cognition of department supervisors regarding training responsibility	0.052	7
D1 Training upon salary, examination, or promotions	0.042	8
D2 Setting of KPI indicators	0.041	9
C2 Employees' learning and action plan before training	0.037	10
B1 Connection between competence demand of different levels and posts of the course	0.035	11
C4 Arrangement of assistants of courses in order to guarantee course execution	0.034	12
B3 Learners' recognition of course objectives before the course	0.032	13
C1 Offering a positive environment for course learning by persons in charge	0.031	14
C3 Complete training record by systematic administrative management	0.026	15
D3 Complete review to continuously improve quality	0.026	16
B5 Lecturers' participation in course design	0.023	17
A5 Complete documents related to training	0.022	18
B4 Department supervisors' participation in course design	0.018	19

Source: Compiled by this study

Total weights mean relative importance at the same level regarding the previous level of the upper level. It refers to the prior vector of the total concern of the total objective. Hence, the sum of all criteria of the next level or prior weights of the plan should be 1. Table 5 shows total relative weights and ranking of the sub-dimensions at the second level. Based on Table 5, the top 5 key factors among the 19 sub-dimensions at the second level are high-ranking supervisors' high degree of support and participation (0.177), active acquisition of TTQS knowledge by persons in charge (0.108), connection between the demands of target learners and the organization (0.105), specific training objectives before training (0.074), and regular training organization experience before introduction of TTQS (0.061). The rankings show that "high-ranking supervisors' support and participation" is the most important factor of organizations when introducing the TTS system. This result matches the findings of many scholars regarding the key factors of ISO 9001 or ISO10015 introduced by enterprises. In addition, recognition of the TTQS knowledge of persons in charge is the second, which matches the sixth indicator "capacity of persons in charge" in the TTQS list of the business version.

5.2. Suggestions

Based on previous results, this study proposes the following suggestions regarding TTQS introduced in international tourist hotels:

(1) High-ranking supervisors must have a high degree of support and participation to result in effective introduction of the training system.

(2) Person in charge of training must enhance personal competence related to TTQS or human resources, which will enhance implementation of the TTQS project, completeness of the training design, and fulfillment of training execution and effectiveness.

(3) The training design should be based on training objectives upon organizational demand. Before training, the members should recognize the training purpose.

(4) When training effectiveness is based on salary, examination, or promotion, it can effectively increase training motivation.

(5) TTQS consultation is a useful resource, and its effective use will accelerate the introduction of the TTQS training system.

References

1. Tourism Bureau, *MOTC Administrative Information System, Monthly Report on Tourist Hotel Operations in Taiwan* (Tourism Bureau, Taiwan, 2012)
2. R.C. Mill and A.M. Morrison, *The Tourism System: A Introduction Text* (Englewood Cliffs, USA, 1985)
3. M. Alazmi and M. Zairi, Total Quality Management & Business Excellence, **14** (2), 199-204 (2003)
4. Y.Y. Lin, Master's thesis, A study of the relationship between employees' educational training and job satisfaction in international tourist hotels in Taiwan (Kaohsiung, Taiwan, 2009)
5. L.L. Wang, Master's thesis, The influence of training on organizational performance: empirical samples from financial industry in Taiwan (Kaohsiung, Taiwan, 2006)
6. C. Bulut and O. Culha, Int. J. Training and Development, **14** (4), 309-322 (2010)
7. Business Training Network, Introduction of project of training quality system, Retrieved from: http://ttqs.evta.gov.tw/ (2013)
8. S.Y. Huang, Master's thesis, KSFs study on TTQS introduced in enterprises (Chia-yi county, Taiwan, 2009)
9. T.S. Huang and J.Y. Cai, East-Asia Rev., **467**, 65-80 (2010)
10. J.S. Lin, Employment Security, **5** (2), 44-51 (2006)
11. A. Aaker, *Strategic Market Management* (New York, USA, 1984)
12. B. Brotherton, E. Heinhuis, K. Miller, et al., Journal of Services Research, 2, 47-78 (2002)
13. Eindt, Jeffcoate, *External Relationships and the Small Business: A Review of Small Business Alliance and Network Research* (Chappell, Mexico, 2002)
14. C.T. Chen, Master's thesis, An empirical research of the key success factors for chain administration— english supplements school industry as an example (Taipei City, Taiwan, 2011)
15. Y.C. Ho, *Practical Marketing Management* (Hwa Tai, Taiwan, 1990)
16. T.L. Saaty and E.H. Forman, *The Hierarchon: A Dictionary of Hierarchies* (Pittsburgh, USA, 1996)
17. W.T. Lin, R.S. Chin, M.R. Huang, and C.C. Huang, *The 43th Annual Meeting of Chinese Society for Quality and 13th National Quality Management Conference* (Hsinchu, Taiwan, 2007)
18. C.W. Yeh and S.T. Tsai, Report on demonstration of IIP international human resource equality system (Workforce Development Agency, Taiwan, 2010)
19. Tourism Bureau, *MOTC Administrative Information System, Monthly Report on Tourist Hotel Operations in Taiwan* (Tourism Bureau, Taiwan, 2013)
20. C.C. Huang, M.C. Lin, S.H. Lin, M.F. Huang, and Y.H. Wang, Innovation and Operational Management, **4** (1), 15-28 (2013)

A study of the correlation between self-confidence and professional achievement of designers

Rain Chen[1,a], Yu-Tzu Yang[2], Li-Chou Chen[3], Chia-Hui Feng[2] and Gen-Ming Guo[4]

[1]*Department of Visual Communication Design, Southern Taiwan University of Science and Technology, Tainan City 710, Taiwan*
[2]*Department of Creative Product Design, Southern Taiwan University of Science and Technology, Tainan City 710, Taiwan*
[3]*Department of Styling and Cosmetology, Tzu Hui Institute of Technology, Pingtung County 926, Taiwan*
[4]*Department of Information Management, Southern Taiwan University of Science and Technology, Tainan City 710, Taiwan*

Abstract: This study mainly investigated which mental state of designers, self-confidence or sense of inferiority, has positive effects on professional design achievement. This study attempted to find if there is correlation between designers' self-confidence or sense of inferiority and their professional achievement. With regard to the tendency of designers' psychological state, the Rosenberg Self-Esteem Scale was used to measure designers' self-confidence, and statistical computations were made based on gathered data. This study used correlation analysis to find if confidence level of 46 seniors of Design Department is relevant to their professional achievement. The results of the study showed that confidence level of designers has a slight correlation to professional achievement. Factors leading to the study's findings may be the small amount of sample analyzed or may be the reason that the Rosenberg Self-Esteem Scale only detected the current level of self-confidence and could not give a proper feedback on designers' self-confidence during the entire semester. In the future, the results of this study can be considered as a pre-test experiment for a more complete study, and it is expected that results of the study can be for design educators' reference.

1 Introduction

With the change of time and the trend of globalization, there has been a growing need for designers to seek cross-domain cooperation. Demands for communication among heterogeneous talents of various types are growing rapidly, and a designer's job has changed from an individual's task to a heterogeneous, cross-field collaboration among multiple persons. With the increasing diversification of the design industry, development of an excellent product must rely on the incorporation of various value chains to achieve marketing success.

When product designers are in the process of creative thinking, brainstorming or discussion with one another to refine ideas is often required. It is also necessary for a designer to acquire sufficient knowledge of the relevant industries. Global design trend has also gradually shifted from practical execution to the phase of design thinking. In Asian design community, Taiwanese corporations tend to

[a] Corresponding author : ocean@stust.edu.tw

DOI: 10.1051/matecconf/201711901024

work on an OEM or ODM basis so that designers' performance of creativity is often subject to budget. Most designers in small and medium-sized businesses in Taiwan do not get duly paid for their professional contributions. One of the reasons for this could be that there are much emphasis on practical skills, hand-drawing techniques, and computer-aided graphics in the designer training while the development of work attitude and character as a designer has been somewhat ignored.

Colleges and universities seem to have based the professional education of product design solely on skills training. Courses with relation to the cultivation of character and self-identity are generally missing in design education. Besides, humanities and arts education, which are closely related to character development, has not received much attention in Taiwan's technological universities, even though personal traits and interpersonal communication are essential to a successful career for designers.

According to studies by other researchers, optimistic and positive characteristics have great impact on designers' professional achievement. Such personality traits are also applied to industrial designers, but they are the most difficult elements to develop in current education system [1]. Jhang has weighed the capability requirements for industrial design and developed a chart, as shown in Figure 1, in which personality traits rank only second to product planning.

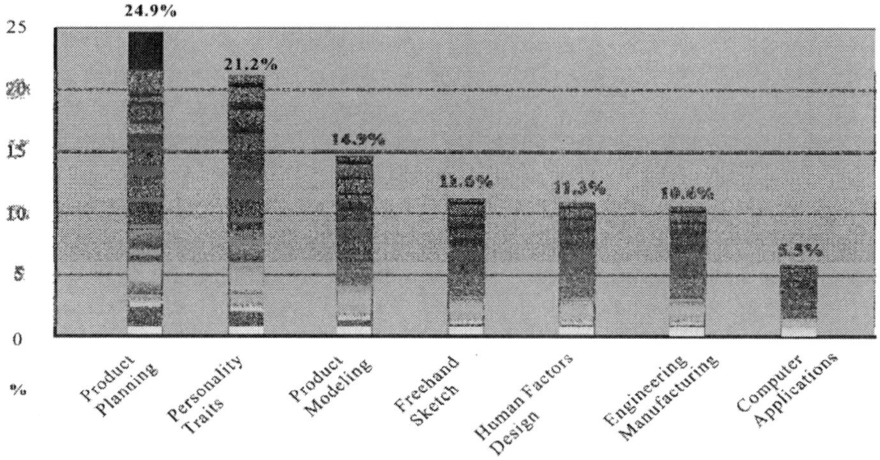

Figure 1. Weights of capability requirements for industrial design.

It is also pointed out in a relevant essay that personality traits are the top requirement for industrial design, followed by the capability of image formation and innovative design. The research results, the essay, and the reality all comply with what scholars have argued: personality traits are the most important quality of designers or people of other professions in modern times. Such traits along with communication skills and expertise and other skills essential for industrial designers are important weapons for competing in the industry [2]. It is stressed in the essay that, in the design profession, personality traits are more important in comparison with design expertise.

Following an argument by Seligman, Alan Carr compared the functions of negative and positive emotions. It is indicated that negative emotions such as fear or anger can serve as a defensive mechanism in evolutionary adaptation while positive emotions help expand the scope of people's concern so that they can sense physical and social environments in a more comprehensive way, accept new ideas, and take new actions, which all lead to a more creative and fruitful outcome [3-4]. According to the study by Tugade & Fredickson, an individual with positive emotions tends to conduct more diversified thinking and is more likely to be inspired in terms of creativity [5].

2 Method

Positive psychology has always been an important concept for psychologists in their study of the connection between people's thinking patterns and their achievements. It has also received much attention in organizations and in the field of human resource studies. In assessment dimensions, "personal proficiency" of designers is most valued by industries, with an average score of 4.53, which is above the level of bearing "significant" importance. "Professional abilities" come next with an average score of 4.33.The average score for competence in 2D software and knowledge in ergonomics and in marketing is between 3.9 and 3.0, indicating "fair" to "significant" importance. The rest of the assessment items all above "significant" level. "Background experience" gets an average score of 3.54, with "certification" given the least concern., The reason could be a lack of professional certification system in industrial design in Taiwan, and consequently, industries pay less attention to junior designers' certification [6].

In the book Get Lucky, author Xuan Liu broached studies related to self-efficacy. After studying patients of phobia in the 1960s, Dr. Albert Bandura, one of the most influential social psychologists in the 20th century, published his theory of self-efficacy in the 1970s. According to Dr. Bandura, self-efficacy stands for "self-confidence concerning whether one can cope with a certain situation." Nowadays, self-efficacy is one of the most popular issues in positive psychology, and its effect has been proven by numerous researches, showing significant association particularly with education, workplace performance, self-management, and exercise and fitness [7].

The topic of this research is "the correlation between self-confidence and professional achievement of designers". It discussed whether a strong correlation between designers' self-confidence and their professional achievement exists. In Change by Design: How Design Thinking Transforms Organizations and Inspires Innovation, Tim Brown suggested that design thinking is an inner exploration of one's true nature [8]. Today, most projects are far too complicated for any individuals, and the go-it-alone style may soon cause a designer to be weeded out. Even in more conventional fields of industrial design and graphic design, teamwork has become normal. However, it takes patience to organize a project team with people from different backgrounds and fields, and it is not easy to have a flair for recognizing people who possess confidence in their professional abilities and are willing to improve themselves. The essay showed which are managers more concerned about: self-confidence or sense of inferiority. Luh indicated that, from corporations' viewpoint, industrial designers should possess proficiency (attitude) and abilities, of which the former encompasses team spirit, quality of work, self-confidence, curiosity, artistic proficiency, patience, and perseverance [9]. In The Design of Everyday Things, Donald A. Norman also suggested that positive thinking is beneficial for creative thinking as it keeps people moving in the face of failure [10].

The aim of the research was to discuss that, in terms of the design profession, which personality trait can provide positive advantages? Self-confidence or sense of inferiority? Due to practical difficulty in defining personality and proficiency and an intention to implement quantified statistics in the research, Rosenberg's scale was used to analyze the level of designers' self-esteem and thereby to measure their self-confidence. The results of the analysis were then cross-matched against their professional achievement.

The focus of this study was on designers' self-confidence and sense of inferiority, which were measured according to the scores they achieved in the survey and the percentage either psychological state made up. According to the research assumption, subjects with higher self-esteem were deemed to be self-confident or super self-confident, subjects with lower self-esteem were deemed to be the ones with stronger sense of inferiority, and subjects with normal self-esteem were deemed to be average. The research attempted to discuss whether designers' self-esteem has an impact on their professional achievement or not.

The research used grades on graduation projects done by 46 seniors of Design Department. Since a graduation project is the ultimate evaluation of a student of Design Department, it is appropriate to use the grades on the projects as a manifestation of professional achievement. Correlation analysis was implemented in the research to cross-match designers' self-esteem (self-confidence and sense of inferiority) against their professional achievement (grades of graduation projects) and probe any correlation in between. It is expected that the outcome of the research can serve as useful information for design educators' reference regarding future curriculum adjustments.

3 Result

Statistical correlation coefficient and correlation analysis were implemented as research instruments, with an attempt to verify the correlation between designers' self-esteem and their professional achievement. Moreover, it was expected to prove that positive correlation exists between self-esteem and professional achievement. Regarding statistics on gender, 18 of the 46 subjects were males (39.1%) and 28 were females (60.9%), as shown in Table 1. The reason to choose college seniors as the research subjects was to examine if there is correlation between their grades on graduation projects and the level of their self-esteem.

Table 1. Statistics on subjects' gender.

	Frequency	Percent	Valid Percent	Cumulative Percent
1 (Male)	18	39.1	39.1	39.1
2 (Female)	28	60.9	60.9	100.0
Total	46	100.0	100.0	--

The subjects' scores on the Rosenberg Self-Esteem Scale and their professional achievement (grades on graduation projects) were collected, as shown in Table 2. The ranges of self-esteem scores were: (I) 10-15 points, indicating a sense of inferiority; (II) 16-25 points, indicating average self-esteem; (III) 26-35 points, indicating self-confidence; (IV) 36-40 points, indicating super self-confidence. The descriptive statistics is shown is Figure 2.

Table 2. Subjects' scores on self-confidence and professional achievement.

subject	Q1	Q2	Q3	Q4	Q5	Q6	Q7	Q8	Q9	Q10	Self-esteem Score	Range	Grade on Grad. Project
1	3	3	1	2	1	3	3	2	1	1	20	II	69
2	4	2	3	2	3	3	3	2	3	3	28	III	71
3	3	3	3	4	3	3	2	2	2	3	28	III	78
4	4	4	2	2	2	2	2	1	1	1	21	II	65
5	3	1	3	2	2	2	1	2	4	4	24	II	65
6	3	3	2	2	2	4	3	2	2	2	25	II	66
7	4	3	4	4	3	4	4	2	4	4	36	IV	67
8	4	3	3	3	3	3	4	3	4	4	34	III	72
9	3	2	3	3	4	3	3	2	3	3	29	III	74
10	4	3	4	3	3	4	3	1	4	4	33	III	74
11	2	2	3	2	2	2	2	2	2	2	21	II	70
12	3	2	4	2	3	3	3	2	3	3	28	III	70
13	3	3	2	3	2	2	2	2	1	1	21	II	67
14	4	3	4	3	2	4	4	1	3	3	31	III	70
15	3	3	4	3	4	3	3	2	3	3	31	III	71
16	3	3	1	2	2	3	4	2	1	1	22	II	74
17	3	3	4	3	4	3	3	2	3	3	31	III	71
18	4	3	3	3	1	4	4	1	4	4	31	III	70
19	3	3	4	3	3	3	3	2	4	4	32	III	71
20	3	3	3	3	3	2	2	2	2	3	26	III	66
21	3	3	3	2	2	3	3	2	2	2	25	II	66
22	4	3	3	2	3	3	2	2	2	2	26	III	78
23	3	3	4	3	3	3	2	2	2	2	27	III	79
24	3	3	3	3	2	2	3	1	2	1	23	II	75
25	3	3	3	3	3	3	3	3	3	3	30	III	71
26	3	3	3	2	3	2	2	4	2	2	26	III	69

27	3	3	3	3	3	3	3	2	3	3	29	III	66
28	3	3	3	3	3	3	2	2	2	3	27	III	71
29	2	3	3	2	2	2	2	1	2	3	22	II	69
30	3	2	3	3	3	2	2	1	4	3	26	III	71
31	3	3	3	3	3	3	2	2	3	4	29	III	67
32	4	3	3	3	3	3	2	2	3	3	29	III	70
33	4	3	4	4	3	3	3	2	2	2	30	III	79
34	4	3	4	3	3	3	4	2	2	2	30	III	75
35	4	3	4	3	4	4	4	2	4	4	36	IV	71
36	3	3	3	3	3	3	3	2	2	2	27	III	71
37	1	4	4	1	4	4	4	4	4	4	34	III	68
38	4	3	3	3	3	2	2	2	3	3	28	III	71
39	4	4	2	3	2	3	2	1	1	2	24	II	65
40	3	3	3	3	3	3	3	2	3	2	28	III	78
41	2	3	3	2	3	3	3	2	2	3	26	III	70
42	3	3	3	3	2	3	3	2	3	3	28	III	66
43	3	3	3	3	3	3	3	2	3	3	29	III	71
44	3	3	3	3	3	3	3	2	2	2	27	III	71
45	4	3	4	3	2	3	3	2	3	3	30	III	75
46	3	3	3	3	3	3	3	2	3	3	29	III	79

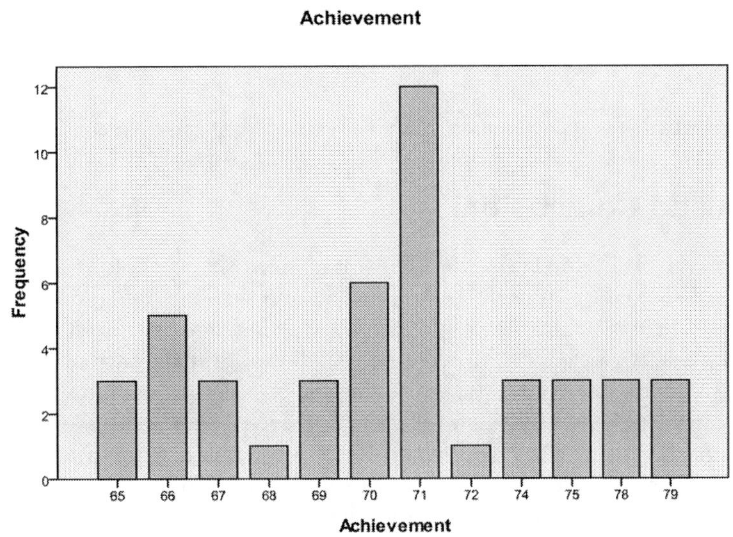

Figure 2. Descriptive statistic on subjects' self-esteem and their professional achievement.

The outcome of the correlation analysis is shown in Table 3 and Figure 3, according to which the Pearson Correlation Coefficient between self-esteem and professional achievement was 0.195. It is therefore revealed that no significant correlation exists between subjects' self-esteem and professional achievement.

Table 3. Correlation analysis between scores on the Rosenberg Scale and professional achievement.

Correlation	Items	Rosenberg	Achievement
Rosenberg	Pearson Correlation	1	.195
	Sig. (2-tailed)		.195
	Sum of Squares and Cross-products	678.370	136.283
	Covariance	15.075	3.029
Achievement	Pearson Correlation	.195	1
	Sig. (2-tailed)	.195	

DOI: 10.1051/matecconf/201711901024

	Sum of Squares and Cross-products	136.283	722.804
	Covariance	3.029	16.062

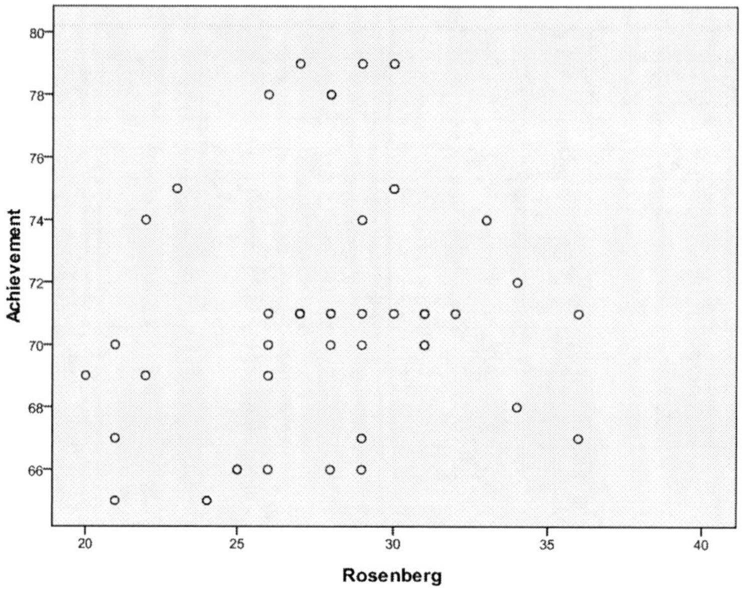

Figure 3. Sample distribution of the Rosenberg Scale and professional achievement.

4 Discussion and Conclusion

Over 100 years ago, William James laid the foundations of psychology in America. He acted on his theories and kept reminding himself to be practical and realistic "pursue the cash value of truth in experiential terms." The cash value he sought for his students does not exist in physical currencies or general achievements and glories, but in what he called the ultimate currency. By ultimate currency, he referred to the end of all means, which is happiness. The rate of depression in the U.S. has increased by ten times since the 1960s, and the average age of first onset is 14.5, compared with 21.5 in the 1960s. According to a study conducted by several universities in the U.S., nearly 45% of college students had learning difficulties due to their suffering from severe depression [11].

Recent studies have revealed that positive emotions work better than negative emotions in terms of effect on the immune system. They can also help improve health. Fredrickson suggested that positive emotions such as happiness, interest, satisfaction, confidence, and love can facilitate actions and broaden people's momentary thought-action repertories, which are of help to building their enduring personal resources and enhancing the sense of happiness [12]. Studies have also shown that individuals with positive mental attitude (greater self-confidence) display more creativity than those with lower self-esteem, which means that positive mental attitude can be predictive of creative performance [13].

However, the outcome of the correlation analysis in the research showed that no significant correlation exists between either high self-esteem or low self-esteem and professional achievement. A possible explanation for such an outcome is that too few samples were available for analysis or that scores of the Rosenberg Self-Esteem Scale represented only designers' self-esteem at the moment instead of their overall confidence throughout the semester. The survey, nevertheless, can be viewed as a pre-test, and more comprehensive studies can be conducted in the future. It is also hoped that the results of such studies can provide useful information for design educators.

Some studies indicated that personality traits of design students are different from those of students of other departments. During training process, designers are asked to show the ability to express themselves. It seems that self-confidence has become one of the important factors in designers' success. Though the results of this research showed that there is no significant correlation between self-esteem and professional achievement, researchers may continue to pay attention to this issue in the future and conduct studies of correlation between designers' level of depression and their professional achievement, and which results are believed to be positively beneficial for the improvement of design education.

Acknowledgements

This work was partially supported by the National Science Council, Taiwan, under the Grant No. MOST 104-2221-E-218-015-MY2.

References

1. C.J. Jhang, R. Chen, T.Y. Sung, C.C. Lin, Social contacts and achievement of design professionals, Psychology, **5** (8), 478-483 (2015)
2. P.H. Kao, A proposal of pre-service self-assessment scale for industrial designers, Thesis of Ling Tung University (2012)
3. A. Cara, *Positive Psychology: The Science of Happiness and Human Strengths*, Routledge (2004)
4. M.E.P. Seligman and M. Csikszentmihalyi, Positive psychology: an introduction, American Psychologist, **55** (1), 5-14 (2000)
5. M.M. Tugade and B. L. Fredrickson, Resilient individuals use positive emotions to bounce back from negative emotional experiences, Journal of Personality and Social Psychology, **86** (2), 320-333 (2004)
6. Y. Hsu, W.C. Chang, and V. Yang, A study on the recruitment and job performance of newly recruited product designers and their implications in design education, The International Journal of Arts Education, **5** (1), 71-109 (2007)
7. X. Liu, *Get Lucky*, Bookzone (2014)
8. T. Brown, *Change by Design: How Design Thinking Transforms Organizations and Inspires Innovation*, HarperBusiness (2009)
9. D.B. Luh, The cultural creation industries plan and the role of industrial design in Taiwan, inter-collaboration in design education and research, KAIST International Design Symposium, Seoul, Korea (2004)
10. D.A. Norman, *The Design of Everyday Things. Revised and Expanded Edition*, Basic Books (2014)
11. B.S. Tal, *Happier: Learn the Secrets to Daily Joy and Lasting Fulfillment*, McGraw-Hill Education (2007)
12. B.L. Fredrickson, The role of positive emotions in positive psychology, American Psychologist, **56** (3), 218-226 (2001)
13. C.H. Su, The study on the relationship between positive psychology and creativity for the university students, Thesis of University of Taipei (2007)

Combining zmet and mec in study of user expectations for ideal catering customer service app

Ton Chaing[1,a], Hsin Rau[1], Wei Jung Shiang[1] and Jon Luen Chiang[2]

[1]*Department of Industrial and Systems Engineering, Chung Yuan Christian University, Taoyuan City 32023, Taiwan*
[2]*Department of Electrical & Electronic Engineering, City University London, Northampton Square, London EC1V 0HB, United Kingdom*

Abstract. The designing and marketing models of general catering customer service APP are made from business owners needs or from commissioned design company's thoughts. The ZMET and MEC technique are used to understand the consumers more properly. Consumers usually will not purchase a product for just the physical content or functionality of the product but will do so because of other reasons, such as what the product presents socially, how it affects their emotions and more In this context, it is necessary to identify the meaning better. This research attempts to understand how customer satisfaction in the context of the catering industry could be understood better by identifying product and company attributes. The catering company can then develop a unique style of their own, so consumers can better understand the ethos of the company, their culture and style. According to consumers' experience, we conclude 5 user expectations of catering customer service APP, including Fans Evaluation, Online-to-Offline Consistency, Timeliness, Smart Order, and Customized Logistics. In this paper, the results of these information show that the user expectations of catering customer services APP will become the effective scientific reference and different patterns of cognitive communities.

1 Introduction

The purpose of this research is to use the ZMET technique along with MEC in order to draw mental diagrams per person and then analyze the mental diagrams in order to develop constructs defining the individual and make recommendations for the catering company. This research makes use of the MEC method and summarizes main constructs from the HVM diagrams that were developed for 12 interviewees. Five key factors are considered for the construct development which is that of Timeliness, Fans Evaluation, Smart Order, Customized Logistics and Online-to-Offline Consistency.

ZMET constructed for different target audiences can help establish the segment's consensus map. This way the marketing group will be able to determine the type of strategy needed to communicate with them. As the ZMET technique allows the audience to tap into their own ideas and develop pictures for just the concepts that they associated with the catering service experience, the researcher is hence able to understand the mental model behind the user opinion construction.

[a] Corresponding author : tchiang.0514@ttri.org.tw

The MEC technique used with the ZEMT technique will be useful for understanding the cognitive hierarchies of consumers when it comes to product consumption. The Means end chain theory MEC has been defined in the research literature on consumer oriented food products as being a more relevant way to understand consumer satisfied designs [1, 2]. Product consumption motives are better understood as the marketer and the product maker will be able to better understand 1) how product attributes are linked to the self-relevant consequences of consumption of the person, their personal life values, goals and more and 2) to identify the hierarchical cognitive structures which present the consumer's mind. This consumer cognitive structure hierarchy will be useful to evaluate and understand how consumers might assess alternative products, why they make a choice to buy a product and why they might choose to buy another product instead. The higher order reasoning constructs from the mental models created from ZMET image collections and the story telling processes are useful to understand salience of some criteria over others [3, 4]. The functional and the psychological consequences of the product and how it connects to meet the consumer goal or value are hence very significant in this study.

1.1 Interview questions

Some of the interview questions to be used are with respect to understanding why a certain image was chosen, why a certain image was chosen over another image, etc.

Interview question I: Why do you select this picture? What kind of feelings and thoughts about the image of catering industry customer service APPs is represented by your selecting this picture?

Interview question II: Would you like us to know what kind of actual benefits or psychological satisfaction this choice brings to you. Why is the choice or construct valuable to you? Is the value in this part concerned with your personal preference? How about the significance of this value to you?

The interview questions are only guidance, the story telling and the mental models are basically encouraged from the consumer end.

2 Consensus and constructs

The number of interviewee consensus with respect is for consumer experience on different constructs below the table 1 shown. The constructs on which there is maximum customer consensus is seen to be that of clean environment and brand reputation. The clean environment consensus indicates that customers connect a favorable experience with a catering company that can project the same. Brand reputation, such as that of the history of catering, the positive experiences that the customer has with the brand, the reach of brand, its marketing, its CSR activities are also considered to be high on the consensus.

Table 1. Consumers Experience of the Catering Services.

Number of Interviewees based on consensus on the Catering Services	Consumers Experience
12	Clean enviroment, Brand reputation
10	Courteous and well trained staff, High rating
9	Green packaging, Polite service, Arrangements, Brand exposure, Food freshness
8	Food taste, Complaint feedback fast, Timely order, Food design
7	Service concept, Service and Product presentation, Knowledge on food practices
5	Unique presentation of food and related services, Discount, Lively ambience
4	Eagemess to serve, Assurance for customers on time and quality, Shared knowledge on food sourcing, Fun service
3	Timely execution, Overall satisfaction, Customer awareness, Fresh orders

Courteous well trained staffs, food taste, service concept, presentation of food, eagerness to service and more are listed under moderate consensus. This collection of data is interesting, as customers are seen to associate clean environment, staff, rating and brand reputation as being their highest expectations with respect to catering. In terms of timely execution, freshness of orders or assurance on time seems to be listed with low consensus. This is a paradox because the most basic elements that a consumer would look for in a catering service are that it should be timely and the products must be fresh, however highest consensus is for brand reputation. Now this need not exactly be a paradox. Brand reputation can be understood to be a more abstract construct or a complex construct. A brand would have good reputation only if the brand is seen to have fulfilled some of the basics such as that of clean and fresh food, timeliness in service and more. So the high consensus rate for constructs such as brand reputation only indicates that the consumers would have taken the most basic constructs as being default requirements for the catering service to fulfill.

3 Experiment of HVM

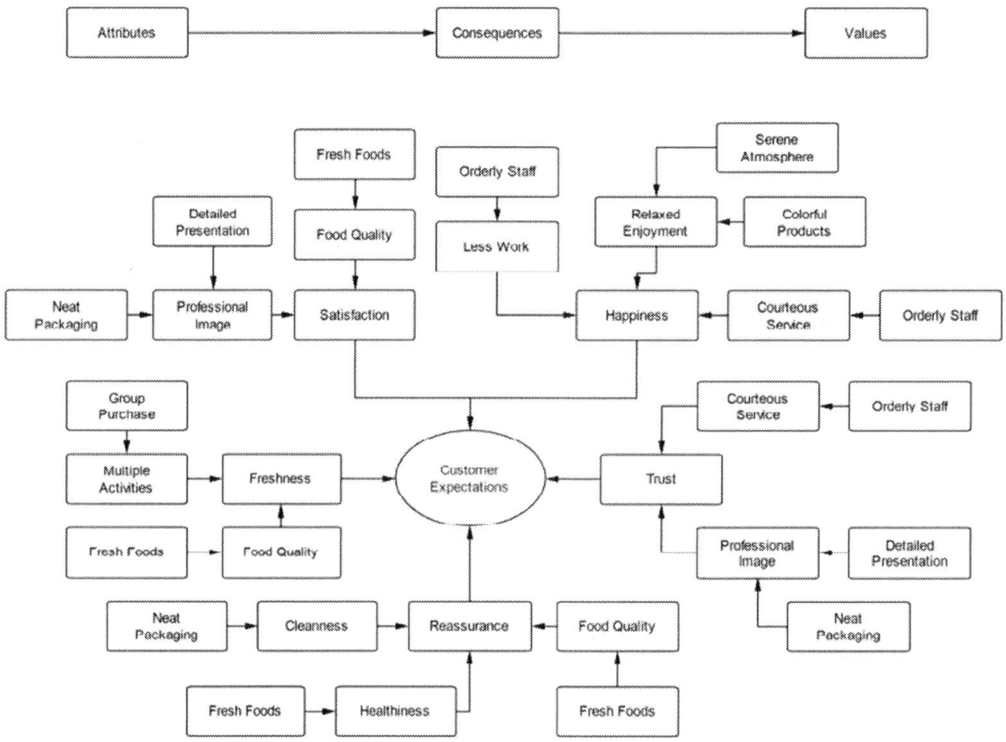

Figure 1. Model of HVM (Hierarchical value map).

The HVM graph has been constructed. The HVM graph is critical to understanding how the attributes, the consequences and the values are connected [5, 6]. Now as noticed in the case of Maslow's hierarchy, the physiological needs have to be satisfied first, so a person might have a need for say food and shelter. Before attaining a second level, the first level has to be satisfied. Now in the case of the figure 1, there are concrete and abstract attributes and these have to be considered before their consequences could be understood. So in the catering services, neat packaging, fresh foods and more are some of the most basic elements. They have to be met irrespective of anything. Secondly, these attributes have consequences which are the more informative end compared to the meta data. These are functional and psychological implications and hence are one order up from just discussing about packaging. The quality of food would have a functional implication. And on the other hand, the way the food looks, its texture, its colour or its aroma would have a psychological implication.

Now using these elements, in the figure 1, the value is understood. The value is at the meta level, where the data state for the app has crossed into the information stage. This is more of the higher end in the catering service, such as the management whose concerns are about value and adding value. Similarly, it can be said that the higher level value element of satisfaction and happiness is guaranteed because of courteous service. Where the service is courteous, then it automatically follows that the person requesting the services is satisfied, he is happier. On the other hand, where the courteous service is lacking, then the higher level values for the customer such as that of satisfaction and happiness is not achieved. In essence this is like a neural network where the firing of neurons at the consequences levels will trigger the value level and where consequences level is not triggered then there will be failure at the (customer) value level.

Marketing would now not look at the basic levels such as that of the constructs, attributes or the consequences, and instead tweaking for successful marketing is now being carried out at the values level. Further iterations in the creation of the graph will fill in any inconsistencies.

4 Customers service App

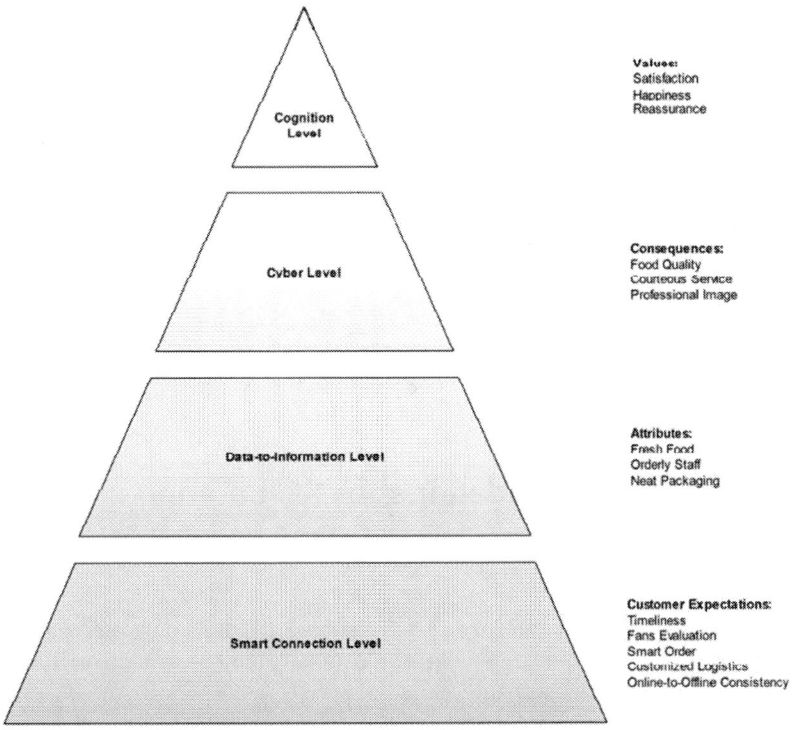

Figure 2. App value proposition process.

The theory of this figure is from Maslow Hierarchy. The Maslow hierarchy of needs or motivation proposed by Abraham Maslow and it has relevance in today management system [7]. It has paved a way to understand about the human psyche and helps in analysing about the people and their requirements. It basically helps in identification of the basic requirements of the people and provides a management style for the people. It also plays an important role in developmental psychology of the people. Maslow uses some the basic tenets of human motivations to develop a robust principle of people management. There are many hierarchical levels to Maslow, this has been divulged in detail in the following. Now similar tenets could be understood in the APP value proposition style.

The first level in the hierarchy is physiological needs. This is the basic requirements that are required for the humans to survive. It discusses in detail about the air, water and metabolic organic requirements of the humans. Apart from this, it also factors in the organic requirements of the human and basic principles of food and shelter. The second stage is the safety requirements of the people. It divulges about the safety requirements both from mental and physical safety. This is a natural security instinct of the people. There are many variables such as finance, personal security. Health and adverse impacts are considered in this analysis. The third hierarchical stage is the love and belonging stage of the people. When the basic tents are met, people aspire for the levels above them. They will want to form groups but at the same time want to affiliate with the people in the next level. This is a basic requirement of wanting to be accepted and valued by others. The people often engage in a profession or hobby to gain recognition. The people want to respect within their peer groups or society around them in order to be recognized.

The esteem levels, inferiority complexes are addressed in these levels. Once these basic tenets are achieved the next stage if the self-actualization state. In this stage, the person starts to perceive and focus on their exact needs and specifications. This is a basic human requirement of wanting to belong to a certain self-requirement that is beyond the scope of the tangible levels. The next level was prescribed by Maslow as people wanting to reach high levels of spirituality. Similar to the levels in Maslow's hierarchy, consider how there are different levels in the case of the app construction. At the base, the consumer oriented constructs, are present which needs to be satisfied for user. These are the basic constructs and similar to the physiological needs, nothing could be analysed further if these are not met. In the figure 2, it shows that these consumer constructs include the elements of timeliness, fans evaluation, smart order, customized logistics, and online-to-offline consistency. The second level is that of the attributes that include the data to information level offering at which point the raw data from the consumer end is already being transformed to information of meaning. At the cyber level, the storage of such information is construed such that at the cognition level the management would be able to understand the values breakdown.

Now there are many criticisms to this Maslow hierarchy of needs. It has been often contended that this is a blanketed approach in dealing with people and that each situation of the individuals is unique. Owing to this, there has been a general contention that this tool can be used to identify basic requirements of the people, and it should be altered to meet the requirements of the individuals in each subjective environment. Thus, this is a broad overview of the Maslow hierarchy that has been used in the office environment for effective people management and also in understanding of the basic psyche of the people.

5 Key factors and recommendations for the catering industry

5.1 Timeliness

Timeliness is a key aspect for the catering service. Some of the consumer participants that were selected only seemed to indicate that they did not choose pictures related to the time aspect only because it was not present in the image bank.

Customers associated service satisfaction with timeliness and consider the choice of brand as having a cause effect on timeliness as in some brands are mot time effective than others. Consumers expect consequences from the timeliness aspect. Some consumers have very strict schedules for the events they want to hire the catering service for and here they would like to see some of the common values identified for all participants in the form of reassurance. Reassurance is the service extended to the customer by means of sharing information on time aspects, information on what the customer could expect and how much time the customer might have to allocate catering services from the time of the order, the time within which a service could be cancelled and deposit returned to the customer etc.

Time based brand messages have to be included by the catering service to remind the customers of how much satisfaction they could have derived from the catering service in the past because of their being very time efficient.

5.2 Fans evaluation

Fans evaluation plays a critical role in how the catering service might be perceived by other potential customers. Now there are some key values identified for the company in the form of reassurance, trust, satisfaction, freshness and happiness. Fans must be given the ability to rate each of these aspects or values for the catering services. In order to increases fan popularity for the catering service, the service must also ensure that they improve on these core elements. Some of these values can be improved on as a form of short term strategy and some of the others could be improved on as a form of long term strategy. For instance, in the context of reassurance, the short term strategy would be to consider how to incorporate freshness and satisfaction. Trust elements could be postponed but nevertheless should form a part of the long term strategy.

5.3 Smart order

Consider the case of the smart ordering system. A smart order system solves many problems for the consumers especially in the case of reducing the time they would take to make the orders. Customers would be able to place smart orders in a clearer way, which would also make the ordering experience very satisfactory in context. Some of the opinions of the customers seem to indicate that they would expect the catering service to be more of a homely one than a professional one. The use of a smart order system could lead to the experience becoming more professional, but care should be taken to ensure that it does not have the rigidity of the automatic systems that most businesses seem to prefer in current times. The smart order system should work such that the catering order placement will trigger the orders for the needed food sourcing, so the food sourcing would be kept as fresh as possible until the last minute. This is a good idea hence to satisfy the needed values of freshness and happiness that the customer participants required.

5.4 Customized logistics

In the interview-story telling process, it was identified that some customers wanted a very elite and professional service, some wanted a homelier service, some wanted a décor or environment that was more similar to a restaurant and others wanted ambience according to their needs at that time. Most of them had said that the service should be fun, and there were many requirements on the service staff, too.

In this context, customized logistics systems would be very useful. A customized logistics procedure is not a modification of an existing procedure but is more of an extension of the existing procedure. The customized logistics system will ensure that any areas of inconsistency or areas of interconnections between suppliers and the end product takers where there are lags would be addressed. This form of customization in logistics would make the brand more commendable.

5.5 Online-to-Offline consistency

Most times the offline presence of the company might not translate in online effects and vice versa. The promises of customer satisfaction made online, the objectives presented and more should be ensured. There should be consistency and only then the brand would be viewed with a little bit of stability. A catering customer service (online) app like the one below would serve to encapsulate the online to offline inconsistencies and hence would be better recommendation for the catering service company.

6 Catering service APP

The app creation is first initiated with a consistent study. A consistent study is a requirement, so as to understand whether the app can improve customer satisfaction [8]. In the figure 3, the primary issue that needs to be investigated is to understand the online-offline inconsistencies that are noticeable in customer feedback. Now based on the target market competitive profile and also the industry-market analysis, it would be necessary for the catering company to formulate an online marketing strategy that would be helpful for them. Recognition becomes an issue, as although the company being represented by a marketing strategy it might still not be recognized well. People searching in social media should know what they look for, so in the context of having both offline and online factors it would be necessary to maintain a consistency. Sometimes, this would mean adopting the same avatars or the same descriptions in marketing throughout, be it online or offline. For instance, in the case of the current situation, some of the attributes that were considered were that of freshness, happiness, satisfaction etc.

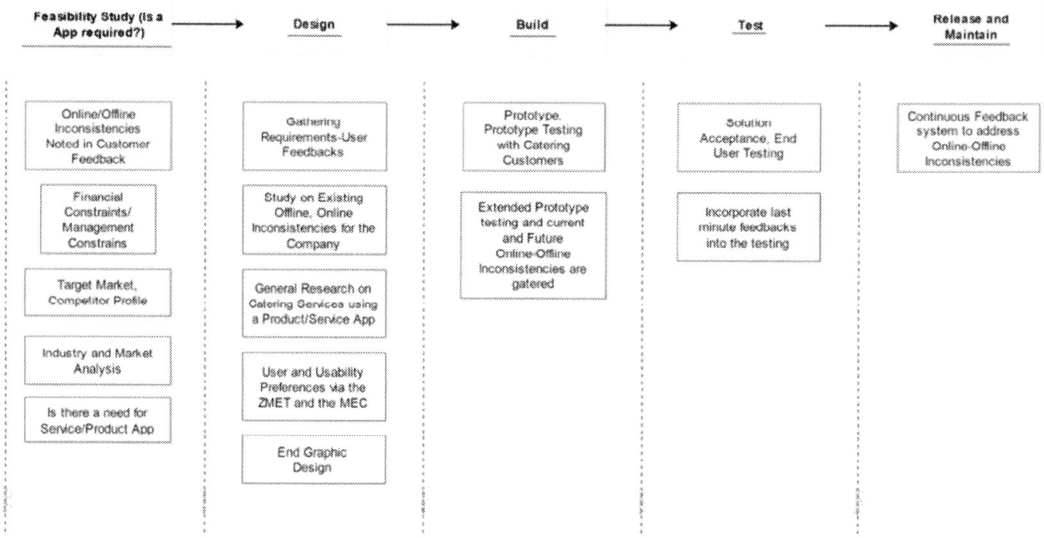

Figure 3. App requirements from design to release.

In developing a form of marketing, it would be necessary to consider each of these attributes is maintained consistently. The product presentation in the store such as that of catalogues which are used to present the form of catering products and services that are provided to the customer might look fresh or fun. But if the social media representation of these products and services do not match the freshness they offer, then this is an issue. Similarly, arguing for the case of adopting the overall tone of the profile, it is critical to remember that the person who searches for the catering company would also be looking it up by the overall profile. So there has to be consistency in creating social media profile to connect with the consumer. A counter argument that would be placed at this point is why a social media profile is needed in the first place, i.e. Is an app necessary? It is, given that more people are beginning use the internet to understand and connect with their brands.

People use multiple platforms in order to find the business that best suits them, they might end up using such platforms as that of messaging for finding their services, or they would use Instagram to look at the pictures of a catering service or they might end up looking at some of the social media talk on a company product and service etc. Now in the development of the app, one of the feasibility concerns that would be introduced is that of how the company is able to connect the different forms of messaging in ways that once again reflect consistency. A catering service cannot just afford to copy past the content that is present in one catering unit onto another. It would be necessary for the catering service to carefully understand the form of audience that would exist for the different messaging services and accordingly tailor make the content, the core essence of the content should however be similar. A consistent tone must be adopted even if specific tailoring has to be done. This will ensure that customers even if they switch different social media platform and search for something they would still be satisfied. Now by ensuring different social media can be connected to with the app, the catering service would be able to find a single point access for its services. It would be able to ensure that all the catering sub units (if there are any) are indeed connected across with just one app. Customers can leave feedbacks, study word of mouth comments and more on the app itself and be informed accordingly.

In feasibility, one of the main concerns as defined earlier is consistency. Consistency in online marketing is more challenges, because most of the social media and marketing media messages are created by the consumers themselves. The marketing media messages are created by the marketing group and then the media message is passed on through the app (as observed in this case), but this media message will not be sent out or passed word for word in the market. The consumer who reads the messages will interpret the marketing message in their own way. Based on how satisfactory they find the service from their personal experiences would be used by the consumer in order to make their own messages. In the context of marketing, this could mean that online marketing by means of the app would never be successful in consistency as is noted. However, to a certain extent the app can deliver consistency and hence deliver satisfaction for the consumer as well.

In the development of the app, prototype testing and end user testing is done, along with the incorporation of a continuous development feedback system. The continuous development feedback system is critical as it ensures that the customers' issues with inconsistency are collected. These issues are then considered as future improvement steps for the app and changes are made accordingly. The consistency check would either be done by working directly with the consumer in the form of feedback or focus groups. This form of continuous feedback system furthermore ensures customer satisfaction as the customer believes they are involved in the system.

7 Discussions

7.1 Maslow Hierarchy of Needs

The main differences and similarities between Maslow's hierarchy of needs theory and Alderfer's need theory are:

The biggest difference between them is that according to Maslow's hierarchy of needs theory, high level needs cannot become incentive factors before low level needs are satisfied. However, according to binary factor theory, hygiene factors and incentive factors can maintain or motivate employees at the same time.

The main differences and similarities between Maslow's hierarchy of needs theory and the Self-determination theory (SDT) are:

STD: The key point focuses on the motivations of personal stimulation and self-determination; in accordance with Maslow's hierarchy of needs, the all behaviors of humans are caused by their needs or demands.

7.2 Smart Connection Level

The ZMET, MEC and HVM in the research are discussions on the demands of network APP consumers, and they are launched to markets after APP value-adding process design is conducted. The most fundamental conditions must be satisfied on the basis of network technology development architecture. In order to reach consumer value HVM, the foundation of APP value-adding shall start with the mart connection level approach of network technology.

7.3 Online-to-Offline consistency

The most fundamental conditions must be satisfied on the basis of network technology development architecture so as to reach the other four factors of consumer values (Fans Evaluation, Timeliness, Smart order, and Customized Logistics). As a result, the APP value-adding process shall start with the Online-to-Offline Consistency of network technology.

8 Conclusions

The research work attempted to understand by means of the ZMET and the MEC technique how a catering service brand is perceived by the consumers using the catering service (or potential new consumers for the catering service) [9]. Using the ZMET data in the form of image selection from the consumers was obtained, and also an interview for story telling was conducted. Based on the data, different key constructs and their numbers of consensus were identified.

The MEC technique application enabled the researcher to identify attributes, values and consequences [10, 11]. The use of this form of technique led to a more holistic understanding of how the consumer will perceive the brand. Based on the data collected, recommendations are made based on the values of reassurance, trust, satisfaction, freshness and happiness. Some of the conclusive units based on which the service industry could hence improve are that of timeliness, fans evaluation, smart order, customized logistics, and online to offline consistency.

A limitation of this research is that in the usual ZMET technique, customers would be allowed to collect their own pictures. The customers (participants) are hence the drivers of the research and not the researchers. However, in this research study, because of the time constraints an image bank was created and the participants choose images from the bank instead. It is recommended that in future research, the participants will be given complete freedom to choose their own images. This will help in unearthing newer issues better.

Using the Maslow hierarchy style modelling helps understand a more abstracted representation of things, it would be necessary for the app developer to go in depth in order to meet customer satisfaction requirements.

Hence, the app can have a more personalized feedback system where the app would be able to connect with each of the customer and then make use of the feedback to set up a batch which would be rolled out in the next patch on the app.

Acknowledgement

In short, since the professional skillful distinction of ZMET and MEC focuses on the discussion about perceptual cognition, the conclusion of discussions about the ideal catering industry customer service image shall also be dominated by the connotations of APP expected by consumers/the sensory and value cognitive dimensions presented by the APP. However, it is suggested that in-depth discussions about functional demands for smart phones like such information system dimensions as the intimacy (downloading speed, easy to search, etc.) of service (response speed, safety and convenience) website (Screen Shot) design could be strengthened in the follow-up research. Besides, the main difficulty in the research method of ZMET lies in the elicitation of main constructs, so constantly improving the elicitation technology and knowledge and making the conclusion more simple and precise shall be the focus worthy of the further research next. Recently, some researchers have started trying to lead this research method to the collection of film studies or film/cartoon development, which will make the effective information obtained during research more abundant; however, with the increasing of the complexity of the based information, it will be more and more difficult in eliciting and mastering the cognition via this method, which shall also be considered in the subsequent research.

References

1. J.A. Gutman, Means-end chain model based on consumer categorization processes, J. mark., **46**, 60-72 (1982)
2. J.C. Olson and T.J. Reynolds, The means-end approach to understanding consumer decision making. In J.C. Olson, & T.J. Reynolds (Eds.), *Understanding consumer decision making: the means-end approach to marketing and advertising strategy*, Mahwah, NJ: Erlbaum, 3-20 (2001)
3. K.G. Grunert and C. Valli, Designer-made meat and dairy products: consumer-led product development, Livestock Production Science, **72**, 83-98 (2001)

4. K.G. Grunert, E. Sørensen, L.B. Johansen, and N.A. Nielsen, Analysing food choice from a means-end perspective, European Advances in Consumer Research, **2**, 366-371 (1995)
5. D.A. Costa, M. Dekker, and W.M.F. Jongen, An overview of means-end theory: potential application in consumer-oriented food product design, Trends food sci. technol., **15** (7), 403-415 (2004)
6. A.L. Biel, *Brand Equity & Advertising: Advertising's Role in Building Strong Brands*, Converting Image Into Equity, 67-82 (1993)
7. M.A. Wahba and L.G. Bridwell, Maslow reconsidered: a review of research on the need hierarchy theory, Organ. Behav. Hum. Perform., **15**, 212-240 (1976)
8. D. Dobni and G. George Zinkham, In search of brand image: a foundation analysis, in advances in consumer research, eds. Marvin E. Goldberg, Gerald Gorn and Richard W. Pollay. Provo, UT: Association for Consumer Research, **17**,100-119 (1990)
9. G.L. Christensen and J.C. Olson, Mapping consumers mental models with ZMET, Psychology & Marketing, **19**, 477-502 (2002)
10. E. Eakin, Penetrating the Mind by Metaphor, The York Times, February 23, B9 (2000)
11. G. Zaltman and L.J. Schuck, Seeing through the Customer's Eyes with Computer Imaging, In Sense & respond: capturing value in the network era, edited by Stephen P. Bradley and Richard L. Nolan. Harvard Business School Press, Boston, MA (1998)
12. G. Zaltman, Consumer research: take a hike, Journal of Consumer Research, **26**,423-428 (2000)
13. G. Zaltman, Metaphorically speaking, Marketing Research, **8**, 13-20 (1996)
14. G. Zaltman, Rethinking market research: Putting people back in, J. Mark. Res., **34**, 424-437 (1997)
15. G. Zaltman, *How Customers Think: Essential Insights into the Mind of the Market*, Harvard Business School Press, Boston, MA (2003)
16. J.J. Collier and C. Malcolm, *Visual Anthropology: Photography as a Research Method*, Revised and expanded edition. Albuquerque, NM: University of New Mexico Press (1986)
17. J. Gutman, Exploring the nature of linkages between consequences and values, Journal of Business Research, **22**, 143-148 (1991)
18. S.M. Kosslyn and O. Koenig, *Wet Mind: The New Cognitive Neuroscience*, NY: The Free Press (1992)
19. M. Blaxton, *Brand Equity & Advertising: Advertising's Role in Building Strong Brands*, Beyond Brand Personality: Building Brand Relationships, 113 (1992)
20. M. De Reuver, H. Bowman, N. Heerschap, and H. Verkasalo, Smartphone measurement: do people use mobile applications as they say they do? *2012 International Conference on Mobile Business*, Paper 2 (2012)

MATEC Web of Conferences **119**, 01026 (2017) DOI: 10.1051/matecconf/201711901026
IMETI 2016

The lifespan of semiconductor patents by assignee and patent characteristics

Shu-Hao Chang[a] and Chin-Yuan Fan

Science and Technology Policy Research and Information Center, National Applied Research Laboratories, Taipei 106, Taiwan

Abstract. Evaluating the values of patents is critical for both managers and investors. Patent maintenance time is an ideal proxy indicator for evaluating commercial patent value. This study investigated the survival rates of patents in the semiconductor industry by using a survival analysis and incorporated a predictive model for patent maintenance times. The technical characteristics of the patents, the assignee factors, and the statuses of the patents were used as the predictive factors for patent maintenance time. The results revealed a small entity size and the number of assignees are the risk factors that increase the likelihood of the maintenance time of a patent decreasing. The litigation status of a patent and the number of its international patent classifications, claims, forward citations, and assignments were determined to be the protection factors that raise the likelihood of its maintenance time extending. Additionally, substantial differences were noted between small entities and nonsmall entities and between being litigated and not in their effects on the survival rates of patents. This study examined the factors that affect the survival rates of patents and provides a reference for the managers and investors of intellectual property rights to use when assessing technical and commercial patent values.

1 Introduction

Since the arrival of the knowledge economy, the governments of advanced countries have paid considerable attention to the economic benefits achieved by owning intellectual properties, which have become one of the key factors of economic development. Patents are a crucial carrier and form of intellectual properties. Therefore, numerous multinational research institutes and companies have tried to measure patent values by using certain tools or methods. Particularly, the method of determining patent values by using the data in patent documents continues to be discussed among scholars [1-4].

Determining patent values is critical for managers, investors, and developers. The most notable indicators of patent values are forward citations [2, 5, 6] and patent maintenance time [7-9]. However, forward citations only indicate that a patent has been cited by another patent. In other words, by measuring the influence of the patent on the technical development of subsequent patents, forward citations indicate the technical value of patents but not the commercial value. Bessen [7] maintained that the correlation between forward citations and commercial patent value is low; in other words,

[a] Corresponding author : shchang@narlabs.org.tw

although a patent may exhibit high technical value, it may not be successfully converted to company profit. The patent maintenance time indicates the validity time of patents. To maintain the validity of their patent claims, patent assignees must not only pay official patent maintenance fees, but also invest a considerable amount of labor force and effort to manage their patents. Therefore, the existence of commercial patent value is typically indicated by the intentions of assignees to continue their payments to maintain their claims [10]. Accordingly, patent maintenance time represents the commercial value of patents. Although no consensus has been formed on how to measure commercial patent value, we maintained that estimating patent maintenance time is a practical means for doing so.

In this study, patent maintenance time was incorporated as the proxy variable for patent values, and a predictive model for patent maintenance time was established on the basis of patent characteristics. Previous studies have reported that the maintenance time of a patent varies considerably according to its technical domains [11]. Therefore, semiconductor technology was selected as the domain of this study to avoid the variability in patent maintenance times. There were two reasons for selecting the semiconductor technology as the domain of this study. First, following the arrival of the Internet of Things, the demand for various sensors and low-power, low-cost semiconductor chips has rapidly increased, and the global semiconductor industry is expected to continue to grow. Second, the patent competition among semiconductor manufacturers is considerably intense; these manufacturers have sought intellectual property protection strategies that involve maximizing their patent portfolios to increase their own competitive advantages [12, 13]. Therefore, semiconductor technology, which is a vital domain for patents, was selected as the target of this study.

This study focused on patent maintenance time to examine its related factors and assess commercial patent value. Recent empirical studies have employed identification models to explore the factors associated with patent maintenance time [9, 14, 15], such as maintenance fees [14-16], technological domains [9, 15, 17], the extents of forward citations [2], and the number of claims [9]. The decision to maintain a patent is typically formulated by the patent assignee. However, few studies have addressed the assignee factors [14]. Practically, patent assignees forfeit their patents for reasons related to the sizes of the assignee companies and their deliberate arrangements in addition to the patent skills of the assignees themselves. Therefore, the sizes and types of assignees were employed in this study to compare academic patents with industrial patents.

Whether patents are assigned and litigated affects their maintenance time. Studies have reported that assigned and litigated patents are more valuable than normal patents [18, 19]. Therefore, these characteristics of patents were included in the model to verify their influence on patent maintenance time. A survival analysis was conducted on patent maintenance time because the patent data in this study were censored data. Patent maintenance fees must be submitted by no later than the 4th, 8th, and 12th years after the date a patent was approved; in other words, each patent can be maintained for a maximum of three times until its 20-year official maintenance time expires. The time interval applied in this study was 12 years. Therefore, whether a patent that was maintained for 12 full years was continued or forfeited was not determined by this study. The information from censored data is incomplete and is typically right censored data. Therefore, conventional regression analysis was inapplicable for this study [20]. Finally, although survival analysis has been applied in previous studies to investigate patent maintenance time [21-24], nearly no study has implemented a survival analysis to explore the characteristics of patents during their approval processes and their statuses after their approval, such as of being assigned and litigated, as well as considered the sizes and types of assignees. This study incorporated a survival analysis to such an effect.

2 Research method

2.1 Patent data sources

This study incorporated the database of the United States Patent and Trademark Office (USPTO). This is a representative patent database because inventors typically simultaneously register their patents in the United States in addition to other countries [25], and US systems are considered widely representative of assignee size. On the definition of the semiconductor technology, the World Intellectual Property Organization (WIPO) established an international patent classification (IPC)–technology concordance table in accordance with the technological domains that patents belong to [26]. The technological domains are divided into a total of 35 types; particularly, the semiconductor technology corresponds to the IPC code of H01L. The maintenance fees of a patent must be submitted by no later than the 4th, 8th, and 12th years after the date the patent was approved; in the period from the 12th year (the third fee submission date) to the 20th year (the year the official patent maintenance time expires), no more maintenance fees are required to be submitted. Therefore, the first 12 years of patent maintenance (2002–2014) were selected as the primary time interval for analysis. In 2014, the third maintenance results of the patents approved in 2002 could be thoroughly examined. A total of 10,148 patents were investigated in this study.

2.2 Survival function estimation methods

For a duration analysis, the duration function is defined as follows. Let $f(t)$ be a continuous probability density of a random available T, where t is a realization of T as the representation of the maintenance. The corresponding cumulative distribution is provided as $F(t) = \int_0^t f(s)ds = \Pr(T \leq t)$. The probability that the spell is of the length of t can then be expressed as $S(t) = \int_t^{\infty} f(s)ds = 1 - F(t) = \Pr(T \geq t)$. S(t) is the survival function. A survival density function can be defined as follows: $s(t) = S'(t) = \frac{d}{dt}S(t) = \frac{d}{dt}[1 - F(t)] = -f(t)$. Because the life duration lasts until t, the probability that it will be terminated at the next time change (Δ) can be defined as Prob $(t \leq T \leq t + \Delta \mid T \geq t)$, which can be characterized by a hazard function:

$$h(t) = \lim_{\Delta t \to 0} \frac{\Pr(t \leq T \leq t + \Delta t | T \geq t)}{\Delta t} = \lim_{\Delta t \to 0} \frac{F(t+\Delta t) - F(t)}{\Delta t S(t)} = \frac{f(t)}{S(t)} \quad (1)$$

The hazard function specifies the instantaneous rate of completion of a duration at $T = t$, conditional upon survival until time t.

There are three approaches to model survival times: nonparametric, parametric and semiparametric. The Kaplan–Meier estimate was employed to clarify the differences among the various groups of patents in their maintenance times. The Kaplan–Meier estimate is a nonparametric estimation method used to calculate the differences among various groups on a survival curve. The formula of the Kaplan–Meier estimate is expressed as follows: $\hat{S}_{KM}(t) = \prod_{t(i)<t} \frac{n(t_i) - d(t_i)}{n(t_i)}$, where $n(t_i)$ represents the number of patents still being maintained during the time $t_{(i)}$ and $d(t_i)$ represents the number of patents no longer maintained during $t_{(i)}$. To determine the factors that affect patent maintenance, this study implemented the Cox proportional hazard model, which is a semiparametric maintenance model. The Cox model does not require a particular probability distribution to represent survival times. The Cox model is as follows:

$$h(t) = h_0(t) \cdot Exp(\beta_1 x_1 + \beta_2 x_2 + \cdots + \beta_k x_k) \quad (2)$$

where $h_0(t)$ is the baseline hazard function and $\beta_1, \cdots \beta_k$ are parameters to be estimated. The remainder of this section further describes the explanatory variables included in the Cox proportional hazard model.

2.3 Explanatory variables

2.3.1 Sizes of assignees

Small assignees are typically divided into three types, namely small companies, individuals, and nonprofit organizations. Studies have indicated that the commercial value of the patents owned by small assignees is generally lower than that of the patents owned by large assignees [7]. Consequently, the long-term maintenance values of the patents owned by small assignees are lower than those of the patents owned by large assignees. Furthermore, because maintaining patents is expensive, small assignees are less capable of affording patent maintenance fees than large assignees are. However, because the patent maintenance fees paid by small assignees are only approximately half of those of large assignees, the payment rates of small assignees are lower than those of large assignees . Consequently, as indicated in some studies, the patent maintenance time of small assignees in the Canadian Intellectual Property Office is longer than that of large assignees [22]. This reveals that the effect of the sizes of assignees on the validity time of patents is undetermined. Therefore, the sizes of patents were differentiated in the aforementioned Cox model to verify their effects on patent maintenance time.

2.3.2 Types of assignees

In the 1980s, the US government passed the Bayh–Dole Act to change the policies that had attributed the research results generated by government-sponsored universities as state-owned, thus causing a considerable increase in patent applications by university authorities [27, 28]. However, university patent assignees differ from enterprise patent assignees regarding their motivations to apply for patents [29], the problems facing them in the patent application process [27], and their means to acquire research and development (R & D) funds [30]. Therefore, the types of patent assignees were differentiated in the Cox model to clarify the difference between academic assignees and nonacademic assignees in their effects on patent maintenance time.

2.3.3 Number of patent assignments

Bessen [7] reported that the validity time for the patents owned by small assignees is shorter than that of the patents owned by large assignees. This is not only because the commercial value of the patents owned by small assignees is lower than that of the patents owned by large assignees, but also possibly because the valuable patents owned by small assignees are acquired by large firms. In other words, a patent with a higher commercial value is assigned more frequently and is valid for a longer duration. Therefore, the number of patent assignments was included in the Cox model.

2.3.4 Litigation

According to previous studies, commercial patent value is indicated by the threat of litigation; that is, the higher the threat of litigation, the greater the monetary value of the patent [31]. Specifically, patents that bear the threat of litigation or have been litigated exhibit high commercial patent value, and assignees are typically strongly inclined to continue paying maintenance fees to protect these patents. Because the research on the effect of litigation on the validity of patents is lacking, litigation was included in the Cox model.

2.3.5 Number of patent assignees and inventors

According to previous studies, technical cooperation effectively improves scientific research efficiency and competence [32, 33]. Technical cooperation is divided into numerous types, such as R & D cooperation, collaborative design, and cooperative production. Patent codevelopment and cooperation, which can be regarded as specific technical cooperative behaviors [34], improve the future citation rates of patents and their influence on subsequent technological development. The size of a patent inventor team can be regarded as the extent of resource investment and commitment toward the patent by an enterprise [35] as well as the intention of the enterprise to increase its resource input to maintain the validity of the patent.

2.3.6 Backward citations

Backward citations are applied to scientific publications and other patents. Particularly, scientific publications are regarded as nonpatent references (NPRs) and indicate the connectivity between the patents and basic science [36]. A larger backward citation indicates a larger amount of technology and knowledge cited by a patent and reflects the scope of knowledge and recognition by inventors or patent examiners on the patent. Therefore, a larger backward citation indicates when a patent is more valuable [37] and will likely be valid for a longer time.

2.3.7 Forward citations

The patents that are cited by other patents are referred to as forward citations [3]. When a patent is cited by other patents, the patent has contributed to further technological development. Forward citations are prevalently used as an index for technical patent value; a larger forward citation indicates a higher technical patent value [5, 6, 37]. Suh [3] incorporated forward citations to examine the relationship between forward patent citation networks and patent prices. This relationship reveals that forward citations do affect technical patent value, which in turn define patent maintenance time.

2.3.8 Number of IPCs

The number of IPCs represents the number of types of registration of a patent and its interdisciplinary characteristic. The number of IPCs implies the technical range of the application of a patent and indicates the scope of the patent [6]. Lerner [38] measured the market values of biotechnological companies by using the number of IPCs. Lerner [38] indicated that the number of IPCs represents the scope of inventions patented by a company and is highly correlated with the market value of the company. Therefore, the number of IPCs was incorporated in the Cox model to explore its relationship with patent maintenance time.

2.3.9 Number of claims

The technical scope and breadth of patent protection can be measured through the number of claims [38], which refers to the number of technical items claimed in a patent. The larger the number of claims in a patent is, the broader the patent is indicated to be. The number of claims affects not only patent values, but also patent maintenance time [24].

3 Data analysis

3.1 Descriptive statistics

Table 1. Descriptive statistics.

Variables	Description	Mean	SD
Time to expiration	Lifespan of the patent	15.57	5.58
Ratio of expired patents (%)	Dummy that equals 1 if the patent has expired and 0 otherwise.	41.71	—
Technical characteristic of the patent			
Number of inventors	Number of inventors of the patent	2.55	1.74

Number of assignees	Number of assignees of the patent	1.01	0.28
Number of backward citations	Number of backward citations of the patent	12.49	15.47
Number of IPCs	Number of IPCs of the patent	5.37	2.58
Number of NPRs	Number of nonpatent references of the patent	1.68	6.25
Number of claims	Number of claims of the patent	16.67	12.38
Assignee factor			
Ratio of small entities (%)	Dummy that equals 1 if the assignee is a small entity and 0 otherwise.	4.73	—
Ratio of university patents (%)	Dummy that equals 1 if the patent is an academic patent and 0 otherwise.	1.01	—
Patent status			
Ratio of litigation (%)	Dummy that equals 1 if the patent has been litigated and 0 otherwise.	0.36	—
Number of forward citations	Number of forward citations of the patent	13.63	23.26
Number of assignments	Number of assignments of the patent	1.53	1.55

The descriptive statistics results are listed in Table 1. In the selected period, 41.71% of patents had already expired because of unpaid annual renewal fees. Regarding the size of patent assignees, 4.73% of patent assignees were small entities. For the types of patents, 0.36% of the patents were litigated, and 1.01% were academic. The results showed that the average expired period of patents was 15.57 years. The descriptive statistics of patent characteristics is also shown in Table 1.

3.2 Cox proportional hazard model

Table 2 shows the results of the Cox proportional hazard model. The following variables were correlated with patent survival: small entity, litigation, university patents, number of inventors, number of assignees, number of forward citations and backward citations of patents, number of IPCs, number of NPRs, number of claims, and number of assignments.

Table 2. Cox proportional hazard results.

Explanatory variables	Hazard ratio: 95% confidence intervals	Significance
Number of inventors	1.006 (0.988–1.024)	0.510
Number of assignees	1.108 (1.001–1.126)	0.048
Number of backward citations	1.000 (0.998–1.002)	0.909
Number of IPCs	0.980 (0.968–0.992)	0.001
Number of NPRs	1.001 (0.995–1.007)	0.798
Number of claims	0.995 (0.992–0.997)	0.000
Small entity	2.588 (2.300–2.914)	0.000
University patents	0.834 (0.633–1.097)	0.194
Litigation	0.233 (0.075–0.722)	0.012
Number of forward citations	0.991 (0.989–0.993)	0.000
Number of assignments	0.850 (0.828–0.872)	0.000
Log likelihood	76329.473	
Chi-squared	573.808***	

Note: *** indicates significance at the 1% level; the total number of observations equals 10,148, of which 4,233 were right-censored.

As the chi-squared value demonstrated, the regression model was significant at the 1% level. The results showed that several patent characteristics have significant effects on the patent maintenance time, such as assignee size (being a small entity), number of assignees, number of IPCs, and number

 DOI: 10.1051/matecconf/201711901026

of claims. Additionally, some of the processing status and development factors of the patents, such as the number of forward citations and assignments and whether the patents in question had been litigated, affected their maintenance time. When the hazard ratio of a factor was higher than 1, it indicated that the factor was a risk factor. As shown in Table 2, small entities and the number of assignees were risk factors, indicating that when the assignees were small entities and the number of assignees was high, the likelihood that the maintenance time of a patent was shortened was increased. The litigation statuses and the numbers of forward citations, IPCs, claims, and assignments were protection factors; when a patent was litigated, and the numbers of its assignments, forward citations, IPCs, and claims were high, its maintenance time was likely to extend. For example, when the other factors remained constant, each additional assignment reduced the survival–risk probability of a patent by 15% (1–0.850=0.150).

3.3 Kaplan–Meier estimate

The factors "small entity" and "litigation" reached the level of significance as shown in Table 2 and were both dummy variables. Therefore, the Kaplan–Meier estimate was employed to verify the survival rate of each group of patents, and the Log–Rank, Wilcoxon, and Tarone–Ware tests were conducted to determine the differences among the groups regarding their survival rates. The Log–Rank test is a statistical measurement method designed to compare two survival curves, in which the data is right skewed and censored. The Wilcoxon and Tarone–Ware methods are alternative measurement methods for comparing survival curves and were also significant. The results of these measurements are shown in Table 3.

Table 3. Tests for comparing survival curves among groups.

Variables	x^2 value		
	Log–Rank	Wilcoxon	Tarone–Ware
Small entity	331.307***	347.771***	340.235***
Litigation	15.177***	14.961***	15.103***

The Kaplan–Meier estimates of survival were graphed on the basis of the size of entity and litigation. Figure 1 presents the patent assignees that were nonsmall entities and survived longer than those that were small entities in the semiconductor field.

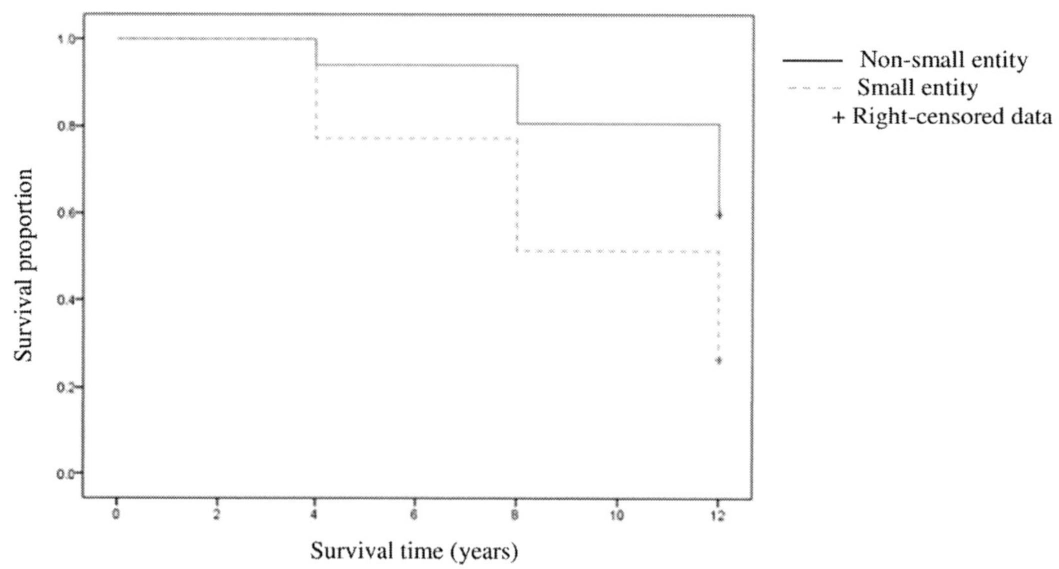

Figure 1. Kaplan-Meier survival estimates, by small entity.

DOI: 10.1051/matecconf/201711901026

As shown in Figure 1, after the first patent maintenance fees were paid (4 years), the survival rate of the small-entity patents was approximately 77%, and that of the nonsmall-entity patents was 94%; after the second patent maintenance fees were paid (8 years), the survival rate of the small-entity patents decreased to approximately 51%, and that of the nonsmall-entity patents was 81%; after the third patent maintenance fees were paid (12 years), the survival rate of the small-entity patents was approximately 27%, and that of the nonsmall-entity patents was 60%. Figure 2 illustrates the comparison of the litigated and nonlitigated patents on their survival rates. The results indicate that litigated patents have higher survival rates than nonlitigated patents do.

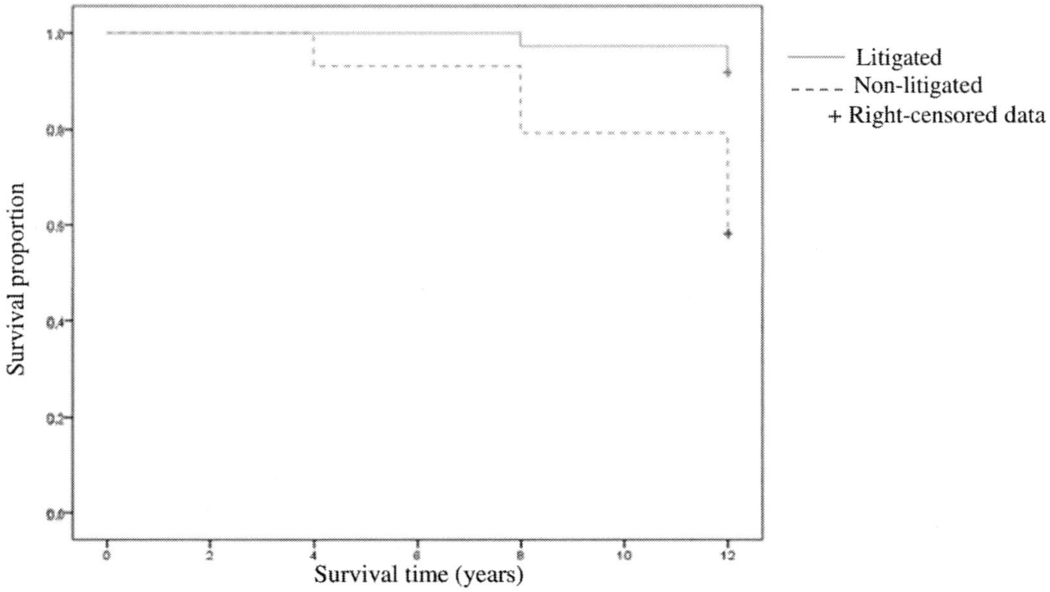

Figure 2. Kaplan-Meier survival estimates, by litigation.

According to Figure 2, after the first patent maintenance fees were paid (4 years), nearly all of the litigated patents remained intact, and the survival rate of the nonlitigated patents was 93%; after the second patent maintenance fees were paid (8 years), the survival rate of the litigated patents decreased to approximately 97%, and that of the nonlitigated patents was 79%; after the third patent maintenance fees were paid (12 years), the survival rate of the litigated patents was approximately 92%, and that of the nonlitigated patents was 58%. Accordingly, after the third patent maintenance fees were paid, a sizable difference was observed between the litigated patents and the nonlitigated patents on their survival rates; more than 90% of the litigated patents remained valid until the 12th year.

4 General discussion

This study investigated the factors that affect the validity time of patents by using a survival analysis. The factors considered in this study were the technical characteristics of patents, comprising the number of IPCs, claims, and backward citations, as well as assignee factors, comprising the sizes and types (academic or nonacademic) of assignees. In addition, this study examined the subsequent development and statuses of the patents, such as their litigation statuses and the numbers of their forward citations and assignments. The differences among the patents of different characteristics of their survival rates were also compared.

Regarding the technical characteristics of patents and the assignee factors, according to the analysis results, small entity sizes and the number of assignees were risk factors; in other words, when the assignees of a patent were small companies and their number was high, the maintenance time of the patent was substantially likely to decrease. This revealed the large-entity-dominant, monopolized, and concentrated characteristics of the semiconductor industry. Previous studies have addressed the semiconductor industry as a capital-intensive industry in which equipment costs account for more than 70% of the capital investment in semiconductor test facilities [39]. For patent skills, this study further verified that the large manufacturing companies were more capable of continuing maintaining their patents than the small companies were, and the patents owned by the large companies were typically more worthy of protecting than those of the small companies were. Additionally, because the semiconductor industry is a technology-intensive industry [40], its high-end technology is commonly controlled by specific companies that hold a considerable monopoly on the industry. Consequently, the technology development alliances in the global semiconductor industry have decreased yearly [41], and research joint ventures have typically been replaced by mergers for companies to acquire necessary technology [42]. Therefore, patents that are valuable and worth protecting have typically been concentrated on one single patent assignee (company).

Some of the technical characteristics examined in this study, namely the numbers of IPCs and claims, increased the probability of extending patent maintenance time. This corresponded to the findings in previous studies. Patent breadth is a critical determining factor for the effectiveness of patent protection [6]. The technical scope and breadth of patent protection can be analyzed by using the numbers of claims and IPCs [39]. Patent applicants can increase their number of claims by writing patent documents and thereby expanding the technical breadth of their patents [43]. Patent assignees can redefine the domain of patents by increasing the number of claims, thus expanding the rights requirements of the patent breadth under certain conditions [7]. The number of claims can be used to explain the operational scope of a patent [44]. Because patents with large numbers of claims are regarded as considerably valuable, they may be frequently litigated [44, 45]. IPCs belong to the technical scope of patent protection (Lerner, 1994); large technical scopes form broad potential values for patents.

No significant differences were observed between the academic patents and the nonacademic patents for their survival rates. This was possibly because the R & D of the semiconductor industry was still concentrated in enterprises. Finally, the subsequent development and statuses of the patents, particularly litigation and the numbers of forward citations and assignments, were observed to significantly affect the survival rates of the patents. Litigation statuses and the number of assignments indicated the commercial value of the patents, and the number of forward citations revealed the technical value of the patents. The patents with higher commercial and technical values were confirmed to be more likely to maintain or extend their validity time.

5 Practical implications

This study examined whether significant differences existed among the patent assignees of different sizes and between the litigated patents and the nonlitigated patents regarding their effects on patent maintenance time. In addition, the survival rates of the patents were separately investigated. The results provide a reference to the Taiwanese government for improving patent maintenance fees. For the existing patent pricing systems, the USPTO currently only varies patent maintenance costs in accordance with the sizes of assignee entities and the time points of patent maintenance. This study addressed a substantial difference between the litigated patents and the nonlitigated patents in their survival rates. Therefore, the prices on these patents can be varied according to different time points of maintenance in the future. In addition, the numbers of IPCs, claims, forward citations, and assignments can be employed as the adjustment factors for maintenance fees.

Patent maintenance time is still an ideal proxy indicator for commercial patent value [7]. According to the patent survival rate estimation results from the survival analysis in this study, each patent exhibited a separate survival rate. A patent value estimation model was established, and the predictive indices for the values and maintenance time of a semiconductor patent, comprising the sizes of its assignees, its litigation statuses, and the numbers of its IPCs, claims, forward citations, and assignments, were incorporated in the analysis. In the future, the values of specific patents should be predicted by using their technical characteristics, assignee factors, and subsequent statuses. The prediction results provide a reference to the managers and investors of intellectual property rights.

6 Limitations and future research directions

Numerous limitations were encountered in this study. First, the patent data employed in this study were from the USPTO database. Although the database is one of the most reliable patent databases worldwide, some semiconductor patents were not registered in the USPTO. This was a limitation in the data source. Second, the USPTO provides preferential fees to microentities in addition to those of general and small entities. Although the microentity fees constitute a considerably small proportion of all the fees provided, the maintenance time and technical characteristics of microentities should be independently discussed to determine whether they differ significantly from those of small entities. Third, because the process of patent maintenance is extremely complex and involves numerous factors, we suggest that future researchers integrate related variables. For example, they can include patent application experience or patent family as one variable. Fourth, the technical characteristics of the patents, assignee factors, and the patent statuses were applied as the determining factors for the survival rates of the patents. However, the decisions to maintain patents are frequently influenced by external factors such as the speed of technical changes, company strategies, and market environments. In this study, patent characteristics were explored from the perspective of the patents themselves; for the assignee factors, only the sizes of the assignees and whether certain patents were academic were considered, but the operations and strategies of the companies were not incorporated in the analysis. If the funds and human resources in future studies are sufficient, then future studies should implement company-level variables, such as the annual revenues and R & D costs of companies to increase the scope of the research and the validity of the survival model.

References

1. M. Reitzig, Res. Policy **32**, 13-26 (2003)
2. T. Schubert, Scientometrics, **88**, 787-804 (2011)
3. J.H. Suh, Technol Anal Strateg, **27**, 485-502 (2015)
4. G. Messinis, Scientometrics, **89**, 813-833 (2011)
5. B. Hall, A. Jaffe, and M. Trajtenberg, Rand J Econ, **36**, 16-38 (2005)
6. D. Harhoff, F Narin, F.M. Scherer, and K. Vopel, Res. Policy, **32**, 1343-1363 (2003)
7. J. Bessen, Res. Policy, **37**, 932-945 (2008)
8. C. Gronqvist, J. Technol. Transfer, **34**, 159-168 (2009)
9. Y.J. Lee and J.D. Lee, Appl Econ, **40**, 229-240 (2008)
10. H. Ernst, World Patent Information, **25**, 233-242 (2003)
11. A. Pakes and M. Simpson, Brookings Pap. Eco. Ac., **Special Issue**, 331-401 (1989)
12. B.H. Hall and R.H. Ziedonis, Rand J Econ, **32**, 101-128 (2001)
13. R.H. Ziedonis, Manage Sci, **50**, 804-820 (2004)
14. J. Kim, Rev Ind Organ, **47**, 195-218 (2015)
15. G. Zhang, X. Lv, and J. Zhou, China Econ Rev, **28**, 37-54 (2014)
16. G. de Rassenfosse and B. van Pottelsberghe de la Potterie, J Econ Surv, **27**, 696-716 (2013)
17. C.J. Serrano, Rand J Econ, **41**, 686-708 (2010)
18. M.S. Meyer and P. Tang, Scientometrics, **70**, 415-440 (2007)
19. N. Ziegler, F. Ruether, M.A. Bader, and O. Gassmann, J. Technol. Transfer, **38**, 930-949 (2013)
20. F.J. Lawless, *Statistical Models and Methods for Lifetime Data*, Hoboken, NJ: John Wiley and Sons (2003)
21. D. Lewensohn, C. Dahlborg, J. Kowalski, and P. Lundin, Res Evaluat, **24**, 197-212 (2015)
22. R. Nikzad, J World Intellect Prop, **14**, 368-382 (2011)
23. R. Svensson, Appl Econ, **45**, 1343-1358 (2013)
24. Y. Xie and D.E. Giles, Appl Econ, **43**, 1375-1384 (2011)
25. S.D. Bass and L.A. Kurgan, Scientometrics, **82**, 217-241 (2010)
26. WIPO, IPC-technology concordance table http://www.wipo.int/ipstats/en/statistics/technology_concordance.html (2016)

27. N. Baldini, Res. Policy, **38**, 1217-1224 (2009)
28. I. Feller and M. Feldman, J. Technol. Transfer, **35**, 597-616 (2010)
29. N. Baldini, R. Grimaldi, and M. Sobrero, Scientometrics, **70**, 333-354 (2007)
30. J.G. Thursby, A.W. Fuller, and M. Thursby, Res. Policy, **38**, 14-25 (2009)
31. Z. Tekic and D. Kukolj, Res Technol Manage, **56**, 18-25 (2013)
32. S. Lee and B. Bozeman, Soc Stud Sci, **35**, 673-702 (2005)
33. F. Rijnsoever, L. Hessels, and R. Vandeberg, Res. Policy, **37**, 1255-1266 (2008)
34. X.P. Lei, Z.Y. Zhao, X. Zhang, D.Z. Chen, M.H. Huang, J. Zheng, R.S. Liu, J. Zhang, and Y.H. Zhao, Scientometrics, **96**, 427-441 (2013)
35. A. Breitzman and P. Thomas, Scientometrics, **103**, 631-647 (2015)
36. Z.Y. Zhao and X.P. Lei, Curr Sci, **104**, 714-720 (2013)
37. J. Suzuki, Res. Policy, **40**, 986-1000 (2011)
38. J. Lerner, Rand J Econ, **25**, 319-333 (1994)
39. C.F. Chien, J.N. Zheng, and Y.J. Lin, J Intell Manuf, **25**, 899-911 (2014)
40. Y. Ho and H. Chiu, Asia Pac J Manage, **30**, 1265-1283 (2013)
41. J.T. Macher, D.C. Mowery, and A.Di Minin, Calif Manage Rev, **50**, 217-242 (2007)
42. K. Gugler and R. Siebert, Rev Econ Stat, **89**, 645-659 (2007)
43. E. Sapsalis, B. van Pottlesberghe de la Potterie, and R. Navon, Res. Policy, **35**, 1631-1645 (2006)
44. M. Reitzig, Res. Policy, **33**, 939-957 (2004)
45. J. Lanjouw and M. Schankerman, Rand J Econ, **32**, 129-151 (2001)

MATEC Web of Conferences **119**, 01027 (2017) DOI: 10.1051/matecconf/201711901027
IMETI 2016

The application and practical benefits of "C theory" in project management

I-Chan Kao[a]

General Education Center, Open University of Kaohsiung, Kaohsiung 812, Taiwan

Abstract. This study regarded adult in-service students who are familiar with project management courses as the subjects and collected data through questionnaires for confirmatory factor analysis and reliability analysis. The aim is to construct a "C Theory" questionnaire scale that encompasses the essence of Chinese management philosophy like decision-making of Taoism, leadership of Legalism, tactics of School of Military Strategists, creativity of Mohism, and coordination of Confucianism. Furthermore, management performance scales were constructed for learning and growth, internal operation, customer satisfaction, and financial control, while the questionnaires and statistical analyses were expected to probe into the impact of "C Theory" on project management performance. This study found that the application and practice of "C Theory" have a high and positive correlation with project management and a significant influence on the improvement of performance. It is therefore suggested that management methods in "C Theory" be appropriately used in project management in order to enhance the efficiency of project management and facilitate the achievement of project management targets.

1 Introduction

1.1 Research motives

Chinese philosophy contains extensive knowledge and profound scholarship with a long history and remarkable impact on human civilization. As early as the Spring and Autumn Period, there was the grand occasion of contention of a hundred schools of thought that were passed down and concentrated into five major scholars: Confucianism, Legalism, Mohism, School of Military Strategists, and Taoism. Many scholars[20-22], [36, 43, 52] pointed out that Chinese style leadership and management philosophy mainly adopt ethics of Confucianism, rule of law of Legalism, universal love of Mohism, power tactics of School of Military Strategists, and letting things take their own course of Taoism as the comprehensive framework. Kao (2014b), Kao (1999), and Tsai and Wu (2013) integrated the previous five Chinese leadership thoughts into "Throne" tactics and leadership philosophy. Lu (2005) pointed out that rule of law of Legalism was similar to X Theory of Douglas McGregor, while "self-discipline and stabilizing others" of Confucianism was similar to Y Theory of Douglas McGregor and Z Theory of William G. Ouchi, and letting things take their own course of Taoism was

[a] Corresponding author : kic0929@ouk.edu.tw

similar to L Theory of William D. Hitt. Furthermore, the Western military academic field spares no effort to utilize the thoughts of School of Military Strategists. Specifically, Master Sun's Art of War has become important material to educate military generals and leaders in the fields of industry, government, and academic circle. Therefore, Chinese leadership and management and that of the West are different in approach, but equally satisfactory in result.

Cheng (1995) observed the differences of the East and the West in terms of cultures, social organizations, philosophical way of thinking, value, and history experiences and concluded "C Theory" from the research results of the Book of Changes, which could embrace all the valuable management and philosophy ideas in the East and the West. Here, "C" refers to the creativity of the Book of Changes of China. Cheng (1995) integrated the principle of mutual promotion of the Five Elements to develop the management model of C Theory (as shown in Figure 1) and employed the principle of mutual promotion and restraint between the Five Elements to elaborate the basic leadership philosophy of the Book of Changes.

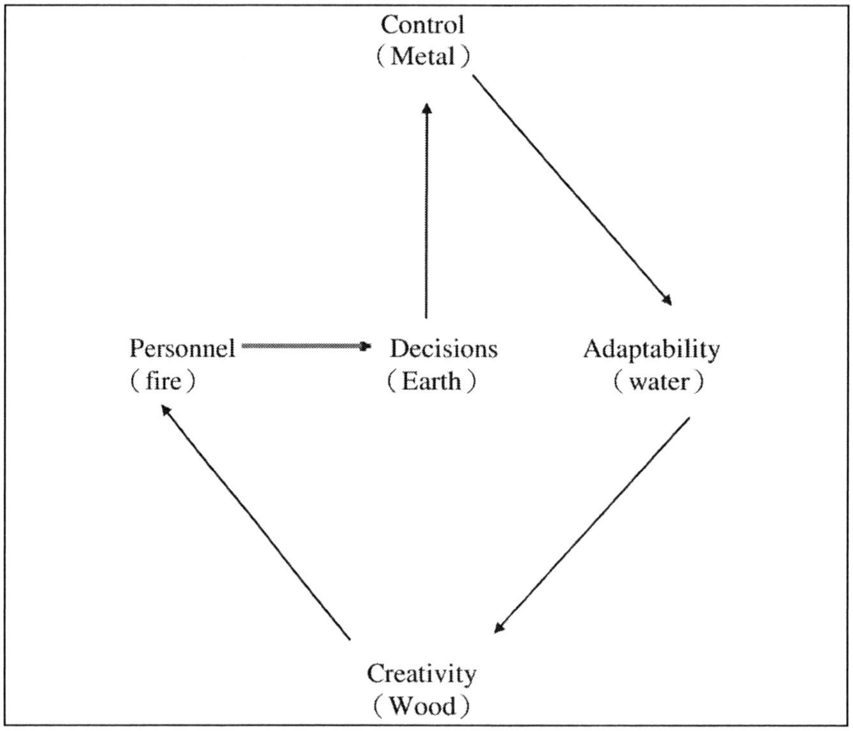

Figure 1. Management Model of C Theory, Source: Cited from Cheng, 38(1995).

According to Figure 1, the "Five Elements" refer to metal, wood, water, fire, and earth. Their key implications are described as follows [5, 46].

1. Metal: It is resolute, fortitude, firm, and flexible and not destroyed after severe training and hammering. It represents "control".
2. Wood: It means prosperous and thriving and stands for "creative plans".
3. Water: It is changing and stands for the "adaptability capacity" of a leader in the face of an ever-changing external environment.
4. Fire: It is strong and vigorous and stands for "personnel" with high morale.
5. Earth: It is thick and sound and boundlessly inclusive. It can create life and guard the center. It stands for "the wisdom of decision-making" of a leader.

Cheng (1995) compared leadership to "gold". Gold stands for strength, firmness, and flexibility, which are the characteristics required for a leader. A good leader shall be like gold rather than iron and shall be loyal, wise, and flexible. Gold surpasses iron and copper, because it is soft, thoroughly tempered, elastic, and resilient. On the contrary, iron rusts and is not plastic, while copper can easily be destroyed. If a leader is like iron or copper, then he/she may be unfaithful and not flexible. Hence, the best leader shall be like pure gold. The order of mutual promotion among the Five Elements is that wood promotes metal, metal promotes water, water promotes wood, wood promotes fire, and fire promotes earth. It implies that decision-making is the core of management. A decision-maker shall have a driving force that must be a power in the center - that is, "earth" in the center. The central power of "earth" shall be deep and sound with the potential of development. In other words, we call it the driving force of decision-making. The wisdom to make decision produces leadership, and one with such wisdom can be a leader. If explained by the principle of the Five Elements, it is "earth promotes metal". "Metal" means being decisive. The strength to cut through metal is a power, or, leadership. The basis of leadership is the wisdom to make decision, or the wisdom of "earth". With the support of vigorous leadership and decision-making, one can conquer "water". One with adaptability capacity knows how to cope with the ever-changing external environment like water and come up with creative plans. If explained by the principle of the Five Elements, it is "water promotes wood". "Wood" means thriving and creative plans. With creative plans, one can combine the wills of members to inspire more potential to form strong, vigorous, and high morale. If explained by the principle of the Five Elements, it is "wood promotes fire". "Fire" has the function to fuse and aggregate. The best decisions can be made by gathering fire and wisdom of all people and giving full play to the role of "fire promotes earth" [53].

Tom Peters & Robert H. Waterman (2005) noted that, "In the future, more than 90% of white-collar workers would face a crisis. All future white-collar works are project works. The same is true for all the works with economic value." A project is not repeatable or routine, but unique. Once a project is completed, it will never be repeated. A project refers to a specific program, recommendation, decision, or matter [44]. According to the Project Management Quality System Guide established by International Standard Organization (ISO) (ISO 10006: 2003), a project is a unique process that is made up by a series of activities with start and end dates, which mutually coordinate and control and which meet the constraint conditions or goals of time, costs, resources, and so on. According to IPMA Competency baseline (ICB), a project refers to the temporary investment and efforts made by an organization to reach a specific goal within a period of time [14]. Project Management Institute (PMI) deemed that a project refers to a temporary effort to create a unique product, service, or result [47]. Based on the above definitions of professional organizations and scholars [23, 30, 32, 35, 37, 41, 44] this study assumed that a project refers to a one-time event, not repeated, with start and end dates, pre-set and clear goals, and limited resources, which shall produce expected deliverables.

The connotation and framework of this study's project management was based on Project Management Quality System Guide (ISO 10006: 2003) established by ISO, IPMA Competency Baseline (ICB), A Guide to the Project Management Body of Knowledge of Project Management Institute (PMI), and Project Management Procedures of British PRINCE2. To develop the methods and strategies of project management, in principle it followed five major procedures: initiation, planning, execution, control, and end [16, 25, 27, 31, 35, 37], [38, 41, 44, 47, 57, 64]. Project management must organize, deploy, and recruit personnel, adopt effective communication and regime development teams, integrate human resources, materials, and financial resources inside and outside an organization, comply with the verification standards to produce deliverables, and meet the needs of the project stakeholders[14, 24, 28, 30], [31, 33], [34, 47, 57, 64].

Modern organizational works have the characteristics of a project. Hence, they can no longer be done by routine management methods. On the contrary, one shall employ the techniques and processes of project management, set and use organizational key performance indicators to evaluate the effect of management, and continue to improve in order to achieve the organization's business objectives. To avoid unsatisfactory performance due to cost overruns, delay, improper range, and poor quality, one shall utilize the methods and strategies of project management, carefully consider the feasibility to initiate a project, rationality of planning, effectiveness of implementation, monitoring accuracy, and completeness, and improve gradually to achieve goals and visions of the organization. However, how can we tell that the goals and visions are reached? It is necessary to apply appropriate assessment methods and indicators to assess whether an organization has reached its desired objectives and visions. Balanced Scorecard (BSC) is a common way used for such a purpose, especially in industry and educational circles. BSC was first proposed by Robert Kaplan and David Norton in 1992. It is a set of strategic management tools to assess the management system and strategy indexation of organizational performance, stressing the spirit of "balance" and effectively converting strategies into action to guide an organization to achieve its goals [18, 65].

The goals and indicators of BSC are derived from the visions and strategies of an organization. The dimensions of evaluation include finance control, customer identity, internal operation, and learning and growth [7, 13]. BSC focuses on balance and seeks a balance between short- and long-term goals, between financial and non-financial metrics, between delayed and advanced assessment indicators, and between internal and external performances of an enterprise. According to Fortune, among the top 1000 enterprises in the U.S., up to 40% have adopted BSC. A Bain & Company survey also pointed out that over 50% of North American companies have adopted BSC as the method to evaluate their internal performances. Harvard Business Review commented that BSC is the most influential strategy management tool since 1975. The strategic and action considerations and management skills of BSC have been stressed and applied by many enterprises and organizations [58-61].

This researcher is a member of teaching staff engaged in in-service education at an adult education institution and a teacher of courses related to project management. Hence, in terms of its research motive, this study regarded adult in-service students with the experience of project management courses as its samples in order to understand their outcomes of the application of C Theory in project management in an organization and the organizational performances in BSC learning and growth, internal operation, customer satisfaction, and finance control. This study learned the influences of the application of "C Theory" in project management by samples with different social variables on organizational performances and the strategies proposed by them to improve the performances of project management, strengthen project management knowledge and system and organizational training to introduce the procedures and methods, and to enhance the project management capacity of members, organizational maturity of project management, and efficiency of practical application.

1.2 Research purposes

Based on the aforementioned research motives and backgrounds, the research goals are described as follows.

1.2.1 To understand the connotation of C Theory and organizational performance.

1.2.2 To develop Application of C Theory and Organizational Performance Questionnaire.

1.2.3 To explore the influence of the application of C Theory in project management on organizational performance.

1.2.4 To propose feasible strategies to improve organizational performance through the application of C Theory in project management based on the research results.

1.3 Research method and process

This study first learned the connotation of C Theory and BSC through a literature review and analysis to develop a first draft of "Questionnaire of Application of C Theory and Organizational Performance". It then distributed the pre-run questionnaire to the adult in-service students with the experience of project management courses to develop the reliability and validity of the questionnaire. It conducted a formal questionnaire survey and collected feedbacks for statistical analysis. The methods of descriptive and inferential statistics helped this study understand the application of C Theory and organizational performances of the subjects during project management. It also learned the differences in the application of C Theory in project management and organizational performance of the subjects with different social variables and analyzed the correlation between the application of C Theory and organizational performance. Lastly, this study shall provide a conclusion and propose feasible strategies to improve organizational performance and efficiency of project management.

DOI: 10.1051/matecconf/201711901027

1.4 Research structure

Figure 2 illustrates the statistical analysis framework of this study. Based on the empirical research in recent years by this researcher [25, 27, 32], [33], this study explored the social variables of gender, age, education, occupation, position, organizational history, and organizational scale and then adopted the decision-making performance of Taoism, control performance of Legalism, tactics performance of School of Military Strategists, creativity performance of Mohism, and coordination performance of Confucianism as the independent variables for the application of C Theory in project management. Lastly, it adopted learning and growth, internal operation, customer satisfaction, and finance control as the dependent variables of performance. Through empirical research and inferential statistical analysis, it explored the relationships among the variables.

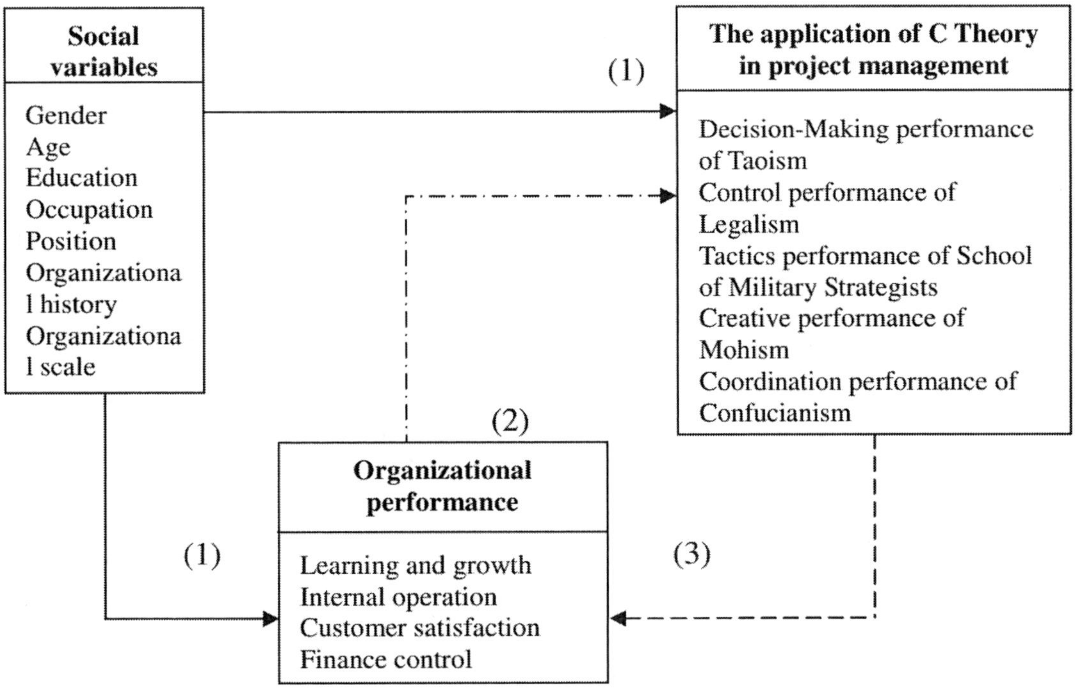

Figure 2. Study Chart, Source: Drawn by this researcher.

Statistical analysis:

(1) T test and One-way ANOVA were adopted to learn the cognitive differences in the application of C Theory in project management and organizational performance of the subjects with different social backgrounds.
(2) Pearson correlation was used to analyze the correlation in the application of C Theory in project management and organizational performance.
(3) Regression analysis was employed to understand the predictability of the application of C Theory in project management for organizational performance.

2 The connotation of C theory and the description of relevant leadership and management

Cheng (2015) proposed C Theory and pointed out in his book on the theory that management factors could be classified into five categories: adjustment, practice of thoughts and ideals, business creation, and creativity transformation. The five categories were the independent and coordinative application of mechanisms, including skills such as operational ability, application of system knowledge, principles and values of principles, and the practice of "Taoism" (the basic principles of formation and transformation) with intelligence. There is no doubt that, as ethical principles, they could also be used in management as guiding principles, because ethics and management are inherently associated in terms of content and functionality. The five categories serve as a reference for the levels of management, which is the transformation of ethical levels. According to the management level structure, C Theory for the first time described how a normative management theory was developed. Cheng (2015) assumed that modern leadership and management meant institutional reform in the face of change and optimized structure. Hence, an appropriate form of leadership was identified and expected. The most important factors for political leadership were listed from historical experience, modern knowledge, and future expectations to discover important experiences and preservation of memories. Learning from competitors and the development of modernization and globalization are equally important.

From the systematic operation of business organizations, the operational framework of the Five Elements shown in Figure 1 could be applied in the system of modern business organizations and the integration and circulation related to management and leadership in order to obtain creative management functions of C Theory. Figure 3 shows the mutual promotion and restriction of each system in an organization. C1 refers to planning and commitment (Chairman of the Board of Directors), C2 to constitution and commands (General Manager, CEO), C3 to competition and cooperation (Marketing Director, Chief Marketing Officer), C4 to creativity and transformation (Product Operation), C5 to communication and coordination (HR), C6 to understanding and communication (Shareholders' Meeting), and C7 to termination and continuity (re-election of Shareholders' Meeting). Based on the operation of the Five Elements management system of an organization shown in Figure 1, icon "→" symbolizes the promotion of one function to another. In other words, Ci → Cj means that Ci produces Cj [5-6]. Thus, C1 representing commitments produces C2 representing commands. Why is this so? In an organization, the commitment to goals leads to the actual participation in commands. Accordingly, when commands enter the market, they become C3 representing competition. What the market needs and produces is C4. C3 follows the needs of customers. When C4 develops in the market, it returns to C5, which is the situation realized through education and development. C5 then needs to return to the beginning to circulate again. The process implies a continuous and mutual promotion relationship, similar to the theory of separation of five powers - namely, wood, metal, water wood, and fire [6, 41, 42].

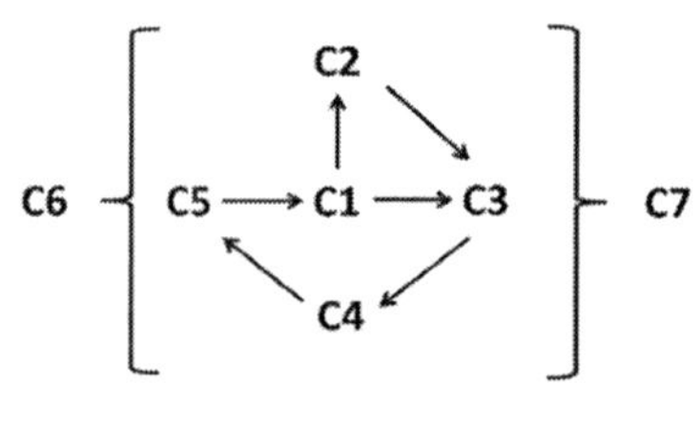

Figure 3. Integration and Circulation of the Creative Management Function of C Theory, Source: Cited from Cheng, 2015, 36.

Cheng (1995) asserted that aside from the integration framework of the Book of Changes, the leadership and management philosophy of C Theory, in terms of leadership and management practices, reflects the decision-making function of "earth" of Taoism, control function of "metal" of Legalism, adaptability capacity of "water" of School of Military Strategists, creativity function of "wood" of Mohism, and coordination function of "fire" of Confucianism, serving as the references and rules of organizational leadership and management. Though each industry promotes different practices in the tasks of project management from initiation, planning, execution, control, to end, the chronological order of all the stages is called "project management procedures". The beginning and end of project management together make up the period of a project. The project management procedures can be classified into six stages: (1) Receipt of a project task; (2) Analysis of the project task; (3) Project planning and action; (4) Distribution of the project task; (5) Execution of the project task; and (6) End of the project task [16, 14], [28-34], [57]. Kao (2014b) believed that after introduction of a leadership philosophy like Confucianism, Legalism, Mohism, School of Military Strategists, and Taoism to the project management and operation of an organization, the feasible strategies include strengthening the professional function of a leader, establishment of the system and regulation of HR management and development, establishment of the authorization system of project management to promptly finish each task, setting up and implementing a project change control system to enhance the efficiency and value of the project, and establishment of a reward and punishment system to enhance the oral and honor of the project members.

Tsai and Wu (2013) pointed out that Confucianism mainly employed the governance of morality, etiquette, and music and developed five major leadership strategies of action by etiquette, governance by morality, consistent cognition and behavior, and govern by doing nothing that goes against nature. Legalism mainly advocated the leadership strategies of reward and punishment, right person for the right position, performance assessment, and letting things take their own course. Mohism stressed hard practice, leading by example, strict self-discipline, virtue, unity, universal love, and strict rules. "According to the rules of Mohism, the murderer shall be sentenced to death, while the person who hurt others shall be punished." This is the way to maintain justice and equality. School of Military Strategists revealed the subjective law of human competition, emphasizing that understanding others and oneself, avoiding the solid and striking the weak, plotting and creating situation, and adaptability capacity could bring about successes. Taoism focused on the leadership strategies of being wise to choose a person for a job, following procedures, letting things take their own course of Taoism, giving authorization to employees, being employ-oriented, recognizing and adopting correct statements, and being inclusive. Kao (2014b) used "Chinese Throne" (see Figure 4) to explain leadership. The left and right handles are like the left and right tokens of a leader and the guarding laws of an organization. The right token is the ethics of Confucianism, while the left is the strategies to govern subordinates of "rules", "strategies", and "trends". The base plate is like the strength of a leader's legs or the military strategies of a strategist. The cushion in the middle balances among the philosophies of Confucianism, Legalism, School of Military Strategists, and Mohism. When the four schools help an organization enter into the right track, the leader can rest assured to implement the ideas of Taoism of "nature and letting things take their own course of Taoism", "following the nature and people", and "adaption to different circumstances".

Figure 4. Leadership Philosophy of Chinese Throne, Source: Kao, 42(2014b).

The aforementioned leadership philosophies of the different Chinese schools were experienced without scientific evidence or objective and universal efficiency. Kang, Chang, Yang, and Huang (2010) deemed that under the influence of traditional Chinese culture, Chinese had a unique perception on the concept of leadership, and the leadership behavior of a leader would reflect some characteristics of Chinese culture. Chen (2014) pointed out that Chinese believed that to complete one thing smoothly, one must need "good timing, geographical convenience, and good human relations". Kang, Su, Chang, and Hsu (2001) believed that both Chinese and Western leadership had the connotations of the trait theory, behavior theory, and contingency theory. However, the behavior model of Chinese contingency leadership was based on "affection", "reason", and "law". A leader considered the respective weight of the leader roles of "sovereign", "parent", and "teacher" and switched among them to realize the true functions of a leader [36]. Chinese often say "being fair and reasonable" or "being reasonable and legal". When talking about "fairness" and "law", "reason" is inseparable, indicating the true leadership philosophy of Chinese: Everything shall be fair, reasonable, and legal [55]. From the perspective of leadership and management, national culture is an important situational factor. Therefore, to understand the ideas or assumptions about Chinese leadership, one must take "culture", "society", and "history" into consideration [63].

No matter for management theory or organization operating practices, the way of thinking of "The West leads the East" has gradually turned into "The West meets the East". The idea of governance by morality advocated by Chinese organizations has gradually attracted the attention of Eastern and Western academia and industry. The interpretation of "the relationship competition and cooperation" with Eastern and Western philosophies could serve as insightful guidance to the competition-cooperation strategy of leadership and management of an organization [3]. Though Chinese leadership philosophy does not stress scientific evidence, it stresses situations, rule by man, and authority and introduces the comprehensive and harmonious beauty to leadership so as to make up for Western leadership philosophy, which focuses on system, law, and power and lacks a cultural atmosphere. In fact, the humanization of science and the scientization of culture might be an alternative solution to Eastern and Western leadership philosophies [49, 52]. Tseng (1987) and Tsai (2008)deemed that Taiwan has long advocated Chinese style management. According to Confucianism, the essence of management was to "human act", while to Legalism, it was "utilitarian act", to Taoism, "nature act", to Mohism, "altruism act", to the Book of Changes, "humanity act", and to Neo-confucianism, "reason act" [39]. Hsu (1992), Chu (1994), Tseng and Liu (1999), Tseng (2002), Huang (1999), et al. all believed that Confucianism advocated "people"-oriented ideas, used appropriate methods to gain the support of subordinates, promoted influencing others by one's own morality, adhered to benevolence, upheld the principle of kindness to the people, and used acceptable ideas and concepts to maintain justice and adaptability composition.

This study integrated the aforementioned description on C Theory and relevant leadership and management, used the operational procedures and characteristics of project management to explore the operation framework and items of the application of C Theory in project management, and summarized five major dimensions of project directors (namely, decision-making performance, leadership performance, adaptability performance, creativity performance, and coordination performance) and 25 management performance items of project directors. The items were adopted in the Scale of Application of C Theory for the "Application of C Theory and Organizational Performance Questionnaire" of this study. See below for a detailed description.

A. Decision-making performance of project directors (decision-making function of Taoism)

A-1. Consideration of the advantages and disadvantages of the organization to make a decision.
A-2. Consideration of the opportunities and risks outside the organization to make a decision.
A-3. Consideration of the instruction of senior management of the organization to make a decision.
A-4. Consideration of the needs of customers to make a decision.
A-5. Consideration of the suggestions of project members to make a decision.

B. Control performance of project directors (control function of Legalism)

B-1. Establish a system and regulation for the implementation of the project.
B-2. Emphasize the power and responsibility performances of project members.
B-3. Implement a reward and punishment system for project members of the project.
B-4. Actively establish the prestige of project directors
B-5. Comprehensive supervision and control of the actions of project members.

C. Adaptability performance of project directors (adaptability function of School of Military Strategists)

C-1. Make execution guidelines for the project according to situations and conditions.
C-2. Understand the needs and characteristics of project members to adjust management approaches.
C-3. Promote the recognition of the visions and goals of the project to project members.
C-4. Promote the cooperation of project members to the instruction of directors.
C-5. Change the control of the project according to changes of internal and external contexts.

D. Creativity performance of project directors (creativity function of Mohism).
D-1. Encourage project members to strengthen project efficiency by creative methods.
D-2. Award given to project members who develop patents.
D-3. Emphasize mutual support and cooperation during the implementation of tasks.
D-4. Enhance efficiency and effectiveness by creative management approaches.
D-5. Actively promote the maximization of project results.
E. Coordination performance of project directors (coordination function of Confucianism)
E-1. Active communication and coordination with project stakeholders.
E-2. Serving as role models for project members.
E-3. Emphasize the moral and ethical performance of project members.
E-4. Promote mutual respect among project stakeholders.
E-5. Promote harmony among project stakeholders.

3. The connotation and relevant description of organizational performance

There are diversified studies on an organizational performance system in the literature. This study regarded the BSC proposed by David P. Norton and Robert S. Kaplan as its assessment system of organizational performance. BSC was derived from the research project co-held in 1990 by David P. Norton, the CEO of Norlan Norton Institute, and Robert S. Kaplan, a professor of Harvard University, which had invited 12 companies to participate. Their team collected and analyzed many cases on a creativity performance assessment system and proposed a performance assessment, BSC, covering four major dimensions: finance, customer, internal processes, and learning and growth [18]. In response to the eagerness to learn BSC, David P. Norton and Robert S. Kaplan published The Balanced Scorecard: Translating Strategy into Action in 1996 [12, 59, 65]. David P. Norton and Robert S. Kaplan believed that in the information age, there were more opportunities to apply BSC creativity strategies than that of performance assessment. Hence, they wrote a second book on BSC: The Strategy-focused Organization: How Balanced Scorecard Companies Thrive in the New Business Environment [13].

After the publication of the research results related to BSC by David P. Norton and Robert S. Kaplan in 1992, BSC had become a hot topic of organizational performance evaluation. Since 2000, non-profit organizations have also introduced BSC and achieved considerable successes. BSC has developed into a widely-used policy management system [2, 48, 60]. BSC makes organizational tasks and strategies more specific, translates them into actions so as to create competitive advantages, employs the linkage of the four dimensions to convert the organizational tasks and strategies into strategic objectives and strategic performance measurement, facilitates an organization to focus on strategic topics in terms of performance evaluation, integrates organizational resources, effectively implements strategies, and accelerates the realization of the visions of an organization. Described as follows, BSC contains four major dimensions: finance, customer, internal processes, and learning and growth [12].

2-1 Financial dimension: Organizational profitability and investment costs. Its assessment indicators include operating income, return on capital, and added value.
2-2 Customer dimension: Other organizations or individuals who have contact with an organization. Its indicators include customer satisfaction, profitability ratios, and increase rate of new customers.
2-3 Process dimension: Its assessment indicators include manufacturing, marketing, after-sales service, product development, and creativity.
2-4 Learning and growth dimension: Its assessment indicators include investments in the growth of skills and information capacity of staff.

BSC was designed and implemented by an organization to enhance its management performance. However, David P. Norton and Robert S. Kaplan (2001) pointed out that BSC could achieve better results in the improvement of governmental or non-profit organizations [48]. As for the establishment of the overall "performance management system" of an organization, BSC also included the visions, goals, and mid- and long-term development plans of an organization [7]. According to the experiences of enterprises that had introduced BSC, BSC was an arduous and time-consuming task. Its difficulty lied in how to find truly proactive management strategies. It did not easily identify unique, "distinct, and excellence elements" superior to the existing ones of an organization [61]. In other words, the competitiveness of an organization could be enhanced by identifying the uniqueness of each of the four dimensions of BSC. Therefore, an organization must re-examine and adjust its organizational structure, processes, and organizational culture and translate its visions and goals into concrete and feasible strategic plans [1, 8, 40]. It is thus necessary to establish a good performance assessment system and appropriate performance assessment indicators to sustain the quality of project management and operations of an organization. Therefore, while introducing BSC to project management, an organization must consider its characteristics, internal and external resources, and measurable indicators inside the organization.

Based on technical ability index 1.06 of IPMA ICB 3.0, Project Management Organization is made up of humans and a relevant basic structure. Project management includes project design and maintenance of appropriate roles, organization, structure, responsibility, and ability. The organization of a project and program is unique and temporary. Its stages or conditions are adapted to conform to the project life cycle. A portfolio project organization is similar to a permanent organization and can be regarded as a permanent sector. Even so, a project organization usually has a shorter circle and changes faster than other permanent organizational sectors [19].

The performance of a project management organization depends on the capacity of all the staff. The capacity of the staff assigned to participate in the project must be tested, and the executive performance of members shall be verified by line management. The optimal operation model is to provide an appropriate way for line managers to discuss with project members together so as to make an individual fulfill a specific role on the team. The knowledge, skills, and experiences of an individual should be reviewed. A line manager shall also consider whether the personality of an individual can match with those of other members. When they do not participate in the selection of team members, line managers must overcome obstacles to develop a project team made up of team members with highly different characteristics [27].

Kao (2012) pointed out that each organization, after a period of operation, evolves into a certain structure that might not necessarily be most suitable for the task and the environment. Different organizational structures have different characteristics. For instance, a centralized and hierarchical structure reduces coordination costs, but hinders communication. Democratic and equal structures encourage communication, but have higher coordination costs. In addition, the size of a team also has an effect on its efficiency, but small organizations tend to have better cohesion, low additional cost, and higher morale. Before a team is formed, organization and structure factors should all be considered. For long-term projects, job satisfaction should also be considered so as to reduce staff turnover.

Based on the above BSC connotation and the organizational characteristics of project management, the organizational performance assessment system of project management includes learning and growth, internal operation, customer satisfaction, and financial control and their items. The twenty items of organizational performance of project management proposed were adopted in the Scale of Application of C Theory from the "Application of C Theory and Organizational Performance Questionnaire" of this study. See below for a detailed description.

A. Learning and growth

A-1. Organization members could actively participate in each advanced study activity.

A-2. Organization members could actively analyze issues to solve and improve them.

A-3. Organization members could actively publish research results and share success experiences.

A-4. The organization could actively encourage staff to innovate to enhance work performance.

A-5. The organization could establish a reward system to openly award excellent staff regularly.

B. Internal operation

B-1. Members could identify the visions, goals, and business strategies of the organization.

B-2. The organization could set up appropriate and feasible business development plans.
B-3. The organization could establish clear rules, regulations, and work norms.
B-4. The organization could arrange tasks among staff according to their expertise and responsibilities.
B-5. The organization could use diversified approaches to enhance work performance.

C. Customer satisfaction

C-1. The organization could establish good channels to communicate with customers.
C-2. The organization could respond positively to the reactions and requirements of customers.
C-3. Customer satisfaction to the organization's services gradually increases.
C-4. Customers' re-visit rate gradually increases.
C-5. Customers actively introduce new customers.

D. Finance control

D-1. The organization could make an accurate budget based on real business needs.
D-2. The organization could establish an appropriate budget implementation, supervision, and evaluation mechanism.
D-3. The organization could assess the efficiency of budget implementation on a regular basis.
D-4. The organization could develop a variety of R&D plans to seek funds.
D-5. ROI gradually increases.

4. Establishment of application of C theory and organizational performance questionnaire

This study summarized and analyzed the relevant literature, edited questions on the application of C Theory and organizational performance, and employed statistical analysis methods like expert validity, configuration validity, and reliability in order to establish the Application of C Theory and Organizational Performance Questionnaire.

4.1 Method to establish the application of C Theory and organizational performance questionnaire

Based on the aforementioned literature discussion and analysis, this study summarized and edited "a first draft of Application of C Theory and Organizational Performance Questionnaire" and invited project management practitioners, university professors, and scholars and experts in the field of project management as auditors of the first draft of the questionnaire. In terms of project management practitioners, it invited 2 project managers and 3 project members. In terms of university professors or scholars in the field of project management, it invited a total of 10 scholars and researchers from National Pingtung University, National Kaohsiung First University of Science and Technology, Republic of China Military Academy, Chang Jung Christian University, Aletheia University, Cheng Shiu University, Tajen University, and Open University of Kaohsiung through the help of Taiwan Project Management Association.

After the audit of the questionnaire's first draft, a pre-test of the questionnaire should be conducted. A total of 160 copies were sent, while 142 valid feedbacks were collected to establish its validity and reliability. For validity, this study adopted Structural Equation Modeling (SEM) to analyze factors. One of the most important characteristics of SEM is to be established on certain theories. In other words, SEM is a statistical technique that verifies the fitness of a priori theoretical model [9]. The content and nature of latent variables should be clearly described at the beginning of the test or have a specific theoretical basis, and the corresponding composition model used to observe variables should be determined. The purpose of analysis was to verify the fitness of the priori factor structure. The procedures could be used to develop test tools as well as the effectiveness of the structure and theoretical framework. Hence, it is also called confirmatory factor analysis (CFA)[10, 62].

As for the "Scale of Application of C Theory", this study referred to project management knowledge systems like ISO10006, IPMA, PMI, and PRINCE2, edited the theoretical framework of the decision function of Taoism, control function of Legalism, adaptability function of School of Military Strategists, creativity function of Mohism, and coordination function of Confucianism of C Theory to the five major dimensions of project directors - decision-making performance, leadership performance, adaptability performance, creativity performance, and coordination performance - and proposed 25 performance items of project directors. As for the "Organizational Performance Scale", this study then referred to the BSC theoretical framework of David P. Norton and Robert S. Kaplan, adapted it to four major dimensions - finance control, customer identity, internal operation, and learning and growth - and proposed 20 assessment items. The above two scales had clear theoretical bases and models. Hence, in terms of the validity analysis of the questionnaire, LISREL was adopted for use of CFA on the structure equation in order to understand the internal factor model. As for reliability, this study adopted SPSS to conduct Cronbach's α analysis to understand the internal consistency of the questions of each level.

4.2 Results of the scales of application of C Theory and organizational performance

The results of expert validity and statistical analysis are described as follows from the two aspects of the Scale of Application of C Theory and the Scale of Organizational Performance.

4.2.1 Scale of application of C Theory

Through the results of expert validity analysis, this study kept the five major dimensions (namely, decision-making performance, control performance, adaptability performance, creativity performance, and coordination performance of project directors) and 25 items in the first draft. Some text of the questions was edited, while some questions were converged. The aforementioned questionnaire was then formed and SEM analysis and Cronbach's α analysis were conducted. The results are as follows.

4.2.1.1 SEM analysis of the scale of application of C Theory

Based on SEM analysis with LISREL, as the scale contained many questions, it could not reflect the relationship road map, but the degrees of freedom (DOF) and chi-square statistics of the overall model were obtained, wherein DOF were 2043 while the chi-square value was 11171.83 and the P value was 0.0, indicating that there were significant differences between the hypothetical model and the observed values. In addition, other adapter indices like NFI, NNFI, and CFI were more than 0.95, indicating that each question and dimension had good structure. The factor loading of the questions of each level is as follows.

(1) The factor loadings from questions A-1 to A-5 of decision-making performance of project directors (decision-making function of Taoism) in order are: 0.53, 0.49, 0.53, 0.57, and 0.53.
(2) The factor loadings from questions B-1 to B-5 of control performance of project directors (control function of Legalism) in order are: 0.51, 0.57, 0.50, 0.53, and 0.50.
(3) The factor loadings from questions C-1 to C-5 of adaptability performance of project directors (adaptability function of School of Military Strategists) in order are: 0.53, 0.58, 0.52, 0.55, and 0.55.
(4) The factor loadings from questions D-1 to D-5 of creativity performance of project directors (creativity function of Mohism) in order are: 0.61, 0.55, 0.64, 0.62, and 0.61.
(5) The factor loadings from questions E-1 to E-5 of coordination performance of project directors (coordination function of Confucianism) in order are: 0.60, 0.53, 0.54, 0.58, and 0.56.

4.2.1.2 Cronbach's α analysis of the scale of application of C Theory

(1) As for the reliability of decision-making performance of project directors (decision-making function of Taoism), α was 0.85, indicating that their decision-making performance had a certain degree of internal consistency. The correlation between each question and the total

questions was bigger than 0.57, indicating that each dimension had high internal consistency.

(2) As for the reliability of control performance of project directors (control function of Legalism), α was 0.86, indicating that their control performance had a certain degree of internal consistency. The correlation between each question and the total questions was bigger than 0.61, indicating that each dimension had high internal consistency.

(3) As for reliability of adaptability performance of project directors (adaptability function of School of Military Strategists), α was 0.86, indicating that their adaptability performance had a certain degree of internal consistency. The correlation between each question and the total questions was bigger than 0.65, indicating that each dimension had high internal consistency.

(4) As for the reliability of creativity performance of project directors (creativity function of Mohism), α was 0.89, indicating that their creativity performance had a certain degree of internal consistency. The correlation between each question and the total questions was bigger than 0.69, indicating that each dimension had high internal consistency.

(5) As for the reliability of coordination performance of project directors (coordination function of Confucianism), α was 0.88, indicating that their coordination performance had a certain degree of internal consistency. The correlation between each question and the total questions was bigger than 0.66, indicating that each dimension had high internal consistency.

Nunnally (1978) believed that α more than 0.7 indicated high reliability cited from [62]. Therefore, from the Cronbach α reliability analysis results of this study, the scale had a high degree of reliability and consistency.

In summary, the five dimensions and their respective questions of the Scale of Application of C Theory were all kept for the formal questionnaire in order to understand the five performances of the application of C Theory: decision-making performance, control performance, adaptability performance, creativity performance, and coordination performance of project directors.

4.2.2 Scale of organizational performance

Through the results of expert validity analysis, this study kept the four major dimensions (namely, learning and growth, internal operation, customer identity, and finance control) and 20 items in the first draft. Some text of the questions was edited, while some questions were converged, and the aforementioned questionnaire was formed. Lastly, SEM analysis and Cronbach's α analysis were conducted. The results are as follows.

4.2.2.1 SEM analysis of the scale of organizational performance

Based on the SEM analysis with LISREL, as the scale contained many questions, it could not reflect the relationship road map, but the degrees of freedom (DOF) and chi-square statistics of the overall model were obtained, wherein DOF were 395 while the chi-square value was 2930.37 and the P value was 0.0, indicating that there were significant differences between the hypothetical model and the observed values. In addition, other adapter indices like NFI, NNFI, and CFI were more than 0.95, indicating that each question and dimension had good structure. The factor loading of the questions of each level is as follows.

(1) The factor loadings from questions A-1 to A-5 of learning and growth in order are: 0.60, 0.62, 0.58, 0.59, and 0.57.

(2) The factor loadings from questions B-1 to B-5 of internal operation in order are: 0.58, 0.61, 0.61, 0.63, and 0.56.

(3) The factor loadings from questions C-1 to C-5 of customer identity in order are: 0.64, 0.63, 0.60, 0.59, and 0.57.

(4) The factor loadings from questions D-1 to D-5 of finance control in order are: 0.58, 0.55, 0.56, 0.56, and 0.51.

4.2.2.2 Cronbach's α analysis of the scale of organizational performance

(1) As for the reliability of learning and growth, α was 0.87, indicating that learning and growth had a certain degree of internal consistency. The correlation between each question and the total questions was bigger than 0.65, indicating that each dimension had high internal consistency.
(2) As for the reliability of internal operation, α was 0.86, indicating that internal operation had a certain degree of internal consistency. The correlation between each question and the total questions was bigger than 0.64, indicating that each dimension had high internal consistency.
(3) As for the reliability of customer satisfaction, α was 0.87, indicating that customer satisfaction had a certain degree of internal consistency. The correlation between each question and the total questions was bigger than 0.68, indicating that each dimension had high internal consistency.
(4) As for the reliability of finance control, α was 0.85, indicating that finance control had a certain degree of internal consistency. The correlation between each question and the total questions was bigger than 0.62, indicating that each dimension had high internal consistency.

Nunnally (1978) believed that α more than 0.7 indicated high reliability cited from [62]. Therefore, from the Cronbach α reliability analysis results of this study, the scale had a high degree of reliability and consistency.

In summary, the four dimensions and their respective questions of the Scale of Organizational Performance were all kept for the formal questionnaire in order to understand the four organizational performances: learning and growth, internal operation, customer identity, and finance control.

5. Research results

After the formal Questionnaire of Application of C Theory and Organizational Performance was completed, the cluster random sampling method was adopted to select 800 subjects to receive the survey, with 713 valid feedbacks collected. After statistical analyses like descriptive statistics, t test, One-way ANOVA, and regression analysis, the results are described as follows.

5.1 Results of descriptive statistics

5.1.1 Gender

Among the valid samples, there were 314 males, accounting for 44.0% of all respondents; while there were 399 females, accounting for 56.0%. The proportion of women was higher than that of men.

5.1.2 Age

Most of the respondents (522) were between 41 and 60 years old, accounting for 73.2% of all respondents, followed by the respondents (109) between 21 and 40 years old, accounting for 15.3%.

5.1.3 Education

Most of them (403) graduated from vocational or senior high schools, accounting for 56.5% of all respondents, followed by college and university graduates (269), accounting for 37.7%.

5.1.4 Occupation

Most of them (350) were engaged in the industry of business and service, accounting for 49.1% of all respondents, followed by civil servants (including those of military, government, education, police, and fire departments) (240), accounting for all samples 33.7%.

5.1.5 Position

Most of them were employees (525), accounting for 73.6% of all respondents, followed by persons in charge (96), accounting for 13.5% of all respondents. In addition, department heads (92) accounted for 12.9%.

5.1.6 Organizational history

Most of the organizations (271) had a history of over 21 years, accounting for 38.0%, followed by those of 5 years (inclusive) (183), accounting for 25.7%. In addition, those organizations between 11 and 20 years (158) accounted for 22.2%, while those between 6 and 10 years (101) accounted for 14.2%.

5.1.7 Organizational scale

Most of the organizations (333) had less than 50 employees, accounting for 46.7%, followed by those (236) of 201 employees, accounting for 33.1%.

5.2 Difference analysis of social background variables in the application of C Theory and organizational performance

5.2.1 Difference analysis of gender

The T test was used to compare the differences between men and women in terms of their application of C Theory and organizational performance. Significant differences were found in terms of adaptability performance, creativity performance, and coordination performance (T value = 2.79, P value = 0.01; T value = 2.68, P value = 0.01; T value = 2.54, P value = 0.01). The values of men were higher than those of women. There were also differences in terms of learning and growth and customer satisfaction (T value = 2.06, P value = 0.04; T value = 2.06, P value = 0.04). The values of men were higher than those of women too.

5.2.2 Difference analysis of age

One-way ANOVA was used to compare the differences among the respondents of different ages in terms of the application of C Theory and organizational performance. Significant differences were found for different ages in adaptability performance, creativity performance, and coordination performance (F value = 2.56, P value = 0.03; F value = 3.05, P value = 0.01; F value = 2.33, P value = 0.04). After further multiple comparisons, it was found that the average factor values of those between 21 and 40 years old, between 41 and 60 years old, and above 61 years old were higher than the values of those younger than 20 years old in terms of adaptability performance, creativity performance, and coordination performance, indicating that the former three age groups had better performances than the latter one in the three aspects.

Based on the statistical results of learning and growth and customer satisfaction (F value = 2.55, P value = 0.04; F value = 3.06, P value = 0.01; F value = 2.32, P value = 0.04), it was also found that the average factor values of those between 21 and 40 years old, between 41 and 60 years old, and above 61 years old were higher than the values of those younger than 20 years old in terms of learning and growth and customer satisfaction, indicating that the former three age groups had better performances than the latter one in the two aspects.

5.2.3 Difference analysis of education

One-way ANOVA was used to compare the differences among the respondents of different education backgrounds. Significant differences were found in decision-making performance, control performance, adaptability performance, creativity performance, coordination performance, and overall project management (F value = 5.04, P value = 0.00; F value = 2.92, P value = 0.03; F value = 3.39, P value = 0.02; F value = 5.46, P value = 0.00; F value = 4.16, P value = 0.00; F value = 5.02, P value = 0.00). There were also significant differences in learning and growth, internal operation, customer satisfaction, finance control, and overall organizational performance (F value = 5.07, P value = 0.00; F value = 9.39, P value = 0.00; F value = 5.22, P value = 0.00; F value = 3.35, P value = 0.02; F value = 4.12, P value = 0.00).

After further multiple comparisons, it was found that the average factor values of those graduated from colleges or universities or graduate institutes were higher than those graduated from senior middle schools or vocational schools in all the dimensions of the application of C Theory and organizational performance, indicating that the former had better cognitive perception than the latter in the aspects. Hence, enhancement of project leaders' education backgrounds and strengthening of in-service learning and growth had significant benefits for enhancing the application of C Theory and organizational performances.

5.2.4 Difference analysis of occupation

One-way ANOVA was used to compare the differences among the respondents of different occupations. Significant differences were found in decision-making performance (F value = 5.8, P value = 0.00), control performance (F value = 5.95, P value = 0.00), adaptability performance (F value = 6.66, P value = 0.00), creativity performance (F value = 6.72, P value = 0.00), coordination performance (F value = 6.72, P value = 0.00), learning and growth (F value = 5.74, P value = 0.00), internal operation (F value = 3.18, P value = 0.01), customer satisfaction (F value = 4.07, P value = 0.00), finance control (F value = 5.36, P value = 0.00), and overall organizational performance (F value = 3.88, P value = 0.00).

From further multiple analysis, it has been found that the average factor values of the respondents engaged in agriculture, forestry, fishery, and animal husbandry were the lowest, while those engaged in business and services or civil servants (including those of military, government, education, police, and fire departments) had high factor average values. The above results show that currently Taiwanese business and services practitioners and civil servants attach great importance to the introduction of project management and organizational performance. Taiwanese industry is currently in the active transition from being manufacturing-oriented to service-oriented, and the government and public sectors are stressing services quality and improvement of administration service satisfaction as their policy objectives. Hence, the perceptions of the respondents in project management performance and organizational performance are significantly enhanced. In addition, the respondents engaged in agriculture, forestry, animal husbandry, fishery, and manufacturing had low scores in the recognition of project management and organizational performance, mainly because the operations of systems and processes of primary and secondary industries are more stable than those of other industries. Hence, the practitioners in the former industries find it more difficult to perceive the improvement of project management and organizational performance.

5.2.5 Difference analysis of position

One-way ANOVA was used to compare the differences among the respondents of different positions. Significant differences were found in decision-making performance (F value = 5.65, P value = 0.00), control performance (F value = 5.94, P value = 0.00), adaptability performance (F value = 6.67, P value = 0.00), creativity performance (F value = 6.65, P value = 0.00), coordination performance (F value = 6.68, P value = 0.00), overall project management (F value = 5.75, P value = 0.00), learning and growth (F value = 3.18, P value = 0.01), internal operation (F value = 4.07, P value = 0.00), customer satisfaction (F value = 5.36, P value = 0.00), finance control (F value = 3.88, P value = 0.00), and overall organizational performance (F value = 3.88, P value = 0.00).

After further multiple comparisons, it was found that the average factor values of persons in charge and department directors were higher than employers in all the dimensions of the application of C Theory and organizational performance, indicating that the former have better cognitive perception than the latter in the aspects. Hence, persons in charge and department directors took full advantages of C Theory, actively introduced project management procedures and methods, strengthened staff's ability of project management, and established team common sense so as to enhance project management and organizational performance.

5.2.6 Difference analysis of organizational history

One-way ANOVA was used to compare the differences among organizations with different history. Significant differences were found in decision-making performance (F value = 11.21, P value = 0.00), control performance (F value = 8.65, P value = 0.00), adaptability performance (F value = 8.57, P value = 0.00), overall project management (F value = 10.53, P value = 0.00), learning and growth (F value = 10.34, P value = 0.00), internal operation (F value = 10.51, P value = 0.00), customer satisfaction (F value = 5.68, P value = 0.00), finance control (F value = 2.57, P value = 0.01), and overall organizational performance (F value = 8.92, P value = 0.00).

After further multiple comparisons, it was found that the average factor values of organizations with a history of over 21 years were higher than those with a history of lower than 5 years (inclusive), between 6 and 10 years, and between 11 and 20 years in all the dimensions of the application of C Theory and organizational performance, indicating that the longer history an organization has, the better project management and organizational performance it has. Hence, skillful application of C Theory in project management, introduction of project management methods and procedures, and enhancement of organizational maturity can enhance operation and management performances and bring about sustainable operation and development opportunities and space.

5.2.7 Difference analysis of organizational scale

One-way ANOVA was used to compare the differences among organizations with different scales. Significant differences were found in decision-making performance (F value = 10.23, P value = 0.00), control performance (F value = 8.72, P value = 0.00), adaptability performance (F value = 8.54, P value = 0.00), creativity performance (F value = 8.43, P value = 0.00), overall project management (F value = 10.53, P value = 0.00), learning and growth (F value = 10.25, P value = 0.00), internal operation (F value = 10.64, P value = 0.00), customer satisfaction (F value = 5.71, P value = 0.00), finance control (F value = 2.58, P value = 0.01), and overall organizational performance (F value = 8.85, P value = 0.00).

After further multiple comparisons, it was found that the average factor values of organizations with over 201 employees were higher than those with lower than 50 employees (inclusive), between 51 and 100 employees, and between 101 and 200 employees in decision, control, adaptability, creativity, and overall project management, and all the dimensions of organizational performance, indicating that the bigger an organization is, the better project management and organizational performance it has. Hence, SMEs could apply C Theory in project management, introduce project management methods and procedures, and enhance organizational maturity so as to enhance operation and management performances.

5.3 Correlation analysis of the application of C Theory and organizational performance

Pearson product moment correlation was adopted to analyze the correlation between project management and organizational performance. Table 1 lists the date related to product moment

 DOI: 10.1051/matecconf/201711901027

correlation of project management and organizational performance. It can be seen that each level and overall dimensions of the application of "C Theory" in project management and organizational performance have high and positive correlation (Correlation coefficients were above .70), indicating that the higher the application of "C Theory" in project management is, the better the organizational performance is.

Table 1. Product Moment Correlation Analysis between Project Management and Organizational Performance.

Correlation Coefficient		Organizational Performance				
		Learning and growth	Internal operation	Customer satisfaction	Finance control	Overall organizational performance
Application of "C Theory" in project management	Decision-making performance	.710***	.725***	.703***	.710***	.762***
	Control performance	.797***	.770***	.778***	.705***	.710***
	Adaptability performance	.844***	.705***	.722***	.785***	.774***
	Creativity performance	.829***	.853***	.856***	.797***	.844***
	Coordination performance	.845***	.714***	.752***	.792***	.760***
	Overall project management	.760***	.760***	.762***	.778***	.799***

*** p<.001

5.4 Regression and predictive power analysis of project management for organizational performance

5.4.1 Predictive power for learning and growth

Table 2. Predictive Power of Project Management in Learning and Growth.

	Unstandardized Coefficient		Standardized Coefficient	t	Significance	Multicollinearity Statistics	
	Estimated value of B	Standard error	Beta distribution			Tolerance	VIF
(Constant)	0.38	0.09		4.00	0.00		
Decision-making performance	0.27	0.05	0.28	5.66	0.00	0.31	3.15
Adaptability performance	0.28	0.06	0.23	3.44	0.00	0.18	4.14
Creativity performance	0.33	0.06	0.36	5.46	0.00	0.21	4.04
Dependent variable: Learning and growth							
Rsquare: 0.65							

Stepwise multiple regression analysis method was adopted to analyze the predictive power of each factor of the application of "C Theory" in project management for learning and growth. Table 2 shows that in the criterion variable of learning and growth, decision-making performance, adaptability performance, and creativity performance reach significant levels (T value = 5.66, P value = 0.00; T value = 3.44, P value = 0.00; T value = 5.46, P value = 0.00), wherein the influence of creativity performance on learning and growth is the biggest, followed by that of decision-making performance. The standardized regression coefficients of the three independent variables are positive, indicating that their influences on learning and growth are positive. The degrees of tolerance are about between 2 and 3. The expansion coefficients of variation are below 5, lower than the evaluation index value of 10, indicating that the issue of multicollinearity among the independent variables is not obvious.

5.4.2 Predictive power for internal operation

Stepwise multiple regression analysis method was adopted to analyze the predictive power of each factor of the application of "C Theory" in project management for internal operation. Table 3 shows that in the criterion variable of internal operation, control performance, adaptability performance, and coordination performance reach significant levels (T value = 4.85, P value = 5.92; T value = 5.92, P value = 0.00; T value = 3.31, P value = 0.00), wherein the influence of adaptability performance on internal operation is the biggest, followed by that of control performance. The standardized regression coefficients of the three independent variables are positive, indicating that their influences on learning and growth are positive. The degrees of tolerance are about between 2 and 3. The expansion coefficients of variation are lower than the evaluation index value of 5 and thus less than the evaluation index value of 10, indicating that the issue of multicollinearity among the independent variables is not obvious.

Table 3. Predictive Power of Project Management in Internal Operation.

	Unstandardized Coefficient		Standardized Coefficient	t	Significance	Multicollinearity Statistics	
	Estimated value of B	Standard error	Beta distribution			Tolerance	VIF
(Constant)	0.33	0.09		3.51	0.00		
Decision-making performance	0.30	0.06	0.30	4.85	0.00	0.23	4.33
Adaptability performance	0.36	0.06	0.38	5.92	0.00	0.21	4.86
Coordination performance	0.27	0.07	0.27	3.31	0.00	0.19	4.34
Dependent variable: Internal operation							
Rsquare: 0.63							

5.4.3 Predictive power for customer satisfaction

Stepwise multiple regression analysis method was adopted to analyze the predictive power of each factor of the application of "C Theory" in project management for customer satisfaction. Table 4 shows that in the criterion variable of customer satisfaction, decision-making performance, adaptability performance, creativity performance, and coordination performance reach significant levels (T value = 5.86, P value = 0.00; T value = 2.57, P value = 0.01; T value = 6.05, P value = 0.00; T value = 6.17, P value = 0.00), wherein the influence of creativity performance on customer satisfaction is the biggest, followed by that of coordination performance. The standardized regression coefficients of the four independent variables are positive, indicating that their influences on learning and growth are positive. The degrees of tolerance are about between 2 and 3. The expansion coefficients of variation are below 5, lower than the evaluation index value of 10, indicating that the issue of multicollinearity among the independent variables is not obvious.

Table 4. Predictive Power of Project Management in Customer Satisfaction.

	Unstandardized Coefficient		Standardized Coefficient	t	Significance	Multicollinearity Statistics	
	Estimated value of B	Standard error	Beta distribution			Tolerance	VIF
(Constant)	0.02	0.07		0.32	0.75		
Decision-making performance	0.27	0.05	0.25	5.86	0.00	0.22	4.65
Adaptability performance	0.13	0.05	0.12	2.57	0.01	0.18	4.77
Creativity performance	0.31	0.05	0.31	6.05	0.00	0.16	4.85
Coordination performance	0.28	0.05	0.28	6.17	0.00	0.19	4.87
Dependent variable: Customer satisfaction							
Rsquare: 0.83							

5.4.4 Predictive power for finance control

Stepwise multiple regression analysis method was adopted to analyze the predictive power of each factor of the application of "C Theory" in project management for finance control. Table 4 shows that in the criterion variable of finance control, decision-making performance, control performance, creativity performance, and coordination performance reach significant levels (T value = 5.42, P value = 0.00; T value = 11.18, P value = 0.00; T value = 4.52, P value = 0.00), wherein the influence of control performance on learning and growth is the biggest, followed by that of decision-making performance. The standardized regression coefficients of the three independent variables are positive, indicating that their influences on learning and growth are positive. The degrees of tolerance are about between 2 and 3. The expansion coefficients of variation are below 5, lower than the evaluation index value of 10, indicating that the issue of multicollinearity among the independent variables is not obvious.

Table 5. Predictive Power of Project Management in Finance Control.

	Unstandardized Coefficient		Standardized Coefficient	t	Significance	Multicollinearity Statistics	
	Estimated value of B	Standard error	Beta distribution			Tolerance	VIF
(Constant)	0.00	0.07		0.05	0.96		
Decision-making performance	0.25	0.04	0.22	5.42	0.00	0.26	3.87
Control performance	0.56	0.05	0.53	11.18	0.00	0.18	4.44
Creativity performance	0.23	0.05	0.21	4.52	0.00	0.19	4.15
Dependent variable: Finance control							
Rsquare: 0.82							

6. Conclusion and suggestions

According to the results of this study, it can be seen that for the differences of social variables and in terms of female, young employees, low education backgrounds, agriculture, forestry, animal husbandry, fishery, and production and manufacturing practitioners, average employees, organizations with short history and middle and small scales, measures like training courses of project management, introduction of the knowledge system, procedures, and methods of project management, enhancement of the function of project management and organizational maturity, that enhancement of the application of "C Theory" in project management shall be adopted to enhance the efficiency and performances of organizations. Furthermore, there is a highly positive correlation between the application of "C Theory" in project management and organizational performance. There are also highly positive correlations among creativity performance, learning and growth, and customer satisfaction, between adaptability performance and internal operation, and between control performance and finance control. In summary, this study suggests that in order to promote organizational performance, during project management the decision-making performance of Taoism, control performance of Legalism, tactics performance of School of Military Strategists, creative performance of Mohism, and coordination performance of Confucianism of "C Theory" can all be fully used to give full play to the benefits of decision-making function, control function, adaptability function, creativity function, and coordination function and to improve the practical benefits of project management so as to enhance organizational performance, such as learning and growth, internal operation, customer satisfaction, and finance control.

References

1. M.H. Chang, The application of balanced scorecard in school performance management, Journal of Educational Research and Development, **123**, 19-25 (2004)
2. S.C. Chang, A probing design of the balanced scorecard application in nonprofit organizations—a case of national vocational high school of commerce, Unpublished Master's Thesis of Chung Yuan University (Taoyuan, Taiwan, 2002)
3. M.J. Chen, Dynamic competition (Taipei, Taiwan, 2010)
4. M.T. Chen, The secret to operation and management in the book of changes (Taipei, Taiwan, 2014)
5. C.Y. Cheng, C Theory—management philosophy of the book of changes (Taipei, Taiwan, 1995)
6. C.Y. Cheng, Chinese business ethics and management based on the book of changes, Chinese Journal of Management, **12** (1), 29-37 (2015)
7. S.T. Chiang, Study on the relationship between teacher efficiency and school efficiency of senior middle schools—A case study of total performance scorecard, Unpublished Master's thesis of National Kaohsiung Normal University (Kaohsiung, Taiwan, 2008)
8. H.C. Chien, Inspiration of balanced scorecard theory to the establishment of the performance assessment indicators of primary and middle schools, Journal of Yu-Da College, **4**, 89-98 (2003)
9. H.C. Chiu, Structural Equation Modeling (SEM): LISREL theories, technologies, and applications (Taipei, Taiwan, 2003)
10. H.C. Chiu, *Quantitative Research and Statistical Analysis*: *Quantitative Research and Statistical Analysis in Social & Behavioral Sciences* (*3rd Edition*) (Taipei, Taiwan, 2008)
11. C.M. Chu, *Management Philosophy of Confucianism*: *Leaders of Utopia* (Taipei, Taiwan, 1994)
12. T.K. Chu (trans), Balanced scorecard (Taipei, Taiwan, 2008)
13. P.D. Norton R.S. Kaplan, *The Strategy-Focused Organization*: *How Balanced Scorecard Companies Thrive in the New Business Environment* (Taipei, Taiwan, 2001)
14. The international project management knowledge system, Editorial Board of Taiwan Project Management Association (2014)
15. C.Y. Hsu, Leadership from the perspective of history (Taipei, Taiwan, 1992)
16. H.Y. Hsu, P.L. Hsiung, Y.F. Chu, Y.H. Chang, M. Fan, C.M. Huang, H.T. Chou, and C.P. Lu, *Basic Project Management Knowledge and Applications in Practice* (*5th Edition*) (New Taipei, Taiwan, 2014)
17. K.K. Huang, Chinese corporate culture and productivity, Applied Psychology Research, **1**, 163-185 (1999)
18. T.H. Huang, Application of basic concepts of balanced scorecard in creative operation of school, Journal of Educational Research and Development, **124**, 52-58 (2004)

19. ICB: IPMA Competence Baseline, http://ipma.ch/resources/ipma-publications/ipma-competence-baseline/ (2009)
20. T.L. Kang, S.H. Chang, L.J. Yang, and S.K. Huang, Development of chinese leadership model: An empirically-based theory, Education Policy Forum, **13** (1), 115-142 (2010)
21. T.L. Kang, K.J. Su, S.H. Chang, and S.C. Hsu, The development of the chinese leadership model and its application in vocational schools, Bulletin of Educational Research, **47**, 133-154 (2001)
22. I.C. Kao, Study on Chinese style contingency leadership-Preliminary structure of "throne" leadership philosophy, Journal of Educational Research and Development, **7**, 45-56(1999)
23. I.C. Kao, *Management of Civic Knowledge: Feasible Strategies to Establish Knowledge Management System of Educational Project of School*, Cited from Wu, Y.M. and Tsai, T.C. (ed.), Civic Urbanology (18-35) (Kaohsiung, Taiwan, 2009)
24. I.C. Kao and Y.S. Chen, A study on model of project management maturity and criteria of assessment, Journal of Project Management, **2** (1), 73-103 (2009)
25. I.C. Kao, Construction and application of the evaluation indications for the practical operation and organizational performance of project management in primary schools, Journal of Research on Elementary Education, **8**, 109-136 (2011)
26. I.C. Kao, Feasible strategies to apply restriction theory, knowledge management, and project management to enhance school performance, Journal of Knowledge Economy, **8** (2), 59-82 (2011)
27. I.C. Kao, The application of the environment analysis and organizational pattern of project management in school organizational development, Journal of Knowledge Economy, **9** (2), 170-199 (2012)
28. I.C. Kao, The research on construction of project management indicators, Storage Management Solutions, **1**, 154-173 (2013)
29. I.C. Kao, Feasible strategies of project management processes, team formation and institutionalized operations applied to school governance, Storage Management Solutions, **1**, 198-229 (2013)
30. I.C. Kao, Application of definition and role analysis of project stakeholders in the human resource development of school project management strategies, Storage Management Solutions, **6**, 158-190 (2013)
31. I.C. Kao, Construction of human resource development strategy of community empowerment by project management, TOC and TTQS, Storage Management Solutions, **2**, 42-65 (2014)
32. I.C. Kao, Control of power and projects by "Sitting on the dragon throne": Feasible school project management strategies based on contingent chinese leadership, Advances in Engineering Education, **5** (4), 38-80 (2014)
33. I.C. Kao, The Correlation between Project Management and Organizational Effectiveness at Taiwan's Public and Private Organizations—Students Who Receive In-Service Continuing Education at the Open University of Kaohsiung as Survey Participants, Advances in Engineering Education, 89-123 (2015)
34. I.C. Kao, Feasible strategies for setting the course of "Community development project management" in adult education institutions, Storage Management Solutions, **2**, 35-62 (2016)
35. H. Kerzner, *Project Management: A System Approach to Planning, Scheduling, and Controlling (9th Edition)* (New York, USA, 2006)
36. C.C. Ko, Study on the correlation among chinese style leadership, organizational culture, and organizational creativity—A case study of SMEs in Taiwan, Unpublished Doctoral thesis, National Changhua University of Education (Changhua, Taiwan, 2012)
37. J. Lewis, *Fundamentals of Project Management* (*2nd Edition*) (New York, USA, 2002)
38. J. Lewis, *Mastering Project Management* (New York, USA, 2008)
39. C.C. Li and H.C. Yu, Master sun's art of war and the book of changes—Decision-making theory and decision-making management (Taipei, Taiwan, 2012)
40. C.W. Liao, Study on the universality of balanced scorecard in school performance assessment, Bulletin of Research on Elementary Education, **12**, 1-17 (2004)
41. Y.J. Lin, *Study on the Book of Changes*, T.J. Pan, H.T. Lai, (ed.) Eastern and Western Philosophy

and Interpretation of Ontology, Kangde Publishing Press October , 63-102(2005)
42. C.T. Liu, Breathtaking heaven, amazing humanity—The first lesson of the book of changes (Taipei, Taiwan, 2015)
43. J.C. Lu, Wisdom of Chinese management philosophy of the pre-qin period–XYZ and L type management, T&D Feixun, **34**, 1-42 (2005)
44. K.M. Lu, Software project management (Taipei, Taiwan, 2009)
45. J.C. Nunnally, *Psychometric Theory* (New York, USA, 1978)
46. C.C. Peng and C. Liu, Study on contingency management theory and the philosophy of the book of changes, Study on the Book of Changes, **79**, 88-91 (2006)
47. PMI, *A Guide to the Project Management Body of Knowledge* (*5th Edition*) (Pennsylvania, USA, 2012)
48. C.W. Shih, Exploration on the performance assessment system of non-profit organizations from the perspective of balanced scorecard—A cast study of student affairs divisions of colleges and universities in Kaohsiung, Unpublished Master's thesis of I-Shou University (Kaohsiung, Taiwan, 2004)
49. R.M. Si, H.E. Yin, C.H. Chen, and K.T. Hsieh, *Read the Books of Changes, Master Management* (Hong Kong, China, 2010)
50. T. Peters and R.H. Waterman, *In Search of Excellence* (Taipei, Taiwan, 2005)
51. C.H. Tsai, Reflection and inspiration of school leadership from the perspective of Chinese style leadership, Journal of Educational Research and Development, **168**, 105-111 (2008)
52. P.T. Tsai and W.Y. Wu, *Leadership Science: Theory, Practice, and Study* (*2nd Edition*) (Kaohsiung, Taiwan, 2013)
53. C.S. Tsao, *"Easy" Mode of Thinking in Systems Thinking and its Reference Value of Modern Management*, Management Science Research, **6**, 26-31 (2009)
54. S.C. Tseng, Modern Chinese style management (Taipei, Taiwan, 1987)
55. S.C. Tseng, *Chinese style management: Management science with Chinese characteristics* (Taipei, Taiwan, 2002)
56. S.C. Tseng and C.C. Liu, Interaction of three ways of management (Taipei, Taiwan, 1999)
57. C.J. Wei, The general project management knowledge system (Taipei, Taiwan, 2013)
58. A.N. Wu, The essence, scope, and integration of balanced scorecard, Accounting Research Monthly, **211**, 45–54 (2003)
59. A.N. Wu, Promotion of strategy and performance management by balanced scorecard, Data Systems Consulting Co., http://www.dsc.com.tw/bsc/bsc_a/bsc_a1.htm (2004)
60. A.N. Wu, Characteristics and functions of balanced scorecard, Data Systems Consulting Co., http://www.dsc.com.tw/bsc/bsc_a/bsc_a2.htm(2004)
61. H. Wu, Balanced scorecard and performance management, Accounting Research Monthly, **12** (2), 35-40 (2000)
62. M.L. Wu, Structural Equation Modeling (SEM)–The application of SIMPLIS (Taipei, Taiwan, 2006)
63. C.F. Yang, *How to Study Chinese* (Taipei, Taiwan, 1996)
64. W.D. Yu, Practical project management (Taipei, Taiwan, 2015)
65. Y. H. Yu and I.P. Chen, Complete guidance on balanced scorecard (Taipei, Taiwan, 2004)

Estimation of peak floor acceleration based on support vector regression and p-wave features

Tzu-Kang Lin[a] and Chang-Han Wu

Department of Civil Engineering, National Chiao Tung University, Hsinchu 300, Taiwan

Abstract. This study proposes a support vector regression (SVR)–based method to evaluate the peak floor acceleration (PFA) in a building subject to a major earthquake. Support vector machine, which constructs a hyperplane in high-dimensional space to solve multivariate problems, is well known for the outstanding performance in classification and regression. Six P-wave parameters obtained from the vertical component of the first three seconds of the acceleration time history of the ground motion, as well as the floor height and the fundamental period of the structure, are adopted as the eight input variables. With these and the support of SVR, the PFA of a specific building can be rapidly estimated to avoid the necessity of installing accelerometers on the building. The SVR model is trained on 2274 representative earthquake records from the Structure Strong Earthquake Monitoring System (SSEMS) of the Central Weather Bureau (CWB) in Taiwan and tested on 757 independent test earthquake records. As demonstrated, the accuracy of the predicted PFA, located within a one-level difference on the seismic intensity scale of Taiwan from the monitored PFA, is 95.51%. The developed algorithm can be integrated into an existing earthquake early warning system (EEWS) to provide comprehensive protection during major earthquakes.

1 Introduction

Two types of EEWS, the regional early warning system and the on-site early warning system, have been adopted in Taiwan. Although an early warning of an earthquake can be announced by the regional early warning system a few seconds before a major earthquake arrives, it requires the P-wave information of several stations and, therefore, it may take approximately 20 to 30 seconds to announce the warning signal. Under this circumstance, regions that are 60 km from the epicenter then belong to the blind space of the early warning system (for the velocity of the S-wave estimated at 3 km/s). On the other hand, as the P-wave information of only one station is required by the on-site early warning to predict the peak ground acceleration (PGA) of an earthquake, the blind space in this instance can be much smaller than the previously described one.

In order to enable earlier earthquake alerts to reduce possible casualties when an earthquake occurs, the National Center for Research on Earthquake Engineering (NCREE) developed an on-site EEWS, which can predict the PGA of the whole time history of the earthquake before the major vibration arrives [1]. The entire operation process is shown in Figure. 1.

However, all current EEWSs focus only on the prediction of the PGA, but the PFA is more

[a] Corresponding author : tklin@nctu.edu.tw

 DOI: 10.1051/matecconf/201711901028

important than the PGA from a structural perspective. As a result, this study proposes an approach to predict the PFA so that the earthquake's influence on a building can be easily assessed. It is expected that an early warning for each floor potentially under threat can be sent out automatically so that efficient evacuation can be conducted to reduce possible casualties.

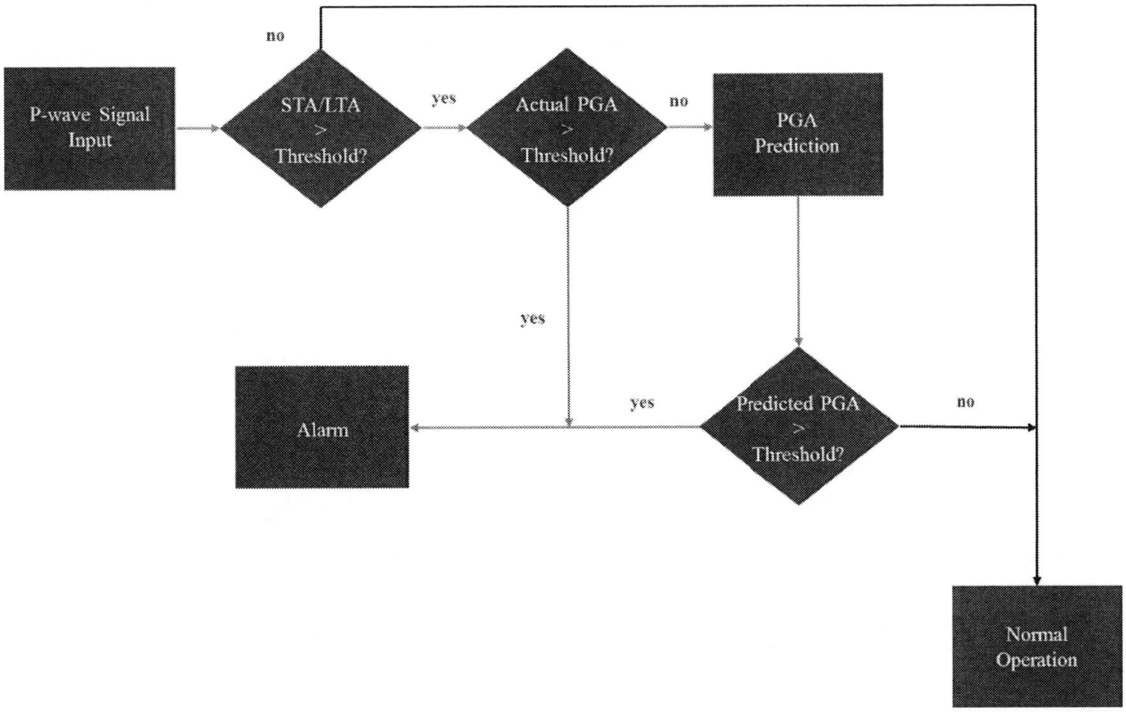

Figure 1. The operation process of an on-site EEWS.

2 The proposed PFA prediction method

2.1 P-wave features

According to Satriano [2], the P-wave features including peak measurements, integral quantities [3], the predominant period, and the average peak can be applied to an EEWS. Since the average peak is similar to the peak measurements, and as it is necessary to set some thresholds in computation, the average peak is not adopted as a parameter for analysis in this study.

Let t_p seconds be the time of observation after a P-wave arrives at the station from the epicenter, and the time that the P-wave arrives at the destination is called P-wave Arrival, which can be determined by a short-term average/long-term average (LTA/STA) method. Three seconds after the P-wave arrives, that is t_p equals 3 seconds, the P-wave features, which are the peak measurement of acceleration (P_a), peak measurement of velocity (P_v), peak measurement of displacement (P_d), effective predominant period (T_c), integral of squared velocity (IV2), and cumulative absolute velocity (CAV), can be quickly estimated from the time history of vertical earthquake acceleration. Of these parameters, peak measurements (P_a, P_v, and P_d) are associated with the PGA, peak ground velocity, and peak ground displacement, respectively, and Kanamori [4] suggested that T_c and IV2 are correlated to the magnitude of an earthquake. Additionally, CAV can be used to detect whether there is a destructive earthquake coming. The T_c, IV2 [5], and CAV [6] can be calculated as:

$$T_c = \frac{2\pi}{\sqrt{r}}, \quad where \ r = \frac{\int_0^{t_p} \dot{u}(t)^2 \, dt}{\int_0^{t_p} u(t)^2 \, dt} \tag{1}$$

$$IV2 = \int_0^{t_p} \dot{u}(t)^2 dt \tag{2}$$

$$CAV = \int_0^{t_p} |\ddot{u}(t)| dt \tag{3}$$

where $u(t)$, $\dot{u}(t)$, and $\ddot{u}(t)$ are the vertical components of the displacement, velocity, and acceleration time histories of ground motion after P-wave Arrival, respectively.

Since the primary earthquake database from the CWB consists of acceleration time-history information, it is necessary to integrate the acceleration to obtain the velocity and displacement for P-wave features. Moreover, a high-pass filter is used to correct for bias shifting at low frequency, where the order and the cutoff frequency of the filter are 2 and 0.075 Hz, respectively.

2.2 Support vector regression

Support vector machine (SVM), a data recognition technique, can be used to generalize and classify the database through image recognition. SVM can be classified as either support vector classification (SVC) or support vector regression (SVR), in accordance with its application purpose. Compared with other analysis methods, the major advantage is that SVR seldom has problems with local minima, overfitting, or underfitting. However, when establishing the SVM model, the amount of training data must be of sufficient representativeness, so that the trained model can be used to classify or regress accurately.

The main concept of SVR is that the original data space can be projected to the hyperspace by using a nonlinear projection function, and the precise outcome can be predicted easily. Consequently, the prediction of PFA for a building is attempted by using SVR [7].

Take a training data set $(x, y) \in R$, where x represents input data, and each set of data can be one or multiple parameters, and y represents the actual value related to the input training data. If the values of input data x have a direct or indirect correlation to the corresponding output value y, then SVR can be employed to find the mathematical relationship between them. The regression function can be expressed as:

$$f(w,b) = w \cdot \phi(x) + b \tag{4}$$

where w is the regression coefficient vector representing the vector of the hyperspace and b is the offset, which represents the distance between the hyperspace and the origin.

If the error between y_i and $f(x_i)$ is less than the error tolerance ε for each x_i, then $f(x_i)$ is approximately equal to y_i. However, as the error of the data is often larger than the error tolerance, which is out of the ε-insensitive area, it is necessary to add ξ_i and ξ_i^* to allow the condition.

The ν-support vector regression problem can be expressed as the following convex optimization problem [8]:

$$\min_{w,b,\xi,\xi^*} \left[\frac{1}{2} \|w\|^2 + C \left(\nu\varepsilon + \frac{1}{m} \sum_{i=1}^{m} (\xi_i + \xi_i^*) \right) \right] \tag{5}$$

subjected to:

$$w \bullet \phi(x_i) + b - y_i \leq \varepsilon + \xi_i$$

$$y_i - w \bullet \phi(x_i) - b \leq \varepsilon + \xi_i^*$$

$$\xi_i, \xi_i^* \geq 0 \text{ for } i = 1, \ldots, m$$

where m is the number of training data, C is a positive constant that determines the degree of penalized loss when a training error occurs, and $\nu\varepsilon$ is a slack variable controlled by the coefficient ν.

By introducing the method of Lagrange multipliers and setting the Lagrange multipliers α_i and β_i, Eq. 5 can be solved by applying the standard quadratic programming algorithm:

$$\max_{\alpha,\beta}\left[-\frac{1}{2}\sum_{i=1}^{m}\sum_{j=1}^{m}(\alpha_i - \beta_i)(\alpha_j - \beta_j)k(x_i, x_j) + \sum_{i=1}^{m} y_i(\alpha_i - \beta_i)\right] \quad (6)$$

subjected to:

$$\sum_{i=1}^{m}(\alpha_i - \beta_i) = 0 \leq \alpha_i$$

$$\beta_i \leq \frac{C}{m} \text{ for } i = 1, 2, \ldots, m$$

$$\sum_{i=1}^{m}(\alpha_i + \beta_i) = 0 \leq C\nu$$

where m is the number of training data and $k(x_i, x_j)$ is the kernel function to yield the inner products in feature space, which is adopted as a radial basis kernel function as follows:

$$k(x_i, x_j) = exp(-x_i - x_j / 2\sigma^2) \quad (7)$$

After the Lagrange multipliers, α_i and β_i, are determined, the parameters w and b can be estimated under the Karush–Kuhn–Tucker complementarity conditions [9]. Therefore, the prediction function can be expressed as:

$$f(x) = \left[\sum_{i=1}^{j}(\alpha_i - \beta_i)k(x, x_i)\right] + b \quad (8)$$

where j is the number of nonzero terms ($\alpha_i - \beta_i$), *i.e.*, the support vectors.

3 Earthquake time-history database

Earthquake data has been measured by the Structure Strong Earthquake Monitoring System (SSEMS) of the CWB since 1992. The system collects original earthquake records from 39 structures in Taiwan, where accelerometers are installed at different heights. The sample frequency of each data set is 200 Hz or 250 Hz, and the resolution is typically 16 bits.

Initially, earthquakes whose magnitudes are larger than $M_L = 6.0$ are chosen. In addition to the six P-wave features used to predict the PFA of a building, the floor height and the fundamental period of a structure are also considered. Consequently, the height of each accelerometer in the structure is checked one by one and the fundamental period of a structure is estimated as one tenth of the structure. The database is carefully filtered, and the data with a time duration of less than three seconds are removed. As a result, 20 stations are finally adopted, and the locations of each structure are shown in Figure. 2. Some 788 sets of earthquake time-history data, namely 3031 points of sample data, are used after the converting for the height of each structure.

Based on the definition of PFA, the peak acceleration values of each accelerometer channel in the whole seismic event are first selected, and then compared with the peak acceleration value of another accelerometer channel at the same height to determine the real PFA. Consequently, for a seismic event, a set of six P-wave features, the fundamental period of a structure, several floor heights, and several real PFAs can be established. The details of the 20 structure stations are given in Table 1.

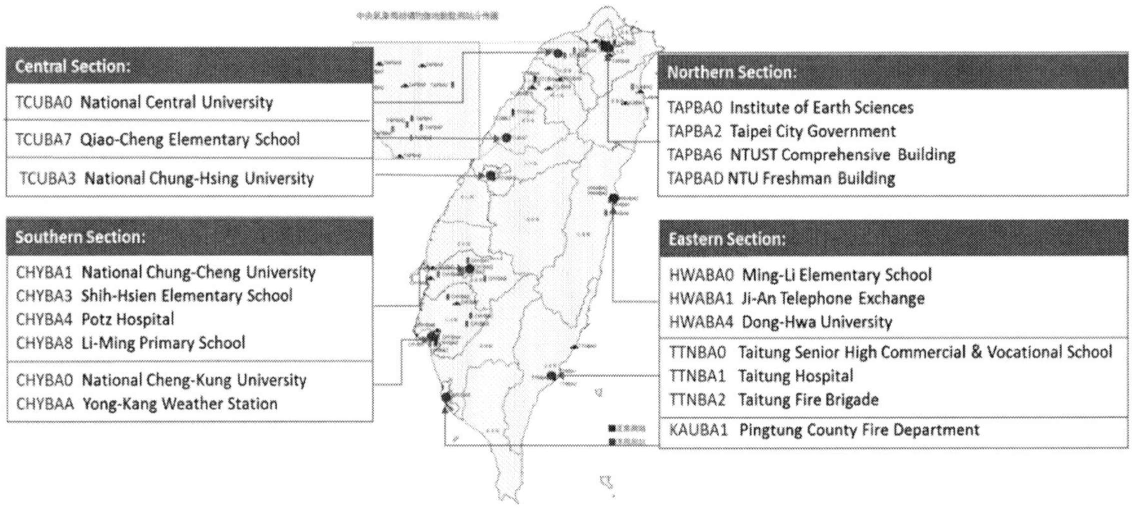

Figure 2. The locations of each station structure.

Table 1. Distribution of sample data.

Station code	Number of floor types	Number of earthquake events	Number of sample data
CHYBA0	3	17	51
CHYBA1	4	84	336
CHYBA3	5	55	275
CHYBA4	5	42	210
CHYBA8	2	35	70
CHYBAA	4	57	228
HWABA0	4	50	200
HWABA1	4	46	184
HWABA4	4	72	288
KAUBA1	4	4	16
TAPBA0	5	16	80
TAPBA2	4	22	88
TAPBA6	4	61	244
TAPBAD	4	32	128
TCUBA0	4	48	192
TCUBA3	3	39	117
TCUBA7	3	12	36
TTNBA0	3	34	102
TTNBA1	3	38	114
TTNBA2	3	24	72
Total		788	3031

4 Analysis result

A total of 2274 points of sample data from the SSEMS in the CWB are used as the representative earthquake data to train the SVR model by means of the eight relevant parameters, namely the six P-wave features, the floor height, and the fundamental period of a structure. On the basis of the correlation between the eight relevant parameters and the real PFA, the predicted PFA can be quickly evaluated. Furthermore, as the floor height and the fundamental period of a structure are fixed, they can be preset before an earthquake occurs.

In this study, a grid search method is used to determine the optimal values of C and σ, where the ranges of C and σ are 2^5 to 2^{15} and 2^{-1} to 2^{-12}, respectively. After searching within the confined range, the optimal values of C and σ can be determined with the smallest mean squared error.

As mentioned above, the data samples are used to train the SVR model, and the reserved testing samples are inputted into the trained SVR model to obtain the predicted PFA. For the purpose of checking the performance of the SVR model, the result is defined as "accurate" if the seismic intensity of the predicted PFA is located within a one-level difference of the monitored PFA, according to the classification criterion of the CWB.

Figure. 3 shows the regression result for the 2274 representative earthquake data. The best values of C and σ are 4096 and 0.0156, respectively, and the squared correlation coefficient is 0.89813. Figure. 4 shows the regression results for the 757 testing earthquake data, and the squared correlation coefficient is 0.319626. The accuracy of the predicted PFA located within a one-level difference of the seismic intensity scale from the real PFA is 95.51%.

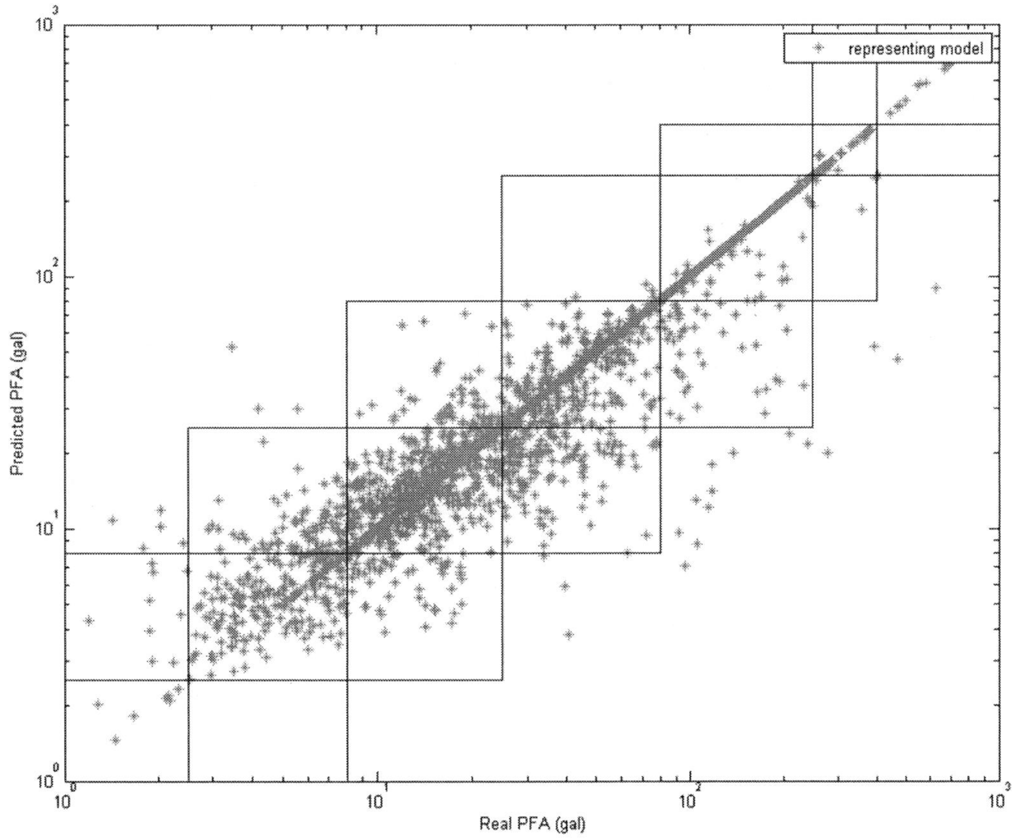

Figure 3. The training results of the representative earthquake records.

 DOI: 10.1051/matecconf/201711901028

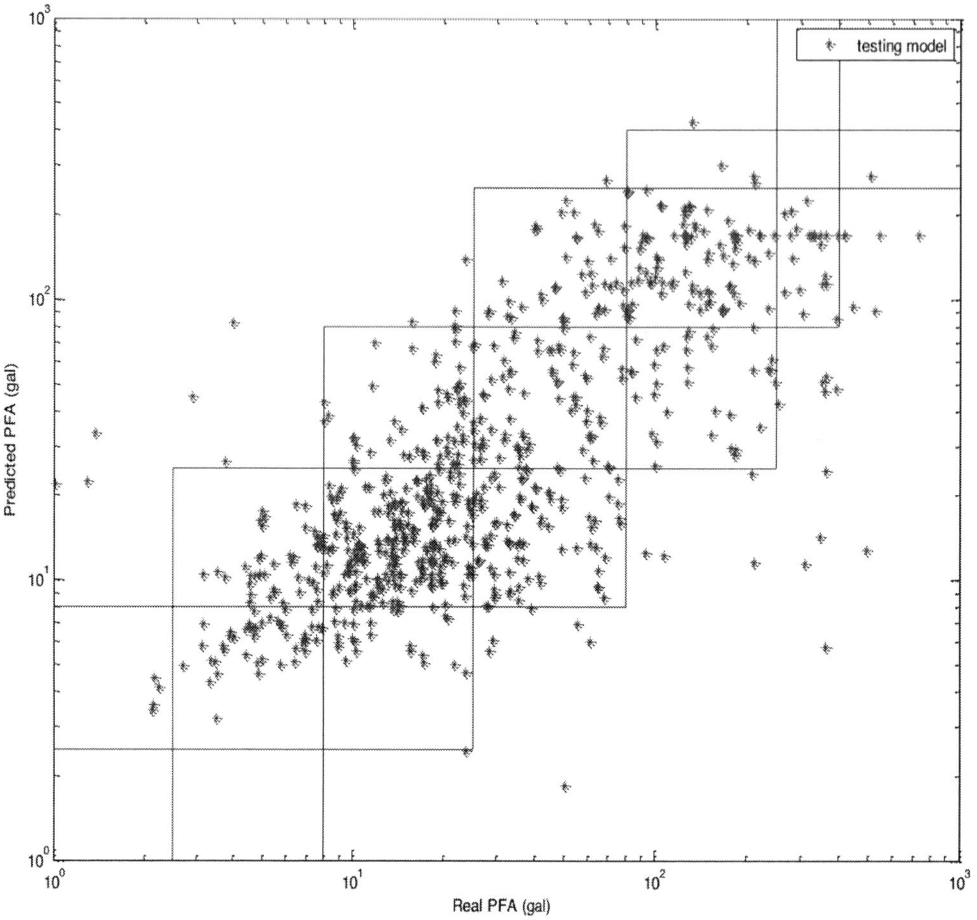

Figure 4. The testing results of the testing earthquake records.

In order to evaluate the practical applicability of this approach more rigorously, three earthquake events are randomly selected from the testing earthquake records, and the corresponding information is shown in Table 2.

Table 2. Details of the three earthquake events.

Earthquake events	Station code	Earthquake occurrence time
1	CHYBA0	2000.06.10
2	TAPBA6	2009.10.03
3	HWABA4	2007.09.06

For these three earthquake events, the seismograms are first examined. A comparison of the PFAs and seismic intensities is also made, and the results are shown in Figures. 5 and 6, Figures. 7 and 8, and Figures. 9 and 10, respectively. As indicated by the bar charts, it is obvious that most of the predictions are certainly larger than their corresponding real PFAs, except in the case of earthquake event 1 at H = 20.95 m, and all the predicted intensities are equivalent to or one-level larger than the real value. Therefore, a conservative estimation of the seismic intensity can be provided by these three earthquake events, and an earthquake early warning can be effectively provided to the residents.

 DOI: 10.1051/matecconf/201711901028

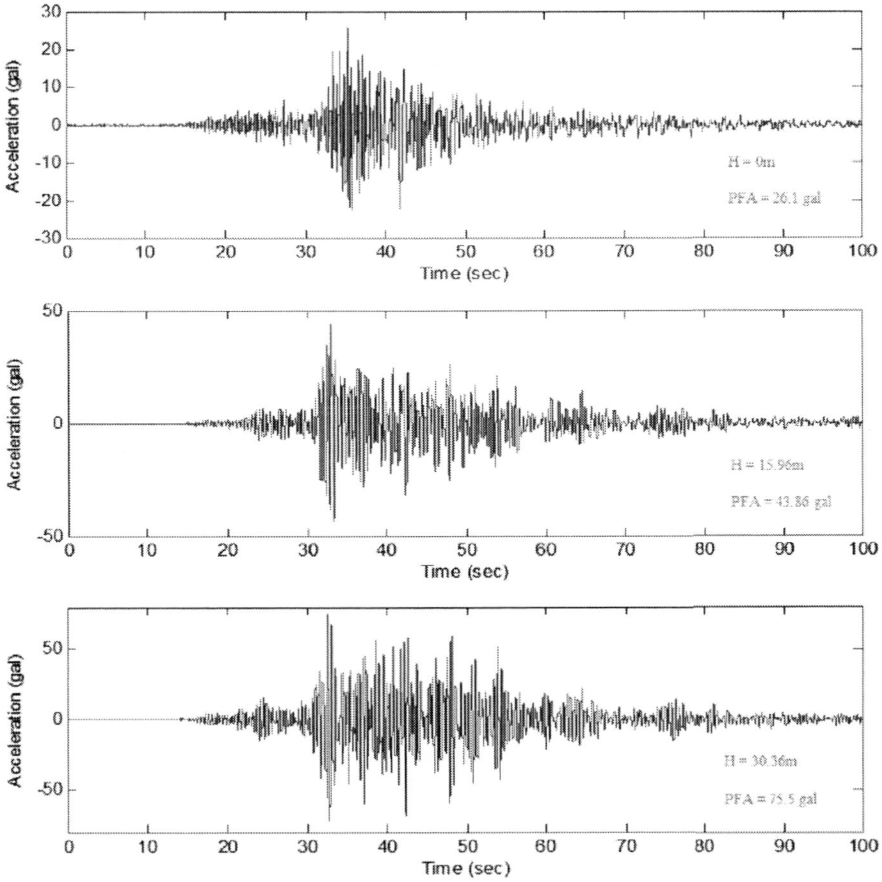

Figure 5. Earthquake event 1: seismograms.

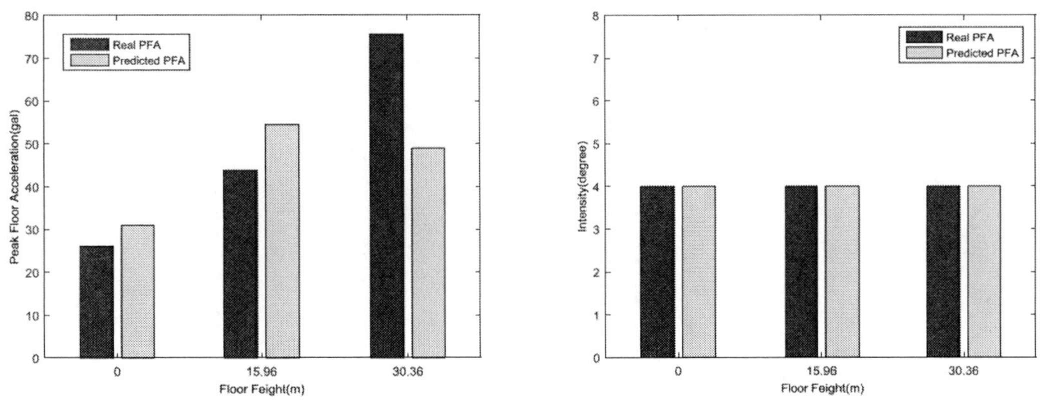

Figure 6. Earthquake event 1: comparison of PFAs and intensities.

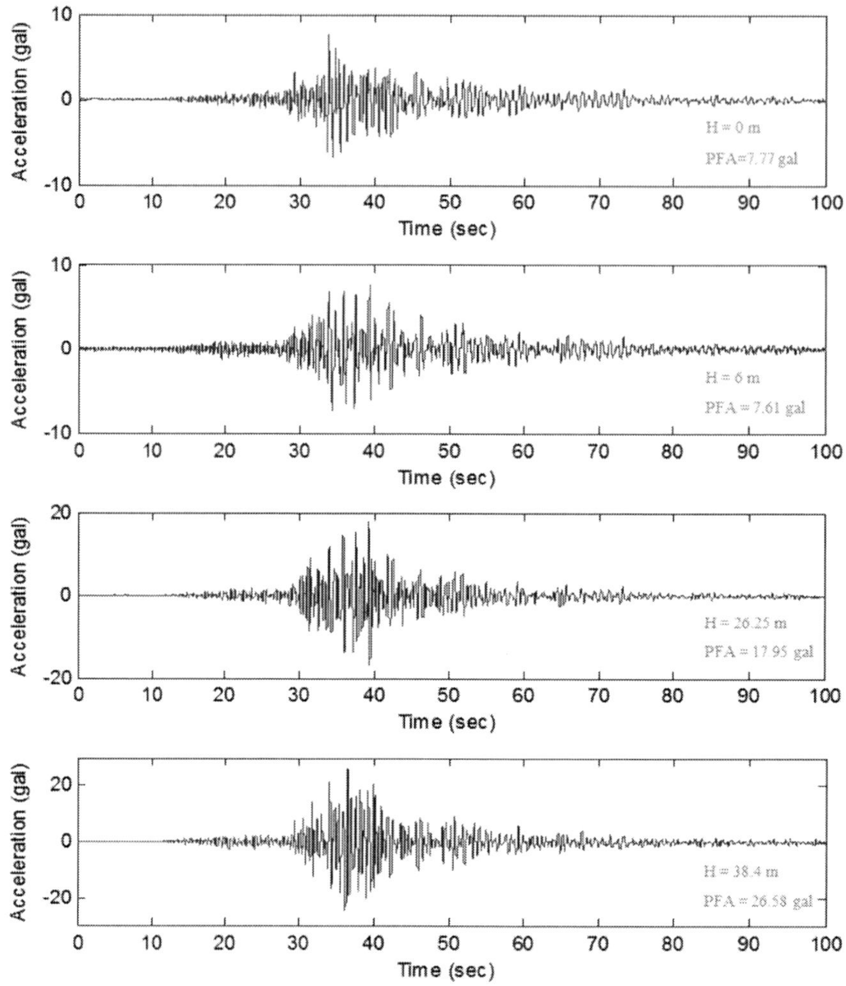

Figure 7. Earthquake event 2: seismograms.

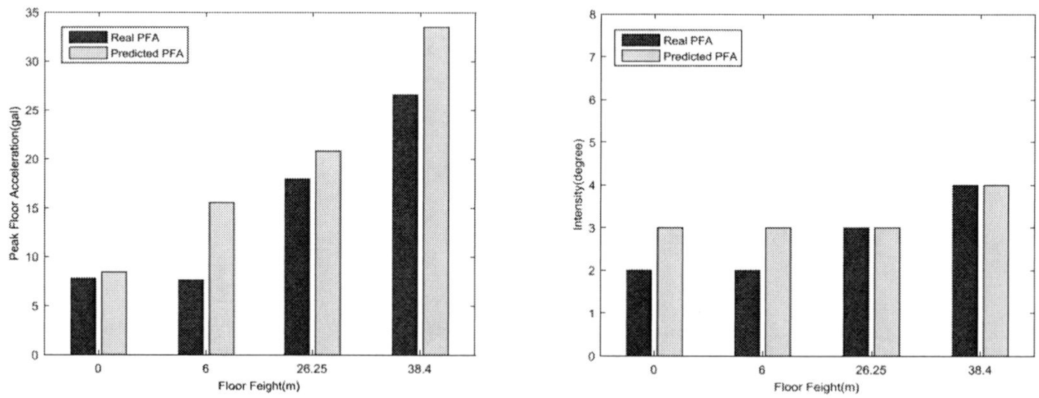

Figure 8. Earthquake event 2: comparison of PFAs and intensities.

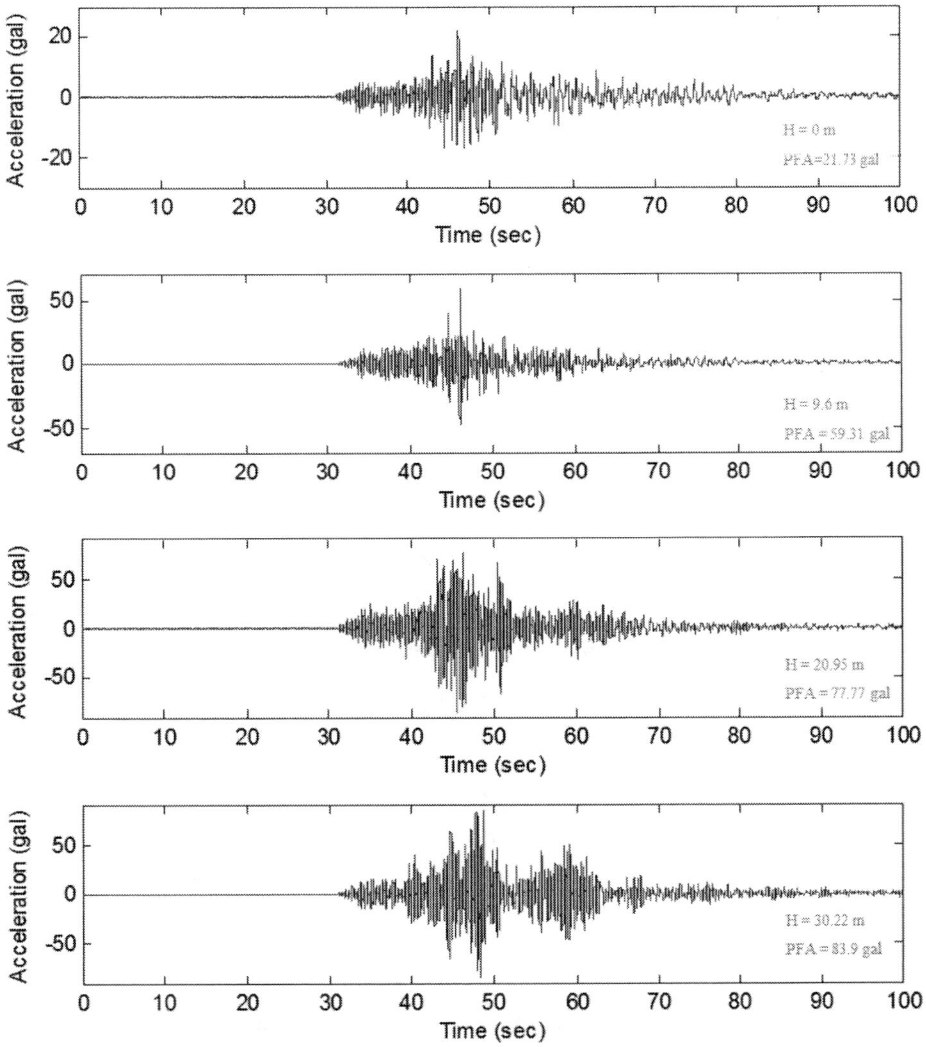

Figure 9. Earthquake event 3: seismograms.

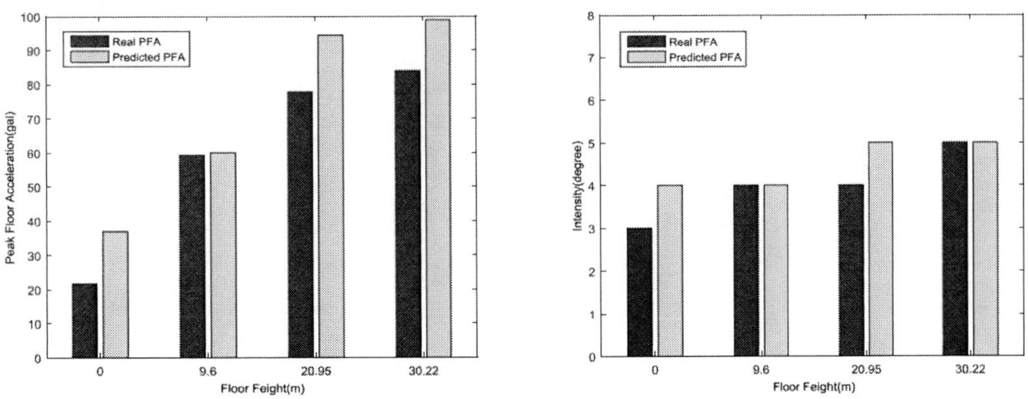

Figure 10. Earthquake event 3: comparison of PFAs and intensities.

5 Summary and conclusion

EEWS have been developed to quickly predict the peak ground acceleration of an impending and to provide early warnings instantly. To enhance the functionality of the existing EEWS system, making precise and rapid predictions of the PFA of a building to allow an evacuation has also become an important issue. In this study, an approach for predicting the PFA from an on-site EEWS was proposed. In addition to using the six P-wave features as characteristic parameters, the floor height and the fundamental period of a structure were also included to estimate the PFA of a building through SVR when an earthquake occurs. For the earthquake time-history database adopted in this study, after testing the earthquake records, which are independent of the representative earthquake records, the accuracy of the predicted PFA located within a one-level difference of the seismic intensity scale from the real PFA is 95.51%. Thus, the accuracy demonstrates that the PFA evaluated by this approach is relatively reliable.

To enhance the performance of the proposed method, more records will be included in the SVR-based method in the near future, especially earthquake records with smaller earthquake magnitudes or records collected from other on-site stations.

References

1. T.Y. Hsu, S.K. Huang, Y.W. Chang, C.H. Kuo, C.M. Lin, T.M. Chang, K.L. Wen, and C.H. Loh, Rapid on-site peak ground acceleration estimation based on support vector regression and P-wave features in Taiwan, SDEE, **49**, 210-217 (2013)
2. C. Satriano, Y.M. Wu, A. Zollo, and H. Kanamori, Earthquake early warning: concepts, methods and physical grounds, SDEE, **31**, 106-118 (2011)
3. M. Böse, T. Heaton, and E. Hauksson, Rapid estimation of earthquake source and ground-motion parameters for earthquake early warning using data from a single three-component broadband or strong-motion sensor, BSSA, **102**, 738-750 (2012)
4. H. Kanamori, Real-time seismology and earthquake damage mitigation, Annu Rev Earth Planet Sci, **33**, 195-214 (2005)
5. G. Festa, A. Zollo, and M. Lancieri, Earthquake magnitude estimation from early radiated energy, Geophys Res Lett, **35** (22) (2008)
6. M. Erdik, Y. Fahjan, O. Ozel, H. Alcik, A. Mert, and M. Gul, Istanbul earthquake rapid response and the early warning syste, B EARTHQ ENG, **1** (1), 157-63 (2003)
7. V.N. Vapnik, *Springer-Verlag*, The nature of wtatistical learning theory (1995)
8. B. Schölkopf, A. Smola, R. Williamson, and P.L. Bartlett, New support vector algorithms, Neural Comput, **12**, 1207-45 (2000)
9. R. Fletcher, *Practical Methods of Optimization*, Second Edition, John Wiley (1987)

Stock prices forecasting based on wavelet neural networks with PSO

Kai-Cheng Wang[1,a], Chi-I Yang[1] and Kuei-Fang Chang[2]

[1]*Ph.D. Program in Mechanical and Aeronautical Engineering, Feng-Chia University, Taichung 40724, Taiwan*
[2]*Department of Applied Mathematics, Feng-Chia University, Taichung 40724, Taiwan*

Abstract. This research examines the forecasting performance of wavelet neural network (WNN) model using published stock data obtained from Financial Times Stock Exchange (FTSE) Taiwan Stock Exchange (TWSE) 50 index, also known as Taiwan Stock Exchange Capitalization Weighted Stock Index (TAIEX) , hereinafter referred to as Taiwan 50. Our WNN model uses particle swarm optimization (PSO) to choose the appropriate initial network values for different companies. The findings come with two advantages. First, the network initial values are automatically selected instead of being a constant. Second, threshold and training data percentage become constant values, because PSO assists with self-adjustment. We can achieve a success rate over 73% without the necessity to manually adjust parameter or create another math model.

1 Introduction

Wavelet neural network (WNN) was first proposed by Zhang and Benveniste in 1992 [1]. It is a special feed-forward network supported by the wavelet theory. Wavelet neural network is not just confined to the analysis of the stock market; it is also widely used in financial derivatives market, gold market and so on [2, 3]. Zhang Qian and Gao Liqun (2002) used wavelet noise filtering technique and least square method to predict price of Denghai Seed industry (stock code: 002041) shares in China [4].

In Taiwan, most investors prefer holding mutual funds over other assets in their investment portfolio. Since the majority of Taiwan's mutual funds are stock-based, there is value in developing a model using opening and closing stock prices. We believe this model should possess qualities such as universalness and ease of use.

The reason why neural networks had been widely adapted is because its distinctive characteristic of simulating the human brain thought process, having the ability to learn and solve problems. Through artificial neural network, parallel computing of the biological neural system can be achieved, allowing high volume data analysis within a short period of time. Now, used across multiple disciplines, stock price prediction is one among the many it's used in. Universalness is the goal for models. In our earlier researches [5], we combined second-order autoregressive mode (AR(2)) with

[a] Corresponding author : gtotony98@gmail.com

wavelet analysis and its relative techniques. Unfortunately, this method was extremely limited by concavity and concavity distribution. However, WNN's network architecture is constructed with many wavelons through combination of weighted wavelons. Therefore, it is naturally unaffected by concavity.

The WNN consists of three parts: parameter; network architecture; and forecasting method. Generally, WNN has five parameters: weight; translation; dilation; threshold; and adjustment. We constructed a model which combines PSO. This method has two advantages. First, network initial values are automatically selected instead of a constant. Second, the threshold and training data percentage become constant values, because after applying PSO, the remaining three of five parameters can self-adjust to the algorithm. Thus, the model presented in this paper has strong operational ease. In 2000, Y. Oussar's research indicated that weight, translation, and dilation parameters shall be decided by the stock price's minimum and maximum values. Experiment results proved, when the initial value choice-making methodology suggested by Y. Oussar was used, not only does the calculation fall into an infinite loop often, the forecasted stock price value also exceeded the government enforced limit-up and limit-down policy, which is non-coherent with Taiwanese laws.

2 Methodology

We collected data of 41 stocks from the Taiwan 50 (Table 1) between 2010/07/15~2010/12/16, 100 trading business days. The daily stock closing price was chosen for input because the closing price reflects all the activities of the index on that day [6]. This means, for each company, a total of 100 stock prices, one from each trading day, is inputted for calculation. During experimentation, 10% of the data points from each stock price data were adopted as the training set. The remaining 90% were used for testing. This 10% was set as the training data percentage parameter. The threshold parameter was 0.01, constant without change. Figure 1 shows the main procedures of our research. The inputs and outputs of each block are further derailed in the material that follows

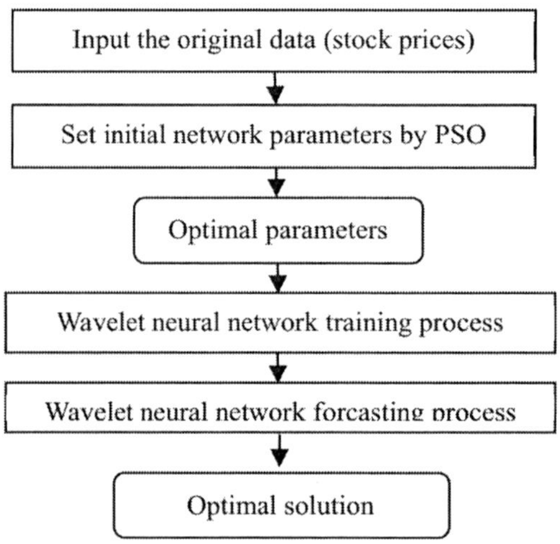

Figure 1. WNN with PSO simple flow chart.

Table 1. 100 times results from 41 Stocks of Taiwan 50.

2010/07/15~2010/12/16	Training data percentage=10% Threshold=10^(-2)		
Stocks (symbol)	PSO(B=C=2) < WNN's RMSE [The number of Success]	WNN's RMSE	PSO(B=C=1.8) <WNN's RMSE [The number of Success]
Yuanta Financial Holdings (2885)	100	1.6988	65

DOI: 10.1051/matecconf/201711901029

Ctbc Financial Holding Co., Ltd.(2891)	100	2.4007	100
China Steel Corporation(2002)	100	0.9933	100
Compal Electronics, Inc.(2324)	100	1.3612	100
Advanced Semiconductor Engineering,Inc.(2311)	98	2.1026	100
AU Optronics Corp.(2409)	100	0.7067	100
Formosa Chemicals & Fibre Corporation(1326)	0	4.6481	0
Taiwan Cement Corp.(1101)	100	0.6244	100
Taiwan Fertilizer Co., Ltd.(1722)	75	2.9549	1
Formosa Plastics Corporation(1301)	99	3.7825	100
Formosa Petrochemical Corp(6505)	100	4.4747	100
Delta Electronics, Inc.(2308)	100	2.1597	100
Taiwan Semiconductor Manufacturing Co., Ltd.(2330)	0	3.7634	99
Taiwan Mobile Co., Ltd.(3045)	0	0.5353	0
Cheng Shin Rubber Ind., Co., Ltd.(2105)	58	1.4546	98
Sinopac Financial Holdings Company Limited(2890)	100	2.2117	100
Mega Financial Holding Company Ltd.(2886)	100	2.3221	100
Lite-On Technology Corp.(2301)	99	2.1014	100
Acer Incorporated(2353)	98	4.4261	100
HTC Corporation(2498)	100	42.1494	100
Asia Cement Corporation(1102)	100	1.9381	100
Siliconware Precision Industries Co., Ltd.(2325)	0	1.0993	0
EVA Airways Corporation(2618)	100	0.6122	100
Cathay Financial Holding Co., Ltd.(2882)	0	0.9139	82
Uni-President Enterprises Corp.(1216)	0	0.5644	0
President Chain Store Corporation(2912)	0	4.4315	0
Fubon Financial Holding Co., Ltd.(2881)	100	1.7556	100
Inotera Memories, Inc.(3474)	0	0.4160	99

Hua Nan Financial Holdings Co., Ltd.(2880)	100	0.9318	100
Asustek Computer Inc.(2357)	100	24.0938	100
China Development Financial Holding Corp.(2883)	100	1.9064	100
Shin Kong Financial Holding Co.,Ltd.(2888)	100	2.2873	100
Far Eas Tone Telecommunications Co., Ltd.(4904)	100	3.1666	100
Quanta Computer Inc.(2382)	98	2.3954	100
Wistron Corporation(3231)	100	1.7876	3
Synnex Technology International Corp.(2347)	0	2.5592	0
MediaTek Inc.(2454)	98	26.6667	2
United Microelectronics Corp.(2303)	1	0.3137	99
Hon Hai Precision Ind. Co., Ltd.(2317)	0	2.3107	0
Foxconn Technology Co., Ltd(2354)	0	6.7135	0
Pou Chen Corporation(9904)	90	0.3341	100
The number of success stocks	28		32
41 stock's success rate	68%		73%

2.1 Practical swarm optimization

In 1995, J. Kennedy and R. C. Eberhart were the first to introduce PSO[7]. Originally, it was for establishing an algorithm simulating social behaviors based on the practical example of simulating behaviors of foraging birds. Later on, it was used for solving optimization problems. The PSO calculation process and parameters used in this research are as follows.

Assuming there is D number of stock prices, each price is a point in the D-dimensional space. Location of the *i*-th point can be presented as:

$$X_i = (x_{i1}, x_{i2}, x_{i3}, \ldots, x_{iD})$$

The best location that the *i*-th price had passed can be recorded as:

$$P_i = (p_{i1}, p_{i2}, p_{i3}, \ldots, p_{iD})$$

The component closest to the target can be found by analysing P_1 through P_D, changing the component to P_{gd}.

The *i*-th price's velocity is recorded as:

$$V_i = (v_{i1}, v_{i2}, v_{i3}, \ldots, v_{iD})$$

The two formulas below simulate the behaviour pattern of the bird flock:

$$V_{id} = \omega \times V_{id} + c_1 \times rand() \times (P_{id} - X_{id}) + c_2 \times Rand() \times (P_{gd} - X_{id}) \quad (1)$$

$$X_{id} = X_{id} + V_{id} \quad (2)$$

Table 2. PSO's parameters.

Parameters	Description
V_{id}	Velocity of particle at ith
ω	Weight. $\omega = 1$.
C_1 and C_2	Acceleration factor. $c_1 = c_2 = 2$
$rand()$ 、 $Rand()$	Random number between 0 and 1
P_{id}	*i*-th best position of particle
X_{id}	*i*-th current position of particle
P_{gd}	Best position of swarm

Based on the data above, (1) was used to update particle velocity. Then, iterated in (2) for the new location.

2.2 Using PSO to optimize the initial values of WNN

Five network parameters were assigned for this WNN: weight parameter w; translation parameter t; dilation parameter d; adjustment parameter $\overline{y}$; and threshold value. The first three parameters, weight parameter w, translation parameter t, and dilation parameter d ,were the optimized parameters. The method of implementation was:

Step 1: The initial values of w, t, and d were assigned based on the parameters developed by Y. Oussar [8], shown below:

w: Set as 0, t: $(M+m)/2$, d: $(M-m)/5$

When m is the original data's minimum value, M is the maximum value. There were 100 initial values for each parameter.

Step 2: The PSO parameters chosen were based on Table 2 [9, 10].

Step 3: A range must be specified when using PSO to optimize the three initial value parameters w, t, and d.

1. Range of weight parameter: $[-m/A, m/A]$.
2. Range of translation parameter: $[m, Bt]$.
3. Range of dilation parameter: $[-Cd, Cd]$.

Various statistics were used to test how "small" the weight needed to be. The experimental results were best when $A = 200$. We have found that the same stands true for the translation parameter and dilation parameter ranges. This was the most significant discovery in this research, the usage of smaller values for t and d after much experimenting, performing favorably in the forecast of stock prices. In this experiment, setting $B = 1.8, C = 1.8$ produced better experimental results compared to our previous works ($B = 2, C = 2$) [11]. However, both B and C have low bounds. By setting $B = 1$ or $C = 1$, WNN fell into infinite loops. We also found that the translation parameter values must all be positive for best performance. The range of PSO parameters cannot be too much beyond the original range, thus the minimum value of stock price data were selected as the left end point, and the right end point assigned was double or less the original t. The range of dilation parameter can be slightly greater than the original parameter. When B and C were assigned the values of 1 and 2.5, the results were extremely poor. When substituted with 1.8 and 2, the results were much more favorable. With the above mentioned, we suggest assigning B and C with values between 1.5 and 2.

Step 4: To obtain the optimized value of w, t, and d, we ran the PSO program loop using the statistics gathered from step 1 through step 3. Objective function is $1/\left(w_i \psi\left(\frac{x_i - t_i}{d_i}\right) + \bar{y} - x_i\right)$. The 100th generation is set to be the stopping generation. Since 100 stock prices were inputted, a corresponding 100 sets of w, t, and d were generated as well, vastly different from Y. Ouusar's experiment, which had identical w, t, and d values. The optimized WNN parameters are $\{w_i\}$, $\{t_i\}$, and $\{d_i\}$.

2.3 Wavelet neural networks

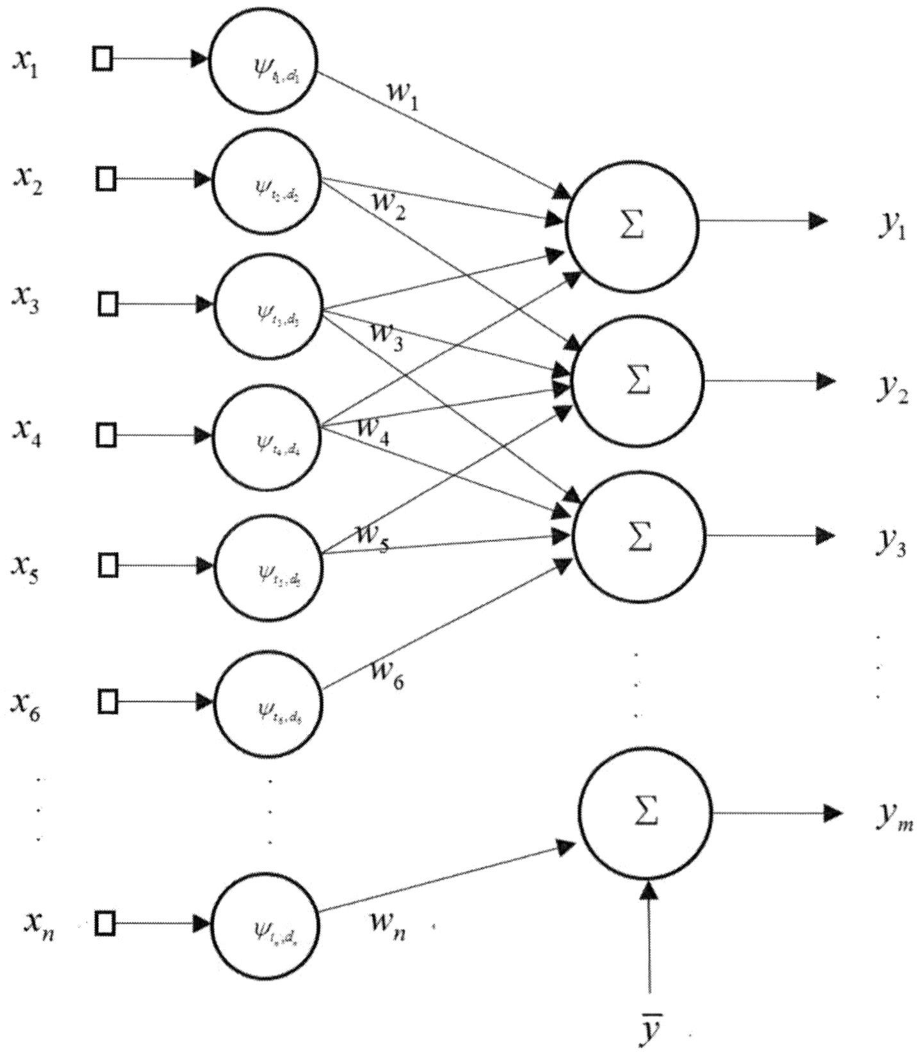

Figure 2. Wavelet Neural Network Structure.

There are many types of artificial neural networks (ANNs). We say that an ANN is a wavelet neural network (WNN) where the activation function is a wavelet [12]. In order to construct a WNN, the number of hidden layers as well as the number of neurons in these hidden layers must be provided. The advantage of the WNN structure proposed in this paper is that neither of these parameters is required (Figure 2). This model uses the Morlet wavelet as the activation function, and there are two main reasons.

First, using wavelet is simpler than using Fourier basis. To understand the first reason, one must first understand the differences between sine/cosine waves and wavelets. Based on the fact that reconstructing signal f from dilations of a fixed sinusoidal function $x \to e^{2\pi ix}$ by $f(\mathrm{x}) = \int_{\mathbb{R}} e^{2\pi i\xi x} f(\xi)d\xi$, Fourier transform of f can be treated as the amount of oscillation $x \to e^{2\pi ix}$ while presenting signal f. However, sine/cosine waves are localized in the frequency domain, but not in time domain. This means that Fourier bases suffer from nonlocality; and nonlocality will interfere with the behavior of Fourier transform. Comparatively, there are more signal representation interference present than in wavelet algorithm. Two issues arise:

Can the use of Fourier transform provide the information needed? The behavior of signal in a time interval (no matter how small it is) influences the "global" behavior of Fourier transform. This means $f(\mathrm{t}), f(\xi)$ cannot both be localized in their respective spaces (Heisenberg's uncertainty principle).

Since sine/cosine waves are infinitely supported in time domain, Fourier bases generally cause high frequency oscillation near the jump point and the endpoints of the interval. The high oscillation causes more time complexity here, than it would in wavelet algorithm, especially, when the signal is stocks.

Luckily, wavelets are localized in both frequency domain and time domain. They can be designed to have desired properties such as near optimal time-frequency localization, compact support, and regularity. Stock signal normally exists as high localized oscillation in time domain. Since, wavelets can be designed to have finitely supported localized oscillation in time domain, using wavelets to represent "bad" signals such as stocks is more advantageous than using Fourier bases. Those are the main reasons why the previous algorithm based on Fourier bases was abandoned.

Secondly, the Morlet wavelet is very useful for the detection of system nonlinearities because of its good support in both frequency and time domain. In practice, the Morlet wavelet is the product of a complex exponential wave and a Gaussian envelope. Based on Figure 3, one can see that the Morlet wavelet enjoys not only advantages of Fourier bases, but also localization in both time and frequency domain, since it is a wavelet.

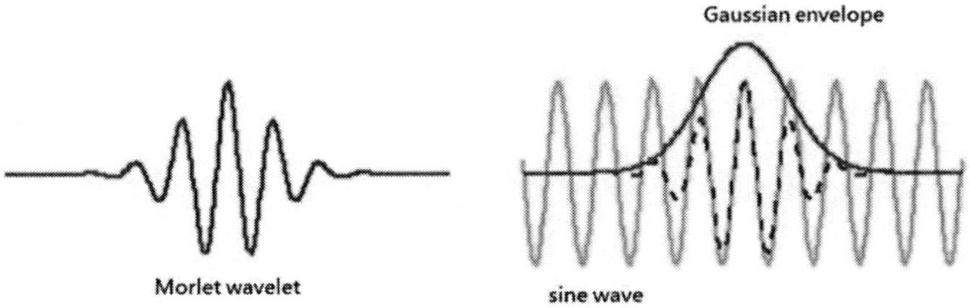

Figure 3. Morlet wavelet, sine wave and Gaussian envelope[16].

Wavelons are basic units of the WNN. The input data was x, with t as the translation parameter and d as the dilation parameter. The Morlet wavelet was used to obtain wavelons.

$$\psi_{t,d}(x) = \psi(\frac{x-t}{d}).$$

$$\psi(x_i) = \sqrt{e} \cdot k_i \cdot e^{-k_i^2/2} \tag{3}$$

where $k_i = (x_i - t_i)/d_i$. The following process describes how a set of data can be used via WNN to generate a forecast. When a set of data $X = (x_1, x_2, x_3, ..., x_n)$ is inputted, it would be translated, using $T = (t_1, t_2, t_3, \cdots, t_n)$, and dilated, using $D = (d_1, d_2, d_3, \cdots, d_n)$, according to the parameters.

$$\frac{x_i - t_i}{d_i}, i = 1, 2, 3, \cdots, n. \tag{4}$$

Each calculated data from (4) would be entered into (3). Linear combinations would be performed with weight vector $W = (w_1, w_2, w_3, \cdots, w_n)$, and values would be calculated by (5) to obtain (6).

$$\psi_{t_i, d_i}(x_i) = \psi(\frac{x_i - t_i}{d_i}), i = 1, 2, 3, \cdots, n \tag{5}$$

$$\sum_{i=1}^{n} w_i \psi(\frac{x_i - t_i}{d_i}) \tag{6}$$

Use (6) and adjustment parameter $\bar{y}$ to obtain the output data y, shown in (7). Adding $\bar{y}$ is required for (7) because of $\int_{-\infty}^{\infty} \psi = 0$. To avoid low convergence rate, adjustment parameter $\bar{y}$ was added.

$$y = \sum_{i=1}^{n} w_i \psi(\frac{x_i - t_i}{d_i}) + \bar{y} \tag{7}$$

2.4 WNN training process

The WNN architecture is composed of weighted linear combinations of many wavelons. Supervised learning method was used to train our WNN, and backpropagation algorithm was used to correct network parameters [6, 13, 14].

The WNN training process is as follows:

Step 1: Input data to the WNN.

Step 2: Five settings were assigned to the WNN, namely the w, t, and d after PSO optimization, as well as the adjustment parameter $\bar{y}$, and the threshold value.

Step 3: Input all data and network parameters and use WNN to calculate the network output values.

Step 4: Calculate the error between the network output value and x using error function E.

$$E = \frac{1}{2} \sum_k (x_k - y_k)^2. \tag{8}$$

Step 5: Determine if the error is within the given threshold value:

1. if yes, proceed to the next step.
2. if not, use the L-M Algorithm to adjust the network parameters and return to Step 3.

Repeat the network output value calculation until the error falls within the given threshold value.

Step 6: Verify whether data still requires training:
1. if yes, return to step 3.
2. if not, the training is complete.

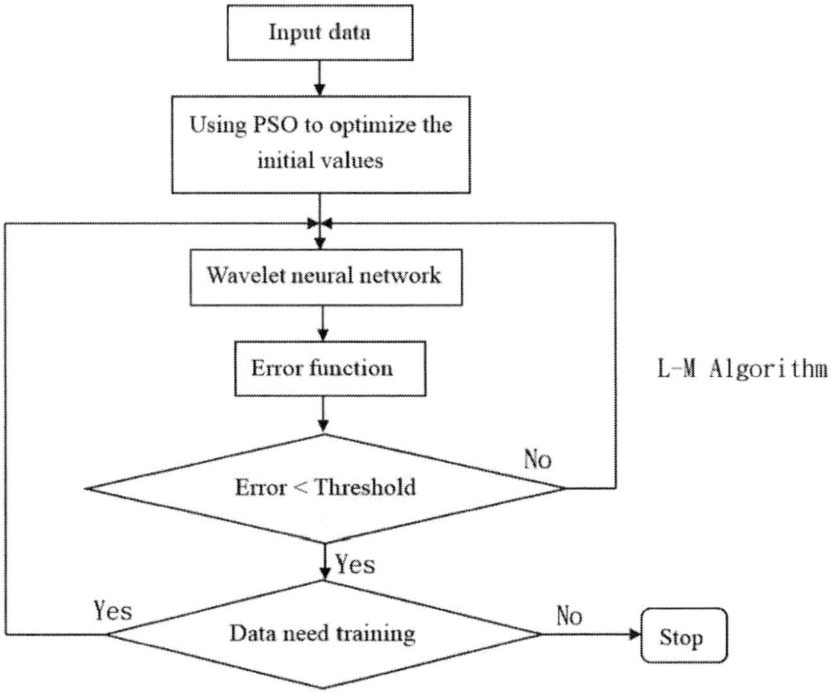

Figure 4. WNN Training Flowchart.

2.5 WNN forecasting process

Our model uses each stock's 100 closing prices to forecast the following 10 days' stock prices. This research used 10% of the training result data for calculation. The chosen 10% were the last 10% of the calculated w, t, and d training results produced. Therefore, the parameters used for calculation were $(w_{90}, w_{91}, \ldots, w_{99})$ 、 $(t_{90}, t_{91}, \ldots, t_{99})$ 、 $(d_{90}, d_{91}, \ldots, d_{99})$ and $\overline{y}$. Applying the parameters listed in (7), the below derived:

$$y = \sum_{90}^{99} w_i \psi\left(\frac{x_i - t_i}{d_i}\right) + \overline{y} \tag{9}$$

This is our WNN forecasting function (9).

$$x_{101} = \sum_{90}^{99} w_i \psi\left(\frac{x_{i+1} - t_i}{d_i}\right) + \overline{y} \tag{10}$$

$$x_{102} = \sum_{90}^{99} w_i \psi\left(\frac{x_{i+2} - t_i}{d_i}\right) + \overline{y} \tag{11}$$

$$x_{110} = \sum_{90}^{99} w_i \psi\left(\frac{x_{i+10} - t_i}{d_i}\right) + \overline{y} \tag{12}$$

Using $x_{101}, x_{102}, \cdots, x_{110}$, stock price of the following 10 days were forecasted.

3 Results and discussions

In Y. Oussar's research, although the initial parameters change along with stock prices taken from different time intervals, but in fact, the parameters w, t, and d remain the same for each data point in the same set of training data. This is not very logical, because it is impossible for all data points to have the same initial value. One disadvantage is the infinite loop that occurs when using the backpropagation algorithm during training. Therefore, the assumption is, if more suitable initial parameters can be found for each data point, the problem should be resolved. The resolution we found was the usage of PSO. Before the initial parameters are assigned for the WNN training process, PSO can analyze each data point, optimizing the corresponding three initial parameters. When this is done, not only does the speed for training increases, the RMSE becomes smaller and the forecast accuracy increases.

In Table 1, we compared our algorithm, which uses PSO optimization, to Y. Oussar's algorithm. The comparison method was running the program 100 times, recording each instance where the RMSE is less than the original RMSE. When the RMSE was less more than 65 times out of 100, we considered using WNN with PSO as better performance than using Y. Oussar's algorithm. For example, when we used China Steel Corporation (2002) to test forecast, the RMSE was consistently less than 0.9933, so it was considered better performing. After running the program using B=C=2, we found 28 companies that had less RMSE more than 65 times. Then, an even smaller range of parameters for were used, B=C=1.8, presenting 30 companies, successfully raising the success rate from 68% to 73%, shown in the last row of Table 1.

This algorithm can not only be used to forecast Taiwanese stocks, but can also be used to forecast other nations' stocks, commodities, and other financial instruments. For testing, we randomly selected three gold index companies: Barrick Gold Corp (ABX); Agnico-Eagle Mines Ltd (AEU); and Anglogold Ashanti Ltd (AU). Data collected were during time interval chosen was March 6, 2014 through August 11, 2014. In Table 3, it shows a success rate of 67%. Then, three large American corporations were randomly chosen: IBM; Facebook; and Apple. Data collected were during time interval February 25, 204 through July 31, 2014 (Table 4). For both algorithms, there were no problems running data of IBM and Facebook.

Table 3. 100 times results from Three Gold Companies.

Company Name	Training data percentage= 10% threshold=10^(-2)	
	WNN with PSO < WNN RMSE [The number of Success]	WNN RMSE
Barrick Gold Corp	100	0.8403
Agnico-Eagle Mines Ltd	0	2.5956
Anglogold Ashanti Ltd	100	1.7455

Table 4. 100 times results from Three American Companies.

Company Stock	Training data percentage= 10% threshold=10^(-2)	
	WNN with PSO < WNN RMSE [The number of Success]	WNN RMSE
IBM	100	3.5754
Facebook	100	7.5791
Apple	One of RMSE = 56.7879	N/A

However, while running the program for Apple, the data produced from this specific experiment was what led us to affirmatively believe that this algorithm is better than Y. Oussar's. When data for Apple was calculated, Y. Oussar's algorithm fell into infinite loops. Looking into the incident further, we found it was because Apple had a stock split of 1:7 on June 9, 2014. The stock price changed from $645.57 to $92.25. Our algorithm, using WNN with PSO, accommodated this drastic change. This is a great advantage. This indicates many future possibilities for this algorithm, such as the possibility of creating an extra step in the model or revising the existing error function to compensate for a stock split. Although, we do point out, in this calculation, there is a dramatic difference in the RMSE value, showing the test results were invalid.

We also realized two major limitations of the algorithm. First, due to the fact that this model was made to have higher adaptability, the accuracy rate is lower. When compared to Ayodele Ariyo Adebiyi's model[15], created a model specific designed to forecast Dell stock, during a ten day interval, our model was not as accurate. In ANN vs. WNN, our model had five lower error values. In ARIMA vs. WNN, our model had only four lower error values. However, if Adebiyi's model was used to forecast the stock price of any other company, the algorithm or the model itself may need complete redesign. Since their model was targeted while our model was not, this error value difference was acceptable and expected (Table 5).

Table 5. Forecasting Value and Error of Three Methods.

Original chart without WNN columns from Adebiyi's paper

DELL (2010/03/01-2010/03/12)							
Sample period	Actual value	Forecasting value			Forecasting error		
		ANN	ARIMA	WNN	ANN	ARIMA	WNN
03/01	13.57	13.16	13.35	13.46	0.03021	0.016212	-0.0081
03/02	13.68	13.55	13.46	13.35	0.0095	0.016082	-0.0239
03/03	13.71	13.7	13.56	13.55	0.00073	0.010941	-0.0116
03/04	13.67	13.55	13.67	13.8	0.00878	0	0.0096
03/05	13.88	13.53	13.78	14.08	0.02522	0.007205	0.0146
03/08	14.01	13.89	13.88	14.06	0.00857	0.009279	0.0035
03/09	14.18	13.92	13.99	14.18	0.01834	0.013399	-0.0002
03/10	14.31	13.85	14.09	14.51	0.03215	0.015374	0.0139
03/11	14.21	14.18	14.2	14.63	0.00211	0.000704	0.0295
03/12	14.26	14.31	14.3	14.93	−0.0035	−0.00281	0.047

The other difficulty is that the range for the PSO parameter is difficult to choose. When the range is too small, the optimal solution may be excluded. When the range is too large, more time is wasted in search of the optimal solution. Therefore, finding a better method of choosing a proper range is an area that needs to be studied.

The limitation of research we had was choosing the time interval for the training data. Dates where significant events that would affect the stock or stock market dramatically had occurred must be avoided. Natural disasters, stock market crashes, or any other significant events such as the Apple stock split will cause the model to fail. We don't have enough manpower to search for all possible events during the 100 days in order to avoid certain time intervals.

4 Conclusion

Our experiment results indicated using smaller range of numbers for the translation and dilation parameters increased performance during forecast. Among some older methods for WNN operations [10,11,13], the network threshold parameters were set as constants, and proper forecast was obtained by adjusting threshold and training data percentage. Our WNN adds PSO and resolves this slightly illogical method. The network initial values of every training data point can be configured by PSO in coordination with WNN, which means the network initial values are "automatically" selected instead of being a constant, and prevents infinite loops during calculation. Threshold and training data percentage become constant values, because PSO provides self-adjustment, which means it is not a true constant during calculation.

From the two advantages mentioned above, our model has higher ease of operation in comparison to the older WNN models, because the end users do not have to set any parameters. The necessary threshold and training data parameters would be chosen for them. If an end user wants to know the following 10 days' stock price, the only data that needs input are the company name and symbol, and the forecast prices will be presented for consideration. In the future, we hope to improve our results by revising the existing training methods or adding additional steps.

Disclosure

The authors declare that there is no conflict of interests regarding the publication of this article.

References

1. Z. Qinghua, IEEE Trans. on Neural Network, **3**, 6 (1992)
2. H. Dongdong and Z. Wenhong, BIFE, 32-36 (2011)
3. H. Zhang and H. Lin, J. Southeast Univ. Nat. Sci., **37** (4), 716-720 (2007)
4. Y. Zhao, Y. Zhang, and C. Qi, PACIIA, **1**, 31-36 (2008)
5. H.R. Ke, K.C. Wang, C.I. Yang, and K.F. Chang, IJAPM, **4** (1), 9-14 (2014)
6. C.J. Lin and H.M. Tsai, Math. Comput. Model., **47** (9-10), 982-996 (2008)
7. J. Kennedy, *Encyclopedia of Machine Learning*, (Online, 2011)
8. Y. Oussar and G. Dreyfus, Neurocomputing, **34** (1), 131-143 (2000)
9. Y. Shi and R.C. Eberhart, IEEE WCCI, 69-73 (1998)
10. Y. Shi and R.C. Eberhart, EP98, 591-600 (1998)
11. C.I. Yang, K.C. Wang, and K.F. Chang, ICDIP, 91590V (2014)
12. J. Zhang, G. Walter, Y. Miao, and W.N.W. Lee, IEEE Trans. Signal Process., **43** (6), 1485-1497 (1995)
13. Y. Wang and X. Cheng, EMEIT, **6**, 3109-3112 (2011)
14. F. Zhang, H. Dai, and D. Tang, J. Appl. Math., 2014 (2014)
15. A.A. Adebiyi, A.O. Adewumi, and C.K. Ayo, J. Appl. Math., 2014 (2014)
16. http://paos.colorado.edu/research/wavelets/wavelet2.html

A study of expertise effects for products with contradictory semantics

Ching-Yi Wang[a]
Department of Industrial Design, Tatung University, Taipei City 104, Taiwan

Abstract. In the design studies, researchers often use the semantic differential method with bipolar adjectives, such as "modern vs. classical" or "simple vs. complex" when investigating the semantics projected by product forms. However, in design practice, some design examples clearly exhibit the simultaneous use of contradictory meanings in product semantics. For example, retro car evokes nostalgia by borrowing characteristics from classical cars. At the same time it exhibits a modern style. However, most studies measure the product semantics mostly by using subjective measurement. There is lack objective measurement for that. In this research, we examined the results of applying the semantic differential method to measure contradiction in product semantics. The results showed that the distributions of semantic differential ratings for the stimuli with contradictory meanings have higher standard deviations. The sensitivity of semantic recognition may depend on participant expertise. The design experts are trained to be good at visual thinking that could easily identify the contradiction semantics between products. In general, successful embedding of contradictory meanings into product forms are based on simple, typical, and rational forms that can display complex, novel, and perceptual images by adding supplementary elements.

1 Introduction

Nowadays, the rapid developments in art and design have created an almost limitless number of items in the same category (for example, chairs in the category of furniture) that have a variety of features and different design styles. Scenes of conflicting furniture arrangement in a space are commonplace, for example, a modern floor lamp next to a classic sofa. How do people react to the scenes of contradiction? In design studies, the most popular method of measuring participants' reactions to objects or images is to apply adjectival descriptions of the semantic differences (STD) and to employ a multidimensional scale (MDS). The studies use questionnaires with a Likert scale, mostly based on the semantics analysis developed by Osgood in 1957 to acquire the participant's subjective responses to stimuli [1]. Different bipolar adjectives (e.g., simple–complicated, plain–luxurious) are tested according to the stimuli and context, in the semantic differential measurement. However, these types of surveys were subjective measurements. Many results have been realized [2-4], and the criticism regarding this method of measurement has arisen for several reasons: (1) that the participants' response may be misguided due to the questionnaire design (2) insufficient reliability and accuracy (3) participants will not answer all of the questions with sufficient care. As the process of acquiring the

[a] Corresponding author : catincar@gmail.com

raw data is not well controlled, the subsequent data analysis may be invalid. In addition, there are many products with both typical and unique visual appearance, such as the "ice lamp" (see Figure 1) which was covered with transparent glass displays warm and cold at the same time. In response to a product within a double semantics, participants generally check a middle option with the hesitant and uncertain attitude (i.e. neutral response) using 7-point bipolar adjective scale. It is difficult to identify their true feelings. In addition, this answer is often considered invalid questionnaires.

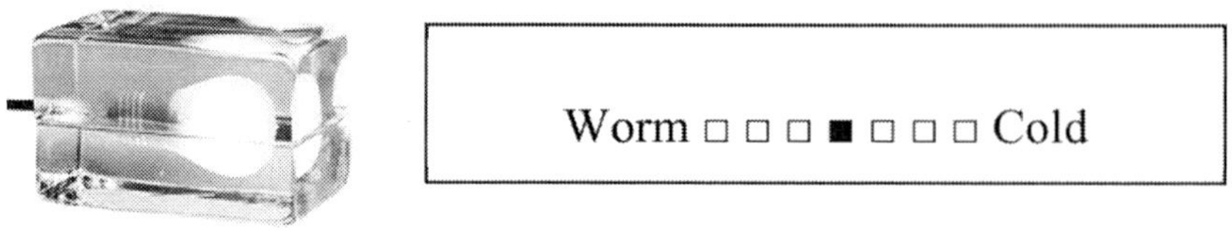

Figure 1. The "ice lamp" product displays a double semantics within warm and cold.

The terms of ambivalence or conjunctive ambiguity are similar with semantic contradiction. The definition of ambivalence includes "a state of mind" when people feel mixed feelings (usually a positive or negative sense) to an object [5]. It can also be seen as "conflict feeling" which aroused by many competitions and assessments of the propositions [6]. It was summarized as "circumplex model" which may explain basic human emotions by two orthogonal dimensions representing "pleasantness" (horizontal, from negative, painful dark to positive, bright) and "activation" (vertical, from introvert to extrovert) [6-9].

Regarding the contradictory semantic study, the method of standard deviation was used to select products which may involve the conflict semantics [2]. They believe that the semantic contradiction would be affected by three factors comprising at odds, contradiction, and insensitivity. They assumed that the measurement results of contradictory semantics may be presented by three types of modes as shown in the Figure 2. Figure 2(a) appears two peaks if participants clearly and respectively perceive the differences across the two semantics. Figure 2(b) shows three peaks if the participants perceive the double semantics because they are unsure about choosing answer. Figure 2(c) displays almost flat distribution if the participants perceive uncertain or unstable semantics due to random guessing. In the preceding three assumptions, they should be appeared a large standard deviation (Std), negative kurtosis (K) value, and the average value (mean) near central (skewness (S) = 0).

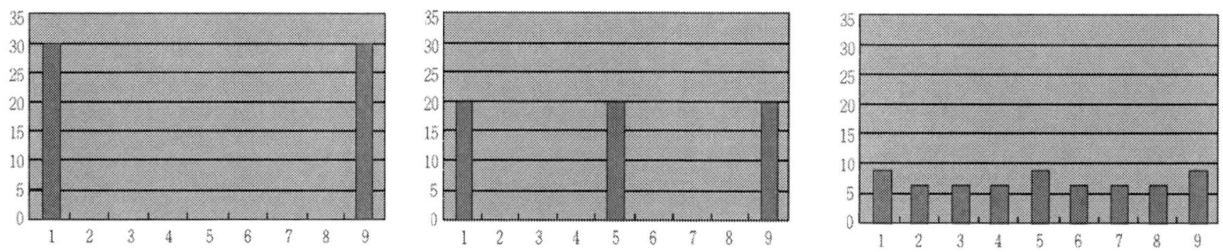

(a) two peaks
Mean=5; Std=4.034
S=0; K=-2.07

(b) three peaks
Mean=5; Std=3.294
S=0; K=-1.526

(c) random guessing
Mean=5; Std=2.681
S=0; K=-1.243

Figure 2. Examples of three conditions of conflict semantic distribution [2].

In the present study, the chair products were as sample to classify using the card sorting. The analysis method using [2] recommended as an indicator to select and define the pictures with and without contradictory semantics. Four dimensions of semantics were used according to the previous studies [11-12], including novel (typical - unique), modern dimension (traditional - modern), perceptual (rational - emotional), and complicated (complicated - simple). We expect to get a reasonable explanation corresponding to contradiction semantics and strengthen this topic for understanding clearly.

2 Methods

2.1 Participants

Sixty participants comprised of 30 experts (11 male, 9 female; mean age=23.6 years) with design training background and 30 novices (9 male, 11 female; mean age=28.6 years, STD=13.65 years) without design training background in the experiment. Different background was used to understand their cognitive differences in response to product works with contradiction semantics.

2.2 Materials

The stimuli included 396 colour pictures of chairs without armrest component (see Figure 3) on a white background. All of the selected pictures acquired via the book 1000 Chairs [12], 500 Chairs [13], and the works by famous designers from Internet. The reason for choosing the chair is that they have various styles and represent one of the most famous products in the design history. These pictures were displayed at a subtended vertical and horizontal visual angle of approximately 45 degrees for better viewing. Each picture was approximately 10 cm high and 10 cm wide and marked a corresponding number at the bottom of the right.

Figure 3. Example picture.

2.3 Procedure

There are two stages in the image clustering, as shown in the Figure 4. In the first stage, the categorization was performed by 6 design experts. The remaining 96 chairs were divided into 4 groups of 24 chairs each. Each dimensions corresponded to one of the following 4 semantics: novel (typical - unique), modern dimension (traditional - modern), perceptual (rational - emotional), and complicated (complicated - simple) [11-12].

In the second stage, each dimension with respect to one of the pairs of adjectives was divided into 9 subgroups corresponding to a 9-point rating scale, as shown in the Table 1. In order to reduce the participants' cognitive loading, each dimension was first divided into 3 groups, representing low, middle and high levels, and then to further divides each of these groups into 3 subgroups, thus arriving at a total of 9 groups. The number of each group was allowed to be uneven or void. After finish card sorting, each participant was asked questions regarding the classified reasons for each group. The overall test took 90 minutes to complete.

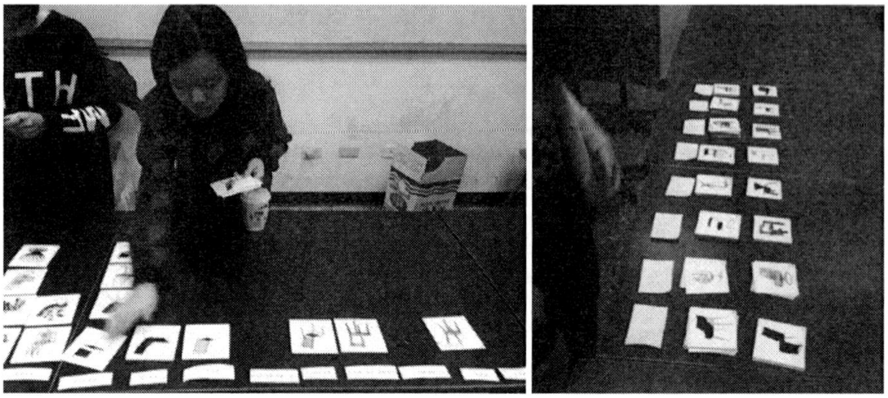

Figure 4. Card sorting: each semantic dimension with respect to one of the pairs of adjectives was divided into 9 subgroups corresponding to a 9-point rating scale

Table 1. The 4 semantic dimensions of 3 groups which represented low, middle and high levels of 9 subgroups.

semantic level	scale	novel	modern	perceptual	complicated
low level	1	highly typical	highly traditional	highly rational	highly complicated
	2	very typical	very traditional	very rational	very complicated
	3	a little typical	a little traditional	a little rational	a little complicated
middle level	4	a little typical and not unique	a little traditional and not modern	a little rational and not emotional	a little complicated and not simple
	5	both typical and unique	Both traditional and modern	both rational and emotional	both complicated and simple
	6	not typical and a little unique	not traditional and a little modern	not rational and a little emotional	not complicated and a little simple
high level	7	a little unique	a little modern	a little emotional	a little simple
	8	very unique	very modern	very emotional	very simple
	9	highly unique	highly modern	highly emotional	highly simple

2.4 Data analysis

Table 2. The representative samples in accordance with different level of semantics.

semantic level / analysis items	low level (range from 1 to 3)	middle level (range from 4 to 6)	high level (range from 7 to 9)
mean (M)	1-3	4-6	7-9
standard deviation (Std)	small	large	small
skewness (S)	positive	negative	negative
kurtosis (K)	positive	negative	positive

The descriptive statistics were first performed to identify a representative sample of 4 semantic dimensions. Table 2 shows the definitions of mean (M), standard deviation (Std), skewness (S), and kurtosis (K) for each level of semantic. The independent-samples T test was performed to analyse the rating scores between different backgrounds which as death with independent variables.

3 Results

3.1 Overall means of four semantic dimensions

Table 3-6 display the card sorting results of chair pictures with the perceptual, novel, modern, and complicated semantics. The middle column exhibits the contradictory semantic with the large standard deviation, and the left and right columns both present the high-level perceptions with small standard deviation to the corresponding to the bipolar adjective pairs.

3.1.1 Perceptual products

Table 3 shows the measurement results of chair with perceptual semantic. The chair (Table 5(a)) were rated the most rational because the seat and back were constructed by vertical and horizontal surfaces and displays rigid outline. Moreover, many participants considered that the chair (Table 5(b)) with large surface-seat and cylindrical metal legs displays soft and rigid combinations, respectively. In addition, the appearance of the chair (Table 5(c)) looks like a female body, such as its headrest, back, and legs respectively symbolize female's head, chest, and legs. Thus, it was rated the most sensual chair.

Table 3. Perceptual dimension (rational-emotional).

	highly rational (small standard deviations)	**both rational and emotional (large standard deviation)**	**highly emotional (small standard deviations)**
	 (a)	 (b)	 (c)
Mean	1.78	5.33	8.12
Std	0.99	1.79	1.04
S	1.32	-0.21	-1.45
K	1.24	-0.31	2.10

Note: M = mean; Std = standard deviation; S = skewness; K = kurtosis.

3.1.2 Novel products

Table 4 shows the measurement results of chair with novel semantic. The wooden chair (Table 3(a)) was rated the most typical. This typical style of chair in the appearance and construction should be structured by a palisading seat/back and four-legged features. Moreover, a typical chair (Table 3(b)) added a back part which comes from another chair as a specially rocking function and made it displays a typical and unique combination. In addition, the unique chair (Table 3(c)) was completely made by a bathtub as its sitting function and confusing participants with its product category, thus they think that it is not a chair.

Table 4. Novel dimension (typical-unique).

	highly typical (small standard deviations)	**both typical and unique (large standard deviation)**	**highly unique (small standard deviations)**
	 (a)	 (b)	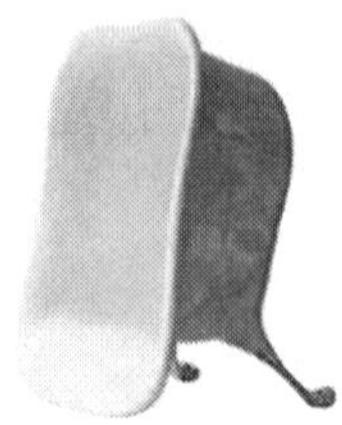 (c)
Mean	1.45	4.75	8.07
Std	0.77	1.94	1.49
S	1.80	-0.04	-2.26
K	2.79	-0.33	5.62

Note: M = mean; Std = standard deviation; S = skewness; K = kurtosis.

3.1.3 Modern products

Table 5 shows the measurement results of chair with modern semantic. The chair (Table 4a) was waved by bamboo and rattan and obviously looks an age-old style. Moreover, the chair which combined with traditional and modern style (Table 4b) has a traditional fencing structure of the seat with the technical metal legs. In addition, the metal chair (Table 4c) was made by an interwoven of bending and netted structure as a main body for supporting sitting function. Most participants believed that it should belong to the highly modern style.

Table 5. Modern dimension (traditional - modern).

	highly traditional (small standard deviations)	**both traditional and modern (large standard deviation)**	**highly modern (small standard deviations)**
	 (a)	 (b)	 (c)
Mean	2.00	5.25	8.08
Std	1.18	2.10	1.18
S	0.96	-0.56	-1.69
K	0.09	-0.53	3.17

Note: M = mean; Std = standard deviation; S = skewness; K = kurtosis.

3.1.4 Complicated products

Table 6 shows the measurement results of chair with complicated semantic. The chair (Table 6a) was almost covered with colourful patches of hexagonal cloth, thus it was rated the most complex type by most participants. Moreover, the structure of chair (Table 6(b)) for its leg was composed by the simple and straight metal lines, but its back was complex. In addition, the chair (Table 5(c)) in the appearance has slender legs, hollow back, and thin seat, overall showing a minimalist form. Its appearance therefore easily seems as a simple type.

Table 6. Complicated dimension (complex-simple).

	highly Complex (small standard deviations)	**both complex and simple (large standard deviation)**	**highly simple (small standard deviations)**
	 (a)	 (b)	 (c)
Mean	2.15	4.72	8.10
Std	1.27	1.85	1.07
S	0.68	0.22	-1.07
K	-0.93	-0.06	0.26

Note: M = mean; Std = standard deviation; S = skewness; K = kurtosis.

3.2 Comparison of cognitive differences between experts and novices in response to the contradictory products

In order to understand the different cognitive models between different backgrounds, we further preformed the T-test to examine the novel, modern, perceptual and complicated dimensions with the contradictory semantic products. According to the T-test analysis results, we found the conditions with significant differences for the contradictory chairs and use the frequency distribution figures to compare the cognitive differences between (average range from 4 to 6 scores) experts and novices, as shown in the Table 7-10.

3.2.1 Perceptual products with contradictory semantics

Table 7 shows the measurement results of perceptual chairs with contradictory semantics. The metal chair (Table 7(a)) was composed of many bending surfaces and three legs. The frequency distribution for the experts concentrated on the left (an average of 3), showing that they consistently agreed that it is the most rational chair. Nevertheless, the frequency distribution for novices presented double peaks, showing that a part of the participants considered that it is emotional style or not. There exist divergent views on this chair between novices.

Moreover, the shape of the chair is similar to the previous one (Table 7(b)), but it is a four-legged chair. The frequency distribution for the experts concentrated on the left (an average of 3.83), presenting that it is more rational. However, the frequency distribution for novices presented double peaks. Because some participants thought that its back legs are slanted and exhibits softness and tenderness.

In addition, the chair was composed by wooden and having high back structure (Table 7(c)). The frequency distribution for the experts focused on the middle, showing that it is a rational and emotional combination. Nevertheless, the frequency distribution for the experts concentrated on the right trend, showing that most novices believed the side contour for this chair presented S curve as a female body.

Table 7. Perceptual chairs with contradictory semantics.

		both rational and emotional		
		 (a)	 (b)	 (c)
experts	Mean	3.00	3.83	5.53
	Std	1.34	1.23	1.11
	S	0.00	0.34	0.73
	K	-1.07	-0.25	-0.05
frequency distribution		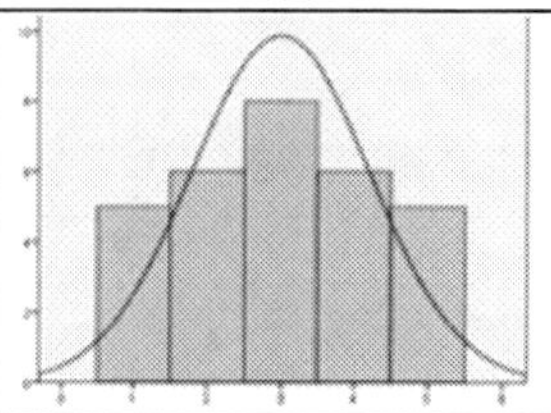	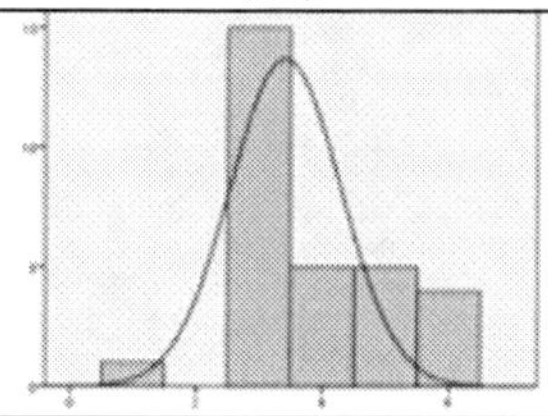	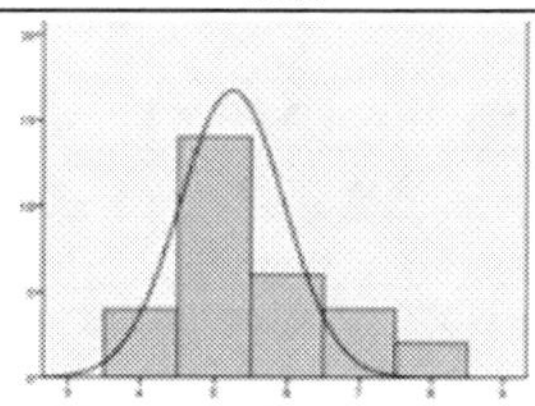
novices	Mean	5.63	5.47	6.93
	Std	2.22	2.10	1.41
	S	-0.27	-0.19	-1.13
	K	-0.68	-0.95	1.62
frequency distribution		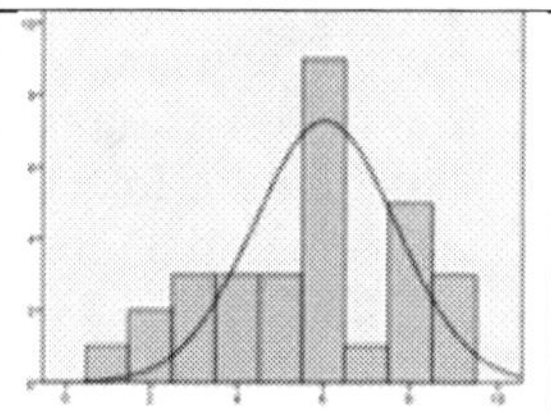	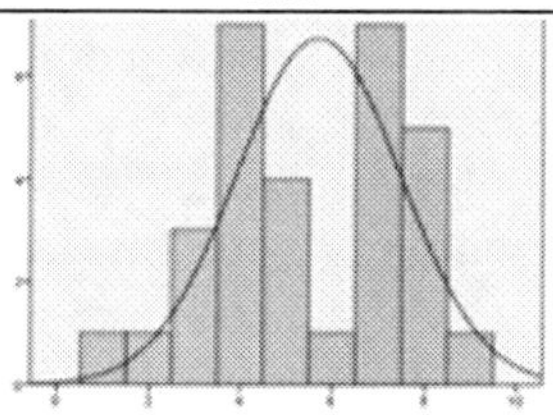	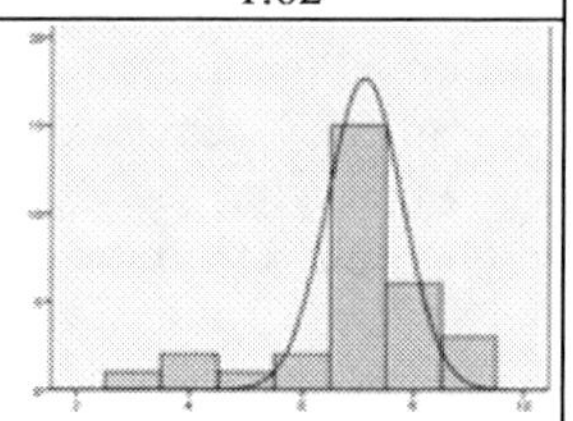
T value		-5.56**	-3.68**	-4.27**

Note: M = mean; Std = standard deviation; S = skewness; K = kurtosis.
*p<.05 and ** p<.01 (2-tailed)

3.2.2 Novel products with contradictory semantics

Table 8 shows the measurement results of novel chairs with contradictory semantics. The structure of this chair (Table 8(a)) was normal, but its seat was composed by hangers. Thus, the frequency distribution for the experts concentrated on the middle, showing it is conflict between typical and unique. Spuriously, most novices had not found its different part. The frequency distribution for novices concentrated on the left, showing that they thought it just a normal chair.

In addition, the chair's leg (Table 8(b)) was an upside down of back which comes from another chair. The frequency distribution for the experts focused on the middle, showing that it is also a conflicting type. However, the frequency distribution for the novices presented double peaks, showing that some participants considered that this chair belongs to typical or unique type.

The chair (Table 8(c)) obviously was two plastic chairs overlapping into one. Therefore, the frequency distribution for the experts focused on the middle, showing that they consistently agreed that this chair is conflict between typical and unique. Nevertheless, the frequency distribution for the novices displayed three peaks, showing that there are strong cognition differences between them.

Table 8. Novel chairs with contradictory semantics.

		both typical and unique		
		 (a)	 (b)	 (c)
experts	Mean	5.43	5.60	5.90
	Std	1.41	1.35	1.09
	S	0.27	0.53	-0.47
	K	-0.62	-0.88	0.29
frequency distribution		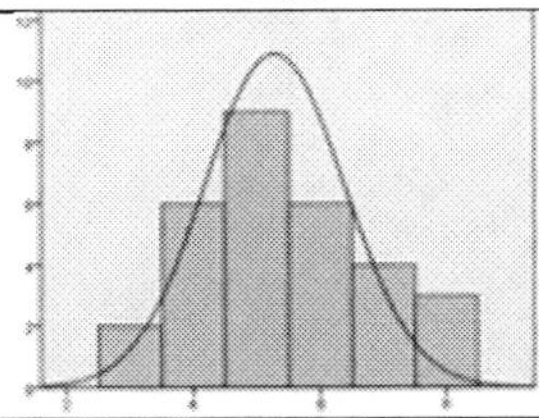	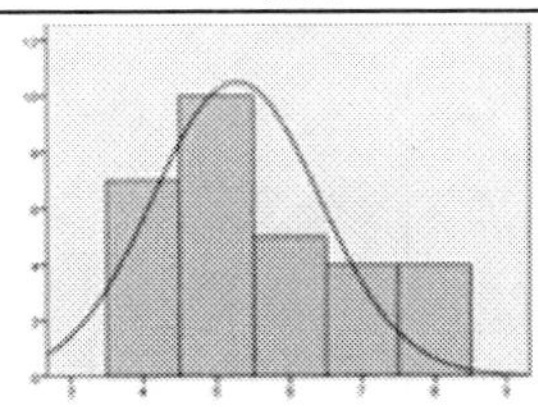	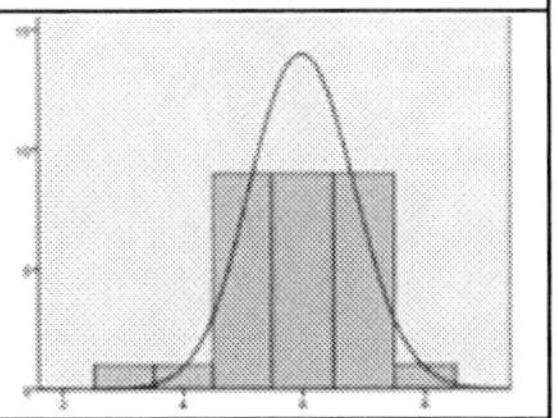
novices	Mean	3.90	4.77	4.97
	Std	1.94	1.59	2.36
	S	0.27	-0.09	-0.01
	K	-0.72	-1.23	-0.59
frequency distribution		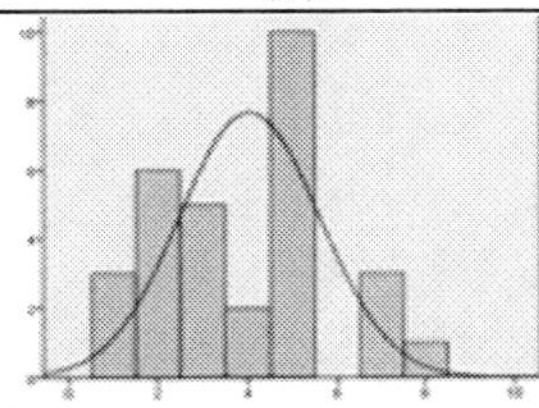	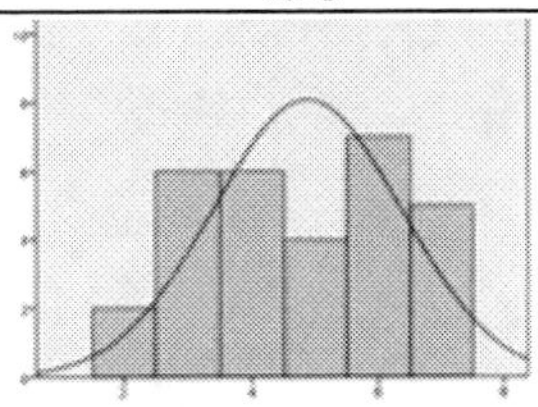	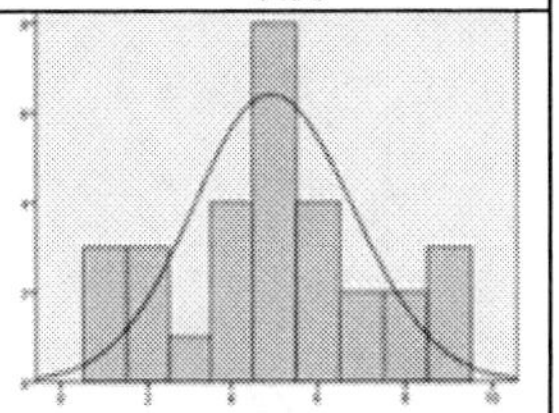
T value		3.5**	2.19*	1.97*

Note: M = mean; Std = standard deviation; S = skewness; K = kurtosis.
*p<.05 and ** p<.01 (2-tailed)

3.2.3 Modern products with contradictory semantics

Table 9 shows the measurement results of modern chairs with contradictory semantics. The chair (Table 9a) was composed with wooden legs and cushion. The frequency distribution for the experts trended to the left (an average of 4.97), showing that this chair was more modern because its white cushion and simple back. However, the frequency distribution for the novices displayed a slightly trended to the right, showing that few novices considered that it still belongs to traditional because it's wooden and four-legged features.

Furthermore, the chair (Table 9b) in the appearance was not different from the classical type, but it is whole made of metal. The frequency distribution for experts presented double peaks. They really read its design history and even know the related design concept from original designer, considering that it is a newly re-designed product that reaches across past and current. They therefore hesitated to choose between tradition and modern. On the contrary, the novices without the design knowledge and product training. Their frequency distribution for the experts concentrated on the left (an average of 3), showing that this chair was undoubtedly considered a traditional type.

Table 9. Modern chairs with contradictory semantics.

		both traditional and modern	
		 (a)	 (b)
experts	Mean	4.97	5.17
	Std	1.07	1.78
	S	-0.66	-0.54
	K	-0.78	-0.41
frequency distribution		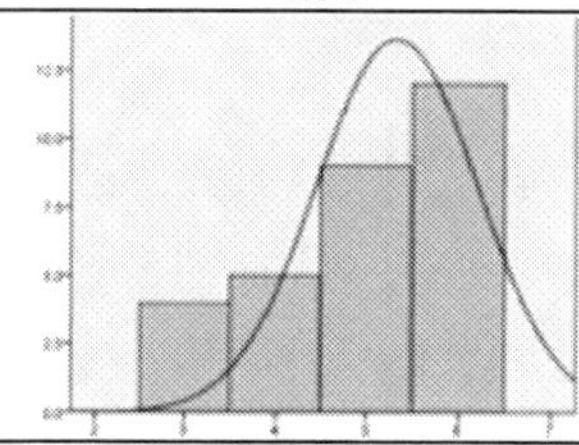	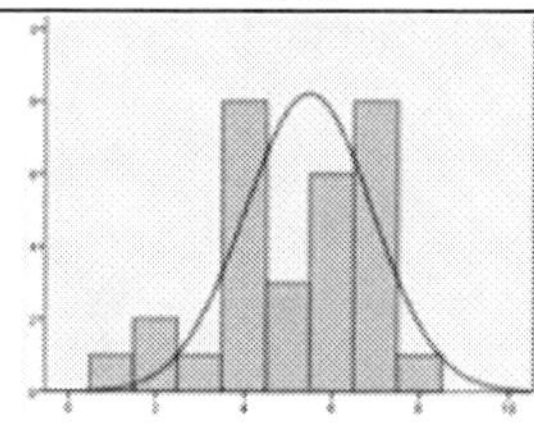
novices	Mean	6.23	2.93
	Std	1.20	2.05
	S	-0.35	0.74
	K	-0.67	-0.78
frequency distribution		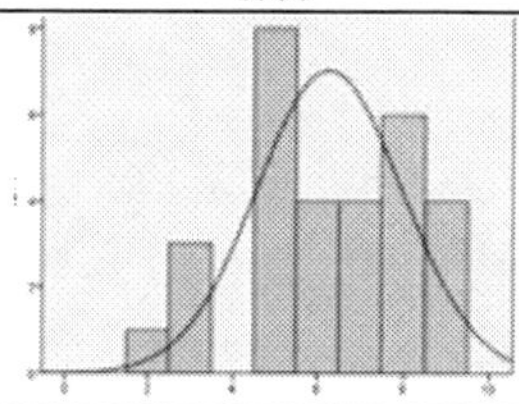	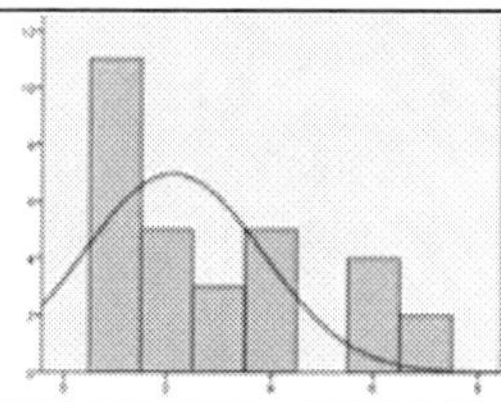
T value		4.50**	-3.09**

Note: M = mean; Std = standard deviation; S = skewness; K = kurtosis.
*p<.05 and ** p<.01 (2-tailed)

3.2.4 Complicated products with contradictory semantics

Table 10 shows the measurement results of complicated chairs with contradictory semantics. This chair (Table 10a) was made of many braided vines. Thus, the frequency distribution for the experts concentrated on the left (an average of 3.87), showing that they did a complex judgment for this chair. However, the frequency distribution for the novices focused on the center, showing that they believed that it involved a simple feature which supported by three-legged.

The chair (Table 10b) was made of metal lines and achieved 2D and 3D mixed views by using a visual illusion. The frequency distribution for the experts concentrated on the center (an average of 5.3). Because the experts would easily switch the different viewings based on their abilities of high imaginations and understand it performs a simple visual principle to create complex viewings. However, the frequency distribution for novices presented double peaks, showing that the some participants considered that this chair is conflict between complex and simple. There exists an opposite opinion for this chair between novices.

Table 10. Complicated chairs with contradictory semantics.

		both complex and simple	
		 (a)	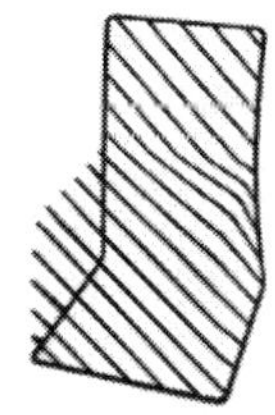 (b)
experts	Mean	1.25	1.09
	Std	0.04	1.76
	S	-0.95	4.45
	K	1.25	1.09
frequency distribution		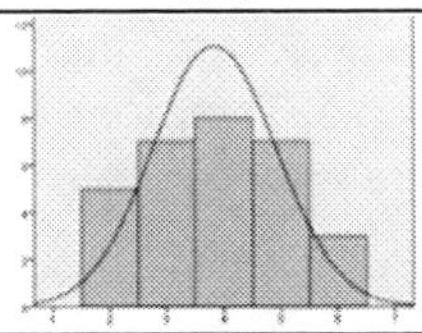	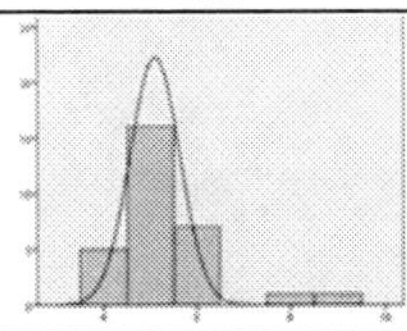
novices	Mean	4.77	6.20
	Std	1.96	1.90
	S	0.85	-0.21
	K	0.42	-0.99
frequency distribution		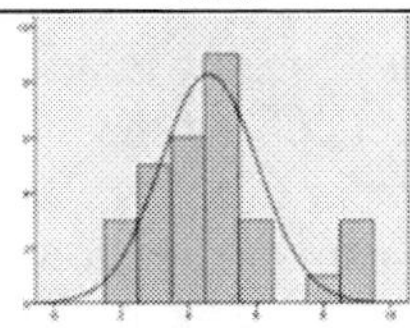	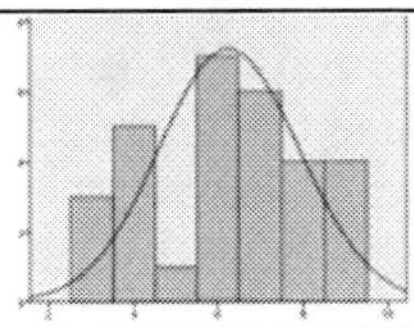
T value		-2.12*	-2.25*

Note: M = mean; Std = standard deviation; S = skewness; K = kurtosis.
*p<.05 and ** p<.01 (2-tailed)

4 Discussion

In this study, we explore the product meaning with semantic contradictory and try to use the standard deviation to find the meanings of contradictory semantics. The results showed that the distributions of semantic differential ratings for the stimuli with contradictory meanings have higher standard deviations.

According to the [2], this suggested that the semantic contradiction would affect by three factors comprising at odds, contradiction, and insensitivity. We concluded that there are several reasons would impact the standard deviation. One explanation may be that the sensitivity of semantic recognition may depend on participant expertise. In general, the designers and artists could identify an artwork's category and style immediately according to their form, structure, material, and composition. Because the participants were not familiar with styling, experts might have been better able to detect the small feature differences. We considered that the experts with design training and background were much more sensitive to them. One the other hand, the novices had less knowledge and experience in distinguishing between objects using subtle visual cues. We considered that the mental mechanism of the imagination process would aggravate the expert's attention at the beginning, but diminish later because of less surprise relative to the others. Another possible reason may be that the artists and designers are trained to be good at visual thinking. To artists and designers, imagination is a key capability that is able to combine a new notion with existing one in novel ways to derive a new image [15]. The expected and unexpected outcomes of arrangements differentiate viewers' reinterpretation, and arouse the new connections of each component to generate a new impression [16].

This concept of combined different components into new image is similar to the tangram. Many educational researchers used the tangram to help student's development of spatial and geometric thinking by various compositions of triangles and portraying things with different angles [17-19]. The activity of assembling a graphic image of a tangram is similar to the endeavor of ready-made chairs (i.e. Table 4b and Table 8a, b, c) that is combined with various materials and turned into something useful. In this study, the experts who are used to design thinking could easily acquire the concept of ready-made chairs. On the contrary, ordinary people experienced difficulty in accepting them as a chair and revealed confusing reaction.

Overall, successful embedding of contradictory meanings into product forms are based on simple, typical, and rational forms that are made to simultaneously exhibit complex, novel, and emotional images by introducing additional elements. The "typical-unique" chairs are likely to have prototypical shapes with additional functions or embedded stories. The "traditional-modern" chairs are based on the traditional shape with additional modern materials. The "rational-emotional" chairs are combined with rigid rational and sensual cures. The "complicated-simple" chairs are achieved through the use of material, texture, and complicated product forms. Therefore, these design methods of operation should enable to arouse more inspirations to designers.

5 Conclusion

Meaning of practices and design theory in the result of this study, semantic dimension can be usually concerned positive meaning and opposite one. For example, the modern dimension "traditional-modern". It still contained argues inside, just like design technique contrast which represented conflict semantic. Designer avoid create boring works, meanwhile, they understand contradictory semantics and help to expend the study domain of practice and theory. However, the studies of contradictory semantics are rare in this field. The measurement of using SD method is nearly subjective. It might be also hard to reveal the real feeling of participants. Therefore, it hopefully can be relative science of brain technology toward objective measurement. The further study will try to discovery the brain reactions and have more reasonable explanation the conflict semantic. It should be understood clearly for this topic.

References

1. C.E. Osgood, G.J. Suci, and P.H. Tannenbaum, *The Measurement of Meaning*, University of Illinois Press (Urbana, IL, 1957)
2. W.K. Hung and L.L. Chen, How to measure product's contradictory semantics, Journal of Design, **15** (4), 41-58 (2010)
3. M.H. Lin, C.Y. Wang, S.K. Cheng, and S.H. Cheng, An event-related potential study of semantic style-match judgments of artistic furniture, International Journal of Psychophysiology, **82**, 188-195 (2011)
4. M.H. Lin, Y.M. Fang, and C.Y. Wang, A preliminary study of applying ERP on users' reactions to web pages with different presentation formats, The Science of Design, **57** (5), 89-98 (2011b)
5. P.L. Gardner, Measuring ambivalence to science, Journal of Research in Science Teaching, **24** (3), 241-247 (1987)
6. S.J. Breckler, A comparison of numerical indexes for measuring attitude ambivalence, Educational and Psychological Measurement, **54**, 350-365 (1994)
7. J.A. Russell, A circumplex model of affect, Journal of Personality and Social Psychology, **39** (6), 1161-1178 (1980)
8. R. Plutchik and H.R. Conte, *Circumplex Models of Personality and Emotions*, American Psychological Association (Washington D.C, 1996)

9. R.E. Thayer, *The Biopsychology of Mood and Activation* (New York: Oxford University Press, 1989)
10. C.M. Whissell, Pleasure and activation revisited: Dimensions underlying semantic responses to fifty randomly selected "emotional" words, Perceptual and Motor Skills, **53**, 871-874 (1981)
11. K.A. Hsiao and L.L. Chen, Fundamental dimensions of affective responses to product shapes, International Journal of Industrial Ergonomics, **36**, 553-564 (2006)
12. P. Hekkert, D. Snelders, and P.C.W. Van Wieringen, Most advanced, yet acceptable: Typicality and novel as joint predictors of aesthetic preference in industrial design, British Journal of Psychology **94**, 111-124 (2003)
13. C. Fiell and P. Fiell, *1000 Chairs* (Kolin. Taschen, New York, USA, 1997)
14. R. Hemachandra, *500 Chairs: Celebrating Traditional and Innovative Designs* (Lark Books, New York, USA, 2008)
15. J. Passmore, *Recent Philosophers: A Supplement to a Hundred Years of Philosophy* (Duckworth Publishing, New York, USA, 1985)
16. M. Suwa, T. Purcell, and J.S. Gero, Macroscopic analysis of design processes based on a scheme for coding designers' cognitive actions, Design Studies, **19**, 455-483 (1998)
17. C.L. Chong and N.M. Siew, Fostering students' creativity through van hiele's 5 phase-based tangram activities, Journal of Education and Learning, **3** (2), 66-80 (2014)
18. C.P. Lin, Y.J. Shao, L.H. Wong, Y.J. Li, and J. Niramitranon, The impact of using synchronous collaborative virtual tangram in children's geometric, Turkish Online Journal of Educational Technology, **10** (2), 250-258 (2011)
19. N.M. Siew, C.L. Chong, and M.R. Abdullah, Facilitating students' geometric thinking through van hiele's phase-based learning using tangram, Journal of Social Sciences, **9** (3), 101-111 (2013)

Assessing the transmission performance of iGMAS real-time data streams

Zhanke He[1,a], Wenda Tang[2], Xuhai Yang[1] and Liming Wang[2]

[1]*National Time Service Center, Chinese Academy of Sciences, Xi'an, Shaanxi 710600, China*
[2]*Xidian University, Xi'an, Shaanxi 710071, China*

Abstract. The use of the BeiDou Navigation Satellite System (BDS) has been increasing in the Asian-Pacific region. Thus, the construction of the international GNSS Monitoring and Assessment System (iGMAS) is accelerating to meet the demanding requirements for high-precision real-time products. Latency and packet loss, two critical data transmission parameters, have great impact on the applications of real-time data streams. This paper aims to propose a Homology-Reverse method to investigate the data transmission performance of the iGMAS stations in local and wide area networks. The proposed method enables not only assessment on receivers but pretesting on network conditions of alternative tracking stations. This paper is expected to provide a reference for the construction of the iGMAS, Continuous Operational Reference System (CORS), and Satellite Based Augmentation Systems (SBAS).

1 Introduction

The rapid development of internet and Global Navigation Satellite System (GNSS) facilitates the widely use of real-time data streams of GNSS are flourishing in various fields, such as navigation and positioning, surveying and mapping, monitoring earthquake, meteorology, transportation, etc. The BeiDou Navigation Satellite System (BDS) indigenously launched by China officially started to serve the Asian-Pacific region since the late 2012. It is now being applied in a wide range of areas and markets with real-time data streams and related products. The international GNSS Monitoring and Assessment System (iGMAS) aims to establish a world-wide GNSS signal tracking network. It is designed to monitor the service performance and signal quality of multi-GNSS by high-precision receivers and high-gain omni-directional antenna. It provides services to customers worldwide [1]. At present, the iGMAS is boosting its pace of construction.

Data latency and packet loss, two critical parameters for transmission, play a key role in the utilization efficiency of real-time data streams. An analysis center always attempts to control the latency as small as possible to provide a more accurate feedback such as clock biases. Many international research institutions in the GNSS field are now exploring the real-time GNSS services with high accuracy. For example, the Jet Propulsion Laboratory (JPL) is the pioneer in this field [2]. The JPL products are transmitted to users with latency within 5 seconds [3-5]. Wuhan University conducted real-time data analysis using PANDA [6, 7]. In their prototype system, the latency of the data streams transmitted from the observation stations to the data center were controlled within 2-5

[a] Corresponding author : hzk@ntsc.ac.cn

seconds after synchronization [3]. The transmission performance of real-time data flows is essential for iGMAS, Continuous Operational Reference System (CORS), and the Satellite Based Augmentation Systems (SBAS), which motivates further study in this paper.

The latency of real-time data streams is composed of network transmission delays, software and hardware processing delays. The former depends on the transmission distance, network type and bandwidth. The hardware processing delays are primarily sourced from the signal processing and output of receivers, while the software processing delay is caused by the data processing and analysis delays of the NTRIP Server, Caster and Client [8]. The packet loss is defined as the ratio of the packets discarded to the total number of the packets sent, or the ratio of the number of packets in the RINEX form to the total number of packets theoretically received. The daily packet number (86400s) was used in this paper.

"Ping method" is commonly used to measure the network latency and packet loss ratio. This involves sending one ping per second between hosts and counting how many times the request is lost. Unfortunately, the "Ping method" for analysing loss is prone to inaccurate measurements due to its statistical limitations. Even in small amounts, packet loss will reduce application throughput, degrade or halt streaming applications, and introduce lags for interactive applications. The Ping method is based on the Internet Control Message Protocol (ICMP) whereas the NTRIP is based on the Transmission Control Protocol (TCP) that includes a connection-oriented service, a reliable data transfer service and a congestion-control mechanism [9]. In addition, the real-time transmission latency and packet loss are different from Ping results because of its various and complicated delay sources. Currently, signal receivers of the four iGMAS systems (BDS, GPS, GLONASS and Galileo) have being improved. Tracking stations are at the stage of site selection and construction and the network types are being chosen. There are notable differences in latency of real-time data output by receivers. The latency will occur during no matter the data transmission from tracking stations to data centers satisfies the requirements or not. Therefore, it is of great importance to study the assessment method of latency and packet loss generated in real-time data streams.

The huge latency of data transmission generated in individual stations may lead to network congestion and thus has a great impact on the use of real-time data streams. This paper proposes a Homology-Reverse method to measure and analyse the transmission latency from domestic tracking stations (see Table 1) to data centers using optimized software [10]. The proposed method is to provide a reference for using the BeiDou real-time data streams.

Table 1. Approximate coordinates of iGMAS domestic tracking stations.

Site	Latitude	Longitude	Altitude (m)
XIA1	N34.369°	E109.221°	449.46
CHU1	N43.791°	E125.443°	273.67
SHA1	N31.099°	E121.200°	21.74
GUA1	N43.471°	E87.177°	2028.82
BJF1	N39.607°	E115.892°	75.52
WUH1	N30.515°	E114.490°	71.24
LHA1	N29.657°	E91.104°	3625.10
KUN1	N25.030°	E102.797°	1988.08

2 Methods and materials

A general assessment strategy is to make a comparison of tested receivers to internationally recognized reference receivers. Real-time data streams are sent out by the NTRIP Server of receivers, retransmitted by the NTRIP Caster in data centers, and decoded by the NTRIP Client in data centers or analysis centers. The data latency and packet loss are then measured. Time synchronization must be maintained between receivers and data center servers. The latency of real-time signals output by receivers can be evaluated in the local area network (LAN). For Wide Area Network (WAN), network conditions between tracking stations and data centers (e.g. network congestion) can be examined by fixing the receivers in tracking stations. Nevertheless, the examination in WAN requires two or more receivers. Thus, its cost is high and the efficiency is low.

The so-called Homology-Reverse method is to establish a recognized receiver and data center in LAN. Real-time data streams sent out from the NTRIP Server of the receiver are retransferred by NTRIP Caster in data centers. After data decoding by the NTRIP Client installed in the receiver or computer, the latency and packet loss are then measured. Time synchronization must be guaranteed among receivers, servers and computers in data center. This proposed method can not only evaluate the performance of receivers being improved but pretest network of tracking stations. Unlike the conventional method in WAN, the proposed method has advantages of low cost and high efficiency.

The "backward" and "forward" latency are the same since the same round-trip paths are taken. For assessing the round-trip latency, a Homology-Returned method can be used by simply exchanging the position of the NTRIP Caster and the NTRIP Client used in the Homology-Reverse method.

As shown in Figure 1, the assessment system consists of GNSS stations, reference stations, data centers, and analysis centers. GNSS stations can be further classified into test stations and base stations. A base station, consisting of receivers, antennae, etc., is able to provide measured data and navigation ephemeris encoded and transmitted in RTCM real-time signals by the NTRIP Server. A test station has receivers or computers on which the NTRIP Client is running. Data centers and analysis centers include data receiving server, data processing server, and time synchronization server. The NTRIP Casters are installed in the data receiving server. Time synchronization server, traced back to UTC (NTSC) provides the time synchronization service for tracking stations and data centers.

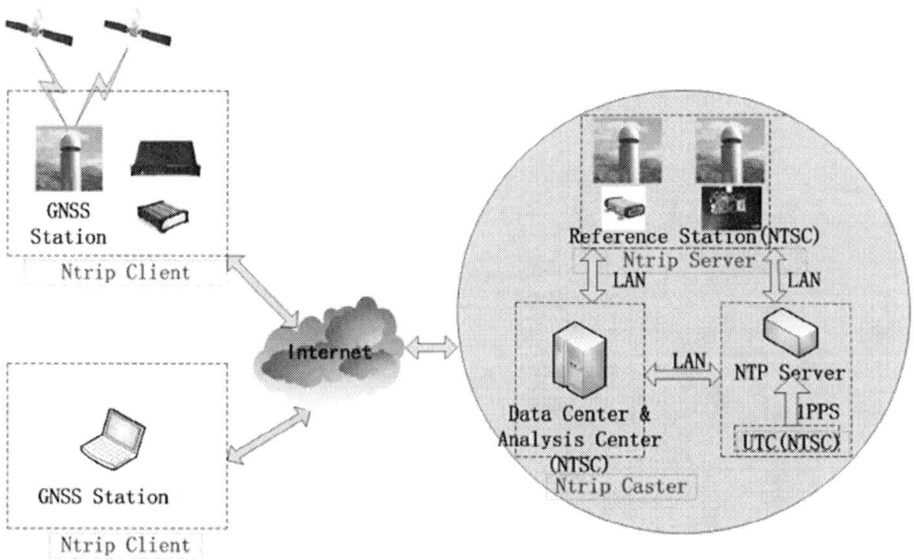

Figure 1. Schematic of the assessment system.

For precise point positioning, the acquisition of update information such as clock biases is affected by many factors [11] (see Figure 2). It involves signal processing and output delays of receivers in tracking stations, encoding delays of NTRIP Server, network transmission delays from stations to data centers, retransmission delays of NTRIP Caster in data centers, decoding delays of NTRIP Client, data analysis delays, delays of update information transmitted from analysis centers to clients, user positioning calculation delays, etc. This study only focuses on the real-time data latency in the data center, including the first five data latency mentioned above.

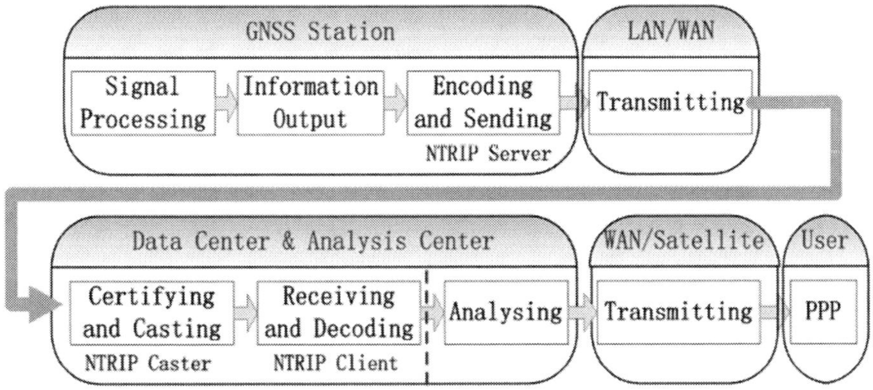

Figure 2. Main sources of Latency.

3 Results and discussion

3.1 Reference receivers

Trimble NetR9 was used as a Reference receiver. The NTRIP Server of receivers sent out RTCM real-time data retransferred by the NTRIP Caster of data centers. Latency was measured after data decoding by the NTRIP Client. Figure 3 shows the measured latency over 24 hours. Table 2 lists the statistical results from the Ping method and the NTRIP software. The real-time data streams output by Reference receivers have a latency of 28-182ms, with an average value of 38.99ms. By adding timestamp on the data packets, the time consumptions of LAN transmission, NTRIP Caster transfer and NTRIP Client decoding were measured, respectively. All the measurements were less than 1ms. Therefore, the latency of real-time data streams mainly caused by the receiver output latency and network transmission conditions.

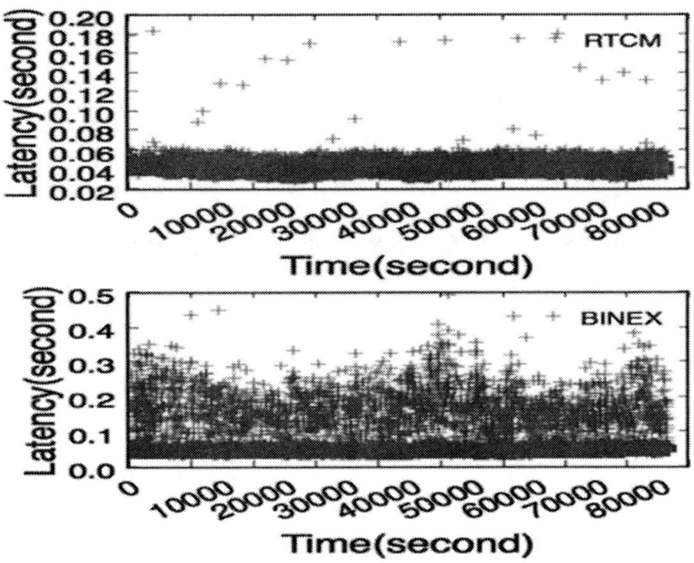

Figure 3. Latency of RTCM and BINEX real-time data streams in LAN.

DOI: 10.1051/matecconf/201711901031

Table 2. Comparison of latency and packet loss from Ping and NTRIP software in LAN.

Types	Min Latency (ms)	Max Latency (ms)	Avg Latency (ms)	Packet loss(s)
Ping(round-trip delay time)	0.21	1.355	0.274	0
NTRIP Software (RTCM)	28	182	38.99	0
NTRIP Software (BINEX)	29	491	46.53	0

On the other hand, BINEX real-time data streams sent out by the NTRIP Caster of Trimble NetR9 were received and decoded by the NTRIP Client [12], transcoded by the BINEX2RTCM, encoded by the NTRIP Server, and finally sent out by the RTCM. The bottom of Figure 3 shows the latency of BINEX. It can be seen that the BINEX latency is much larger than the RTCM latency. In addition, for a receiver with four systems, under the current constellation distribution for each navigation system, the BINEX data quantity is 1.19kB per second whereas RTCM data is 0.47 kB per second. The total differences for one day can reach 60.75MB. The application of the RTCM real-time data streams greatly reduces the Network transmission pressure.

3.2 Test stations

Generally, the latency and packet loss measured based on the proposed assessment method are consistent with the practical measurements. However, the practically measured latency is a slightly larger than that from the proposed method. The XIA1 station used UB370 (OEM plate), self-developed NTRIP Server, Data center and analysis center in LAN. The data latency of the tested receiver was in basic agreement with the Reference receiver, as shown in the top of Figure 4 (A represents assessment results). For the BJF1 station, the measured mean latency of data streams is 1s greater than the assessed results. This means the tested receiver has relatively large data latency. Therefore, the proposed assessment method can be used to determine the source of data latency. If the assessed latency was the same as the measured results, the latency is expected to be from a network that should be checked to satisfy the station requirements. If not as shown in the bottom of Figure 4, the latency should come from the receiver.

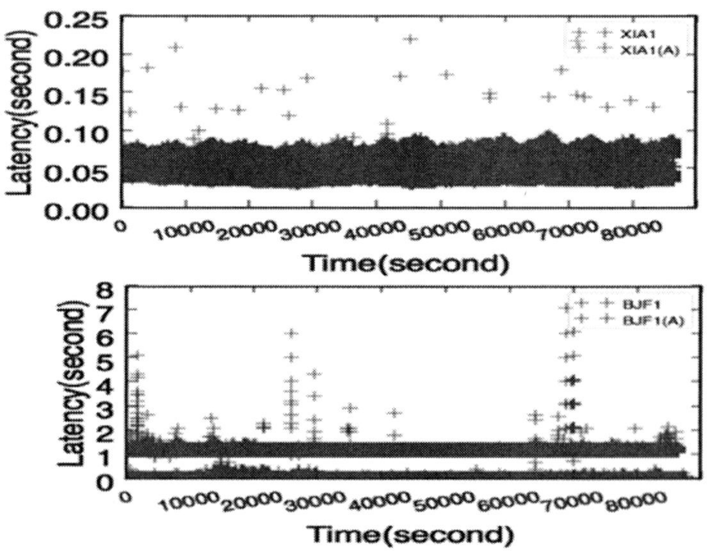

Figure 4. Comparison of measured and evaluated real-time data latency in the XIA1 and BJF1stations.

 DOI: 10.1051/matecconf/201711901031

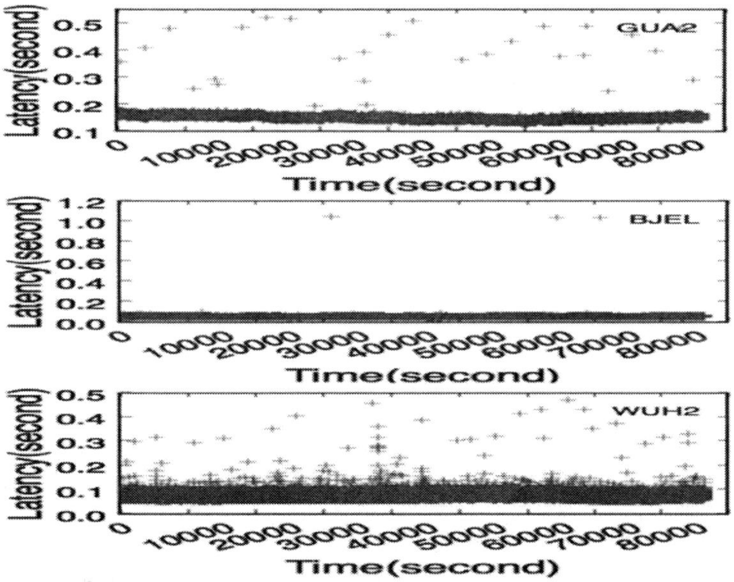

Figure 5. Real-time data latency in the GUA2, BJEL and WUH2 (SDH) stations.

Table 3. Latency and packet loss from Ping in WAN.

Sites	Min Latency (ms)	Max Latency (ms)	Avg Latency (ms)	Packet loss(s)
XIA1	0.17	11.96	0.365	33
GUA2	65.54	135.25	66.71	41
BJEL	13.27	127.59	16.74	4
BJF1	19.4	361.57	45.86	235
WUH2	24.78	266.16	31.34	1

Table 4. Latency and packet loss from NTRIP software in WAN.

Sites	Min Latency (ms)	Max Latency (ms)	Avg Latency (ms)	Packet loss(s)
XIA1	43	218	67.74	0
GUA2	124	519	150.2	0
BJEL	35	1041	45.9	0
BJF1	1027	5988	1072	0
WUH2	45	467	65.75	0

Synchronous Digital Hierarchy (SDH) Network is safe, reliable and has low latency but costly. System builders should think twice before choosing SDH Network (MSTP) or Virtual Private Network (VPN). The WUH2 station used the SDH Network (MSTP) to connect data center and analysis center. Its latency is illustrated in Figure 5. In addition, both BJEL and GUA2 stations from

China Science and Technology Network (CSTNET) were used. Their latency and packet loss agree well with special lines. Table 3 and 4 list the results from Ping and NTRIP software. Network conditions play a key role in the latency and packet losses. Several common internets have the same amount of latency and packet loss as the special line. This is against the fact that common networks have larger latency and packet loss than special lines.

3.3 iGMAS stations

The data center and analysis center deploys the China Science and Technology Network. Assessments were conducted at eight iGMAS tracking stations (XIA1, BJF1, SHA1, KUN1, SHA1, WUH1, GUA1 and LHA1) and two overseas tracking stations (CNYR and ZHON). The measurement work lasted one week and each day had similar results. Figure 6～10 show the measured results one sample day. Each tracking station provides broadband at a speed no less than 2Mbps. The domestic network operators include the China Science and Technology Network, the China Unicom Network (CUNET) and the China Telecom Network (CTNET). The latency and packet losses measured from Ping and NTRIP software are shown in Table 5 and 6.

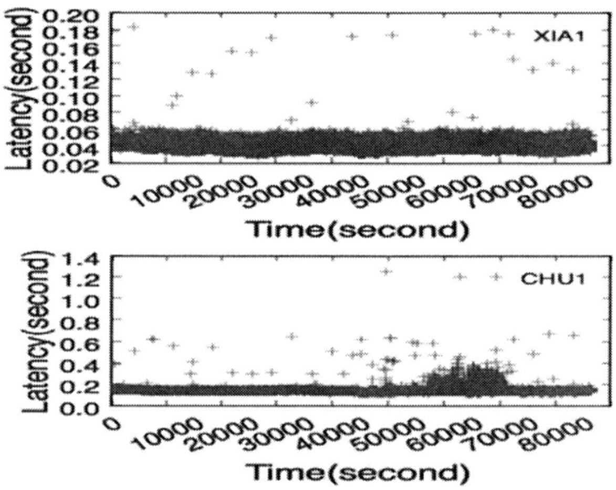

Figure 6. Real-time data latency in the XIA1 and CHU1 stations.

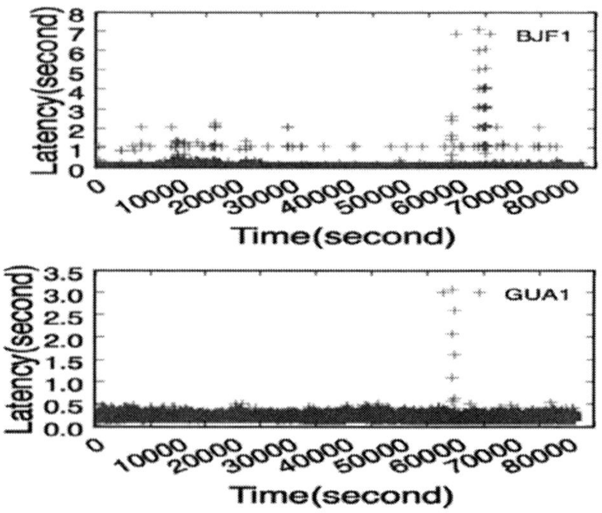

Figure 7. Real-time data latency in the BJF1 and GUA1 stations.

DOI: 10.1051/matecconf/201711901031

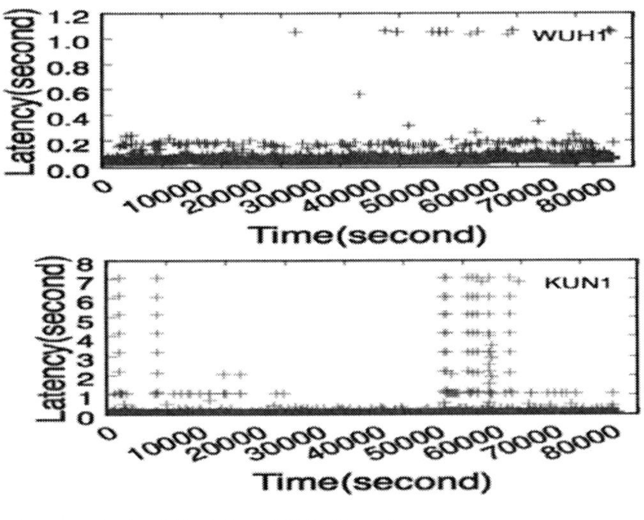

Figure 8. Real-time data latency in the WUH1 and KUN1 stations.

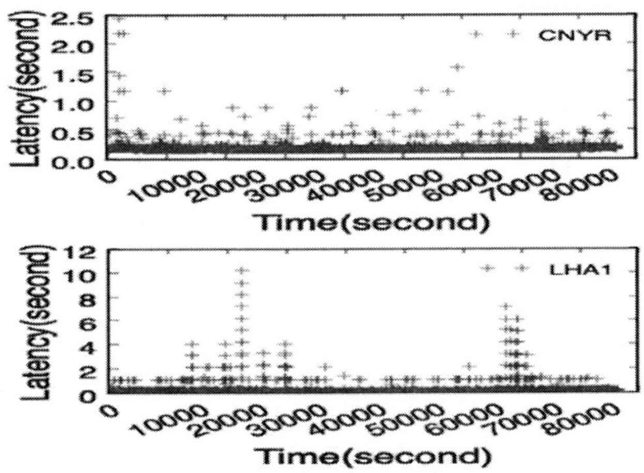

Figure 9. Real-time data latency in the CNYR and LHA1 stations.

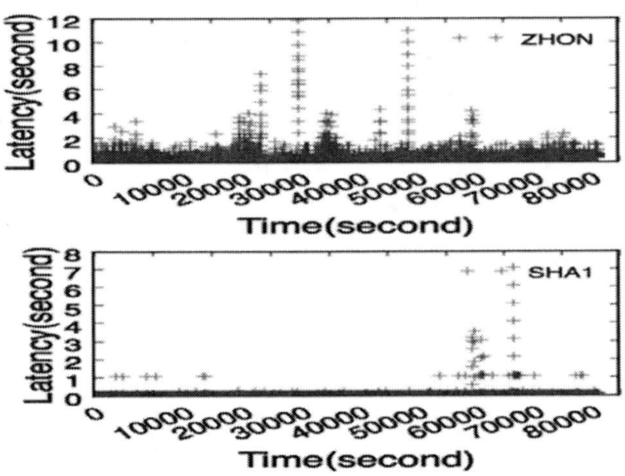

Figure 10. Real-time data latency in the ZHON and SHA1 stations.

No packet loss occurs in all the domestic stations except KUN1 and LHA1. Despite network congestion at some stations, 99.62% of the latency is controlled within 1s and 99.82% within 2s. The overseas station ZHON has no packet loss whereas CNYR does. However, more than 99.40% of the latency in the two stations was controlled within 1s. It is observed that some stations have network congestion and latency as high as 10s. The reason should be further studied.

The short-term instable characteristics of clock biases bring a certain error after a short-term extrapolation. The extrapolation duration should be reduced as much as possible. Therefore, the waiting time for estimating the clock biases should be short enough considering timeliness. However, to ensure a sufficient amount of measurements for each assessment, reasonable estimation waiting time must be set to maintain the stability and high precision of the estimation and prediction of clock biases. The majority of the data from the iGMAS domestic network arrives at the data center and analysis center within 1 seconds of its BeiDou time-tag. Data with latency less than 1 second are used in computing the BeiDou clocks.

Table 5. Latency and packet loss from ping of tested iGMAS stations.

Sites	Min Latency (ms)	Max Latency (ms)	Avg Latency (ms)	Packet loss(s)
XIA1(A)	0.17	11.96	0.365	33
CHU1(A)	32.81	698.2	41.26	55
SHA1(A)	37.85	63.06	42.23	19
GUA1(A)	66.84	132.68	67.97	43
BJF1(A)	19.4	361.57	45.86	235
WUH1(A)	38.03	175.09	40.94	157
LHA1(A)	69.56	381.86	75.14	23
KUN1(A)	69.55	843.59	71.06	230
ZHON(A)	42.68	73.95	43.49	32
CNYR(A)	233.9	341.03	245.2	23

Table 6. Latency and packet loss from NTRIP software tested iGMAS stations.

Sites	Min Latency (ms)	Max Latency (ms)	Avg Latncy (ms)	Packet loss(s)
XIA1(A)	43	218	67.74	0
CHU1(A)	111	1254	138.4	0
SHA1(A)	48	7067	62.58	0
GUA1(A)	125	3065	154.2	0
BJF1(A)	45	3996	66.77	0
WUH1(A)	41	3054	53.04	0
LHA1(A)	65	10180	85.93	84
KUN1(A)	66	7080	82.84	41
ZHON(A)	313	11792	362.7	0
CNYR(A)	156	2437	170	173

4 Conclusion

According to the measured and evaluated results, the RTCM transmission of real-time data streams significantly reduces the network transmission pressure. Some stations have the same amount of latency and packet loss by using either internet or special lines. For the current internet conditions, packet loss can be eliminated for real-time data streams. The latency of 99.4% of data packets is controlled within 1s. The use of internet can satisfy the requirements of the iGMAS, CORS, and SBAS for improving data transmission performance. In the NTRIP software, the encryption scheme enables the enhancement of safety and privacy as well as optimization of operation costs. The findings in this paper offer a reference for the use of the BeiDou real-time data streams as well as the facility construction.

Acknowledgements

This work was supported by the National Natural Science Foundation (11173026), the international GNSS Monitoring and Assessment System (iGMAS), the West Light Foundation of the Chinese Academy of Sciences (2013YB10), and the Fundamental Research Funds for the Central Universities.

References

1. W.H. Jiao, Q. Ding, J.W. Li, X.C. Lu, L.P. Feng, J.Q. Ma, and G. Chen, Monitoring and assessment of GNSS open services, Journal of Navigation, **64**, S19-S29 (2011)
2. R.J. Muellerschoen, W.I. Bertiger, and M.F. Lough, Results of an internet-based dual-frequency global differential GPS system, *Proceedings of the Iain World Congress & Annual Meeting of the Institute of Navigation*, 796-802 (San Diego, USA, 2000)
3. C. Shi, Y.D. Lou, W.W. Song, S.F. Gu, C.J. Geng, W.T. Yi, and Y.Y. Liu, A wide area real-time differential GPS prototype system in China and result analysis, Survey Review, **43** (322), 351-360 (2011)
4. R.J. Muellerschoen, A. Reichert, D. Kuang, M. Heflin, W. Bertiger, and Y. Bar-Sever, Orbit determination with NASA's high accuracy real-time global differential GPS system, *Proceedings of the 14th International Technical Meeting of the Satellite Division of the Institute of Navigation (ION GPS 2001)*, 2294-2303 (Salt Lake City, USA, 2001)
5. M.O. Kechine, C.C.J.M. Tiberius, and H.V.D. Marel, An experimental performance analysis of real-time kinematic positioning with NASA's Internet-Based Global Differential GPS, GPS solutions, **8** (1), 9-22 (2004)
6. J.N. Liu and M.R. Ge, PANDA software and its preliminary result of positioning and orbit determination, Wuhan University Journal of Natural Sciences, **8** (2), 603-609 (2003)
7. C. Shi, Q.L. Zhao, J.H. Geng, Y.D. Lou, M.R. Ge, and J.N. Liu, Recent development of PANDA software in GNSS data processing, *Proc. SPIE. International Conference on Earth Observation Data Processing and Analysis (ICEODPA)* (eds. D.R. Li, J.Y. Gong, and H.Y Wu), **7285**, 72851S1-1 (Wuhan, China, 2008)
8. G. Weber, D. Dettmering, H. Gebhard, and R. Kalafus, Networked transport of RTCM via internet protocol (Ntrip) … IP-Streaming for Real-Time GNSS Applications, *Proceedings of the 18th International Technical Meeting of the Satellite Division of the Institute of Navigation (ION GNSS 2005)*, 2243-2247 (Long Beach, USA, 2005)
9. J.F. Kurose and K.W. Ross, *Computer Networking. 6th edition*, Pearson Education (New Jersey, USA, 2012)
10. Z.K. He, W.D. Tang, X.H. Yang, L.M. Wang, and J.H. Liu, Use of NTRIP for optimizing the decoding algorithm for real-time data streams, Sensors, **14** (10), 18878-18885 (2014)
11. M. Li, L.Z. Qu, Q.L. Zhao, J. Guo, X. Su, and X.T. Li, Precise point positioning with the BeiDou navigation satellite system, Sensors, **14** (1), 927-943 (2014)
12. R. Galas and W. Kohler, A binary exchange format for GPS data, Physics and Chemistry of the Earth Part A Solid Earth and Geodesy, **26**, 645-648 (2001)

Design of the crashworthy structure of an urban aircraft

Bairong Shang, Jingwu He[a], Yuexi Xiong and Zhe Hu

School of Aeronautic Science and Engineering, Beihang University, Beijing, China

Abstract. With the development of general aviation, the urban aircraft is around the corner. The urban aircraft with composite is considered as an ultralight vehicle and the crashworthiness is of vital importance for such an ultralight aircraft. Composites are being widely and increasingly used in the aerospace industry because of their advantages that include the high specific strength and stiffness over traditional metallic materials. Besides, composites have the potential for absorbing the energy in a crash event. The crashworthiness of the cockpit section is analyzed in this paper and some modifications in the subfloor have been made to improve the survivability of the pilot. Advances in commercial softwares have enabled engineers to simulate crash events. The three-dimensional structure model is established by use of CATIA software and the crash process is simulated by MSC/DYTRAN. By comparing the crashworthiness of composite structures, reliable basis is provided for the design of a safe and sound structure.

1 Introduction

The crashworthiness is the ability to protect the safety of crew members and passengers during an aircraft crash accident and the aircraft accidents are severe threats to the life safety of people. With the development of the civil aviation scientific technology, more and more attention is attracted by the crashworthiness of civil aircraft [1]. The airplane's crash is an uncertain process and the true process of impact is also very complicated, therefore the reasonable crash condition and crash environment is the precondition to acquire an accurate structural response [2].Two methods are mainly used in the research on crashworthiness of aircraft structure, the drop impact test and the finite element simulation. The drop impact test needs a lot of time and costs. Moreover, the accuracy and reliability of test data depend much on sensor elements and data collection methods. However, the crash simulation method can greatly shorten the research cycle and reduce the costs. So a reasonable crash simulation will raise the design efficiency and give designers a comprehensive understanding of the characteristics of the crash.

An increasing number of research institutions have studied the crashworthiness of the fight vehicle. Back in the 1970s, Federal Aviation Administration and National Aeronautics and Space Administration started the drop impact test of aircrafts, including small aircraft and large aircraft supported by industry departments of the United States. Meanwhile, extensive research and analysis

[a] Corresponding author : hejingwu@buaa.edu.cn

on the dynamic characteristics of aircraft crash and the safety were carried out [3]. In 1999, NASA Langley Research Center conducted a crash test of the Sikorsky ACAP helicopter and simulated the crash process by the finite element method, aiming at verifying the validity of the finite element simulation. The simulation was by use of the software MSC.Dytran and the results of the simulation and crash test were similar. So it proved that the finite element simulation method is feasible in crash resistance design [4].

In the past 50 years, composites have been used in the aircraft structure gradually and a great deal of work about the failure behavior of composites has been done. In the 1990s, the European Union organized a project entitled "CRASHWORTHINESS" which acquired the crash test data of a full-scale fuselage. Then a series study of composite structure about crash resistance were carried on, which were emphasized on carbon fiber reinforced composite material characteristics and failure mechanism [5-7].

The study of the crashworthy structure mainly contains seat systems, landing gear systems and fuselage structure [8,9]. The method of the crashworthy structure design is different among the light fixed-wing aircraft, the helicopter and the large and medium-sized aircraft [10]. Due to constrained space under the floor in the light fixed-wing aircraft and helicopter, the subfloor of energy absorption is widely used to dissipate the kinetic energy during the impact. However, it is focused on the struts and reinforced bulkheads in the large and medium-sized aircraft and the reinforced bulkheads are the important structure to absorb the kinetic energy. The cross-sections are respectively shown in Figure 1 and Figure 2.

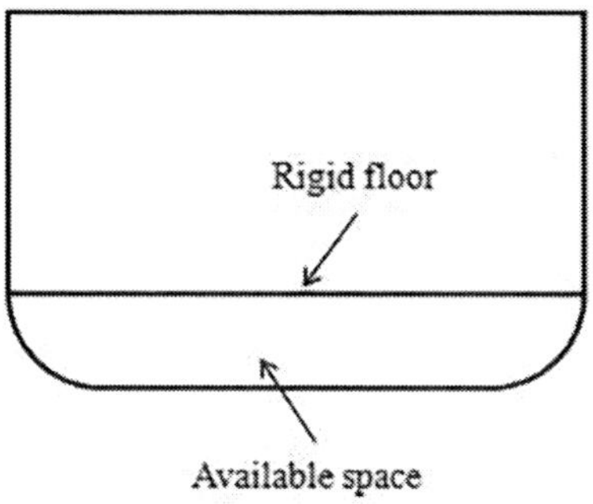

Figure 1. Layout of the light fixed-wing aircraft and helicopter.

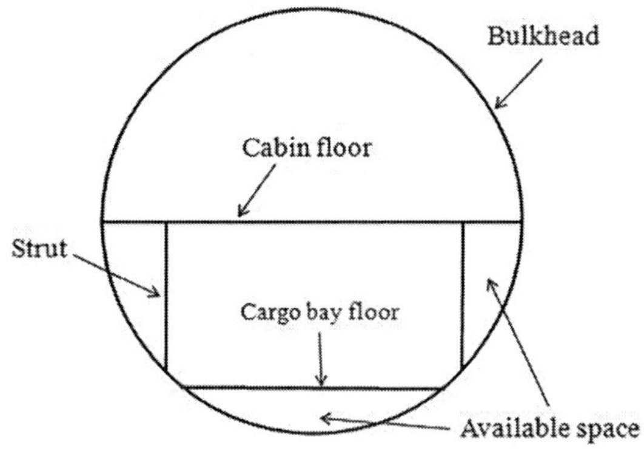

Figure 2. Layout of the large and medium-sized aircraft.

2 Energy analysis during the crash process

In the process of aircraft crash, by the law of conservation of energy

$$\frac{1}{2}M(V_0^2 - V_1^2) = U_G + U_F \tag{1}$$

Where M is the quality of the aircraft, V_0 is the initial velocity when contact the ground, V_1 is the velocity when fuselage rebound, U_G is the energy absorbed by the ground and U_F is the energy absorbed by the fuselage structure.

U_F will meet the equation

$$U_F = U_{cabin} + U_{subfloor} \tag{2}$$

Where U_{cabin} is the energy absorbed during the cabin destruction and $U_{subfloor}$ is the energy absorbed during the subfloor destruction

According to the equation (1) and equation (2) , U_{cabin} will meet the equation

$$U_{cabin} = [\frac{1}{2}M(V_0^2 - V_1^2) - U_G] - U_{subfloor} \quad (3)$$

In order to maintain people's survival space and reduce overload, the equation indicates:

1) Try to reduce the impact velocity and it requires that the pilot have to control the descent velocity and attitude reasonably . But this is not the emphasis in this paper.
2) Try to increase the extent of the damage of subfloor structure which helps to absorb the kinetic energy as much as possible

3 Model of the unmodified cockpit section

Because of the development of general aviation and the urban traffic jam, the urban aircraft is around the corner. The security must be the priority for the ultralight aircraft. A vertical drop of the cockpit section, which crashes vertically to impact at 8 m/s is simulated in this paper. The cockpit section is approximately 1200mm in length and 800mm in maximum width. The structure of the aircraft is mainly composite structure and the three-dimensional structure model is established in the software CATIA, shown in Figure 3 and Figure 4. The main structure of the cockpit section are the skin, the bulkheads, the stringers and the floor. The crossbeams and the bulkheads, supporting the floor panel are integrated together. There are stiffeners below the floor panel as well. In order to attain enough stiffness and strength and reduce weight, Carbon Fiber Reinforced Plastic (CFRP) is chosen and T300 carbon fabric is used in the whole cockpit section.

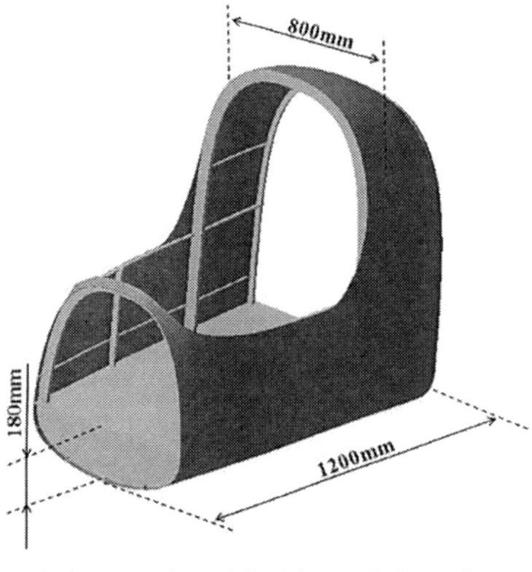

Figure 3. Structural model of the cockpit section.

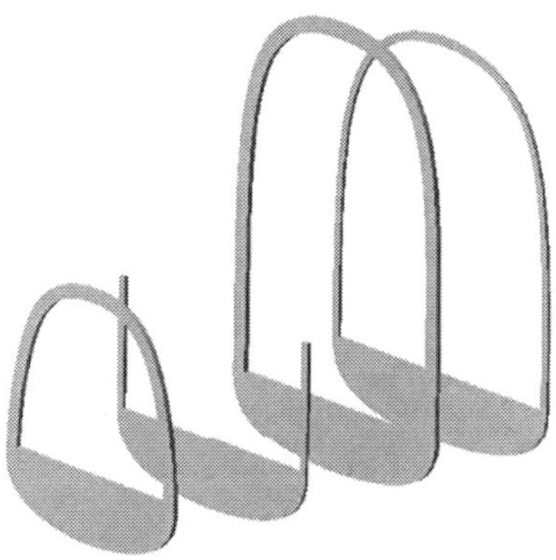

Figure 4. Structural model of bulkheads.

The unmodified cockpit section is simplified and the finite model is established. First ,the mesh is completed in the software Hypermesh and then the model is imported into the software Patran where the loads, the boundary conditions and the output requirements are set up. At last, the simulation is operated in the solver Dytran. According to the standards about the crashworthiness, the vertical impact velocity is set as 8m/s and the ground is simplified as a rigid plane. The direction of the impact is vertical to the rigid plane. Furthermore, the pilot and the seat are simplified as a mass element which is connected to the floor and the connections in the model are regarded as rigid connections. The finite model is shown in Figure 5, Figure 6 and Figure7. The floor panel is a composite sandwich construction with a foam core and T300 carbon fabric face sheets, shown in Figure 8. The property of T300 and other materials is shown in Table 1 and Table 2.The whole model is meshed with shell elements and consists of 33209 nodes, 32837 elements and one concentrated mass element. The explicit algorithms for structural dynamic analysis is adopted to solve the impact problem. The structural mass of the cockpit section is 11.94kg.There are a few contact constraints in the finite model, in consequence it will cost several hours to complete the simulation.

Table 1. Material property data of T300.

ρ (g/cm^3)	ν	G_{12} (GPa)	E_{11} (GPa)	E_{22} (GPa)	X_T (MPa)	X_C (MPa)	Y_T (MPa)	Y_C (MPa)	S (MPa)
1.57	0.33	5.0	135	9.75	1560	1226	80	218	89.9

Table 2. Other material property data used in the finite model.

Material	ρ (g/cm^3)	E (MPa)	ν	σ_y (MPa)	E_h (MPa)	ε_{psf}
Rohacell foam	0.045	13.78	0.3	0.62	0.37	0.8
Foam 1	0.052	75	0.38	1.6		0.03
0°/90° E-glass	0.24	18495	0.13	85	813	0.01
±45° E-glass	0.24	10360	0.47	63	813	0.01

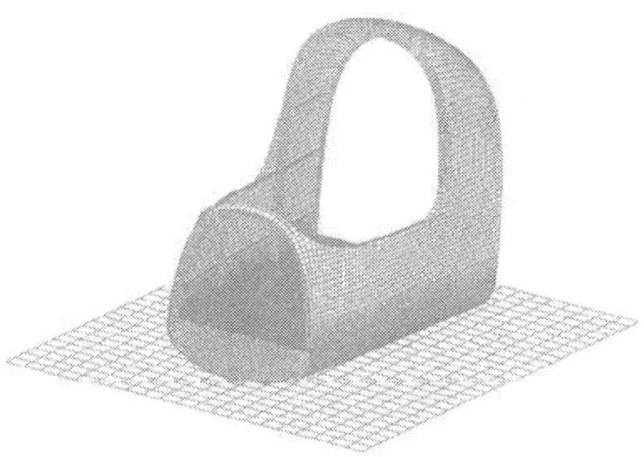

Figure 5. Finite model of the unmodified cockpit section.

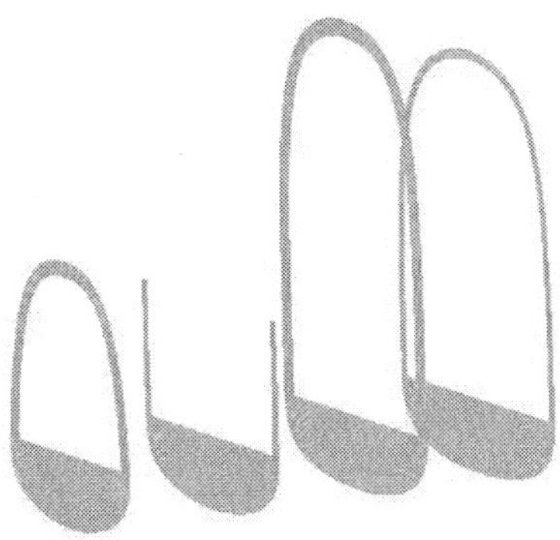

Figure 6. Finite model of the bulkheads.

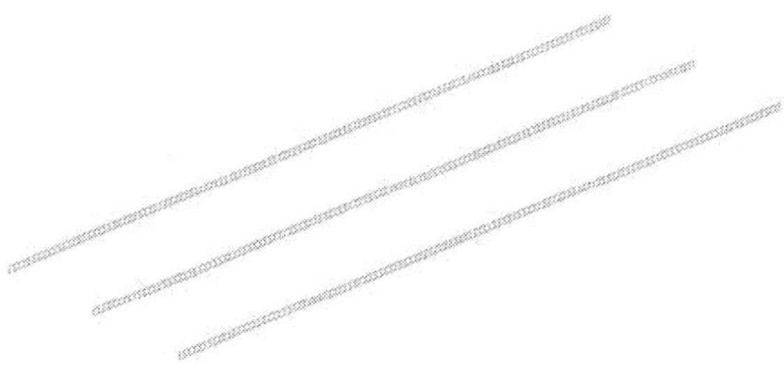

Figure 7. Finite model of the floor stiffeners.

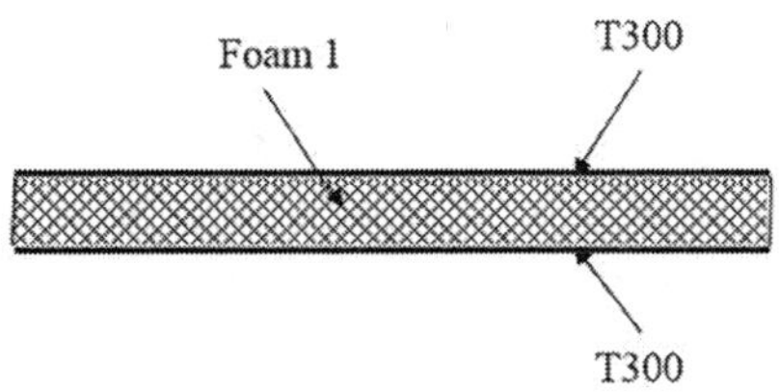

Figure 8. Schematic drawing of the floor panel.

4 Simulation results of the unmodified cockpit section

The peak value of acceleration, appearing at the time 2.2ms is 364.6g, shown in Figure 10. The peak acceleration appears at the time when the cockpit section just contacts to the rigid plane which indicates that the strength and stiffness of the structure are too high. Then with the destruction of the structure and the energy absorption, the acceleration decreases gradually. From Figure 11, it is easy to figure out that the cockpit section rebounds at 6.5ms. As can be seen from Figure 9, only a little part

 DOI: 10.1051/matecconf/201711901032

has deformed and damaged, and the whole section maintains the intact structure. Although the survival space is enough, the peak acceleration is as high as 364.6g. The overload does sever harm to people's body, so it requires that the stiffness and strength have to be reduced so that the structure will damage further and the kinetic energy will be absorbed step by step, which will contribute to reduce the peak acceleration.

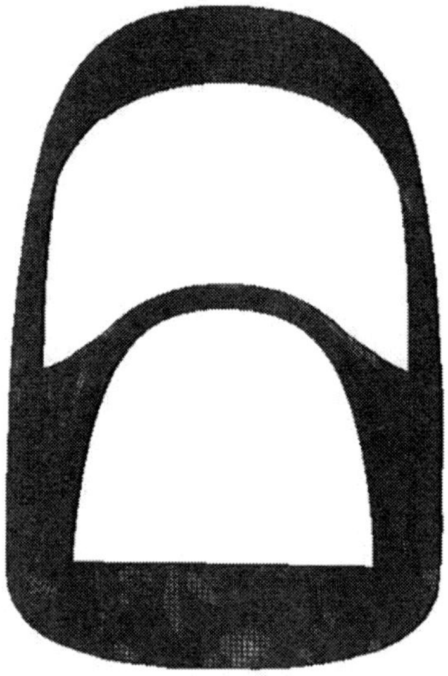

Figure 9. Deformation of the unmodified cockpit section at 6.5ms.

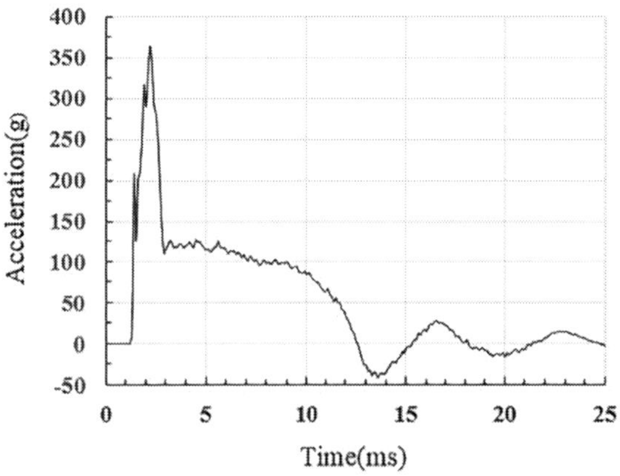

Figure 10. Time history of acceleration.

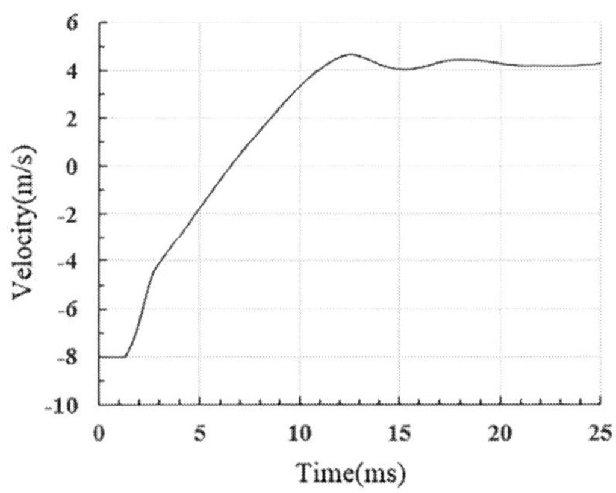

Figure 11. Time history of velocity.

5 Improvement of the subfloor

The Rohacell foam that has good energy-absorbing performance is added at the bottom and the bulkheads and some auxiliary structure are added to support the floor. In addition, the beams supporting the floor panel have also been modified.The finite models are shown in Figure 14 , Figure 15 and Figure 16. In order to make the upper structure of the cockpit section stiff enough, the top part of the skin, the bulkheads and the auxiliary structure are carbon fiber reinforced plastic structure and the lower part are glass fiber reinforced plastic structure which is easier to damage and absorb the energy. The material distribution is shown in Figure 12 and Figure 13 and T300 and E-galss fabric are used in the model. The structural mass of the cockpit section is 10.72kg.

DOI: 10.1051/matecconf/201711901032

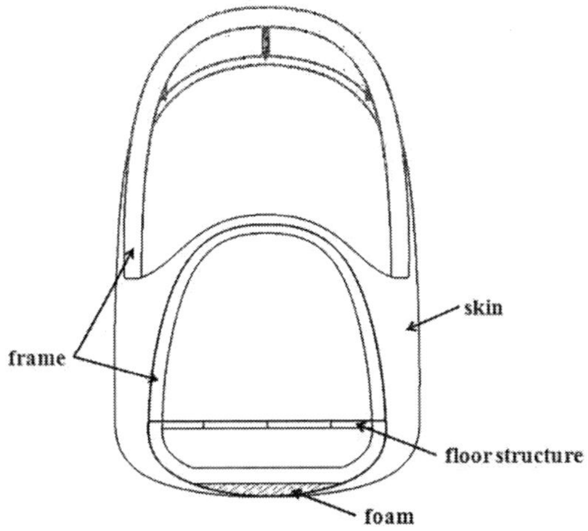

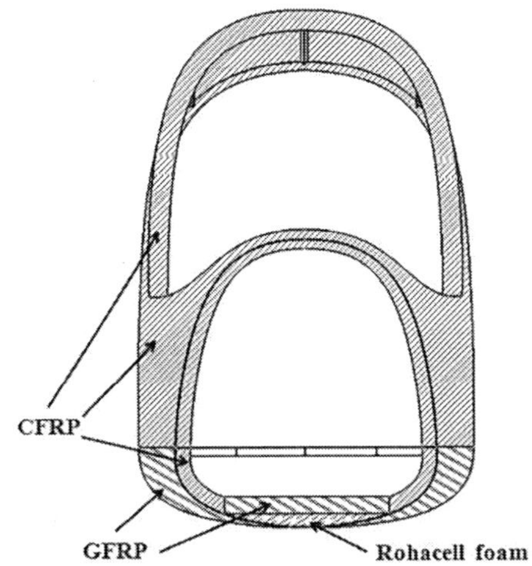

Figure 12. Front view of the modified cockpit section. **Figure 13.**Schematic drawing of material distribution.

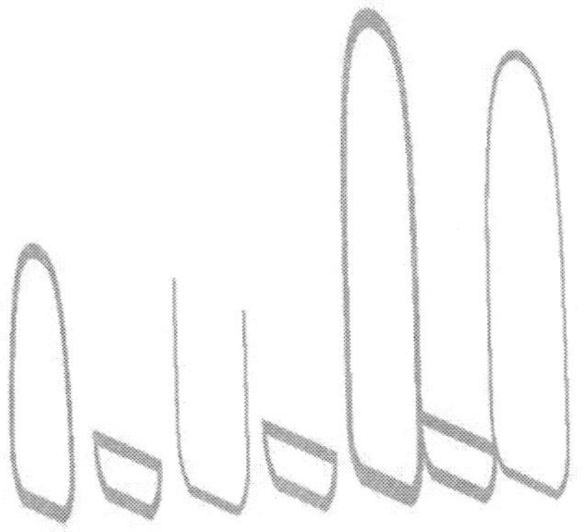

Figure 14. Finite model of the modified bulkheads and the auxiliary structure.

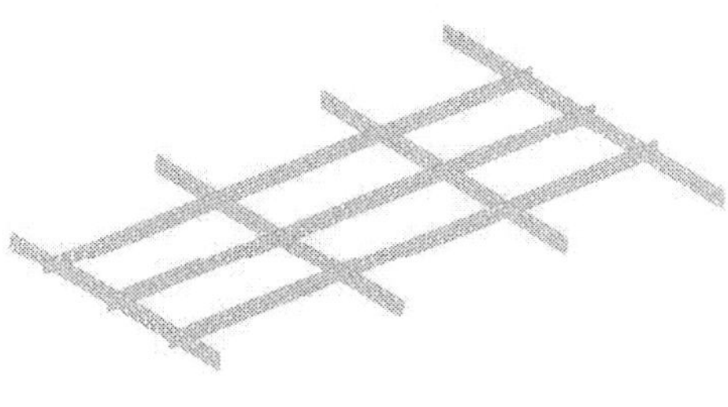

Figure 15. Finite model of the foam. **Figure 16.** Beams of the floor structure.

As can be seen from Figure 17, the subfloor has damaged badly and been crushing. For the sake of reducing the acceleration when the cockpit section contacts the ground, firstly the added foam at the bottom is compressed to absorb the energy. Secondly, the plastic deformation and the destruction occurs in the glass fiber reinforced plastic. Finally, the subfloor with CFRP is destroyed gradually. This multi-level buffering structure makes the process of energy absorption slow and steady. The peak acceleration is about 55.2g, shown in Figure 18. It can been seen from Figure 19 that the cockpit section rebounds at nearly 28.9ms and remains at the velocity of 2.6m/s at 60ms.

Compared with the result of the unmodified cockpit section, the peak acceleration obviously decreased by 84.9% and the acceleration remains in a stable range, shown in Figure 20. In addition, the upper cockpit section which has a little deformation and destruction maintains an enough survival space. Therefore the chance of the pilot's survival has been greatly increased. The mass of the modified cockpit section is 1.22kg less than the unmodified one ,which will conrtibute to improve the flight performance.

Figure 17. Deformation of the modified cockpit section at 28.9ms.

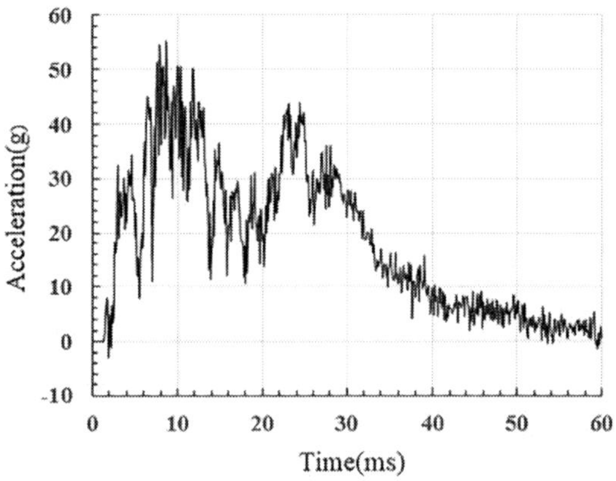

Figure 18. Time history of acceleration for the modified cockpit section.

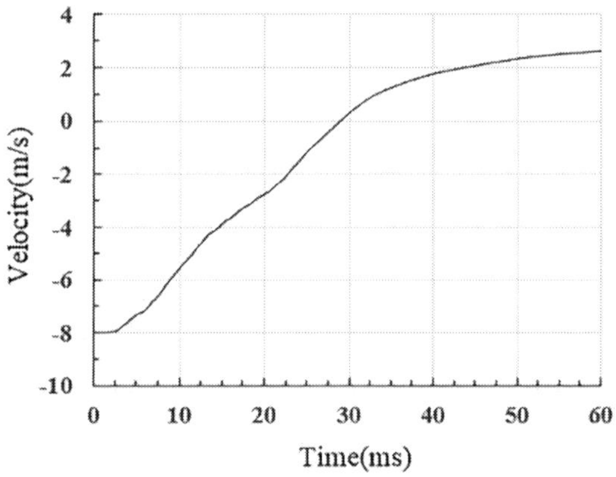

Figure 19. Time history of velocity for the modified cockpit section.

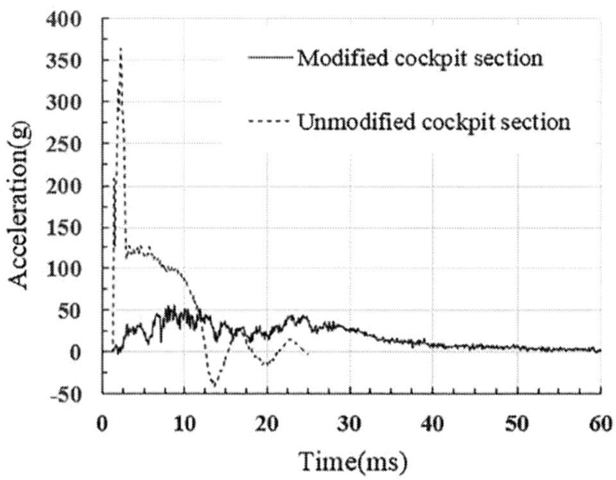

Figure 20. Comparison of acceleration histories.

6 Summary

The crashworthiness of aircraft is of vital importance and a crash simulation of the composite cockpit section is simulated in the transient dynamic finite element software MSC.Dytran. Because of the high specific strength and specific stiffness, T300 carbon fabric is used in the structure. But the stiffness and strength of the unmodified cockpit section is so high that the high impact loads transmits to the occupant. In order to limit the loads transmitted to the seat, the Rohacell foam and E-glass fabric are used at the bottom of the modified cockpit section. The progressive failure of the subfloor makes the kinetic energy absorbed step by step and reduces the loads transmitted to the seat. Compared with the peak acceleration of the unmodified cockpit section, the peak acceleration of the modified one is reduced by 80% and the mass reduced by 10.2%. The Rohacell foam and E-glass help to absorb the energy during the crash and the progressive failure of the subfloor contributes to make the acceleration remain in a stable range. The design concept for the subfloor of the ultralight, introduced in this paper has certain significance.

References

1. H. Zhang, Design guideline for crash resistance of the general aircraft (2009)
2. Z.Y. Feng, X.M. Zhang, H.L. Mou, and T. C. Zou, MSTAE, **32**, 353 (2013)
3. E.L. Fasanella and K.E. Jackson, *Proceedings of the 44th AIAA Structures, Structural Dynamics and Materials Conference* (2003)
4. E.L. Fasanella, K.E. Jackson, and K.H. Lyle, J. Am. Helicopter Soc., **47**, 156-168 (2002)
5. G. Labeas and T. Kermanidis, FFEMS, **26**, 449-457 (2003)
6. E. Deletombe and D. Delsart, Aerosp. Sci. Technol., **4**,189-199 (2000)
7. J.F.M. Wiggenraad, D. Santoro, F. Lepage, C. Kindervater, and H.C. Mañez, *The 57th Annual Forum of the American Helicopter Society* (2001)
8. A.O. Bolukbasi and D.H. Laananen, J. Aircraft, **21**, 512-519 (1984)
9. K.H. Lyle, K.E. Jackson, and E.L. Fasanella, J. Aircraft, **39**, 142-147 (2002)
10. J.D. Cronkhite and V.L. Berry, NASA Contractor Report No.3603 (1982)

Development of an MDO platform for aero-structural interactions

Zhe Hu[a], Yuexi Xiong, Jingwu He and Yang Pan
School of Aeronautical Science and Engineering, Beijing University of Aeronautics and Astronautics, Beijing, China

Abstract. An interactive MDO platform for multivariate and high coupling aero-structural interactions is presented in this paper. Programs based on RSM and GA are utilized as optimization solver, which is connected with CAD/CAE software by a friendly GUI. To generate and modify CAD models in the analysis and optimization process, CATIA Automation Technique is adopted to create a CAD model generator. During the analysis process, high-fidelity models in both disciplines are used. For aircraft wing MDO problems, optimal wing shape and structural configuration can be obtained using this platform in the preliminary design phase. While the current platform is aiming at solving Aero-Structural problems in the preliminary design phase of UAV components, it also provides a reference solution for other related MDO problems. Two cases are provided to verify the validity and practicability of this platform. Besides the design ideas and techniques, this paper also demonstrates its further development capabilities.

Nomenclature

C_l	=	Lift coefficient
$C_l^{Baseline}$	=	Lift coefficient of the initial design
C_d	=	Drag coefficient
$C_d^{Baseline}$	=	Drag coefficient of the initial design
L/D	=	Lift to drag ratio
Λ	=	Sweep angle
λ	=	Taper ratio
U_{max}	=	Maximum displacement of the wing
M	=	Structural Mass
rsb	=	Thickness of the rear spar flanges
fsb	=	Thickness of the front spar flanges
pt	=	Thickness of the spar plates
st	=	Thickness of the skin

[a] Corresponding author: huzhebuaa@163.com

1 Introduction

With the continuous development of the Computer Aided Engineering (CAE) Technology, high precision analysis methods based on Computational Fluid Dynamics (CFD) and Computational Structure Mechanics (CSM) can be applied during the preliminary design phase of the Unmanned Aerial Vehicles (UAV) to obtain more precise simulation results [1]. Wing, as important part of an UAV, is closely related to the aero-structural interaction issues. And the design of a wing is a comprehensive issue connected with various disciplines. Recent years, researchers have been working on the development of Multidisciplinary Design Optimization (MDO) frameworks and platforms in which high precision modeling, analysis techniques are tightly integrated [2-5]. Therefore, it is evident that a platform with friendly GUI which can help to deal with MDO problems that considering aero-structural interactions is essential for aircraft designers and engineers. The MDO platform is consist of the modeling, analysis, optimization modules and the data exchanger. Equipped with a graphic user interface, the platform is able to provide various options for modeling, analysis and optimization, and can seek for an optimal configuration for preliminary wing design issues.

2 CAD modeling module

Parametric modeling is essential to be applied in the pre-processing stage. Especially in the optimization stages, design variables related to parametric modeling should be modified to generate new design points. Therefore, on one hand, parametric modeling with a graphic interface can help to generate, modify, and check the model in a more efficient way. On the other hand, its property makes it possible to be easily applied in the optimization algorithms. Computer Aided Three-dimensional Interactive Application (CATIA), known as one of the most common used Modeling software in aviation industry, provides a programming technique named CATIA Automation to create and modify models in form of VBSripts. The generation process of the model is demonstrated in Figure 1.

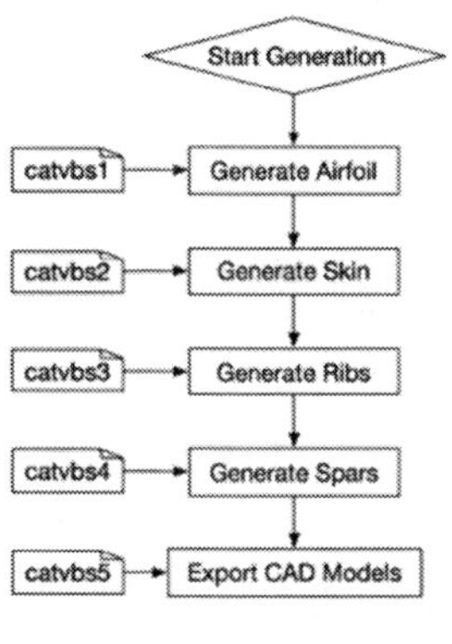

Figure 1. CAD Model Generation Process.

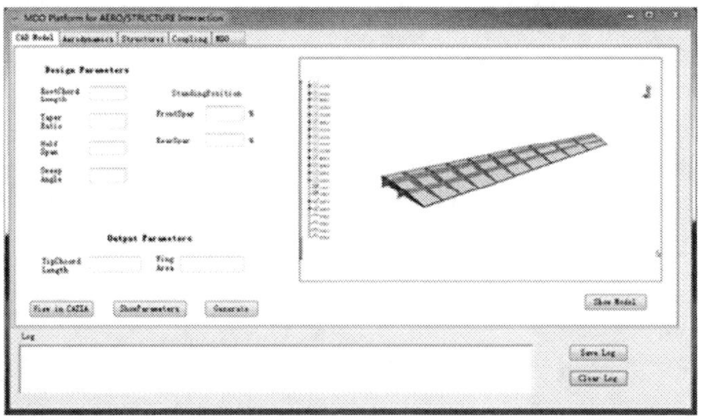

Figure 2. CAD Modeling Module.

Specifically, the generation form of the CAD Model is in a sequential way in which the skin, ribs, spars are generated in order. Different kinds of wing structures can be generated by modular scripts to expand the model database. The CAD parametric modeling process is aimed to generate models for both aerodynamics and structural analysis processes. The interface of CAD modeling module is shown in Figure 2. Wing shape parameters can be quickly modified to generate CAD models through the GUI. In addition, a log Window is provided to record and show the important information and user operations, and it can help to check the input and intermediate process information when error occurs.

3 Analysis module

The analysis workflow is shown in Figure 3. The analysis process begins with the aerodynamic mesh generation. After the aerodynamic analysis, pressure data generated by interpolation method can be obtained after the aerodynamic analysis process. Python script is applied to extract the load data from '.dat' file and create an '. inp' file for ABAQUS in which the finite elements information, material properties, boundary conditions and aerodynamic loads are recorded. Once the analysis process is completed, the analysis results for each discipline can be obtained from the interface.

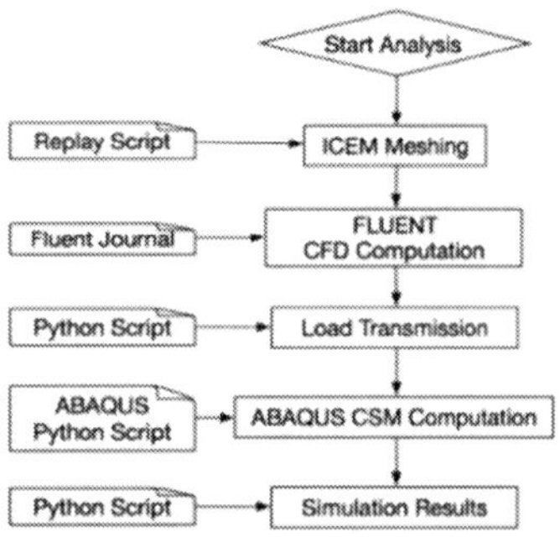

Figure 3. Analysis Workflow.

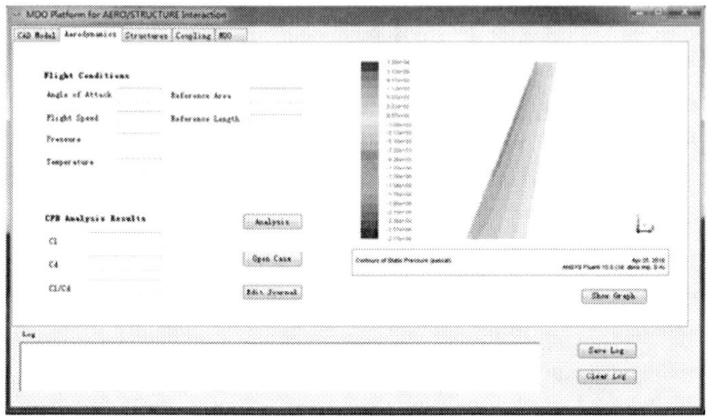

Figure 4. Aerodynamic Analysis Module.

3.1 Aerodynamic analysis

Aerodynamic analysis method based on CFD is able to compute the aerodynamic performance through a high precession way. In the interface of aerodynamic analysis module shown in Figure 4, flight conditions can be modified through the GUI to obtain the aerodynamic performance of the wing.

During the analysis process, Spalart-Allmaras viscosity model is adopted to simulate the turbulence of the flow field. Flux Type of Roe-FDS and Implicit formulation are chosen as the solution method [6]. A wing mesh for CFD computation is shown in Figure 5. It is generated through an ICEM replay script and has the O type grid topology. The automatic aerodynamic analysis process is driven by a '. journal' script of FLUENT. When the analysis process is completed, a python script is used to extract the flight performance parameters and generate an '. inp' file for the following structural analysis.

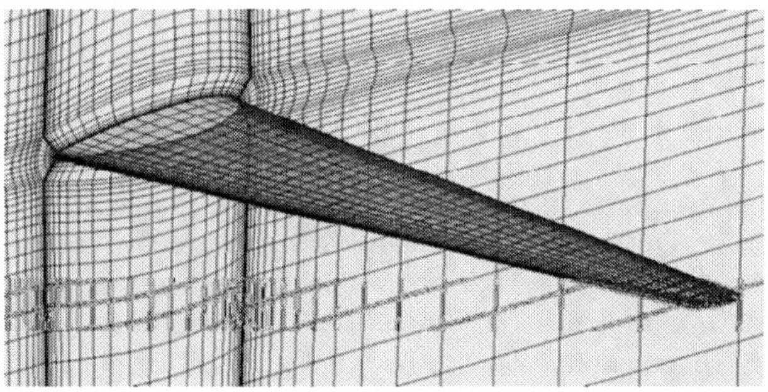

Figure 5. Aerodynamic Analysis Module.

3.2 Structural analysis

Patran is often chosen as the structural analyzer for wing MDO problems [7-9]. In this paper, in order to create CSM model more efficiently, ABAQUS is selected as the structural pre-processer and analyzer. The structural analysis simulation is based on the finite element method (FEM). The interface of structural analysis module is shown in Figure 6. Structural parameters and material properties can be modified through the interface to get the maximum stress and displacement on each component. In addition, stress and displacement nephograms, structural mass and other simulation results can be obtained. Moreover, the platform can also provide more detailed analysis information such as the analysis database and the original project files.

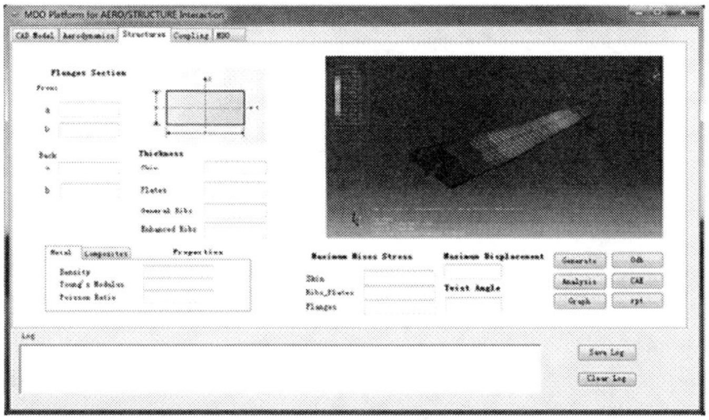

Figure 6. Structural Analysis Module.

4 Aero-structural coupling

Aero-structural interactions have a notable influence on the aerodynamic performance of the wing. The deformation and its impacts on external flow should be taken into consideration. The max displacement on a wing tip is chosen as a characteristic variable. The changing tendency of the maximum displacement on a wing tip is shown in Figure 7.

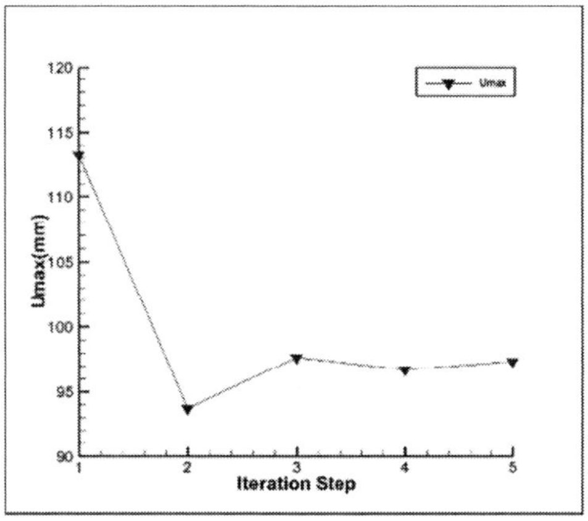

Figure 7. Variation of U_{max} with Iteration Step.

The solution is converged after 5 iterations according to the convergence criterion demonstrated as below:

$$\left|\frac{U_{\max}^{i}-U_{\max}^{i-1}}{U_{\max}^{i}}\right|<0.01 \tag{1}$$

Where U_{max}^{i} is the maximum displacement of the wing tip in the ith iteration step. A series of case have been tested to come to a conclusion that the computation results will be converged after 3-5 times of iterations. The impact of the coupling effect on lift coefficient (Cl) is shown in Figure 8.

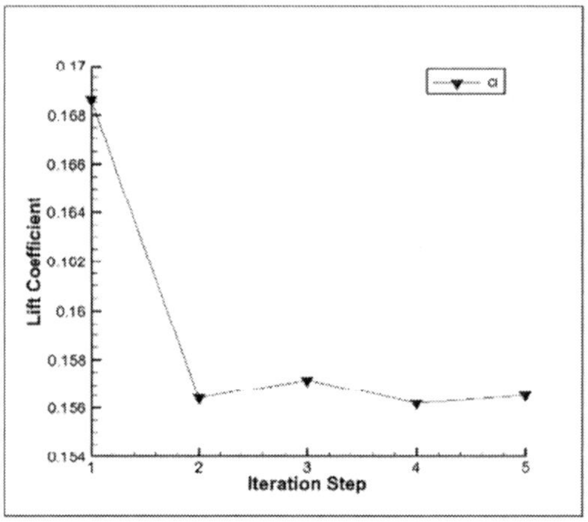

Figure 8. Variation of Cl with Iteration Step.

Compared with the case of a rigid wing, the lift coefficient decreases about 7%. Therefore, for the optimization of aerodynamic performance, it is of great significance to consider the effects of aero-structural interactions in order to establish a more actual and precise simulation model.

5 Multidisciplinary design optimization

5.1 MDO Architecture

MDO is a research field which has been growing rapidly as the analysis and approximation techniques are developed. Wing design is a representative issue since the interactions of disciplines like aerodynamics and structural theory have to be taken into account. Sharing many similarities with Collaborative Optimization (CO) [10], the MDO architecture integrated in the platform is demonstrated in Figure 9.

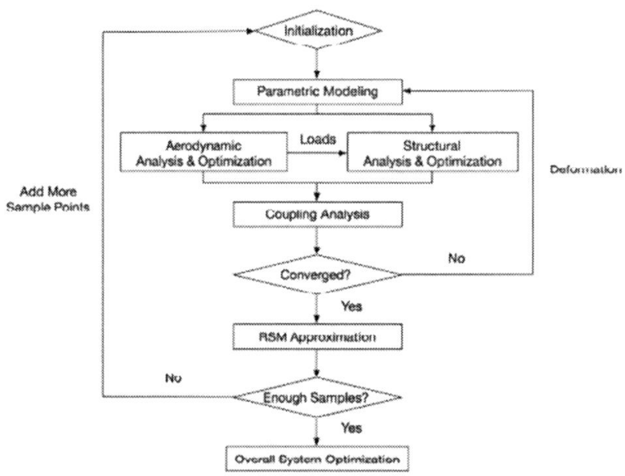

Figure 9. MDO Process.

The optimization process has two levels. One is the overall system optimization in which the global design variables that can influence the optimization process in all subsystems are defined, the other is the subsystem optimization in which local design variables are determined for each discipline after the initialization of the global design. After the optimization processes in all subsystems are finished, the coupling analysis process is executed to obtain the objective variables. Since the optimization cost without approximation techniques is huge, Respond Surface Method (RSM) [3] is integrated in the platform to establish the system level surrogate model and reduce the optimization time.

6 Case study

6.1 Case 1: Sequential optimization

The wing in case 1 flies at the speed of Ma 0.4, and its AOA is 2°. Semi-coupling method is applied in the optimization process of case 1. Specifically, the aerodynamic load is defined as the decoupling variable between the two disciplines, and the impacts on the external flow field due to structural deformation have not been taken into consideration during the optimization process. The aerodynamic optimization is performed in the first phase. Then the optimal wing shape is confirmed and the structural optimization and coupling analysis are executed in the second phase. The design target of case 1 is to maximize the lift to drag ratio during the aerodynamic optimization process and minimize the structural weight of the wing in the structural optimization phase. Six structural design variables are selected through the Design of Experiment (DOE). The mathematic model of the optimization problem is described as follows:

$$\left.\begin{array}{ll} Maximize: & L/D \\ Subject\ \text{to:} & C_l \ge C_l^{Baseline},\ X_{aero}^{\min} \le X_{aero} \le X_{aero}^{\max} \end{array}\right\} Aerodynamic\ Optimization$$
$$\left.\begin{array}{ll} Minimize: & M \\ Subject\ \text{to:} & \begin{array}{l} \sigma_{\max} \le \sigma_{allowable} \\ u_{\max} \le u_{allowable} \end{array},\ X_{struc}^{\min} \le X_{struc} \le X_{struc}^{\max} \end{array}\right\} Structural\ Optimization \quad (2)$$

Where X_{aero} is the aerodynamic design vector which is constrained by the design limits. It consists of the sweep angle and taper ratio of the wing. And the X_{struc} is the structural design vector which includes the skin thickness, spar section parameters and the thickness of the spar plates. $\sigma_{allowable}$ represents the allowable stress of the structure and $u_{allowable}$ stands for the allowable structural displacement. RSM are both applied for each discipline to create approximation models after the precision test. Fifty-eight design points generated by Optimal Latin Hypercube [11] are used to create RS surrogate model for aerodynamic analysis while 50 design samples generated by Optimal Latin Hypercube Design are applied to create RS surrogate model for structural analysis. The results of the prediction accuracy evaluation for respond surface models are shown in Table 1.

Table 1. Prediction Accuracy Evaluation for RSM in Case 1.

Items	R^2	RMSE
L/D	0.985	0.025
M	0.999	0.009

During the optimization process, Sequential Quadratic Programming (NLPQL) [12] is selected as the optimization algorithm and integrated in the platform. The final results are shown in Table 2.

Table 2. Optimization Results of Case 1.

Items	Variables Constraints	Initial Design	Optimal Design(RSM)	Optimal Design(Actual)	Variation(Initial vs Optimal)	Error(RSM vs Actual)
Λ	[15,25]	25	20.29	N/A	-18.84%	N/A
λ	[0.25,0.35]	0.3	0.25	N/A	-28.6%	N/A
rsb	[2,3.5]	3mm	2.02mm	N/A	-32.6%	N/A

fsb	[2,3.5]	3mm	3.49mm	N/A	+16.3%	N/A
st	[1,2]	1mm	1.1mm	N/A	+10%	N/A
pt	[2,4]	3mm	3.51mm	N/A	+17%	N/A
C_l	[0.1705,∞]	0.1705	0.1887	0.1832*	+3.9%	N/A
C_d	N/A	0.0222	0.0211	0.0225*	-2.1%	N/A
L/D	N/A	7.65	8.54	8.14*	+6.4%	N/A
M	N/A	18.62kg	20.11kg	19.93kg	+7.03%	0.9%

*Coupling analysis has been performed.

Compared with the baseline design, the lift to drag ratio of the optimal design increases about 0.83%. However, the structural weight increases about 7.03%. The optimization results illustrate that the sequential optimization process from the aerodynamics discipline to the structures discipline is failed to find an optimal design point satisfying the design targets of all the disciplines.

6.2 Case 2: Multidisciplinary design optimization

Tight-coupling method is applied in case 2. Specifically, during the whole optimization process, aerodynamic load and the structural deformation are both taken into account and defined as the decoupling variables. The initial design conditions are the same as case 1. That means the influence of the aerodynamic loads and the impacts on external fluid flow caused by the structural deformation have been considered in the optimization process. The optimization workflow is shown in Figure 9. In this case, the optimization process is divided into 2 levels. One is the overall system level in which global design vectors are determined. The other is the subsystem level which includes the aerodynamic analysis (Note: No variables are defined to optimize in aerodynamic subsystem) and the structural optimization. The mathematic model of the optimization problem is described as follows:

System Level Optimization

$$\left.\begin{array}{ll} \textit{Maximize}: & F(X_{sys}) = W_1 * C_{mass} + W_2 * C_{l/d} \\ \textit{Subject to}: & C = \left|Y_{sys1} - Y_{aero}\right| + \left|Y_{sys2} - Y_{struc}\right| = 0 \end{array}\right\} \quad (3)$$

Where X_{sys} represents the design vector of the overall system level. The consistency constraint C ensures the supplementary vector Y_{sys1}(or Y_{sys2}) is equal to the coupling vector Y_{aero}(or Y_{struc}). Weighting method is applied in the system level optimization to convert the multi-objective optimization problem to a single objective optimization problem. Also in the equation above, W_1 is the weight coefficient of the structural mass and W_2 is the weight coefficient of the lift to drag ratio. In this case, they are both set as 0.5. C_{mass} represents the decrease coefficient of structural weight while $C_{l/d}$ stands for the increase coefficient of lift to drag ratio, both of which are determined as follows:

$$\left\{\begin{array}{l} C_{mass} = \dfrac{M_{Baseline} - M_{opt}}{M_{Baseline}} \\ C_{l/d} = \dfrac{L/D_{opt} - L/D_{Baseline}}{L/D_{Baseline}} \end{array}\right. \quad (4)$$

The mathematic model of subsystem optimization (Structural Discipline) can be described as follows:

$$\begin{aligned} &\textit{Minimize}: && M \\ &\textit{Subject to}: && \begin{matrix}\sigma_{\max} \le \sigma_{allowable} \\ u_{\max} \le u_{allowable}\end{matrix}, \quad X_{struc}^{\min} \le X_{struc} \le X_{struc}^{\max} \end{aligned} \tag{5}$$

When the optimization process in all subsystems are completed, the coupling analysis will be performed. Then statistical data of single design point is obtained. In the overall system level, 48 design points are used to generate the RS model to perform the overall system optimization. The prediction accuracy evaluation for RSM is shown in Table 3 and precision of the RS Model is high enough to replace the original CAE analysis process.

Table 3. Prediction Accuracy Evaluation for RSM in Case 2.

Items	R^2	RMSE
L/D	0.998	0.012
M	0.999	0.002

Finally, the optimization results based on the 2-level MDO architecture is illustrated in Table 4. The L/D of the optimal wing has increased about 7.5% and the structure weight has decreased about 5.9%. The optimization results demonstrate that the optimal wing shows better performance in both disciplines, while the prediction error of the surrogate model is less than 2% which proves the validity and practicability of the MDO platform. Compared to the results of case 1, the MDO architecture is more capable to seek the global optimal design.

Table 4. Optimization Results of Case 2.

Items	Variables Constraints	Initial Design	Optimal Design(RSM)	Optimal Design(Actual)	Variation(Initial vs Optimal)	Error(RSM vs Actual)
Λ	[15,25]	25	15.95	N/A	-36.2%	N/A
λ	[0.25,0.35]	0.3	0.252	N/A	-28.0%	N/A
rsb	[2,3.5]	3mm	2mm	N/A	-33.3%	N/A
fsb	[2,3.5]	3mm	3.15mm	N/A	+5%	N/A
st	[1,2]	1mm	1mm	N/A	0%	N/A
pt	[2,4]	3mm	2.63mm	N/A	+12.6%	N/A
C_l	[0.1705,∞]	0.1705	0.1889	0.1865	+9.4%	1.2%
C_d	N/A	0.0222	0.0229	0.0227	-2.25%	0.87%
L/D	N/A	7.65	8.251	8.225	7.5%	0.32%
M	N/A	18.62kg	17.63kg	17.52kg	-5.9%	0.62%

7 Further development possibilities

The platform specifically designed for Wing MDO problems is able to seek an optimal wing configuration and provides modules for computation tests. The MDO platform has integrated the MDO architecture, high-precision simulation methods, approximation techniques and optimization algorithms by scripting language and packages all the modules through the GUI. The concept of modular design makes it a potential platform that can allow third-party users and programmers to quickly customize and enhance its functions. The architectures of different modules in this platform is illustrated in Figure 10. Owing to the concept of modular design and the standardization of this platform, the benefits are showed as follows.

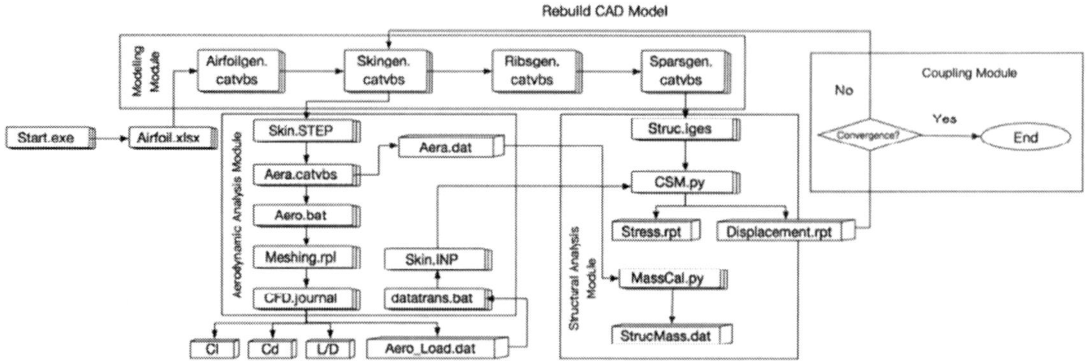

Figure 10. Architectures of different modules of the platform.

Firstly, various VBScripts are integrated in the CAD generator to create different inner structures. The wing structure database can be extended by integrating more CAD generation scripts in a efficient way. Secondly, because of the portability and interchangeability of the scripting language, the secondary development environment can be simple and neat. Third-party optimization algorithms and approximation techniques can be integrated by adding a simple corresponding script. Lastly, owing to the universal storage form of simulation data, it is easier for different post-processing tools to exact data from the database.

Original designed for wing MDO issues, the platform is being developed for more possibilities, allowing more disciplines, MDO architectures, optimization algorithms to be integrated in the future development. In terms of user interaction and data management, the platform will be improved to simplify the setup process and make the data storage normalized.

References

1. T.W. Simpson, V. Toropov, V. Balabanov, et al., *12th AIAA/ISSMO multidisciplinary analysis and optimization conference*, **5**, 10-12 (2008)
2. J. Martins, C. Marriage, *Aiaa/asme/asce/ahs/asc Structures, Structural Dynamics, and Materials Conference* (2007)
3. R.H. Myers and D.C. Montgomery, Technometrics, **38** (3), 284-286 (1996)
4. R. Paiva, *Aiaa/issmo Multidisciplinary Analysis and Optimization Conference* (2008)
5. L. Attanasi, A.D. Marco, F. Nicolosi, et al., *Aiaa Applied Aerodynamics Conference* (2015)
6. P. Spalart and S. Allmaras, La Recherche Aérospatiale, **439** (1), 5-21 (1992)
7. T. Long, L. Liu, S. Zhou, et al., *Aiaa/issmo Multidisciplinary Analysis and Optimization Conference* (2008)
8. H. Jie and Y. Xiongqing, Comput Aid Eng., **22** (1), 16-20 (2013)
9. H. Jie, R. Wang, et al., TNUAA, **44** (4), 458-463 (2012)
10. R.D. Braun, A.A. Moore, and I.M. Kroo, *6th AIAA/USAF/NASA/ISSMO Symposium on Multidisciplinary Analysis and Optimization* (2010)
11. J.S. Park, J Stat Plan Infer, **39** (1), 95-111 (1994)
12. K. Schittkowski, Ann Oper Res., **5** (1), 485-500 (1986)

MATEC Web of Conferences **119**, 01034 (2017) DOI: 10.1051/matecconf/201711901034
IMETI 2016

The critical success factors of building smart green city logistics business in Fuzhou

Shih-Ming Ou[a], Qiaoping Mei, Li Yang and Dongling Chen

Department of Logistics Management, Fuzhou University of International Studies and Trade, Fujian, China

Abstract. This paper intends to find the critical success factors for smart green city logistics business in Fuzhou. We first investigate and collect the factors that impact the operations of smart green city logistics business. In order to acquire the experts' assessments on the influence between each two factors, we invite the managers who serve in the top five logistics companies in Fujian and conduct expect questionnaire investigation. By using DEMATEL method, we explore the causal effect among factors. Our findings show that the government investments and green power development are the two most influential factors among all.

1 Introduction

IBM proposed in 2009, to build a future-oriented concept of "smart logistics". Under this concept the supply chain is advanced, interconnected and intelligent. Through sensors, RFID tags, actuators, GPS as well as other devices and systems generate the general idea of "real-time intelligent supply chain". Smart logistics emphasizes on building a virtual logistics dynamic information management system.

With rapid rise of e-commerce, consumers' demands are stirred and largely increasing in the size of number of goods that results in highly development of logistics business. For sustainable development, green logistics concept has been spread worldwide. However, green technology and management still do not be adopted by most of the logistics companies in Fuzhou.

This study aims to explore the critical success factors for developing smart green city logistics business. To achieve the research objective, we first construct a hierarchy framework with the aspects of product, customer, operations management and finance as well as their rules. Then DEMATEL (Decision Making Trial and Evaluation Laboratory) method is employed to find the crucial factors.

The rest parts of this paper are organized as follows. Methodology that we employ in this paper is introduced in section two. In section three, we conduct an empirical study. The last part is the conclusion of this paper.

2 Methodology

DEMATEL (Decision Making Trial and Evaluation Laboratory) was first developed by A. Duval, E. Fontela and A. Gabusl in 1970. It can be used to solve the world's complex problems, such as agriculture, race, environmental protection, trade, health care, etc.([1],[3],[4],[8],[9])

[a] Corresponding author : benao13@gmail.com

After establishing the aspects and rules that affect the system, acquiring the assessments from experts, as well, DEMATEL quantifies and compares the factors each other, and then clarifies the causal diagram.([2],[5]) Mathematically, the operational procedure can be presented as follows.[11]
Step 1: Build hierarchy structure.
Step 2: Acquire experts' assessments.
Step 3: Construct initial direct-relation matrix.

$$X=\begin{bmatrix} 0 & x_{12} & \cdots & x_{1n} \\ x_{21} & 0 & \cdots & x_{2n} \\ \vdots & \vdots & \ddots & \vdots \\ x_{n1} & x_{n2} & \cdots & 0 \end{bmatrix} \tag{1}$$

where x_{ij}, i = 1, 2, …, n; j = 1, 2, …, n, denote the influence of factor i to factor j.
Step 4: Normalize direct-relation matrix

$$S=\frac{1}{\max\limits_{1\le i\le n}\sum_{j=1}^{n} a_{ij}} \tag{2}$$

$$N=X\times S \tag{3}$$

Step 5: Compute total relation matrix

$$T=N\ (I-N)^{-1} \tag{4}$$

Step 6: Plot and analyze causal-effect diagram

3 Empirical study

Table 1. Aspects and rules.

Aspect	Rule	Explanation
Product	Usage of green energy and carbon reduction	Pursuing environmental protection and energy saving as well as carbon reduction as the main value proposition.
Consumer	Environmentalist	The customers who are willing to pay more for the environmental sustainability.
	Support from Government	Assistance from government could improve the operations of smart green city logistics business.
	Appling information technology to logistics business	Appling information technology to logistics business for more efficient customer relationships maintaining.
Operations Management	Logistics service quality	Improvement of logistics service quality.
	Infrastructure	R&D platform and facility construction.
	laws	Legalization, standardization and institutionalization of logistics industry.
Finance	Logistics costs	Logistics cost control.
	Company investments	Expenditure on developing smart green logistics business.

 DOI: 10.1051/matecconf/201711901034

We conducted expert questionnaire during the period from September to December in 2015 to acquire the assessments of the managers who work in the major green logistics companies in Fuzhou. After rearranging data collected from experts, we have the aspects and rules as shown in Table 1.

We then processed the experts' assessments by using DEMATEL. The causal-effect diagram is presented as follows：

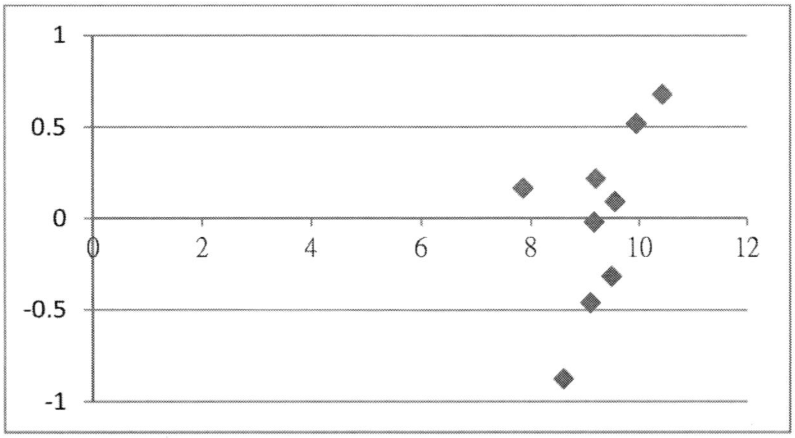

Figure 1. Causal-effect diagram.

Observing Figure 1 "Causal-effect diagram", the two factors of "Financial support from Government" and "Usage of green energy and carbon reduction" are crucial factors on developing smart green city logistics business.

4 Conclusions

In this study, we aim at finding the key success factors on developing smart green city logistics business. We acquire experts' assessments from main logistics companies in Fuzhou and then employ DEMATEL method to explore and visualize the causal relationships among factors. Our findings show that "Financial support from Government" and "Usage of green energy and carbon reduction" are the two key factors. That is, as long as the Government can give vigorously promote investment, it will facilitate the development of smart green city logistics business.

References

1. B. Chang, C.W. Chang, and C.H. Wu, Expert Syst Appl, **38**, 3 (2011)
2. C.Y. Hsu, K.T. Chen, and G.H. Tzeng, Int J Fuzzy Syst, **9**, 4 (2007)
3. J. Jassbi, F. Mohamadnejad, and H. Nasrollahzadeh, Expert Syst Appl, **38**, 5 (2011)
4. J.L. Yang and G.H. Tzeng, Expert Syst Appl, **38**, 3 (2011)
5. M.S. Kuo, Expert Syst Appl, **38** (2011)
6. Q. Zhou, W. Huang, and Y. Zhang, Safety Sci, **49**, 2 (2011)
7. S. Opricovic and G. H. Tzeng, Int J Uncertain Fuzz, **11**, 5 (2003)
8. T.F. Golob and A.C. Regan, Transport Res C- Emer, **9**, 2 (2001)
9. W.H. Tsai and W.C. Chou, Expert Syst Appl, **36**, 2 (2009)
10. W.S. Lee, A.Y. Huang, Y.Y. Chang, and C.M. Cheng, Expert Syst Appl, **38** (2011)
11. W.W. Wu and Y.T. Lee, Expert Syst Appl, **32**, 2 (2007)
12. W.W. Wu, Expert Syst Appl, **35** (2008)

Planning for sustainable development of tourism in Songkou ancient town

Hong-Jun Ou[a]

Department of fine arts and design, Fuzhou University of International Studies and Trade, Fuzhou City, Fujian Province, China

Abstract. This study is aimed to investigate strategies for urban landscape planning and ecotourism development of Songkou Town to present distinctive cultural landscape. Questionnaire is designed in terms of economy, environment and life. Based on results of questionnaire survey, interview theme and outline are developed and designed and individual in-depth interviews are performed. Besides, through comparison with interview emphasis of report forms in related literatures, we intend to propose optimal policies on urban and rural landscape and ecotourism of Songkou town in order to provide a reference for local government and future researchers.

1 Introduction

Ancient towns have gradually become new travel destinations for modern people because of their special geographical environment, distinctive architecture style and important historical value. Most ancient towns are located beside beautiful mountains and clear waters, as products of mutual integration of natural environment and human development through thousands of years. Along with rapid development of tourism, many ancient towns are seriously destroyed during exploitation. Therefore, with Songkou ancient town as an example, based on analysis of characteristics of ancient town tourism, this paper tries to propose a concept of "eco-culture as the main axis and sustainable development as the idea" to establish new rural tourism criteria of ancient towns.

This paper is mainly divided into three parts to describe and record current cultural landscape planning and sustainable development situations of Songkou ancient town. The first part introduces history and current development of this ancient town. The second part discusses and analyzes cultural landscape cases in progress. The third part puts forward several directions based on planning of sustainable development idea of the ancient town. Finally, the author makes generalization and summarization to propose possible strategies for cultural-ecological tourism and sustainable development of Songkou ancient town.

2 History and current development of Songkou ancient town

Songkou, called Songyang in the past, is situated at southwest of Yongtai, upstream of Dazhang River. Historically, it is at the junction of four prefectures and five counties, acting as the distributing centre

[a] Corresponding author : 2968981426@qq.com

of products from Youxi, Dehua, Xianyou, Putian, and etc. By virtue of such geographical advantage, ancient Songkou wharf was once an important location for merchandises from various parts of China. A flourishing ancient town is gradually formed through tourism and business.

Songkou is the only famous Chinese historical and cultural town in Fuzhou. It is also the hometown of Yuangan Zhang, a patriotic ci writer in the Southern Song Dynasty. Songkou ancient town is the political, economic and cultural center and the important transportation hub of southwestern Yongtai. Within the township, there are more than 60 well-preserved folk houses of Ming and Qing dynasties. In various villages of this town, you can find more than one 100 well-preserved folk houses of Ming and Qing dynasties.

On October 14, 2008, Songkou town was formally named as "famous Chinese historical and cultural town" by State Administration of Cultural Heritage and Ministry of Housing and Urban-Rural Development, and became the third town of Fujian Province and the only one of Fuzhou City that owns this title. For promoting development of ecological and cultural tourism, in 2014, Oustudio of Taiwan was invited to "feel the pulse" of this old and new town.

Figure 1. Memorial archway of Songkou town.

3 Several cases of architectural and cultural landscapes

3.1 Crane-shaped road

As a part of Longkoucuo district, crane-shaped road has been always a popular scenic spot of Songkou due to its unique shape. After establishment at the end of Song Dynasty, multiple parts of this road have been damaged. Some segments have been occupied by surrounding new and old buildings. Some other segments are repaired by descendants, and lose original primitive style due to replacement of cob wall with red bricks and cements. Before renovation, this road was once blocked. In the phase-I tourism line program of Taiwan Oustudio,[b] crane-shaped road is an important point. During initial repair, Zheng's family of Longkou also paid much attention to re-construction of this road. Consequently, this road becomes the only project of the town that is lead by the government as well as constructed and supervised by local family.

To achieve better effect of cob wall punning during re-construction, many craftsmen who have retired for many years make a lot of attempts under the guidance of Oustudio to test soils of different origins and proportions. The renovated crane-shaped road presents three-segment façade of local conventional low-rise building: lower segment of stone bench-table, middle segment of loess and upper segment of grey- tile roof. The completed crane-shaped road becomes a landscape that

[b] Oustudio was established in 2001, with Guocang Liu as the design director. It is set up in ancient capital of Taiwan-Tainan City.

combines old and new styles and is not bitty any more. Crane-shape road is associated with ancestral directives and culture of Zheng's family of Longkou. Therefore, conceptual design of re-construction should obtain sufficient respect from Zheng's family. Legend of crane-shaped road passed on from generation to generation locally detailedly introduces implication of each structure of this road.[c] Finally, the repaired crane-shaped road is also respected during renovation. Meanwhile, people can see the traces of changes in local residential building forms at different historical stages.[d]

Figure 2. Crane-shaped road[e].

3.2 Movie Temple

Movie Temple, adjacent to Songkou town government, is the common conception and action of town government, community, local people and Oustudio. It not only functions as the faith center for revitalization of local tradition, but also be endowed with new public functions of new era. It has functions of cinema, auditorium and temple.

Bi-directional experience place is created to achieve cooperation between government and community. In terms of historical development, functions of Movie Temple change with times, which lead to contradiction between government and community in cognition of use of this building in the contemporary era. Movie Temple was once the biggest Dacheng palace in Yongtai County. The main hall faces the north. Since the Cultural Revolution period, this temple has been used as cinema and auditorium. People enter it from the north. The stage is situated in the south. In recent years, people of rural areas generally move out, leading to rising of cities and decline of countryside. Under such context, this cinema and auditorium becomes an idle dangerous building. However, people who have moved out to cities succeed in business, give older generation sufficient economic resources and initiate donation. They plan to dismantle cinema and re-build original temple. The government also has the idea of reviving this place but their direction of plan is different from the community. The government hopes to create a new community public place for current and future uses.

Oustudio utilizes the space plan of "one place, two attributes" to break the deadlock between the government and the community. They transform the base of this dangerous building into a temple

[c] Cob wall of the crane-shaped road represents skin layer of crane neck. Therefore, soil road in the center represents esophagus.

[d] Citation: Homeland (2015)

[e] It is originally expected to construct pure cob wall. However, current soil ramming technique is not able to achieve such a great bending degree. Therefore, the biggest wall of crane-shaped road is repaired via both bricking and soil paste techniques.

desired by common people and a safe public place expected by the government. New public place functions are set in the new Movie Temple. The second floor inside the bricked façade in the north is used as a meeting space for artistic activities, leisure, playing and singing. [f]

Figure 3. Internal and exterior of the Movie Temple.

4 Investigation into the idea of sustainable development of Songkou ancient town

As a Millennium town, Songkou ancient town has rich tourism resources. In consideration of the principle of "no destruction to the natural and human landscape", design of sustainable development is especially significant. With regard to this argument, several planning principles are proposed:

4.1 Sustainability of tourism product and economy

To achieve sustainable development of tourist area, the prerequisite is long-term and stable development of economy. On the road of sustainable development, Songkou ancient town should firstly enhance economic benefits. Development and design of tourism product is the key to enhance economic benefits. Under the leadership of Oustudio team, a model point is found on Zhi Street, which leads to the ancient ferry.[g] In this composite space, fabrics, wood products, bamboo knitting articles, rattan plaited articles, desserts and local products are gradually dug out. More than six kinds of local agricultural and cultural products are sold in this store.

Besides historical and cultural background, product development of Songkou ancient town should also be introduced with new design ideas to satisfy tourists' thirst for knowledge as well as increase sense of participation and experience. Also, development of various leisure and tourism products should aim to increase tourists' length of stay, promote consumption chain and improve economic income. It is essential to constantly deepen tourism product structure and enhance vital force of products, so as to achieve long-term stable development of tourism economy of this ancient town as well as provide economic guarantee for social sustainability and ecological sustainability.

4.2 Sustainability of integrative development of ancient town tourism and cultural creative industry

Tourism is not only a kind of creative and aesthetic economy but also a kind of experience economy. Tourism development of ancient town is based on rich cultural resources and emphasizes outstanding

[f] Citation: Taiwan Architecture (2015)

[g] For a two-floor store along the Zhi Street of Songkou ancient ferry, Oustudio makes reconstruction and interior decoration. They build a composite space with functions of cafe, accommodation and retail, and name it as "Open Songkou". This store was opened in October, 2015.

originality and feature. Cultural creative industry needs to dig out, integrate and stimulate the spirit of ancient-town cultural tourism resources, in order to motivate the core competitive force, apply such competiveness into development of cultural creative products and create better products for satisfying consumers' demands. Therefore, sustainable development of ancient town tourism must form its characteristic and distinctive industrial layout through integrative and interactive development of human environment, technology and work division of cultural creativity so as to avoid similarity to other ancient towns.

4.3 Sustainability of ecological culture

Ecological culture is extension and innovation of culture under certain environment. It mainly advocates harmonious coexistence of human and nature as well as green lifestyle and civilized human morality so that people can actually understand, adore, protect and enjoy the nature. As a kind of emerging culture, ecological culture calls for a conceptual system of harmonious coexistence of human and nature.

To provide theoretical basis of sustainable development, ecological culture that utilizes scientific theory can realize coordinating development of tourists and environment. Ancient town tourism is a carrier of ecological culture. Like Songkou town, all ancient towns bears strong cultural ambience and sufficiently exhibits harmonious coexistence of human and nature. In current tourism bloom, common people more focus on short-term economic benefits due to the effect of market economy. As a consequence, many ancient towns are destroyed.

Ancient town is a special utilizable resource. It has characteristics such as irreplaceability, regeneration, manufacturing, recycle and public. Irrespective of the creator and the preserver, ancient town is historical and cultural heritage and thus considered as common wealth jointly created by human beings. Therefore, in development of ancient town tourism, each tourist and each tourism developer should establish the idea of ecological culture so that tourism resource of ancient town and promotion of sustainable tourism development can have profound significance. [1, 2]

Figure 4. Rattan plaited bags and cases from Yongtai and Songkou.

5 Conclusion

To sum up, it is suggested to promote cognition direction of ancient town tourism development and combine with development and construction plan of the government, in order to sufficiently develop the area and achieve sustainable activation of the area. Through research of strategies for developing ecological tourism, conservative and restrained development pattern is recommended. Excessive public utilities will compromise original comfort of the space. Original simple and distinctive style can only be restored through subtraction technique.

In addition, this study suggests Songkou ancient town to integrate human resources and local activities. Festival activities can be used to drive legend of local culture so as to achieve win-win situation of local characteristic business and economic industry as well as strengthen centripetal force of local residents. In this process, ecological cultural tourism concept and local organization can be combined to create employment opportunities for grassroots and enhance additional value of ecological tourism industry. It is also recommended to integrate special landscapes and folk activities of Songkou ancient town to form a sustainably-developed eco-museum strategy and concept as well as increase consciousness of local culture and living environment. It is expected to make "the industry, the officials, the academic circle, the researchers" and the local residents jointly participate in the construction, thus to improve quality of local living environment, attract more tourists enthusiastic for ecological tourism, as well as increase quality of ecological tourism "production" enterprise and revenue of residents. "Cultural and ecological museum" will be used as the concept and blueprint for promoting cultural ecological tourism development and local development of Songkou ancient town.

References

1. Y. Li , Journal of Tourist Market, **79**, 77 (2010)
2. H. Zhao and J. Yu , On Economic Problems, **50**, 55 (2015)

Analysis of hospital interior air quality audits

Lee-Kuo Lin[a], Yan-Sin Cai and Yao-Chien Tsai

Department of Civil Engineering, National Taipei University of Technology, Taipei 106, Taiwan

Abstract. In general, people spent more than 80~90% of living time in the indoor every day, human health and indoor environmental quality are closely related. The hospital has a complex and unique environmental characteristics, medical personnel and patients are prolonged exposed to risk factors in a variety of environments. Therefore, the merits of indoor air quality in the hospital, not only has a threat to the health of medical personnel and patients, but also will directly affect the quality and efficiency of health care work. A regular monitoring can, improve and maintain a well of indoor air quality, thus ensuring the safety maintenance of medical personnel and patients in hospital, it has become an important issue for hospital. This study has literatures review to collate and analyse that are related issues with indoor air quality. Then measures the indoor air quality test with direct-reading instruments. In selected hospital of this study were field-tested, then use the measured data in the field, discussion and analysis of the causes of air pollutants and the establishment of the sensing area of pollutants Concentration empirical mode.

1 Introduction

In recent years, as environmental awareness promotion, indoor air quality and harm to people's health issues more attention, Taiwan is the world's second country indoor air quality legislation. People per day on average, about 80~90% time in indoor environments, Indoor environmental quality and human health related, poor indoor air quality can lead to sick building syndrome, legionellosis and other various diseases of indoor environmental quality is at stake.

Hospitals have complex and special environmental characteristics, indoor air quality requirements are also relatively more stringent, health care workers are exposed to different risk factors for a long time in the environment of, so the quality of indoor air quality, Will threaten the health of medical staff and the quality of care and efficiency. In addition, the long stay patient in the hospital, Because of its body has been weak, poor immune system, More vulnerable to the effects of poor indoor air quality, Therefore, how to monitor, on a regular basis to improve and maintain good indoor air quality, And ensuring the safety of health care workers and hospital patients maintain hospital has become an important topic.

Implementation of this research purposes is to discussion on indoor air quality for hospitals, research purposes include the following:

(1) On the actual hospital indoor air quality testing, understanding hospital indoor air (CO_2), (HCHO), (PM_{10}) levels.

[a] Corresponding author : lklin@ntut.edu.tw

(2) Collection and analysis of relevant literature, medical home indoor air quality may be affected by factors, and the effect type, characteristics, source ,through data analysis to analyze of causes of air pollution.
(3) Through the characteristics of the highly correlated pollutants and environmental factors, Using data analysis, Establish concentrations of pollutants in the area experience formula , Assessing the degree of pollution, determine whether the three pollutants in hospitals out of indoor air quality standard.

2 Main contents

2.1 Indoor air quality and health

According to the Bulletin of the Republic of China Executive Yuan's Environmental Protection Department the purpose of "indoor air quality management Act," definition of article III of the "indoor" refers to the use of the public buildings of closed or semi-closed spaces, and the public means of transport to take the space. "Indoor air quality" refers to the presence in indoor air after long-term exposure is sufficient to directly or indirectly to prejudice the health or living conditions of substance, Include carbon dioxide, carbon monoxide, formaldehyde and total volatile organic compounds, bacteria, fungi, particle size less than or equal to ten Micron particulate matter (PM_{10}), particle size less than or equal to the second. Five microns particulate matter ($PM_{2.5}$) and ozone and other substances designated by the central competent authority notice concentrations, air humidity and temperature.

World in view of indoor air pollution for the public health impact or potential threats, according to their architectural characteristics and lifestyles of States, invested a lot of resources for research, laws are in place and management programmers, and so on. Overall, the poor air quality on human health hazard in severity can be divided into three different classes (Figure1), First for the sick building syndrome, the second for building-related illness, Third is through research and reasoning are aware of many air pollutants have carcinogenic properties, long time stimulation can affect the heart and blood vessels and chronic respiratory diseases.

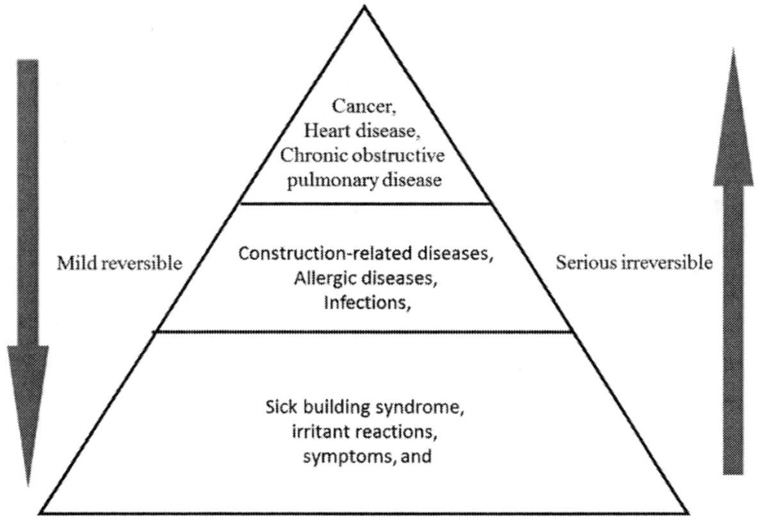

Figure 1. Indoor air quality and health hazards.

2.2 Indoor air pollutants

The national institute for occupational safe and health, investigation and analysis of indoor air quality, sources of indoor air pollutants include outdoor and indoor human activities, indoor air conditioning systems, building materials, equipment and supplies, and indoor sources of organic matter, such as the six, pollutants and sources is shown in table1.

Table 1. Indoor environmental pollution source and pollutant

Pollution Source		Pollutant
Outdoor air	Stationary sources	SO_2, O_3,Dust,CO,Hydrocarbon,Other toxic substances
	Mobile sources	CO,Dust,Nitrogen oxides, Oxysulfide, Lead
	Soil	Radon, Microorganism（Bacteriumgerm, Fungifungus.）
Indoor human activities	Human activities	CO_2, Vapor, Bad smell,Microorganism
	Smoke	Dust,CO,CO_2, Ammonia,Nitrogen oxides, Hydrocarbon, Volatile organic contaminants,Carcinogenic compounds
	Spray	Fluorocarbon, Bad smell,Volatile organic contaminants
Air conditioning system	Air handling unit	Fungifungus,Allergen ,Bacteriumgerm , Bad smell
	Air conditioner line	Dust,Fiber,Fungifungus,Allergen, Bacteriumgerm
Building materials	Concrete, stone	Radon
	Inorganic mineral board, Plywood, Carpetrug, Furniture	HCHO,Volatile organic contaminants
	Insulating materials, Insulation or fireproofing materials	HCHO, Glass fiber,Asbestos,Volatile organic contaminants
Equipment and supplies	Burning appliances	CO_2,CO, Nitrogen oxide, Hydrocarbons,Dust
	Service machines (Copier,Clean machine)	O_3, Dust,Bacteriumgerm, Fungifungus
	Other	Stationery solvent, Fungicide,Pesticides,Volatile organic contaminant
Indoor organic material	Perishable foods	Fungifungus, Bad smell
	Plants and flowers	Pollen, Fungifungus
	Waste	Ammonia,Allergen, Bacteriumgerm

This study only CO_2, HCHO and PM_{10}, three projects, Learn about indoor air pollutants will help to judge the sources of pollutants and control, improve indoor air quality and human health, the following will be described in detail for three pollutants.

(1) CO_2

CO_2 is a colorless, odorless and nonirritating gas, not toxic, Low concentrations are harmless to the human body, and High concentration has an impact on human health. Sources of CO_2 produced by human breathing as the main source. Concentration is usually with density and ventilation efficiency has great relevance. CO_2 concentration on human physiological effects as shown in table2.

Table 2. Effects of CO_2 concentration on the body

Concentration (ppm)	Physiological responses
≦600	No
600~1000	Occasional headaches, Lethargy, Sultry
1000~10000	Respiratory and circulatory organs and brain function is affected
10000~30000	Breathing speeds up, face the warm feeling
30000~40000	Ringing in the ears, Headache, Increased blood pressure
40000~60000	Vascular dilatation of the skin, Nausea, Vomiting
70000~80000	Mental confusion, Difficulty in breathing
80000~100000	Mental confusion, Convulsions, Respiratory arrest
100000~200000	Central nervous system disorders, Life-threatening

(2) HCHO

HCHO is a transparent and pungent gas, its extensive use in man-made sheet metal, plastic flooring, synthetic materials, coatings and adhesive. Table3 is Effects of HCHO concentration on the body past animal and human epidemiological and toxicological data confirmed HCHO exposure, In addition to causing stimulation of the eyes, skin, throat, and is one of the chemicals that cause human cancer, International Agency for Research of Cancer was HCHO in 2006 as a definite human carcinogen. Indoor HCHO main source is dominated by furniture or use of a product, such as furniture or wood products on the surface of the adhesive, wallpapers, cleaning agents, or electronic devices is one of the sources. In addition, indoor decoration materials such as Styrofoam, insulation, adhesive materials, fabrics, carpets and floor layout with HCHO, And thus, HCHO levels to determine the health of the test building materials building materials one of the most important projects.

Table 3. Effects of HCHO concentration on the body

Concentration (ppm)	Effect
0.1~0.3	Minimum trigger amount
0.8	Bad smell
1.0~2.0	Micro stimulation
2.0~3.0	Eye, nose and throat stimulation
4.0~5.0	Mucous membrane irritation involuntary tears
10.0~20.0	Severe burning sensation, cough can only be tolerated for a few minutes
50.0~100.0	Seriously hurt in 5-10 minutes

(3) PM_{10}

Suspended particulates are tiny particles floating in the air, and the severity of harm to human health depends on the particle size, When the particle size is smaller, easier to enter the body of tracheal or Broncho alveolar region. Particles smaller than 2.5 μm particles, known as $PM_{2.5}$, Deposition in the lungs of the efficiency is the highest, and deposit resistance will harm human respiratory function, Allergic rhinitis, asthma, chronic obstructive pulmonary disease and other diseases, Effects of Suspended particles concentration on the body as shown in table4.

Table 4. Effects of Suspended particles concentration on the body

Concentration (mg/m^3)	Effect	Notes
0.15	Increase the frequency of asthma	Increased mortality when is above $0.1 mg/m^3$
0.23	To reduce the FEV1.0	
0.35	Influence of pulmonary function of chronic disease patients	
0.35	Increase in respiratory tract of patients with chronic bronchitis symptoms	

2.3 Data analysis and discussion

Detection in all test points in 8 hours, Living factors of temperature and relative humidity, Chemical factors on CO_2 and HCHO and physical factors in aerosols of size less than or equal to 10 microns (PM_{10}), choose the average regression analyses.

Table 5. Correlation between chemical pollution pointer-CO_2

CO_2 Pointer	High positive correlation	Moderate positive correlation	High negative correlation	Moderate negative correlation
Indoor temperature	0.846			
Outdoor temperature	0.828			
Indoor PM_{10}		0.590		
Outdoor PM_{10}		0.429		

(1) CO_2 concentrations influence factor analysis

CO_2 concentrations affect indoor pollutant concentrations of important factors and correlation of factors, to detect the average data using statistical methods to further some factor of its relevance, as shown in table5. CO_2 show in the pointer in the statistical results to the indoor temperature, outdoor temperature Max.

(2) HCHO concentrations influence factor analysis

Establishment of HCHO concentrations affect indoor pollutant concentrations of important factors and correlation of factors, to detect the average data using statistical methods to further some factor of its relevance, as shown in table6. HCHO statistical outcome, the pointer appears to be indoor humidity maximum.

Table 6. Correlation between chemical pollution pointer-HCHO

HCHO Pointer	High positive correlation	Moderate positive correlation	High negative correlation	Moderate negative correlation
Indoor humidity	0.729			
Outdoor humidity		0.628		

(3) PM_{10} concentrations influence factor analysis

PM_{10} concentrations affect indoor pollutant concentrations of important factors and correlation of factors, to detect the average data using statistical methods to further some factor of its relevance, as shown in table7. For PM_{10} statistical outcome, the pointer displays indoor temperature, outdoor $PM_{2.5}$ PM_{10}, affected the most. For PM_{10} statistical outcome, the pointer displays indoor temperature, outdoor $PM_{2.5}$ PM_{10}, affected the most.

Table 7. Correlation between physical pollution pointer-PM_{10}

PM_{10} Pointer	High positive correlation	Moderate positive correlation	High negative correlation	Moderate negative correlation
Indoor temperature	0.760			
Outdoor temperature		0.634		
Indoor humidity				-0.504
Outdoor humidity				-0.594
CO_2		0.590		
Outdoor PM_{10}	0.887			
Outdoor $PM_{2.5}$	0.841			

(4) Regression model analysis

Regression analysis on optimization of inspection data, the detection zone can be obtained CO_2, HCHO, PM_{10} the respective regression model are as follows:

(i) Pharmacies:

$CO_2 = 364.671 + 2.304 \times$ (Number of people) - $0.004 \times$ (Number of people)^2 + $2.4 \times 10 - 6 \times$ (Number of people)^3

HCHO = 0.189 - 7.417 / (Indoor humidity)

$PM_{10} = 7.039 \times$ (Number of people) 0.348

(ii) Draw blood:

$CO_2 = 656.771 + 1.650 \times$ (Number of people) + $0.003 \times$ (Number of people)^2

HCHO = $1.3 \times 10 - 7 \times$ (Indoor humidity) 3.321

$PM_{10} = 48.782 - 0.059 \times$ (Number of people) + $0.002 \times$ (Number of people)^2

(iii) Registration area:

$CO_2 = -294.806 + 13.280 \times$ (Number of people) - $0.046 \times$ (Number of people)^2 + $5.4 \times 10 - 5 \times$ (Number of people)^3

HCHO = 0.322 - 13.331 / (Indoor humidity)

$PM_{10} = -55.823 + 1.030 \times$ (Number of people)

3 Conclusions and suggestions

Environmental awareness promotion, has attracted more attention to the health hazards and indoor air quality, Indoor air quality testing equipment in specific medical institutes carry out onsite inspection, to understand indoor air quality changes and features, Conclusions are as follows:

(1) Determine whether the pollutants in hospitals out of indoor air quality standard, Pharmacies are in compliance with the standard, Blood area are HCHO of five days does not meet standards and PM_{10} of one day does not meet standards, And the rest are in compliance with, Registered Area are CO_2 of two days does not meet standards and HCHO of one day does not meet standards and PM10 of five day does not meet standards, And the rest are in compliance with.

(2) Through the collection and analysis of data to know, CO_2 concentrations, and number, temperature, opening and closing with a high degree of correlation, HCHO concentrations, climate, and construction, opening and closing a high degree of association, PM_{10} concentrations, and construction, cleaning, temperature, opening, closing, outdoor $PM_{2.5}$, PM_{10} high outdoor Association

Indoor air quality testing and inspection data analysis, There are still inadequacies can be used as research recommendations:

(1) Recommends that the follow-up could be explored in addition to medical institutions place, the change of indoor air quality and characteristics, and hospitals to compare indoor air quality analysis.

(2) Can continue to discuss the improvement of indoor air quality benefits.

References

1. Executive Yuan Environmental Protection Agency, Indoor air quality maintenance management specialist staff training materials, Environmental training of personnel of the Executive Yuan EPD (2014)
2. Gold pass stick, Analysis of indoor carbon dioxide concentrations detected - in Changhua County National Primary and Secondary Computer Classroom Case, Large Yeh University, Thesis (2011)
3. S. Q. Luo, and B. X. Chen, Comparative study of management mechanism of countries indoor environmental quality (including air quality), Ministry of the Interior Building Research Institute, Self-study (2006)
4. Xie Yun stiles, Consider the elaboration of localization of indoor air quality indicator of health risk assessment,Taipei University of Technology, Thesis (2003)
5. Y. Y. Li, Research office space Indoor Air Quality Control Strategies, National Cheng Kung University, PhD thesis (2004)
6. S. Hui Zhen, Z. M. Jiang, and L. J. Zhang, Effects of the use of the air conditioning system of the air quality in the business district of the office, NSC / EPD Technical Cooperation Suhui Zhen, Jiang Zheming, Lijun Zhang plans (2000)
7. M. X. Li, Kaohsiung metropolitan area SO2, O3 and PM10 to assess the impact on human health of the, Chaoyang University of Technology, Thesis (2010)
8. Y. T., Architects study indoor air quality - A Case Study in Taichung, Feng Chia University, Thesis (2011)

A study on the development of the "Taiwan project management competence scale"

I-Chan Kao[a]

General Education Center, Open University of Kaohsiung, Kaohsiung 812, Taiwan

Abstract. On the basis of the 46 competence elements in the International Project Management Association (IPMA) ICB Version 3.0 and the PMP competence guidelines in PMBOK® Guide -- Fifth Edition by the Project Management Institute, this study formulates a "Preliminary Draft of the Taiwan Project Management Competence Scale". After expert review, a pretest was developed for the Taiwan Project Management Competence Scale. With adult students in adult education institutions as the respondents, a pretest was conducted in order to collect data. Confirmatory factor analysis, and reliability analysis were then conducted to construct the "Taiwan Project Management Competence Scale", which covers the four dimensions of project management of cognitive and conceptual competences, behavioral and technical competences, contextual competences, and moral and ethical competences. It is intended that this scale could be used as reference in future investigations into and measurement of project management competence in Taiwan, thus, facilitating the diagnosis, improvement, and refinement of performance in terms of project management competence.

1 Introduction

1.1 Study motives

According to the IPMA competence Baseline (ICB 3.0), as formulated by the International Project Management Association (IPMA), "Project" refers to temporary investment and efforts of an organization within a certain period of time in order to achieve special targets [31]; in addition, the American Project Management Institute (PMI) suggests that "Project" refers to temporary efforts to create a unique product, service, or result [29]. Based on all the definitions of project management, as made by international professional organizations (IPMA & PMI) and project management scholars [5, 12, 15], [18-26], [34,37], this research concludes that project management refers to a kind of managerial art created by the application of effective management methods and systematic management processes to integrate humans and resources in order to achieve tasks under the drive of limited resource conditions, and reach the established and unique goal visions.

According to the IPMA competence Baseline, as formulated by IPMA, Kao (2013a) established the contents of the Taiwan National Competence Baseline (TNCB) as the main body structure, and

[a] Corresponding author : kic0929@ouk.edu.tw

then, integrated the knowledge system and management processes of ISO, PMI, and PRINCE2 of England in order to put forward the organizational roles, tasks, management structure, and management processes of project management; the establishment of "Project Management Indicators" includes project initiation, integration, planning, and management; project scope planning and management; project schedule planning and management; project cost planning and management; project quality planning and management; project human resources planning and management; project communication planning and management; project risk planning and management; and project procurement planning and management, for a total of 9 levels of project management with 66 indicators. Kao(2011a) established 82 school project management indicators regarding the ten dimensions of the integration management of school projects, scope management, schedule management, cost management, quality management, human resources management, communication management, risk management, procurement management, and organization maturity degree. Kao (2014) established initiation and integration management, scope management, schedule management, cost management, quality management, human resources management, communication management, risk management, procurement management of community construction and execution, maturity degree of project management, and other methods and strategies, with community construction project management as the title, and a total of 82 indicators of are established. However, even with the norms and reference standards of project management indicators, do the managers of the execution of a project have the project management competence to effectively perform the work and tasks of the project?

As an old saying goes "Human efforts is the decisive factor", meaning even if the connotation and standard of the indicators of project management are understood, if the competence of the execution of project management is not owned, the tasks of the project may not be performed properly. Therefore, this study intends to explore the connotation and projects of Taiwanese project management competence, develop a Taiwanese project management competence performance scale through empirical investigation of the current project management situations, and examine cultural characteristics through ICB 3.0 of IPMA and by referencing PMI competence specifications, which will measure managers' degrees of management competence in project execution, measure the training courses for strengthening project management competence and planning project management professional development, as well as the basis for the formulation of a development strategy for project management personnel in order to cultivate talents with project management competence and give play to the functions of human resources management of proper talents, proper arrangements, and proper use to promote the realization of project management goals, and this is the motivation of this study.

1.2 Study purposes

Based on the foregoing statement of study motivations and study background, the purposes of this study are, as follows:

1.2.1 To understand the connotation of project management competence.

1.2.2 To establish a questionnaire scale for project management competence.

1.2.3 To propose a Taiwanese project management competence performance scale according to the results of this study, which shall act as reference for future investigations into and measurements of project management competence in Taiwan, thus, facilitating the diagnosis, counseling, improvement, and refinement of performance in terms of project management competence.

1.3 Study method and process

First, through literature review and analysis, this study integrates the connotation of project management competence and constructs the "Preliminary Draft of a Questionnaire on Taiwan's Project Management Competence Performance", where experts from industry, government, and academia are invited to construct expert validity, and then, a total of 328 persons from industry, government, and academia are extracted as the study samples in order to implement a questionnaire survey of Taiwan's project management competence performance; the validity of the recovered questionnaire data is analyzed with Structural Equation Modeling (SEM), and Cronbach's α reliability analysis and other statistical methods are adopted to establish Taiwan's project management competence performance scale.

2 Exploration of project management competence

2.1 Operation process of project management

The operation system of project management includes: setting project management goals, planning operational process, operation of methods, construction of team, allocation and use of resources, a platform base of information and knowledge management, establishment of the system, etc., which are all required to integrate project business resources and labor, and complete set tasks, in order to achieve the project goals. In terms of the connotation of project management processes, many scholars and professional groups in the field of project management[2],[3], [6-8], [10], [11, 13, 16],[17], [19], [21-23], [25], [26, 29, 32], [33] have their own unique theoretical and practical understanding, thus, this research integrates the various statements to put forward the operation connotation of project management processes, including setting project goals, initiating, planning, executing, controlling, and closing.

2.1.1 Setting of project goals

Project goals shall be determined by the project initiator, and must be combined with the operational strategy goals of the organization. "Achieve Project Goals", refers to the final deliverables of the project, which are accepted by the customer and all project parties are satisfied with the performance of the project team. The setting of project goals must meet the following five principles: Specific, Measurable, Achievable, Realistic, and Time - the first letters of the five words above are connected to form SMART, that is to say, the setting of project management goals shall conform to SMART principles [9, 19, 34].

2.1.2 Project management process

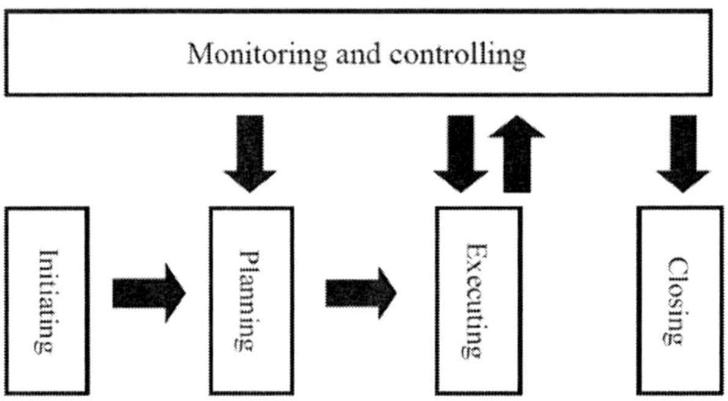

Figure 1. Project management process diagram, Source, Editorial Committee of Taiwan Project Management Association, , 21(2014).

The sequence of project management work at each stage is also called the Project Management Process, which includes the five processes of Initiating, Planning, Executing, Controlling, and Closing on the basis of the ***International Project Management Knowledge System*** (detailed in Figure 1), as compiled by the Editorial Committee of Taiwan Project Management Association, and ***A Guide to Project Management Body of Knowledge Fourth Edition*** (detailed in Figure 2), as compiled by the American Project Management Institute (PMI) [9, 28].

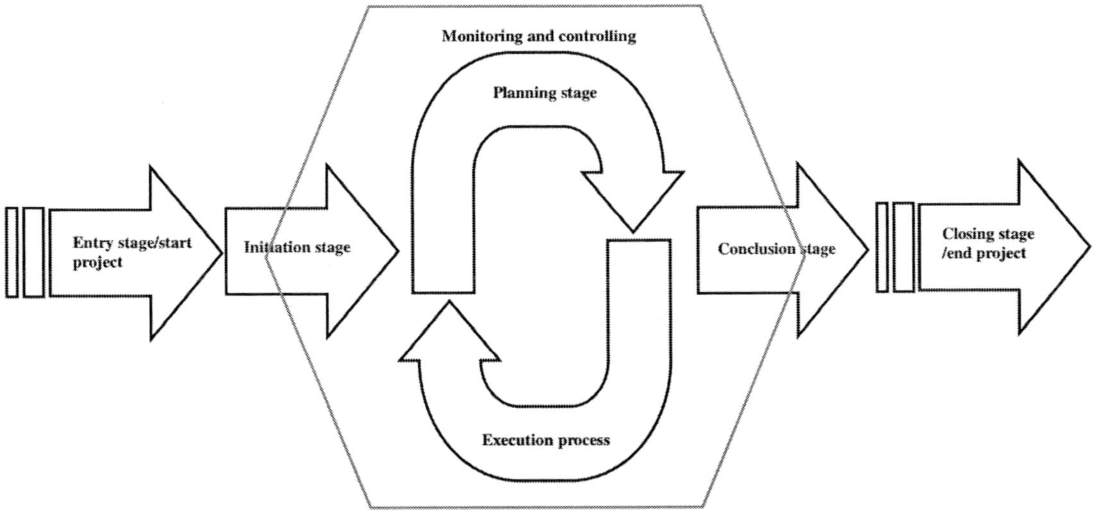

Figure 2. Project management process diagram, Source, PMI, 40(2008).

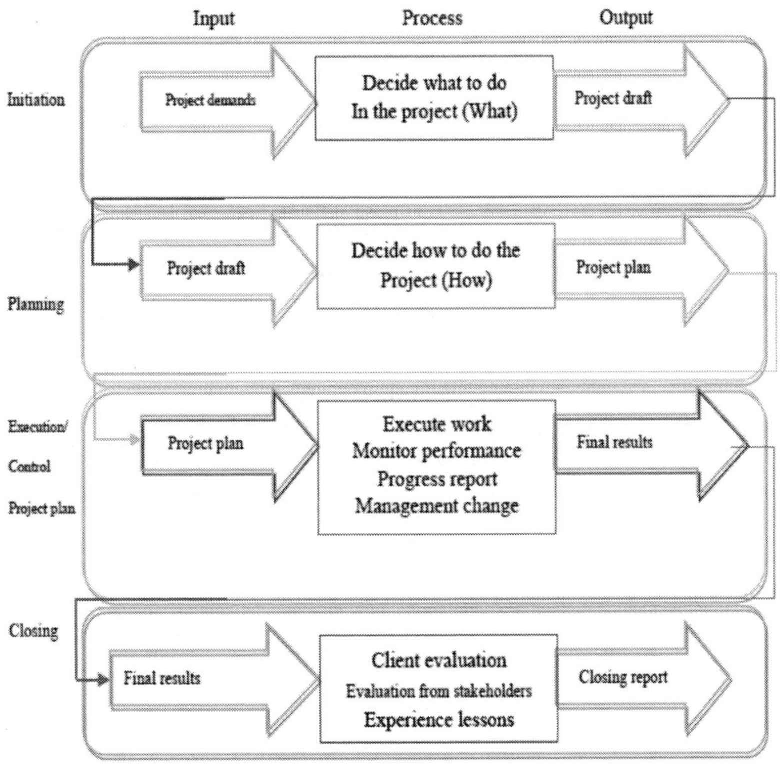

Figure 3. Project management process, Source: Kao, 207(2013b).

The connotation of the project management process, as shown in Figure 3, can be preliminarily summed up by project management activities that follow the Process of Input→Process→Output

(IPO), in combination with the five stage activities of initiating, planning, executing, controlling, and closing. In the initiating stage, the main operational task is to determine the steps of the project: project requirements are the input source basis of project initiation, while the project draft is the output of documents in the initiation stage; in the planning stage, project draft- output in the initiating stage becomes the input source basis for the planning and operation of the project, and the project plans are the output documents of planning and operation; in the executing/controlling stage, project plan-the output of the planning stage becomes the input source basis for executing project works, monitoring performance, progress reports, and change of management projects, and the final deliverables are the output documents, products, or services of executing and controlling a project; in the closing stage, final deliverables-the output of executing/controlling stage becomes the input source basis for the project closing operation, including customer's evaluation, evaluations of related personnel, experience, and lessons, and the close-out report is the output document of project closing. It can be known from the above IPO operational processes that, the operational mode of the project management process is to closely connect previous operations with the latter stages of project management, in other words, project management has obvious stage connections and operational systems [19].

2.2 Connotation of project management competence

McClelland (1973) suggested the measurement of competency to replace intelligence, which acts as the prediction basis for personal success. The connotation of competence includes knowledge, skills, and emotions, with outside behaviors as the measurement basis; in addition, it can be divided into individual's "possession" of ability, referring to possessed knowledge, emotions, skills, or "presented" ability, meaning actual performance of behaviors and successful performance of certain tasks [14]. As defined by Cheng, Lin, and Cheng (2009), competence refers to "the ability to use knowledge and the attitude and skills to perform tasks, meaning every behavior, motivation, and knowledge in relation to the success of works" quoted from, [36]. Spencer and Spencer (1993) put forward the Iceberg Model, (like an iceberg floating on the surface of the ocean), where the proposed competence includes 5 elements in terms of skills, knowledge, self-concept, traits, and motives; where skills and knowledge are more explicit, are easily trained, developed, and changed, which belong to superficial competence; and self-concept, traits, and motives, which are relatively implicit and not easily changed, and they belong to implicit competence (just like an iceberg hidden under the water). Project management competence, as explored in this study, refers to the explicit competence of knowledge and skills possessed by project management personnel.

In accordance with the ICB 3.0 competence baseline, as issued by IPMA (2009), a total of 46 project management competence elements, including 20 technical competence elements, 15 behavior competence elements, and 11 situation competence elements, shall act as the reference standard for the world to join IPMA member organizations, as well as the promotion of project management education training and project management competence certification. The connotation of ICB 3.0 competence baseline is, as follows.

2.2.1 20 technical competence elements for project management

1.01 Project management success: refers to the identification of interested parties towards project management results.

1.02 Interested parties: defined as direct or indirect personnel related to the project, as well as the priority of their interest in the project, and their opinions on the significance of the project are clarified.

1.03 Project requirements and objectives: refers to meeting the requirements and objectives of interested parties and customers.

1.04 Risk and opportunity: they refer to qualitative/quantitative risk management and opportunity evaluations in the process of planning and executing a project.

1.05 Quality: it refers to quality policy, objectives, and responsibilities determining what level the project should achieve in order to satisfy customers.
1.06 Project organization is a group of people: the arrangements of authority, relationships, and responsibilities conforming to enterprises, as well as the functional processes and related infrastructure.
1.07 Teamwork: it refers to the promotion of the project team to work harmoniously together in order to reach a common project objective.
1.08 Problem resolution: it refers to the use of many different methods to solve problems related to the project.
1.09 Project structures: it refers to the decomposition of project works into tasks, work packages, and activities in order to ensure no activity is leaked during the execution of a project.
1.10 Scope and deliverables: they refer to clarifying project scope and deliverables.
1.11 Time and project phases: they refer to the sequence of the project work package/activity, project period estimation, process arrangements, establishing a project deadline, and monitoring its implementation in a timely manner.
1.12 Resources: they refer to ensuring personnel have the necessary techniques, behaviors, and situational competence, and providing sufficient information, tools, and training in order to perform a task successfully.
1.13 Cost and finance: they refer to the assessment of the sum of all the activity costs since project initiation.
1.14 Procurement and contract: they refer to identifying potential suppliers and price inquiries for bidding and tendering to select suppliers, determining the best suppliers, and signing and managing contracts.
1.15 Changes: it refers to the changes of project plans or contract clauses after the occurrence of unpredicted incidents.
1.16 Control and reports: they refer to the evaluation of actual project progress and performance, and providing reports after comparing project baselines.
1.17 Information and documentation: they refer to the data and information conforming to organization policies and project management.
1.18 Communication: it refers to accurately conveying the correct information to all parties in order that they can understand and meet expectations.
1.19 Start-up: it refers to analyzing project feasibility, clarifying concepts, and signing letters of authorization for the project.
1.20 Close-out: it refers to the evaluation of project results and gained experience after the project is completed, and documenting the experience for future project reference.

2.2.2 15 behavior competence elements of project management

2.01 Leadership: it refers to exerting effective guidance in order that a team can understand and accept the guidance trend; it also refers to providing guidance and motivating others to achieve project goals.
2.02 Engagement and motivation: the project manager must understand the skills, experience, personal attitude, environment, and intrinsic motives of each member in order to encourage each member, and make project members be devoted to the project work; obtain the project and identification of internal staff and relevant persons regarding the project.
2.03 Self-control: it refers to handling daily work, changing requests and responding to pressure situation with a systematic and trained method.
2.04 Assertiveness: it refers to reaching a consensus through debate and argument in the common goals, and stating own views with authority.

2.05 Relaxation: it refers to the ability to relieve tension in tough times so as to restore the spirit of team.

2.06 Openness: it refers to allowing people to express their opinions and suggestions freely, accepting knowledge and experience of others and avoiding discrimination in discussing issues.
2.07 Creativity: it refers to the ability to blend all kinds of knowledge and promote team members to think and handle matters with original and imaginative methods.
2.08 Results orientation: it refers to making team focused on key objectives so as to make the interested parties of the project get the best results.
2.09 Efficiency: it refers to the ability to effectively use time cost and resources cost so as to produce agreed deliverables and realize the expectations of interested parties.
2.10 Consultation: it refers to the exchange of disagreed opinions on project topic, and the openness of disagreed opinions so as to lead the resolution agreed mutually.
2.11 Negotiation: it refers to the resolution of different opinions of each party of a project or program in a public manner in order to achieve a mutually satisfactory solution program.
2.12 Conflict and crisis: refers to collecting information, weighing and selecting programs, working towards positive and synergetic solutions in the shortest possible time in face of conflict and crisis, and solving project and contract negotiations with calmness, self-control, and friendliness.
2.13 Reliability: it refers to making others trust you, and the realization of the quality delivery of results within the planned time.
2.14 Values appreciation: it refers to knowing about the inherent essence of others, understanding the different values among project members, and communicating and accepting other views.
2.15 Ethics: it refers to the acceptance of behaviors of team members through ethics communication of legal basis.

2.2.3 11 situation competence elements of project management

3.01 Project orientation: it refers to managing a project management job to make the operations become more efficient, and be beneficial to market competition.
3.02 Program orientation: it refers to integrating specific and related projects, with the application of policy to program management and the development of program management abilities, in order to achieve strategy goals.
3.03 Portfolio orientation: it refers to the knowledge/experience of integrating projects not necessarily associated in order to control program coordination and optimize budget and resources.
3.04 Project, program, and portfolio implementation: they refer the ability of the organization to continuously improve projects, enlarge projects, combine projects, and be engaged in change management, in order to determine best practices in line with the baseline.
3.05 Permanent organization: it refers to continuously providing resources/facilities/products to a permanent organization (such as, accounting and personnel departments), and establishing friendly relations.
3.06 Business: it refers to project management through industry, business, or profession to provide products or services in order to ensure project execution results conform to the requirements of the operation (such as law, finance, economy, human resources, sales and marketing, information technology).
3.07 Systems, products, and technology: they insure that the systems, products, and technologies used inside and outside the organization are all managed as projects, from the development of conception to production to distribution.
3.08 Personnel management: it refers to the planning, recruitment, selection, training, retention, performance evaluations, and motives and abilities of a project team to meet the human resource management needs of the project.
3.09 Health, security, safety, and environment: they refer to ensuring all activities in the project are planned and implemented under the conditions of health, security, safety, and environmental protection.
3.10 Finance: it refers to providing the information required by the project's financial management department, and managing and controlling all finances.

3.11 Legal: it refers to the planning and execution of a project in accordance with laws (such as, no claims caused by infringement, breach of contract, or legal review before signing the contract).

PMI (2012) points out that, effective project management requires a project manager to have the following competences, in addition to specific domains skills and general management expertise: (1) Knowledge: it refers to the project manager's knowledge of the processes and operations of project management; (2)Performance: it refers to the project manager's ability to execute and complete all project matters through management knowledge; (3)Personal ability: it refers to the behavior performance of a project manager in the execution of related projects or activities. Personal effectiveness traits involved in project management include attitude, core personality, and the ability to lead and guide teams to achieve goals and balance project limits. In addition, the project manager has the responsibility to meet the following requirements, including meeting the requirements of tasks, teams, and individuals, organizing and developing strategies and links within the project team, and understanding and applying best practice knowledge, tools, and techniques, thus, strengthening their knowledge management, performance management, goal leadership, and project management. PMI (2012) also points out that a project manager completes tasks through the project team and other interested parties, thus, an effective project manager must balance moral, interpersonal interactions, and conceptual skills to assist in the analysis of situations, and conduct appropriate interactions with other interested parties, including: Leadership, Team building, Motivation, Communication, Influencing, Decision making, Political and cultural awareness, Negotiation, Trust building, Conflict management, and Coaching.

2.3 Connotation of project management competence performance scale

A summarization of the above mentioned contents is, as follows: there are five processes for project management competence: initiating, planning, executing, controlling, and closing, which provide the structure for competence in project management, including cognitive and skill performances, in combination with attitude, moral ethics, and project management standards. This study has a total of 30 cognitive competence project management performances, 30 technical competence performances, 10 attitude competence performances, and 10 moral ethics competence performances, for a total of 80 items in the performance scale. The items of the project management competence performance scale are, as follows:

2.3.1 Cognitive competence performance of project management

2.3.1.1 Cognitive competence performance of initiating a project

01. I know the processes and implementation methods for initiating a project.
02. I know the principles and implementation methods for setting project goals.
03. I know the items and methods for the proposal of a project concept book (draft).
04. I know the process and implementation methods for project feasibility analysis.
05. I know the items and methods for the proposal of a project authorization letter.

2.3.1.2 Cognitive competence performance of project planning

01. I know the processes and methods for the proposal of a project plan.
02. I know the planning processes and methods for a project schedule.
03. I know the planning processes and methods for a project budget.
04. I know the planning processes and methods for project quality.
05. I know the planning processes and methods for project scope.
06. I know the planning processes and methods for project risk.
07. I know the planning processes and methods for a project's human resources.

08. I know the planning processes and methods for equipment, tools, and materials of a project.
09. I know the planning processes and methods for project procurement and bidding.
10. I know the planning processes and methods for communication with a project's interested parties.

2.3.1.3 Cognitive competence performance of project controlling

01. I know the processes and methods for the control and management of a project schedule.
02. I know the processes and methods for the control and management of a project budget.
03. I know the processes and methods for the control and management of project quality.
04. I know the processes and methods for the control and management of project scope.
05. I know the processes and methods for the control and management of project risk.
06. I know the processes and methods for the control and management of a project's human resources.
07. I know the processes and methods for the control and management of project equipment and materials.
08. I know the processes and methods for the control and management of project procurement and bidding.
09. I know the processes and methods for effective control and management of a project's interested parties.
10. I know the processes and methods for controlling and managing the related laws and contracts of a project.

2.3.1.4 Cognitive Competence Performance of Project Closing

01. I know the processes and methods of closing a project.
02. I know the processes and methods for closing a project's contract.
03. I know the processes and methods for closing the administrative affairs of a project.
04. I know the processes and methods for holding a project review meeting.
05. I know the processes and methods for managing the knowledge and experience contained in project documents.

2.3.2 Technical competence performance of project management

2.3.2.1 Technical competence performance of project initiation

01. I will initiate projects with benefits.
02. I will set project goals.
03. I will propose a project concept book (draft).
04. I will operate the tools and techniques used for project feasibility analysis.
05. I will propose a project authorization letter.

2.3.2.2 Technical competence performance of project planning

01. I will propose a project plan.
02. I will operate the tools and techniques used for project schedule planning.
03. I will operate the tools and techniques used for project budget planning.
04. I will operate the tools and techniques used for project quality planning.
05. I will operate the tools and techniques used for project scope planning.
06. I will operate the tools and techniques used for project risk planning.
07. I will operate the tools and techniques used for project human resource planning.
08. I will operate the tools and techniques used for project equipment and materials planning.

09. I will operate the tools and techniques used for project procurement and bid planning.
10. I will operate the tools and techniques used for communication planning with a project's interested parties.

2.3.2.3 Technical competence performance of project controlling and management

01. I will control and manage the project schedule.
02. I will control and manage the project budget.
03. I will control and manage project quality.
04. I will control and manage project scope.
05. I will control and manage project risk.
06. I will control and manage the project's human resources.
07. I will control and manage the project's equipment and materials.
08. I will control and manage the procurement bidding of a program.
09. I will effectively control and manage the interested parties of the project.
10. I will control and manage the related laws and contracts of the project.

2.3.2.4 Technical competence performance of project closing

01. I will operate the tools and techniques of project closing.
02. I will operate the tools and techniques of closing the project's contracts.
03. I will operate the tools and techniques of closing the project's administrative affairs.
04. I will hold project review meetings.
05. I will manage the knowledge and experience documents of the project.

2.3.3 Attitude competence performance of project

01. I will actively study and solve the problems of the project.
02. I will actively apply innovative methods to improve project performance.
03. I will have great confidence in the executed project.
04. I will learn and grow continuously.
05. I can effectively control and manage my emotions and pressure.
06. I am willing to cooperate with team members.
07. I am willing to share my creations and opinions with team members.
08. I agree with the objectives and aspirations of the project.
09. I will try to promote the success of the project.
10. I will actively engage in the development of a project management career.

2.3.4 Moral ethnics competence performance of project

01. I will observe laws and ethical norms.
02. I will fulfill my responsibilities and obligations.
03. I will observe promises and agreements.
04. I will respect the cultural differences between team members.
05. I will keep program secrets confidential.
06. I will cooperate with project limits and hypothesis state.
07. I will treat my colleagues with equality and mutual benefit.
08. I will promote team harmony.
09. I will actively maintain the security of the project's environment.
10. I will actively promote the welfare and benefit of the project team.

3 Establishment of project management competence performance scale

Taking Taiwan's government, industry, and academic fields as the study subjects, this study establishes a project management competence performance scale, conducts comprehensive analysis and sorting of related literatures, and compiles the items for a project management competence performance scale questionnaire, and then, expert validity, construct validity, reliability, and other statistical analysis methods are adopted in order to establish the project management competence performance scale of this study.

3.1 Methods for the establishment of project management performance and the organization of performance questionnaire

Through the foregoing literature reviews and analysis, this study summarizes and edits the preliminary draft of a project management competence performance questionnaire scale, and then, invites project management practitioners, university (college) professors, and study scholars and experts in the field of project management as the auditing experts of the preliminary draft of the study questionnaire. In terms of project management practitioners, 2 high-level senior project supervisors from public sector institutions, and 3 high-level senior project supervisors from industry are invited as the auditing experts for the preliminary draft of the questionnaire scale in this study; secondly, in terms of university (college) professors or scholars in the field of project management, with the assistance of Taiwan's project management institutes and organizations, for a total of 10 scholars and experts from Taiwan's National Pingtung University, National Kaohsiung First University of Science and Technology, Republic of China Military Academy, Chang Jung Christian University, Aletheia University, Cheng Shiu University, Tajen University, and the Open University of Kaohsiung are invited as experts to audit the preliminary draft of questionnaire in this study.

After the preliminary draft questionnaire is approved by the experts, pretesting is conducted, with 350 pretested questionnaires sent out, 328 valid questionnaires are recovered, and validity and reliability are established based on the data of the recovered questionnaires. The validity of this study shall be analyzed by the Structural Equation Modeling (SEM) of the LISREL software application. One of the most important characteristics of SEM Analysis is that it must be based on certain theoretical basis, meaning that SEM is a kind of statistical technique used to test the applicableness of priori theoretical model [4]. The researchers must have very clear instructions or specific theoretical basis for the contents and natures of the potential variables before measurement, and have an established early decision composition mode for each corresponding observation variable. The implementation of analysis is to test the applicableness of the proposed factor structure; this procedure can be used to test structure effectiveness during the development of measurement tools, and can also be used to test the theoretical framework, thus, it is also known as confirmatory factor analysis (cfa) [3, 35].

The four dimensions of cognitive competence, technical competence, attitude competence, and moral ethics competence for project management are compiled in the questionnaire "Taiwan Project Management Competence Performance Scale", as developed in this study, by referring to ISO, IPMA, PMI, British PRINCE2, and other project competence baseline and knowledge systems, and a total of 80 project management competence performance items are listed. As the scale of this study has a definite theoretical basis and model, confirmatory factor analysis of Structural Equation Modeling is carried out by LISREL software regarding the validity analysis of the questionnaire in this study, in order to understand the factor model of its internal structure; Cronbach's α is carried out by SPSS software for reliability analysis of this study, in order to understand the internal consistency of the items at each level.

3.2 Results of the establishment of Taiwan management project competence performance

Through the results of expert validity analysis in this study, four dimensions and 80 items in the preliminary draft of the questionnaire are retained, among them, some words in the items are modified and some items are combined to form the questionnaire contents mentioned above. Then SEM analysis and Cronbach's α are conducted on the questionnaire contents, as shown in the following:

3.2.1 SEM analysis

Through SEM analysis with LISREL software, the obtained results and relation path diagram cannot be presented here due to too many items in the scale. However, the degree of freedom and chi-square statistics of the entire model are listed from the analysis report, with the degree of freedom of 2043, chi-square value of 11171.83, and P value of 0.0, which indicates that there exists obvious difference between the hypothesis model and the observed values. In addition, other adaptation indicators of NFI, NNFI, and CFI are all more than 0.95, which indicates that each item and dimension is well structured. The factor loading of each item at each level is, as follows:

(1) In terms of the factor loading of each item of cognitive competence performance for project initiation, where the factor loadings from Item 01 to Item 05 are, respectively, 0.58, 0.56, 0.54, 0.55, and 0.57.
(2) In terms of the factor loading of each item of cognitive competence performance for project planning, the factor loadings from Item 01 to Item 10 are, respectively, 0.53, 0.50, 0.51, 0.57, 0.50, 0.51, 0.52, 0.54, 0.52, and 0.50.
(3) In terms of the factor loading of each item of cognitive competence performance for project control and management, the factor loadings from Item 01 to Item 10 are, respectively, 0.52, 0.55, 0.53, 0.58, 0.55, 0.52, 0.54, 0.53, 0.57, and 0.60.
(4) In terms of the factor loading of each item of cognitive competence performance for project closing, the factor loadings from Item 01 to Item 05 are, respectively, 0.62, 0.61, 0.62, 0.60, and 0.64.
(5) In terms of the factor loading of each item of technical competence performance for project initiation, the factor loadings from Item 01 to Item 05 are, respectively, 0.57, 0.58, 0.55, 0.56, and 0.54.
(6) In terms of the factor loading of each item of technical competence performance for project planning, the factor loadings from Item 01 to Item 10 are, respectively, 0.52, 0.51, 0.52, 0.56, 0.50, 0.51, 0.50, 0.53, 0.51, and 0.52.
(7) In terms of the factor loading of each item of technical competence performance for project control and management, the factor loadings from Item 01 to Item 10 are, respectively, 0.53, 0.52, 0.55, 0.57, 0.56, 0.53, 0.54, 0.53, 0.56, and 0.58.
(8) In terms of the factor loading of each item of technical competence performance for project closing, the factor loadings from Item 01 to Item 05 are, respectively, 0.58, 0.60, 0.61, 0.57, and 0.62.
(9) In terms of the factor loading of each item of attitude competence performance of the project, the factor loadings from Item 01 to Item 10 are, respectively, 0.61, 0.58, 0.60, 0.58, 0.63, 0.60, 0.57, 0.60, 0.62, and 0.60.
(10) In terms of the factor loading of each item of moral ethic competence performance of the project, the factor loadings from Item 01 to Item 10 are, respectively, 0.54, 0.55, 0.52, 0.57, 0.55, 0.56, 0.58, 0.55, 0.58, and 0.54.

3.2.2 Cronbach's α analysis

(1) In terms of the reliability of the overall dimension of the cognitive competence performance of project initiation, the dimension α value is 0.86, which indicates the decision making

performance of the project supervisor has a certain degree of internal consistency. In addition, the correlation value of each item, as well as the overall construct item, are all greater than 0.64, which indicates that the internal consistency of each dimension is relatively high.

(2) In terms of the reliability of the overall dimension of the cognitive competence performance of project planning, the dimension α value is 0.85, which indicates the control and management performance of the project supervisor has a certain degree of internal consistency. In addition, the correlation value of each item, as well as the overall construct item, are all greater than 0.58, which indicates that the internal consistency of each dimension is relatively high.

(3) In terms of the reliability of the overall dimension of the cognitive competence performance of project control and management, the dimension α value is 0.86, which indicates the emergency response performance of the project supervisor has a certain degree of internal consistency. In addition, the correlation value of each item, as well as the overall construct item, are all greater than 0.63, which indicates that the internal consistency of each dimension is relatively high.

(4) In terms of the reliability of the overall dimension of the cognitive competence performance of project closing, the dimension α value is 0.88, which indicates the cost management business among the project work has a certain degree of internal consistency. In addition, the correlation value of each item, as well as the overall construct item, are all greater than 0.69, which indicates that the internal consistency of each dimension is relatively high.

(5) In terms of the reliability of the overall dimension of the cognitive competence performance of project initiation, the dimension α value is 0.87, which indicates the decision making performance of the project supervisor has a certain degree of internal consistency. In addition, the correlation value of each item, as well as the overall construct item, are all greater than 0.66, which indicates that the internal consistency of each dimension is relatively high.

(6) In terms of the reliability of the overall dimension of the cognitive competence performance of project planning, the dimension α value is 0.86, which indicates the control and management performance of the project supervisor has a certain degree of internal consistency. In addition, the correlation value of each item, as well as the overall construct item, are all greater than 0.64, which indicates that the internal consistency of each dimension is relatively high.

(7) In terms of the reliability of the overall dimension of the cognitive competence performance of project control and management, the dimension α value is 0.86, which indicates the emergency response performance of the project supervisor has a certain degree of internal consistency. In addition, the correlation value of each item, as well as the overall construct item, are all greater than 0.64, which indicates that the internal consistency of each dimension is relatively high.

(8) In terms of the reliability of the overall dimension of the cognitive competence performance of project closing, the dimension α value is 0.87, which indicates the cost management business among the project work has a certain degree of internal consistency. In addition, the correlation value of each item, as well as the overall construct item, are all greater than 0.68, which indicates that the internal consistency of each dimension is relatively high.

(9) In terms of the reliability of the overall dimension of attitude competence performance of a project, the dimension α value is 0.88, which indicates the emergency response performance of the project supervisor has a certain degree of internal consistency. In addition, the correlation value of each item, as well as the overall construct item, are all greater than 0.69, which indicates that the internal consistency of each dimension is relatively high.

(10) In terms of the reliability of the overall dimension of moral ethic competence performance of a project, the dimension α value is 0.88, which indicates the cost management business among the project work has a certain degree of internal consistency. In addition, the correlation value of each item, as well as the overall construct item, are all greater than 0.68, which indicates that the internal consistency of each dimension is relatively high.

To sum up, the four dimensions and items of Taiwan's project management competence performance scale in this study can all be retained, which can be the measurement used for the general situations of competence performance in industry, government, and academic fields in Taiwan, as well as a reference indicator for the evaluation of project management personnel performance, human resources management, and proposing talent development strategies.

4 Conclusion

A total of 80 items for the "Taiwan Project Management Competence Performance Scale" are established, including 30 items of "Project Management Cognitive competence Performance", with 5 items of cognitive competence performance of project initiation, 10 items of cognitive competence performance of project planning, 10 items of cognitive competence performance of project control and management, and 5 items of cognitive competence performance of project closing in terms of the connotation; secondly, it includes 30 items of "Project Management Technical Competence Performance", with 5 items of technical competence performance for project initiation, 10 items of technical competence performance of project planning, 10 items of technical competence performance of project control and management, and 5 items of technical competence performance of project closing in terms of connotation; thirdly, it includes 10 items of "Attitude Competence Performance of Project"; fourthly, it includes 10 items of "Moral Ethic Competence Performance of Project".

The results of this study shall be used for subsequent performance measurements of project management personnel, as well as a reference indicator for the development of project management competence and career development. It can be used in IPMA and PMI international project management communities, for the promotion of project management education and training all over the world, and for the certification of Project Management Professional (PMP). The study results can be integrated into the auditing and certification of project management competence to gradually import a knowledge system of project management, including best practices experience, and a competence baseline for project management organization and operation of all fields. A complete operational mechanism is constructed for project management in order to improve the maturity of project management organization and operation, and shape a project management culture for organizations, which has deep and wide functions and values for strengthening the benefits of project management.

References

1. S.C. Cheng, P.C. Lin, and Y.C. Cheng, Research into the development of college student competence scale, Psychological Testing, **56**, 3, 397-430 (2009)
2. A.P. Chi, Modern project management (Beijing, China, 2005)
3. H.C. Chiu, *Quantitative Research and Statistical Analysis*: *Quantitative Research and Statistical Analysis in Social & Behavioral Sciences* (*3rd Edition*) (Taipei, Taiwan, 2008)
4. H.C. Chiu, Structural Equation Modeling, SEM: LISREL theories, technologies, and applications (Taipei, Taiwan, 2003)
5. W.W. Chung, *Eleven Project Management Course* (*Newly Revised Edition*) (New Taipei, Taiwan, 2014)
6. D.I. Cleland and L.R. Ireland, *Project Management-strategic Design and Implementation* (New York, USA, 2005)
7. H.S. Cooke and K. Tate, *The McGraw-Hill 36-hour Project Management Course* (New York, USA, 2005)
8. R.C. Cooke, *Just Enough Project Management* (New York, USA, 2005)
9. The international project management knowledge system, Editorial Committee of Taiwan Project Management Association (Kaohsiung, Taiwan, 2014)
10. G.R. Hearkens, *Project Management* (New York, USA, 2002)
11. Hsu, *Textbook of Modern Project Management* (New Taipei, Taiwan, 2004)
12. H.Y. Hsu, P.L. Hsiung, Y.F. Chu, Y.H. Chang, M. Fan, C.M. Huang, H.T. Chou, and C.P. Lu, *Basic Project Management Knowledge and Applications in Practice* (*5th Edition*) (New Taipei, Taiwan, 2014)
13. ICB: IPMA Competence Baseline, IPMA, http://ipma.ch/resources/ipma-publications/ ipma-competence-baseline/ (2009)
14. T.L. Kang, Competence-based vocational training theory and curriculum development, Employment and Training, **15** (6), 3-9 (1997)

15. I.C. Kao and Y.S. Chen, A study on model of project management maturity and criteria of assessment, Journal of Project Management, **2** (1), 73-103 (2009)
16. I.C. Kao, Construction and application of the evaluation indications for the practical operation and organizational performance of project management in primary schools, Journal of Research on Elementary Education, **8**, 109-136 (2011)
17. I.C. Kao, Feasible strategies for the application of theory of constraints, knowledge management and project management for the improvement of school's performance, Journal of Knowledge Economy, **8** (2), 59-82 (2011)
18. I.C. Kao, The research on construction of project management indicators, Storage Management Solutions, **1**, 154-173 (2013)
19. I.C. Kao, Feasible strategies of project management processes, team formation and institutionalized operations applied to school governance, Storage Management Solutions, **1**, 198-229 (2013)
20. I.C. Kao, Construction of human resource development strategy of community empowerment by project management, TOC and TTQS, Storage Management Solutions, **2**, 42-65 (2014)
21. H. Kerzner, *Project Management: A System Approach to Planning, Scheduling, and Controlling (9th Edition)* (New York, USA, 2006)
22. M.J. Kuan, Training materials for APMP project management teachers (Taoyuan, Taiwan, 2009)
23. J. Lewis, *Fundamentals of Project Management (2nd Edition)* (New York, USA, 2002)
24. J. Lewis, *Mastering Project Management* (New York, USA, 2008)
25. H.H. Lin, M.H. Huang, and W.L. Wang, Software project management (Taipei, Taiwan, 2005)
26. K.M. Lu, Software project management (Taipei, Taiwan, 2009)
27. D.C. McClelland, Testing for competence rather than for intelligence, American Psychologist, **28** (1), 1-24 (1973)
28. *A Guide to the Project Management Body of Knowledge (4th Edition)*, PMI (Pennsylvania, USA, 2008)
29. *A Guide to the Project Management Body of Knowledge (5th Edition)*, PMI (Pennsylvania, USA, 2012)
30. L.M. Spencer and S.M. Spencer, *Competence at Work: Models for Superior Performance* (New York, USA, 1993)
31. Selection of cases of management quality certification of international items and C class interview, Taiwan Project Management Association (Beijing, China, 2012)
32. J.K. Ting, Project management—Project thought and key to management (Taipei, Taiwan, 2004)
33. I.P. Tobis, M. Tobis, *Managing Multiple Projects* (New York, USA, 2003)
34. C.J. Wei, The General project management knowledge system (Taipei, Taiwan, 2013)
35. M.L. Wu, Structural Equation Modeling (SEM) –The application of SIMPLIS (Taipei, Taiwan, 2006)
36. S.C. Wu, Development of "Teacher competence scale", Psychological Testing, **60** (2), 397-428 (2013)
37. W.D. Yu, Practical project management (Taipei, Taiwan, 2015)

A study on the impact of the occupational performance of teachers in adult education institutions on instructional satisfaction

I-Chan Kao[a]

General Education Center, Open University of Kaohsiung, Kaohsiung 812, Taiwan

Abstract. This study treated adult in-service students of adult education institutions as the subjects and collected data through a questionnaire survey. This study conducted confirmatory factor analysis, and reliability analysis in order to develop questionnaire on "Occupational Performance of Teachers in Adult Education Institutions" and "Instructional Satisfaction in Adult Education Institutions". The collected data were analyzed to discuss the occupational performance and instructional satisfaction in adult education institutions, as well as the correlation between the two and their predictability. It is found that occupational performance and instructional satisfaction in adult education institutions are highly positively correlated and greatly predictable. It is therefore suggested that teachers in adult education institutions work actively to deliver their occupational performance so as to enhance instructional satisfaction and help students to achieve their learning objectives in an effective manner.

1 Introduction

1.1 Research motives

Adult students refer to learners who are above 18 years old, including post-adolescents, the middle-aged, and the elderly. They have more diverse and complex social backgrounds than do adolescent learners, and their reasons for education activities are very different from that of young students whose purpose is relatively simple, mainly for obtaining knowledge and skills and gaining a diploma and degree. With different social environments and physical and mental attributes and needs, adult learners participate in education activities in order to not just get a diploma and degree, but also to strengthen professional skills, enrich life knowledge and related skills, solve problems in life and work, pursue lifelong learning, make friends, expand interpersonal network, look for business partners, explore for customers, obtain a student identity for social welfare, kill time, etc. Compared with adolescent students, adult learners are confronted with gradually aging physiological functions, and so their abilities to absorb and respond to knowledge and skills are obviously poorer than those of young students. Nevertheless, according to ten years of experience for researchers in adult education, the older adult learners are, the better learning motivation they will gain and the more time and energy they will put into learning.

[a] Corresponding author : kic0929@ouk.edu.tw

To achieve teaching objectives, it is necessary for teachers engaged in adult education to understand the adult students' learning attributes, so as to design effective course contents, teaching methods, learning guidance, and assessment methods to help them to achieve their learning objectives. Nicknamed as Father of Adult Education, Knowles (1990) put forward seven steps of andragogy for helping adults to learn effectively: create a good learning atmosphere, build a joint planning mechanism (teachers and students design the training contents and implementation methods together), diagnose learning needs, set up plans and objectives, build a model of learning experience (think about methods for teaching and learning), guide learning activities with appropriate technology and teaching materials, assess learning outcomes, and re-diagnose new learning needs.

Teachers in adult education institutions should consider the following important enlightenments when planning learning activities: first, teachers should be learner-oriented, inspire learners on learning needs and responsibility, and encourage adults to establish self-directed learning habits; second, effective learning should be combined with adults' life experience and development tasks, so that the learners can have stronger motivation and better learning outcomes; third, to achieve effective learning, learners must participate in the design of learning activities, including goal setting, needs recognition, design of learning styles, etc.; fourth, the teacher plays a role as a learning partner, providing a learning environment and support for adults' learning; fifth and lastly, the adult education model is a dynamic process - that is to say, the seven principles form an alternate cycle. For instance, creating a good learning atmosphere not only is the first step, but also runs through the whole teaching process, just like the rest of the principles [18, 52].

If teachers in adult education institutions could grasp the following principles in the teaching process, then it will be helpful for maximizing teaching effects: understand attributes of adult learners, enhance adults' learning motivation, arrange appropriate learning environment, build a relaxing and comfortable learning atmosphere, set up learning groups, utilize adults' experience as important learning resources, adopt problem-centered learning structure, and employ evaluation methods suitable for adult learning; as for the curriculum planning of adult education, teachers should take into account adult learners' learning motivation, design flexible study hours catering to different student groups, and arrange convenient learning places to facilitate study of adult learners [1, 21, 44].

While primary and secondary school teachers have formal training programs for the implementation of professional teacher training, there is no formal training program for teachers in higher education institutions and in adult education institutions. There are obvious differences between adult education and child or teenager education. In addition to the minor differences between basic educational philosophies and principles, the interactive relationship between teachers and adult students, the needs and purpose of adults participating in the education activities, learning places, learning time, learning atmosphere, course topics and design, arrangement and implementation of teaching activities and so on are totally diverse from the case of adolescent students. In other words, teachers should not teach adult students in the same way they teach children or teenagers. Instead, they should plan curriculum and teaching according to learning features and needs of adult students, based on the principles of adaptive and differentiating teaching. Therefore, taking adult students participating in adult education activities as subjects, from the viewpoint of adult students, this study aims to understand the connotation of the occupational performance of teachers in adult education institutions, and whether they can meet the learning objectives and needs of adult students.

1.2 Research purposes

Based on the foregoing research motives, the research objectives are as follows.

1.2.1 To explore the connotation of the occupational performance of teachers and instructional satisfaction in adult education institutions.

1.2.2 To design questionnaires on the occupational performance of teachers and instructional satisfaction in adult education institutions.

1.2.3 To explore the influence that the occupational performance of teachers exerts on students' learning satisfaction in adult education institutions.

1.2.4 To draw a conclusion and propose suggestions according to the study results on how teachers in adult education institutions should deliver their occupational performance effectively so as to improve the instructional satisfaction of adult learners.

1.3 Research methods and process

Based on the literature review and analysis, this study first explored the connotation of the occupational performance of teachers and instructional satisfaction in adult education institutions, so as to design the first draft of the questionnaire on "Occupational Performance of Teachers in Adult Education Institutions" and "Instructional Satisfaction in Adult Education Institutions". Next, it took adult students participating in teaching activities of adult education institutions (including open university, community university, workers' university, university for the elderly, tribal university, and continuing education classes) as subjects to pilot test the questionnaire in order to improve the reliability and validity of the questionnaire. Afterwards, the formal questionnaire was designed to conduct the formal survey; after the survey, the collected data were analyzed with the method of descriptive statistics and inferential statistics to understand the subjects' attitudes toward teachers' occupational performance and instructional satisfaction, to realize the difference among subjects of different social variables on satisfaction with teachers' occupational performance and instructional satisfaction, and to analyze and discuss the correlation between teachers' occupational performance and instructional satisfaction. Lastly, this study draws a conclusion and puts forward feasible suggestions on how to improve teachers' occupational performance and instructional satisfaction.

1.4 Research structure

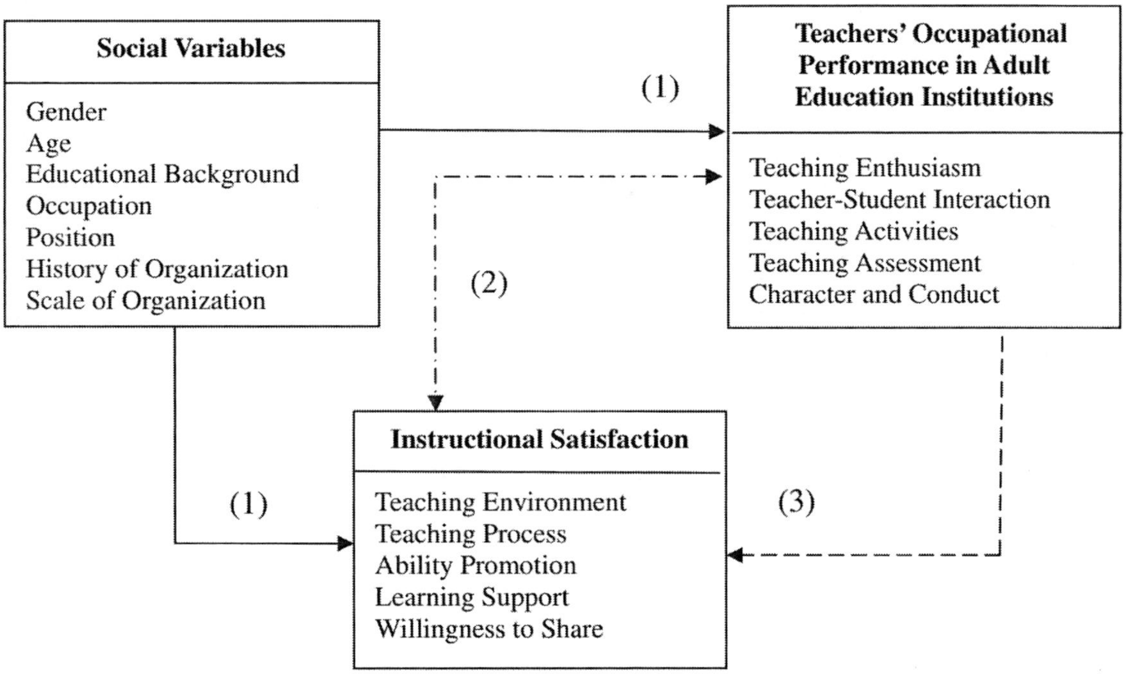

Figure 1. Statistical Analysis Structure, Source: Drawn by the Author.

As shown in Figure 1, the statistical analysis structure of this study includes exploring social variables, such as the adult student respondents' gender, age, educational background, occupation, positions, history of organization, and scale of organization. The study then took occupational performance of teachers in adult education institutions, such as teaching enthusiasm, teacher-student interaction, teaching activities, teaching assessment, and teachers' character and conduct, as independent variables.

Finally, this study took teaching environment, teaching process, ability promotion, learning support, willingness to share, and other learning satisfaction feelings as dependent variables. Through empirical research and inferential statistical analysis, this study explored the relation among the variables.

Statistical Analysis:

(1) With T test and One-way ANOVA, to explore how different social background affects the occupational performance of teachers in adult education institutions and their cognition and feelings of instructional satisfaction.
(2) With Pearson product-moment correlation, to explore the correlation and degree between the occupational performance and cognition and feelings of teachers' instructional satisfaction in adult education institutions.
(3) With regression analysis, to find out the predictive power of the occupational performance on cognition and feelings of teachers' instructional satisfaction in adult education institutions.

2 Connotation of teachers' competency in adult education institutions and related research

Competency is derived from McClelland (1973) replacing intelligence with competency measurement as the prediction basis of personal success. The connotation of competency includes knowledge, skills, and cordiality, with external behaviors as the basis of measurement. In addition, it can also be distinguished as an individual's "possessed" ability, referring to possessing knowledge, emotions and skills, or abilities to "perform" - that is, successful implementation of a task with actual behavioral performance [24]. Cheng, Lin, and Cheng (2009) defined competency as the "ability to use knowledge, attitude and skills to perform work tasks, which was all the behavior, motivates and knowledge related to work success or failure" cited from [54]. Spencer and Spencer (1993) proposed the Iceberg Model, regarding competency being composed of five elements: skills, knowledge, self-concept, traits, and motives, among which, skills and knowledge are more explicit and easier to be trained, developed, and changed, similar to the part of the iceberg above the surface of the ocean, and thus they exhibit surface competency; while self-concept, traits, and motivations are relatively implicit and cannot easily be changed, like the part of the iceberg hidden under the surface, and thus they present implicit competency.

Shulman (1986) held that the required competency of the teaching profession includes 7 aspects: content knowledge, general pedagogical knowledge, curriculum knowledge, pedagogical content knowledge, knowledge of learners and their characteristics, knowledge of educational contexts, as well as knowledge of educational goals, purposes, values, philosophy, and history. Royer & Feldman (1984) stressed that the most important thing for a successful teacher is that he/she must genuinely love teaching, be patient, have acute awareness of students' needs, fully prepare for lectures, etc. The teacher must possess knowledge that the student is learning and developing his/her professional knowledge and further organize the knowledge and convey it to students; at the same time, the teacher must be capable of classroom management and adopt a multi-dimensional evaluation of students; in addition, the teacher must develop his/her latent traits, showing a professional image [16, 22, 43].

The competency requirements for teachers in adult education institution have common features with those for traditional teachers. Nonetheless, considering the adult students' diverse social background and learning needs, there should also be some special requirements, and when reflected on the quality of competency, there should be comprehensive quality merging together with ideology and morality, scientific and cultural knowledge, teaching activity characteristics, level of physical and mental development, ability, and personality traits, embodied in knowledge, ability, body and mind, character, methods, and so on. One should also emphasize the professional characteristics of teachers in adult education institutions, such as a good command of professional skills and rich work experience [20].

From the perspective of the course of its development, the teaching profession has advanced from the organizational development stage to the professional development stage. With special characteristics of professional development, teachers in adult education institutions are geared for professional development. Teachers' professional development has gone through the external professional standard construction to individual independent development and is accompanied by improvements in various education systems and the development of a safeguard mechanism. Despite the variety of adult education objects, contents, and forms, adult education can still be evaluated from the following six aspects [5, 30]: (1) use of professional knowledge and skills, (2) possession of service concept and work ethic, (3) undertaking long-term professional training, (4) continuously seeking professional education for professional development, (5) entitlement to effective professional autonomy; and (6) forming a strong professional organization.

Professional teachers in adult education institutions are familiar with a series of procedures, such as plans, teaching, assessments, etc., and they can choose different teaching contents and methods according to different teaching objects so as to effectively manage the class, attract students' attention, and smoothly carry out teaching activities. They should also take appropriate actions in uncertain and complex teaching situations in order to successfully complete teaching tasks [13, 26]. With adults as their educational objects - who have a higher level of psychological maturity and a richer life experience - in order to interact and communicate with adult learners and give guidance in their learning, teachers in adult education institutions must have extensive knowledge in various fields, so as to meet the interests and developing needs of adult learners and to win admiration from adult learners. In addition, character, words and deeds, ideology, and moral sentiment of the teachers in adult education institutions all directly or imperceptibly influence their students. During work and interaction with students, teachers' excellent qualities, such as dedication, responsibility, enthusiasm, kindness, humility, self-discipline, democracy, justice, integrity, and willpower, are not only conducive to the teaching activities and development of students' body and mind, but also help teachers to act as a role model and play an exemplary role for adult students [10].

This study integrates other studies and discussions of relevant researchers [1, 4, 6, 15, 17, 19, 22], [23, 28, 31, 35, 38, 45], [46, 49], [54-56] on adult education and competency of teachers. This paper summarizes adult education institution teachers' competency into 5 aspects: teaching enthusiasm, teacher-student interaction, teaching activities, teaching assessment, and character and conduct, with a total of 25 competency items that are illustrated as follows.

A. Teaching Enthusiasm

A-1. Energetic and vigorous
A-2. Devoted to teaching
A-3. Confident about teaching
A-4. Aware of students' learning needs
A-5. Innovative and motivated in the teaching profession

B. Teacher-Student Interaction

B-1. Cheerful and humorous
B-2. Able to create an interactive atmosphere in class
B-3. Able to arouse students' learning motivation
B-4. Open to students' suggestions and criticism
B-5. Caring about and guiding students in life and study

C. Teaching Activities

C-1. Fully prepared for teaching materials
C-2. Logical and certain about teaching objectives
C-3. Clear and articulate
C-4. Able to quote extensively and give examples
C-5. Vivacious and funny in teaching activities

D. Teaching Assessment

D-1. Coordinate teaching assessment with teaching objectives
D-2. Design appropriate assessment activities according to students' traits
D-3. Adopt multi-dimensional assessment methods
D-4. Able to assess and grade fairly
D-5. Abel to improve teaching according to assessment results

E. Character and Conduct

E-1. Observe disciplines and fulfill responsibilities
E-2. Sett an example with own conduct
E-3. Virtuous and speaking friendly
E-4. Truthful in speech and firm in action
E-5. Friendly and amiable

3 Connotation of student satisfaction and related research

Chang (2010) explained satisfaction as an internal state when individual motives (physical or psychological) drive behaviors to achieve the pursued goal, which are psychological feelings when the individual's desire is satisfied. Webster's unabridged dictionary (1997) defines satisfaction as a kind of feeling and state of mind after the desire fades away. Tough (1982) pointed out that student satisfaction is a kind of feeling or attitude towards learning activities, which results from learners' love for the learning activity, or in the process of learning their desire and needs are satisfied. Student satisfaction therefore can be used to explain the motives and results of learners to participate in the learning activities. Flammger (1991) defined student satisfaction as the fulfillment of demands and needs, as well as a feeling of enjoyment and satisfaction. Spence & Evans (1956) believed that everyone in different life phases have different learning needs and desires; if these needs and desires were realized, he/she would be satisfied; otherwise, he/she would feel dissatisfied and stop learning.

Lam & Wong's (1974) research suggested that if the learning content is in accordance with students' interests or needs, then it will improve student satisfaction; they also noted that the more intimate the teacher is to students and the more informal the interactions are between the teacher and students, the more satisfied the students feel about the teacher. Lin (1999) argued that the factors affecting student satisfaction are comprehensive. In addition to students' personal factors, teachers and the school can also exert an effect. His research summarized the factors affecting student satisfaction into the following aspects: (1) students' personal factors: students' background (such as gender, age, and educational background), personality, interests, inclination, participation motives, learning interests, experience, family background, self-concept, self-esteem, etc. are related to student satisfaction, (2) teachers' factors: teachers' professional intelligence, traits, teaching strategies, teaching behaviors, teaching methods, teaching contents, attitude toward students, teaching attitude, enthusiasm, etc. affect student satisfaction, (3) school's factors: school location, school-located downtown environment, learning environment, teaching facilities, administrative arrangement, transportation, etc. also affect student satisfaction.

Based on an opinion survey of learners at Evergreen School in Kaohsiung, Pai (1984) concluded that student satisfaction included 8 aspects: class places, curriculum, teachers' attitude in class, teaching methods, course contents, professional knowledge, expressive skills, and service attitude of school administrative staff. Tsai (1991) in his study of student satisfaction and its related factors of elderly learners divided students' satisfaction into two aspects: Evergreen School and students' learning outcomes; the aspect of Evergreen School includes course contents, teachers, hardware facilities and administrative measures; the aspect of students' learning outcomes includes entertainment, self-fulfillment, interpersonal relationship, and social adjustment.

Chen (1997) conducted a study on National Open University students and divided student satisfaction into teaching methods, course contents, learning outcomes, teacher-student interaction, peer relationship and support, etc. Conducting research on university extension education, Wang (1999) categorized educational service quality into five aspects: administrative support, physical facilities, curriculum design and teaching materials selection, teachers' qualities, and interactions. In his study of teacher-student interaction and student satisfaction in National Open University, Lin (2000) divided satisfaction into institutional support, upgrading education, teaching materials, and peer relationship. Chen (2003) explored student satisfaction of students from National Open University from five aspects: learning environment, administrative quality, teachers' quality and teaching attitudes, curriculum, and interpersonal relationship. Kao (2014) probed into student satisfaction of adult learners about digital learning from five aspects: digital learning environment, teaching performance, ability improvement, learning resources support, and willingness to participate and share.

Based on the above review of student satisfaction variables, this paper takes five aspects to examine student satisfaction: teaching environment, teaching process, ability promotion, learning support, and willingness to share, for a total of 25 items of student satisfaction that are illustrated as follows.

A. Teaching Environment

A-1. I am satisfied with the lighting equipment in the classroom.
A-2. I am satisfied with ventilation and air-conditioning in the classroom.
A-3. I am satisfied with the desks and chairs in the classroom, because they are comfortable.
A-4. The teaching aids are in good condition and can satisfy my learning needs.
A-5. The network and digital teaching system is easy to use and satisfies my learning needs.

B. Teaching Process

B-1. I am satisfied with the teaching materials and lecture notes.
B-2. I am satisfied with the teaching schedule.
B-3. I am satisfied with the teacher's teaching attitude.
B-4. I am satisfied with the teaching methods.
B-5. I am satisfied with the teaching assessment.

C. Ability Promotion

C-1. I become more interested in learning.
C-2. The teacher gives me great help in life and work.
C-3. The teacher enables me to control my schedule of life and work.
C-4. The teacher teaches me better methods in learning and work.
C-5. The teacher makes me improve my learning and work performance.

D. Learning Support

D-1. The teacher's encouragement and support are the source of my learning motives.
D-2. The learning climate created by the teacher helps me in learning.
D-3. The teacher knows my learning status well and gives me timely tutorial.
D-4. Emergency equipment is in good maintenance.
D-5. Abnormal situations are under effective and quick control.

E. Willingness to Share

E-1. I am willing to introduce the school's services to others.
E-2. I have introduced the school's services to others.
E-3. If others seek advice from me, I will give priority to introducing my school.
E-4. I will tell others about the advantages of my school.
E-5. I feel proud when introducing my school to others.

4 Design of questionnaires on the occupational performance of teachers and instructional satisfaction in adult education institutions

The questionnaires on the occupational performance of teachers and on the instructional satisfaction in adult education institutions are designed based on a comprehensive analysis of relevant literature. Questions for the questionnaire were then designed, and after specialist validity, construct validity and reliability, and use of statistical analysis methods, the questionnaire was finalized.

4.1 Methods used in the design of the questionnaire

Through a literature review and analysis, this study summarized and edited the first draft of the questionnaires on the occupational performance of teachers and on instructional satisfaction in adult education institutions. The questionnaire adopts a five-point Likert Scale: "strongly agree" for 5 points, "agree" for 4 points, "medium" for 3 points, "disagree" for 2 points, and "strongly disagree" for 1 point. Adult education practitioners, scholars, and experts who conduct studies on adult education were invited as reviewers of the first draft of the questionnaire. In terms of adult education practitioners, two senior teachers were invited from each of the adult education institutions of National Open University, Community University, University of Aboriginal Tribes in Kaohsiung, and Workers' University, for a total number of 8; as for scholars and experts conducting studies on adult education, a total of 7 scholars and experts were invited from Open University of Kaohsiung, National Pingtung University, National Kaohsiung Normal University, National Chung Cheng University, National Chi Nan University, Chang Jung Christian University, and Aletheia University, for a total of 15 experts who reviewed the first draft of the questionnaire.

After review of the first draft by the experts, the pilot test of the questionnaire was carried out. A total of 200 questionnaires were handed out and 186 were collected, and based on the collected questionnaire data, its validity and reliability were established. The validity of this study is based on the factor analysis of Structural Equation Modeling (SEM) by LISREL software. One of the most important characteristics of SEM analysis is that it must be based on certain theories - SEM is a statistical technique used to verify the applicability of a prior theoretical model [11]. Researchers must give a clear explanation about the content and character of the latent variables at the beginning of the measurement, or they must provide a specific theoretical basis and have set up the model of corresponding observed variables earlier. Analysis of the test is for proving the applicability of the factor structure proposed at an early phase. In addition to improving the measurement equipment, this procedure can be used to test the effectiveness of the structure and also can be used in the verification of the theoretical framework. Thus, SEM analysis is also known as confirmatory factor analysis (CFA) [12, 53].

Questionnaires on the occupational performance of teachers and instructional satisfaction in adult education institutions have a strong theoretical foundation and model. Therefore, the validity analysis of the questionnaire adopts LISREL software to conduct SEM confirmatory factor analysis, so as to understand the factor model of internal construct. The reliability analysis of this study carried out Cronbach's α analysis with SPSS software, so as to understand the internal consistency at all levels.

4.2 Results of questionnaires on teachers' occupational performance and student satisfaction in adult education institutions

The questionnaires were designed to cover the contents through expert validity analysis, after reserving 10 parts of the first draft, issuing a total of 50 questions, correcting some words in the questions, and consolidating some questions,. After SEM analysis and Cronbach's α analysis of the questionnaire, the results are as follows.

4.2.1 Questionnaires on teachers' occupational performance in adult education institutions

The questionnaires were designed to cover the contents through expert validity analysis, after reserving 5 parts of the first draft, issuing a total of 25 questions on teaching enthusiasm, teacher-student interaction, teaching activities, teaching assessment, character and conduct, correcting some words in the questions, and consolidating some questions,. After SEM analysis and Cronbach's α analysis of the questionnaire, the results are as follows.

4.2.1.1 SEM Analysis of questionnaires on teachers' occupational performance in adult education institutions

Through SEM analysis with LISREL software and with a large number of questions, it is impossible to present all the results and a relational graph. However, the analysis report lists the degrees of freedom and chi-square statistics of the overall model, among which the degrees of freedom are 2043, chi-square value is 11171.83, and P value is 0.0, indicating that there are differences between the hypothesis model and the observed value. In addition, the other goodness-of-fit index NFI, NNFI, and CFI are all over 0.95, indicating the good condition of each question and structural dimension. Factor loadings for the questions in each part are as follows.

(1) Factor loadings for each question in the part of teaching enthusiasm, from A-1 to A-5, are 0.53, 0.57, 0.53, 0.49, and 0.53.
(2) Factor loadings for each question in the part of teacher-student interaction, from B-1 to B-5, are 0.55, 0.64, 0.61, 0.62, and 0.61.
(3) Factor loadings for each question in the part of teaching activities, from C-1 to C-5, are 0.53, 0.58, 0.52, 0.55, and 0.55.
(4) Factor loadings for each question in the part of teaching assessment, from D-1 to D-5, are 0.57, 0.52, 0.51, 0.53, and 0.50.
(5) Factor loadings for each question in the part of character and conduct, from E-1 to E-5, are 0.60, 0.54, 0.58, 0.53, and 0.56.

4.2.1.2 Cronbach's α analysis

(1) For reliability in the overall teaching enthusiasm, the α value is 0.85, indicating that there is a certain degree of internal consistency for teachers' teaching enthusiasm in adult education institutions. In addition, the correlation values of each question and for all questions in the dimension are above 0.56, showing that there is high internal consistency in each dimension.
(2) For reliability in the overall student-teacher interaction, the α value is 0.86, indicating that there is a certain degree of internal consistency for student-teacher interaction in adult education institutions. In addition, the correlation values of each question and for all questions in the dimension are above 0.60, showing that there is high internal consistency in each dimension.
(3) For reliability in the overall teaching activities, the α value is 0.86, indicating that there is a certain degree of internal consistency for teaching activities in adult education institutions. In addition, the correlation values of each question and for all questions in the dimension are above 0.63, showing that there is high internal consistency in each dimension.
(4) For reliability in the overall teaching assessment, the α value is 0.85, indicating that there is a certain degree of internal consistency for teaching assessment in adult education institutions. In addition, the correlation values of each question and for all questions in the dimension are above 0.57, showing that there is high internal consistency in each dimension.
(5) For reliability in the overall character and conduct, the α value is 0.86, indicating that there is a certain degree of internal consistency for character and conduct of teachers in adult education institutions. In addition, the correlation values of each question and for all questions in the dimension are above 0.60, showing that there is high internal consistency in each dimension.

Nunnally (1978) believed that an α value above 0.7 shows high reliability cited from [53]. Therefore, based on the Cronbach's α analysis on reliability, the questionnaire has high reliability and consistency.

To summarize, this study retained 5 dimensions and questions of the questionnaire on the occupational performance of teachers in adult education institutions. This questionnaire was then used as the formal survey questionnaire to explore the overview of the occupational performance of teachers in adult education institutions in teaching enthusiasm, teacher-student interaction, teaching activities, teaching assessment, and character and conduct.

4.2.2 Questionnaire on student satisfaction in adult education institutions

The questionnaires were designed to cover the contents through expert validity analysis, after reserving 5 parts of the first draft, issuing a total of 25 questions on teaching environment, teaching process, ability promotion, learning support, and willingness to share, correcting some words in the questions, and consolidating some questions. After SEM analysis and Cronbach's α analysis of the questionnaire, the results are as follows.

4.2.2.1 SEM analysis of questionnaires on instructional satisfaction in adult education institutions

Through SEM analysis with LISREL software and with a large number of questions, it is impossible to present all the results and a relational graph. However, the analysis report lists the degrees of freedom and chi-square statistics of the overall model, among which the degrees of freedom are 395, chi-square value is 2930.37, and P value is 0.0, indicating that there are differences between the hypothesis model and the observed value. In addition, the other goodness-of-fit index NFI, NNFI, and CFI are all over 0.95, indicating the good condition of each questions and structural dimensions. Factor loadings for questions in each part are as follows.

(1) Factor loadings for each question in the part of teaching environment, from A-1 to A-5, are 0.56, 0.56, 0.58, 0.55, and 0.51.
(2) Factor loadings for each question in the part of teaching process, from B-1 to B-5, are 0.61, 0.58, 0.61, 0.56, and 0.63.
(3)Factor loadings for each question in the part of ability promotion, from C-1 to C-5, are 0.53, 0.56, 0.52, 0.54, and 0.55.
(4) Factor loadings for each question in the part of learning support, from D-1 to D-5, are 0.58, 0.59, 0.60, 0.62, and 0.57.
(5) Factor loadings for each question in the part of willingness to share, from E-1 to E-5, are 0.57, 0.60, 0.55, 0.58, and 0.56.

4.2.2.2 Cronbach's α analysis on questionnaire on instructional satisfaction in adult education institutions

(1) For reliability in the overall teaching environment, the α value is 0.87, indicating that there is a certain degree of internal consistency for teaching environment in adult education institutions. In addition, the correlation values of each question and for all questions in the dimension are above 0.66, showing that there is high internal consistency in each dimension.
(2) For reliability in the overall teaching process, the α value is 0.86, indicating that there is a certain degree of internal consistency for teaching process in adult education institutions. In addition, the correlation values of each question and for all questions in the dimension are above 0.64, showing that there is high internal consistency in each dimension.
(3) For reliability in the overall ability promotion, the α value is 0.83, indicating that there is a certain degree of internal consistency for ability promotion in adult education institutions. In addition, the correlation values of each question and for all questions in the dimension are above 0.61, showing that there is high internal consistency in each dimension.
(4) For reliability in the overall learning support, the α value is 0.85, indicating that there is a certain degree of internal consistency for learning support in adult education institutions. In addition, the correlation values of each question and for all questions in the dimension are above 0.63, showing that there is high internal consistency in each dimension.
(5) For reliability in the overall willingness to share, the α value is 0.85, indicating that there is a certain degree of internal consistency for willingness to share in adult education institutions. In addition, the correlation values of each question and for all questions in the dimension are above 0.64, showing that there is high internal consistency in each dimension.

Nunnally (1978) believed that an α value above 0.7 shows high reliability cited from [53]. Therefore, based on the Cronbach's α analysis on reliability, the questionnaire has high reliability and consistency.

To summarize, this study retained 5 dimensions and questions for the questionnaire on the instructional satisfaction in adult education institutions. This questionnaire was used as the formal survey questionnaire to explore the overview of cognition and feelings of adult learners about teaching environment, teaching process, ability promotion, learning support, and willingness to share.

5 Research results

After completing the design of the formal Questionnaires on Occupational Performance of Teachers and on the Instructional Satisfaction in Adult Education Institutions, this study adopted the cluster random sampling method, extracting 800 subjects to conduct the questionnaire survey and collecting 674 effective questionnaires. The study then conducted descriptive statistics, T test, One-way ANOVA, regression analysis, etc. The analysis results are stated below.

5.1 Descriptive statistics

5.1.1 Gender

Among the effective samples, 301 are male (44.7%); 373 are female (55.3%).

5.1.2 Age

The majority of the respondents are ages 21 to 40, for a total of 344 people (51.0%); the second largest group is ages 41 to 60, for a total of 278 people (41.2%).

5.1.3 Educational background

The majority of the respondents are high school graduates, with a total number of 328 people (48.7%); followed by graduates of junior college or university, for a total of 222 people (32.9%).

5.1.4 Occupation

The majority of the respondents work in industrial and commercial services, with a total of 345 people (51.2%); followed by civil servants (including soldiers, government workers, teachers, police, firefighters, etc.), for a total of 216 people (32.0%).

5.1.5 Positions

Most of the samples are employees, for a total of 502 people (74.5%); followed by department managers, for a total of 86 people (12.8%); in addition, 86 of the respondents are heads of their company (12.8%).

5.1.6 History of organization

The majority of 264 people are from organizations with over 21 years of history (39.2%); followed by 176 people from organizations with a history of less than 5 years (26.1%); in addition, 150 people are from organizations with a history of 11 to 20 years (22.3%); while 84 people are from organizations with a history of 6 to 10 years (12.5%).

5.1.7 Scale of organization

The majority of the sample, 309, are from organizations with less than 50 employees (45.8%); followed by 226 people from organizations with over 201 employees (33.5%).

5.2 Current situation of the occupational performance of teachers and instructional satisfaction of adult learners in adult education institutions

Tables 1 and 2 lists the statistics of the average number and standard deviation at all levels concerning the current situation of teachers' occupational performance and adult learners' student satisfaction in adult education institutions. Table 1 shows that the average number for the performance of teachers in adult education institutions is above 4.50, indicating high performance. Table 2 shows that the average number for adult learners' student satisfaction is above 4.50, indicating a high degree of student satisfaction.

Table 1. Average Number and Standard Deviation of Teachers' Performance in Adult Education Institutions.

Items of Teachers' Performance in Adult Education Institutions	**Average Number**	**Standard Deviation**	**Number of Questions**
Teaching Enthusiasm	4.63	0.56	5
Teacher-Student Interaction	4.67	0.58	5
Teaching Activities	4.66	0.55	5
Teaching Assessment	4.62	0.59	5
Character and Conduct	4.67	0.55	5
Overall Performance	4.65	0.56	25

N=674

Table 2. Average Number and Standard Deviation of Adult Learners' Student Satisfaction.

Standard Deviation of Adult Learners' student satisfaction	Average Number	Standard Deviation	Number of Questions
Teaching Environment	4.58	0.61	5
Teaching Process	4.65	0.57	5
Ability Promotion	4.61	0.59	5
Learning Support	4.58	0.61	5
Willingness to Share	4.65	0.60	5
Overall student satisfaction	4.61	0.59	25

N=674

5.3 Variance analysis of teachers' occupational performance and instructional satisfaction in adult education institutions with social background variables

5.3.1 Variance analysis on gender

Comparing between males' and females' occupational performance and student satisfaction in adult education institutions with the T test shows that they have significant differences in satisfaction about teaching enthusiasm (T=2.44, P=0.01), teacher-student interaction (T=2.75, P=0.01), teaching activities (T=2.06, P=0.04), teaching assessment (T=2.29, P=0.02), character and conduct (T=2.45, P=0.01), and overall performance of teachers (T=2.52, P=0.01), with males having a higher average than females for each factor. There are also differences in ability promotion (T=2.12, P=0.03), learning support (T=2.26, P=0.02), and overall student satisfaction (T=2.01, P=0.04), with males having a higher average than females for each factor.

5.3.2 Variance analysis on age

Using One-way ANOVA to compare student satisfaction of different age groups in adult education institutions, the statistics suggest that there are significant differences among different age groups in their feelings for teaching enthusiasm (F=2.52, P=0.03), teacher-student interaction (F=3.04, P=0.01), and teaching assessment (F=2.32, P=0.04). Further multiple comparisons and analysis show that the age group from 41 to 60 has profound feelings about teaching enthusiasm, teacher-student interaction, teaching assessment, and other performances and has the highest average among all the age groups, indicating that this age group has a better cognitive feeling in teachers' occupational performance.

After further multiple comparisons and analysis on different cognitions of student satisfaction about ability promotion (F=2.52, P =0.04), learning support (F=2.30, P=0.04), and willingness to share (F=3.04, P=0.01), it is found that learners aged between 41 and 60 have the deepest feelings for ability promotion, learning support, and willingness to share, and that they also have a higher average in each factor than any other age groups, indicating that learners of this age group have a better cognitive feeling in student satisfaction.

5.3.3 Variance analysis on educational background

Using One-way ANOVA to compare student satisfaction of learners in adult education institutions with different educational background, the statistics suggest that there are significant differences in their feelings for teaching enthusiasm (F=2.50, P=0.03) and teacher-student interaction (F=2.42, P=0.03). Further multiple comparisons and analysis show that graduates from junior high school and primary school have profound feelings about teaching enthusiasm and teacher-student interaction, and that they have a higher average than graduates of other educational levels, indicating that graduates from junior high school and primary school have a better cognitive feeling in teachers' occupational performance.

After further multiple comparisons and analysis on different cognitions of student satisfaction about ability promotion (F=2.52, P =0.04) and willingness to share (F=3.02, P=0.02), it is found that graduates from junior high school and primary school have the deepest feelings for ability promotion and willingness to share, and that they also have a higher average in each factor than any other graduate groups, indicating that graduates from junior high school and primary school have a better cognitive feeling in student satisfaction.

5.3.4 Variance analysis on occupation

Using One-way ANOVA to compare student satisfaction of learners in adult education institutions with different occupations, the statistics suggest that there are significant differences in their feelings for teaching enthusiasm (F=4.63, P=0.001), teacher-student interaction (F=2.92, P=0.02), teaching activities (F=4.41, P=0.002), teaching assessment (F=4.28, P=0.002), character and conduct (F=3.05, P=0.02), and teachers' performance (F=4.04, P=0.003). Further multiple comparisons and analysis show that learners from the manufacturing industry and civil servants (including soldiers, government workers, teachers, police, firefighters, etc.) have a significantly higher average than learners working in information technology industry in teaching enthusiasm and teaching assessment. Learners from the manufacturing industry have a significantly higher average than learners working in the information technology industry in teaching activities and teachers' overall performance. Therefore, learners working in the information technology industry have a lower satisfaction about teachers' occupational performance in adult education institutions.

After further multiple comparisons and analysis on different cognitions of student satisfaction about teaching environment (F=6.36, P =0.000), teaching process (F=5.54, P=0.000), learning support (F=4.93, P=0.001), willingness to share (F=4.17, P=0.002), and overall student satisfaction (F=4.91, P=0.001), it is found that learners from the manufacturing industry and civil servants have the deepest feelings for teaching environment, teaching process, learning support, willingness to share, and overall student satisfaction, and that they also have a higher average in each factor than learners from the industrial and commercial services and information technology industry. The results also indicate that learners from the manufacturing industry and civil servants have a higher student satisfaction, while learners from the industrial and commercial services and information technology industry have a lower satisfaction.

5.3.5 Variance analysis on positions

Using One-way ANOVA to compare student satisfaction of adult learners in different positions, the statistics suggest that there are significant differences in their feelings for teaching enthusiasm (F=2.52, P=0.04), teaching activities (F=2.42, P=0.04), and teachers' overall performance (F=2.41, P=0.04). Further multiple comparisons and analysis show that department managers and heads of their company have a higher average in teaching enthusiasm, teaching activities, and teachers' overall performance than average staff, indicating that the average staff have less profound feelings about teachers' occupational performance.

After further multiple comparisons and analysis on different cognitions about teaching environment (F=2.52, P=0.04), learning support (F=2.42, P =0.04), willingness to share (F=2.52, P=0.04), and overall student satisfaction (F=2.50, P=0.04), it is found that department managers and the heads of their company have a significantly higher average than the average staff in teaching environment and willingness to share the deepest feelings for ability promotion, learning support, and willingness to share. This indicates that the average staff have lower cognition about student satisfaction.

5.3.6 Variance analysis on organization history

Using One-way ANOVA to compare feelings of adult learners working in organizations with different history about teachers' performance and student satisfaction different positions, the statistics suggest that there are significant differences in their feelings for teaching enthusiasm (F=3.15, P=0.03), teacher-student interaction (F=4.25, P=0.006), teaching activities (F=2.76, P=0.04), teaching assessment (F=2.86, P=0.04), and teachers' overall performance (F=3.15, P=0.03). Further multiple comparisons and analysis show that learners from organizations with a history above 21 years have an obviously higher average than learners from organizations with a history of less than 5 years in teaching enthusiasm, teacher-student interaction, and teachers' overall performance. This indicates that learners from organizations with a history of less than 5 years have less profound feelings about teachers' occupational performance.

After further multiple comparisons and analysis on different cognition about teaching environment (F=6.68, P=0.000) and overall student satisfaction (F=3.06, P=0.04), it is found that learners from organizations with a history between 6 and 10 years and a history above 21 have the most profound feelings about teaching environment and overall student satisfaction, and that they have a higher average than learners from organizations with a history between 11 and 20 years, indicating that learners from organizations with a history between 11 and 20 years have lower cognition about student satisfaction.

5.3.7 Variance analysis on organization scale

Using One-way ANOVA to compare feelings of adult learners working in organizations with different scale about teachers' performance and student satisfaction of different positions, the statistics suggest that there are significant differences in their feelings for teaching enthusiasm (F=2.82, P=0.04), teacher-student interaction (F=5.04, P=0.002), teaching activities (F=2.67, P=0.04), character and conduct (F=2.88, P=0.04), and teachers' overall performance (F=3.14, P=0.03). Further multiple comparisons and analysis show that learners from organizations with more than 201 employees have an obviously higher average than learners from organizations with less than 50 employees and organizations with employee numbers between 101 and 200 in teacher-student interaction and teachers' overall performance. This indicates that learners from small- and medium-sized organizations with less than 50 employees and organizations with employee numbers between 101 and 200 have less profound feelings about teachers' occupational performance.

After further multiple comparisons and analysis on different cognitions about learning support (F=3.14, P=0.04) and overall student satisfaction (F=3.06, P=0.04), it is found that there is no significant difference in cognition about student satisfaction among learners from organizations with different scales.

5.4 Analysis on item management performance and organizational performance

After Person product-moment correlation analysis, Table 3 lists the statistics of the correlation coefficients of the factors in teachers' occupational performance and learners' student satisfaction in adult education institutions. The results indicate that there is a highly positive correlation (correlation coefficients are all above 70) between factors and the overall condition for teachers' occupational performance and learners' student satisfaction - that is, the better teachers' occupational performance is, the higher the degree of student satisfaction is.

Table 3. Person Product-Moment Correlation Analysis on Project Management Performance and Organizational Performance.

Correlation Coefficients		Factors of student satisfaction: Teaching Environment	Teaching Process	Ability Promotion	Learning Support	Willingness to Share	Student Satisfaction
Factors of Teachers' Performance in Adult Education Institutions	Teaching Enthusiasm	.733***	.828***	.780***	.818***	.769***	.837***
	Teacher-Student Interaction	.733***	.793***	.762***	.801***	.714***	.811***
	Teaching Activities	.746***	.817***	.767***	.797***	.737***	.824***
	Teaching Assessment	.771***	.847***	.795***	.827***	.742***	.849***
	Character and Conduct	.758***	.835***	.758***	.780***	.738***	.825***
	Teachers' Performance	.787***	.867***	.813***	.847***	.778***	.872***

***$p<.001$ N=674

5.5 Regression analysis on the predictive power of teachers' occupational performance on student satisfaction

5.5.1 Predictive power analysis on teaching environment

Table 4. Predictive Power Analysis of Teachers' Performance Factors on Teaching Environment.

	Unstandardized Coefficient		Standardized Coefficient	t	Significance	Multicollinearity Statistics	
	Estimated value of B	Standard error	Beta distribution			Tolerance	VIF
(Constant)	0.38	0.08		4.10	0.00		
Teachers' Performance	0.28	0.06	0.28	5.68	0.00	0.31	3.15
Teacher-Student Interaction	0.27	0.05	0.23	3.65	0.00	0.18	4.14
Dependent variable: Teaching Environment							
Rsquare :0.63							

This study used step-wise multiple regression to analyze the effects of teachers' performance factors on teaching environment. As shown in Table 4, with teaching environment as the criterion variable, teacher-student interaction (T=5.68, P=0.00) and overall teachers' performance (T=3.65, P=0.00) reached a significant level (T=3.44, P=0.00; T=5.46, P=0.00). Teacher-student interaction exerts the largest effect on teaching environment, and the next is overall teachers' performance. The standardized regression coefficients of the two independent variables are positive, signifying that they have positive effects on teaching environment. In the aspect of tolerance, the variance inflation factor values are between 2 and 3, which are under 5 and less than the evaluation index value of 10, indicating little multicollinearity of regression equation.

5.5.2 Predictive power analysis on teaching process

This study used step-wise multiple regression to analyze the effects of teachers' performance factors on teaching process. As shown in Table 5, with teaching process as the criterion variable, among overall teachers' performance (T=5.67, P=0.00), teacher-student interaction (T=3.64; P=0.00), and teaching activities (T=3.58, P=0.00), overall teachers' performance exerts the largest effect on teaching process, and the next is the teacher-student interaction. The standardized regression coefficients of the three independent variables are positive, signifying that they have positive effects on teaching process. In the aspect of tolerance, the variance inflation factor values are between 2 and 3, which are under 5 and less than the evaluation index value of 10, indicating little multicollinearity of regression equation.

Table 5. Predictive Power Analysis of Teachers' Performance Factors on Teaching Process.

	Unstandardized Coefficient		Standardized Coefficient	t	Significance	Multicollinearity Statistics	
	Estimated value of B	Standard error	Beta distribution			Tolerance	VIF
(Constant)	0.38	0.09		4.20	0.00		
Teachers' Performance	0.28	0.05	0.28	5.67	0.00	0.32	3.16
Teacher-Student Interaction	0.27	0.06	0.23	3.64	0.00	0.20	4.15
Teaching Activities	0.21	0.04	0.21	3.58	0.00	0.20	4.16
Dependent variable: Teaching Process							
Rsquare :0.66							

5.5.3 Predictive power analysis on ability promotion

This study used step-wise multiple regression to analyze the effects of teachers' performance factors on ability promotion. As shown in Table 6, with ability promotion as the criterion variable, among overall teachers' performance (T=5.85, P=0.00), teacher-student interaction (T=3.85; P=0.00), teaching activities (T=3.58, P=0.00), and character and conduct (T=3.21, P=0.00), overall teachers' performance exerts the largest effects on ability promotion, and the next is the teacher-student interaction. The standardized regression coefficients of the four independent variables are positive, signifying that they have positive effects on ability promotion. In the aspect of tolerance, the variance inflation factor values are between 2 and 3, which are under 5 and less than the evaluation index value of 10, indicating little multicollinearity of regression equation.

 DOI: 10.1051/matecconf/201711901038

Table 6. Predictive Power Analysis of Teachers' Performance Factors on Ability Promotion.

	Unstandardized Coefficient		Standardized Coefficient	t	Significance	Multicollinearity Statistics	
	Estimated value of B	Standard error	Beta distribution			Tolerance	VIF
(Constant)	0.38	0.08		4.00	0.00		
Teachers' Performance	0.33	0.05	0.28	5.85	0.00	0.32	3.16
Teacher-Student Interaction	0.28	0.06	0.23	3.85	0.00	0.23	4.15
Teaching Activities	0.22	0.04	0.21	3.58	0.00	0.21	4.17
Character and Conduct	0.21	0.04	0.21	3.21	0.00	0.20	4.18
Dependent variable: Ability Promotion							
Rsquare :0.67							

5.5.4 Predictive power analysis on learning support

This study used step-wise multiple regression to analyze the effects of teachers' performance factors on learning support. As shown in Table 7, with learning support as the criterion variable, among overall teachers' performance (T=3.55, P=0.00), teaching enthusiasm (T=4.45; P=0.00), and teaching assessment (T=3.79, P=0.00), teaching enthusiasm exerts the largest effects on learning support, and the next is overall teachers' performance. The standardized regression coefficients of the three independent variables are positive, signifying that they have positive effects on ability promotion. In the aspect of tolerance, the variance inflation factor values are between 2 and 3, which are under 5 and less than the evaluation index value of 10, indicating little multicollinearity of regression equation.

Table 7. Predictive Power Analysis of Teachers' Performance Factors on Learning Support.

	Unstandardized Coefficient		Standardized Coefficient	t	Significance	Multicollinearity Statistics	
	Estimated value of B	Standard error	Beta distribution			Tolerance	VIF
(Constant)	0.38	0.09		4.63	0.00		
Teachers' Performance	0.28	0.05	0.23	3.55	0.00	0.23	4.12
Teaching Enthusiasm	0.33	0.06	0.28	4.45	0.00	0.21	3.15
Teaching Assessment	0.22	0.04	0.21	3.79	0.00	0.24	4.13
Dependent variable: Learning Support							
Rsquare :0.70							

5.5.5 Predictive power analysis on willingness to share

This study used step-wise multiple regression to analyze the effects of teachers' performance factors on willingness to share. As shown in Table 8, with willingness to share as the criterion variable, among overall teachers' performance (T=7.39, P=0.00), teacher-student interaction (T=4.11, P=0.00), and teaching enthusiasm (T=3.88; P=0.00), overall teachers' performance exerts the largest effects on willingness to share, and the next is teacher-student interaction. The standardized regression coefficients of the three independent variables are positive, signifying that they have positive effects on willingness to share. In the aspect of tolerance, the variance inflation factor values are between 2 and 3, which are under 5 and less than the evaluation index value of 10, indicating little multicollinearity of regression equation.

Table 8. Predictive Power Analysis of Teachers' Performance Factors on Willingness to Share.

	Unstandardized Coefficient		Standardized Coefficient	t	Significance	Multicollinearity Statistics	
	Estimated value of B	Standard error	Beta distribution			Tolerance	VIF
(Constant)	0.38	0.08		4.70	0.00		
Teachers' Performance	0.33	0.06	0.26	7.39	0.00	0.22	3.12
Teacher-Student Interaction	0.24	0.05	0.23	4.11	0.00	0.23	4.11
Teaching Enthusiasm	0.22	0.04	0.21	3.88	0.00	0.24	4.13
Dependent variable: Willingness to Share							
Rsquare :0.63							

5.5.6 Predictive power analysis on overall student satisfaction

This study used step-wise multiple regression to analyze the effects of teachers' performance factors on overall student satisfaction. As shown in Table 9, with overall student satisfaction as the criterion variable, among overall teachers' performance (T=9.71, P=0.00), teacher-student interaction (T=5.85, P=0.00), and teaching activities (T=3.87; P=0.00), overall teachers' performance exerts the largest effects on overall student satisfaction, and the next is teacher-student interaction. The standardized regression coefficients of the three independent variables are positive, signifying that they have positive effects on overall student satisfaction. In the aspect of tolerance, the variance inflation factor values are between 2 and 3, which are under 5 and less than the evaluation index value of 10, indicating little multicollinearity of regression equation.

Table 9. Predictive Power Analysis of Teachers' Performance Factors on Overall Student Satisfaction.

	Unstandardized Coefficient		Standardized Coefficient	t	Significance	Multicollinearity Statistics	
	Estimated value of B	Standard error	Beta distribution			Tolerance	VIF
(Constant)	0.37	0.08		4.22	0.00		
Teachers' Performance	0.34	0.07	0.26	9.71	0.00	0.22	3.12
Teaching Enthusiasm	0.27	0.05	0.25	5.85	0.00	0.23	3.32
Teaching Assessment	0.22	0.04	0.21	3.87	0.00	0.24	4.13
Dependent variable: overall student satisfaction							
Rsquare :0.73							

6 Conclusion and suggestions

According to the results herein, in terms of different social variables for teachers' performance in adult education institutions, this study suggests that teachers should pay attention to female learners, young and senior learners, learners with medium and higher degrees, learners working in the industrial and commercial services industry and information technology industry, learners working in organizations, learners working in organizations with a shorter history, and learners working in small- and medium-sized organizations. Teachers should learn about students' learning needs and objectives, design adaptive curriculum and teaching activities, and strengthen teaching enthusiasm, teacher-student interaction, teaching activities, teaching assessment, and character and conduct, so as to improve student satisfaction over the teaching environment, teaching process, ability promotion, learning support, and willingness to share. There is an obvious positive correlation between teachers' occupational performance and student satisfaction in adult education institutions, while at the same time overall teachers' occupational performance exerts positive predictive power on all aspects of student satisfaction. Therefore, it is indeed a challenge for teachers in adult education institutions to teach adult learners from various and complex social backgrounds and to effectively tailor the teaching methods to meet each learner's learning needs.

This study believes that teachers in adult education institutions can play a role like a "ferryman", encouraging, assisting, and leading adult learners to steer their ship of learning to become a competent captain roaming in the sea of knowledge. Teachers should instruct adult learners to set up learning objectives in a self-directed manner and to obtain the knowledge and skills that they need. In conclusion, teachers in adult education institutions should participate in learning activities to improve their competency, strengthen performance analysis and application, train design patterns, improve teaching methods, strengthen human resource development, study adult education theories, principles of learning outcomes assessment, and information and knowledge on industry overview, train skills in needs analysis, motivation and reward theory teaching, teacher-student interaction, effective listening and communication, and problem analysis and solving, improve their ability at creating a teaching atmosphere, and lastly compile and present teaching materials and information feedback. Teachers in adult education institutions should develop enthusiasm for teaching, patience, willingness to share, positive thinking, and inclusiveness, display good performance in enthusiasm, interaction with students, teaching assessment, and character and conduct, and adopt adaptive and diverse teaching methods, so as to improve adult learners' satisfaction towards the teaching environment, teaching process, ability promotion, learning support, and willingness to share and to maintain a sustainable management of adult education institutions and achieve the educational objective of lifelong learning.

References

1. Effective adult education, Adult Education Institute of Republic of China and Graduate School of Adult and continuing education of National Chung Cheng University (Taipei, Taiwan, 1995)
2. Professionalization of adult education, Adult Education Institute of Republic of China (Taipei, Taiwan, 1995)
3. C.H. Chang, Modern psychology (Taipei, Taiwan, 2010)
4. C.S. Chang, Criteria and training pattern for primary and middle school teachers, Special Research Report of National Science Committee of Executive Yuan (Taitung, Taiwan, 2000)
5. J. Chang, Study on professionalization of adult education and professional training strategies of teachers in adult education institutions, Vocational Education Forum, **10**, 32-34 (2007)
6. T.J. Chang, W.H. Wu, T.C. Tseng, T.C. Hsu, P.L. Hung, M.S. Wang, W.H. Chen, M.H. Tseng, Y.J. Chang, C.C. Tseng, C.M. Huang, S.S. Pai, and C. T. Li, *Teaching Profession Development System of Middle School Teachers* (Taipei, Taiwan, 2004)
7. K.E. Chen, Study on relation among learning styles, student satisfaction, and learning achievement, Unpublished Master's Thesis of National Taiwan Normal University (Taipei, Taiwan, 1997)
8. S.L. Chen, Study on relation among goal-orientation, student satisfaction, and learning performance: A case study of kaohsiung learning guidance center of national open university, Unpublished Master's Thesis of National Sun Yat-sen University (Kaohsiung, Taiwan, 2003)
9. S.C. Cheng, P.C. Lin, and Y.C. Cheng, Study on development of competence evaluation on

college students, Psychological Testing, **56** (3), 397-430 (2009)
10. C.Y. Chi and W.W. Chien, Study on training strategies of teachers in adult education institutions of our country, China Adult Education, **4**, 84-85 (2005)
11. H.C. Chiu, Structural Equation Modeling, SEM: LISREL theories, technologies, and applications (Taipei, Taiwan, 2003)
12. H.C. Chiu, *Quantitative Research and Statistical Analysis: Quantitative Research and Statistical Analysis in Social & Behavioral Sciences (3rd Edition)* (Taipei, Taiwan, 2008)
13. C.C. Chung, Professionalization of teaching: ideology, system, and subjects, Educational Research, **2**, 14-16 (2001)
14. D.M. Flammger, Nontraditional students and postsecondary school satisfaction, ERIC Document Reproduction Service (1991)
15. A. Gilis, M. Clement, L. Laga, and P. Pauwels, Establishing a competence profile for the role of student-centred teachers in higher education in Belgium, Research in Higher Education, **49** (6), 531-554 (2008)
16. E.G. Hall and H.L. Jones, *Competency-based Education: A Process for the Improvement of Education* (Englewood Cliffs, USA, 1976)
17. F.T. Ho and J.Y. Lo, Education reform and teacher professionalization, Teacher Education Institute of Republic of China, (1-30) (Taipei, Taiwan, 1992)
18. M.C. Hu, Theory and model of adult education (Taipei, Taiwan, 1998)
19. M.C. Hu, Modernization and professionalization of adult education (Taipei, Taiwan, 2001)
20. C. Huang, Investigation and analysis on current situation of professional quality of teachers in adult education institutions: from evaluation survey of adult learners of higher education institutions, Journal of Higher Correspondence Education(Philosophy and Social Sciences Edition), **22** (5), 10-13 (2009)
21. F.S. Huang, Adult psychology and learning (Taipei, Taiwan, 1995)
22. H. Huntly, Teachers' work: beginning teachers' conceptions of competence, The Australian Educational Researcher, **35** (1), 125-145 (2008)
23. L. Ingvarson, Professional development as the pursuit of professional standards: The standard based professional development system, Teaching and Teacher Education, **14** (1), 127-140 (1998)
24. T.L. Kang, Theory and curriculum development of ability-oriented professional training, Employment and Training, **15** (6), 3-9 (1997)
25. I.C. Kao, Study on the relationship between e-learning service quality of schools and satisfaction with learning effectiveness, Annual Quality Congress Transactions, **2**, 114-141 (2014)
26. I.C. Kao, Feasible strategies for setting the course of "Community Development Project Management" in Adult Education Institutions, Storage Management Solutions, **2**, 35-62 (2016)
27. M. Knowles, *The Adult Learner: A Neglected Species* (Houston, USA, 1990)
28. B. Koster, J. Dengerink, F. Korthagen, and M. Lunenberg, Teacher educators working on their own professional development: Goals, activities and outcomes of a project for the professional development of teacher educators, Teachers and Teaching: Theory and Practice, **14** (5), 567-587 (2008)
29. Y.L. Lam and A. Wong, Attendance regularity of adult learners: An examination of content and structural factors, Adult Education, **14**, 2 (1974)
30. T. Lei, Study of characteristics of professionalized development of teachers in adult education institution*s*, Continuing Education Research, **3**, 9-10 (2011)
31. H.H. Lin, Study on adult teaching evaluation index construction and application of community college teachers, Unpublished Master's thesis of National Chung Cheng University (Taipei, Taiwan, 2005)
32. M.C. Lin, Study on teacher-student interaction and student satisfaction of Open University, Unpublished Master's thesis of National Kaohsiung Normal University (Kaohsiung, Taiwan, 2000)
33. P.W. Lin, Study on student satisfaction about professional program in engineering of comprehensive senior high school, Academic Journal of Nankang Vocational High School, **17**, 217-240 (1999)

34. D.C. McClelland, Testing for competence rather than for intelligence, American Psychologist, **28** (1), 1-24 (1973)
35. Ministry of Education., UCAN College Employment Platform, retrieved from https://ucan.moe.edu.tw/ (2016)
36. J.C. Nunnally, *Psychometric Theory* (New York, USA, 1978)
37. H.H. Pai, Kaohsiung evergreen school opinion survey on community education for the elderly (Kaohsiung, Taiwan, 1984)
38. H.L. Pan, L.Y. Wang, M.F. Chien, C.L. Sun, S.C. Chang, H.H. Chang, S.H. Chen, S.M. Chen, and P.J. Tsai, Development of professional competence index of national primary and middle school teachers, Education Research Information, **12** (4), 129-168 (2004)
39. J. Royer and R. Feldman, *Educational Psychology: Application and Theory* (New York, USA, 1984)
40. L.S. Shulman, Those who understand: Knowledge growth in teaching, Educational Research, **15** (2), 4-14 (1986)
41. R.B. Spence and L.H. Evans, Dropouts in adult education, Adult Education, **6** (4), 221-225 (1956)
42. L.M. Spencer and S.M. Spencer, *Competence at Work: Models for Superior Performance* (New York, USA,1993)
43. Y. Steinert, Mapping the teacher's role: The value of defining core competencies for teaching, Medical Teacher, **31**, 371-372 (2009)
44. J.H. Su, Effective adult learning, T&D Fashion, **34**, 1-7 (2005)
45. C.L. Sun, Future policy and practice of teacher education (Taipei, Taiwan, 2010)
46. Y.T. Sung, T.Y. Chao, and Y.P. Chen, Preparation of situational judgment tests in the teaching context, Annual Academic Seminar on Testing of China Academy of Testing (Taipei, Taiwan, 2009)
47. A. Tough, *Intentional Change* (Chicago, USA, 1982)
48. M.Y. Tsai, Study on student satisfaction and related factors of the elderly learners: A case study of Taipei Evergreen School, Unpublished Master's Thesis of Soochow University (Taipei, Taiwan, 1991)
49. C.K. Wang, Teaching profession update (Taipei, Taiwan, 2016)
50. P.C. Wang, Study of the relationship between consumer motivation, service quality and post-purchase behavior: A case study of university extension education, Unpublished Master's Thesis of National Chung Cheng University (Chiayi, Taiwan, 1999)
51. Webster's unabridged dictionary, http://www.merriam-webster.com/netdict.htm (1997)
52. H.C. Wei, Create effective adult learning: theory and practice, T&D Fashion, **161**, 1-18 (2013)
53. M.L. Wu, Structural Equation Modeling (SEM) –The application of SIMPLIS (Taipei, Taiwan, 2006)
54. S.C. Wu, Development of "Teachers' performance evaluation form," Psychological Testing, **60** (2), 397-428 (2013)
55. S.K. Yang, C.L. Huang, and C.W. Yao, International comparison of teachers' quality and reform measures, Taiwan Association for Sociology of Education (Taitung, Taiwan, 2011)
56. Y.H. Ying, Practice and exploration of network-based teaching and management of adult education, China Educational Technology, **272**, 60-62 (2009)

The crisis early warning of the quality of supply chain based on rough set&feature weighted support vector machine

Xiu-lian Hu[a], Dan Liu and Qi Jiang

Department of Logistics, Fuzhou University of International Studies and Trade, Fuzhou 350202, Fujian, China

Abstract. A Rough Set&Feature Weighted Support Vector Machine(RS-FWSVM) model is proposed for the quality of supply chain crisis early-warning,which aims at some problems of the quality of supply chain. This model combines the advantages of the RS and FWSVM,which can get classification per-formances by changing the weights of different linear functions in the feature space. Application process of this model to the crisis early warning of SCQ is researched, which can help enable chain enterprises to identify crises in the process of operations and to predict possible crises.

1 Introduction

The complicated structure and the global extension of the supply chain have exacerbated the chain instability, not only the supply chain faces more and more risks, but also the supply chain crisis break out repeatedly. All above make the the quality of supply chain crisis early-warning management become an extremely important subject. So enterprises must realize that it is very important to adopt the early-warning management of the crisis of the quality of supply chain,do a good job of warning and preventing the crisis in advance, and then take effective measures to avoid that.

In 1993, Japan's famous supplier of semiconductor materials Sumitomo Chemical plant got exploded, which posed a serious threat to the global supply [1]. The international financial crisis caused by the U.S.subprime crisis in 2009 putted great pressure on enterprises of our country. In 2015, the explosions at Tianjin Port brought the huge impact on global supply chain, which leaded to the electronics industry supply disruptions.

Frequent crisis shows to enterprises that the quality of supply chain crisis is needed. The quality of supply chain crisis early-warning is one of the important components of supply chain management.

2 RS-FWSVM Model

The RS (Rough Set) theory, introduced by Pawlak [2], is a new mathematical tool to deal with vagueness and uncertainty.

SVM (Support Vector Machine), suggested by Cortes and Vapnik [3], evolves from optimal hyperplane under nonlinearly separable situation. It has special advantages in solving problems of small samples, nonlinearity, and high dimension mode recognition.

[a] Corresponding author : 563354676@qq.com

The capability of fault-tolerance and generalization of RS is poor and it can deal only with quantized data . But this is an advantage of SVM. However, SVM can not single out redundant knowledge, which is the advantage of RS. So the RS-SVM model is proposed with both advantages and disadvantages of RS and SVM by some scholars.

Y. L.Wu [4], et. al. built logistics risk evaluation model based on RS and SVM. They carried on application study combining with examples by identifying and classifying the risk information set.X.L.Zhang [5], et. al. proposed a partner selection model based on RS and SVM.The core point of this model was using rough sets theory to pick out the important attributes, and then SVM was used for customer classification. This method that didn't affect classification performance can reduce the data dimensions and the complexity in classification process.J.F.Wang [6], et. al. presented a evaluation model of RS and SVM. They concluded that the accuracy of RS and SVM model in this article would be improved from 98.6% to 100% by comparison with applying the SVM directly.

The literatures mentioned above use RS to find out the features of reduced sets by retaining the core attributes to improve the ability of classification of SVM. But the influence of the importance of attributes on the classification failed to be taken into account. But in many cases, features that are not completely relative or even complete-ly irrelative to the problem in the used data, it may affect the classification performances of SVM [7].

In order to solve this problem, the Rough Set--Feature Weighted Support Vector Machine (RS-FWSVM)model is put forward So, it is required to impose different weights to different features.

Let S is the information system of RS,and S =(U, A, V, F), $P \subseteq A$, U/ind(P)=$\{X_1, X_2, X_3, \cdots, X_n\}$, the information quantity [8] of P is

$$I(P)=\sum_{i=1}^{n} p(X_i)p(X_i^c)=\sum_{i=1}^{n}\frac{|X_i|}{|U|}\left[1-\frac{|X_i|}{|U|}\right]=1-\frac{1}{|U|^2}\sum_{i=1}^{n}|X_i|^2 \tag{1}$$

The $|X|$ indicates the base of X, $|X_i|/|U|$ is the probability of the equivalence class X_i in U.

The importance of attribute a(a $\subseteq$ A) in A is

$$Sig_{A-\{a\}}(a)=I(A)-I(A-\{a\}) \tag{2}$$

The SVM that bases on feature weighted kernel function is called FW-SVM [8].Feature weighted kernel function K_p is defined as

$$k_p(x_i, x_j)=k(x_i^T P, x_j^T P) \tag{3}$$

The P=diag（W_1, W_2, ..., W_n）is the matrix of a linear transformation, also known as feature weighted matrix. Then FWSVM classification model is

$$\begin{cases} \max \sum_{i=1}^{l} a_i - \frac{1}{2}\sum_{i=1}^{l}\sum_{j=1}^{l}\alpha_i\alpha_j y_i y_j k(x_i^T P, x_j^T P) \\ s.t \quad 0 \le a_i \le C, i=1,\cdots,l \\ \sum_{i=1}^{l} a_i y_i = 0 \end{cases} \tag{4}$$

It's decision function is

DOI: 10.1051/matecconf/201711901039

$$f(x)=\mathrm{sgn}(\sum_{i=1}^{l} a_i\, y_i \mathrm{k}(x_i{}^{\mathrm{T}}\mathrm{P} \cdot x^{\mathrm{T}}\mathrm{P}) + b) \tag{5}$$

3 Crisis early warning of the quality of supply chain based on RS-FWSVM

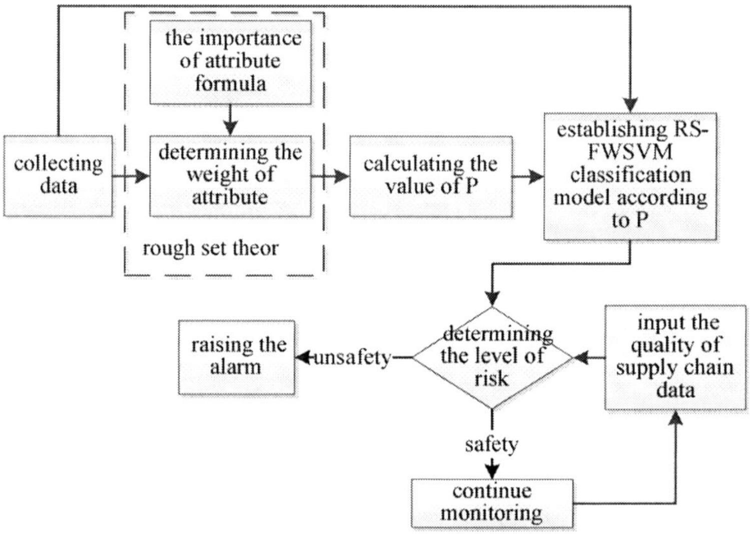

Figure 1. The application process of the crises early warning of the quality of supply chain

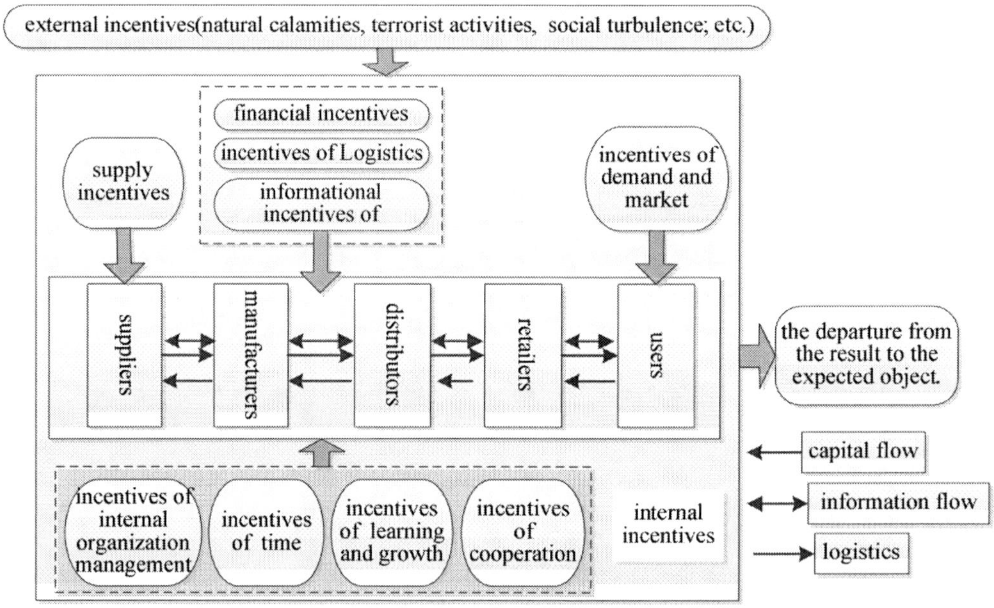

Figure 2. The schematic of incentives of the crises of the quality of supply chain.

The application process of RS-FWSVM in crisis early warning of the quality of supply chain is shown in Fig. 1. Firstly, designing questionnaire according to the index system for the crisis early warning of the quality of supply chain and collecting the data. Secondly, determining the weight of indicators using with the rough set theory and the computation formula of the importance of attribute,and obtaining the value of P. Finally, establishing RS-FWSVM classification model according to the collected data and P.

As long as relevant data of the supply chain to be assessed are collected, the crises level of it can be known with the model. If the output is safety, the system will keep monitoring the quality of supply chain, whereas an alarm is raised. Risky supply chain must take action.

3.1 The Incentives of the crisis of the quality of supply chain and the evaluation index analysis

The incentives of the crises of the quality of supply chain are the most di-rect reason for supply chain crises. From the source of the crises,which can be grouped into external and internal incentives (Figure 2).

According to the incentives for the crises, relevant references and the investigation in the enterprises ,a set of original evaluation index system $C=\{a_1, a_2, \cdots, a_m\}$is designed.

3.2 Data collection

This investigation altogether provides questionnaire 110, the actual recovery of 86 copies, of which 84 valid questionnaires, for an effective questionnaire returns-ratio is 76.4%. Data from the questionnaires is collected and analyzed.

3.3 Weight determination with rough set theory

Taking data collected in 84 questionnaires as sample, the weights of indicators of the crises of the quality of supply chain is determined by equation 2(see Table 1).

TABLE 1. the weights and means of indicators of the crises of the quality of supply chain

The Incentives of the Crisis	No.	Indicators	$Sig_{A-\{a_i\}}(a_i)$	Means
external incentives	a1	nature disaster	0.00765	3.11
	a2	macro-economic situation	0.00510	3.76
	a3	political stability and government intervention	0.00538	3.5
	a4	the impact degree of industry	0.00538	3.75
financial incentives	a5	turnover of inventory	0.01694	3.87
	a6	turnover of account receivable	0.01204	3.7
	a7	return on equity	0.01204	3.59
	a8	sales growth	0.01286	3.47
	a9	quick ratio	0.01204	3.93
	a10	asset-liability ratio	0.01286	3.31
informational incentives of	a11	the timely information transfer rate among node enterprises	0.05591	3.2
	a12	the accurately information transfer rate among node enterprises	0.0351	3.1
	a13	the frequency of informational communication among node enterprises	0.02489	3.3
incentives of logistics	a14	not strictly inventory control	0.03920	3.63
	a15	risk in transit	0.01857	3.3
	a16	order disposal failure	0.02224	2.91
supply incentives	a17	the suppliers's on time delivery(OTD)	0.01061	2.8
	a18	the suppliers's quality of products	0.01673	2.8
	a19	key suppliers's bankruptcy and losses	0.03428	3.17

The Incentives of the Crisis	No.	Indicators	$Sig_{A-\{a_i\}}(a_i)$	Means
incentives of internal organization management	a20	the supply chain structure rationality	0.00530	3.51
	a21	the supply chain structure stability	0.00708	3.61
	a22	the influence and control ability of the core enterprise	0.00326	3.11
	a23	emergency response system of supply-chain risk	0.01326	3.47
	a24	overall labour productivity	0.00208	3.37
incentives of time	a25	processing time of customer complaint	0.02000	2.84
	a26	customer response time	0.01715	3.21
incentives of learning and growth	a27	income ratio of new product(service)	0.04488	3.53
	a28	training expenditures ratio	0.02325	4.27
incentives of cooperation	a29	the equitable degree of the value distribution among node enterprises	0.03960	3.49
	a30	the coordination of supply chain culture	0.01674	3.46
	a31	strategic alignment among node enterprises	0.01143	3.41
	a32	the coordination among node enterprises	0.02245	3.23
incentives of demand and market	a33	core enterprise's market share	0.00408	3.4
	a34	core enterprise's customer satisfaction	0.00735	3.19
	a35	customer-keeping ratio	0.01959	3.1
	a36	inexactly market forecast	0.00571	3.27
	a37	perfect order fill rate	0.00571	2.91

3.4 The construction of the crisis early warning of the quality of supply chain based on RS-FWSVM model

A method of multi-class classification based on FWSVM is put forward. Then FWSVM classifiers are used to conduct classification on crisis level of the quality of supply chain. Where I means safety, II means mildly early-warning, III means moderately early-warning and IV means heavy early-warning.First, Classifier 1 can divide the data into two groups: I IIand III IV, then Classifier 2 sets III and IV apart from each other, at last, Classifier 2 sets I and II apart from each other(see Fig.3).

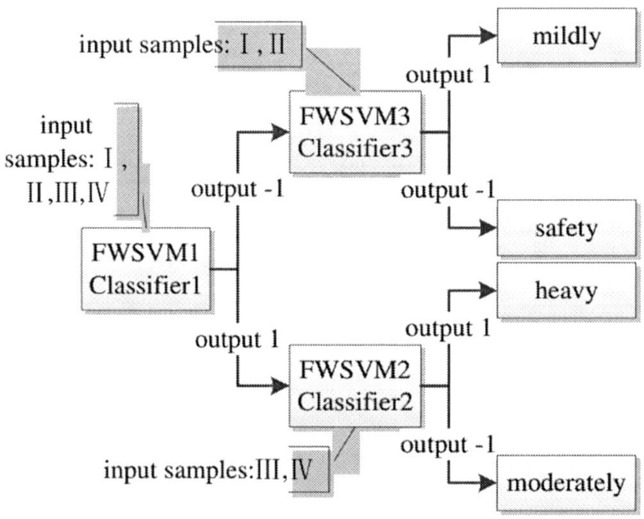

Figure 3. the multi-class classification based on FWSVM.

DOI: 10.1051/matecconf/201711901039

63 training samples and 21 test samples are used to the FWSVM1 Classifier 1. The LIBSVM software is also used to classify the samples. The information about the FWSVM1and SVM1, as shown in Fig.4 and Fig.5.

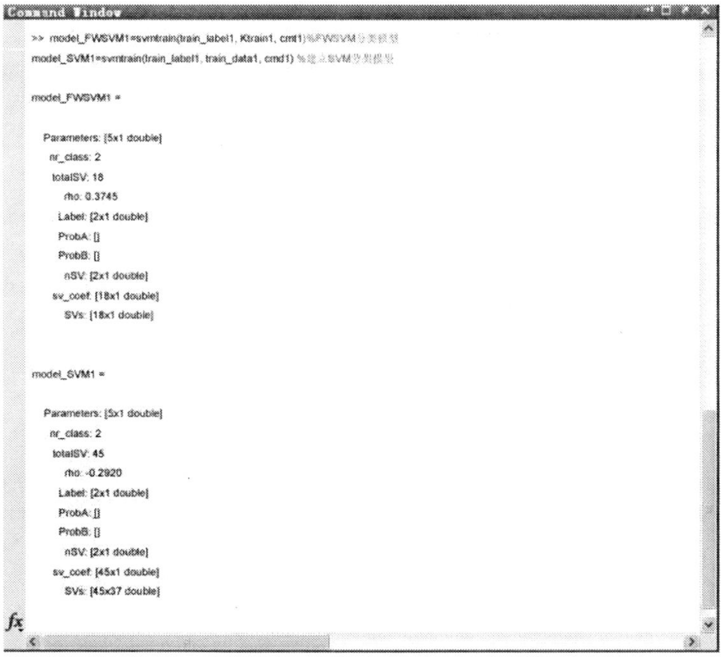

Figure.4 The parameter information about FWSVM1and SVM1

Figure.5 The prediction of FWSVM1

Similarly, additional information about FWSVM and SVM can be calculated. Table 2 is a comparison table of FWSVM and SVM.

Table 2. Comparison table of FWSVM and SVM.

Classifier	Test Samples	Actual Crisis Level	C-SVM Output	classification accuracy of C-SVM	C-SVM Support Vectors Number	FWSVM Output	Classification Accuracy of FWSVM	FWSVM Support Vectors Number
classifier 1	sc2,sc5,sc6,sc4,sc7,sc8,sc10,sc12,sc31,sc33,sc38,sc39,sc42,sc46,sc53,sc56,sc58,sc59,sc61,sc62,sc63	II, IV, III, I, I, I, III, II, III, II, I, II, I, II, I, II, III, IV, III, II, II	-1,1,1,-1,-1,-1,1,1,1,-1,-1,1,-1,-1,1,1,1,1,-1,-1	80.95%	45	-1,1,1,-1,-1,-1,1,-1,1,-1,-1,-1,-1,-1,1,1,1,1,-1,-1	95.24%	18
classifier 2	sc31,sc58,sc59,sc61,sc80,sc81,sc84,	III, III, IV, III, III, IV, III	-1,1,1,-1,-1,1,-1	85.71%	21	-1,-1,1,-1,-1,1,-1,	100%	15
classifier 3	sc33,sc38,sc39,sc42,sc46,sc53,sc56,sc62,sc63	II, I, II, I, II, I, II, II, II,	1,-1,1,-1,-1,-1,1,-1,1	77.78%	17	1,-1,1,-1,-1,-1,1,1,1	88.89%	13

The predict results suggest that the generalization ability of FWSV is superior to C-SVM. It's forecast accuracy is more accurate and than C-SVM,and the number of support vectors is decreased. This could mean that,by compressing geometric space,it has the performance of distinguishing attribute weight,which can improve it's classification capacity.

4 An application example on the crisis early warning of the quality of supply chain based on RS-FWSVM

An automobile industry supply chain ,where is marked as A ,and a textile business supply chain, where is marked as B,are selected to make some empirical analysis. Through investigation and researchnand according to RS-FWSVM model , then the crisis levelof the quality of supply chain is determined.

Data and information is obtained through conducting a number of high-level interviews with A and B. The final data of evaluating indexes as shown in table 3.

Table 3. The evaluation data of the quality of supply chain

Evaluation Indicators	A	B	Mean Value	Evaluation Indicators	A	B	Mean Value
nature disaster	6	2	3.11	the supply chain structure rationality	2	6	3.51
macro-economic situation	5	4	3.76	the supply chain structure stability	2	1	3.61
political stability and government intervention	2	4	3.5	the influence and control ability of the core enterprise	2	3	3.11
the impact degree of industry	5	4	3.75	emergency response system of supply-chain risk	2	4	3.47
turnover of inventory	2	3	3.87	overall labour productivity	2	4	3.37
turnover of account receivable	2	4	3.7	processing time of customer complaint	2	3	2.84
return on equity	3	3	3.59	customer response time	2	3	3.21
sales growth	4	3	3.47	income ratio of new product(service)	2	3	3.53
quick ratio	3	4	3.93	training expenditures ratio	4	6	4.27
asset-liability ratio	5	4	3.31	the equitable degree of the value distribution among node enterprises	2	3	3.49

Evaluation Indicators	A	B	Mean Value	Evaluation Indicators	A	B	Mean Value
the timely information transfer rate among node enterprises	1	3	3.2	the coordination of supply chain culture	2	3	3.46
the accurately information transfer rate among node enterprises	2	3	3.1	strategic alignment among node enterprises	2	4	3.41
the frequency of informational communication among node enterprises	2	3	3.3	the coordination among node enterprises	2	3	3.23
not strictly inventory control	2	4	3.63	core enterprise's market share	6	3	3.4
risk in transit	2	2	3.3	core enterprise's customer satisfaction	3	3	3.19
order disposal failure	2	1	2.91	customer-keeping ratio	3	2	3.1
the suppliers's on time delivery(OTD)	1	4	2.8	inexactly market forecast	4	3	3.27
the suppliers's quality of products	1	3	2.8	perfect order fill rate	3	4	2.91
key suppliers's bankruptcy and losses	1	2	3.17				

Inputting the data from A and B into FWSVM1 classifier, the output is -1 and 1 respectively, indicating that the crisis level of A is safety or mildly early-warning and B is moderately early-warning or heavy early-warning. When the data from A fed into the FWSVM3 classifier, the result is 1, so the quality of A's supply chain in the mild warning state.Then inputting the data from B into FWSVM2 classifier, the output is -1, so company B is moderately early-warning. The management personnel of the company A and B confirms that the results are consistent with actual situation , that the method and model has a certain practicality and feasibility.

According to the former research on preventing and controlling the crisis of the quality of the supply chain, A should adopt measures to focus in the incentives, paying close attention to deviate from normal levels of early warning indicators. B should take measure to monitor and prevent crisis, to monitor crisis incentives, and take corresponding measures to reduce crisis level of the quality of the supply chain.

Although the crisis level of the quality of the supply chain of B company is not high, can not rule out the individual indicators in a high position, such as the timely information transfer rate among node enterprises, not strictly inventory control, order disposal failure, training expenditures ratio and other indicators. They should be promptly and efficiently controlled by the preventive measures.Otherwise, it will lead to supply risks and damages the quality of the supply chain. Which will bring a huge economic loss to these supply chain enterprises. B company can establish an information sharing platform with suppliers to strengthen the communication among supply chain members, and heighten operation efficiency.

At the same time, supply chain enterprises can establish the information system to record the root cause of the crisis and propose solutions. We can effectively avoid the occurrence of the supply chain quality crisis by settitng up the mechanism of preventing, pre-alarming pre-controlling of crisis. The crisis can't be avoided but can be controlled. Effective crisis prevention can reduce the possible risks. Positive crisis reaction can minimize the loss the crisis may cause.

5 Conclusion

RS-FWSVM model is put forward based on the analysis of the advantages and disadvantages of the RS and SVM, and the application flow of this model in the crisis early warning of the quality of the supply chain is studied.The reduced index system of the quality of the supply chain and index weights are obtained based on the rough set,Which simplifying the sample space under the condition of good recognition capability of samples. The improved method takes into account the differences resulting from the classification of indexes and gives different weights to them. It is in conformity with authentic situations.

Acknowledgement

This paper is supported by young and middle-aged teachers education scientific research project in fujian province(No.JAS160713) and project at the university level(No. FWX15012) .

References

1. X.D. Tang, The strategy analysis of the bullwhip effect in product delivery in the supply chain, China Economist, **5**, 46-47 (2010)
2. Z. Pawlak, Rough sets, International Journal of Computer and Information Sciences, **11**, 341-356 (1982)
3. C. Cortes and V. Vapnik, Support Vector net-works, Machine Learning, **20**, 273-297 (1995)
4. Y.L. Wu, M.L. Liu, and X.Q. Li, A research on risk evaluation of express logistics service based on RSSVM, Logistics Sci-Tech, **37**, 11-14 (2014)
5. X.L. Zhang, D.P. Wang, and X.Z. Zhao, Dynamic supply chain partner selection research based on RS and SVM, Mathematics in Practice and Theory, **44**, 1-6 (2014)
6. J.F. Wang, X.Y. Sheng, X.Q. Zhai, and L.J. Feng, Capability evaluation of coal mine disaster emergency logistics based on rough set and support vector machine, Safety in Coal Mines, **45**, 237-239 (2014)
7. B. Sun, S.J. Song, and C. Wu, A new algorithm of support vector machine based on weighted feature, in *Proc. 8th Conf. International Conference on Machine Learning and Cybernetics*, Baoding, China, 1616-1620 (2009)
8. W.X. Zhang and G.F. Fang, Uncertain decision-making based on rough set, Beijing, China: Tsinghua University Press, 23 (2006)

Research on the path of synergetic development between logistics industry and the leading industry in Fujian province

Qiaoping Mei[a], Qi Jiang and Li Yang

Department of Logistics, Fuzhou University of International Studies and Trade, Fuzhou 350202, Fujian, China

Abstract. Logistics industry is closely related to other industry, the synergetic development between logistics industry and leading industry is benefit for leading industry to reduce costs,improve efficiency and promote industrial upgrading. This book analyzes the association of logistics industry and leading industry on the basis of the input-output tables of Fujian and proposes advises to promote their synergetic development based on the calculation.

1 Introduction

During the period of 11th Five-Year Plan Fujian adjust the industrial structure and initially form large-scale industrial clusters taking electronics, machinery and petrochemical as leading industry. The added value of the leading industry in 2010 is 220.432 billion yuan, increase 26.3% and elevate 14.9 percent over the previous year, accounting for 36.4% of the above-scale industrial added value, The three leading industries realize a total profit of 45.832 billion yuan, up 56.9% over the previous year and faster than the average level of the province , Contribution rate to the profit growth of above scale industries is 37.8%.Obviously,the economic development of Fujian mainly depends on the leading industries. As a Productive service industry, Logistics industry is the foundation of the national economy with the effect of creating jobs, promoting production, stimulating consumption and promoting industrial restructuring, accelerating the transformation of the mode of economic development and enhance the competitiveness of the national economy, logistics services an link for others, the efficiency and effectiveness of logistics operations is an important manifestation of the competitiveness of leading industry. Domestic scholars have many research about the logistics industry and related industries, for example Li Guanlin (2001) [1] analyzes the input-output of China's logistics industry based on 1997 statistics, Jun Ruan et al. (2007) [2] uses the input-output method to analyze the modern logistics industry of Fujian Province, and pointes out the strategy of further development of the logistics , Wang Yan (2010) [3] summarizes the characteristics and trend of China's logistics industry from the nature of the logistics industry.

This paper studies the association analysis between logistics industry and the leading industry of Fujian province based on the 2007 statistics, and put forward countermeasures about promoting the linkage development of logistics industry and the leading industry in Fujian province according to the analysis results, which is the innovation of this paper. In addition, the three leading industry plays a great contribution to the development of economy, the three leading industries of Fujian province have been formed, but the development of the logistics industry still need coordination, so it has great

[a] Corresponding author : 460903976@qq.com

theoretical and practical significance to study the linkage development path of logistics industry and the leading industry of Fujian province.

2 Effect analysis between logistics industry and leading industry in Fujian

Industrial linkage refers to technical economic ties of the connection link of various inputs and output products between industries, industrial correlation analysis is called input-output analysis method (referred to input-output method).Input-output analysis analyzes the input-output relationships of the various parts of the economic system (as an industrial sector production units or consumption units, industries, products, etc.)

Based on the 2007 input-output tables of Fujian Province , this paper merges the railway transportation industry, road transportation industry, maritime transportation industry, air transportation industry, handling and other transportation services, etc. into logistics, While the communications equipment manufacturing, computer manufacturing, electronic components manufacturing industry, home audio-visual equipment are consolidated into electronic information manufacturing, metal processing machinery manufacturing, other general equipment manufacturing, rail transportation equipment manufacturing, automobile manufacturing, ships and floating equipment manufacturing, electrical machinery manufacturing are regarded as machinery and equipment manufacturing.

Oil and natural gas industry, petroleum and nuclear fuel processing industry, special chemical products manufacturing, chemical fiber manufacturing are merged into petrochemical industry. Then we calculated on the basis of the divisions.

2.1 Direct consumption coefficient analysis

In the basic flow table, provided xij is the total output in the production process, $a_{ij} = \frac{x_{ij}}{X_J}$ is called direct consumption coefficient, which represents producing unit product of j industrial sector need value of direct product consumption of the i industrial sector. Suppose the national economy have n sectors, the direct consumption coefficient of the industries a_{ij} is composed by n×n order matrix, which we denoted A.

Table 1. Direct consumption coefficient between Logistics and leading industries of Fujian Province in 2007.

	Electronic information industry	**Equipment manufacturing**	**Petrochemical industry**	**Logistics industry**
direct consumption coefficient of Logistics to other industries	0.013726	0.015120	0.016095	0.056631
direct consumption coefficient of other industries to Logistics	0.036509	0.042261	0.027066	0.056631

According to the input-output tables of Fujian Province in 2007.

The data in Table 1 shows that the logistics industry needs to consume itself 566.31 yuan,petrochemical 160.95 yuan,machinery and equipment manufacturing 151.20 yuan, electronic information industry 137.26 yuan,when the total output of the logistics industry increase ten thousand yuan,indicating the development of the logistics industry has a very significant effect in promoting leading industries.The total output of electronics, machinery, petrochemicals respectively consume 365.09,422.61,270.66 when they increase ten thousand yuan,in fact, the direct consumption of each industry on logistics is logistics costs that they produce production units product.Obviously the logistics costs of equipment manufacturing is the highest, this means that logistics costs account for a large proportion of the cost of the equipment manufacturing, the main reason is that logistics is one of the basic indemnification for normal production of equipment manufacturing enterprises, and it is an important part throughout the entire process of production and sales of businesses.The logistics in procurement, production, sales, recycling and waste disposal of equipment manufacturing are difficult to coordinate,resulting in a significant cost of waste.The logistics costs of electronic information industry take second place, the reason is that the product value of the electronic information manufacturing is relatively high, the logistics requirements is also higher, also the logistics of electronics manufacturing starts relatively late, which indicating the province should improve the information degree of logistics and improve internal,external information systems.Promoting the upgrading of the logistics management level of Information manufacturing will have a very important significance by using the specialization of logistics.

2.2 Influence coefficient and sensitivity coefficient analysis

We have an industry that is the extent of the impact of other industries called sensitivity, and the impact of an industry level is called the influence of other industries. Productive activities of an industrial sector will inevitably affect and be affected other productive activities by correlation between industrial sectors.We call the extent of the impact of a industry by other industries sensitivity,and call the impact of an industry on other industries influence power.

According to Leontief inverse matrix, let S_i denote the Sensitivity coefficient of i industry sector, T_j indicate the influence coefficient of j industry sector, then:

$$S_i = \frac{\sum_{i=1}^{n} \overline{b_{ij}}}{\frac{1}{n}\sum_{i=1}^{n}\sum_{j=1}^{n} \overline{b_{ij}}} \quad (i, j = 1, 2, \cdots, n)$$

Industry sensitivity reflects the sensing capability of a industry on various industrial sectors of the national economy. This sensing capability represent the direct and indirect pull capability of the national economy to the industry.The higher the industry sensitivity and sensitivity coefficient are, the greater the pull effect from the national economy,the industry will have the nature of Basic industry and bottleneck industry.

Industrial influence reflects the influence ability of the change of final product of an industry on the total output of the entire national economy. This influence ability manifest the direct and indirect promotion of capacity of the industry on national economic development.The higher the influence and influence coefficient are,the greater driving force the industry to the national economic development will be,developing these industries will make "multiplier" effect on economic growth.

Table 2. Influence coefficient and sensitivity coefficient between logistics and the leading industries of Fujian in 2007.

	Electronic information industry	Equipment manufacturing	Petrochemical industry	Logistics
influence coefficient	1.717459	0.586473	0.864848	1.44181
sensitivity coefficient	1.676967	0.655711	1.354258	1.70247

According to the input-output tables of Fujian Province in 2007.

Table 2 shows the sensitivity coefficient of logistics is 1.70247, indicating Pull effect of the leading industries on logistics in Fujian is large. influence coefficient of logistics is only 1.44181, higher than the average level of society, so logistics has great function in promoting development of regional economic,We should give priority to develop it.We can known that sensitivity coefficient of logistics to leading industries is greater than industrial influence by contrast,namely the pull of the development of leading industries to logistics is greater than the promotion logistics to leading industries,so Fujian should better develop leading industries and take some initiative measures.

2.3 Intermediate demand rate and intermediate inputs rate analysis

Intermediate demand rate is the proportion of an industry's intermediate demand to total demand, the higher the intermediate demand rate is,the industry will more have the nature of providing the intermediate product.Intermediate input rate is the proportion of an industry's intermediate inputs to the total investment, reflecting the size added value of industry.

Table 3. Intermediate input rate and Intermediate demand rate between industry and leading industries of Fujian in 2005.

	Electronic information industry	Equipment manufacturing	Petrochemical industry	Logistics
Intermediate input rate	0.507531	0.257896	0.316928	0.257087
Intermediate demand rate	0.466352	0.224018	0.572154	0.692625

According to the input-output tables of Fujian Province in 2007.

The data in Table 3 shows that the intermediate input rate of industry in Fujian to leading industries is 25.7087%,which indicated that intermediate demand rate of production of the logistics to other industrial product is not high, and the proportion of added value created by logistics is higher.So it is obviously to find that the logistics of Fujian plays an important basic role in the process of regional economic development.Intermediate demand rate is 69.2625%, it indicate that logistics of Fujian has a characteristics of being a strong basic industry, that is to say that as a service industry,logistics is to provide services for the development of leading industries.The development of various industries highly need logistics to provide services for them.

3 Path analysis of promoting interactive development of logistics and leading industries of Fujian

Through the input-output analysis,we can see the logistics in Fujian has a certain degree of development, logistics and leading industry contact closely,the development of leading industry needs coordination of logistics, in turn leading industry will promote the development of logistics.Leading Industry make outstanding contributions to the economic development of Fujian, measures to promote interactive development of modern logistics and leading industries are as follows:

3.1 Common path of interactive development of the logistics industry and Leading Industry in Fujian

3.1.1 Develop the third-party logistics of three leading industries vigorously

Electronic information industry is a technology and capital intensive industry with various kinds of product and fast upgrading. Production and distribution of product require a higher level of logistics services, while the supply chain of electronic information products require a higher level of supply logistics of raw materials,such as sorting, packaging,reasonable transportation stowage, accuracy and timeliness warehousing, etc. , which resulted in high logistics cost of electronic information industry.

Logistics is one of the basic indemnity to normal production of equipment manufacturing enterprises which is an important part of the entire production and sale process throughout business.Procurement, production, sales, recycling and waste disposal of equipment manufacturing industry are difficult to coordinate effective, resulting in tremendous waste of cost.Direct consumption coefficient of logistics of Fujian to petrochemical industry is relatively low and their correlation is not great,the reason is that development logistics outsourcing of domestic petrochemical enterprises is slow,most enterprises not only are engaged in production, sales business, but also has their own fleet, warehouse, which restrict the development of third party logistics.

Today, most companies have a serious shortage of resources, improving resource utilization become a prerequisite for business survival. As market competition become intensified and social division is refined,for related businesses of leading industry,using third-party logistics services can reduce costs, improve customer level, increase flexibility, improve production efficiency, make business owner focusing on and improve the professional level of logistics, thus developing third-party logistics is an inevitable trend for related enterprises of three leading industries.

3.1.2 Enhance the ability of logistics enterprises

Developing the leading industries of Fujian should enhance the ability of logistics enterprises.The direct consumption coefficient of logistics in Fujian is the highest, indicating its low operational efficiency,therefore strive to enhance the capabilities of logistics enterprises in Fujian is necessary.First, we should guide and support the logistics companies update logistics management concepts, and actively penetrate the professional field, extending logistics services; improve service function, and innovative service models by high quality, efficient, low-cost objectives and careful studying the production and operation mode of businesses and logistics requirements;promote the innovation of applied technology and mechanization, automation, standardization of logistics facilities,constantly optimize the service and information network to improve service levels.Second, focus on fostering some competitive large-scale logistics enterprises of higher level of service.Merger and restructure small businesses in the industry with the principle of optimal allocation of resources,expand the concentration ratio of logistics market and improve operational efficiency of the logistics.Third, encourage and guide the logistics enterprises in Fujian to participate in the comprehensive assessment of the national A-class logistics enterprises.Implement brand strategy to enhance the operation and management level of logistics enterprise.Fourth, encourage and guide logistics enterprises to actively expand the international logistics channels, carry out high-end logistics service based on multinational business, expanding service area, and gradually provide logistics information planning ,management consulting and other value-added services in order to better serve for the leading industries.

3.2 Specific path of interactive development between Logistics and Leading industries in Fujian

3.2.1 Electronic information manufacturing enterprise remain to identify core competitiveness and choose service providers correctly.

Sensitivity coefficient of logistics on electronic information manufacturing is less than industry influence, namely the promotion of logistics on electronic information manufacturing is greater than the pull of electronic information manufacturing on logistics, so Fujian should better develop electronic information manufacturing,the development level of logistics largely depends on the aggregation and release of logistics demand of manufacturing.In recent years,logistics of manufacturing has achieved rapid development,concepts of modern logistics and management techniques gradually popularize,but it is still a weak link of the development of service industry.So we must pay attention to synergetic development of manufacturing and logistics,manufacturing must identify their core competencies and objectives and conditions of logistics outsourcing,choose the right logistics service providers.

DOI: 10.1051/matecconf/201711901040

3.2.2 Improve application level of information systems of equipment manufacturers and related logistics businesses.

Partition coefficient reflect the product,s proportion and flow logistics sectors, machinery and equipment manufacturing is only 4.2427%, indicating that there are 4.2427% logistics service flows to machinery and equipment manufacturing,so we should expand investment of machinery and equipment manufacturing in logistics.The tools that produced for other services industries by machinery and equipment manufacturing have large universality and pertinence, there is no uniform standard that the same type of device will has large difference because of different manufacturers, the characteristic of mechanized equipment can not be destined to have a great amount, so mechanical parts of mechanical equipment are difficult to form the bulk, which lead to high logistics requirements, the corresponding logistics companies can build logistics information system to improve their reaction capacity and provide better services for enterprises.In addition to the establishment of the system by business between the network and the supplier or the customer has no specific business integration planning, Information level of enterprise in the area of product development,materials management,financial management,office automation is high,while it is low in area of supply chain management, E-commerce, customer relationship management,which is the reason that prevent interactive development between logistics and machinery and equipment manufacturing.

3.2.3 Provide customized logistics solutions by improving service capacity of logistics enterprises

Correlation between logistics and petrochemical industry is not large, mainly because the development of logistics enterprises in petrochemical Industry is slow, wihch is unable to meet the logistics needs of petrochemical enterprises, the development of logistics outsourcing of petrochemical enterprises in Fujian is slow, most businesses are both engaged in the production, sales, and has its own fleet, warehouse, restricted the pace of development of this third-party logistics industry.While self-operate logistics consume investment of enterprises, but some had to operate logistics themselves from the safety and security point.So logistics enterprises should provide customized logistics services according to the requirements of petrochemical enterprises, and constantly improve their service capacity to contribute to the development of petrochemical industry.

References

1. G. Li, Input-output analysis of China logistics, Chinese Circulation Economy, **6**, 15-18 (2001)
2. J. Ruan, Z.Y. Zheng, Input-output analysis of modern logistics in Fujian, Statistics & Information Forum, **3**, 72-76 (2007)
3. Y. Wang, Industrial nature and trend of China, s Logistics-Based on 5 point and 12 sectors input-output table, Port Economy, **5**, 31-33 (2010)
4. M. Liao, Input-output and its extension analysis, Beijing, China: The Capital Economic and Trade University press, 30-50 (2009)

Research on the effect of Fujian free trade zone to Xiamen international shipping center

Li Zhang[1,a], Jie Zhu[2] and Dan Liu[3]

[1]*Department of Logistics, Fuzhou University of International Studies and Trade, Fuzhou, China*
[2]*Department of Scientific Research, Fuzhou University of International Studies and Trade, Fuzhou, China*
[3]*Department of Logistics, Fuzhou University of International Studies and Trade, Fuzhou, China*

Abstract. China free trade zone is an important carrier of the Belt and Road to achieve the internal and external linkage. There are four free trade zones in china. They are in Shanghai, Guangdong, Tianjin and Fujian. Fujian free trade zone supplies greater openness and favorable policies for Xiamen international shipping center, and dispenses with the formalities, reduces shipping costs, and realizes the territory of the commissioner of policies. So it will be in favor of FDI and benefited for employment-generating. It will boost the economy and increase international trade. Firstly this paper found out the influence factors of Fujian free trade zone to Xiamen international shipping center through the investigation. Then index system of these influencing factors was designed. And then the index system was designed into a questionnaire. Relevant experts filled the questionnaire. Then the main influence factors and secondary influence factors were been found. The weights of each index were got by using the method of analytic hierarchy. At last, development strategies of Xiamen international shipping center were been proposed.

1 Introduction

Prime Minister Keqiang Li proposed that set a free trade zone in Fujian in2014. Fujian free trade zone has a total area of 118.04 square meters including Fuzhou, Pingtan, Xiamen three places. With the introduction of preferential policies in Fujian Province, Fujian free trade zone will attract more consumer groups and more foreign goods to the Fujian free trade zone and transit. International Trade and financial flows will be freer in Fujian free trade zone [1]. Xiamen international shipping center will get support for the volume of trade, national policy, infrastructure, financial services and other aspects.

Fujian free trade zone has played a huge role in supporting Xiamen international shipping center. This paper focuses on the impact of Fujian free trade zone on Xiamen international shipping center. Firstly, the index system of Fujian free trade zone to Xiamen international shipping center was established. Three first level indicators were constructed. They are shipping level, shipping services and shipping ecology. Three first level indicators contain 17 second grade indexes. Secondly, the index system was

[a] Corresponding author : 415613585@qq.com

set up into a questionnaire which the experts filled in. The data was collected by expert survey. With the AHP method of data processing, the weights of the indicators were got. Finally, suggestions on the construction of Xiamen international shipping center were put forward.

2 Methodology

2.1 Analytic hierarchy process theory

Analytic hierarchy process is a kind of hierarchical weight decision analysis method which applies network system theory and multi objective comprehensive evaluation method. In this paper, the AHP hierarchy model is built in a progressive system structure with M layer. There is only one factor in the top level. The factors of the next layer belong to the upper layer. The different factors in the same layer are parallel. There is no direct relationship between the two adjacent factors [2]. In any one of the index system, the meaning of each factor is different and the social problems described by the index system are different. So the total index values are a relation of weighted sum, but not simply add the various indicators of the system factors. Specific models such as formula1:

$$S = \sum_{i=1}^{n} W_i f_i(I_i) \quad i = 1,2,\ldots\ldots, \ n \tag{1}$$

Among them, $f_i(I_i)$ is a function of I_i. W_i represents the weight of each indicator.

2.2 Steps of analytic hierarchy process

2.2.1 Structure judgment matrix

The importance degree of each factor can be got by the comparison between two factors of each level. According to the degree of importance, the level of each factor can be rated. The comparison result is the judgment matrix. Saaty gave 9 importance levels and their assignment as shown in Table 1. B indicates the judgment matrix b_{ij} indicates the comparison between the two factors. b_{ij} is identified based on expert opinion and system analyst repeated studies.

Table 1. The meaning of importance scale.

quantized value	comparison between the two factors
1	equally important
3	slightly important
5	extremely important
7	Strongly important
9	crucial importance
2，4，6，8	The middle value of two adjacent assignment

2.2.2 Calculating Weighting Vector

For the constructing judgment matrix, the eigenvectors which matched largest eigenvalueis are calculated. The relative weight of each factor in the index system is obtained after normalized the eigenvectors. Take solving the judgment matrix M as an example:

$$HW = \lambda W \tag{2}$$

Among them, M represents the judgment matrix, W represents the eigenvectors. λ_{max} and W are solved out and W is normalized. Then the weights of each factor are got. Weight indicates the relative importance of a low level to a higher level.

2.2.3 Consistency check

When applying analytic hierarchy process, it is very important to keep the consistency of judgment thinking. If the judgment matrix satisfies the formula 3, 4 and 5 , the judgment matrix is consistent.

$$b_{ii} = 1 \tag{3}$$

$$b_{ji} = 1/b_{ij} \tag{4}$$

$$b_{ij} = b_{ik} / b_{jk} \tag{5}$$

Among them, i, j, $k = 1,2,\ldots\ldots, n$

The consistency index of judgment matrix:

$$\text{CI} = \frac{\lambda_{max} - n}{n-1} \tag{6}$$

The greater the value of the consistency index CI is, the more inconsistent the judgment matrix is. The smaller the value of the consistency index CI is, the more consistent the judgment matrix is. For multi-step judgment matrix, it can be tested by used the average random consistency index RI. The average random consistency index which is gotten by 1-10 step matrix calculation after 1000 times is given in the table 2.

Table 2. The average random consistency index.

n	1	2	3	4	5	6	7	8	9	10
RI	0	0	0.58	0.90	1.12	1.24	1.32	1.41	1.46	1.49

When $n \le 2$, judgment matrix is always consistent. The ratio of judgment matrix consistency index CI and average random consistency index RI is called the random consistency ratio CR.

$$CR = \frac{CI}{RI} \tag{7}$$

When $CR < 0.1$, it is believed that the consistency of judgment matrix can be accepted. When $CR > 0.1$, the consistency cannot be accepted. It is necessary to modify the judgment matrix in order to achieve acceptable consistency. In order to make the evaluation result of the index system is correct, the consistency check of each judgment matrix must be carried out in the analytic hierarchy process.

3 Data analysis and results

3.1 Impact index system of Fujian Free Trade Zone on Xiamen international shipping center

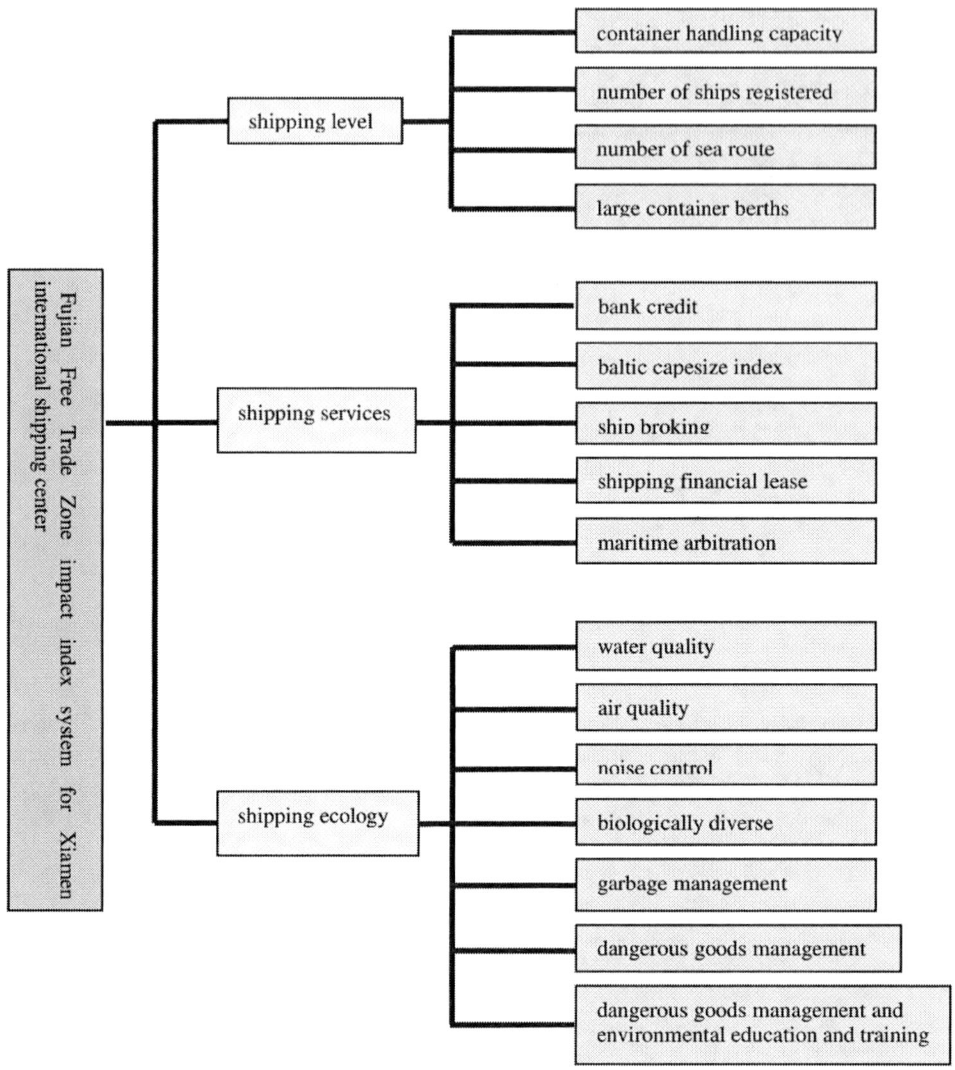

Figure 1. Fujian Free Trade Zone impact index system for Xiamen international shipping center.

Xiamen international shipping center covers an area of 24.41 square kilometers. It is Xiamen port hub, shipping and trading center and service enterprise gathering center. The geographical position of Xiamen port is excellent. It is port of navigation routes which the north and south route to the Yangtze River Delta and the Pearl River Delta must pass through in our country. The west coast of Taiwan Strait can be arrived direct from Xiamen port. Compared with other ports, Xiamen port spents less time, navigates the shortest route distance and costs the lowest money. The combination of 200 thousand ton container ship and 150 thousand ton container ship can full tidal and two-way navigate in the whole channel of Xiamen port. At the same time, it can meet the two-way navigation requirement of the 200 thousand ton container ships which have a draught of 15.5 metres.

According to "the fourth generation international shipping center" innovation mode, the competitiveness of international shipping center has been assessed and evaluated. Based on the Global international Shipping-Center Index, this paper established a two level index system to study the influence of Fujian Free Trade Zone on Xiamen international shipping center. There are 3 one class indexes. And one class index sets target of 16 2 class. One class indexes contain shipping level, shipping services and shipping ecology. Shipping level sets target of 4 2 class. They are container handling capacity, number of ships registered, number of sea route and large container berths. Shipping services set target of 5 2 class. They are bank credit, baltic capesize index, ship broking, shipping financial lease and maritime arbitration. Shipping ecology sets target of 7 2 class. They are water quality, air quality, noise control, biologically diverse, garbage management, dangerous goods management and environmental education and training. It is shown in figure 1.

3.2 Applying the method of AHP to confirm the index weight

According to impact index system of Fujian Free Trade Zone on Xiamen international shipping center which is constructed above, expert questionnaire was compiled. Based on the 9 importance rating method given by Saaty, experts gave each index scoring. After getting the experimental data, several judgment matrixes were constructed. Judgment matrixes were calculated by using the yaaph software of analytic hierarchy process, and the weight of each index is obtained. Then whether the judgment matrix satisfied the consistency index was tested. The judgment matrix and weight of one class indexes are shown in table 3.

Table 3. The judgment matrix and weight of one class indexes.

index	Shipping level	Shipping services	Shipping ecology	weight W_i
Shipping level	1.0000	2	3	0.5396
Shipping services	0.5000	1.0000	2	0.2970
Shipping ecology	0.3333	0.5000	1.0000	0.1634

The consistency ratio of the one class index judgment matrix is 0.0088. So the judgment matrix passed the consistency tests. We can draw a conclusion from the obtained weight values that shipping level is the biggest impact influence of Fujian Free Trade Zone on Xiamen international shipping center, followed by the shipping services, and finally the shipping ecology. The judgment matrix and weight of shipping level index are shown in table 4.

Table 4. The judgment matrix and weight of shipping level index.

index	container handling capacity	number of ships registered	number of sea route	large container berths	weight W_i
container handling capacity	1.0000	3.0000	4.0000	2.0000	0.4747
number of ships registered	0.3333	1.0000	3.0000	3.0000	0.2551
number of sea route	0.2500	0.5000	1.0000	0.5000	0.1072
large container berths	0.5000	0.3333	2.0000	1.0000	0.1630

Table 5. The judgment matrix and weight of shipping services index.

index	bank credit	baltic capesize index	ship broking	shipping financial lease	maritime arbitration	weight W_i
bank credit	1.0000	0.2500	1.0000	2.0000	0.3333	0.1160
baltic capesize index	4.0000	1.0000	4.0000	4.0000	2.0000	0.4379
ship broking	1.0000	0.2500	1.0000	2.0000	0.5000	0.1258
shipping financial lease	0.5000	0.2500	0.5000	1.0000	0.5000	0.0830
maritime arbitration	3.0000	0.5000	2.0000	2.0000	1.0000	0.2374

The consistency ratio of the shipping level index judgment matrix is 0.0871. So the judgment matrix passed the consistency tests. We can draw a conclusion from the obtained weight values that container handling capacity is the biggest impact influence in the shipping level index, followed by number of ships registered, large container berths, and finally number of sea route. The judgment matrix and weight of shipping services index are shown in table 5.

The consistency ratio of the shipping services index judgment matrix is 0.0268. So the judgment matrix passed the consistency tests. We can draw a conclusion from the obtained weight values that baltic capesize index is the biggest impact influence in the shipping services index, followed by maritime arbitration, ship broking, bank credit ,and finally shipping financial lease. The judgment matrix and weight of shipping ecology index are shown in table 6.

Table 6. The judgment matrix and weight of shipping ecology index.

index	water quality	air quality	noise control	biologically diverse	garbage management	dangerous goods management	environmental education and training	weight W_i
water quality	1.0000	3.0000	4.0000	3.0000	2.0000	2.0000	5.0000	0.3041
air quality	0.3333	1.0000	2.0000	0.5000	0.3333	0.5000	0.5000	0.0712
noise control	0.2500	0.5000	1.0000	0.3333	0.3333	0.3333	1.0000	0.0551
biologically diverse	0.3333	2.0000	3.0000	1.0000	0.5000	0.3333	2.0000	0.1121
garbage management	0.5000	3.0000	3.0000	2.0000	1.0000	0.5000	3.0000	0.1723
dangerous goods management	0.5000	2.0000	3.0000	3.0000	2.0000	1.0000	4.0000	0.2189
environmental education and training	0.2000	2.0000	1.0000	0.5000	0.3333	0.2500	1.0000	0.0662

The consistency ratio of the shipping ecology index judgment matrix is 0.0406. So the judgment matrix passed the consistency tests. We can draw a conclusion from the obtained weight values that water quality is the biggest impact influence in the shipping ecology index, followed by dangerous goods management, garbage management, biologically diverse, air quality, environmental education and training, and finally noise control.

4 Conclusion

This paper established the impact index system of Fujian free trade zone to Xiamen international shipping center. It can be seen that the important impact factors of Fujian Free Trade Zone on the Xiamen international shipping center are container handling capacity, number of ships registered and baltic capesize index.

Acknowledgement

Funding from Fujian province education scientific research project of the education of young teachers (NO: JAS150703) supported this research.

References

1. S. Liu and Y. Zhu, International Trade, **7**, 87-89 (2015)
2. Q. Zhang, Logistics technology, **34**, 11-14 (2015)

MATEC Web of Conferences **119**, 01042 (2017) DOI: 10.1051/matecconf/201711901042
IMETI 2016

A method of curvature mode difference and wavelet transform to damage detection on space grid structure

Li-Mei Zhang[1,a], Xiang-kun Yang[2] and Wei-ran Liu[1]

[1]*School of Civil Engineering, Hebei University of Science and Technology, Shijiazhuang, Hebei Province, 050018, China*
[2]*NORENDAR International, Shijiazhuang, Hebei Province, China*

Abstract. A method combined curvature mode difference and wavelet transformation was used to determine the damage location of space gird structure. This method considered the curvature mode difference of pre and post damage as the wavelet transform analysis signal. Moreover, these signals were analyzed by wavelet transform using wavelet function db3. As the wavelet transform coefficient occur sudden change, the damage locations of links will be obtained. In this paper, square pyramid space grid structure, orthogonal space truss structure, three way latticed grid and honeycomb triangular pyramid grid structure will be analyzed using this method to identify single damage and multi-damage location. The solutions from numerical simulations show that this method can take effective identification of single damage location and multi-damage location of square pyramid space grid structure by the first mode condition. And this method has versatility to detect damages from different space grid structures.

1 Introduction

Curvature mode (Li 2002, Chen 2007, Ding 2007, VALSAN 2008) is one of the mode parameters, and curvature mode often expressed the signals of dynamics testing. But the signals of curvature mode are continuous, and the influence of noise must be considered. So, these signals are not very sensitive to structural damage. However, the method of wavelet transformation can show the local features of signals and remove noise. Therefore the wavelet transformation was used to discrete these continuous curvature mode signals and eliminate noise. According to these features, the wavelet transformation can be used to detect structural damages (Law 2005, Zhang 2006, Yan 2012). Sun Zengshou (2005) use the method of curvature mode and wavelet transform to identify damage, considered the difference of wavelet transform coefficient of pre and post damage as injury index; this method also was used by Yin Zhixiang (2009)to detect the damage location of spatial latticed structure; Xu Wenjie(2008) used CWT and DWT wavelet mode to test the damage location of spatial latticed structure; Zhang Yue(2010) use the method of wavelet analysis and curvature mode to detect damage of square pyramid grid structure.

Through analysis, the curvature modes difference from pre and post damage were considered as importing signals which were analyzed by db3 wavelet in this paper. This method was used to detect

[a] Corresponding author : zhanglimei168@126.com

the damage location of square pyramid grid structure. The solutions show that this method can locate single and multi damage location more accurately.

2 Theory of damage detection

2.1 Theory of curvature mode

Link damage often occur in local structure, which can change structural local parameter, for example stiffness EI, structural section area, etc.. The change of EI can cause curvature mode change and these changes from pre and post damage can express the damage index to detect damage location(Wang 2012).

Structure vibration equation is written as:

$$M\ddot{v} + C\dot{v} + Kv = f(t) \tag{1}$$

While, M, C, K respectively represent matrix of mass, damp and stiffness, v is vibration displacement, $f(t)$ is the structural exciting load. According to mode theory, the solution of Eq.(1) can be expressed a s:

$$v = \sum_{i=1}^{n} \varphi_i q_i(t) = \Phi q \tag{2}$$

While, ϕ_i is displacement modal vibration shape and$q_i(t)$ is modal coordinates, $\Phi$is modal matrix. Curvature modes are often obtained by difference approximate calculation and can be written as:

$$\varphi_i'' = (\varphi_{i-1} - 2\varphi_i + \varphi_{i+1}) / (l_{i-1,i} l_{i,i+1}) \tag{3}$$

While, subscript i express the ith measuring point; $l_{i\text{-}1,i}$ express distance between measuring point$i$-1and$i$. Then pre and post damage of curvature mode difference $\Delta\phi_{ki}''$ can be expressed as:

$$\Delta\varphi_{ki}'' = \varphi_{ki}''^{u} - \varphi_{ki}''^{d} \tag{4}$$

While, k express the kth mode; $\Delta\phi_{ki}''$ is curvature mode difference of node i with the kth mode; $\phi_{ki}''^{d}$ is curvature mode difference of node i after structural damage; $\phi_{ki}''^{u}$ is curvature mode difference of node i with intact structure.

2.2 Theory of wavelet transformation

The purpose of wavelet transformation is to analysis the response signals by wavelet function. Through analysis, some mutation information was obtained and these information can determine the situation of this structure. In this paper, the curvature mode difference $\Delta\phi_{ki}''$ was considered as input signal S . These signals were analyzed by specific wavelet function $\varphi(t)$ which have two basic parameters scale a and translation distance b . Considering structural discrete characteristics, the wavelet transform coefficients C adopt discrete wavelet transform (DWT). Through analysis, the wavelet transform coefficients C will be gained and this coefficient will be written as follows:

$$C = (W_\varphi S)(a,b) = \langle S, \phi_{a,b} \rangle = \frac{1}{\sqrt{a}} \int_{-\infty}^{+\infty} S\phi\left[\frac{t-b}{a}\right] dt, \quad a > 0 \tag{5}$$

While, $C = \langle W_{\varphi}S \rangle(a,b)$ is wavelet transform coefficient (Zhang 2012), "$\langle\ \rangle$" means inner product. This equation show that the minimal changes of wavelet function $\varphi(t)$ can lead to larger variety of wavelet coefficient C, signal S is variable that change by damage. Therefore, the bigger the wavelet transform coefficient C means the larger the mutation degree of signal, namely damage location. So, with different wavelet function $\varphi(t)$, the analyzing signals will show huge difference. The db3 wavelet belongs to Daubechies group was selected in this paper because of its properties of fast transformation, better orthogonality and vanishing moment(Li 2012).

3 Numerical simulations of four type space grid structures

3.1 Model and damage conditions of four type space grid structures

Element link8 was selected to simulate bars in software ANSYS. Its elastic modulus $E = 2.06e11Pa$, poisson ratio $\mu = 0.3$, section area of upper and lower chord is 0.00005 m^2, web member is 0.000025 m^2, this truss bearing uniform load 1 kN/m^2, using surrounding support and constraint in bottom chord nodes. Square pyramid space grid truss shows in figure 1(unit mm)and this structure has 326 nodes, 1200 bars. Orthogonal space truss structure shows in figure 2 (unit mm)and it has 348 nodes, 1139 bars. Three way latticed grid structure shows in figure 3 (unit mm)and it has 656 nodes, 3096 bars. Honeycomb triangular pyramid grid structure shows in figure 4(unit mm) and it has 390 nodes, 1110 bars. Damage degree and damage conditions list in table 1. Single damage and multi damages all exit. Damage express link element damage and damage was obtained by the reduction of section area. The damage degree of upper chord was 10%, 30% and 50% respectively and other damage degree all considered 10%.

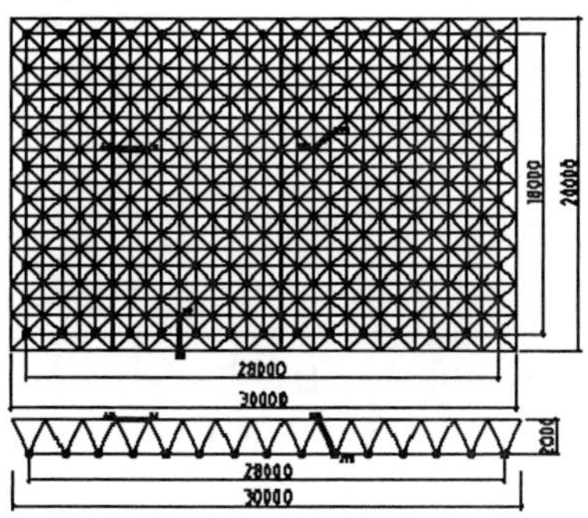

Figure 1. Square pyramid space grid structure.

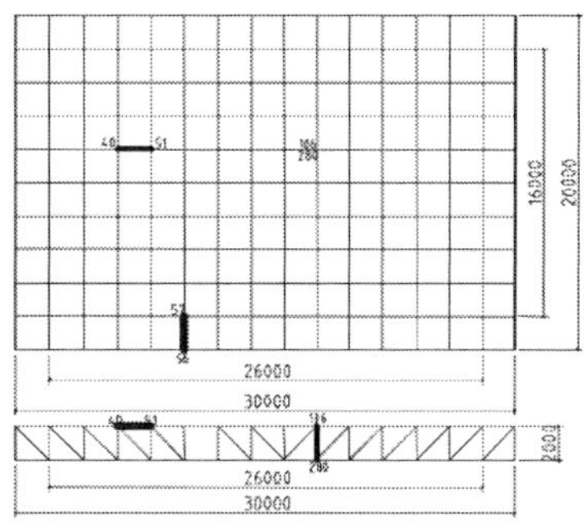

Figure 2. Orthogonal space grid structure.

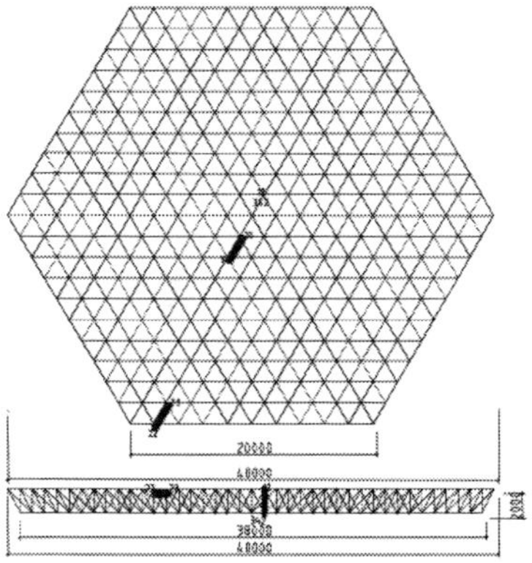

Figure 3. Three way latticed grid structure.

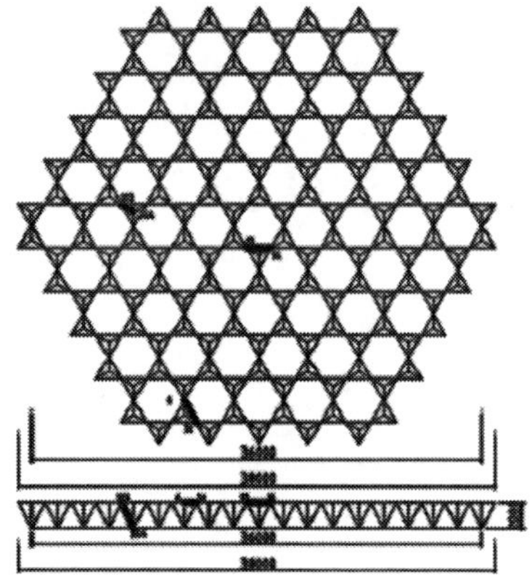

Figure 4. Honeycomb triangular pyramid grid structure.

Table 1. Damage conditions of four type space grid structures.

Structure type	Damage Condition	Damage Member（Node Number）	Damage Degree	Member Type
square pyramid space grid structure	1	51（56—57）	10%	Upper Chord
	2	51（56—57）	30%	Upper Chord
	3	51（56—57）	50%	Upper Chord
	4	985（106—273）	10%	Web Chord
	5	200（40—51）	10%	Middle Mhord
	6	51、200、985	10%	—
orthogonal space truss structure	1	51（56—57）	10%	Upper Chord
	2	51（56—57）	30%	Upper Chord
	3	51（56—57）	50%	Upper Chord
	4	746（106—280）	10%	Web Chord
	5	200（40—51）	10%	Middle Mhord
	6	51、200、746	10%	—
three way latticed grid structure	1	21（22—23）	10%	Upper Chord
	2	21（22—23）	30%	Upper Chord
	3	21（22—23）	50%	Upper Chord
	4	9（9—10）	10%	Middle Mhord
	5	1853（12—342）	10%	Web Chord
	6	9、21、1853	10%	—
honeycomb triangular pyramid grid structure	1	82（31—6）	10%	Upper Chord
	2	82（31—6）	30%	Upper Chord
	3	82（31—6）	50%	Upper Chord
	4	9（13—14）	10%	Middle Chord
	5	1065（183—364）	10%	Web Chord
	6	9、82、1065	10%	—

3.2 DWT coefficient of four type space grid structures

Calculating the first-order curvature mode difference of above condition, then the first scale DWT coefficients of wavelet db3 were obtained and the results were shown in figure 5 to figure 8.

From Fig.5 to fig.8 can be seen, ①Wavelet transform coefficients (DWT) were sensitive to the square pyramid space grid structure, orthogonal space truss structure and three way latticed grid structure, but the honeycomb triangular pyramid grid structure has poor sensitivity to DWT coefficient. ②More great mutations in node of damage member with single damage condition and multi-damage condition were illustrated in these figures. Based on these mutations, the damages can be determined and single damage can be judged easily. ③As damage degree increase, the value of DWT coefficient also increase, so that the size of damage degree can be pre-judgment. ④This method is closely related to the damage location. Damage on web chord and upper chord near edge are sensitive to DWT coefficient, while damage on middle chord in the structure was insensitive relatively.

(a) Condition 1

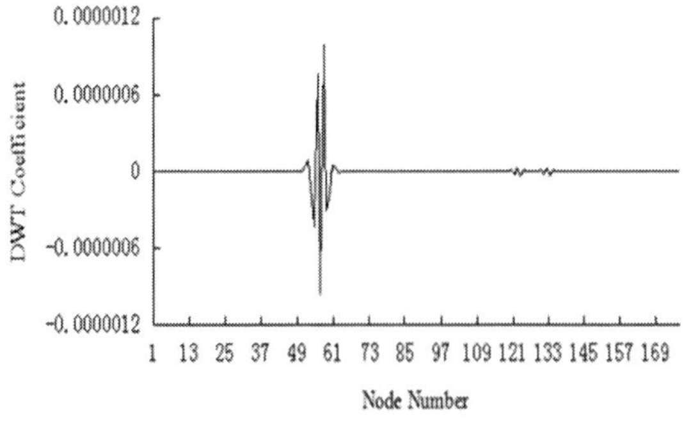

(b) Condition 2

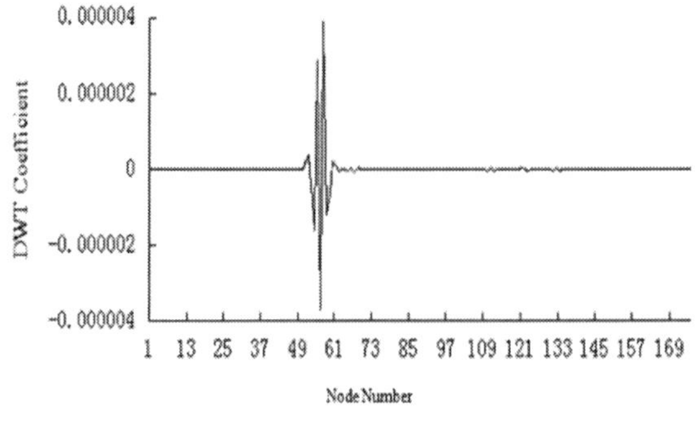

 DOI: 10.1051/matecconf/201711901042

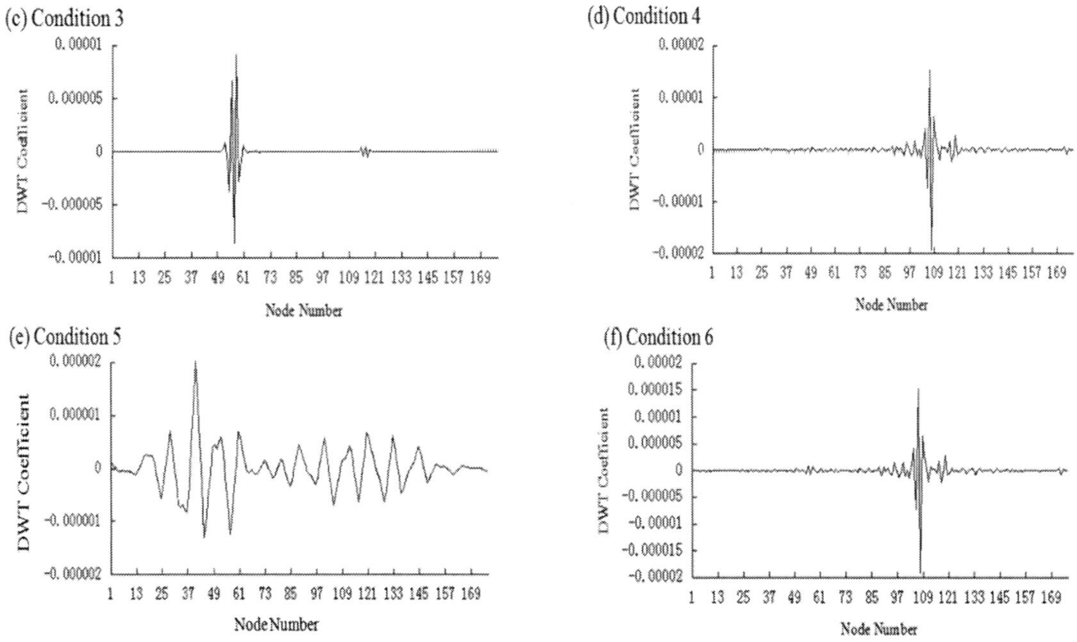

Figure 5. DWT Coefficients of square pyramid space grid structure.

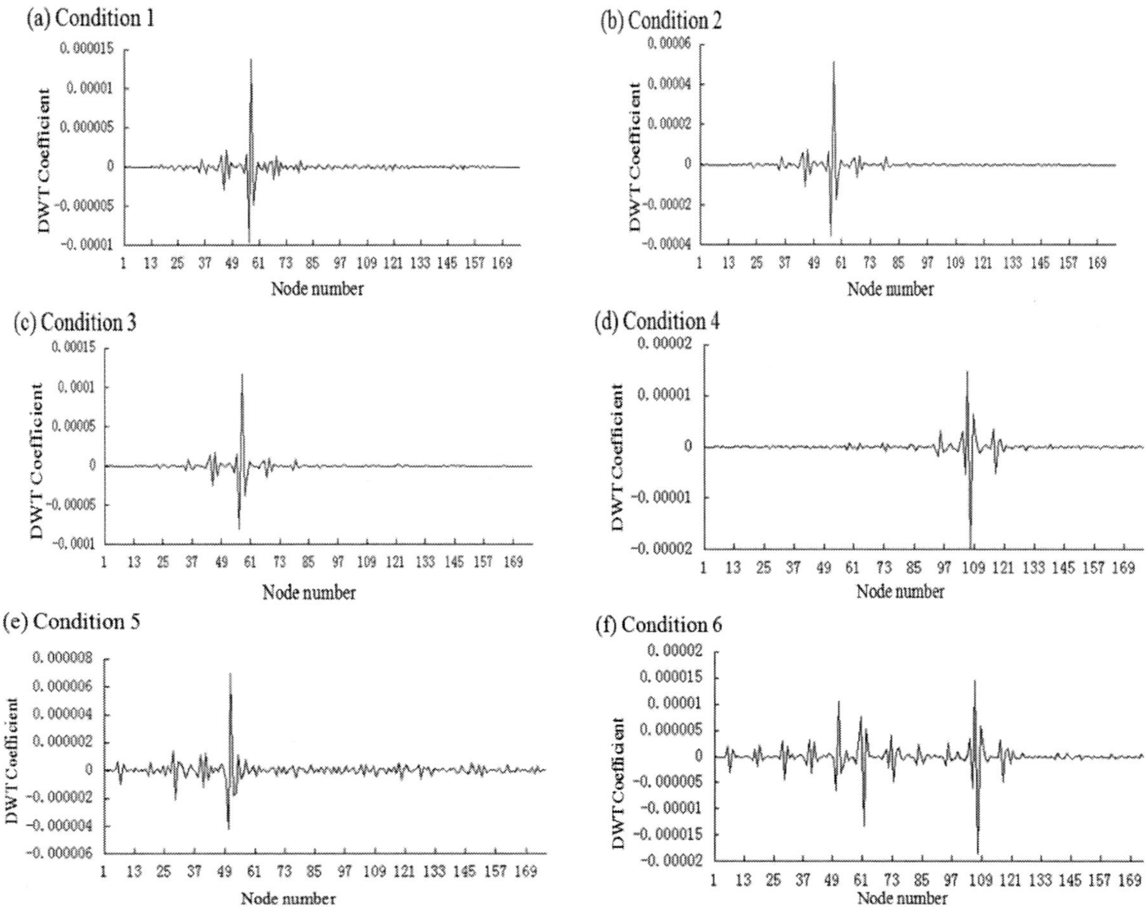

Figure 6. DWT Coefficients of orthogonal space truss structure.

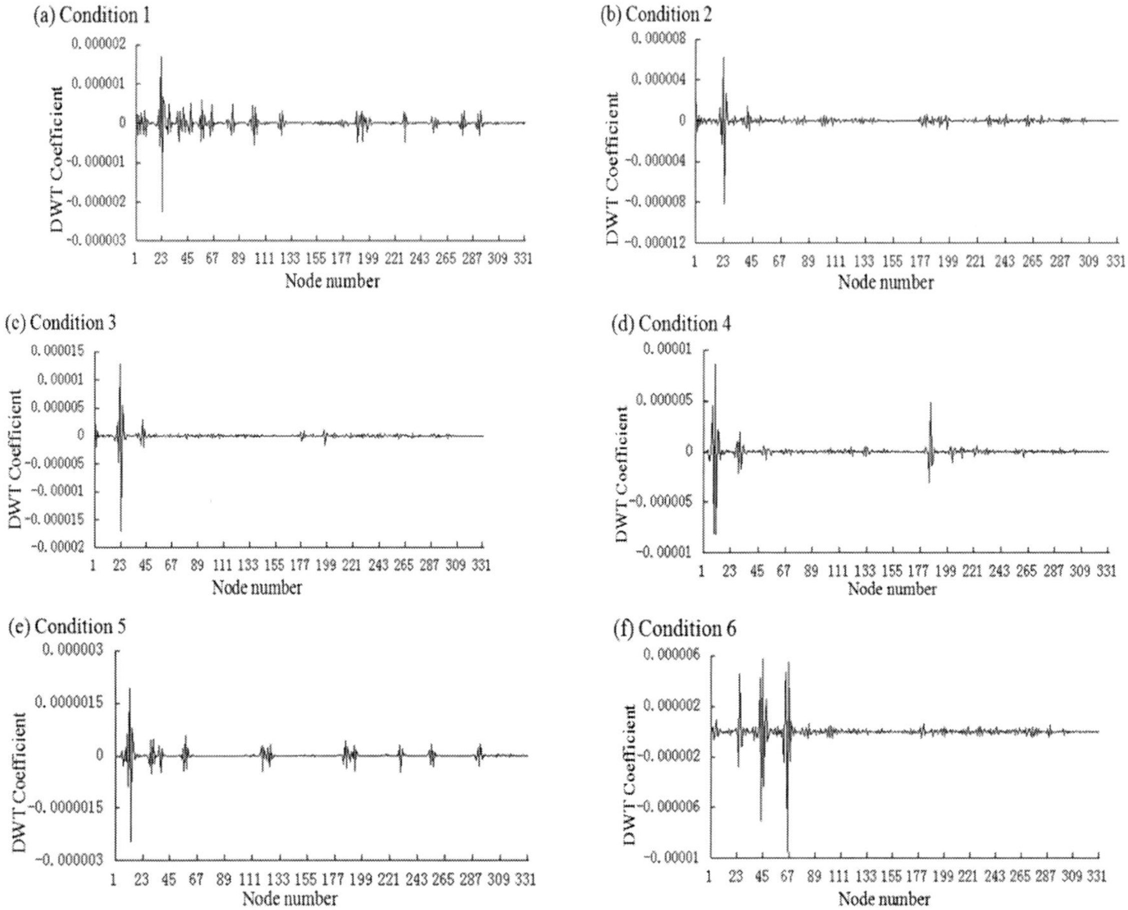

Figure 7. DWT Coefficients of three way latticed grid structure.

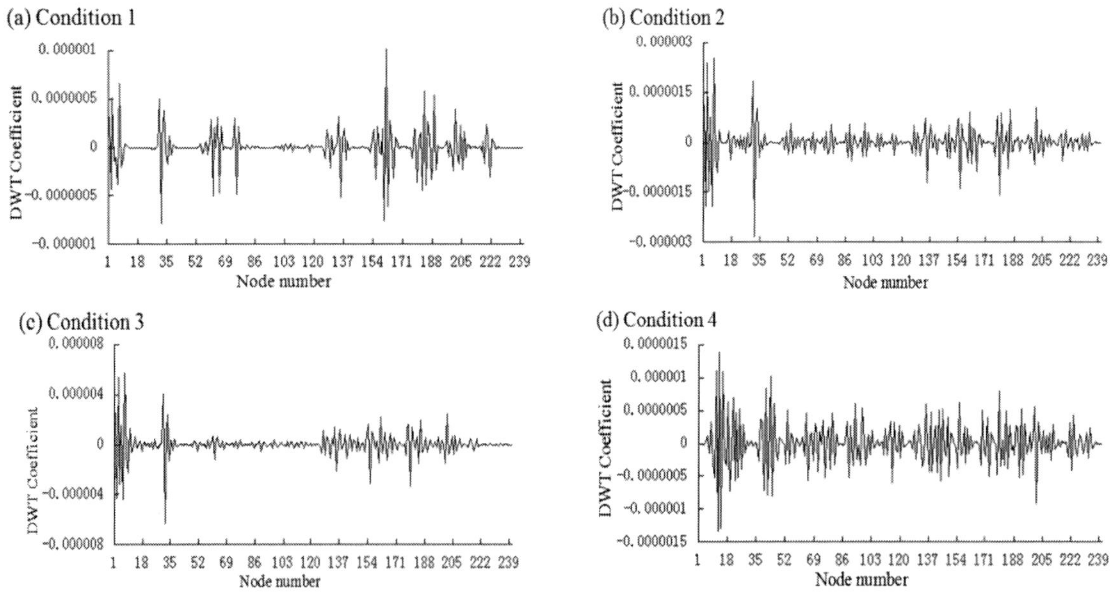

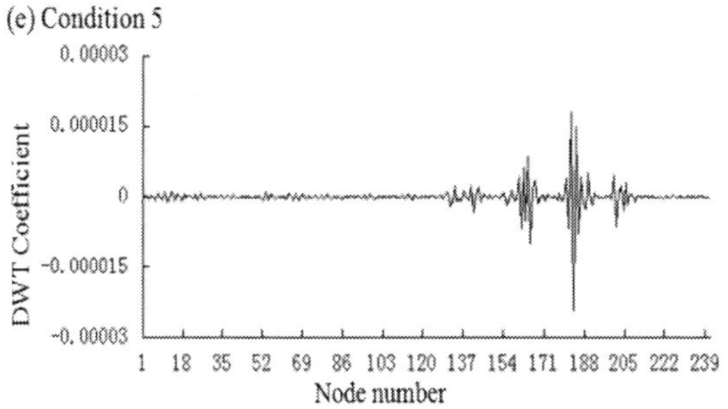

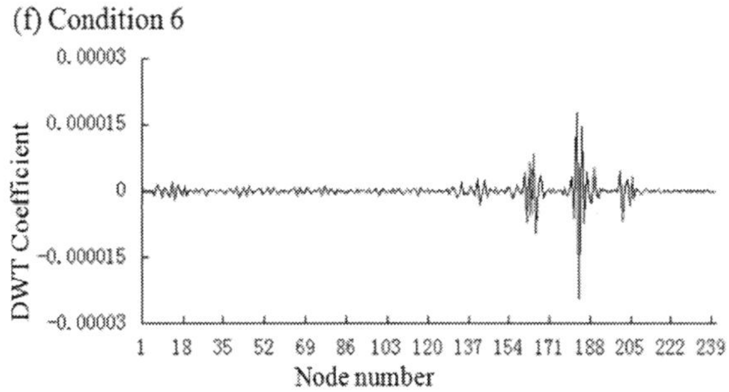

Figure 8. DWT Coefficients of honeycomb triangular pyramid grid structure.

4 Conclusion

This paper use a method that based on curvature mode difference and wavelet transform to judge damage location of four space grid structures. Damage expressed by reducing member elastic modulus and damage members including chord and web member, the structural curvature mode differences of pre and post damage were considered as signal which were used to do DWT by db3 wavelet. And the nodes corresponding to the maximum wavelet transform coefficient response represent damage location. The results of numerical simulation show that:

(1) The db3 wavelet transform coefficients of curvature mode difference from first mode shape of pre and post damage were used to judge damage location of square pyramid space grid structure. Through analysis, this method can obtain better identification effect in both single damage and multi-damage condition of four different space grid structures.

(2) Based on the DWT mutations, the damages' location can be determined. Damage on web chord and upper chord near edge are sensitive to DWT coefficient, while damage on middle chord was insensitive. So the links in different location have different identification results.

(3) The damage detection results show that for the same link with different damage degree, the greater degree of injury, the greater mutation value in damage node.

Acknowledgement

This work appreciates very much the Natural Science Foundation of Hebei Province (No. 2014208135) and Hebei province science and technology support project (No. 15275429).

References

1. G.Y. Li and H.W. Zheng, Measurement & Diagnosis, Journal of Vibration, **22**, 136 (2002)
2. J. Chen and F. Xiong, Journal of Wuhan University of Technology, **29**, 99 (2007)
3. Y. Ding, Y. Song, and Z.X. Li, Journal of Tianjin University, **40**, 726 (2007)
4. S.P. Valsan and K.S. Swarup, Electric Power Systems Research, **78**, 547 (2008)
5. S.S. Law, X.Y. Li, and X.Q. Zhu, Engineering Structures, **27**, 1339 (2005)
6. X.B. Zhang. Shanxi Architecture, **32**, 58(2006)
7. P. Yan, H.S. Li, and J.P. Ge, Journal of Vibration and Shock, **31**, 121 (2012)
8. Z.S. Sun, J.G. Han, and W.X. Ren, Earthquake Engineering and Engineering Vibration, **25**, 44 (2005)
9. Z.X. Yin, W.J. Xu, and L.Q. Mao, Earthquake Engineering and Engineering Vibration, **29**, 164 (2009)
10. W.J. Xu, Thesis for master degree of Liaoning Technical University (to be published, 2008)

11. Y. Zhang, Thesis for master degree of Hebei University of science and technology (to be published, 2010)
12. Z. Wang, Sichuan Building Science, **38**, 97 (2012)
13. D.F. Zhang, China Machine Press, (to be published, 2012)
14. Y.M. Li, B. Zhou, and W.J. Zhang, Spatial Structures, **18**, 24 (2012)

Assessment of soil-structure interaction on a 51-story building from the spectral ratio of earthquake recordings

Kun-Sung Liu[a]

Department of Civil Engineering & Hazard Mitigation Research Center, Kao Yuan University, Kaohsiung City 82151, Taiwan

Abstract. The soil-structure interaction (SSI) can significantly alter the characteristics of recorded motions in buildings. The ratio of Fourier amplitude spectrum of the top-story accelerations to that of the foundation accelerations permits the identification of the natural frequency of the fixed-base building. In this study, records of the Chi-Chi earthquake and the 1226 Hengchun earthquake doublet from the structural array in a 51-story highrise building are used to obtain the dynamic characteristics of the buildings by the transfer function (TF) method. As a result, the acceleration amplification of seismic excitation on the 47th storey of buildings is 4.24, in the horizontal component, from the Chi-Chi earthquake data greater than those of the 1226 Hengchun earthquake doublet with a value of 2.82 and 2.06, respectively. In addition, from the spectral ratio of the accelerations (47th floor/basement), together with the Fourier Amplitude Spectrum (FAS) of the 47th floor and basement accelerations, it is noted that the peaks of the 47th floor FAS and the spectral ratio appear to coincide with each other from the records of 1226 Hengchun earthquake doublet, suggesting that there is no significant SSI effects in both the longitudinal and transverse directions.

1 Introduction

Taiwan is located in the Circum-Pacific seismic belt. The Philippine Sea plate and the Eurasian plate collide with each other around Taiwan. Therefore, it is not surprising that large earthquakes have repeatedly hit the Taiwan [1]. Safeguarding life and property from destructive effects of earthquakes is a major concern of the people there. Since the most widespread damaging effects of earthquakes are caused by strong ground shaking, effective reduction of life and property losses from strong earthquakes requires conscientious application of earthquake resistant building codes and implementation of appropriate retrofit measures. Ideally, implementation of these mitigation measures should be based on instrumental recordings of strong earthquakes, for such data are crucial for improving earthquake resistant design of buildings [2]. The inclusion of SSI phenomena in the seismic analysis and design of structures has been addressed in seismic code provisions, including the FEMA450 document [3].

The effects of SSI on the seismic response of structures had been seriously taken into account since the 1971 SanFernando earthquake and the beginning of nuclear plant construction in the earthquake-prone California. In the 1980s and 1990s, many studies investigated the effects of SSI on

[a] Corresponding author : lk.sung99@msa.hinet.net

special structures [4]. Two schemes of SSI are widely recognized for structures founded on or in soils: inertial and kinematic interaction. Inertial interaction results from inertia forces produced by seismic accelerations of the structure. Inertia forces increase the base shear and overturning moments acting on the foundation, resulting in relative displacements between the free field and foundation [3]. Kinematic interaction, on the other hand, occurs due to the inability of the relatively rigid foundation to conform to the ground motion, causing deviation of the foundation motion from the free-field motion. This is particularly accentuated by incoherent ground motion, wave inclination, and foundation embedment effects [5]. A detailed discussion on SSI effects and analysis techniques is presented by [6].

In general, SSI can decrease the base shear, lateral forces, and overturning moments experienced by a structure during earthquakes. However, it increases lateral displacements and secondary forces associated with P-delta effects [7]. Therefore, SSI effects should be considered during the design process [3]. Aviles and Perez-Rocha [8] investigated site effects and SSI during the Mexico earthquake of 1985. They found that interaction effects were larger for tall and slender structures than for short and squat structures of the same period, and these effects decreased as the foundation depth increased. Crouse and Ramirez [9] identified SSI effects and nonlinear site response as the main reasons for the differences observed in three sets of earthquake records for two buildings in the Jensen filtration plant during the Northridge main shock and aftershock motions.

Full-scale monitoring can provide valuable data for realistic evaluation of the dynamic behaviors of tall buildings under earthquake excitations. Analyses of records obtained in buildings during strong earthquakes and ambient vibrations from building have shown the natural frequencies and damping ratio may vary with increasing amplitude of shaking [10-17]. At present reliable field measurements of earthquake effects on tall buildings are still limited [3]. Fortunately, the Central Weather Bureau (CWB) of Taiwan has installed 60 digital strong-motion array systems in buildings and other structures for engineering and scientific purposes since 1992, as part of its Taiwan Strong Motion Instrumentation Program (TSMIP) [18]. It is remarkable that these digital strong-motion arrays have capabilities to record building responses not only due to strong motions but also weak ambient vibrations, owing to a flexible gain selection of the amplifier-filter unit at 1, 10, or 100 [2].

In this study, seismic data recorded in the structural array of a 51-story highrise building in Kaohsiung City are analyzed with several vibration parameters, including its transfer functions and natural frequencies. In addition, the spectral ratio of the accelerations (47th floor / basement), together with the Fourier Amplitude Spectrum (FAS) of the 47th floor and basement accelerations excited by the M7.6 Chi-Chi earthquake and the 1226 Hengchun earthquake doublet, are used to assess the SSI effects in both the longitudinal and transverse directions.

2 Seismic data

The KBA0 Building is located in Kaohsiung. The building has 51 stories with a total height of 222.1 m at the roof, as shown in Figure 1. A typical floor plan of the KBA0 Building has a rectangular shape, with a length of 42.5m in the NS (i.e. x-x or longitudinal) direction and a width of 34.5m in the EW (i.e. y-y or transverse) direction. A total of 26 accelerometers are placed at the basement, 1st, 15th, 30th and 47th floor. These floors are at height of –19.5m, 0.0m, 57.6m, 119.2m, and 191.7m relative to the ground surface, respectively (Figure 1).

A major earthquake of M_w magnitude 7.6 took place in central Taiwan at 17:47:16 UT of 20 September 1999. The epicenter was located near the town of Chi-Chi, after which the event was named. This was the largest inland earthquake to strike Taiwan in the twentieth century. The CWB located its epicenter at 120.82° E and 23.85° N, with a focal depth of 8 km [1]. In addition, on 26 December 2006, two closely linked earthquakes, both ML 7.0, occurred only 8 minutes apart at 12:26:21 and 12:34:15 UT, respectively, in southwestern offshore of Taiwan near Pingtung. The epicenters of three earthquakes and location of the KBA0 building array station, as shown in Figure 2 and given in Table 1, were selected for this study. Records of the Chi-Chi earthquake and the 1226 Hengchun earthquake doublet from the structural array in the KBA0 Building are shown in Figures 3,

4, and 5, respectively. The accelerograms of channel number 1, 4, 7, 10, 13, 15, 17, 19, 21 and 24 are plotted for the accelerometers installed in the NS component and channel number 2, 5, 8, 11, 14, 16, 18, 20, 22 and 25 are those in the EW component as well as channel number 3, 6, 9, 12, 23 and 26 are those in the UD component. The largest peak acceleration of 101 cm/sec^2 was recorded in Channel 21, located at the 47th floor of the KBA0 Building as shown in Figure 3.

Table 1. The earthquake parameters used in this study.

No	Date	Time (UTC)	Lat. (N)	Lon. (E)	Depth (km)	Peak Acc (1F)			M_L	M_w	Epicentral Distance(km)
						X(gal)	Y(gal)	Z(gal)			
1	1999/09/20	17:47:15.85	23.853	120.816	08.00	20.9.	24.1	12.0	7.30	7.60	144.2
2	2006/12/26	12:26:21.00	21.687	120.555	44.11	17.7	19.2	18.2	6.96	7.00	108.8
3	2006/12/26	12:34:15.13	21.970	120.420	50.22	36.3	40.5	37.8	6.99	6.90	75.3

Remark: The moment magnitude (M_w) adopted from the Harvard CMT.

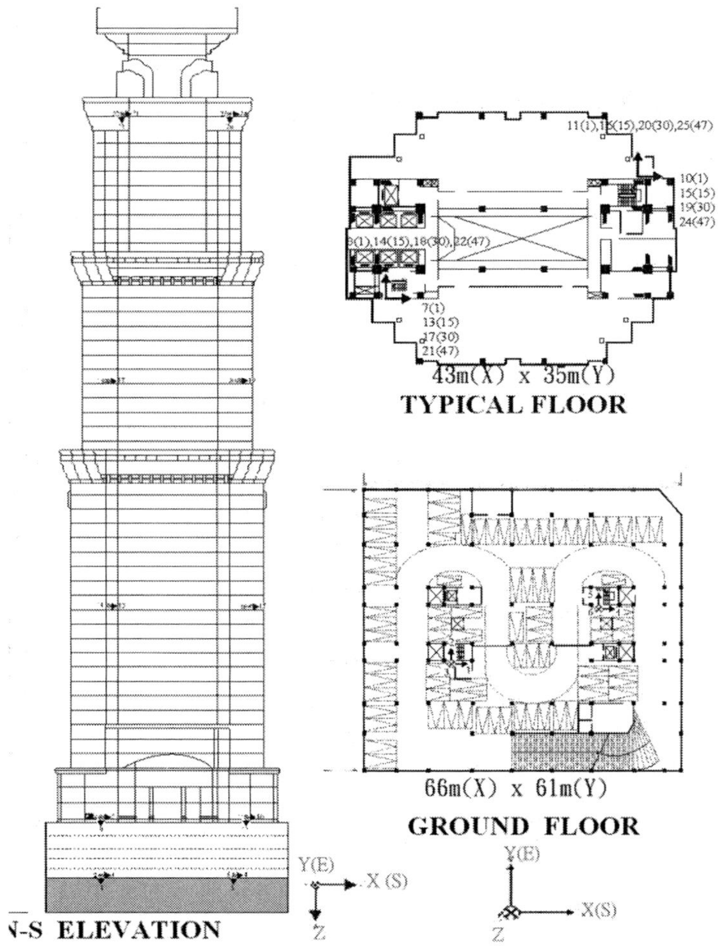

Figure 1. Configurations of the KBA0 Building. Left: NS elevation profile; Top right: a typical floor plan; Lower right: the ground floor plans. The accelerometer components are indicated by arrows.

DOI: 10.1051/matecconf/201711901043

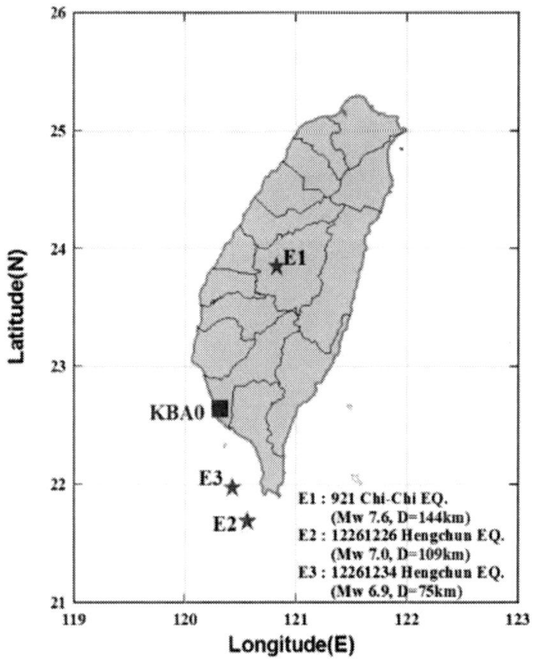

Figure 2. Location of the KBA0 building array station and the three epicenters of earthquakes used in this study.

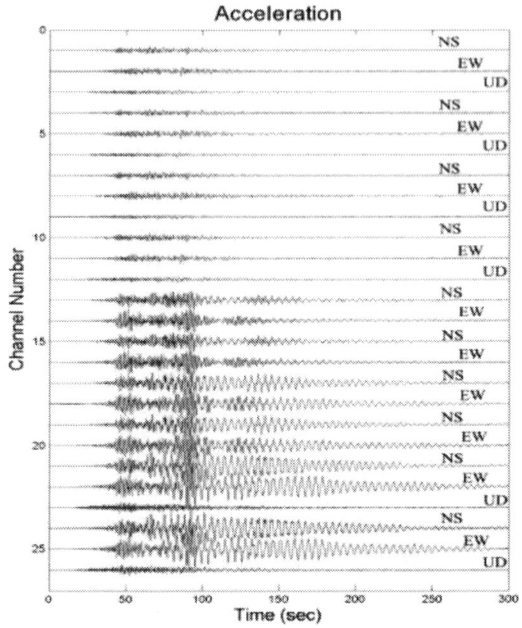

Figure 3. Acceleration time histories of the Chi-Chi earthquake, as recorded by the structural array in the KBA0 Building.

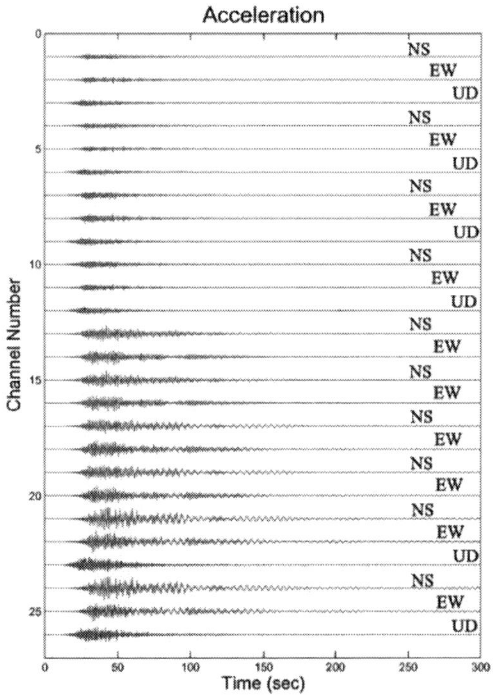

Figure 4. Acceleration time histories of the Hengchun Mw7.0 earthquake, as recorded by the structural array in the KBA0 Building.

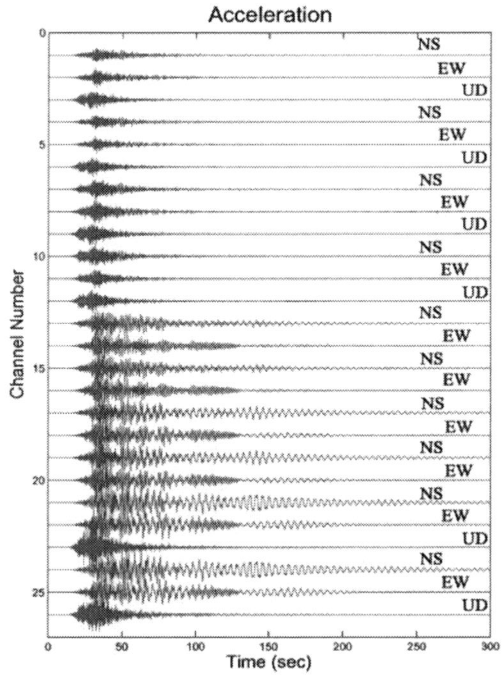

Figure 5. Acceleration time histories of the Hengchun Mw6.9 earthquake, as recorded by the structural array in the KBA0 Building.

3 Methodology

3.1 Parameters of vibration of the KBA0 building

The measured acceleration data are used to obtain the dynamic characteristics of the buildings. Different sets of modal parameters can be determined from the recorded time histories by several system identification methods, either in frequency domain or time domain. Most commonly used methods are based in frequency domain [19]. In order to identify the natural frequencies of the KBA0 Building, the transfer function (TF) method are used. Details of the TF method can be found in [20-22], respectively.

3.2 Identification of soil-structure interaction in the KBA0 building

The soil-structure interaction (SSI) can significantly alter the characteristics of recorded motions in buildings. The dominant frequency of a building subject to SSI will be smaller than that of a fixed-base building. The ratio of Fourier amplitude spectrum of the top-story accelerations to that of the foundation accelerations permits the identification of the natural frequency of the fixed-base building [13,23].

4 Results and discussion

4.1 Naturaltural frequencies of vibration of the KBA0 building

In the present study, only the first five modes of the building response are considered since higher modes are of little significance in overall building responses [24]. In the following the recorded data in the KBA0 Building from three earthquakes are used to identify the natural frequencies of the 1st, 2nd, 3rd, 4th, and 5th modes of vibration in the longitudinal and transverse directions.

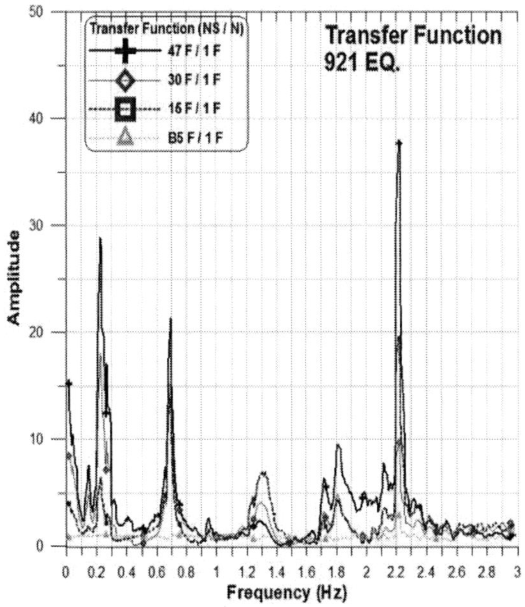

Figure 6. Transfer functions of the KBA0 Building at the north-side in longitudinal direction from the Chi-Chi earthquake.

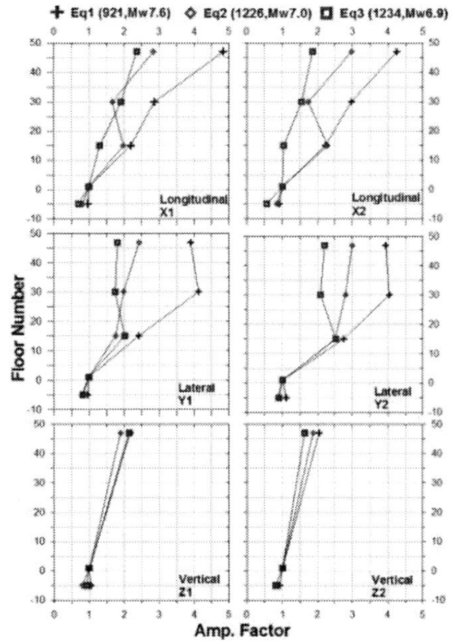

Figure 7. Acceleration amplification of seismic excitation on the different story of KBA0 building. The top, middle and bottom panels in Figure 7 are shown the longitudinal, lateral, and vertical direction,

respectively.

The natural frequencies of the first five translational modes in each direction evaluated by using the TF method from the observed acceleration data are summarized in Table 2. Figures 6 show the longitudinal transfer functions in the frequency domain as obtained from the Chi-Chi earthquake data by the north-side sensors in the KBA0 Building. By examining the observed transfer functions, the natural frequencies of the 1st, 2nd, 3rd 4th, and 5th longitudinal modes of the KBA0 Building are found to be 0.238, 0.697, 1.284, 1.834 and 2.207 Hz, respectively. Likewise, the corresponding results for the transverse component are 0.220, 0.715, 1.290, 1.797, and 2.293 Hz, respectively.

Taking the KBA0 Building as a steel structure with eccentrically braced frame, the corresponding fundamental mode period according to the building code can be calculated as follows [25]:

$$T = 0.070h_n^{3/4} \tag{1}$$

Where h_n is the height of building from the ground. The calculated natural frequencies for the first five modes are 0.248, 0.745, 1.242, 1.738, and 2.235 Hz, respectively. It is found that these frequencies conducted from the code formula are closer to those identified from actual data.

In addition, the building array with 26-channel accelerograms recorded from the three earthquakes has been analyzed to study the acceleration amplification of seismic excitation on the 47 story of KBA0 building. The results as shown in Figure 7 show the amplification factor is 4.24, in the horizontal component, from the Chi-Chi earthquake data greater than those of the 1226 Hengchun earthquake doublet with a value of 2.82 and 2.06, respectively.

Table 2. Observed modal frequencies *f* (Hz) in the longitudinal (L) directions and transverse (T) directions of the KBA0 Building.

Earthquake	Mode	Longitudinal (L) X1(Hz)	Longitudinal (L) X2(Hz)	Transverse (T) Y1(Hz)	Transverse (T) Y2(Hz)
EQ1	f_1	0.238	0.238	0.220	0.220
	f_2	0.697	0.697	0.715	0.715
	f_3	1.284	1.290	1.290	1.290
	f_4	1.834	1.834	1.797	1.797
	f_5	2.207	2.201	2.293	2.293
EQ2	f_1	0.238	0.238	0.232	0.232
	f_2	0.702	0.702	0.702	0.702
	f_3	1.270	1.270	1.331	1.331
	f_4	1.752	1.752	1.752	1.752
	f_5	2.069	2.069	2.191	2.191
EQ3	f_1	0.238	0.232	0.226	0.226
	f_2	0.696	0.696	0.708	0.708
	f_3	1.276	1.276	1.294	1.294
	f_4	1.794	1.794	1.825	1.825
	f_5	2.045	2.045	2.197	2.197

4.2 Identification of soil-structure interaction in the KBA0 building

It has been demonstrated that the fundamental frequency of the building may vary with the intensity of excitation force. The change in frequency may be attributed to one or more of the following causes: 1. Change in building effective stiffness, which could be due to the cracking of RC sections on the tension side and/or the disengagement of stiffness-contributing nonstructural elements; 2. Change of foundation stiffness due to softening in the soil at larger strains and the resulting effect of soil-structure interaction; and 3. Structural damage [14-15,17].

In order to understand the characteristics of response spectra between the three earthquakes and torsional predominant frequency, the Fourier amplitude spectrum of the basement, 1st, 15th, 30th and 47th floor in the longitudinal direction (X1) are plot in the Figure 8. The top, middle and bottom panels in Figure 8 are shown the chi-chi earthquake, Hengchun M7.0 earthquake, and Hengchun M6.9 earthquake, respectively. In addition, the Fourier amplitude spectra of the three earthquakes in the 47th floor are plot in the Figure 9. The top, middle and bottom panels in Figure 9 are shown the longitudinal, lateral, and vertical direction, respectively. From the Figures 8 and 9, we can find the following results: (1) it shows that amplitude spectrum decrease with increasing frequency, especially for that of the chi-chi earthquake. It probably due to that the difference of vary for frequency content decay with distance. (2) The high story (47F) has high Fourier spectrum while the frequency is lower than 0.8 Hz among the three earthquakes, especially for the horizontal component and chi-chi earthquake. (3) The torsional predominant frequency is 0.960 Hz after comparing the recording from the different site in the transverse direction of same floor.

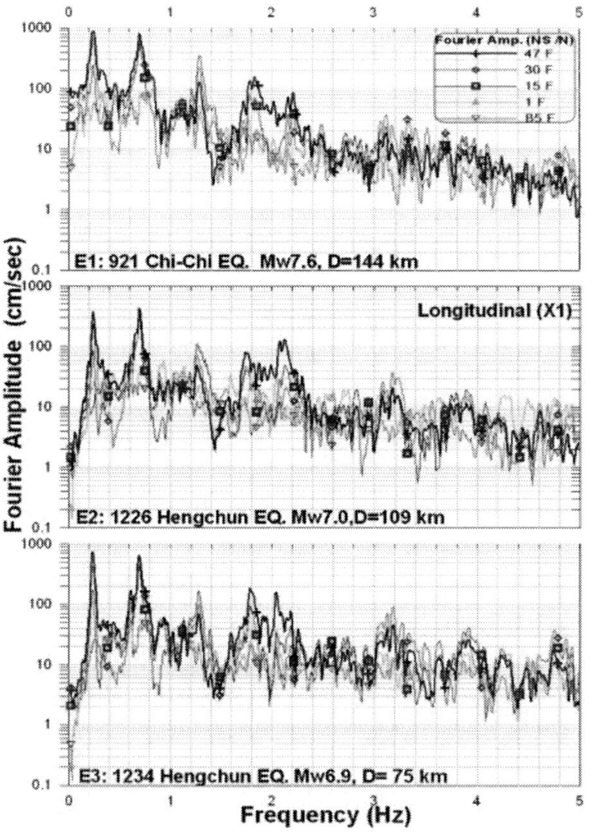

Figure 8. Fourier amplitude spectrum of the basement, 1st, 15th, 30th and 47th floor in the longitudinal direction (X1). The top, middle and bottom panels in Figure 8 are shown the chi-chi earthquake, Hengchun M7.0 earthquake, and Hengchun M6.9 earthquake, respectively.

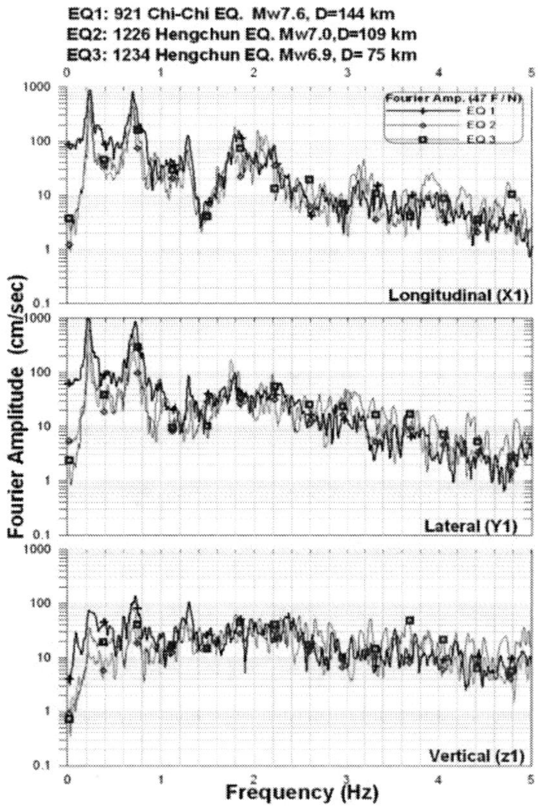

Figure 9. Fourier amplitude spectra of the three earthquakes in the 47th floor. The top, middle and bottom panels in Figure 9 are shown the longitudinal, lateral, and vertical direction, respectively.

In order to determine if the frequency reduction in the earthquake was caused by soil-structure interaction, we calculated spectral ratio for the three earthquakes. The spectral ratio of the accelerations (47th floor/ basement), together with the Fourier Amplitude Spectrum (FAS) of the 47th floor and basement accelerations are shown in Figures 10 and 11 for the Hengchun M7.0 earthquake in the longitudinal and transverse directions, respectively. It is noted that the peaks of the 47th floor FAS and the spectral ratio appear to coincide with each other from the records of 1226 Hengchun earthquake doublet, suggesting that there is no significant SSI effects in both the longitudinal and transverse directions.

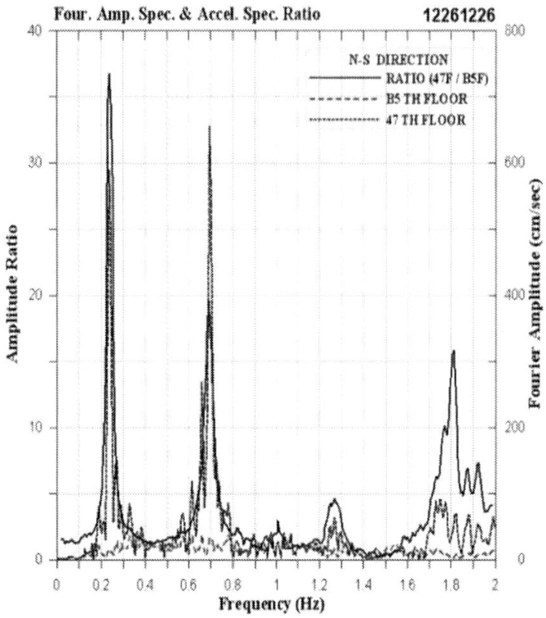

Figure 10. Fourier amplitude spectra of acceleration and spectral ratio of accelerations of 47th floor and foundation of KBA0 in the longitudinal (north-south) direction for the Hengchun M7.0 earthquake.

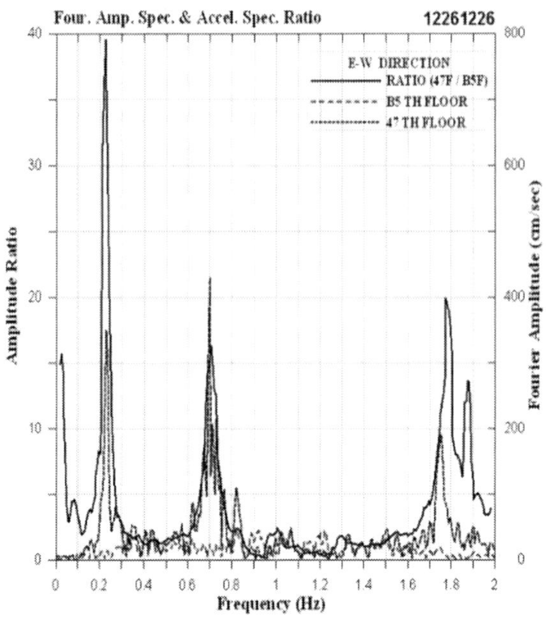

Figure 11. Fourier amplitude spectra of acceleration and spectral ratio of accelerations of 47th floor and foundation of KBA0 in the transverse (east-west) direction for the Hengchun M7.0 earthquake.

5 Conclusion

The present study investigates the dynamic characteristics of a 51-story high-rise building by the transfer function (TF) method as well as assesses the SSI effects. Based on above results and discussion, we can draw the following conclusions:

1. Based on the observed transfer functions from the records of the Chi-Chi earthquake, the natural frequencies of the 1st, 2nd, 3rd 4th, and 5th longitudinal modes of the KBA0 Building are found to be 0.238, 0.697, 1.284, 1.834 and 2.207 Hz, respectively. Likewise, the corresponding results for the transverse component are 0.220, 0.715, 1.290, 1.797, and 2.293 Hz, respectively. It is found that these frequencies conducted from the code formula are closer to those identified from actual data.
2. From the building array with 26-channel accelerograms recorded from the three earthquakes, the acceleration amplification of seismic excitation on the 47 storey of KBA0 building is 4.24, in the horizontal component, from the Chi-Chi earthquake data greater than those of the 1226 Hengchun earthquake doublet with a value of 2.82 and 2.06, respectively.
3. From the spectral ratio of the accelerations (47F/ basement), along with the FAS of the 47-story and basement accelerations for the 1226 Hengchun earthquake doublet Chi-Chi earthquake, it is noted that the peaks of the 47th floor FAS and the spectral ratio seem to coincide with each other, suggesting that there is no significant SSI effects in both the longitudinal and transverse directions.

Acknowledgement

The author thanks the Central Weather Bureau of Taiwan, ROC for providing excellent seismic data for the present study. This research was supported by the Ministry of Science and Technology of the Republic of China under Grant No. MOST 104-2116-M-244-001.

References

1. T.C. Shin and T.L. Teng, An overview of the 1999 Chi-Chi, Taiwan, Earthquake, Bull. Seis. Soc. Am., **91** (5), 895-913 (2001)
2. K.S. Liu and Y.B. Tsai, Observed natural frequencies, damping ratios and mode shapes of vibration of a 30-story building excited by a major earthquake and typhoon, Earthquake Spectra, **26** (2), 371-397 (2010)
3. A. Turan, S.D. Hinchberger, E. Naggar, and M. Hesham, Seismic soil–structure interaction in buildings on stiff clay with embedded basement stories, Canadian Geotechnical Journal, **50**, 858-873 (2013)
4. C.C. Spyrakos, I.A. Koutromanos, and Ch.A. Maniatakis, Seismic response of base-isolated buildings including soil–structure interaction, Soil Dynamics and Earthquake Engineering, **29**, 658-668 (2009)
5. J.P. Stewart, G.L. Fenves, and R.B. Seed, Seismic soil–structure interaction in buildings. I: Analytical method, Journal of Geotechnical and Geoenvironmental Engineering, **125** (1), 26-37 (1999)
6. J.J. Johnson, Soil–structure interaction, In: W.F. Chen and C. Scawthorn, editors, *Earthquake Engineering Handbook*, *Part 2*, Boca Raton, FL: CRC Press (2003)
7. FEMA 450. NEHRP recommended provisions for seismic regulations for new buildings and other structures (2003 edition), Building Seismic Safety Council (BSSC), Washington DC, (2003)
8. J. Aviles and L.E. Perez-Rocha, Site effects and soil-structure interaction in the Valley of Mexico, Journal of Soil Dynamics and Earthquake Engineering, **17**, 29-39 (1997)
9. C.B. Crouse and J.C. Ramirez, Soil-structure interaction and site response at the Jensen Filtration Plant during the 1994 Northridge, California, mainshock and aftershocks, Bull. Seis. Soc. Am., **93** (2), 546-556 (2003)
10. ATC, *Guidelines for Using Strong-Motion Data and Shakemaps in Postearthquake Response*, ATC-54 Report, Applied Technology Council,Redwood City, California, (2005)
11. J.F. Clinton and T.H. Heaton, Is there a linear building transfer function for small excitation? *EOS* Trans. Am. Geophys. Union 84 _46_, Fall Meet. Suppl., Abstract, S51E-0102 (2003)
12. D.A. Foutch, The vibrational characteristics of a twelve-storey steel frame building, Earthquake Eng. Struct. Dyn., **6**, 265-294 (1978)
13. M.D. Kohler, P.M. Davis, and E. Safak, Earthquake and ambient vibration monitoring of the steel-frame UCLA factor building., Earthquake Spectra, **21**, 1-22 (2005)
14. Y. Li and S.T. Mau., Learning from recorded earthquake motion of buildings, Journal of Structural Engineering, **123**, 62-69 (1997)
15. D. Skolnik, Y. Lei, E. Yu, and J.W. Wallace, Identification, model updating, and response prediction of a 15-story steel frame building, Earthquake Spectra, **22**, 781-802 (2006)
16. M.D. Trifunac, Comparisons between ambient and forced vibration experiments, Earthquake Eng. Struct. Dyn., **1**, 133-150 (1972)
17. E. Yu, D. Skolnik, D.H. Whang, and J.W. Wallace, Forced vibration testing of a vour-story reinforced concrete building utilizing the nees@ucla mobile field laboratory, Earthquake Spectra, **24**, 969-995 (2008)
18. K.S. Liu, T.C. Shin, and Y.B. Tsai, A free field strong motion network in Taiwan: TSMIP, TAO, **10** (2), 377-396 (1999)
19. C.H. Loh and H.M. Lin, Application of off-line and on-line identification techniques to building seismic response data, Earthquake Engineering and Structural Dynamics, **25**, 269-290 (1996)
20. G.W. Housner, Design spectrum, In *Earthquake Engineering*, (R. L. Weigel, Ed.), Prentice-Hall, Englewood Cliffs, New Jersey (1970)
21. G.H. McVerry, Structural identification in the frequency domain from earthquake records, Earthquake Engineering and Structural Dynamics, **8**, 161-180 (1980)
22. P.C. Jennings, An introduction to the earthquake response of structures, In *International Handbook of Earthquake and Engineering Seismology*, (William Lee, Ed.), Academic Press, (2002)

23. E. Safak, Detection and identification of soil-structure interaction in buildings from vibration recordings, J. Struct. Eng., **121**, 899-906 (1995)
24. Y. Li and S.T. Mau., Anatomy of the loma prieta earthquake records of two steel buildings using mimo system identification, *Proc., Conf. Engrg. Mech.*, ASCE, New York, 689-693 (1992)
25. K.C. Tsai, C.H. Loh, C.H. Yeh, and C.M. Su, Design examples and the recommended modifications on the tentative seismic force provisions for building structures in Taiwan, Architecture & Building Institute, Ministry of Interior Research Project Report, Taipei (2001)

Hand tool handle design based on hand measurements

Ching-yi Wang[1,a] and Deng-chuan Cai[2]

[1]*Department of Industrial Design, Tatung University, Taipei City 104, Taiwan*
[2]*National Yunlin University of Science and Technology, Douliou, Yunlin City 64002, Taiwan*

Abstract. This study provides a tool handle design guideline based on measurements of hand gripping diameter. Anthropometric measurements on hands of 60 participants were collected with a caliper. The results in the current study concluded the guidelines for handle design, including: (1) a definition of handle design was depended on the users' hand length and hand breadth, (2) different gender would affect the size of tool handle, (3) the handle length at least should be 100mm, (4) the cross-section handle of elliptical shape was better for gripping, and (5) the best ratio of handle cross-section with width and length should be 1:1.25. These findings can serve as references for tool handle design, and improve comfort for users.

1 Introduction

Correct tool design is important for preventing upper-extremity musculoskeletal disorders. Considering the ergonomics of a hand-tool, in addition to its main function, the most important part is the tool's handle. Tool handle design research has been previously limited to the determination of the diameters of cylindrical handles to increase performance and comfort for reducing the chances of CTD (i.e., blisters, inflamed skin, cramped muscles, etc.) [1-4]. Many studies have researched the topic of tool handle design to define the optimal size and shape of a tool handle. Most of the studies have focused on cylindrical [5-7] or elliptical [8-9] shapes of the handles to provide guidelines for determining the optimal diameters to increase finger-force comfort, exertion, and the contact area. According to the previous studies on the related designs of the ergonomic handle, a hand tool design principles are summarized as following:

- Handle diameter: a grip design should consider the hand action when holding the grip [10]. A handle with a diameter of 30 to 40mm is most suitable for a gun or electric tool [5]. Moreover, the handle suitable for both male and female was a cylindrical handle with a diameter of around 40cm [8]. Some studies suggested that the handle diameter should be in a range of 30 to 45 mm [7]. However, some studies argued that the diameter should be in a range of 25 to 50mm [6]. In addition, a suitable handle diameter should be 19.7% of the hand length [11]. The handle shape should be an elliptical shape with a width and length ratio of 1: 1.25 that can be used for the operation of push force and pull force [8].
- Handle length: the minimum handle length should be 100 to 125mm [12]. The palm width of 95% of workers is less than 100mm [7]. Moreover, the average of hand breadth for the males was 87mm, and that for the females was 78mm [13].

[a] Corresponding author : catincar@gmail.com

• Handle slope: the pistol-shaped handle should be tilted at 78° [10]. Some studies found that users prefer a slightly curved hammer handle of 10° to replace the straight grip [14]. A hammer handle can reduce wrist damage when it is tilted 20° to 40° [15]. For the hand tool with a normal linear handle, the grip central line and the forearm support shaft form a degree of 110° [16].

The purpose of this study was used anthropometric measurements to develop a handles that fits human-hand and suitable for the most people. The findings can serve as references for the industry and hand tool designers in the design of handles.

2 Methods

2.1 Hand anthropometric measurements

Sixty adult participants (30 male, 30 female; mean age=31.2 y; SD=4.4 y) took part in the experiment. Hand anthropometric measurements were obtained with the participant's right-hand using a caliper.

2.2 Definition of idealized gripping gesture

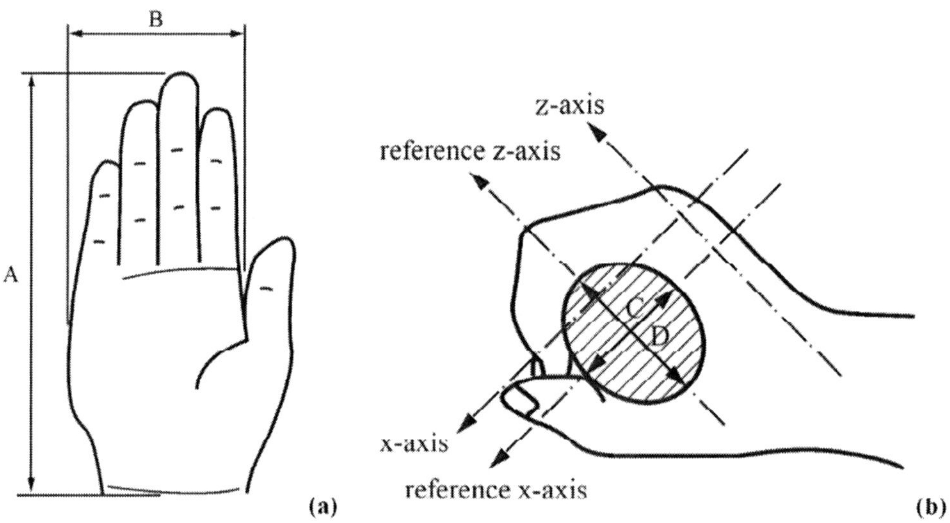

Figure 1. The definitions of the dimensional measurement of hand: (a) A, B, C, and D dimensions were measured with a caliper. (b) The grip gesture was determined by the biodynamic hand coordinate system defined in ISO 8727 [17].

Table 1. Definitions of hand measurements.

item	definition
A hand length	distance from top of the middle finger to the distal crease of the wrist.
B hand breadth	maximum hand breath where the fingers join the palm.
C grip breath inside width diameter	inside hand elliptical diameter with the length of major axis measured at grip breadth
D grip breath inside length diameter	inside hand elliptical diameter with the length of minor axis measured at grip breadth

Figure 1 and Table 1 shows the definitions of the dimensional measurement of hand. The A-D dimensions were measured by using a caliper, as shown in the Fig.1a. In order to obtain the consistency of the gripping gesture, the palm and fingers of the hand needs to be position of the force distribution. The grip gesture were determined by the biodynamic hand coordinate system defined in

ISO 8727 [17] is included in Fig. 1b. The z-axis is passes proximally through the third metacarpal bone when gripping gesture. The x-axis of the system is approximately normal to the palm of the hand, projecting anteriorly from the origin when the hand lies open in the normal anatomical position, i.e. palms facing forward. The axis parallel to the x and z-axis and passing through the center of the handle circle were used as the reference x and z-axis.

3 Results and discussions

3.1 Hand anthropometric data

Table 2 presents the results of anthropometric data. The overall means of hand length was 177.9mm (187.9mm for males and 167.9mm for females). In addition, the average of hand breadth (four fingers) was 79.4mm (83.6mm for males and 75.2mm for females). The measurement results of hand in the current study were close to the anthropometric measurements of Taiwanese aged between 25 to 34 years old [18], suggesting that the anthropometric data in the present study can represent a design reference for related tool manufacturing industries in Taiwan.

Table 2. The results of hand dimensions (unit: mm) (standard deviation of the mean in parentheses).

item		This study		Wang et al. (2002)
male	A hand length	187.9	(7.9)	183
	B hand breadth (four fingers)	83.6	(4.8)	86
	C grip breath inside width diameter	44.5	(5.3)	-
	D grip breath inside length diameter	35.6	(4.9)	-
female	A hand length	167.9	(6.6)	167
	B hand breadth (four fingers)	75.2	(6.2)	75
	C grip breath inside width diameter	32.2	(4.5)	-
	D grip breath inside length diameter	25.7	(5.0)	-
all	A hand length	177.9	(12.4)	175
	B hand breadth (four fingers)	79.4	(6.9)	80.5
	C grip breath inside width diameter	38.3	(7.9)	-
	D grip breath inside length diameter	30.7	(7.0)	-

3.2 Handle length

The handle length was determined by the hand breadth. In the current study, the result of hand breadth (four fingers) was consistent with the findings which suggested that approximately 95% of workers were below 100mm [7]. According to the result of hand breadth, this study therefore considered that the handle length at least should be 100mm. Moreover, the average of hand breadth for the males of American was 87mm, and that for the females was 78mm [13]. It was inferred that the hand breadth of Taiwanese was slightly shorter than that of American.

3.3 Handle diameter

Table 3. The ratio of major-to-minor diameter of handle dimensions.

item	**handle diameter of elliptical grip***	
	This study	Cochran & Riley (1986)
male	1:1.25	-
female	1:1.25	-
all	1:1.25	1:1.25

Table 3 illustrates the results of handle diameters for males and females. The grip breath inside width-to-length diameter for all participants was 30.7-to-38.3mm (35.6-to-44.5mm for males and 25.7-to-32.2mm for females). The overall griping diameter of handle in the present study was consistent with the [5], which suggested that handle diameter should be in the range of 30 to 40 mm. However, this suggested range of handle diameter was small for males, and large for females. Because of the hand length for females in the griping was smaller than for the males. Moreover, this study agreed the the diameter should be in a wider range of 25 to 50mm [6]. It was more suitable for most people, but not clearly for the specific tool handle (i.e. power tool, peeler, or saw). No matter what a tool design should consider the different sizes of handle for different gender used.

3.4 Handle cross-section

The best cross-section of a handle for gripping was elliptical shape. The ratio of cross-section with width and length diameter for males and females were both 1: 1.25, as shown in the Table 3. According to the [8] recommended that the ratio of width and length diameter was 1: 1.25 (28mm for the width diameter and 35mm for the length diameter), suggesting that it is the best ratio for gripping. Moreover, a handle cross-section with the width-to-length diameter for gripping was 17.3-to-21.5% of the hand length (18-to-23.7% for males and 15.3-to-19.2% for females), as shown in the Table 4. The overall results were close to the findings of [11], that the diameter of a suitable handle was 19.7% of hand length. However, this suggested value for a handle was large for females because their hand length was too small that did not enough to completely hold. Thus, a great size of handle cross-section should be depended on the hand length of users.

Table 4. The percentage of handle dimensions to hand length.

item	handle diameter		
	This study		Kong & Lowe (2005)
	width diameter*	length diameter**	
male	18%	23.7%	-
female	15.3%	19.2%	-
all	17.3%	21.5%	19.7%

*width diameter =D grip breath inside width diameter / A hand length
**length diameter =C grip breath inside length diameter / A hand length

4 Conclusion

This study presented a guideline for handle design based on hand anthropometric measurements. The findings in the current study are important for developing a comfortable tool handle to develop and manufacture a best shaped tool handle for users. Thus, our results could provide a handle reference for constructing a new tool handle design.

References

1. J. Marsot and L. Claudon, Design and ergonomics: methods for integrating ergonomics at hand tool design stage, International Journal of Occupational Safety and Ergonomics, **10** (1), 13-23 (2004)
2. P.G. Dempsey, R.W. McGorry, and N.V. O'Brien, The effect of work height, workpiece orientation, gender, and screwdriver type on productivity and wrist deviation, International Journal of Industrial Ergonomics, **33**, 339-346 (2004)
3. L.F.M. Kuijt-Evers, J.W.R. Twisk, L. Groenesteijn, M.P.de Looze, and P. Vink, Identifying predictors of comfort and discomfort in using hand tools, Ergonomics, **48** (6), 692-702 (2005)

4. L.F.M. Kuijt-Evers, T. Bosch, M.A. Huysmans, M.P.de Looze, and P. Vink, Association between objective and subjective measurements of comfort and discomfort in hand tools, Applied Ergonomics, **38**, 643-654 (2007)
5. The Eastman Kodak Company, *Kodak's Ergonomics Design for People at Work* (Canada: John Wiley & Sons, 2004)
6. A. Mital and W. Karwowski, *Workspace, Equipment and Tool Design* (Elsevier, Amsterdam, Netherlands, 1991)
7. V. Putz-Anderson, *Cumulative Trauma Disorders: a Manual for Musculoskeletal Diseases of the Upper Limb* (Taylor and Francis, London, England, 1988)
8. D. Cochran and M. Riley, An evaluation of knife handle guarding, Human Factors, **28** (3), 295-301 (1986)
9. N.J. Seo and T.J. Armstrong, Effect of elliptical handle shape on grasping strategies, grip force distribution, and twisting ability, Ergonomics **54** (10), 961-970 (2011)
10. T. Fraser, *Ergonomics Principles in the Design of Hand Tools: Occupational Safety and Health Series* (International Labour Office, Geneva, Switzerland, 1980)
11. Y.K. Kong and B.D. Lowe, Optimal cylindrical handle diameter for grip force tasks, International Journal of Industrial Ergonomics, **35**, 495-507 (2005)
12. S. Konz, *Work Design: Industrial Ergonomics* (Grid Publishing Inc, Columbus, OH, 1983)
13. A.R. Tilley, *The Measure of Man and Woman: Human Factors in Design* (Henry Dreyfuss Associates, Wiley, New York, USA, 2002)
14. R. Krohn and S. Konz, Bent hammer handles, *Proceedings of the Human Factors Society 26th Annual Metting*, Santa Monica, CA: Human Factors Society, 413-417 (1982)
15. R. Schoenmarklin and W. Marras, Effects of handle angle and work orientation on hammering: II. Muscle fatigue and subjective ratings of body discomfort, Human Factors, **31** (4), 413-420 (1989)
16. K. Kroemer, H. Kroemer, and K. Kroemer-Elbert, *Ergonomics: How to Design for Ease and Efficiency* (Prentice Hall, Englewood Cliffs, NJ, 1994)
17. ISO 8727. Mechanical vibration and shock – human exposure –biodynamic coordinate systems, International Organization for Standardization, Geneva, Switzerland (1997)
18. M.J. Wang, M.Y. Wang, and Y.C. Lin, *The Anthropometric Database Manual of Taiwan* (Hsinchu: Ergonomics Society of Taiwan, 2002)

Creating a smart application system to provide a beneficial maintenance service for elderly drivers

Sebin Jung[1,a], Hongying Meng[1] and Shengfeng Qin[2]

[1]*Department of Electronic and Computer Engineering, Brunel University London, Uxbridge, UB8 3PH, UK*
[2]*Northumbria School of Design, City Campus East, Northumbria University, Newcastle upon Tyne, NE1 8ST, UK*

Abstract. As overall population ages, elderly drivers have become a larger percentage of the driving population. With this trend, a lot of vehicle systems have been improved for elderly's safety and convenience using different advanced technologies. However, elderly drivers have often paid more money than other drivers in a car-repair shop due to their lack of knowledge about vehicle systems with modern technologies. Given this fact, developing a tool to diminish this disadvantage and to help elderly drivers maintain their cars with confidence and with minimal cost is necessary. Therefore, this research work mainly focuses on suggesting a system concept on user-interface application, which is connected to a smart phone or a tablet to provide beneficial services anywhere. For the research outcome, diverse research activities – surveys, interviews with small focus groups, observations of the focus groups, and discussions – has been conducted to understand the elderly driver's difficulties and behaviours regarding vehicle maintenance, to investigate what specific problems make them uncomfortable in repair shops, and to demonstrate how new system-concepts could be developed for the elderly. Furthermore, we conclude that adequate system-concepts for the elderly would offer elderly drivers convenient vehicle repair and maintenance and provide them a confident driving experience.

1 Introduction

In recent decades, elderly drivers have become a larger percentage of the driving population as the aging population increases worldwide. A report compiled for the National Highway Traffic Safety Administration (NHTSA) says that 36.8 million licensed drivers were age 65 and older in the United States in 2013 [15]. By 2020, there will be more than 40 million drivers on the road in this age group. Also, in the UK, the IAM Road Safety forecasts that the number of over 65 drivers will increase from 9.5 million in 2004 to 10.4 million in 2011, and to 12.7 million in 2021 [8]. This shows that for the automotive industry, the elderly group is not only becoming more influential in deciding what car to buy, but should also be considered as requiring important services. However, the elderly have not yet taken any beneficial services from the industry despite being a large portion of the driving population. One issue is particularly prevalent in the vehicle repair and maintenance service: Elderly drivers have often paid more money than younger drivers in car-repair garages due to their lack of knowledge of current vehicle systems with modern technologies. Although a lot of vehicle systems have become

[a] Corresponding author : sebin.jung@brunel.ac.uk

more technically sophisticated for a driver's safety and convenience by incorporating different high technologies, such a complicated system is difficult for elderly drivers to understand and repair. According to the Ripoff Report [18], "National auto service centers company over charges for parts, elderly people most, excessive unneeded work, parts being installed when not needed. The company over charges excessively for auto parts 4.2 over what they pay."

Given this fact, developing a tool to diminish this disadvantage and to help elderly drivers maintain their cars with confidence and with minimal cost is necessary. Although some tools have been created for vehicle repairs, they only provide guidelines via an application or book, and they are not helpful enough for vehicle owners to appropriately repair their vehicles without the help of experts. Drivers should have guidelines to identify a specific problematic component and a system to provide a solution for problematic parts. Additionally, although many automotive companies have updated vehicle system networks to help customers easily use and maintain them, it is still difficult for the elderly drivers to identify vehicle errors and repair their vehicles by themselves. Moreover, even though this is a regular issue, many scholars and developers have not considered training or helping them to avoid potential damage nor have they offered convenient vehicle repair and maintenance services. As a result, there is no product for elderly drivers to benefit from vehicle repair and maintenance, and they can continuously suffer from the same issue. Therefore, first, this paper will investigate challenges for the elderly with modern vehicle systems based on their body conditions and study how detecting vehicle failure modes are complicated and difficult to understand. It will also emphasize the significance of a system concept for the elderly by analyzing current systems and suggesting possible system developments for the elderly based on focus-group survey and interviews. Finally, it will provide proper system feature-concepts for the elderly throughout the focus-group evaluation.

2 Challenges for the elderly using modern vehicle systems

The elderly population face many problems due to their declining body conditions, such as hearing loss, vision and cognitive impairments, late reaction, and other chronic illnesses. Elderly users could also experience differing familiarity in using and understanding high technologies, and their usability could be influenced by the technologies. These barriers can affect how they react to technological systems. Thus, understanding their capabilities and limitations will help scholars and developers create usable application systems for elderly users.

Here are some general problems associated with the elderly that we should address.

Firstly, some eye conditions can interfere with the elderly's ability to focus their peripheral vision and can cause difficulty seeing in the dark or blurred vision and lowered recognition of colour contrast [5]. Poorly sized and coloured components of the vehicle systems can be very frustrating to view. With imperfect visual acuity, they have difficulties in detecting and judging the location of objects correctly [5]. Secondly, decreased hearing is a key symptom of ageing. Sometimes, elderly people experience difficulty hearing higher pitched sounds and separating particular sounds from important auditory cues in the vehicle systems. Thirdly, elderly users may experience reduced dexterity and motor control, making precise system control and accurate movement on the vehicle systems very difficult [21]. Thus, when vehicle systems show abnormal signal patterns due to vehicle system errors, elderly drivers could have a very difficult time responding to the signals.

In addition to sensory problems, elderly drivers often undergo cognitive decline with a loss of short-term memory, becoming more easily distracted, and declining in analytical and communication power, and reading skills [17]. Elderly users might also experience difficulties understanding compound and large amounts of text, and following long and complicated instructions [3]. If a lot of unfamiliar information is displayed on the vehicle system, they could have a higher chance of getting lost when trying to manage the system. Thus, they often become confused and feel helpless when they encounter error messages from vehicle systems. This confusion and helplessness can affect their confidence in using current vehicle systems, and limit their ability to use different systems in the vehicle.

Elderly drivers do not only face a physical and age-related impairment. The different age groups can also pose unique challenges for elderly drivers since they have less experience with modern technology than other age groups [9]. Lack of experience with particular interfaces and systems in the vehicle can make them more difficult to interpret. An owners-manual for user's tutorials is also difficult for the elderly to understand due to a lot of unfamiliar jargon. Thus, elderly drivers can be more fearful to use vehicle systems with modern technology [22].

3 Analysis of current vehicle system modes for detecting vehicle symptoms

Due to age-related conditions and less experience with modern technology, there are many possibilities for this age group to have more financial losses in repairing and maintaining current vehicle systems. With rapidly improved technologies in vehicle electronics, control, and software contents for vehicle diagnostic systems, using and understanding systems has become even more difficult for the elderly users.

To have an accurate and easier detection method in vehicle symptoms, many studies [1, 4, 6, 12, 13, 14, 16, 20, 22] have been introduced in vehicle diagnostic system concepts and networks. However, we still need to find better ways to diagnose vehicle failure modes.

For example, Pattada and Signh [12] introduced a basic diagnostic process based on the Fault Detection and Isolation (FDI) model. In their study, FDI is defined as a subfield of control engineering that concerns itself with monitoring a system, identifying when a fault has occurred, and pinpointing the type of fault and its location. Also, the fault-model-based interactive service procedure tool is discussed as a guidance role in this study. The tool uses a fault model, and is built using service-procedures information, historical repair data, and engineering inputs. This fault model captures dependencies between multiple symptoms, such as Diagnostic Trouble Codes (DTC), vehicle-operational parameters, customer complaints, and technician-test outcomes and relates them to probable faults. In the service bay, this tool employs the fault model and vehicle symptoms to provide a ranked list of suspect components, related repairs, and the next steps necessary to further isolate the fault. In effect, as the technician collects and inputs more symptoms or test outcomes with the tool, it isolates the fault to specific components and recommends further actions until the fault is isolated to one specific component, and then recommends the correct repair. When examining the fault-model-based diagnostic system, this study pointed out one problem. An expert technician who is familiar with service procedures can infer faults and failure modes from DTCs and PIDs and perform correct repairs with a very high probability. But, a less-experienced technician can perform service procedures that lead to a lower probability of correct first-time fixes and increased repeat visits. It would also lead to increased troubleshooting time and increased labour costs. This means that the drawback on this system should be fixed to reduce human errors and a new and supplementary system to appropriately mange the system errors from the fault model inside vehicles should be developed.

Also, Palai [1] mentioned a limitation of existing fault management: Typically, fault-detection techniques are limited to determine whether some fault has occurred or not. The problems associated with the existing practice are that for any given fault, multiple Diagnostic Trouble Codes (DTCs) are registered by multiple Electric Control Units (ECUs), from the central module to the entire system, which results in confusion for service technicians. It does not focus on correctness of the data received from the network nor monitor the root cause of such faults over various vehicle-manoeuvring conditions. This leads to increased time to repair a given fault. Therefore, he stated that a vehicle-level system that supports fault management and a suitable vehicle network should be created.

Given this fact, electric and computer systems play a key role in controlling the fault modes and should properly communicate with diverse modules in the vehicle. Also, such a current system mode could make it difficult to detect and diagnose vehicle errors for users. Therefore, we need to explore and refine better ways to provide a tool for non-experts like elderly drivers, as well as service technicians, allowing them to easily, quickly, and effectively repair and maintain vehicles.

4 A current assistive device for vehicle repair and maintenance

I-Mechanic is the first Augmented Reality application from Inglobe Technology [7] based on an innovative computer vision 3D tracking software for smartphones to support people in ordinary maintenance of vehicles anytime and anywhere. Car engines and equipment can be a totally unknown universe for most of us. Even if a person with significant driving experience is driving in a new car, the person might not know where the most relevant parts in the engine are exactly located. With this mobile application, anyone can contextually access the instructions required to maintain their car, where and when it is necessary. Also, this application can provide a person with useful information on nearby auto repair and parts stores with the opportunity to reach the person in a more effective and contextual way.

This application must be sufficiently good to show people exactly where vehicle errors are located by showing problematic components in the vehicle with a blue mark via a smartphone display. However, it shows only one way to fix the vehicle errors although many causes can exist within one component error. Furthermore, the function of the detection-system application is limited to engine detection, and it only works properly when drivers open the front hood of the vehicle. In addition, it only provides closest auto repair shops without repair and maintenance cost details. Thus, when an elderly driver goes to the store that the system indicates, they may still pay more than they expected, thus repeating the current issue. Therefore, a smart assistive tool should be designed and developed which can detect vehicle errors easily and effectively, indicate them with more effective designs than the current one, and offer diverse ways to repair the errors.

5 Research methods

5.1 Questionnaire study/survey for identifying initial system concepts

Since smartphone ownership among older adults has risen in recent years, their convenient vehicle repair and maintenance service via a smartphone display could make them more easily accessible at an affordable cost. Thus, we decided to design and develop application system concepts considering the physical and mental conditions of the elderly.

To propose an adequate application system concept for the elderly to have a usable vehicle repair and maintenance system, surveys and interviews were conducted with elderly focus group. These short research activities were conducted in Seoul, Korea where many elderly drivers actively drive around the city area. For the first activity, the survey was conducted with 50 participants aged 60-75 with at least 20 years of driving experience and age-related health issues, such as decreased eyesight, lowered memory, and slower reaction times. The survey questions asked about their specific situations when they have vehicle system errors and visit repair and maintenance stores. Based on the short survey questions, interviews were also conducted with the participants. The fifty interviewees were divided into 3 groups: The age of 60-65 (15 persons), 65-70 (15 persons), and 70-75 (20 persons) and the interview session had been taking for 3 hours. Once the survey and interviews were returned, the resulting data were analysed (see Table 1).

From the survey data, we learned that elderly drivers tend to go to auto repair stores at an average of one to three times per month and have vehicle errors predominantly in the infotainment systems. They have difficulty easily recognizing error signals. Particularly, in the vehicle infotainment system, a lapse in recognizing visual signs is often reported in other studies. According to a report from Parker, Reason, Manstead, and Stradling about driving errors, elderly respondents reported more lapses than other full-age-range samples of drivers in the recognition of the visual signs [2]. This fits in with what is known about the declining cognitive capacities of older people and the increasing likelihood of problems with attention. Villalba, Kirk, and Stamatiadis also stated in their study that head-on crashes requiring a driver's visual recognition and search capabilities occurred most in the elderly group [11]. Also, 83% of elderly participants feel afraid when vehicle errors are occurred. The main reason is that they are worried about the money required to pay for repairing vehicle errors. They also said that since a lot of high technologies have been applied to current vehicle systems, they thought that repairing the vehicle errors would be more expensive than it had been in the past. Some elderly people said that technicians in auto repair stores could overcharge them for repairing services more than other age groups because they have less experience with modern technology in the vehicle and merely know about current vehicle system modes. Also, from the last survey question, 78% of the elderly participants feel that they pay more than they expected and they stated they have difficulty finding a trustworthy repair shop to fix their vehicles for reasonable prices.

Furthermore, the elderly group could not understand technical jargon for the vehicle systems and most of the words are English-based. Thus, they often get lost when technicians are explaining vehicle system errors. Based on the analysis of the survey and interview results, current repair and maintenance services are not effective and beneficial enough for the elderly and need to be improved.

After analysing the survey data of the elderly experiences, the focus group interviews were conducted to establish a repair and maintenance service system for them and the focus group was asked if they had suggestions for an application system that would be more beneficial for their driving life.

Table 1. Survey data from 50 elderly drivers.

Q1	How often do you go to auto repair stores per month?	
	① 0	0
	② 1-3 times	35
	③ 3-6 times	12
	④ Over 6 times	3
Q2	What kind of vehicle error have you had?	
	① Engine	15
	② Infotainment	30
	③ Seating system	4
	④ Lighting Systems	1
Q3	Could you catch error signals in the vehicle well?	
	① Yes	18
	② No	32
Q4	If yes, do you feel afraid when you have errors from your vehicle?	
	① Yes	15
	② No	3
Q4-1	If yes, what makes you nervous from the situation?	
	① Doesn't know why and where the errors are occurred.	4
	② Would pay a lot of money for the repair	6
	③ Where to go for repairing my vehicle	3
	④ Above all	2
Q5	Could you understand when an auto-technician explains your vehicle conditions?	
	① Yes	3
	② No	47
Q6	Do you think that you have paid more than you expected?	
	① Yes	39
	② No	11

5.2 Focus group study on confirming and enriching the initial system concepts

Based on the research, identical application concepts specifically for the elderly could be considered with diverse suggestions from the elderly focus group. Here are some application concept proposals in both design and technological aspects based on the previous research data gathered.

5.2.1 Design aspect

When designing a proper system for the elderly group, there are some elements that should be considered to help them use it without errors. Also, design and technology should be well-collaborated to satisfy and support many elderly drivers. Thus, during the interview session with the focus group, we asked the elderly interviewees what design elements should be considered for the repair and maintenance service application. From the interview results, all elderly drivers complained about text and icon sizes in the current applications. Due to their failing eyesight, their sizes are not big enough to read, so they often can't easily recognize functional menu options in the applications. If the text sizes and icon images enlarged, they would easily perceive the options in the application and would react quicker when another option needs to be processed. Moreover, those with arthritis and swollen fingers can have limited usability to communicate with the applications. A growing number of applications offer numerous multi-finger gestures for new touchscreen interfaces. Those with upper limb and mobility problem could face difficulties in performing different gestures like touch buttons, and controlling sliding functionalities.

According to Ramussem, Pejtersen, & Goodstein[10], "Information presented by the system should be offered in such a way that it can be easily and unambiguously understood without any additional help". So, the iconic images should be carefully simplified with representative images for the elderly.

5.2.2 Technological aspect

Identical application concepts considering diverse technologies for the elderly could be developed with feasible system ideas.

Here are some technical feature concepts based on previous research data.

First, indicating vehicle failure errors throughout the application can allow the elderly driver to easily find problematic parts in the vehicle. The interviewees stated that they have had many experiences when they don't know where the vehicle errors are occurred. So, they just followed a technician's directions, and sometimes required many more repair parts than they thought.

And, a technical function showing closest repair stores with prices of different vehicle components should be created to help the elderly repair their vehicle for a reasonable price. The main complaint from the elderly drivers in the repair and maintenance service is that they pay more than they expected. Most interviewees said that if they have a common repair price list with each vehicle component in the different vehicle models, they could have a chance to negotiate the repair price and select the repair shops within their budget. Thus, detailing the closest stores with comparative repair prices can help the elderly drivers avoid wasting their money on the service.

Lastly, an instruction voice and sound would be helpful for elderly drivers with vision problems. The elderly interviewees indicated that they could not understand current vehicle system manuals because most of the contents are described with small text size and are difficult to find. Thus, most elderly drivers have merely read the manuals although a lot of vehicle information is explained in it. So, if the application has a function to explain each current component audibly, they could understand the vehicle parts more easily and have more knowledge about vehicle systems.

5.2.3 Concept recommendations for the elderly drivers

Based on the previous proposal, we conducted the application concept development for elderly drivers. In this development process, suggested design and technological aspects were developed with diverse concepts. To determine whether the application concepts would be beneficial to elderly drivers, we asked the focus group to evaluate them by rating them on a scale of 1 to 5, with 1 being the minimum and 5 being the maximum. We asked them to rate the concepts using the following criterion: ease of use, frequency, and whether it was a good match for its intended user group. Furthermore, another interview session was conducted to evaluate the usefulness of the application concepts. The focus group was asked about how often each person would use the application system if it existed and if they had suggestions for other design concepts that would be more beneficial to aid with vehicle repair and maintenance.

6 Results

The data was collected from subjective concept ratings from the elderly focus group and the focus-group suggestions from the interview. From this data, we finalized fine-design concepts for elderly drivers that have the highest ratings in terms of their ease of use, frequency, and whether they are a good match for elderly drivers.

6.1 Designing set of application concepts

To be a properly defined application design concept that is marketable as a future application design for elderly drivers, the sets of design concepts considered their physical conditions and minimized drawbacks of previously proposed design systems.

First, indicating vehicle failure errors were designed with cues on the problematic part, such as a red marking, a red marking with a blinking effect, and a red marking with text description because a blue marking used in the current application is not easily recognizable and the color was not effective enough to draw their attention to the problematic parts.

Second, showing closest stores on the map was provided with two options: showing the store map with different prices and indicating a store with the cheapest price based on the price comparison.

Third, a voice instruction concept was systemized with two functions. With one, all instructions are explained with a voice for all the options and with the other, only a warning sound and voice instructions indicating a problematic part and a way of repair and maintenance were suggested.

6.2 Evaluation of the system concepts and finalized application concepts

After designing the application concepts, the focus group rated them based on previously defined criterion. The rating results are also in Table 2.

Table 2. Designed system concept evaluation.

Proposed Systems for The elderly	**Potential Design Concepts**	**Evaluation**			
		Ease of use	Frequency	Fit for the elderly	Total
Indicating vehicle errors	Red marking	2	3	2	7
	Red marking with a blinking effect	4	3	4	11
	Red marking with text description	4	5	5	14
Informing auto repair shops	Providing a shop on the map with different prices	4	3	3	10
	Providing a shop on the map in the cheapest price	4	4	5	13
Sound function	Voice function for all vehicle components	3	3	4	10
	Voice function for a problematic part with instruction	4	5	5	14

After rating each concept from the focus group, the final preferred application concepts were suggested. From the final systems, we can see how each concept has affected the focus group conditions. The final concepts are shown in the Table 3.

First, indicating vehicle failure error with a red marking with a test description was selected by the focus group. The elderly focus group said that they mostly want to know what kind of problem their vehicle has. If the application shows a red marking with text on a problematic part, they would easily recognize where to repair. However, if there is only a marking with a colour or blinking cue, sometimes they would be confused or misunderstand where the problematic part is since many vehicle systems are connected with each other. Also, elderly focus group suggests a different colour marking on the problematic part based on a level of severity for that part. For example, a red marking would indicate that a vehicle owner should repair the system as soon as possible and a yellow marking means that there is a symptom on a certain part that requires attention at some point.

Second, providing an auto repair shop with the cheapest price was more preferable for elderly drivers because they don't want to check all the stores for different prices. Particularly, selecting and showing the cheapest repair price would be very effective for elderly drivers with diminished memories and vision. Several elderly people said that the closest shops with different prices might give them more opportunity to select a shop. However, if they check all the prices in the stores, they stated feeling dizzy and too stressed to choose. Also, the focus group proposed that proving the shop's pros and cons with the cheapest price would be more beneficial for them when selecting a repair shop.

Table 3. Designed system concept evaluation

Proposed systems for the elderly	Finalized design concepts	Feature
Indicating vehicle errors	Red marking with text description	
Informing auto repair shops	Providing a shop on the map in the cheapest price	
Sound function	Voice function for a problematic part with instruction	

For the sound function, a warning sound on a vehicle error and explaining how to repair the erroneous parts was chosen from the focus group. They said that if they have too much information for the part, they can not remember everything due to their memory deficits. Knowing only basic, essential information would be enough for them. They suggested that a voice explanation should be operated with Stop, Rewind, Fast Forward, and Start options because they need to carefully check if they miss the name of problematic parts, where the parts are located, and how to repair the parts. Also, the elderly focus group said that if both voice explanation and text are used at the same time, they would feel more confident to use the system and would be able to understand difficult jargon for vehicle systems.

7 Conclusion

In this paper, we analysed the drawbacks of existing vehicle detection modes and application systems and highlighted the significance of system design based on a particular user group, i.e. the elderly. We also surveyed and interviewed 50 elderly people to identify their repair and maintenance difficulties and gain insight into what type of application systems would be most beneficial to them. Finally, we proposed and developed essential application concepts that the focus group needs and that could mitigate issues the elderly currently have and provide a convenient and enjoyable maintenance experience. However, a limitation of our study was that the application concepts for the elderly group were not fully tested with an actual vehicle system. The finalized system concepts could be changed from actual tests. In future work, we would test how the finalized concept works for the elderly group.

References

1. D. Palai, SAE Int., **6**, 261 (2013)
2. D. Parker, J.T. Reason, A.S. R. Manstead, and S.G. Stradling, Ergo. **38**, 1036 (1995)
3. D. Williams, M.A. Ul Alam, S.I. Ahamed, and W. Chu., QSIC, **373** (2013)
4. E. Armengaud, A. Tengg, R. Lieger, and E. Kristen, SAE Int., **1**, 1044 (2011)
5. F. Nunes, *Int. Con. On Per. Tech. Rel. to Ass. Environ.*, **49**, 8 (2010)
6. G. Abusamra, R. Carpenter, and S. Kiselewich, SAE Int. Conf., **7** (2004)
7. Inglobe Technologies, Augmented Reality (2013), Retrieved from http://www.armedia.it/
8. IAM Policy and Research, *IAM Driving Road Safety* (IAM House, London, 2010)
9. J. Durick, T. Robertson, M. Brereton, F. Vetere, and B. Nansen, *Proceedings of the 25th Australian Computer-Human Interaction Conference: Augmentation, Application, Innovation, Collaboration* (ACM, New York, 2013)
10. J. Rasmussem, A.M. Pejtersen, and L.P. Goodstein, *Cognitive Systems Engineering* (Wiley & Sons, Inc., New York, 1994)
11. J. Villalba, A. Kirk, and N. Stamatiadis, SAE Int., **1**, 367 (2001)
12. K. Pattada and S. Singh, SAE Int., **1**, 749 (2011)
13. M. Mateos, P. Robin, S. Sauvage, V. Joloboff, G. Madhusudan, Y. Bennani, P. Dague, and L. Bourgeols, SAE Int. Con., **5** (2002)
14. M.R. Busse, A. Israeli, and F. Zettelmeyer, *Repairing the Damage: The Effect of Price Expectations on Auto-Repair Price Quotes* (National Bureau of Economic Research, Cambridge, MA, 2013)
15. National Highway Traffic Safety Administration, *Traffic Safety Facts: Older Population* (NHTSA, Washington, 2014)
16. P. Peti, R. Obermaisser, A. Ademaj, and H. Kopetz, *Parallel and Distributed Processing Symposium* (PDPS, Atlanta, 2005)
17. R. Leung, J. McGrenere, and P. Graf, *Behav. & Infor.* Tech., **30**, 5 (2011)
18. Ripoff Report, National-Auto-Service-Centers (2015), Retrieved from http://www.ripoffreport.com/r/National-Auto-Service-Centers/Port-Richey-Hudson-Florida-34668/National-Auto-Service-Centers-Company-over-charges-for-parts-elderly-people-most-excessi-231387
19. S. Sayago, D. Sloan, and J. Blat, *Inter. with Com.*, **23**, 543 (2011)
20. T. Robertson, SAE Int., **1**, 1978 (2014)
21. W3C Web Accessibility Initiative, W3 Org. (2013), Retrieved from http://www.w3.org/WAI/older-users/
22. Y. Zhang, G.W. Gantt Jr., M. Rychlinski, R. Edwards, J. Correia, and C. Wolf, *International Conference on Prognostics and Health Management* (ICPHM, Denver, 2008)

A backtracking evolutionary algorithm for power systems

Ji-Pyng Chiou[1,a], Chung-Fu Chang[2] and Chin-Ju Li[1]

[1]*Department of Electrical Engineering, Ming Chi University of Technology, New Taipei City 24301, Taiwan*
[2]*Department of Electrical Engineering, WuFeng University, Chiayi County 62153, Taiwan*

Abstract. This paper presents a backtracking variable scaling hybrid differential evolution, called backtracking VSHDE, for solving the optimal network reconfiguration problems for power loss reduction in distribution systems. The concepts of the backtracking, variable scaling factor, migrating, accelerated, and boundary control mechanism are embedded in the original differential evolution (DE) to form the backtracking VSHDE. The concepts of the backtracking and boundary control mechanism can increase the population diversity. And, according to the convergence property of the population, the scaling factor is adjusted based on the 1/5 success rule of the evolution strategies (ESs). A larger population size must be used in the evolutionary algorithms (EAs) to maintain the population diversity. To overcome this drawback, two operations, acceleration operation and migrating operation, are embedded into the proposed method. The feeder reconfiguration of distribution systems is modelled as an optimization problem which aims at achieving the minimum loss subject to voltage and current constraints. So, the proper system topology that reduces the power loss according to a load pattern is an important issue. Mathematically, the network reconfiguration system is a nonlinear programming problem with integer variables. One three-feeder network reconfiguration system from the literature is researched by the proposed backtracking VSHDE method and simulated annealing (SA). Numerical results show that the perfrmance of the proposed method outperformed the SA method.

1 Introduction

Recently, many mathematic algorithms have used to solve various industrial systems. Among mathematical algorithms, the evolutionary algorithms (EAs) have been widespread interest based on advantages including ease to use, robust, and so on. EAs are a class of stochastic search and optimization methods that include genetic algorithms (GAs), evolutionary programming (EP), evolution strategies (ESs), genetic programming (GP), and their variants [1]. These algorithms, based on the principles of natural biological evolution, have received considerable and increasing interest over the past decade. EAs operate on a population of potential solutions, applying the principle of survival of the fittest to produce successively better approximations to a solution. EAs are robust and suitable for effectively obtaining optima, and with a smaller probability than other algorithms of falling into local optima. The methodology of the EAs has been discussed in many literatures [2-9], and EAs are also successfully used in various industrial applications [10-15].

[a] Corresponding author : jipyng@mail.mcut.edu.tw

Variant methods of EAs were proposed to increase the convergence speed and the capability of global search during development of EAs. Hybrid differential evolution (HDE) [16-17] is a EAs' variant method that is one of the most excellent EAs. HDE is a stochastic search and optimization method. The fittest of an offspring competes one-to-one with that of the corresponding parent, which is different from the other evolutionary algorithms (EAs). This one-to-one competition gives rise to a faster convergence rate. However, this faster convergence also leads to a higher probability of obtaining a local optimum because the diversity of the population descends faster during the solution process. To overcome this drawback, migrating operation and accelerated operation act as a trade-off operator for the diversity of population and convergence property in HDE. Migrating operation maintain the diversity of population, which guarantees a high probability of obtaining the global optimum. And, accelerated operation is used to accelerate convergence. However, a fixed scaling factor is used in HDE. Using a smaller scaling factor, HDE becomes increasingly robust. However, much computational time should be expanded to evaluate the objective function. HDE with a larger scaling factor should results generally fall into local solution or misconvergence. Lin et al. [16] used a random number that its value is between zero and one as a scaling factor. However, a random scaling factor could not guarantee the fast convergence. The selection of mutation operator is also a very important issue in HDE. The proper mutation operator can accelerate to search out the global solution [14]. However, the selection of mutation operator is problem-dependence. In HDE, the proper mutation operator is not easy to select. To overcome the drawback of the fixed and random scaling factor and alleviate the problem of selection of mutation operator in HDE, the concept of variable scaling factor is used in variable scaling hybrid differential evolution (VSHDE) method [18-19]. The rule of updating scaling factor based on the 1/5 success rule of the evolution strategies (ESs) [1, 20] is used to adjust the scaling factor. The 1/5 success rule emerged as a conclusion of the process of optimizing convergence rate of two functions (the so-called corridor mode and sphere model [1, 20-21]). To increase the population diversity, the concept of backtracking is used in the VSHDE method (called backtracking VSHDE method).

Power systems across the world have been recently deregulated, which resulted in competition among the power sellers. This resulted in increased complexity of distribution networks. Further the distribution system planner and operator are facing new challenges that put emphasis on optimal planning and operation of the distribution system. Reduction of power loss, improvement of power quality, cost minimization and load balance among the branches are important factors for better operation of the distribution system. These can be accomplished by optimal network reconfiguration [22]. Distribution systems consist of groups of interconnected radial circuits. The configuration may be varied via switching operations to transfer loads among the feeders. Two types of switches are used in primary distribution systems. There are normally closed switches (sectionalizing switches) and normally open switches (tie switches). Those two types of switches are designed for both protection and configuration management. Network reconfiguration is the process of changing the topology of distribution systems by altering the open/closed status of switches. Because there are many candidate-switching combinations in the distribution system, network reconfiguration is a complicated combinatorial, non-differentiable constrained optimization process aimed at finding optimal operation of the distribution system. Optimal feeder reconfiguration has been investigated over decades with heuristic and meta-heuristic techniques for single and multi-objectives in order to find optimal planning and operation of the distribution system [22-28]. A multi-objective invasive weed optimization algorithm is proposed by Sudha et al. [22] to solve the optimal network reconfiguration. While solving optimal network reconfiguration of the radial distribution system, minimization of active power loss, maximum node voltage deviation, number of switching operations and the load balancing index are considered as the objectives simultaneously. Huang et al. [23] proposed parallel undirected spanning tree-based genetic algorithm (PSTGA) to solve the network reconfiguration in parallel. Liu et al. [24] proposed an improved type of immune algorithm to ensure that the algorithm can quickly converge to the global optimal solution to improve the efficiency of the algorithm solving and solution accuracy. Nguyen et al. [25] proposed a reconfiguration methodology based on a cuckoo search algorithm (CSA) for minimizing active power loss and the maximizing voltage magnitude.

Syahputra et al. [26] proposed particle swarm optimization (PSO) algorithm based multi-objective optimization for reconfiguration of radial distribution network with the presence of distributed energy resources (DER). The benefits of DER integration in distribution system are reducing power losses, improving voltage profiles and load factors, eliminating system upgrades, and reducing environmental impacts. Sureshkumar et al. [27] presented the use of differential evolution method to reduce the losses and balancing the loads in the radial distribution network. Iswarya et al. [28] deals with reconfiguration of distribution system to minimize the power loss and considering reliability.

In this study, a backtracking VSHDE for solving the network reconfiguration of distribution systems is proposed. Here, 1/5 success rule of evolution strategies (ESs) [20-21] is used to adjust the scaling factor to accelerate searching out the global solution. In addition, the concept of backtracking used in the proposed method to increase the population diversity. One three-feeder distribution system from the literature is solved by the proposed method is sloved respectively by the proposed method and SA. From the computational results, it is observed that the convergence property of the bracktracking VSHDE method is better than that of the SA method.

2 Backtracking VSHDE algorithm

The backtracking VSHDE method is briefly described in the following.

Step 1. Initialization

The initial population is chosen randomly and would attempt to cover the entire parameter space uniformly as (1).

$$\mathbf{Z}_i^0 = \mathbf{Z}_{i,\min} + \sigma_i \cdot (\mathbf{Z}_{i,\max} - \mathbf{Z}_{i,\min}),\ i = 1,\ldots,N_p \tag{1}$$

Where $\sigma_i \in (0,1]$ is a random number, $\mathbf{Z}_{i\min}$ and $\mathbf{Z}_{i\max}$ are the lower bounds and upper bounds of i th individual, respectively. The initial process can produce N_p individuals of $\mathbf{Z}_i^0$ randomly.

In addition, the initial historical population is also generated using equation (1) to maintain the population diversity.

Step 2. Backtracking Operation [29]

Increasing the population diversity to increase the probability of searching out the global solution is the aim of the backtracking operation. Backtracking operation has two strategies. First, two random numbers between 0 and 1 are generated. Second, if the first random number is less than the second random number, then current population is replaced with the historical population and re-permuting the every individual's position randomly in the population. Otherwise, if the first random number is greater or equal than the second random number, then current population is only re-permuting the individual's position in the historical population.

Step 3. Mutation operation

The essential ingredient in the mutation operation is the difference vector. Each individual pair in a population at the G-th generation defines a difference vector $\mathbf{D}_{jk}$ as

$$\mathbf{D}_{jk} = \mathbf{Z}_j^G - \mathbf{Z}_k^G \tag{2}$$

The mutation process at the G-th generation begins by randomly selecting either two or four population individuals $\mathbf{Z}_j^G$, $\mathbf{Z}_k^G$, $\mathbf{Z}_l^G$, and $\mathbf{Z}_m^G$ for any $j, k, l,$ and m. These four individuals are then combined to form a difference vector $\mathbf{D}_{jklm}$ as

$$\mathbf{D}_{jklm} = \left(\mathbf{Z}_j^G - \mathbf{Z}_k^G\right) + \left(\mathbf{Z}_l^G - \mathbf{Z}_m^G\right) \tag{3}$$

A mutant vector is then generated based on the present individual in the mutation process by

$$\hat{\mathbf{Z}}_i^{G+1} = \mathbf{Z}_p^G + F \times \mathbf{D}_{jklm},\ i = 1, 2, \cdots, N_p \tag{4}$$

Where F is the scaling factor. Furthermore, $j, k, l,$ and m are randomly selected.

The concept of variable scaling factor used in the proposed method is to overcome the drawback of the fixed and random scaling factor. The rule of updating a scaling factor based on the 1/5 success rule of the ESs is used to adjust the scaling factor. The 1/5 success rule emerged as a conclusion of the process of optimizing convergence rate of two function(the so-called corridor mode and sphere model) [1, 20-21].The scaling factor should be updated in every q iterations as follow:

$$F^{t+1} = \begin{cases} c_d \times F^t & \text{if } p_s^t < \dfrac{1}{5} \\ c_j \times F^t & \text{if } p_s^t > \dfrac{1}{5} \\ F^t & \text{if } p_s^t = \dfrac{1}{5} \end{cases} \tag{5}$$

Where p_s^t is the frequency of successful mutations measured. The initial value of the scaling factor, F is set to 2 [30-31]. The factors of $c_d = 0.82$ and $c_j = 1/0.82$ [1] are used for adjustment, which should be taken place for every q iterations. When the migrating operation performed or the scaling factor is too small to find the better solution, the scaling factor is reset as follow:

$$F = 1 - \frac{iter}{itermax} \tag{6}$$

Where *iter* and *itermax* are the number of current iteration and the maximum iteration, respectively.

Step 4. Crossover operation

The perturbed individual of $\hat{\mathbf{Z}}_i^{G+1}$ and the present individual of $\mathbf{Z}_i^G$ are chosen by a binomial distribution to progress the crossover operation to generate the offspring. Each gene of i-th individual is reproduced from the mutant vectors $\hat{\mathbf{Z}}_i^{G+1} = \left[\hat{\mathbf{Z}}_{1i}^{G+1}, \hat{\mathbf{Z}}_{2i}^{G+1}, \cdots, \hat{\mathbf{Z}}_{ni}^{G+1}\right]$ and the present individual $\mathbf{Z}_i^G = \left[\mathbf{Z}_{1i}^G, \mathbf{Z}_{2i}^G, \cdots, \mathbf{Z}_{ni}^G\right]$. That

$$\hat{Z}_{gi}^{G+1} = \begin{cases} Z_{gi}^G & \text{if a random number} > C_r \\ \hat{Z}_{gi}^{G+1} & \text{otherwise} \end{cases} \tag{7}$$

Where $i = 1, \cdots, N_p$; $g = 1, \cdots, n$; and the crossover factor $C_r \in [0, 1]$ is assigned by the user.

Step 5. Boundary Control Mechanism

When the gene of the child is less than the lower bound, two random numbers between 0 and 1 are generated. If the first random number is less than the second random number, the gene of child set to lower bound, otherwise, the gene of child is chosen randomly in an attempt to cover the entire parameter space uniformly. Similarly, when the gene of the child is greater than the upper bound, two random numbers between 0 and 1 are generated. If the first random number is less than the second random number, the gene of child set to upper bound, otherwise, the gene of child is chosen randomly in an attempt to cover the entire parameter space uniformly.

Step 6. Estimation and selection

The evaluation function of a child is one-to-one competed to that of its parent. This competition means that the parent is replaced by its child if the fitness of the child is better than that of its parent. On the other hand, the parent is retained in the next generation if the fitness of the child is worse than that of its parent, i.e.

$$\mathbf{Z}_i^{G+1} = \arg\min\left\{f\left(\mathbf{Z}_i^G\right), f\left(\hat{\mathbf{Z}}_i^{G+1}\right)\right\} \tag{8}$$

$$\mathbf{Z}_b^{G+1} = \arg\min\left\{f\left(\mathbf{Z}_i^{G+1}\right)\right\} \tag{9}$$

Where *arg min* means the argument of the minimum.

Step 7. Migrating operation if necessary

In order to effectively enhance the investigation of the search space and reduce the choice pressure of a small population, a migration operation is introduced to regenerate a new diverse population of individuals. The new population is yielded based on the best individual $\mathbf{Z}_b^{G+1}$. The g-th gene of the i-th individual is as follows:

$$Z_{ig}^{G+1} = \begin{cases} Z_{bg}^{G+1} + \sigma_i \times \left(Z_{g\min} - Z_{bg}^{G+1}\right) & if\ \delta < \dfrac{Z_{bg}^{G+1} - Z_{g\min}}{Z_{g\max} - Z_{g\min}} \\ Z_{bg}^{G+1} + \sigma_i \times \left(Z_{g\min} - Z_{bg}^{G+1}\right) & otherwise \end{cases} \tag{10}$$

Where σ_i and δ are randomly generated numbers uniformly distributed in the range of [0,1]; $i = 1, \cdots, N_p$; $g = 1, \cdots, n$.

The migrating operation is executed only if a measure fails to match the desired tolerance of population diversity. The measure is defined as follows:

$$\varepsilon = \sum_{i=1, i\neq b}^{N_p} \sum_{g=1}^{n} \frac{\eta_Z}{n \times \left(N_p - 1\right)} < \varepsilon_1 \tag{11}$$

Where

$$\eta_Z = \begin{cases} 0 & if\ \varepsilon_2 < \left|\dfrac{Z_{gi}^{G+1} - Z_{bi}^{G+1}}{Z_{bi}^{G+1}}\right| \\ 1 & otherwise \end{cases} \tag{12}$$

Parameter $\varepsilon_1, \varepsilon_2 \in [0, 1]$ express the desired tolerance for the population diversity and the gene diversity with respect to the best individual.

Step 8. Accelerated operation if necessary

When the best individual in the present generation cannot be improved any longer by the mutation and crossover operations, a decent method is then employed to push the present best individual towards attaining a better point. The accelerated phase is expressed as follows:

$$\mathbf{Z}_b^{G+1} = \begin{cases} \mathbf{Z}_b^{G+1} & \text{if } f\left(\mathbf{Z}_b^{G+1}\right) < f\left(\mathbf{Z}_b^{G}\right) \\ \mathbf{Z}_b^{G+1} - \alpha \nabla f & \text{otherwise} \end{cases} \tag{13}$$

Where $\mathbf{Z}_b^G$ denotes the best individual, as obtained from equation (9). The step size $\alpha \in (0, 1]$ in (13) is determined by the descent property. Initially, α is set to one to obtain the new individual.

Step 9. Repeat step 2 to step 8 until the maximum iteration quantity or the desired fitness is accomplished.

The main calculation procedure of the backtracking VSHDE method is as shown in Fig. 1.

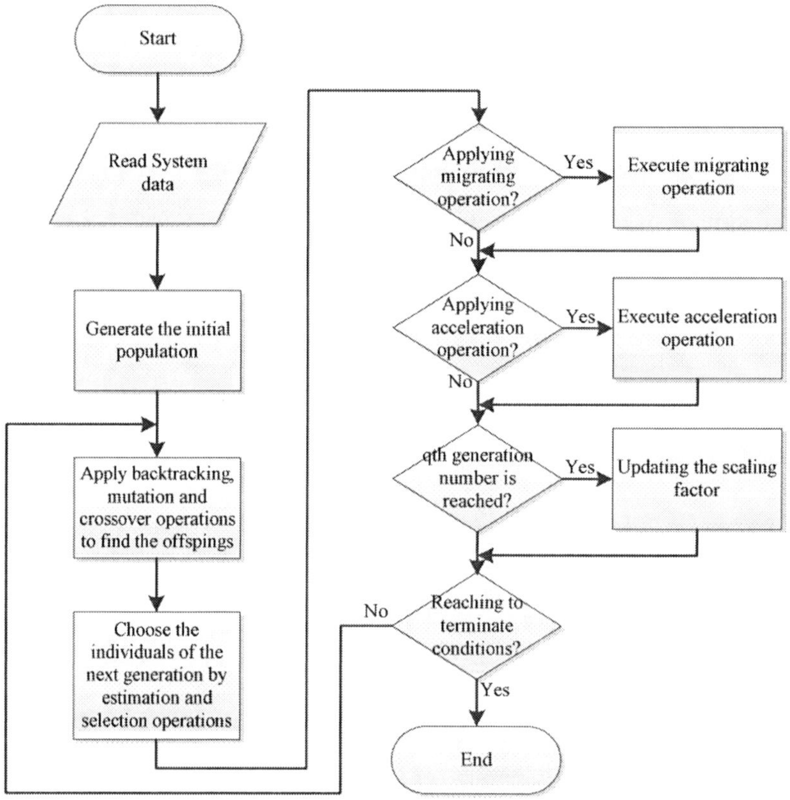

Figure 1. Main calculation procedures of the backtracking VSHDE method

3 Problem formulation

This network reconfiguration problem aims to minimize the system power loss, subject to operating constraints under a certain load pattern. The mathematical model of the problem can be expressed as follows:

$$\min f = \min \left(P_{T,Loss} \right) \tag{14}$$

Where $P_{T,Loss}$ is the total real power loss of the system.

The voltage magnitude at each bus must maintain within its limits. The current on each branch has to lie within its capacity rating. These constraints are expressed as follows

$$V_{\min} \le |V_i| \le V_{\max} \tag{15}$$

$$|I_i| \le I_{i,\max} \tag{16}$$

Where

$|V_i|$ voltage magnitude of bus i

$V_{\min}$, $V_{\max}$ bus minimum and maximum voltage limits

$|I_i|$, $I_{i,\max}$ current magnitude and maximum current limit of branch i

A set of simplified feeder-line flow formulations is employed. Considering the single-line diagram depicted in Fig.2, the following set of recursive equations is used for power flow computation [32-33].

$$P_{i+1} = P_i - P_{Li+1} - R_{i,i+1} \cdot (P_i^2 + Q_i^2) / |V_i|^2 \tag{17}$$

$$Q_{i+1} = Q_i - Q_{Li+1} - X_{i,i+1} \cdot (P_i^2 + Q_i^2) / |V_i|^2 \tag{18}$$

$$|V_{i+1}|^2 = |V_i|^2 - 2(R_{i,i+1} \cdot P_i + X_{i,i+1} \cdot Q_i) + (R_{i,i+1}^2 + X_{i,i+1}^2) \frac{(P_i^2 + Q_i^2)}{|V_i|^2} \tag{19}$$

Where P_i and Q_i are the real and reactive powers flowing out of bus i , and P_{Li} and Q_{Li} are the real and reactive load powers at bus i . The resistance and reactance of the line section between buses i and $i+1$ are denoted by $R_{i,i+1}$ and $X_{i,i+1}$, respectively. The power loss of the line section connecting between buses i and $i+1$ may be computed as

$$P_{Loss}(i,i+1) = R_{i,i+1} \cdot \frac{P_i^2 + Q_i^2}{|V_i|^2} \tag{20}$$

The total power loss of the feeder, $P_{F,Loss}$, may then be determined by summing up the losses of the all line sections of the feeder. Which is given by

$$P_{F,Loss} = \sum_{i=0}^{n-1} P_{Loss}(i,i+1) \tag{21}$$

and the total system power loss, $P_{T,Loss}$, is the sum of power losses of all feeders in the system.

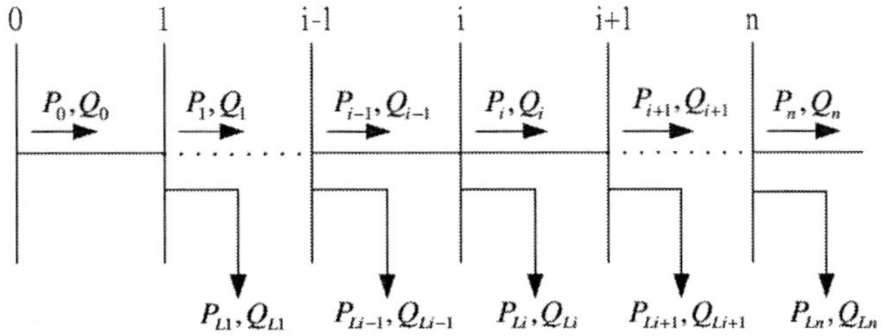

Figure 2. Single-line diagram of a main feeder.

4 Application of the proposed method

Implementation of the problem begins from the parameter encoding. A tie switch and some sectionalizing switches with the feeders form a loop. A certain switch of each loop is then selected to open to make the loop become radial and the selected switch naturally become a tie switch. The network reconfiguration problem is identical to the problem of selection of an appropriate tie switch for each loop so that the power loss can be minimized. A coding scheme that recognizes the tie switch positions is proposed. The total number of tie switches is kept constant regardless of the change in the system's topology or the tie switches' positions. Fig. 3 shows an individual composing of wherein the tie switch (TS) positions. Different switches from a loop are respectively selected to cut as a tie switch to decide its associated fitness value to determine a feasible solution (radial configuration) with minimum loss. The fitness function to be maximized is defined as follows

$$\max f = \min P_{T,Loss} = \min \sum_{k=1}^{M} \left(\sum_{i=0}^{n_k - 1} P_{Loss}(i, i+1) \right) \tag{22}$$

Where M is the total feeder number of the system and n_k is the total section number of feeder k.

TS No.(1)	TS No.(2)	· · ·	TS No.(n)

Figure 3. An individual composing of n tie switches.

One three-feeder distribution system from the literature is investigated and the results are used to compare the performance of the proposed backtracking VSHDE method with the SA method. The Fortran SA [34] algorithm solver accessed from http://www-aig.jpl.nasa.gov/public/home/decoste /HTMLS/NN/glopt/glopt.html#sa_codes, is used to solve the network reconfiguration of distribution systems. The SA solver recommended some setting factors for a user. The recommended factors are used for solving network reconfiguration of distribution systems. For comparison, SA package is rewritten by Matlab software.

Example:

The three-feeder distribution system [35] as shown in Fig. 4 Input data of this example system are shown in Table 1. The system consists of 3 feeders, 13 normally closed switches, and 3 normally open switches. The system load is assumed to be constant and $S_{base} = 100MVA$. The setting factors used in backtracking VSHDE to solve this example are as follows. The population size, N_p, is set to 5. The maximum generation, *itermax*, is set to 50. The crossover factor, C_R, is set to 0.5. Two tolerances, ε and ε_1, used in the migrating operation are both set to 0.1. The scaling factor is updating in every 10 iterations. Six strategies of mutation operation are respectively used to solve this example. To verify the performance of the proposed algorithm, this example was repeatedly solved one hundred times. Table 2 shows the best computational results. The best solutions obtained by the proposed method are same as that obtained from [36]. Number of the successful runs to achieve the best solution as [36] is 63, 64, 61, 79, 62 and 72 for six different strategies of mutation operation used in backtracking VSHDE, respectively. From the computational results, the selection of the mutation operator in backtracking VSHDE does not affect the performance of the proposed method significantly. The SA is also applied to solve this problem. The number of times to which the best solution obtained by the SA is 35. From the above discussion, it could be concluded that the performance of the proposed backtracking VSHDE method is better than the SA method.

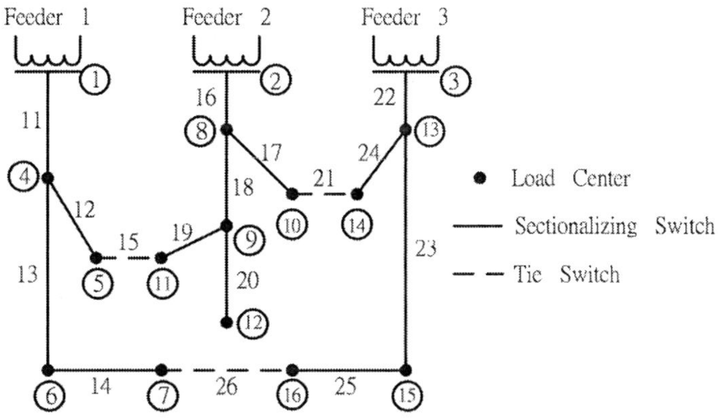

Figure 4. A three-feeder distribution system for example.

Table 1. Input data for example.

Bus to bus	Section resistance $(P.U.)$	Section reactance $(P.U.)$	End bus real load (MW)	End bus reactive load $(MVAR)$	End bus fixed capacitor $(MVAR)$
1-4	0.075	0.1	2.0	1.6	
4-5	0.08	0.11	3.0	1.5	1.1
4-6	0.09	0.18	2.0	0.8	1.2
6-7	0.04	0.04	1.5	1.2	
2-8	0.11	0.11	4.0	2.7	
8-9	0.08	0.11	5.0	3.0	1.2
8-10	0.11	0.11	1.0	0.9	
9-11	0.11	0.11	0.6	0.1	0.6
9-12	0.08	0.11	4.5	2.0	3.7
3-13	0.11	0.11	1.0	0.9	
13-14	0.09	0.12	1.0	0.7	1.8
13-15	0.08	0.11	1.0	0.9	
15-16	0.04	0.04	2.1	1.0	1.8
5-11	0.04	0.04			
10-14	0.04	0.04			
7-16	0.09	0.12			

Table 2. Numerical results of example.

Main items	Original configuration	After reconfiguration
Tie switches	15, 21, 26	19, 17, 26
Power loss (KW)	511.4	466.1
Voltage Magnitude $(P.U.)$	$V_{\max} = 1.0000$ (Bus 1, 2, 3)	$V_{\max} = 1.0000$ (Bus 1, 2, 3)
	$V_{\min} = 0.9693$ (Bus 12)	$V_{\min} = 0.9716$ (Bus 12)
Number of tie switches changed	2	
Loss reduction	8.86%	

5 Conclusion

Two heuristic methods including backtracking VSHDE and SA for solving the network reconfiguration of distribution systems have been described in this work. The bactracking VSHDE method utilized the 1/5 success rule of the evolution strategies (ESs) to adjust the scaling factor to accelerate searching out the global solution. The variable scaling factor is used to overcome the drawback of fixed and random scaling factor used in HDE. And, the concept of backtracking is used in this paper to increasing the population diversity to increase the possibility of searching out the global solution. The computational results obtained of sloving one three-feeder distribution system from the literature is investigated. From the computational results of solving one three-feeder distribution system from the literature showed the performance of the backtracking VSHDE method is better than those obtained by the simulated annealing (SA).

Acknowledgment

Financial research support from the Ministry of Science and Technology of the R. O. C. under grant MOST 104-2221-E-131-017 is greatly appreciated.

References

1. Z. Michalewicz, *Genetic Algorithms + Data Structures = Evolution Programs 3rd ed* (New York, 1999)
2. I. Zelinka, A survey on evolutionary algorithms dynamics and its complexity - Mutual relations, past, present and future, Swarm and Evolutionary Computation, **25** (1), 2-14 (2015)
3. B. Kazimipour and X. Li, A.K. Qin, A review of population initialization techniques for evolutionary algorithms, *Proceedings of IEEE Congress on Evolutionary Computation*, 2585-2592 (2014)
4. M. Crepinsek, S.H. Liu, and M. Mernik, Exploration and exploitation in evolutionary algorithms: A survey, ACM Computing Surveys, **45** (3), 35 (2013)
5. J. Prakash, P.K. Singh, Partitional algorithms for hard clustering using evolutionary and swarm intelligence methods: a survey, Advances in Intelligent Systems and Computing, **2**, 515-528 (2013)
6. B. Li, J. Li, K. Tang, and X. Yao, Many-objective evolutionary algorithms: a survey, ACM Computing Surveys, **48** (1), A10 (2015)
7. H. Ishibuchi, H. Masuda, Y. Tanigaki, and Y. Nojima, Review of coevolutionary developments of evolutionary multi-objective and many-objective algorithms and test problems, IEEE Symposium on Computational Intelligence in Multicriteria Decision-Making (MCDM), 178-184 (2014)
8. V.L. Vachhani, V.K. Dabhi, and H.B. Prajapati, Survey of multi objective evolutionary algorithms, IEEE International Conference on Circuit, Power and Computing Technologies (2015)
9. C. Von Lücken, B. Barán, and C. Brizuela, A survey on multi-objective evolutionary algorithms for many-objective problems, Computational Optimization and Applications, **58** (3), 707-756 (2014)
10. R.V. Devi, S.S. Sathya, and M.S. Coumar, Evolutionary algorithms for de novo drug design-a survey, Applied Soft Computing Journal, **27**, 543-552 (2015)
11. P.M. Pradhan and G. Panda, Comparative performance analysis of evolutionary algorithm based parameter optimization in cognitive radio engine: a survey, Ad Hoc Networks, **17**, 129-146 (2014)
12. M. Gen and L. Lin, Multiobjective evolutionary algorithm for manufacturing scheduling problems: State-of-the-art survey, Journal of Intelligent Manufacturing, **25** (5), 849-866 (2014)
13. S. Li, L. Kang, and X.M. Zhao, A survey on evolutionary algorithm based hybrid intelligence in bioinformatics, Bio Med Research International, **2014** (2014)
14. D.G.N. Rani and S. Rajaram, A survey on B*-Tree-based evolutionary algorithms for VLSI floorplanning optimisation, International Journal of Computer Applications in Technology, **48** (4), 281-287 (2013)
15. A. Ponsich, A.L. Jaimes, and C.A.C. Coello, A survey on multiobjective evolutionary algorithms for the solution of the portfolio optimization problem and other finance and economics applications, IEEE Transactions on Evolutionary Computation, **17** (3), 321-344 (2013)

16. Y.C. Lin, K.S. Hwang, and F.S. Wang, Plant scheduling and planning using mixed-integer hybrid differential evolution with multiplier updating, *Proceedings of the IEEE Conference on Evolutionary Computation*, **1**, 593-600 (2000)
17. J.P. Chiou and F.S. Wang, Hybrid method of evolutionary algorithms for static and dynamic optimization problems with application to fed-batch fermentation process, Computers & Chemical Engineering, **23**, 1277-1291 (1999)
18. J.P. Chiou, C.F. Chang, and C.T. Su, Variable scaling hybrid differential evolution for solving network reconfiguration of distribution systems, IEEE Transactions on Power Systems, **20** (2), 668-674 (2005)
19. J.P. Chiou, Variable scaling hybrid differential evolution for large-scale economic dispatch problems, Electric Power Systems Research, **77**, 212-218 (2007)
20. T. Back, F. Hoffmeister, and H.P. Schwefel, A survey of evolution strategies, *Proc. of 4th Int. Conf. Genetic Algorithms*, 2-9 (1991)
21. T. Back and H.P. Schwefel, An overview of evolutionary algorithms for parameter optimization, Evol. Comput., **1**, 1-23 (1993)
22. D. Sudha Rani, N. Subrahmanyam, and M. Sydulu, Multi-objective invasive weed optimization-An application to optimal network reconfiguration in radial distribution systems, International Journal of Electrical Power and Energy Systems, **73**, 932-942 (2015)
23. H. Huang, J. Gu, and C. Fang, Application of undirected spanning tree-based parallel genetic algorithm in distributed network reconfiguration, Dianli Xitong Zidonghua/Automation of Electric Power Systems, **39** (14), 89-96 (2015)
24. F.C. Liu, W. Xu, G. Zhang, W.Z. Wang, and Z.Y. Li, Distribution network reconfiguration based on immune clonal selection algorithm, Environment, *Proc. of 3rd International Conference on Frontier of Energy and Environment Engineering*, 657-660 (2015)
25. T.T. Nguyen and A.V. Truong, Distribution network reconfiguration for power loss minimization and voltage profile improvement using cuckoo search algorithm, International Journal of Electrical Power and Energy Systems, **68**, 233-242 (2015)
26. R. Syahputra, I. Robandi, and M. Ashari, PSO based multi-objective optimization for reconfiguration of radial distribution network, International Journal of Applied Engineering Research, **10** (6), 14573-14586 (2015)
27. K. Sureshkumar nad P. Vijayakumar, Distribution network reconfiguration for loss minimisation using differential evolution algorithm, ARPN Journal of Engineering and Applied Sciences, **10** (7), 2861-2866 (2015)
28. R. Iswarya and R.M. Sasiraja, Network reconfiguration of distribution system in presence of harmonic load using expert system approach, International Journal of Applied Engineering Research, **10** (55), 1961-1966 (2015)
29. P. Civicioglu, Backtracking Search Optimization Algorithm for numerical optimization problems, Applied Mathematics and Computation, **219**, 8121-8144 (2013)
30. R. Storn and K.V. Price, Minimizing the real functions of the ICEC '96 contest by differential evolution, IEEE Conference on Evolutionary Computation, 842-844 (1996)
31. K.V. Price, Differential evolution vs. functions of the 2nd ICEC, IEEE Conference on Evolutionary Computation, 153-157 (1997)
32. C.T. Su and C.C. Tsai, A new fuzzy reasoning approach to optimum capacitor allocation for primary distribution systems, *Proc. IEEE on Industrial Technology Conf.*, 237-241 (1996)
33. C.T. Su, C.S. Lee, and C.S. Ho, Optimal selection of capacitors in distribution systems, *Proc. IEEE Power Tech. Conf.*, 301 (1999)
34. B. Goffe, Global optimization of statistical functions with simulated annealing, Journal of Economics, **60** (12), 65-100 (1994)
35. S. Civanlar, J.J. Grainger, H. Yin, and S.S.H. Lee, Distribution feeder reconfiguration for loss reduction, IEEE Trans. Power Delivery, **3**, 1217-1223 (1988)
36. H.C. Cheng and C.C. Kou, Network reconfiguration in distribution systems using simulated annealing, Electric Power System Research, **29**, 227-238 (1994)

MATEC Web of Conferences **119**, 01047 (2017) DOI: 10.1051/matecconf/201711901047
IMETI 2016

A hybrid Big Bang-Big Crunch algorithm for energy optimization on heterogeneous distributed system

Yan Kang[1,a], Hao Li[2], Chunhui Wang[1] and Li Dai[1]

[1]*Department of Software Engineering, School of Software, Yunanan University, Kunming, 650091, China*
[2]*Department of Network Engineering, School of Software, Yunanan University, Kunming, 650091, China*

Abstract. A hybrid algorithm is presented for global optimization of the energy consumption rather than makespan of the system. The cost profile is optimized by extending the execution time of the tasks on dynamic voltage scalable processing elements in embedded environment with meeting the execution constraints. The modified Big Bang Big Crunch (BBBC) method improves the scheduling efficiency of the tasks by simulating one of the theories of the evolution of the universe. BBBC algorithm generates disorder data points inspired by energy dissipation procedure of the Big Bang phase, and moves those data points to a single representative data point inspired by a center of mass cost approach in the Big Crunch phase. The Logistic and Sinusoidal chaotic maps are investigated and utilized to improve the movement step of the BBBC algorithm. The proposed hybrid algorithm is tested on several benchmark data sets and its performance is compared with those of Ant Colony Optimization and HEFT strategies. The simulation experiment shows that the presented evolutionary optimization algorithm is robust and suitable for energy saving.

1 Introduction

Energy efficiency has become an important issue of the distributed system especially the embedded system which is composed of processing elements (PES) normally powered by batteries. There exist slacks among the system since the process ability of the processor is usually not fully utilized. The processors that enabled dynamic voltage scaling (DVS) strategy switches off gated clocks for unused circuit parts during idle periods and then reduce the power consumption. The power consumption is at least a quadratic function of the supply voltage (hence CPU speed) in the cased of function executed. It was shown in [1,2,3] that further energy consumption can be reduced by decreasing process frequencies during the voltage selection. Investigations of DVS processors have shown that the power consumption can be decreased by up to 10 times after using voltage scales when executing real-life applications [4].

DVS results in the different processing elements characteristics if utilize the slacks to extend the execution time of the processors. DVS decreases the energy consumption to extend the processing time of the tasks that have the slacks between the deadline and its completion time as slowly as possible in the case of the slacks. Yao. [5] Presented one of the earliest theoretical models for DVS, and presented an $O(n^3)$ algorithm for obtaining an optimal schedule and computing a characterization of the minimum-energy DVS schedule. The schedule gave a main benchmark for referencing other scheduling

[a] Corresponding author : kangyang@ynu.edu.cn

algorithms in both theoretical and simulation aspects under the assumption of convexity of the power consumption function.

There are static and dynamic techniques to using dynamic voltage scaling to decide appropriate operating voltage/speed. Static techniques take advantage of offline parameters, such as periods and worst-case execution cycles while dynamic techniques (based on slack reclamation) exploit early completions of tasks to further decrease the speed and energy consumption. Here, we investigate static one. Dynamic methods [6,7] normally reduce the speed based on these predicted values and achieve more energy saving than static ones. But the interest in static techniques is still high since static approaches can be enhanced to develop dynamic ones.

Three important co-synthesis steps are: a) Mapping: Determining the assignment of computational tasks to PEs and data transfers to communication links (CLs), b) Scheduling: Determining the execution order (sequencing) of tasks mapped to PEs and communications to CLs, and c) Evaluation: Determining the quality of the implementation candidate (timing feasibility, cost, power, area, etc.).

Pillai and Shin [8] computed a single optimal speed off-line and obtained the minimal speed to make a task set schedulable without changing it under Earliest Deadline First, and proposed a near-optimal method under Rate Monotonous. Without fixing the processor speed, Saewong and Rajkumar [9] assumed that the speed of the processor can be changed continuously in a given range and presented an algorithm to find the optimal speed value based on fixed priority assignments. Based on the task parameters, some methods in [10] were proposed to decide statically processor speed and assign different speed to every task before system execution. A more general scheme in [11] chose the speed switching instants more freely which typically occur at the activation/deadline of some job.

Classical programming approaches such as Non-linear programming (NLP), Linear programming (LP) and Dynamic programming methods are utilized to solve NP hard problem at the cost of increased dimensionality and depends on mathematical formula of the cost function [12]. Heuristics algorithms as the recent advances in AI techniques are often used as complementary approach to solve complex optimization problems such as NP hard problem. Heuristic techniques pave the way towards finding quality solutions to the NP hard problem during tolerable time from the inspiration of natural behavior of various social/natural procedures. And the heuristic methods are easy to transform and adapt according to specific problem. Genetic algorithms [13], Swarm intelligence[14], Bacteria foraging, ant colony search techniques are population based and stochastic in nature, and are utilized to find global/near global solutions to optimization problems.

As another population based algorithm, Big Bang and Big Crunch (BBBC) developed by Erol and Eksin [15] are inspired by the concept of universal evolution. BB-BC method has been proved to outperform genetic algorithm for benchmark test functions [16]. Big Bang-Big Crunch (BBBC) algorithm, one of the most recent heuristic optimization strategies, is utilized to solve optimization problem by simulating the Big Bang phase and Big Crunch phase during the evolution process of the universe. According to the Big Bang and Big Crunch theory, disorder and randomness are generated by energy dissipation during the Big Bang phase, and randomly distributed particles are drawn into an order during the Big Crunch phase. Based on the good performance in exploiting the local optimization solution during the BBBC evolution process, new solution is obtained as the influx of random solutions by means of a center of mass or minimal cost approach. The search strategy of cannot explore the global optimization effectively despite. The studies about BBBC optimization are very few as considering the premature convergence and the convergence speed of BBBC approach. This paper aims to improve the optimization performance of BBBC optimization by combining the self-adaptive operator and competitive strategy.

This paper is organized as follows: The next section briefly describes the problem model which includes task model and energy model. In Section 3, hybrid algorithm based on BBBC is presented while its performance is studied in section 4.

2 Task model and energy model

The energy optimization problem of the tasks can be divided into two problems: the task scheduling problem and the DVS problem. The task scheduling problem is to order a set of tasks and assign them to different processors of a distributed system for accomplishing an application. The heterogeneous of the system implies the processing time of each task may not be the same on different processors.

The task scheduling problem can be formulated as follows. There is an application consisting of n tasks at_i (i=1, 2,...,m) which processes on a set of processors ap_j (j=1,2,...,n)according to a predetermined order. Each task at_i is ordered by its priority and processed on a specific processor ap_j according to the earliest ending time. The task scheduling problem is thus to determine the routing sequence and the assignment of the tasks on the processors so that precedence constraint among tasks is satisfied. In this paper, the tasks and their executing constraints are depicted as the nodes and the directed edges of a Directed Acyclic Graph (DAG). There is an entry task and an exit task to imply the start and end of all the tasks. The directed edge implies starting point (the processor task) must complete before the beginning of its end point (successor task). When the task and its successor task are assigned on different processors, a communication time will be needed for transferring the data between two tasks. For simplification, the communication time is assumed the same among the whole communication network.

The power consumption is at least a quadratic function of the supply voltage. Here the static power is ignored and dynamic power is studied and reduced by extend the execution time of the tasks based on its slack. The relation between dynamic power dissipation and supply voltage is computed as Eq.1

$$P_v = C_f \cdot V_s^2 \cdot fc \;\; P_v = C_f \cdot V_s^2 \cdot fc \tag{1}$$

where P_y, C_f, V_s, fc are the dynamic power, the workload capacitance, the supply voltage and the clock frequency, respectively.

And fc is computed according to Eq. 2

$$fc = k \cdot {(V_s - V_k)^2}/{V_{ds}} \tag{2}$$

where k is a constant, and V_k is the threshold voltage. The rate between the clock frequency and the threshold voltage is a constant.

According to Eq. 1 and 2, the power consumption of a processor is obtained according to Eq. 3

$$E = P_{dv} \cdot t = C_{ef} \cdot V_{ds}^2 \cdot c \tag{3}$$

Normally, the goal in an energy optimization of task scheduling problem is to find a feasible schedule with the minimum energy consumption. For a feasible schedule of the task scheduling problem, the execution time of all the tasks sequence fulfilling the precedence constraint are adjusted for reducing the energy consumption. Several symbols are introduced as follows:

- ec_i: the energy consumption of the task at_i
- dt_i: the deadline of the task at_i
- et_{ij}: the execution time of task at_i which is assigned on processor ap_j
- prt_i: the priority of task at_i
- ef_i: the earliest finish time of task at_i
- lf_i: the latest finish time of task at_i
- ppt_i: the directly predecessor of task at_i
- spt_i: the successor of task at_i on the processor
- st_i: the directly successor of task at_i
- slt_i: the slack of task at_i
- ot_i: is the order task at_i in the schedule
- pt_i: the processor which execute task at_i
- Sch_k: kth schedule which is composed of ot_i and pr_i

- v_{imax}: the maximum voltage of task at_i
- v_{is}: the scaling voltage of task at_i
- EC_k: is the initial energy consumption of kth schedule
- AEC_i: is the adjusted energy consumption of kth schedule after using DVS strategy
- MS_i: the makespan of the schedule, i.e., the finish time of the last task at_m

Consider the feasibility of the solution in real time systems, the execution of the task cannot be interrupted, and the finish time of each task should be less than its deadline as $ef_i <= dt_i$.

3 Hybrid BBBC algorithm

3.1 BBBC algorithm

BBBC optimization includes Big Bang procedure and Big Crunch procedure. The Big Bang procedure is a random phase that simulates the energy dissipation in nature, one of the theories of the evolution of the universe. The following Big Crunch procedure is a contraction phase that calculates a center of mass for the population viewed as gravitational attraction [17]. In this way, the disorder solutions are generated over the entire search, and then the solutions convergence to a local or global optimum point iteratively. Big Bang and Big Crunch phases are firstly obtained random candidates distributed over the center of mass or the minimum candidate, and then compute the center of the candidates. The pseudocode of BBBC is given as follows:

Begin
k=1, random candidates are generated as the kth generation.
While stopping criteria not met **Do**
  Pc, Center of mass of candidates, is obtained by Big crunch procedure.
  k+1th generation is generated around Pc by Big bang procedure.
  Evaluate k+1th generation.
  k++.
End while
End

3.2 Population initialization

The scheduling situation of all the tasks is obtained according to a routing pattern and an ordering pattern alternatively. The first pattern demonstrates the assignment of the tasks, and the second pattern guarantees the execution order of the tasks on the heterogeneous processors. A candidate is represented as a schedule, Sch={OT,PT} that each task and processor is assigned a fixed ID. A random solution is obtained as the combination of two random orders of (1,2,..,m) and (1,2,..,n) respectively. The solutions to task scheduling problem can be represented as a population which is composed by a set of alternatives or candidates. At the beginning of the evolution, some list algorithms as HEFT are utilized in a hybrid way to generating various initial solutions with relative good quality [18]. The HEFT algorithm [19] is known as good trade off between schedule length and search effort as its time complexity is low as (O(nlogn+(e+n)p)).

To generate the initially promising schedule, the tasks are assigned onto processor with minimum earliest finish time. The insertion technique is used to insert the task into the earliest available time on its assigned processor. The remaining tasks are selected according to the descending order of the priority.

3.3 Objective function computation

For a feasible schedule of the task scheduling problem, the objective of the problem is to minimize the energy consumption. The energy consumption is obtained by extending the execution time of the tasks based on using DVS strategy on their slacks. The slack of task at_i, slt_i, equals difference between ef_i and

lf_i. Lf_m of the last task at_m is initialized as its deadline, and lf_i (i=1,2,..,*m 1*) is calculated recursively from the last task as shown in Eq.4:

$$lf_i = \min\{lf_{st_i} - et_{ij}, lf_{spt_{ij}} - et_{ij}\}, j \in pt_i \tag{4}$$

Here method for minimizing total dynamic energy dissipation are based on variable voltage systems (nearly continuous range of possible supply voltages), that can be adjusted to multi voltage systems (small and limited number of potential supply voltages) easily [20]. AEC_i of task ta_i is obtained by the optimization strategy according to the PE power consumption and supply voltage optimization as Eq.5:

$$AEC_i(et_{ij} + \Delta t) = EC_i(t) \cdot vt_{is}^2 / vt_{i\max}^2 \tag{5}$$

where $\Delta t = \min\{slt_i\} / C$,C is a constant, and v_{is} is computed as Eq.6:

$$V_{is} = V_{it} + V_{ic}/2 * d_i + \sqrt{(V_{it} + V_{ic}/2 * d_i)^2 - V_{it}^2} \tag{6}$$

where $d_i = (et_{ij} + \Delta t)/et_{ij}$, V_{ic} is a constant and is computed as Eq.7:

$$V_{ic} = (V_{i\max} - V_{it})^2 / V_{i\max} \tag{7}$$

The task with high energy saving is adjusted firstly according the slack with satisfying deadline and precedence constraints. The procedure of extending execution time of the task is repeated until no slack existing.

3.4 Modified BC phase

The mass is calculated as the inverse of the energy consumption and makespan value respectively as the population is divided into two groups. With viewing ot_i as the position of task at_i, the convergence operator of the Big Crunch phase generates the center of mass (Pco_i, Pcp_i). $Pco_i(E)$ is computed according to Eq. 8:

$$Pco_i(E) = (\sum_{j=1}^{Num} ot_i / E_j) / (\sum_{j=1}^{Num} 1 / E_j), i = 1, \cdots, m \tag{8}$$

where *Num* is the population size of the feasible solutions.

$Pcp_i(E)$ is computed as Eq. 9:

$$Pcp_i(E) = (\sum_{j=1}^{G} pt_i / E_j) /(\sum_{j=1}^{G} 1 / E_j), i = 1, \cdots, m \tag{9}$$

where *E* equals makespan and energy consumption respectively for each group. The fit candidates with minimum makespan and energy consumption respectively are also viewed as the starting points in the Big Bang phase.

3.5 Modified BB phase

New shcedule (ot_i, pt_i) is randomly generated around the starting point obtained by BC phase.
Firstly pt_i is computed as Eq. 10:

$$pt_i = (Pcp_i + \alpha_i \cdot \gamma_i) \bmod m, i = 1, \cdots, n \tag{10}$$

where α_i is a random real less than 1. r_i is the standard deviation of a normal distribution, and is computed as Eq. 11:

$$\gamma_i = speed1 \cdot (n-1) / iter \quad (11)$$

where *iter* represents the number of iterations, *speed*1is an parameter to control convergence speed of the search.

Secondly the set of ot_i is obtained by in ascending order to sorting pt_i which is computed as Eq.12:

$$pt_i = (Pcp_i + \alpha_i \cdot \beta_i) \bmod n, i = 1, \cdots, n \quad (12)$$

where β_i is computed as Eq.13:

$$\beta_i = speed_i \cdot (pt_{\max} - pt_{\min}) / iter \quad (13)$$

where $pt_{\max}$ and $pt_{\min}$ are the upper and lower bounds of the variable pt_i, and $speed_i$ is an parameter acting as *speed*. It is noted that the set of ot_i maybe infeasible as it disobeys the precedence constraint. An adjust strategy need to recalculated pt_i recursively from the task as Eq.14:

$$pt_i = \max_{j \in ppt_i} \{ pt_j + cc_i, pt_i \}, i = 1, \cdots, n \quad (14)$$

where cc_i is a constance equal to 1.

3.6 Chaos based BB phase

Meanwhile, to guarantee the diversity of the initial solution, a chaos strategy is applied to randomly order the tasks and generate their assigned machines. The non repetition is obtained by mathematics formula defined by chaos strategy that has been successfully applied in various research areas as physics and engineering [21]. Chaotic based functions have utilized to help heuristic algorithms such as imperialistic competitive algorithm [22], harmony method [23], charged system search strategy [24] to trade off between exploration and exploitation. Chaos strategy is incorporated with BBBC algorithm to improve the diversity among candidates and away from premature convergence. In this paper, logistic and sinusoidal chaotic map function are utilized to replace parameter cc_i and $speed_i$ in Eq. 13, and 14 respectively as Eq. 15 and 16.

$$cc_{i+1} = a1 \cdot cc_i (1 - cc_i) \quad (15)$$

$$speed_{i+1} = \alpha 2 \cdot speed_i^2 \sin(\pi speed_i) \quad (16)$$

where $a1$ is a constant equals to 4, *cc* is initialized as a real number between 0 and 1, and cannot equals 0, 0.25, 0.5, 0.75, and 1. *A*2 is a constant equals to 2.3, *speed* is initialized as 0.7.

The Hybrid BBBC (HBBBC) algorithm is described as follows:

Input: A set of tasks to be assigned on heterogeneous processors.

Output: A task schedule with adjusted execution voltage.

1: the population composed of *Num* candidates is initialized by HEFT and random strategy;
2: compute the makespan, slack energy consumption of each candidate;
3: new energy consumption of each candidate is obtained by using Eq. 5;
4: **while** the stopping criteria are not met **do**
5: the feasible candidates are selected as each task satisfying the precedence and deadline constraint;
6: the center of mass (Pco_i, Pcp_i) is computed as Eq. 8 and 9 with *E=MS*;
7: the center of mass (Pco_i, Pcp_i) is computed as Eq. 8 and 9 with *E=EC*;
8: new candidates are generated by Eq. 10 and 12 with the starting point (Pco_i, Pcp_i);
9: the adjust strategy described as Eq.14 guarantees the feasibility of the candidates;

DOI: 10.1051/matecconf/201711901047

10: Chaos BB strategy changes candidates by modifying the parameters in Eq. 13 and 14;
11: compute the makespan and energy consumption of each candidate;
12: new energy consumption of each candidate is obtained by using Eq. 5;
13: **end while**

4 Experiment and Result

The extensive experiments are implemented to evaluate HBBBC algorithm which was programmed in java and run on Intel Core TM processors with 2 GHz speed with 4GB of memory. HEFT scheduling algorithm that proved to perform well for the task scheduling problem and our previously proposed ACO [25] are also implemented for verifying our algorithm, the energy saving of HBBBC is enabled by the exploitation of the DVS technique.

4.1 Experimental settings

Each method is run 10 times with different initial control variables for 100 generations. The best results of each run are presented. The performance of HBBBC was thoroughly evaluated with two test cases: randomly and benchmarks as three real world parallel applications: Fast Fourier Transformation [26], the Ludecomposition [27] and the Laplace equation solver [28]. A random tasking scheduling problem generator is designed to generate various task graphs with different characteristics.

Specifically, m is generated between 10 and 200. The ratio of the computational and communication value of the tasks is randomly generated as less 4 or larger than 1 that represents the application with larger computation or communication costs respectively. The maximum and minimum execution time of each task is defined to describe difference between the fastest processor and the slowest processor in a given system.

4.2 Results

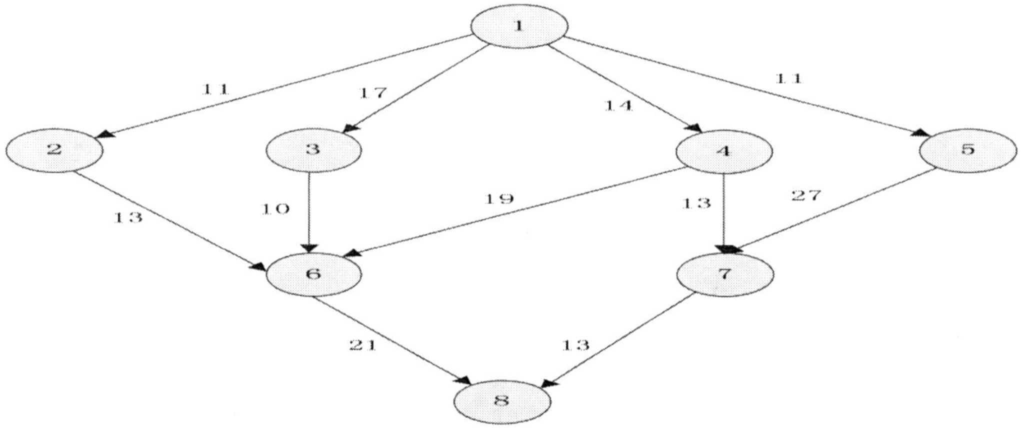

Figure 1. Example of a task graph with 8 tasks.

Table 1. Computation cost of Fig 1.

at	1	2	3	4	5	6	7	8
P1	11(1)	10(17)	9(23)	11(24)	15(35)	12(37)	10(61)	11(39)
P2	13(21)	15(24)	12(31)	16(27)	11(41)	9(43)	14(23)	15(41)
P3	9(3)	11(2)	14(2)	10(20)	19(37)	5(25)	13(66)	10(40)

Here, test firstly is implemented on a benchmark set. Figure 1 demonstrates a DAG with nine tasks and 13 edges labelled the communication times that remain the same on the all communication edges. Figure 2 and Figure 4 show the schedules of the tasks in Figure 1 by HEFT and HBBBC strategies on three processors respectively based on the computation cost (energy) described in Table 1. Figure 3 and Figure 5 show the energy saving by DVS strategy based on schedules shown in Figure 2 and Figure 4. If the deadline is extended to 150, the energy consumption is decreased from the 2855.0 to 393.19257 by using the DVS strategy on the schedule obtained by HEFT strategy. And the energy consumption is decreased from 1723.0 to 152.40613 by using the DVS strategy on the schedule obtained by HBBBC strategy.

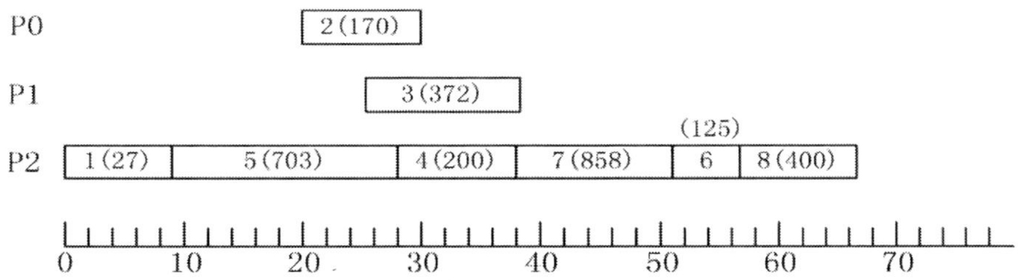

Figure 2. The schedule of Fig. 1 by HEFT.

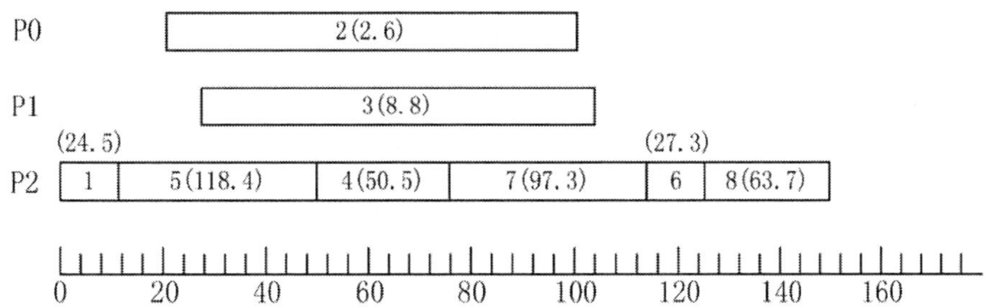

Figure 3. The energy saving of the schedule in Fig 2.

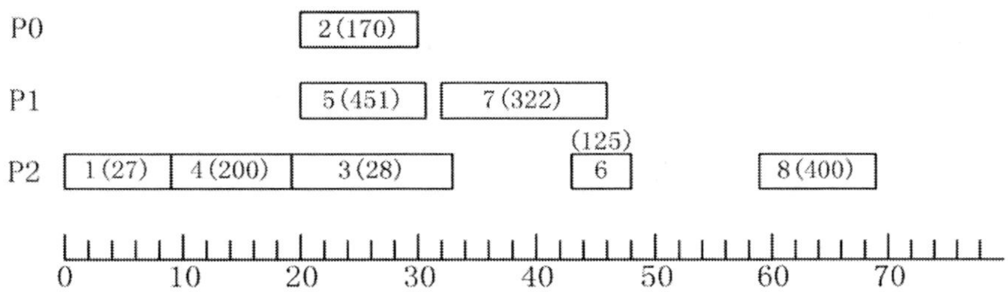

Figure 4. The schedule of Fig 1 by HBBBC.

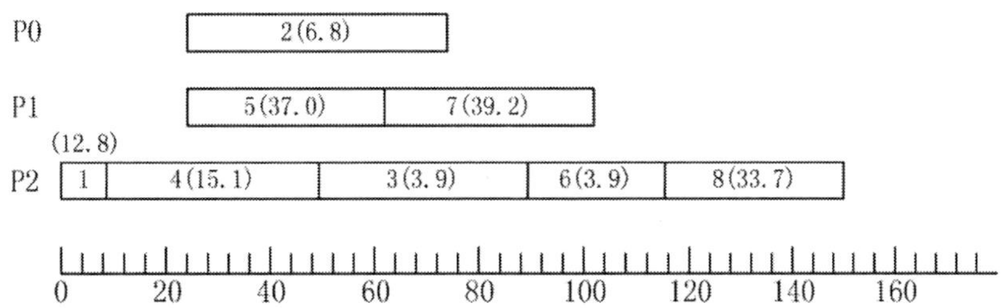

Figure 5. The energy saving of the schedule in Fig 4.

Table 2 presents the energy saving results obtained from our extensive comparative evaluation study among HBBBC, HEFT and our previously proposed ACO algorithms. Simulations have been conducted using the simulator we developed as part of this study as HEFT is energy unconscious and is not directly applicable to DVS problem.

Table 2. Comparative results

Algorithm over DAG set	HBBBC over ACO	HBBBBC over HEFT
FFT	21%	22%
Laplace	15%	25%
LU	13%	19%
Random	20%	26%

5 Conclusion

Since most traditional algorithms only focus on makespan, we presented HBBBC to investigate the energy consumption and task scheduling altogether. The energy will be saved by adjusting slack and execution time obtained by schedule algorithm. Our algorithm hybridizes HBBBC algorithm and dynamic scaling strategy to minimize the energy consumption of the tasks on the heterogeneous distributed mobile system. An adjust strategy is used to guarantee the solutions meeting the precedence constraints, and the feasible solutions are selected to generate around the centric mass in BC phase. The centric mass is treated as the starting point for generate new random solutions around it in BB phase. In essence, the energy saving of dynamic scaling strategy is incorporated with the scheduling strategy based on BBBC and chaos strategy. Chaos map functions are applied to trade off the exploration and exploitation. The computational results on a set of random and specific instances testify the performance of HBBBC. Our study provides promising results showing the significance and potential of hybridizing the task scheduling and dynamic scaling strategy in the reduction of energy consumption.

References

1. T. Ishihara and H. Yasuura, Voltage scheduling problem for dynamically variable voltage processors, *Proceedings of International Symposium on Low Power Electronics and Design*, 197-202 (1998)
2. A. Manzak and C. Chakrabarti, Variable voltage task scheduling for minimizing energy or minimizing power, *Islped Proceedings of the International Symposium on Low Power Electronics and Design*, **6** (2), 3239-3242 (2000)
3. M.S. Al Hashimi, Considering power variations of DVS processing elements for energy minimisation in distributed systems, International Symposium on System Synthesis, 250-255 (2001)
4. T. Burd, T. Pering, A. Stratakos, and R. Brodersen, A dynamic voltage scaled microprocessor system, IEEE International Solid state Circuits Conference, **35** (11), 1571-1580 (2000)
5. F. Yao, A. Demers, and S. Shenker, A scheduling model for reduced CPU energy, Foundations of Computer Science Annual Symposium, **23** (25), 374-382 (1995)
6. A. Qadi, S. Goddard, and S. Farritor, A dynamic voltage scaling algorithm for sporadic tasks, IEEE Real time Systems Symposium, 52-62 (2004)
7. C. Scordino and G. Lipari, A resource reservation algorithm for power aware scheduling of periodic and aperiodic real time tasks, Computers IEEE Transactions, **55** (12), 1509-1522 (2007)
8. P. Pillai and K. Shin, Real time dynamic voltage scaling for low power embedded operating systems, Acm Sigops Operating Systems Review, **35** (5), 89-102 (2001)
9. S. Saewong and R. Rajkumar, Practical voltage scaling for fixed priority RT systems, IEEE Real time and Embedded Technology and Applications Symposium, 106-114 (2010)

10. Y. Liu and A. Mok, An integrated approach for applying dynamic voltage scaling to hard real time systems, IEEE Real time and Embedded Technology and Applications Symposium, 116-123 (2003)
11. H. Aydin, V. Devadas, and D. Zhu, System level energy management for periodic realtime tasks, *Proceedings of IEEE 27th Real Time Systems Symposium*, 313-322 (2006)
12. N. Sinha, R. Chakrabarti, and P. Chattopadhyay, Evolutionary programming techniques for economic load dispatch, IEEE Transactions on Evolutionary Computation, **7** (1), 83-94 (2003)
13. V. Bilolikar, K. Jain, and M. Sharma, An adaptive crossover genetic algorithm with simulated annealing for multi mode resource constrained project scheduling with discounted cash flows, International Journal of Operational Research, **25** (1), 28-46 (2016)
14. A. Meng, Z. Li, H. Yin, S. Chen, and Z. Guo, Accelerating particle swarm optimization using crisscross search, Information Sciences, **329**, 52-72 (2015)
15. O. Erol and I. Eksin, A new optimization method: big bang-big crunch, Advances in Engineering Software, **37** (2), 106-111 (2006)
16. B. Alatas, Uniform big bang-chaotic big crunch optimization, Communications in Nonlinear Science and Numerical Simulation, September, **16** (9), 3696-3703 (2011)
17. C. Rao and G. Yesuratnam, Big bang and big crunch (BB BC) and firefly optimization (FFO): application and comparison to optimal power flow with continuous and discrete control variables, American Journal of Nursing, **104** (9), 26-26 (2004)
18. H. Arabnejad and J. Barbosa, List scheduling algorithm for heterogeneous systems by an optimistic cost table, IEEE Transactions on Parallel and Distributed Systems, **25** (3), 682-694 (2014)
19. H. Topcuouglu, S. Hariri, and M.Y. Wu, Performance effective and low complexity task scheduling for heterogeneous computing, IEEE Trans. Parallel Dist. Systems, **13** (3), 260-274 (2002)
20. M. Schmitz and B. Al, Iterative schedule optimization for voltage scalable distributed embedded systems, Acm Trans. Embedded Comput. Syst, **3** (1), 182-217 (2004)
21. M. Tavazoei and M. Haeri, Comparison of different one dimensional maps as chaotic search pattern in chaos optimization algorithms, Applied Mathematics and Computation, **187** (2), 1076-1085 (2007)
22. S. Talatahari, B. Azar, R. Sheikholeslami, and A. Gandomi, Mperialist competitive algorithm combined with chaos for global optimization, Communications in Nonlinear Science & Numerical Simulation, **17** (3), 1312-1319 (2012)
23. Q. Pan, L. Wang, and L. Gao, A chaotic harmony search algorithm for the flow shop scheduling problem with limited buffers, Applied Soft Computing, **11** (8), 5270-5280 (2011)
24. S. Talatahari, A. Kaveh, and R. Sheikholeslami, An efficient charged system search using chaos for global optimization problems, International Journal of Optimization in Civil Engineering, **1** (2), 305-325 (2011)
25. Y. Kang, An ant colony system for dynamic voltage scaling problem in heterogeous system, Lecture Notes in Electrical Engineering, **277**, 73-81 (2013)
26. R.E. Lord, J.S. Kowalik, and S.P. Kumar, Solving linear algebraic equations on an mimd computer, J. ACM, **30** (1), 103-117 (1983)
27. T.H. Cormen, C.E. Leiserson, and R.L. Rivest, *Introduction to Algorithms*, MIT Press (1990)
28. M.Y. Wu and D.D. Gajski, Hypertool: a programming aid for message passing systems, IEEE Trans. Parallel and Distributed Systems, **1** (3), 330-343 (1990)

Multi-stage forging on torx-pin screw with high torque

Gow-Yi Tzou[1], *Dyi-Cheng* Chen[2] and *Shih-Hsien* Lin[2,a]

[1]*Department of Mechanical and Automation Engineering, Chung-Chou University of Science and Technology, Chang-hua County 510, Taiwan*
[2]*Department of Industrial Education and Technology, National Chang-Hua University of Education, Chang-hua County 500, Taiwan*

Abstract. The study assumes constant shear friction to simulate multi-stage forging on torx-pin screw with high torque using FEM commercial software. Because the screw head has a plum needle shape with the effect of high torque, the 3D simulation model is used to explore the effective stress, the effective strain, the velocity field, and the forging force in each stage. The forming pass schedule considers three stages to obtain the product, and uses the Deform-3D software to explore the feasibility of forming pass schedule. Finally, the realistic manufacturing of torx-pin screw can be carried out to compare with the simulation results to verify the acceptance of FEM simulation. The maximum dimension error between simulation and experiment is around 2.45%, thus the feasibility of FEM simulation to die design for manufacturing the torx-pin screw can be verified.

1 Introduction

Screw & nut products are categorized into the fastener. Screw & nut can be divided into two main types. One is the threaded fastener, including screw, bolt, stud and nut; and the other is the non-threaded fastener, including rivet, washer and hook. In general, screws are indicated the threaded products with smaller diameters, such as hex bolt, square-head bolt, foundation bolt, T-shape bolt, etc. How to manufacture the fasteners with low cost is still very important. Generally speaking, when selecting appropriate tool designs to manufacture the fasteners, it needs to consider whether these methods satisfies the production conditions, such as the number of stages, the forging force for the individual stage and whole forging process, the diameter of wires, the dies and the length of front/rear punch. Additionally, the simulation of tool design needs to know distribution of effective strain and stress for final products, and the forging force for each stage also needs to be controlled within a proper range. Li and Oh el al.[1-2] think that the design ways of multi-stage tool design are generally depended on subjective working experience and technologies of designers, but there still have insufficient experiences or undistributed considerations. For the manufacture of product, the most important issue is that how to use the lowest costs including money and time to produce the best products. Therefore, in order to shorten the lead time of the tool designs, reducing the cost of dies and using proper tool designs for machines are needed. Of course it is necessary to look for effective methods. In recent years, due to accelerating computation and lowering cost of computers, it makes a significant advancement of design and development in metal forming using Finite Element Method

[a] Corresponding author : fitrc@yahoo.com.tw

(FEM), especially when it produces a more precise, rapid and stable simulation tool. Oh et al. [2] use the FEM DEFORM software to discover the damage causes of die and the die elastic deformation phenomenon during the cold forging.

Shih et al. [3] and Hsia [4] use the simulation analysis software, except for greatly increasing the efficiency of making dies, shortening the production cycle of new products and increasing the lifespan of dies, it also can provide new-coming designers with the prompt understanding of tool design, the analysis of defect and the improvement of simulation. Therefore, it can save massive time in pre-process schedule to further upgrade the design level of dies. Lange et al. [5] applies various computer-aid simulation analyses (NASTRAN, ANSYS, DEFORM, ABAQUS, I-DEAS) to analyze the stress distribution on cold-forging die to predict the damage positions of the die, as well as conducted experiments to investigate the impact of the die curve fillet dimension and incline angle, surface treatment used and the hardness of various materials and thermal treatment on the service life of die. Nagao et al. [6] use the finite element software DEFORM to analyze the car forging parts and compare the optimal designed stress ring of die, to understand the pressure phenomenon forced on die, in order to increase the service life of die. Knoerr et al. [7] use the integrated computer-aid simulation analysis system to calculate the force for the cold forging die, in order to investigate the problem points of over-force damage or fatigue damage. Wang et al. [8] make improvements in the defective phenomenon, such as folding that caused by the flowing of blank. They adopt the feature-based method with focusing on the location and timing of occurring defects to complete these feature requirements for these problems in early stages before redesigning the tool designs, also fix these features to avoid occurring the defective phenomenon with initiating feature requirements that caused by the material flowing in follow-up forming stages again. Shih et al. [9] and Sun et al. [10] explore that when the bolt is forging and forming, its tensile strength would be influenced by the crack of forging stream line. Therefore, the simulation Simufact software is used to know the cause of cracking the forging stream line, changing tool designs to reduce the number of cracked forging stream lines, and then improving bolt finished products' tensile strength. Chang et al. [11] aims at analyzing the influence of the maximum principal stress on Tungsten Carbide Steel die core in an extrusion die which caused the crack of die core, and then adjusts the dies assembly method in order to improve the service life of die.

This study mainly investigates that how to plan the forming pass schedule how many stages can be used to obtain the product, and uses the Deform-3D software to explore the feasibility of forming pass schedule. Finally, the realistic manufacturing of plum needle screw can be carried out to compare with the simulation results to verify the acceptance of FEM simulation.

2 Finite element method

Finite Element Method (FEM) is a numerical method to analyze product parts or assembly systems, in order to make sure of the entire product development cycle and function's consistency, its main spirit is still focusing on changing the questions with limitless freedom to be finite freedom, and then sort out the required approximation with using the numerical method. FEM function is to design the geometric model, after carrying out the finite element analysis to understand the structural response (deformation, stress, temperature, etc.), and the calculated results can be showed in graphs and diagrams. If in the design compared the calculated stress value and tolerance or maximum value, when it failed the design requirement, then the structure needs to re-design and re-analyze until reaching the acceptable design values.

FEM's applications of the plastic forming can be divided into the elastic FEM and rigid-plastic FEM, and FEM is the most common method to analyze the plastic deformation for materials. According to the work patterns and geometric shape of objects, tools and materials are segmented into several elements (triangle, tetragon, tetrahedron or hexahedron), so as to solve the stiffness matrix for individual element, then reconsider the material stress and strain model to get the structural equations, and cooperate with the stress and strain relations and boundary condition to compute the force, deformation, stress and strain distribution by computers.

DOI: 10.1051/matecconf/201711901048

(1) Elastic-plastic Model

For the elastic-plastic model, assuming the deformation of materials including the recoverable and unrecoverable plastic deformation, which can provide the material flow information, including stress distribution, forming force, residual stress and strain distribution; therefore, its applicable range is comprehensive; however, in terms of the numeric computation, since varied material flow types between elastic and plastic deformation, it needs to inspect whether the deformation of every element is in the elastic or plastic range or not, thus the increment of deformation time should be very small, which is the economic defect for this computing method.

(2) Rigid-plastic Model

For the rigid-plastic materials, assuming it is an ideal material ignoring its elastic properties, thus certainly simplify the solving process when carrying out the analysis, thus the material stress distribution is mainly dependent on the strain when deforming. Even the rigid-plastic mode analysis couldn't obtain the distribution of residual stress, but it could be applied to the metal forming analysis, which could allow bigger increment for the deformation time; as a result, its computation is better.

2.1 DEFORM-2D (3D) software

It is mainly in an integrated environment to integrate with the modeling, forming, heat conducting and property of forming equipment to carry out the simulation analysis. In addition, it is applied for the cold, hot and warm forming with providing valuable engineering analytical data, such as material flow, die filling, forming force, die stress, grain flow, metal microstructure and defect yielding status, etc. These complicated 3D parts and die are the target to be treated. Also, it obtains the effective stress, effective strain, velocity field and forming force of every stage for the forging part, and greatly increases the service life of die. The simulation of the forging part is almost as same as the actual product. Among many finite element simulation analysis software, DEFORM-2D (3D) is the most optimal one of carrying out the simulation of fastener forging, thus this study is applied DEFORM-3D to conduct the die design and improvement of the high-torque torx-pin screw.

2.2 Simulations of torx-pin screw

Because the torx-pin head is not axial symmetrical, Deform-2D is unable to simulate it. It has to use the DEFORM-3D to implement an advanced and precise simulation, and the stages of its forming schedule are shown in Fig. 1, and wire rod (ϕ 7.75 X 22.30 mm, the material is AISI4140) is needed 3 stages to conduct the product, where the part of torx-pin head at the 3rd stage has the most difficulty to compete, as shown in Fig. 2. At the 1st stage, a forward extrusion method will be used to reduce the outer diameter of material from ϕ 7.75 mm to ϕ 5.20 mm, at the 2nd stage it is pre-forming the round head, and at the 3rd stage the workpiece at the 2nd stage can be formed to the required dimension, as well as the torx-pin head can be formed at the 3rd stage.

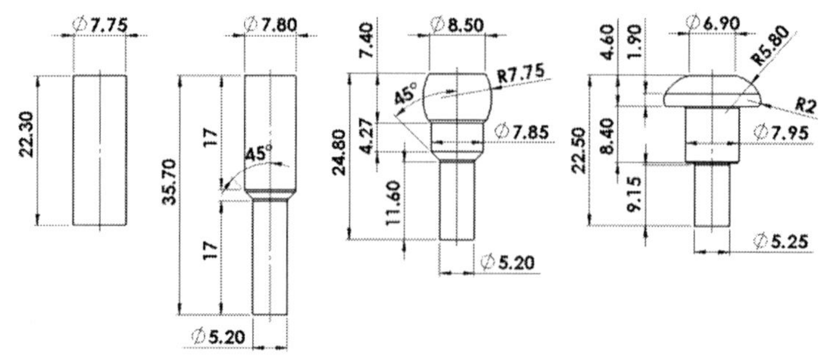

Figure 1. Forming schedule for each stage.

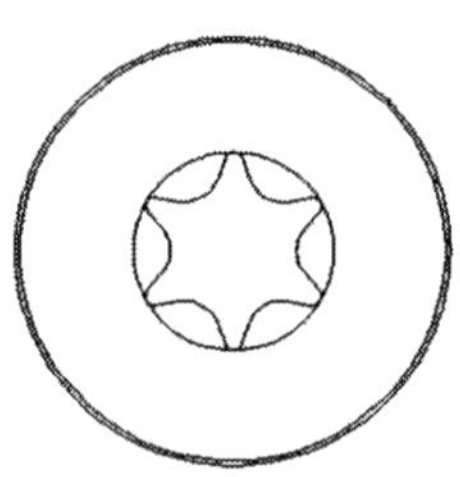

Figure 2. Torx-pin head at the 3rd stage.

 DOI: 10.1051/matecconf/201711901048

2.2.1 Mesh convergence analysis

The main function of the mesh convergence analysis is to find the optimal mesh numbers to obtain more precise analytic results with using the 1st stage as the analytic subject. As shown in Fig. 3 and Table 1, the analysis mesh numbers are between 50000 to 130000, and according to the forging force to observe the force change in every increase of 10,000 mesh and capture the most stable value of force, its error should be controlled within ±1%. These results also can be showed by the convergence error, and the least error would be used to conduct the analysis for each stage. The least error obtained from the mesh convergence analysis that is 0.979%, thus Model 8 should be adopted, and the mesh numbers used in this 3D research would be 120,000.

Convergence Conditions can be wrote as:

$$\text{Convergence error } (\%) = \frac{\text{Result}_N - \text{Result}_{N-1}}{\text{Result}_N} * 100\% \quad (1)$$

where

Result $_N$: current process; Result $_{N-1}$: previous process

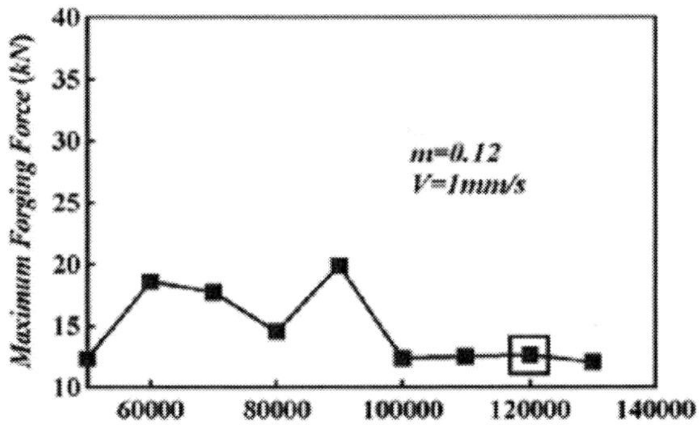

Figure 3. Mesh convergence analysis diagram for forging force.

Table 1. Mesh convergence analysis.

Model	1	2	3	4	5	6	7	8	9
Mesh	50000	60000	70000	80000	90000	100000	110000	120000	130000
Force(kN)	12.338	18.589	17.739	14.535	19.918	12.353	12.488	12.611	12.027
Error (%)		33.625	-4.792	-22.045	27.026	-61.24	1.078	0.979	-4.853

2.2.2 Pre-process setting

The actual material of the torx-pin screw is AISI4140 with higher hardness; the flow stress is used from the data base in the software. With a view to understanding the convergent effect of mesh numbers, the mesh convergence analysis is considered the mesh number from 50,000 to 130,000; the optimal mesh number is discovered to be 120,000 meshes. The simulation conditions in this study is shown in Table 2; the frictional factor sets at m=0.12 according to the defaulted cold forming in Deform software, and the punch velocity is assumed to be 1mm/s due to the cold forming process, as well as after using SolidWorks to make drawing, then imports the geometry of 3D drawing into the DEFORM-3D to conduct the analytic research.

Table 2. Analytic parameters for 3D simulation.

Workpiece	AISI4140
Young's Modulus (*E*)	206.754GPa
Poisson's Ratio (υ)	0.3
Flow stress	Defaulted

 DOI: 10.1051/matecconf/201711901048

Object type	Plastic
Number of Elements	120,000
Front die	WC
Velocity (V_o)	1mm/s
Frictional factor (m)	0.12
Rear die	WC
Velocity (V_o)	0mm/s
Frictional factor (m)	0.12

2.3 Die design

This study uses SolidWorks to design individual stage die, as shown in Fig. 4, after drawing the files would be saved as the STL format, which would conveniently read in the DREFORM software. In this study, designing the 3D upper die torx-pin punch, the rear punch also added to make the workpiece to be easily kicked out from the die. Moreover all dies are made of Tungsten Carbide.

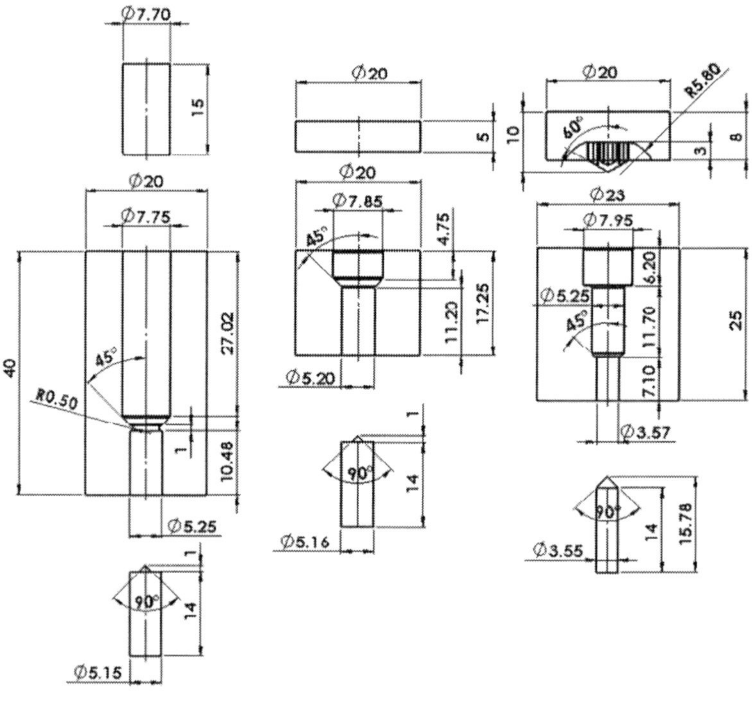

Figure 4. Die design diagram for each stage.

3 Results and discussions

3.1 Forming simulation analysis at the 1st stage

After completely designing the die, input it into DEFORM-3D to conduct the analysis simulation, and complete the positioning process of the die and workpiece during pre-process. In Fig. 5, it shows the 1st stage die forging, and Table 1 is used as the setting reference for each parameter. After undergoing the mesh convergence analysis, the obtained mesh numbers is 120,000, the friction factors for the upper and bottom dies is 0.12, the cold forging analysis is adopted, thus disregards the heat conducting, the front punch velocity is 1mm/s, and the die is made of Tungsten Carbide.

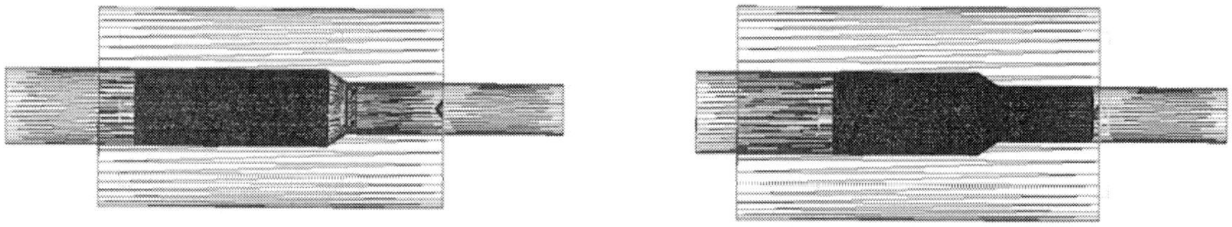

(a) Before forming (b) After forming

Figure 5. Die assembly diagram of the 1st stage before and after forging.

After completing the pre-process setting, it starts to conduct the simulation analysis. After finishing the analysis, with post-treatment, it can be known about the effective stress, strain, velocity field, and forging force. Fig. 6(a) is the effective stress diagram, and from this diagram, it is noted that the stress is mainly centralized on the connecting position of the shrinking outer diameter, which is similar to the extrusion forging, thus the maximum effective stress could be reached to 1590MPa. Fig. 6(b) is the effective strain diagram, the strain is occurred on the small diameter, and the maximum effctive strain is 4.97mm/mm. Fig. 6(c) is the velocity field distribution diagram, all materials would be flew downwardly, which the flow rate is rapid on the small diameter, and the highest flow rate is 2.99mm/sec. The maximum forging force could be reached to 97.03kN, as showen in Fig. 7.

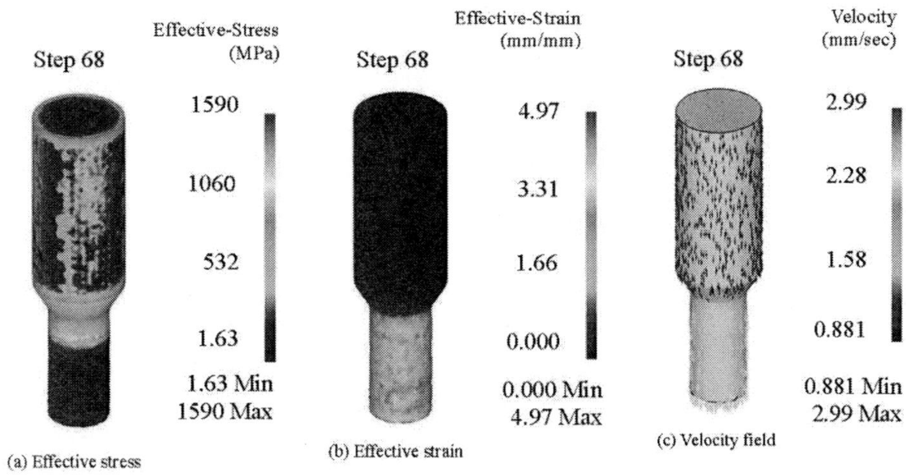

Figure 6. Effective stress, effective strain, and velocity field at the 1st stage.

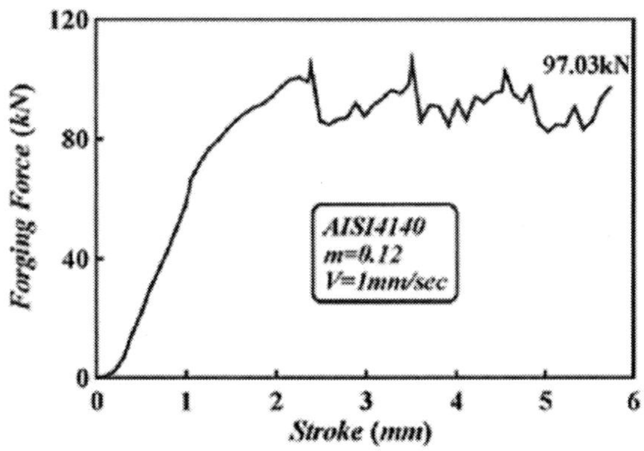

Figure 7. Forging force at the 1st stage.

DOI: 10.1051/matecconf/201711901048

3.2 Forming simulation analysis at the 2nd stage

Using the workpiece completed at the 1st stage forging to conduct the 2nd stage forging, replacing the die with the design die at the 2nd stage, after positioning the workpiece and die, setting parameters, as shown in the Fig. 8, which is the die assembly diagram of the 2nd stage before and after forging formation.

(a) Before forming (b) After forming

Figure 8. Die assembly diagram of the 2nd stage before and after forging.

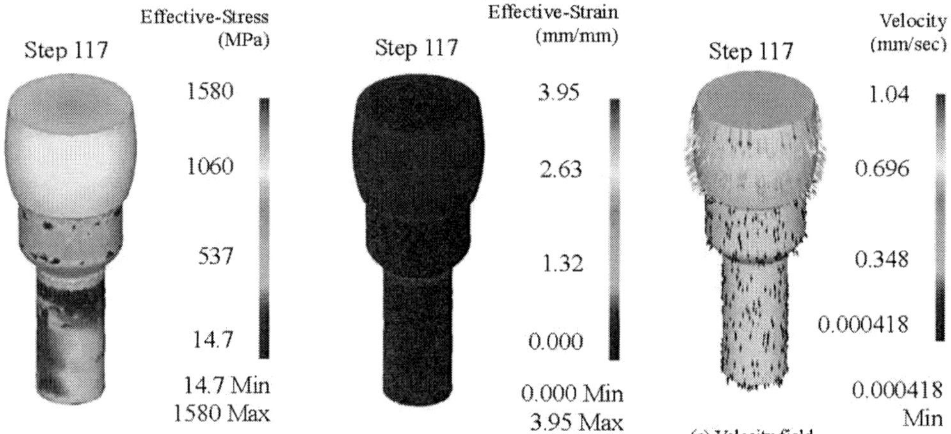

(a) Effective stress (b) Effective strain (c) Velocity field

Figure 9. Effective stress, effective strain, and velocity field at the 2nd stage.

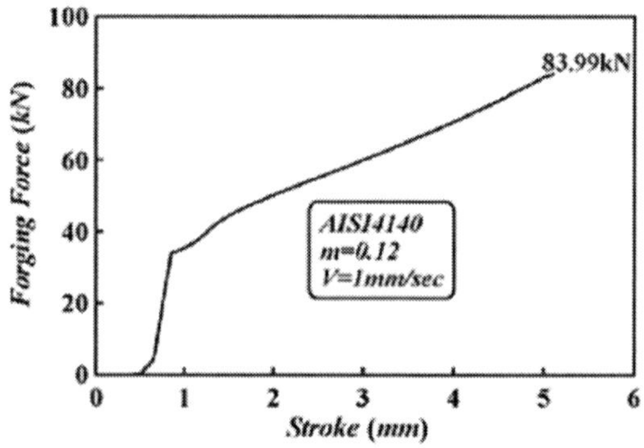

Figure 10. Forging force at the 2nd stage.

The 2nd stage is mainly used to make the pre-form process for the top at the 3rd stage, after completing the simulation analysis at the 2nd stage, it would obtain the effective stress, effective strain,

DOI: 10.1051/matecconf/201711901048

velocity field, and forging force. Fig. 9(a) is the effective stress diagram, and from this diagram it is noted the stress is mainly centralized on the top round head, the maximum effective stress could be reached to 1580MPa, the effective strain diagram is shown in Fig. 9(b), the maximum sffective strain is occurred on the top round head, and its maximum strain is 3.95mm/mm. The velocity field distribution diagram is shown in Fig. 9(c), the materials are flew downward, and flew outward on the round head part, and the material flow status is very even. As shown in Fig. 10, the maximum forging force could be reached to 83.99kN.

3.3 Forming simulation analysis at the 3rd stage

Using the workpiece completed at the 2nd stage to conduct the 3rd stage forging, replacing the die with the design die at the 3rd stage, after positioning the forging piece and die, setting parameters, as shown in the Fig. 11, which is the die assembly diagram at the 3rd stage before and after forging.

(a) Before formation (b) After formation

Figure 11. Die assembly diagram at the 3rd stage before and after forging.

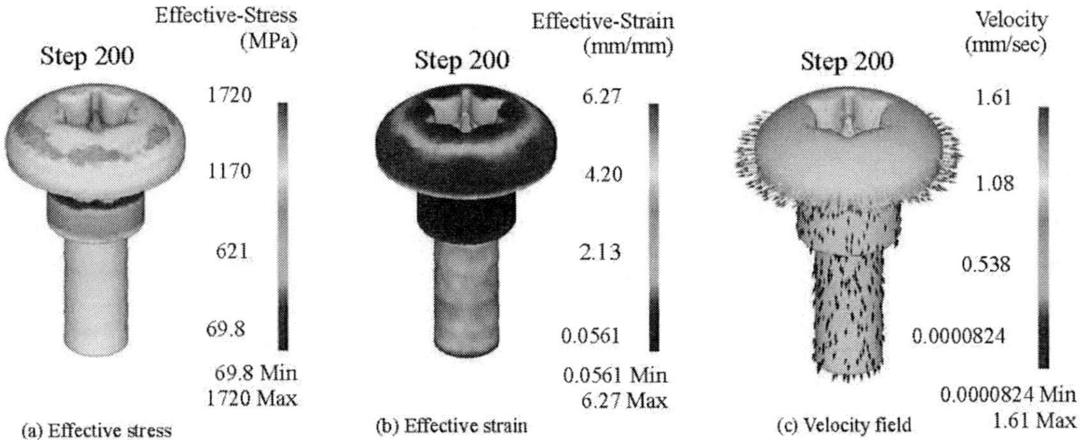

(a) Effective stress (b) Effective strain (c) Velocity field

Figure 12. Effective stress, effective strain, and velocity field at the 3rd stage.

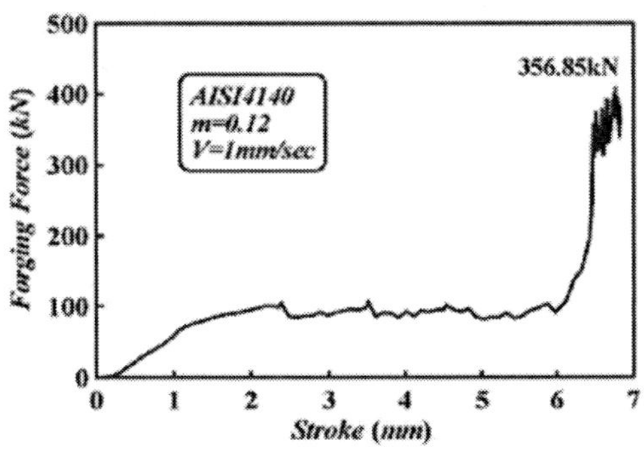

Figure 13. Forging force at the 3rd stage.

DOI: 10.1051/matecconf/201711901048

After completing the 3rd stage finite element simulation, it can be known about the effective stress, effective strain, velocity field, and forging force, as shown in Fig. 12(a), its effective stress is higher than the previous 2 stages, where the stress is significantly centralized on the torx-pin head, which the maximum effective stress could be reached to 172 MPa. Fig. 12(b) is the effective strain diagram, the maximum effective strain is 6.27mm/mm, and centralized on the torx-pin head bottom. Fig. 12(c) is the velocity field distribution diagram, the highest flow velocity is on the round head, and the highest flow velocity is 1.61mm/sec. The maximum forging force could be reached to 100kN, as shown in Fig. 13, and the force is suddenly raised in the process of torx-pin head forming, the maximum value has reached to 356.85 kN.

3.4 Forging force for each stage

The forging force for each stage can be summarized in Fig. 14 and Table 3, the forging force is very even at first in these 3 stages; however it needs to form the torx-pin head and presses the round head to the required dimension, thus the force is suddenly raised at the 3rd stage. From Table 3, the total forging force is 537.83 kN which does not exceed the rate forging force of machinery, 600 kN.

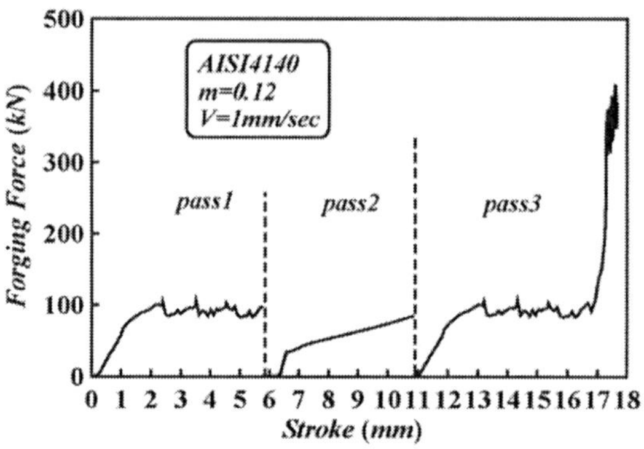

Figure 14. Forging force for each stage.

Table 3. Summary of forging force for each stage.

Stage Number	Forging Force (kN)
Pass 1	97.03
Pass 2	83.99
Pass 3	356.85
Total forging force	537.83

3.5 Experimental verification

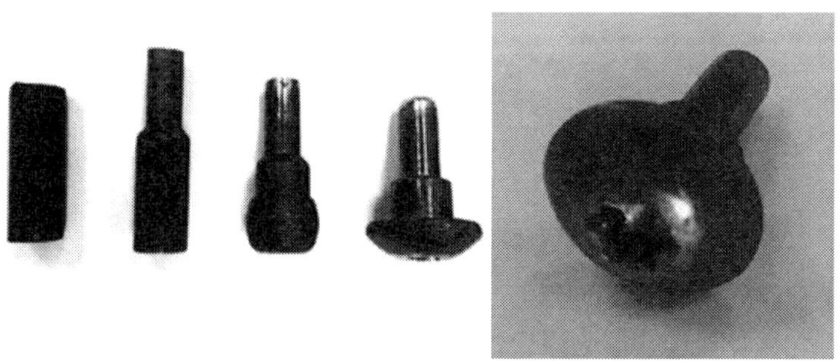

Figure 15. Forging parts for each stage

According to the FEM simulation to check the feasibility of dies, the dies for each stage can be made to perform the realistic experiments to get the forging parts for each stage, as shown in Fig. 15.

Table 4. Comparisons of FEM result and experiment.

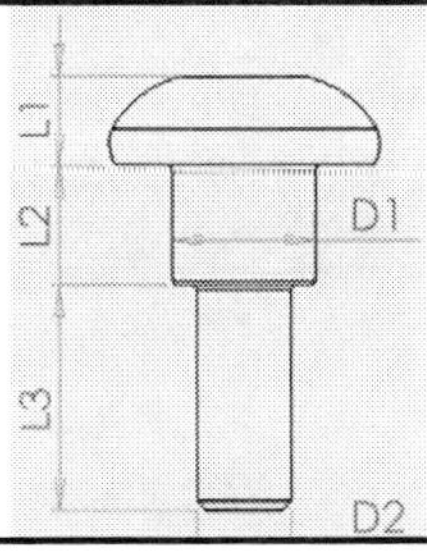

Position measurement	Experiment (mm)	Simulation (mm)	Error (%)
D1	7.95	7.943	0.09
D2	5.3	5.246	-1.03
L1	4.6	4.49	-2.45
L2	6.3	6.23	-1.12
L3	12.2	12.32	0.97

The dimensions of product between FEM result and experiment can be compared in Table 4. The maximum error on L1 position is 2.45%, the FEM simulation is in agreement with the experiment. Therefore, the FEM simulation model proposed can be acceptable for the trox-pin screw manufacture.

4 Conclusions

In this study, SolidWorks and DEFORM-3D have been to implement the finite element simulation analysis, it shows that DEFORM-3D can improve the die design to reduce the die cost and design time. It also can obtain the effective stress, effective strain, velocity field, and forging force under various forming conditions for each stage . Lead time and technology development time of product design would be greatly reduced, as well as lowers the expensive on-site testing costs so as to increase the benefit of die design, reduce material cost, and increase company's economic benefit. Therefore FEM simulation can actually apply to the multi-stage forging. Moreover, it uses the dimension simulated by DEFORM-3D to compare the dimension of actual product, and the biggest error for L1 is 2.45% which is very close to the actual dimension of product. The applicability of this simulation can be proved, thus this study can be a reference for the die design of the fastener industry.

References

1. G. Li, J.T. Jinn, W.T. Wu, and S.I. Oh, Recent development and applications of three-dimensional finite element modelling in bulk forming process, Journal of Materials Processing Technology, **113**, 40-45 (2001)
2. S.I. Oh, W.T. Wu, and K. Arimoto, Recent development in process simulation for forming process, Journal of Materials Processing Technology, **111**, 2-9 (2001)
3. C.W. Shih, G.Y. Tzou, and K.H. Chang, Multi-stage forming analysis and experiment of flange sleeve, Journal of Chinese Society of Mechanical Engineers, **35** (3), 243-250 (2014)
4. S.Y. Hsia, Optimization of microextrusion preforming using taguchi method, Mathematical Problem in Engineering, Article ID 305797 (2013)
5. K. Lange, A. Hettig, and M. Knoerr, Increasing tool life in cold forging through advanced design and tool manufacturing technique, Journal of Materials Processing Technology, **35**, 495-512 (1992)
6. Y. Nagao, M. Knoerr, and T. Altan, Improvement of tool life in cold forging of complex automotive parts, Journal of Materials Processing Technology, **46**, 73-85 (1994)
7. M. Knoerr, K. Lange, and T. Altan, Fatigue failure of cold forging tooling: causes and possible solutions through fatigue analysis, Journal of Materials Processing Technology, **45**, 57-71 (1994)
8. J.L. Wang, M.W. Fu, and J.Q. Ran, Analysis and avoidance of flow-induced defects in meso-forming process: simulation and experiment, International Journal of Advanced Manufacturing Technology, **68**, 1551-1564 (2013)
9. C.W. Shih, G.Y. Tzou, and K.H. Chang, Effects of forging stream line on tensile strength using FEM simulation, Advances in Mechanical Engineering, Article ID 216246 (2014)

10. M.C. Sun, G.Y. Tzou, and L.A. Zheng, Processing animation simulation and FEM analysis of multi-stage cold forging of stainless automotive battery fastener, Indian Journal of Engineering & Materials Science, **20**, 219-224 (2013)
11. K.H. Chang, C.W. Shih, and G.Y. Tzou, Defect improvement of extrusion dies using combination of FEM stress analysis with Taguchi method, Transactions of the Canadian Society Mechanical Engineering, **39** (3), 729-738 (2015)

Reliability analysis of imperfect coverage systems with a shared warm standby

Pei-Chang Li[a], Hong-Jie Yuan and Jie Lan
School of Reliability and System Engineering, Beihang university, Beijing 100191, China

Abstract. Redundancy technique is commonly applied to satisfy the reliability requirements of fault-tolerant systems. Warm standby, a compromise between hot standby and cold standby in term of power consumption and recovery time, has attracted wide attention over the past several decades. However, the existing reliability analysis methods for warm standby system with imperfect coverage are difficult to deal with some cases, such as non-exponential time-to-failure distributions for the system components and the systems with shared standbys. In this paper, a new approach based on step function and impulse function is proposed to overcome the limitations of the existing approaches. The reliability of a system including shared standbys is deduced considering two kinds of imperfect fault coverage models, which contain Element Level Coverage (ELC) and Fault Level Coverage (FLC). The proposed approach can applicable to any type of time-to-failure distributions for the system components subject to imperfect fault coverage. A case study is presented to illustrate the applications and advantages.

1 Introduction

Standby Redundancy is an important concept in improving the reliability of systems [1]. It is universally accepted that fault tolerant and safety critical systems cannot achieve intended reliability without employing standby [2]. Typically, the standby techniques can be classified into hot standby sparing, cold standby sparing and warm standby sparing [3]. Cold standby sparing is used when the energy consumption is of concern, while hot standby is applied when the recovery time is of vital importance. Warm standby is a compromise between cold standby and hot standby in terms of power consumption and recovery time [4-6]. Warm spares have different failure rates or failure distributions before and after they are used to replace a faulty component, which may fail randomly. Therefore, the failure behavior of the warm standby component is time-dependent [7]. Due to the complex failure behavior of the warm standby system, it has attracted extensive attention [8-10]. Existing approaches for modeling and analyzing the reliability of warm-standby systems are mostly Markov-based methods, simulation-based methods, and combinatorial methods. The Markov-based methods suffer from the well-known state space explosion problem [11] and are typically limited to exponential time-to-failure distributions. The simulation-based methods, for instance, Monte-Carlo simulations, are usually involve long computational time, especially when high accuracy results are required [12]. A combinatorial approach was proposed by Liu et al. [13], which enumerates all minimal cut sets or sequences and then applies the inclusion-exclusion formula to assess the reliability of warm standby

[a] Corresponding author : lipeichang@buaa.edu.cn

system. However, the inclusion-exclusion expansion makes the complexity of the method exponential. Another combinatorial approach based on sequential binary decision diagram (SBDD) has recently been proposed to evaluate the standby systems [14-15].

In redundancy standby systems, automatic recovery mechanisms are generally designed to make the system continue to function correctly even in the presence of faults or errors. If a component fails without being detected or isolated due to the failure of system recovery mechanism, it may produce incorrect result which can lead to an overall system failure [16]. This behavior is known as IFC [17-19]. System reliability can't improve unlimitedly with the increase of standby components when IFC is considered [20]. There are two conceptual models for system reliability subject to IFC: fault coverage can be modeled as a function of the number of faults that the system has experienced, which is Element Level Coverage (ELC), or it can be assumed that a particular coverage value can be associated with each redundant element in the system, which is Fault Level Coverage (FLC) [21]. The simple and efficient algorithm (SEA) is a well-known approach used to incorporate the IPC into combinatorial models for the system reliability analysis [22-23]. However, it is not applicable to dynamic systems with time or sequential dependence.

It is a difficult and challenging task to reliability analysis of WSP systems subject to IFC due to the sequence-dependent failure component behavior. Some researchers have studied the reliability of the WSP system subject to FLC [24-27]. However, their work is restricted to cases where the failure time of each system component obeys an exponential distribution. When several components have a shared warm standby, the process to system reliability analysis will be more complex.

In this work, a new approach is proposed for the reliability analysis of systems with a shared warm standby subject to IPC. The reliability of a system including a shared standby is assessed considering ELC and FLC. The time-to-failure distributions of system components are not limited in the paper.

The rest of this paper is organized as follows. Section 2 presents the fundamentals of warm standby system and IPC. Section 3 presents the suggested approach, which introduces Heaviside step function and Impulse function to descript the dynamic failure behaviors of warm standby system subject to IPC. An algorithm is proposed to deduce the failure probability density function and evaluate the reliability of system with a shared warm spare. Case studies are provided in Section 4 to illustrate the approach. Section 5 summarizes the paper.

2 Preliminaries

2.1 Warm standby system

Warm standby sparing is a fault-tolerant design technique widely used to improve system reliability. Warm spares have different failure rates or failure distributions before and after they are used to replace a faulty component. In practice, some important components usually equip a shared warm standby due to expensive cost. For example, a series system is composed of component A and B. In order to improve the system reliability, a warm standby S is equipped for A and B to share considering the cost, the fault tree of which is shown as Figure 1.

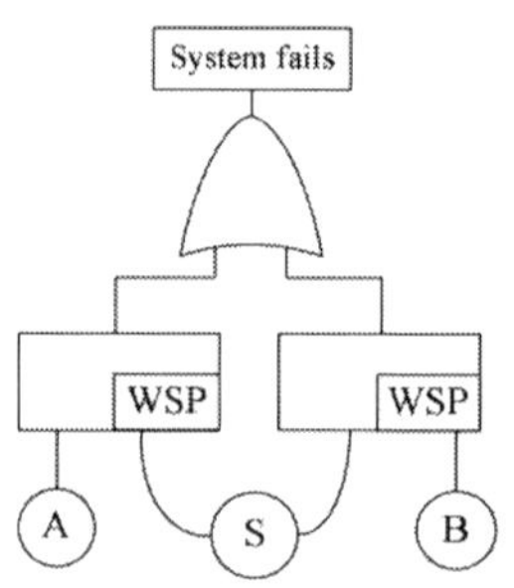

Figure 1. Fault tree of a shared warm standby system.

DOI: 10.1051/matecconf/201711901049

Once A or B fails, S is used to replace the faulty component. However, sometimes the warm standby fails to turn into working state due to the failure of monitoring, locating and switching devices. This refers to fault coverage problem, which will be detailed below.

2.2 Imperfect fault coverage (IPC)

Fault coverage problem has attracted extensive attention in fault tolerant system. In general ， fault coverage is divided into perfect fault coverage(PFC) and imperfect fault coverage (IPC). PFC means that whether the system can continue to work or not depends on the remaining structure of the system, not the faulty component, while the IPC behavior is caused by the imperfect coverage mechanisms [28]. The general IPC model is shown in Figure 2 [23].

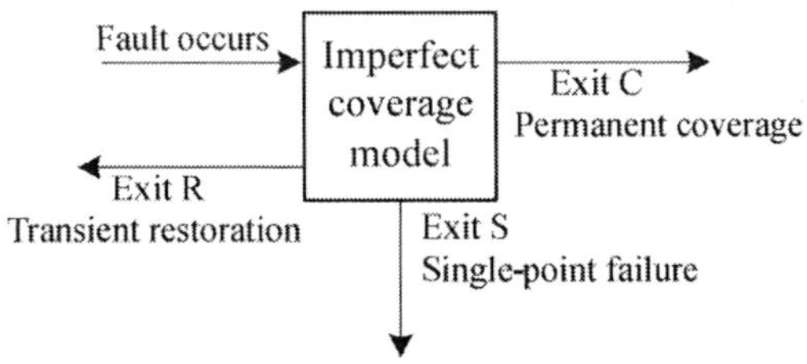

Figure 2. General structure of imperfect coverage model.

When a fault occurs, there are three possible outcomes corresponding to transient restoration R, permanent coverage C and single-point failure S. Exist R represents the failure is transient and causes no change to the state, while Exit C denotes the failure is permanent and causes only outage of offending component. And Exit S means the occurrence of the fault that causes the total system to fail.

The probabilities r_0 , c_0 and s_0 , respectively represents the occurrence probability of R exit, C exit and S exit, and $r_0 + c_0 + s_0 = 1$. In this paper, the probability r_0 is assumed to be zero, which means that the transient restoration is not considered as a fault.

3 The proposed method

3.1 Heaviside step function and Impulse function

(1) Heaviside step function
The Heaviside step function $u(t-\tau)$ is conventionally defined as [29]

$$u(t-\tau)=\begin{cases} 0 & if\ t<\tau \\ \dfrac{1}{2} & if\ t=\tau \\ 1 & if\ t>\tau \end{cases} \tag{1}$$

where t and τ are nonnegative real numbers. The discontinuity of the Heaviside step function occurs at $t=\tau$. In the warm spare structure, the Heaviside step function $u(t-\tau)$ can be used to denote that a component fails at time t after the time instant τ .
(2) The impulse function
The impulse function is used to express a variable taken on a specific value and defined as [29]

$$\delta(t-\tau)=0 \quad for\ t \neq \tau \text{ , and } \int_{-\infty}^{+\infty} \delta(t-\tau)dt = 1 \tag{2}$$

Because all the variables in warm standby system belong to the set of nonnegative real numbers, the integral in (2) can be transformed as

$$\int_0^{+\infty} \delta(t-\tau)dt = 1 \tag{3}$$

Besides, an important property of the impulse function will be used extensively in this paper, shown as follows.

$$\int_0^{+\infty} f(t)\delta(t-\tau)dt = f(\tau) \tag{4}$$

3.2 The conditional PDF of warm standby

As shown in Figure 1, component A and B respectively possess a time-to-failure probability density function (PDF), $f_A(a)$ and $f_B(b)$. Let $\lambda_\alpha(s)$ and $\lambda_S(s)$ respectively denote the failure rates of warm standby S in spare state and working state. It is often assumed that $\lambda_\alpha(s)$ is equal to $\alpha\lambda_S(s)$, where α takes values between 0 and 1.

Therefore, the conditional failure rate $\lambda_{S|A,B}(s|a,b)$ of S can be denoted as

$$\lambda_{S|A,B}(s|a,b) = u(a-s)u(b-s)\alpha\lambda_S(s) + [u(s-a)u(b-a) + u(s-b)u(a-b)]\lambda_S(s) \tag{5}$$

The formula (5) consist of 2 terms, and the former means the failure rate of warm standby S in spare state while the latter indicates the failure rate in working state.

According to reliability theory, the relationships among failure rate $\lambda(t)$, PDF $f(t)$ and reliability $R(t)$ are shown as follow.

$$f(t) = \lambda(t)\cdot\exp[-\int_0^t \lambda(t)dt] \tag{6}$$

$$R(t) = \exp[-\int_0^t \lambda(t)dt] \tag{7}$$

Using the formula (6), the conditional PDF $f_{S|A,B}(s|a,b)$ of S is

$$f_{S|A,B}(s|a,b) = \lambda_{S|A,B}(s|a,b)\exp[-\int_0^s \lambda_{S|A,B}(t|a,b)dt] \tag{8}$$

We assume that the components A, B and S don't fail at the same time. Utilizing the property of Heaviside step function and formula (7), the conditional PDF of S can be simplified as follows

$$\begin{aligned} f_{S|A,B}(s|a,b) &= u(a-s)u(b-s)\alpha\lambda_S(s)\exp[-\int_0^s \alpha\lambda_S(t)dt] + u(s-a)u(b-a)\lambda_S(s)\exp[-\int_0^a \alpha\lambda_S(t)dt - \int_a^s \lambda_S(t)dt] + \\ &\quad u(s-b)u(a-b)\lambda_S(s)\exp[-\int_0^b \alpha\lambda_S(t)dt - \int_b^s \lambda_S(t)dt] \\ &= u(a-s)u(b-s)\alpha f_S(s)[R_S(s)]^{\alpha-1} + u(s-a)u(b-a)f_S(s)[R_S(a)]^{\alpha-1} + u(s-b)u(a-b)f_S(s)[R_S(b)]^{\alpha-1} \end{aligned} \tag{9}$$

where $R_S(t)$ means the reliability of warm standby S in time t .

3.3 Reliability evaluate subject to IPC

Reliability evaluate is important for reliability analysis. Considering the IFC of warm standby systems, FLC model and ELC model are used to analyze system reliability.

DOI: 10.1051/matecconf/201711901049

3.3.1 FLC model

In FLC model, a coverage parameter c is used to represent the probability of success in detecting and isolating the faulty component. In this case, the conditional PDF of output X is

$$\begin{aligned} f_{X|A,B,S}(x|a,b,s) = & [(1-c)u(b-a)u(s-a)+cu(a-b)u(s-a)+u(a-s)u(b-a)]\delta(x-a)+ \\ & [(1-c)u(a-b)u(s-b)+cu(b-a)u(s-b)+u(b-s)u(a-b)]\delta(x-b)+ \\ & [c\cdot u(s-a)u(b-s)+c\cdot u(s-b)u(a-s)]\delta(x-s) \end{aligned} \tag{10}$$

Where the first term denotes that the state of X is equal to the state of A in three cases: S fails to replace the faulty A; S succeeds to replace the faulty B, then A fails; and S fails first, then A fails. The second term states that the state of X is equal to the state of B, which also includes three cases similar to the cases of first term. While the last term indicates that when S succeeds to replace the faulty A or B, and then S fails, the state of X is equal to the state of S.

It is noticed that $f_{X|A,B,S}(x|a,b,s)$ is a proper distribution function, which can be proved by (1) and (10), shown as follows.

$$\int_0^{+\infty} f_{X|A,B,S}(x|a,b,s)dx = 1 \tag{11}$$

Therefore, the joint failure probability density function of X is

$$f_{ABSX}(a,b,s,x) = f_{X|A,B,S}(x|a,b,s)f_{S|A,B}(s|a,b)f_A(a)f_B(b) \tag{12}$$

Here the failures of A and B are considered to be independent on each other. Using formula (9) and (10), we can get

$$\begin{aligned} f_{ABSX}(a,b,s,x) = \{ & [(1-c)u(b-a)u(s-a)+c\cdot u(a-b)u(s-a)+u(a-s)u(b-a)]\delta(x-a)+ \\ & [(1-c)u(a-b)u(s-b)+c\cdot u(b-a)u(s-b)+u(b-s)u(a-b)]\delta(x-b)+ \\ & [c\cdot u(s-a)u(b-s)+c\cdot u(s-b)u(a-s)]\delta(x-s)\}\cdot\{u(a-s)u(b-s)\alpha[R_S(s)]^{\alpha-1}+ \\ & u(s-a)u(b-a)[R_S(a)]^{\alpha-1}+u(s-b)u(a-b)[R_S(b)]^{\alpha-1}\}f_S(s)f_A(a)f_B(b) \end{aligned} \tag{13}$$

The formula (13) can be simplified as

$$\begin{aligned} f_{ABSX}(a,b,s,x) = & \{(1-c)u(b-a)u(s-a)[R_S(a)]^{\alpha-1}+c\cdot u(a-b)u(s-a)[R_S(b)]^{\alpha-1}+ \\ & u(a-s)u(b-a)\alpha[R_S(s)]^{\alpha-1}\}f_S(s)f_A(a)f_B(b)\delta(x-a)+ \\ & \{(1-c)u(a-b)u(s-b)[R_S(b)]^{\alpha-1}+c\cdot u(b-a)u(s-b)[R_S(a)]^{\alpha-1}+ \\ & u(b-s)u(a-b)\alpha[R_S(s)]^{\alpha-1}\}f_S(s)f_A(a)f_B(b)\delta(x-b)+ \\ & \{c\cdot u(s-a)u(b-s)[R_S(a)]^{\alpha-1}+c\cdot u(s-b)u(a-s)[R_S(b)]^{\alpha-1}\}f_S(s)f_A(a)f_B(b)\delta(x-s) \end{aligned} \tag{14}$$

To integrate $f_{ABSX}(a,b,s,x)$ with respect to a, b, and s from 0 to $+\infty$, the marginal failure PDF of X can be gotten

$$f_X(x) = \int_0^{+\infty}\int_0^{+\infty}\int_0^{+\infty} f_{ABSX}(a,b,s,x)dadbds \tag{15}$$

The $f_X(x)$ of formula (15) can be the sum of 8 terms

The first term is

$$f_1 = \int_0^{+\infty}\int_0^{+\infty}\int_0^{+\infty}(1-c)u(b-a)u(s-a)f_S(s)[R_S(a)]^{\alpha-1}f_A(a)f_B(b)\delta(x-a)dadbds \tag{16}$$

Using the properties of the Heaviside step function and Impulse function, f_1 can transformed into

$$\begin{aligned} f_1 &= \int_0^{+\infty}\int_0^{+\infty}(1-c)u(b-x)u(s-x)f_S(s)[R_S(x)]^{\alpha-1}f_A(x)f_B(b)dbds \\ &= (1-c)[R_S(x)]^{\alpha-1}f_A(x)\int_x^{+\infty}f_B(b)db\int_x^{+\infty}f_S(s)ds \\ &= (1-c)f_A(x)R_B(x)[R_S(x)]^{\alpha} \end{aligned} \tag{17}$$

The second term of $f_X(x)$ is

$$\begin{aligned} f_2 &= \int_0^{+\infty}\int_0^{+\infty}\int_0^{+\infty}c\cdot u(a-b)u(s-a)f_S(s)[R_S(b)]^{\alpha-1}f_A(a)f_B(b)\delta(x-a)dadbds \\ &= \int_0^{+\infty}\int_0^{+\infty}c\cdot u(x-b)u(s-x)f_S(s)[R_S(b)]^{\alpha-1}f_A(x)f_B(b)dbds \\ &= cf_A(x)\int_0^{x}f_B(b)[R_S(b)]^{\alpha-1}db\int_x^{+\infty}f_S(s)dc \\ &= cf_A(x)R_S(x)\int_0^{x}f_B(b)[R_S(b)]^{\alpha-1}db \end{aligned} \tag{18}$$

Using a similar approach made to compute the first term and second term, the other terms of $f_X(x)$ are

$$f_3 = f_A(x)R_B(x)\{1-[R_S(x)]^{\alpha}\} \tag{19}$$

$$f_4 = (1-c)R_A(x)f_B(x)[R_S(x)]^{\alpha} \tag{20}$$

$$f_5 = cf_B(x)R_S(x)\int_0^{x}f_A(a)[R_S(a)]^{\alpha-1}da \tag{21}$$

$$f_6 = R_A(x)f_B(x)\{1-[R_S(x)]^{\alpha}\} \tag{22}$$

$$f_7 = cR_B(x)f_S(x)\int_0^{x}f_A(a)[R_S(a)]^{\alpha-1}da \tag{23}$$

$$f_8 = cR_A(x)f_S(x)\int_0^{x}f_B(b)[R_S(b)]^{\alpha-1}db \tag{24}$$

Thus,

$$\begin{aligned} f_X(x) &= f_1+f_2+f_3+f_4+f_5+f_6+f_7+f_8 \\ &= [f_A(x)R_B(x)+R_A(x)f_B(x)]\{1-c[R_S(x)]^{\alpha}\}+c[f_A(x)R_S(x)+R_A(x)f_S(x)]\int_0^{x}f_B(b)[R_S(b)]^{\alpha-1}db+ \\ & c[f_B(x)R_S(x)+R_B(x)f_S(x)]\int_0^{x}f_A(a)[R_S(a)]^{\alpha-1}da \end{aligned} \tag{25}$$

Especially, when $f_A(x)=\lambda_1 e^{-\lambda_1 x}$, $f_B(x)=\lambda_2 e^{-\lambda_2 x}$, $f_S(x)=\lambda_3 e^{-\lambda_3 x}$, $f_X(x)$ turns to

$$\begin{aligned} f_X(x) &= (\lambda_1+\lambda_2)e^{-(\lambda_1+\lambda_2)x}+\frac{c(\lambda_1+\lambda_3)\lambda_2}{\lambda_2+(\alpha-1)\lambda_3}e^{-(\lambda_1+\lambda_3)x}+\frac{c(\lambda_2+\lambda_3)\lambda_1}{\lambda_1+(\alpha-1)\lambda_3}e^{-(\lambda_2+\lambda_3)x}- \\ & c(\lambda_1+\lambda_2+\alpha\lambda_3)[\frac{\lambda_1}{\lambda_1+(\alpha-1)\lambda_3}+\frac{\lambda_2}{\lambda_2+(\alpha-1)\lambda_3}]e^{-(\lambda_1+\lambda_2+\alpha\lambda_3)x} \end{aligned} \tag{26}$$

Therefore, the reliability of system is

$$\begin{aligned} R_X(x) &= \int_x^{+\infty}f_X(x)dx \\ &= e^{-(\lambda_1+\lambda_2)x}+\frac{c\lambda_1}{\lambda_1+(\alpha-1)\lambda_3}[e^{-(\lambda_2+\lambda_3)x}-e^{-(\lambda_1+\lambda_2+\alpha\lambda_3)x}]+\frac{c\lambda_2}{\lambda_2+(\alpha-1)\lambda_3}[e^{-(\lambda_1+\lambda_3)x}-e^{-(\lambda_1+\lambda_2+\alpha\lambda_3)x}] \end{aligned} \tag{27}$$

3.3.2 ELC model

In ELC model, every element has a particular coverage value, which is the probability of being covered when it fails. Here, we assume that the coverage probability of A is c_1 while B is c_2. In this case, the conditional PDF of output X is

$$\begin{aligned} f_{X|A,B,S}(x|a,b,s) = {} & [(1-c_1)u(b-a)u(s-a)+c_2u(a-b)u(s-a)+u(a-s)u(b-a)]\delta(x-a)+ \\ & [(1-c_2)u(a-b)u(s-b)+c_1u(b-a)u(s-b)+u(b-s)u(a-b)]\delta(x-b)+ \\ & [c_1\cdot u(s-a)u(b-s)+c_2\cdot u(s-b)u(a-s)]\delta(x-s) \end{aligned} \tag{28}$$

Using a similar approach to the IFC model, the failure PDF of X can be obtained

$$\begin{aligned} f_X(x) = {} & f_A(x)R_B(x)\{1-c_1[R_S(x)]^\alpha\}+R_A(x)f_B(x)\{1-c_2[R_S(x)]^\alpha\}+c_2[f_A(x)R_S(x)+ \\ & R_A(x)f_S(x)]\int_0^x f_B(b)[R_S(b)]^{\alpha-1}db+c_1[f_B(x)R_S(x)+R_B(x)f_S(x)]\int_0^x f_A(a)[R_S(a)]^{\alpha-1}da \end{aligned} \tag{29}$$

Especially, when $f_A(x)=\lambda_1 e^{-\lambda_1 x}$, $f_B(x)=\lambda_2 e^{-\lambda_2 x}$, $f_S(x)=\lambda_3 e^{-\lambda_3 x}$, $f_X(x)$ turns to

$$\begin{aligned} f_X(x) = {} & (\lambda_1+\lambda_2)e^{-(\lambda_1+\lambda_2)x}+\frac{c_1\lambda_1}{\lambda_1+(\alpha-1)\lambda_3}[(\lambda_2+\lambda_3)e^{-(\lambda_2+\lambda_3)x}-(\lambda_1+\lambda_2+\alpha\lambda_3)e^{-(\lambda_1+\lambda_2+\alpha\lambda_3)x}]+ \\ & \frac{c_2\lambda_2}{\lambda_2+(\alpha-1)\lambda_3}[(\lambda_1+\lambda_3)e^{-(\lambda_1+\lambda_3)x}-(\lambda_1+\lambda_2+\alpha\lambda_3)e^{-(\lambda_1+\lambda_2+\alpha\lambda_3)x}] \end{aligned} \tag{30}$$

In this case, the reliability of system is

$$\begin{aligned} R_X(x) &= \int_x^{+\infty} f_X(x)dx \\ &= e^{-(\lambda_1+\lambda_2)x}+\frac{c_1\lambda_1}{\lambda_1+(\alpha-1)\lambda_3}[e^{-(\lambda_2+\lambda_3)x}-e^{-(\lambda_1+\lambda_2+\alpha\lambda_3)x}]+\frac{c_2\lambda_2}{\lambda_2+(\alpha-1)\lambda_3}[e^{-(\lambda_1+\lambda_3)x}-e^{-(\lambda_1+\lambda_2+\alpha\lambda_3)x}] \end{aligned} \tag{31}$$

4 Case study

Taking a system including a shared warm standby element for example, the dynamic fault tree of the system is shown in Figure 3. The system consists of elements A, B, S, C, D, E, and the of dormancy coefficient α of warm standby element S is 0.5. Considering the time-to-failures of the system components obey exponential distributions and Weibull distributions, the system reliability is evaluated by the proposed method.

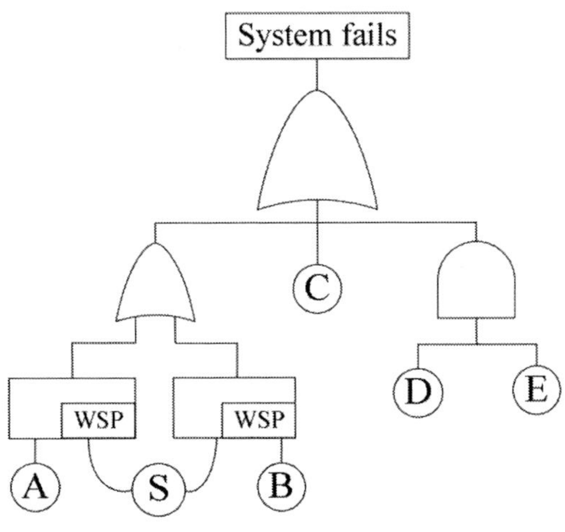

Figure 3. Dynamic fault tree of the example system.

4.1 Exponential distributions

When the time-to-failures of system components follow exponential distributions, the parameters of components are presented at Table 1.

Table 1. parameter values of system components.

Component	A	B	S	C	D	E
Parameter Value (λ_i / h^{-1})	3×10^{-4}	4×10^{-4}	2×10^{-4}	10^{-4}	2.5×10^{-4}	2.8×10^{-4}

Seeming the shared warm standby structure as a new component X, the system reliability $R(t)$ can be obtained

$$R(t) = R_X(t) \cdot R_C(t) \cdot \{1-[1-R_D(t)] \cdot [1-R_E(t)]\} \tag{32}$$

Considering the fault coverage of warm standby structure, the PFC, FLC, ELC models are discussed in this paper. The PFC model equals to the coverage parameter being 1. We suppose that the system coverage parameter equals to 0.85 in FLC model while the coverage parameter of element A equals to 0.95 and element B equals to 0.9 in ELC model.

Utilizing the MATLAB software, the system reliability curve changing over time is shown in Figure 4.

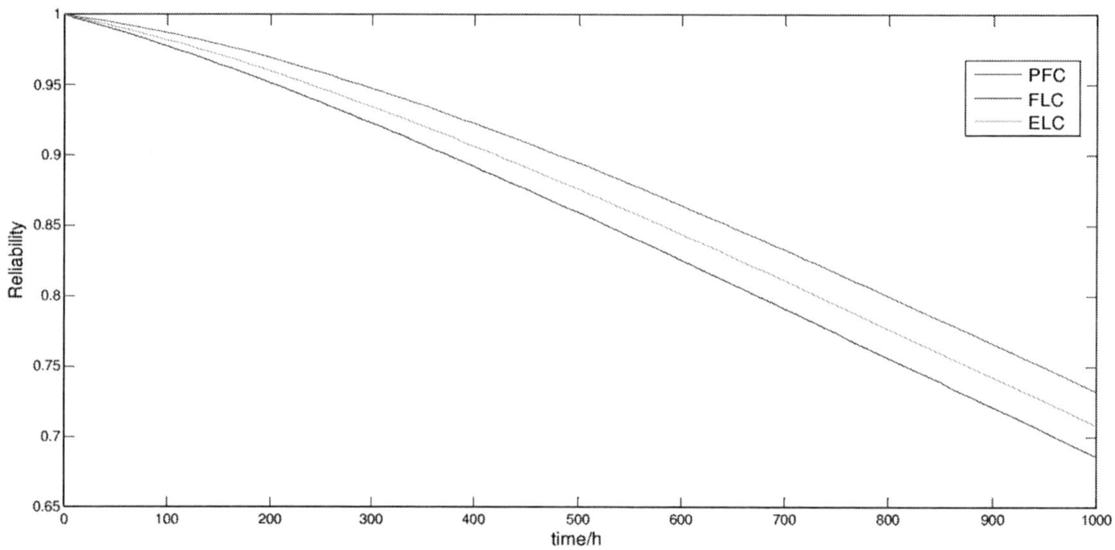

Figure 4. System reliability curve changes over time.

Table 2. Partial derivatives with respect to failure rate of each element

	$\frac{\partial R}{\partial \lambda_A}$	$\frac{\partial R}{\partial \lambda_B}$	$\frac{\partial R}{\partial \lambda_S}$	$\frac{\partial R}{\partial \lambda_C}$	$\frac{\partial R}{\partial \lambda_D}$	$\frac{\partial R}{\partial \lambda_E}$
PFC $(c=1)$	-189	-208	-186	-136	-23	-20
FLC $(c=0.85)$	-222	-215	-173	-129	-22	-19
ELC $(c_A=0.95, c_B=0.9)$	-195	-238	-171	-122	-21	-18

In the mission time $t = 800h$, the system reliability in PFC, FLC, ELC models is 0.800, 0.756 and 0.777, respectively. For the mission time $t = 800h$, partial derivatives of system reliability function are taken with respect to failure rate of each element, the results are shown as Table 2.

From the Table 2, the decrease of failure rates of element A, B and C are more useful for improving the system reliability. The importance of the elements impacting on the system reliability is related to the fault coverage models.

4.2 Weibull distributions

When the time-to-failures of system components follow Weibull distributions, the parameters of components are shown at Table 3.

Table 3. parameter values of system components.

component	A	B	S	C	D	E
Scale parameter value η (h)	1.2×10^3	1.1×10^3	1.5×10^3	5×10^3	1.8×10^3	2×10^3
Shape parameter value m	3	3	2.5	1.2	2	2

Using a similar approach to the case of exponential distributions, the system reliability with elements obeying Weibull distributions can be obtained, shown in Figure 5.

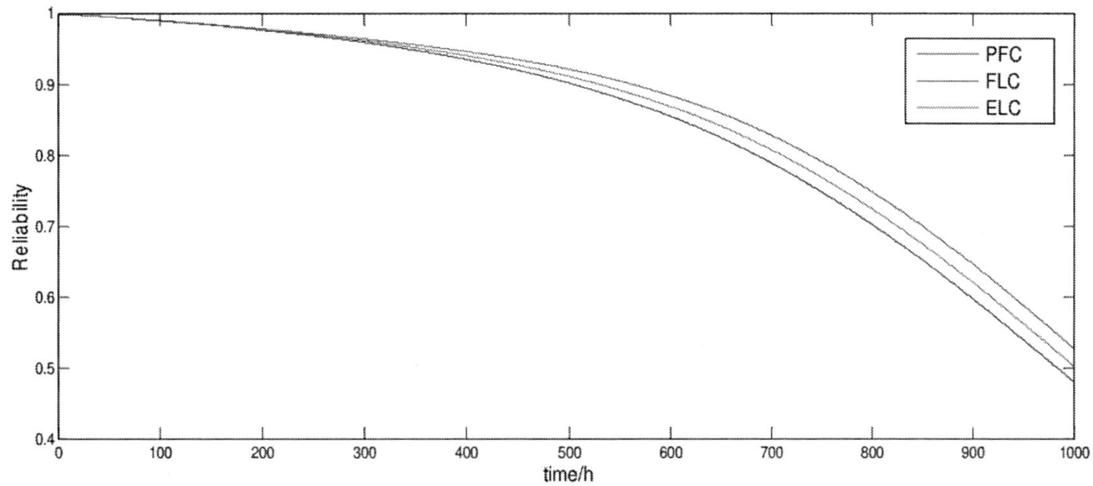

Figure 5. System reliability curve changes over time in the case of Weibull distributions.

In the mission time $t = 800h$, the system reliability changes over the coverage parameters in FLC model and ELC model are shown in Figure 6 and Figure 7. Seeming the system reliability as a function with respect to the scale parameters of elements, the mission reliability changing over the scale parameter of each component is presented in Figure 8.

From Figure 8, the change of the system reliability over the scale parameters of A, B and S is more pronounced. Therefore, the key to enhancing the reliability of the system is the developments of components A, B and S. Besides, the increasing of coverage parameter has an important significance on security operation of warm standby system.

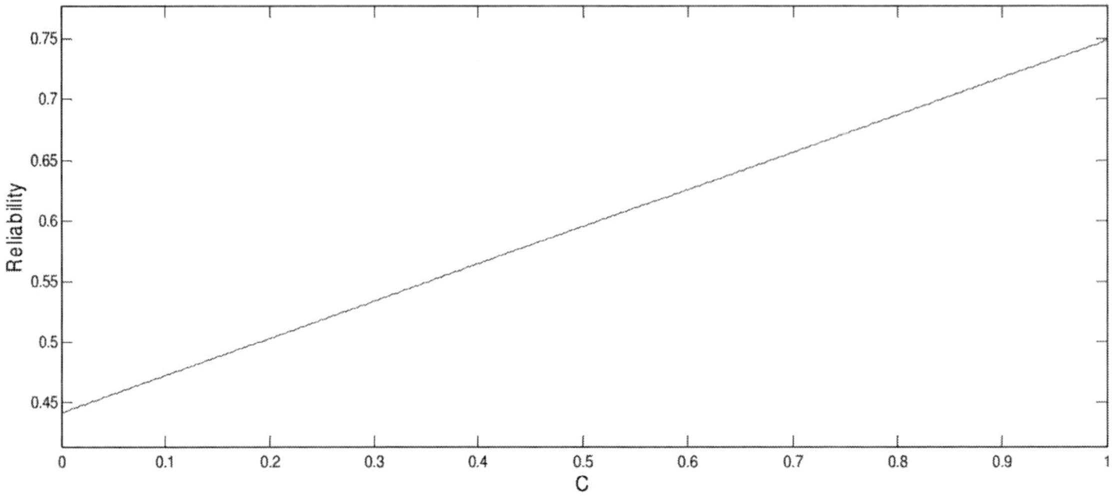

Figure 6. System reliability changes over the coverage parameter in FLC model.

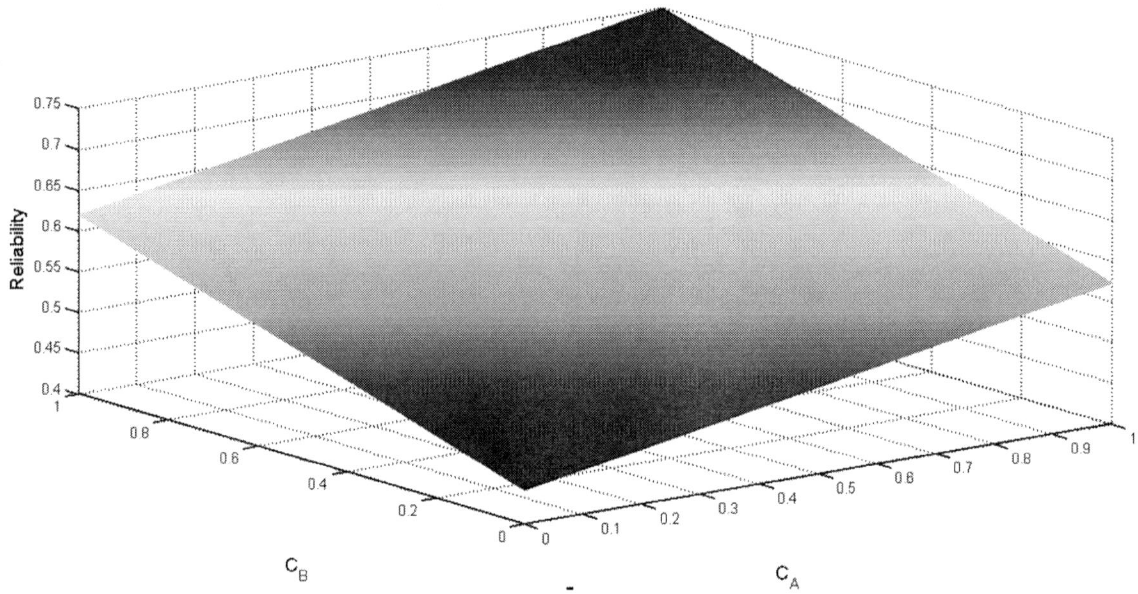

Figure 7. System reliability changes over the coverage parameters in ELC model.

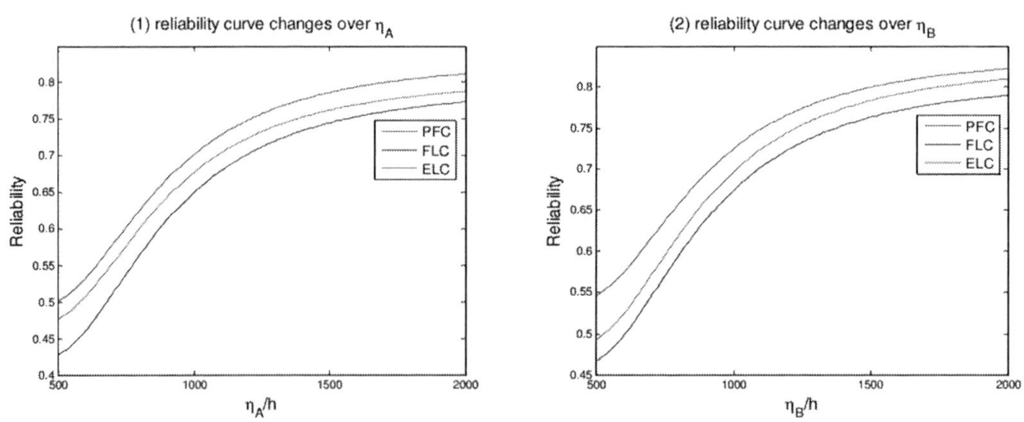

DOI: 10.1051/matecconf/201711901049

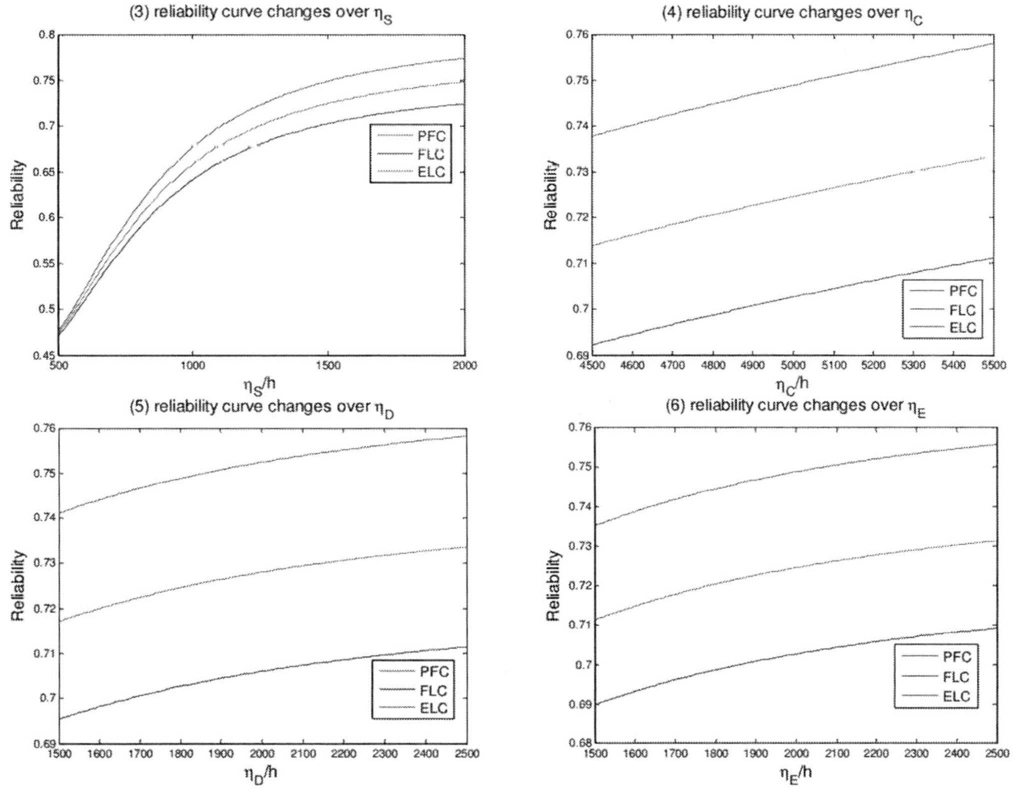

Figure 8. Mission reliability changing over the scale parameter of each component.

5 Conclusion

This paper proposes a new approach to the system reliability analysis with shared warm standby element. The step function and impulse function are introduced to describe the failure characteristic of warm standbys. Two kinds of imperfect fault coverage models: Element Level Coverage (ELC) and Fault Level Coverage (FLC) are considered in the system reliability evaluation. The proposed approach has no limitation on the type of time-to-failure distributions for the system components. In the mission time, the reliability of warm standby system changing over the time-to-failure parameters of each element and the coverage parameters in different fault coverage models are obtained.

Acknowledgements

This work was supported by the National Defense Basic Scientific Research program of China under the number 61325102. The authors are thankful to the editor and reviewers for their valuable comments that help us improve the quality of the article.

References

1. W. Kuo, R.V. Prasad, A.F. Tillman, et al., Microchem. J., **32** (10), 911-911 (2001)
2. S.V. Amari, J.B. Dugan, and R.B. Misra, IEEE Trans. Rel., **48** (3), 275-284 (1999)
3. O. Tannous, L. Xing, P. Rui, et al., IEEE Trans. Rel., **49** (2), 1261-1265 (2009)
4. S. Amari, G Dill, et al., RAMS, 417-422 (2009)
5. J. She and M.G. Pecht, IEEE Trans. Rel., **41**(1), 72-75 (1992)
6. S. Eryilmaz, IEEE Trans. Rel., **62**(2), 499-503 (2013)

7. W.Y. Yun and H.C. Ji, Reliab. Eng. Syst. Saf., **95** (8), 880-886 (2010)
8. S.V. Amari, H. Pham, and R.B. Misra, IEEE Trans. Rel., **61** (4),1007-1018 (2012)
9. K. Geismar, C. Enevold, L.K. Sørensen, et al., Eur. J. Oper. Res., **201** (3), 821-827 (2010)
10. G. Levitin and S.V. Amari, Reliab. Eng. Syst. Saf., **95** (4), 396-401 (2010)
11. T. Zhang, M. Xie, M. Horigome, et al., Reliab. Eng. Syst. Saf., **91** (4), 381-387 (2006)
12. W. Long, T.L. Zhang, and F.Y. Lu, J. Mer. Mar. Nat. Sci., **53**, 49-54 (2002)
13. D. Liu, C. Zhang, W. Xing, et al., HPCC, 755-765 (2007)
14. L. Xing, O. Tannous, and J.B. Dugan, IEEE Trans. Syst, Man, Cybern., Syst., **42** (3), 715-726 (2012)
15. O. Tannous, L. Xing, J.B. Dugan, et al., RAMS, 24-27 (2011)
16. I. Dobson, B.A. Carreras, V.E. Lynch, et al., J. Nonlinear. Sci., **17** (2), 026103-026113 (2007)
17. W. Bouricius, W.C. Carter, and P. Schneider, ACM, 295-309 (1969)
18. T.F. Arnold, IEEE Trans. Comput., **C-22** (3), 251-254 (1973)
19. L. Xing, IEEE Trans. Rel., **56** (1), 58-68 (2007)
20. A. Myers, IEEE Trans. Rel., **57** (2), 349-354 (2008)
21. A. Myers, *Complex System Reliability* (Springer, London, 2010)
22. T. Arnold, IEEE Trans. Comput., **22** (3), 325-339 (1973)
23. V.S. Amari, J.B. Dugan, R.B. Misra, et al., IEEE Trans. Rel., **48** (3), 267-274 (1999)
24. J.C. Ke, Z.L. Su, et al., Simul. Model. Pract. Th., **18** (3), 338-347 (2010)
25. Y. L. Hsu, S.L. Lee, and J.C. Ke, Math. Compute. Simul., **79** (7), 2227-2239 (2009)
26. S.L. Lee, J.C. Ke, and Y.L. Hsu, Int. J. Syst. Sci., **40** (11), 1149-1159 (2009)
27. Y.L. Hsu, S.L. Lee, and J.C. Ke, Simul. Model. Pract. Th., **16** (3), 353-367(2008)
28. S.V. Amari, H. Pham, G. Dill, et al., IEEE Trans. Rel., **53** (4), 567–575 (2005)
29. J. Spanier and K.B. Oldham, *An Atlas of Functions* (Springer-Verlag, New York, 2009)

Estimating reliability of degraded system based on the probability density evolution with multi-parameter

Ge Jiang[a], Hongjie Yuan and Hailong Zhang

School of Reliability and System Engineering, Beijing University of Aeronautics and Astronautics, Beijing 100191, China

Abstract. System degradation was usually caused by multiple-parameter degradation. The assessment result of system reliability by universal generating function was low accurate when compared with the Monte Carlo simulation. And the probability density function of the system output performance cannot be got. So the reliability assessment method based on the probability density evolution with multi-parameter was presented for complexly degraded system. Firstly, the system output function was founded according to the transitive relation between component parameters and the system output performance. Then, the probability density evolution equation based on the probability conservation principle and the system output function was established. Furthermore, probability distribution characteristics of the system output performance was obtained by solving differential equation. Finally, the reliability of the degraded system was estimated. This method did not need to discrete the performance parameters and can establish continuous probability density function of the system output performance with high calculation efficiency and low cost. Numerical example shows that this method is applicable to evaluate the reliability of multi-parameter degraded system.

1 Introduction

For high reliability and long life products, it is difficult to obtain failure life time through life test and accelerated life test. The performance degradation data can provide important information for reliability assessment. Therefore the system reliability assessment based on degradation data becomes an important direction in the field of reliability research[1-2]. The performance of the system depends on the performance parameters of its components. According to the degradation law of the main parts and the function relationship between the performance of system and the parameters of various components, Monte Carlo simulation method can provide the degradation law of the performance of system and evaluate system reliability[3]. But this method need the simulation calculation on each discrete time point and large amount of calculation. So the literatures[4-8] put forward the universal generating function method. The method discreted the input parameter and combined the discrete state of the input parameters. Finally, the universal generating function algorithm deduced the reliability that meet the requirements of system output performance. But the precision of the system reliability evaluated by the universal generating function was related with the discrete degree of parameter. And

[a] Corresponding author : jungle@buaa.edu.cn

with the increasing of the number of discrete state, the precision significantly improves. Because the discrete state combinations increase like geometric model with the increase of the number of discrete state, this method makes the calculation very complex[9] and can not get the probability density function of the system output performance. This paper proposes a system reliability assessment method based on multiple-parameter density evolution. This method can deduce the continuous probability density function of the system output performance parameters and calculate the system reliability. Compared with the Monte Carlo method, this method needs a small amount of calculation, but it provides higher precision than the universal generating function method.

2 Universal generating function method

The universal generating function method defines the contact between the possible value of the random parameters and the corresponding probability. It relates the state of the component performance parameter with the state probability of the system performance parameter. According to the system universal generating function computed by universal generating function algorithms, system reliability is obtained.

System degradation belongs to multiple random degradation process. The distribution characteristics of the degraded component performance parameters change with time. Degraded performance parameters obey normal distribution $X \sim N(\mu, \sigma^2)$ at the given moment. For discrete random variable, Universal generating function discretes the distribution for J states $\{x_1, x_2, ..., x_j\}$, The corresponding probability of each state is

$$q_j = \Pr\{X = x_j\} \text{ and } \sum_{j=1}^{J} q_j = 1 \tag{1}$$

At the time t, universal generating function of the component performance parameters is expressed as

$$U(z) = \sum_{j=1}^{J} q_j \bullet z^{x_j} \tag{2}$$

When there are n performance parameters degenerating, the parameter vectors are $X = [X_1, X_2, ..., X_n]$ and system output function is $G(X,t)$。At the time t, the universal formula of the system universal generating function is

$$U_{G(X,t)}(z) = \Omega(U_1(z), U_2(z), ..., U_n(z)) = \sum_{i_1=1}^{I_1} \cdots \sum_{i_n=1}^{I_n} q_{1i_1} \cdots q_{ni_n} \bullet z^{g(x_{1i_1}, x_{2i_2}, ..., x_{ni_n})} \tag{3}$$

among them, $g(\bullet)$ is the system output performance value, $I_1, I_2, ..., I_n$ are different discrete states of n system performance degraded parameters at the time t.
Particularly when n=2, the system universal generating function is

$$U_{G(X,t)}(z) = \Omega(U_1(z), U_2(z)) = \sum_{i=1}^{I} \sum_{j=1}^{J} q_{1i} \bullet q_{2j} \bullet z^{g(x_{1i}, x_{2j})} \tag{4}$$

When the degradation threshold of system output performance parameters is D, system reliability is

$$R(t) = \sum_{g(x_{1i}, x_{2j}) \geq D} q_{1i} \bullet q_{2j} \tag{5}$$

The literature[7] defined two special operators for the system output function: π operator and σ operator.The output functions of the system with parallel structure is

$$g(a,b) = \pi(\mathrm{a},\mathrm{b}) = \mathrm{a}+\mathrm{b} \tag{6}$$

The output functions of the system with tandem structure is

$$\mathrm{g}(\mathrm{a},\mathrm{b}) = \sigma_2(\mathrm{a},\mathrm{b}) = \frac{a \bullet b}{a+b} \tag{7}$$

According to the literature[9], the more the discrete states of component performance parameters, the higher the accuracy of the assessment. But the increase in discrete states will lead the number of the performance parameters state combinations increase like the geometric growth. When the system degraded performance parameters are more and the parameters are not the same type of distribution, it will bring huge computational burden.

3 Probability density evolution method

3.1 Establishing evolution equations

Probability density evolution method evolved from static and dynamic stochastic systems. This method based on systems physics equations researched the law of probability evolution from system multivariate stochastic parameters to the system output response and its theory is based on probability conservation principle. The existing random factors don't disappear and new random factor isn't added in conservative random system. The evolution of the physical state of the random system leads to the migration of probability and causes the evolution of the probability density. Establish the probability density evolution equation as follows [10-12], assuming that the system output function is

$$Z(t) = G(X,t) \tag{8}$$

$Z(t)$ is the system output function and random variables of system are $\mathrm{X}=[\mathrm{X}_1,\mathrm{X}_2,...,\mathrm{X}_n]$.Augmented system $(Z(t), X)$ constitutes a conservative stochastic system. Joint probability density function of $(Z(t), X)$ is $p_{ZX}(z,x,t)$.

According to probability conservation principle, probability density evolution equation can be established [13, 14]:

$$\frac{\partial p_{ZX}(z,x,t)}{\partial t} + \sum_{i=1}^{m} \dot{Z}_i(x,t) \frac{\partial p_{ZX}(z,x,t)}{\partial z_i} = 0 \tag{9}$$

Formula (9) is general probability density evolution equations. When m=1, the general probability density evolution equations are

$$\frac{\partial p_{ZX}(z,x,t)}{\partial t} + \dot{Z}(x,t) \frac{\partial p_{ZX}(z,x,t)}{\partial z} = 0 \tag{10}$$

Typically, the boundary conditions of equation (9) is

$$p_{ZX}\left(z,x,t\right)\Big|_{z_i\to\pm\infty}=0,\ i=1,2,\cdots,m \tag{11}$$

The initial condition is

$$p_{ZX}\left(z,x,t\right)\Big|_{t=t_0}=\delta(z-z_0)p_X(x) \tag{12}$$

For partial differential equation (10), when $\dot{z}(x,t)$ is known, according to the initial and boundary conditions the analytical solution is

$$p_{ZX}\left(z,x,t\right)=\delta[z-z(x,t)]p_X\left(x\right) \tag{13}$$

$p_X\left(x\right)$ is the joint probability density of random parameters. $\delta(\bullet)$ is Dirac function. When using one-dimensional normal distribution density function δ sequence, Dirac function can be approximated by a normal distribution.

$$\delta(z-z(x))=\frac{1}{\sqrt{2\pi}\lambda}\exp[-\frac{1}{2\lambda^2}(z-z(x))^2] \tag{14}$$

λ above is a custom amount in Dirac function. When λ is close to zero, the above equation is theoretically established. Therefore, the following equation can be used to solve the probability density function of system output.

$$p_Z(z,t)=\frac{1}{\sqrt{2\pi}\lambda}\int_{\Omega_x}\exp[-\frac{1}{2\lambda^2}(\frac{z-z(x,t)}{z_{\max}})^2]\bullet\frac{1}{z_{\max}}\bullet p_X(x)dx \tag{15}$$

It is complex to directly solve equation (15). According to the method of multi-dimensional space selected point [15], the approximate calculation formula is:

$$p_Z(z,t)=\frac{1}{N_{sel}}\sum_{i=1}^{N_{sel}}\frac{1}{\sqrt{2\pi}\lambda}\exp[-\frac{1}{2\lambda^2}(z-G(x_i,t))^2]\bullet V_{x_i}\bullet\prod_{j=1}^{n}p_j(x_i) \tag{16}$$

N_{sel} is the total number of selected points.

x_i is the coordinates of the selected point.

$p_j\left(x_i\right)$ is the probability density of performance parameters X_j at the selected point.

V_{x_i} is the volume of the region the point represents.

3.2 System reliability assessment

According to the described method above, the probability density function of system output performance can be established at any time t. Assume that the system performance levels required for normal work is D, the system reliability at the time t is

$$R(t)=\Pr\left\{G(t)\ge D\right\}=\int_D^{+\infty}p_Z(z,t)dg \tag{17}$$

4 Comparison with Monte Carlo simulation

Taking an example of the function of system output performance given in the literature [9], $G(X,t)$ is the parameters of output performance.

$$G(X,t)=\frac{1}{4}\left(\left(\sin(X_1(t)-3)\right)\cdot(X_2(t)-1)+(X_3(t)-1)^2\right)-1 \tag{18}$$

Suppose at time of $t=t_0$, $X_1(t_0)$ and $X_3(t_0)$ obey uniform distribution $U[0,10]$, $X_2(t_0)$ obeys uniform distribution $U[6,16]$.The level of performance required by the system is 20. Using Monte Carlo simulation and the method proposed in this paper estimate the system reliability. And the compared result is shown in Table 1.

Table 1. Compared with Monte Carlo.

$G(X,t_0)$	20	The number of calculations
Monte Carlo	0.9837	100000
This method	0.9865	135
The error	0.28%	--

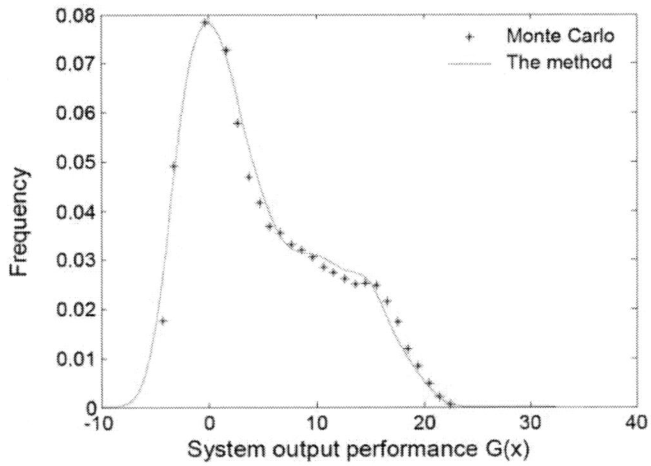

Figure 1. The probability density curve.

Figure 1 is the probability density curve of system output. As is shown in Figure 1, the calculation results based on the multi-parameter probability density evolution and Monte Carlo simulation are very close. Compared with Monte Carlo simulation results, when obtaining the same precision of system output performance parameters probability density function and system reliability, the computational efficiency is nearly 750 times of Monte Carlo simulation. Obviously, the system reliability assessment based on multi-parameter probability density evolution can not only ensure the accuracy of the assessment but also reduce the amount of calculation.

5 Example of system reliability assessment

If the system performance output function is

$$G(X,t)=\frac{X_1(t)\cdot X_2(t)}{X_1(t)+X_2(t)} \tag{19}$$

$X_1(t)$ is performance parameter of the component 1 and its deterioration law is in line with the Wiener process. Its characteristic parameters are

$$\mu_1: \ 6.0-2\times10^{-5}\bullet t$$

$$\sigma_1: \ 0.2+3.0\times10^{-6}\bullet t$$

$X_2(t)$ is performance parameter of the component 2 and its deterioration law is in line with the Wiener process. Its characteristic parameters are

$$\mu_2: \ 5.0-1\times10^{-5}\bullet t$$

$$\sigma_2: \ 0.3+1.5.0\times10^{-6}\bullet t$$

Table 2. Contrast of results among the three method.

Time/h	Monte Carlo Simulation	Probability Density evolution	Universal Generating Function（3 discrete states）	Universal Generating Function（7 discrete states）
0	0.99921	0.999293	1	0.999585
5000	0.997045	0.99663	1	0.99533
10000	0.991045	0.989008	0.974829	0.990157
15000	0.975943	0.971918	0.974829	0.973631
20000	0.94636	0.939747	0.866516	0.93574
25000	0.895974	0.887415	0.866516	0.86028
30000	0.821404	0.812742	0.758204	0.835575
35000	0.724797	0.718108	0.733032	0.739338
40000	0.612443	0.610291	0.733032	0.572939
45000	0.495858	0.498538	0.733032	0.572901
50000	0.384628	0.391953	0.241796	0.349516
55000	0.287996	0.297411	0.241796	0.349139
60000	0.208795	0.218657	0.241796	0.147861
65000	0.146107	0.156497	0.241796	0.147861
70000	0.099876	0.109638	0.133484	0.123155
75000	0.066361	0.075663	0.025171	0.041518
80000	0.04402	0.051825	0.025171	0.041518
85000	0.028085	0.035552	0.025171	0.03914
90000	0.018184	0.024694	0.025171	0.036762
95000	0.011826	0.017586	0.025171	0.007465
100000	0.007314	0.01301	0.025171	0.007465

The time-varying probability density curve of the system output performance parameters based on multi-parameter probability density evolution method is shown in Figure 2. Table 2 lists the calculated results based on different methods and the calculation accuracy comparison is shown in Figure 3. The result of the system reliability assessment method based on multi-parameter probability density evolution is closest to the result of Monte Carlo simulation through comparison. Therefore, the method presented in this paper is suitable for assessing the reliability of multi-parameter degraded system.

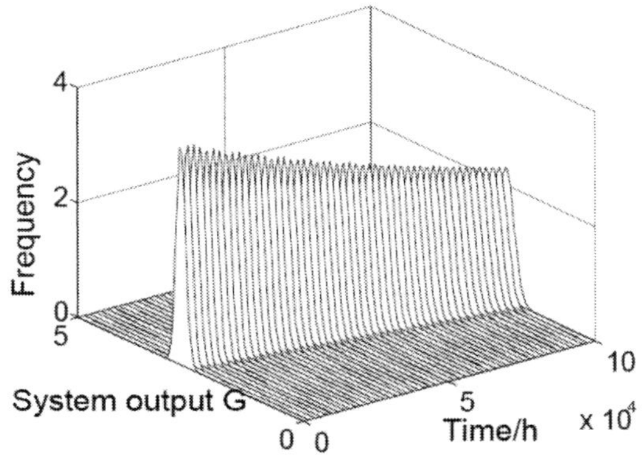

Figure 2. Time-varying probability density curve of the system output.

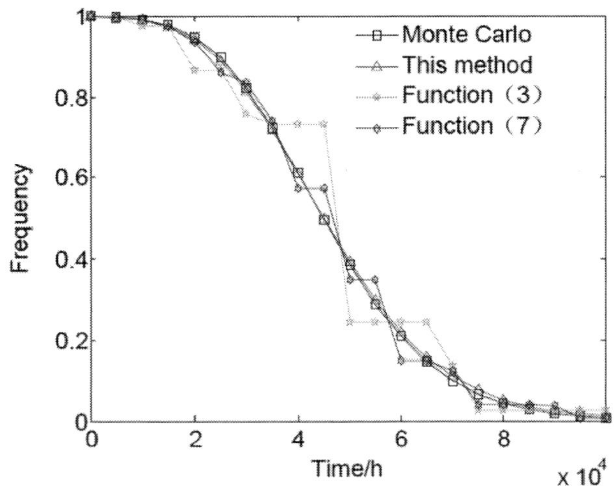

Figure 3. Contrast of reliability curve.

6 Conclusion

This paper presents a system reliability assessment method based on multi-parameter probability density evolution for complex parameter degraded system. This method can deduce the continuous probability density function of the system output performance and calculate the system reliability. This method is more accuracy than universal generating function method. Compared with Monte Carlo simulation, the assessment method proposed in this paper reduces the computational cost in the case to achieve the same accuracy and it provides an efficient way for system reliability assessment.

References

1. J. Tang, T.S. Su, Naval Research Logistics, **55** (3), 265-276 (2008)
2. K.A. Doksum and A. Hoyland, Technometrics, **34** (1), 74-82 (1992)
3. S.K. Zeng, T.D. Zhao, and J.G. Zhang. Tutorial of system reliability design and analysis, Beijing (2001)

4. G. Levitin, A. Lisnianski, H. Ben-Haim, et al., IEEE Transactions on Reliability, **47** (2), 165-172 (1998)
5. A. Azadeh, B. Maleki Shoja, and S. Ghanei, Reliab. Eng. Syst. Safety, **136**, 62-74 (2015)
6. M. Nourelfath, E. Chatelet, N. Nahas, Reliab. Eng. Syst. Safety, **103**, 51-60 (2012)
7. G. Levitin, IEEE Trans. Reliab., **50** (4), 380-388 (2001)
8. V. Ebrahimipour, M. Sheikhalishahi, and B. Maleki Shoja, IEEE, 255-239 (2010)
9. C. Li, Research on reliability analysis and optimization based on the multi-state system theory[D], Changsha(2010)
10. I. Elishakoff, Y.J. Ren, and M. Shinozuka, J. Eng. Mech., **122** (6), 559-565 (1996)
11. R.N. Mantegna, J. Stat. Phys., **70** (3/4), 721-736 (1993)
12. J.B. Chen and J. Li, Probabilist. Eng. Mech., **20** (1), 33-44 (2005)
13. J. Li and J. Chen, Acta Mechanica Sinica, **35** (6), 716-722 (2003) (in Chinese)
14. J. Li and J.B. Chen, Adv Mech, **40** (2), 170-188 (2010) (in Chinese)
15. K.T. Fang and Y. Wang, *Number Theoretic Methods in Statistics* (London,1994)
16. J. Deng, Int J Solids Struct, **43**, 3255-3291 (2006)

Particle swarm optimization based networked control system design with uncertainty

Ti-Hung Chen[1,a], Ming-Feng Yeh[2] and Wen-Yo Lee[1]

[1] *Department of Computer Information and Network Engineering, Lunghwa University of Science and Technology, Taoyuan City 33306, Taiwan*
[2] *Department of Electrical Engineering, Lunghwa University of Science and Technology, Taoyuan City 33306, Taiwan*

Abstract. This study discusses the design of state feedback control for networked control system (NCS) with uncertainty, where the feedback controller is used to solve the latency and the packet dropout caused by NCS. The conventional design approach generally adopts Lyapunov stability theorem to analyse the stability of NCSs and then Lyapunov-Krasovskii stability criterion is solved to obtain the control gain matrix. Rather than using complex Lyapunov-Krasovskii stability criterion and solving difficult linear matrix inequality (LMI) problem in the control design, this study attempts to utilize particle swarm optimization (PSO) algorithm to search an optimal feedback control gain for the NCS. Finally, the computer simulation is illustrated the proposed control strategy and demonstrated the control performance.

1 Introduction

Nowadays, increasing attention has been paid to design of networked control systems (NCSs) in which the components (e.g., sensors, actuators, and controllers) are spatially distributed and connected via a share communication network [1-10]. In NCSs, data is transmitted in atomic unit called data packets. To transmit data packets, the system states and output signals are sampled by the sensors, transmitted to the controllers over the network, and finally the controller signals are transmitted to the actuators by the data packets. The actuators would control the controlled plants according to the receiving data packets.

Two major challenges in NCSs are the effects of "latency" and "packet dropout" on the system performance [11]. The former is caused by propagation, transmission, router, processing, computer, etc., and the latter is resulted from network congestion, multi-path fading, faulty network drivers, etc. To treat the mentioned effects, several methods have been proposed. For example, [2, 8] adopted H_∞ control theory to stabilize NCSs and analysed the system robustness. [3] applied state feedback control method to treat the control problem of NCSs. [4] applied Takagi-Sugeno (T-S) fuzzy control technique to guarantee the stability of NCSs.

While the state feedback control applying to NCSs, the conventional approach generally adopts Lyapunov stability theorem to analyse the stability of NCSs. Then solving Lyapunov-Krasovskii

[a] Corresponding author : cth.chen@gmail.com

stability criterion, resulted from Lyapunov stability theorem, could attain the control gain matrix [2, 8]. Generally speaking, Lyapunov-Krasovskii stability criterion is characterized in terms of a linear matrix inequality (LMI) problem or an eigenvalue problem [12]. However, solving LMI or eigenvalue problem is difficult and generally obtains a feasible solution.

Particle swarm optimization (PSO) algorithm, proposed by Kennedy and Eberhart in 1995 [13], was inspired by the social behavior of bird flocking and fish schooling. It works by maintaining a swarm of particles that move around in the search space influenced by the improvements discovered by the other particles. One important advantage of PSO algorithm is that it does not use the gradient of the problem to be optimized. Therefore, PSO algorithm can be readily employed for a host of optimization problems and has received considerable research interest [14-17]. On the other hand, owing to its simplicity of implementation and ability to quickly converge to a reasonably good solution [18], PSO algorithm has been successfully applied in solving many real-world optimization problems.

With the help of randomly searching ability of PSO algorithm, this study attempts to utilize PSO algorithm to search the optimal control gain matrix for the NCS with uncertainty, instead of solving complex LMI or eigenvalue problem. The computer simulation is illustrated the proposed control strategy and demonstrated the corresponding control performance.

This study is organized as follows: The problem formulation of NCS is given in section 2. Section 3 presents the control scheme of PSO based NCS. In section 4, computer simulation will be given to illustrate the control performance of the proposed control strategy. Section 5 concludes the paper.

2 Problem formulations

Figure 1 illustrates a class of the networked control system which is consisted of the following five parts: a sensor, a controller, a zero-order-hold (ZOH), a communication network channel, and a controlled plant [18]. In the figure, the ZOH plays a role of an actuator. Consider the following controlled plant with uncertainty [8, 19]:

$$\dot{\mathbf{x}}(t) = \mathbf{A}\mathbf{x}(t) + \Delta\mathbf{A}\mathbf{x}(t) + \mathbf{B}\mathbf{u}(t) + \boldsymbol{\omega}(t) \tag{1}$$

where $\mathbf{x}(t) = [x(t), \dot{x}(t), ..., x^{(n)}(t)] \in \Re^n$ is the state vector, $\mathbf{u}(t) \in \Re^m$ is the control signal, $\boldsymbol{\omega}(t) \in \Re^n$

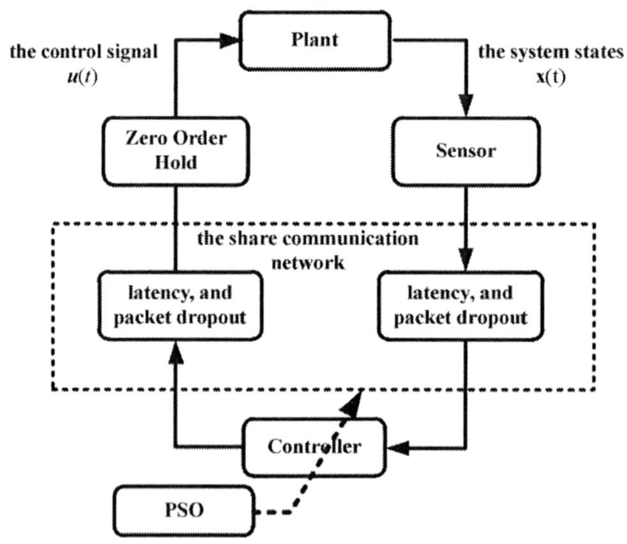

Figure 1. The framework of PSO-NCS.

DOI: 10.1051/matecconf/201711901051

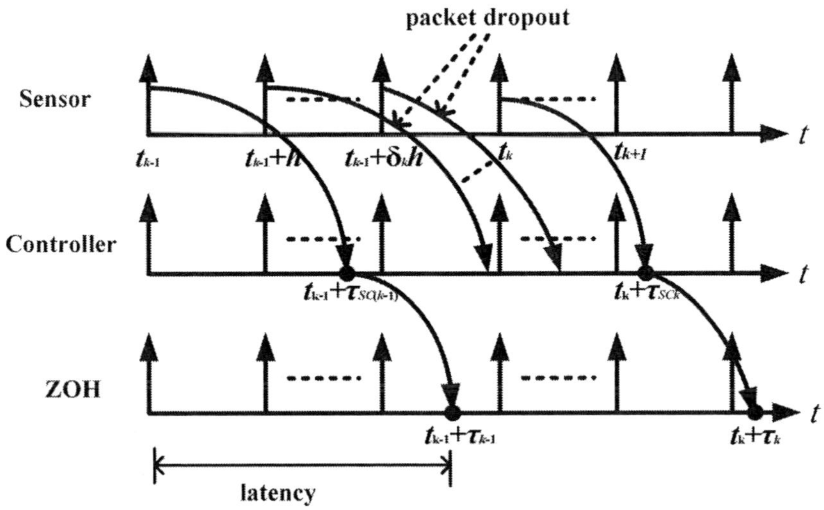

Figure 2. Latency and packet dropout of NCS.

is the unknown external disturbance, $\mathbf{A} \in \Re^{n \times n}$ and $\mathbf{B} \in \Re^{n \times m}$ are the known state matrices, and $\Delta\mathbf{A} \in \Re^{n \times n}$ is the uncertainty state matrix. The initial state vector is given as $\mathbf{x}(t_0) = \mathbf{x}_0$. Without loss of generality, we make the following assumptions [8, 19]:

Assumption 1: $(\mathbf{A}, \mathbf{B})$ is controllable pair.

Assumption 2: The external disturbance $\boldsymbol{\omega}(t)$ is bounded. That is, $\boldsymbol{\omega}(t) \in L_2[0, t_f]$, $\forall t_f \in [0, \infty)$.

Assumption 3: $\Delta\mathbf{A}$ is unknown, but bounded.

In the NCSs, the controller is connected with the sensor (or ZOH) via a share communication network channel (consisting of propagation, transmission, router, processing, computer, storage delays, etc.). The use of a share communication network will lead to latency. In addition, network congestion, multi-path fading, and faulty network drivers would cause the packet dropout. Latency and packet dropout might be potential sources to instability and poor performance of NCSs.

Figure 2 shows latency and packet dropout of NCSs. Because of latency and packet dropout, packets are transmitted from the sensor to the controller spend τ_{sck}, and from the controller to the actuator spend τ_{czk}. Define the total latency as $\tau_k = \tau_{sck} + \tau_{czk}$ at every sampling instant t_k and make the following assumptions for the total latency and packet dropout [2, 8, 20]:

Assumption 4: The total latency τ_k is bounded.

Assumption 5: The packet dropout number η_k is bounded as $0 \le \eta_k \le \eta_{\max}$, where $\eta_{\max}$ is upper bound of η_k.

Here, we adopt the state feedback control into inhibit the effect of the latency and the packet dropout such that the state vector $\mathbf{x}(t)$ of NCS would converge to zero, that is $\lim_{t \to \infty} \|\mathbf{x}(t)\| = 0$. In the state feedback control scheme, the controller is designed as follows:

$$\mathbf{u}(t) = \mathbf{K}\mathbf{x}(t) \tag{2}$$

where $\mathbf{K}$ is the control gain matrix and is obtained by solving the Lyapunov-Krasovskii stability criterion. Solving the stability criterion generally requires calculating the complex LMI problem. It is a difficult task sometimes. Rather than using LMI solver, this study attempts to apply PSO algorithm to obtain the control gain matrix $\mathbf{K}$.

3 PSO based NCS design

PSO algorithm uses a simple mechanism that imitates their swarm behaviors to guide the particles to search for globally optimal solutions. In PSO algorithm, each potential solution is regarded as a particle associated with two vectors, i.e., the velocity vector and the position vector. During the evolutionary process, the trajectory of each individual is adjusted by dynamically altering the velocity of each particle, according to its own flying experience (*pBest*) and the flying experience of the other particles (*gBest*) in the search space. With such a search mechanism, the particles have a tendency to fly towards better search area over the course of search process.

In past, the optimal control parameters used in NCSs are obtained by solving the Lyapunov-Krasovskii stability criterion. Such a solving process is complex and generally obtains a feasible solution. In addition, the network communication quality is also unpredictable. Thus, it is impossible to describe latency and packet dropout of NCSs accurately. Sequentially, the gradient of system state cannot obtain. With the help of randomly searching ability of PSO algorithm, this study applies PSO algorithm to search the optimal control gain matrix $\mathbf{K}$, instead of solving the complex Lyapunov-Krasovskii stability criterion. The following procedure can be used for searching the optimal control gain matrix with PSO algorithm.

Step 1: Initialize the swarm by assigning a random position $\mathbf{K}_i^o$ and the corresponding velocity vector $\mathbf{v}_i^o$ to each particle, where $i = 1,2,\ldots, N$, and N represents the size of the swarm. Also determine the maximum number of generations J_{max}.

Step 2: Evaluate the fitness value for each particle. In this study, the fitness function is defined as

$$F(\mathbf{K}_i^j) = f(\mathbf{x}(t), \mathbf{u}(t)) = f(\mathbf{x}(t), \mathbf{K}_i^j\mathbf{x}(t)) \tag{3}$$

where $\mathbf{K}_i^j$ denotes the *i*th particle for the *j*th generation.

Step 3: Update the best position for each particle $\mathbf{K}_i^{pBest}$ and the global best position $\mathbf{K}^{gBest}$ by (4) and (5), respectively.

$$\mathbf{K}_i^{pBest} = \arg \max_{h=0,1,\ldots,j} F(\mathbf{K}_i^h) \tag{4}$$

$$\mathbf{K}^{gBest} = \arg \max_{\substack{h=0,1,\ldots,j; \\ i=1,2,\ldots,N}} F(\mathbf{K}_i^h) \tag{5}$$

where $\mathbf{K}_i^{pBest}$ represents the position with the best fitness found so far for the *i*th particle and $\mathbf{K}^{gBest}$ is the best position discovered by the swarm.

Step 4: Update the position and the corresponding velocity for each particle using

$$\mathbf{v}_i^{j+1} = w \cdot \mathbf{v}_i^j + c_1 \cdot \phi_1 \cdot (\mathbf{K}_i^{pBest} - \mathbf{K}_i^j) + c_2 \cdot \phi_2 \cdot (\mathbf{K}^{gBest} - \mathbf{K}_i^j) \tag{6}$$

$$\mathbf{K}_i^{j+1} = \mathbf{K}_i^j + \mathbf{v}_i^{j+1} \tag{7}$$

where w is the inertia factor, c_1 is the "self-confidence" factor, c_2 is the "swarm confidence" factor, and ϕ_1 and $\phi_2 \cdot$ are two uniformly distributed random numbers independently generated within [0, 1].

Step 5: Repeat Steps 2-4 until a stopping criterion is fulfilled (e.g., the maximum number of generations or the goal is reached). And, $\mathbf{K}^{gBest}$ is the optimal control gain matrix.

 DOI: 10.1051/matecconf/201711901051

4 Simulation result

In order to demonstrate the performance of the proposed control strategy, an inverted pendulum control system [8, 21] is used verify it. Let $x_1(t)$ be the angular of the pendulum with respect to the vertical line and $u(t)$ be the applied the control signal. Define $\mathbf{x}(t)=[x_1(t),\dot{x}_1(t)]^T=[x_1(t),x_2(t)]^T$, the dynamic equations of the inverted pendulum system can be described as follows:

$$\begin{aligned}\dot{x}_1(t)&=x_2(t)\\ \dot{x}_2(t)&=\frac{g\sin(x_1)-(m_p l x_2^2\sin(x_1)\cos(x_1)/M)}{(4l/3)-(m_p l\cos^2(x_1)/M)}+\frac{\cos(x_1)u/M}{(4l/3)-(m_p l\cos^2(x_1)/M)}+\omega(t)\end{aligned} \tag{8}$$

where m_c is the mass of the cart, m_p is the mass of the pole, $M = m_c + m_p$, $g = 9.8\ m/s^2$ is the acceleration due to gravity, and l is the half length of the pole. In this study, the parameters are set as $m_c = 1\ kg$, $m_p = 0.1\ kg$, $l = 0.5\ m$, and the sampling period as 0.001sec. In addition, the unknown external disturbance $\omega(t)$ is a uniformly distributed random number independently generated within [−0.1, 0.1] and the initial state as $\mathbf{x}(0) = [\pi/6,\ 0]^T$. The Jacobian linearized model around the equilibrium $\mathbf{x}(0) = [0,\ 0]^T$ can be obtained directly from the dynamic equations (8), which yields

$$\begin{bmatrix}\dot{x}_1(t)\\ \dot{x}_2(t)\end{bmatrix}=\begin{bmatrix}0 & 1\\ \dfrac{g}{(4l/3)-m_p l/M} & 0\end{bmatrix}\begin{bmatrix}x_1(t)\\ x_2(t)\end{bmatrix}+\Delta\mathbf{A}\begin{bmatrix}x_1(t)\\ x_2(t)\end{bmatrix}+\begin{bmatrix}0\\ \dfrac{a}{(4l/3)-m_p l/M}\end{bmatrix}u+\begin{bmatrix}0\\ \omega(t)\end{bmatrix} \tag{9}$$

where $\Delta\mathbf{A}$ is be regarded as the uncertainty terms due to latency, packet dropout and linearization error. In the simulation, each entry in $\Delta\mathbf{A}$ is a uniform random number between −0.1 and 0.1.

In the simulations, three PSO variants, i.e., the standard PSO (Std. PSO), PSO with a linearly varying inertia weight (PSO-LVIW, [22, 23]), and PSO with time-varying acceleration coefficients (PSO-TVAC, [24]), were tested using the same swarm size of 100. Each algorithm was run 50 independent trials with the same maximum number of generations $J_{max} = 100$ for each trial and their worst, mean and best results are used in the comparison. Shi and Eberhart [22, 23] suggested that when w is medium ($0.8 < w < 1.2$), the PSO will have the best chance to find the global minimum, but also takes a moderate number of generations. Besides, as mentioned in Kennedy and Eberhart [13], a recommended choice for the factors c_1 and c_2 is a fixed value of 2 since it on average makes the weights for "social" and "cognition" parts to be 1. This setting also has been adopted by many other researchers. With the help of the above suggestions, the parameters used for the standard PSO is set as $w = 1$ and $c_1 = c_2 = 2$. As suggested in Shi and Eberhart [23], the inertia weight for PSO-LVIW was set to change from 0.9 to 0.4 over the generations. Besides, the boundary values of acceleration coefficients for HPSO-TVAC were set as $c_{1,0} = 2.5$, $c_{1,f} = 0.5$, $c_{2,0} = 0.5$, and $c_{2,f} = 2.5$, which are the best ranges for c_1 and c_2 suggested by Ratnaweera [24]. In the simulation, the PSO based NCS is also compared with the real coded genetic algorithm (RCGA) based NCS, where RCGA makes use of uniform arithmetical crossover with the crossover rate of 0.8 and uniform mutation with mutation rate of 0.1 [25]. Finally, the fitness function is defined as

$$F(\mathbf{K}_i^j)=\int_0^T(x_1^2(t)+0.0005u^2(t))dt \tag{10}$$

where $T = 5$ in the simulation.

All the programs coded by Matlab version R2014b were executed by a personal computer with Intel® Core™ i5-3470 CPU @ 3.20/3.60 GHz processor, 16.0-GB RAM, and Windows 7 with service pack 1 operating system.

The best, worse and mean fitness of the best-of-run solutions for each algorithm are presented in Table 1, where Std. Dev. represents the corresponding standard deviation. In addition, the optimal control gain matrix associated with the best fitness for each algorithm is given in the last column of Table 1. Since the best fitness values obtained by different algorithm are almost the same, their responses, including the curve of control signal $u(t)$ and the trajectories of $x_1(t)$ and $x_2(t)$, are close to each other. Figures 3 and 4 therefore only illustrates the trajectories of $x_1(t)$ and $x_2(t)$, respectively, obtained by the standard PSO algorithm, where the solid and dotted lines represents the trajectories corresponding to the best and worst fitness, respectively. The corresponding control signal $u(t)$ is shown in Figure 5. The simulation results show that the proposed control strategy implemented by each evolution algorithm could perform good stability and inhibit effectively the effect of "latency" and "data packet dropout".

Table 1. Performance comparison on fitness value and the optimal gain.

	Fitness				Optimal gain
	Best	Worst	Mean	Std. Dev.	
Std. PSO	6.0482	6.6321	6.2341	0.1654	[-54.0255, -8.3900]
PSO-LVIW	6.0488	6.8441	6.1844	0.1469	[-52.3876, -8.0424]
PSO-TVAC	6.0471	6.7676	6.2315	0.2023	[-54.4973, -8.4962]
RCGA	6.0469	NaN	(6.0469)*	(0.0000)*	[-53.9474, -8.3808]

*Without regarding a failed run.

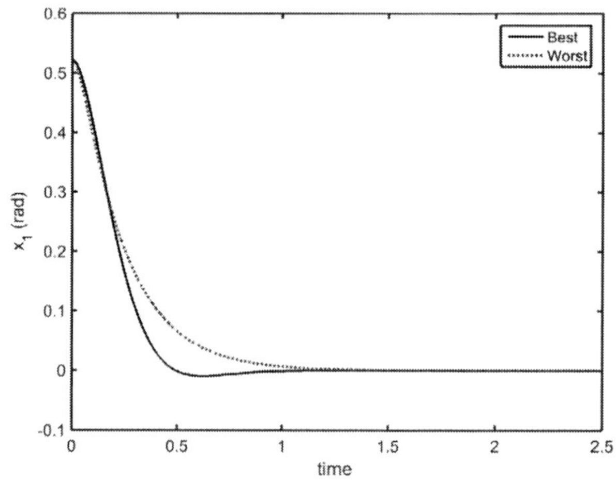

Figure 3. The trajectory of $x_1(t)$.

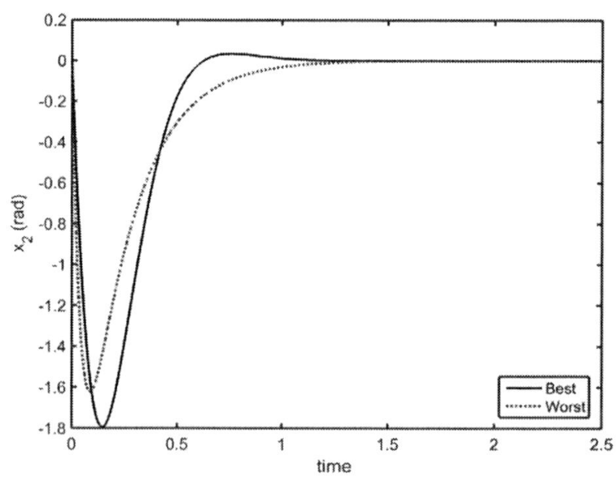

Figure 4. The trajectory of x2(t).

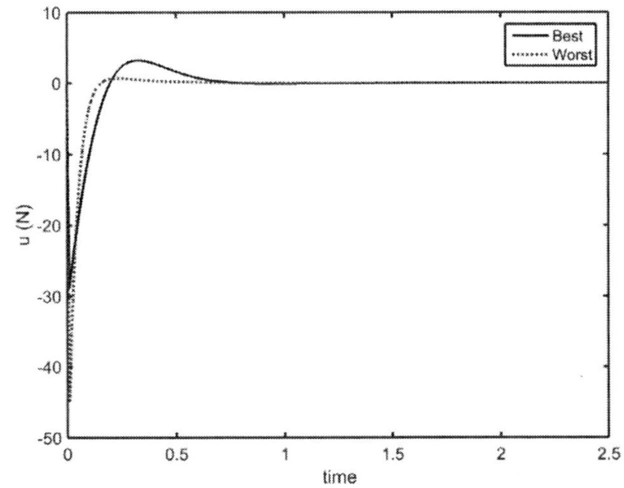

Figure 5. The control signal u(t).

In order to compare the convergence speeds and reliabilities (success performance) of different algorithms, this study selects a threshold value of 6.6 to determine whether a trial (run) is successful or not. The success of an algorithm means that the algorithm can result in a fitness value no worse than the prespecified threshold, with the number of generations less than the corresponding maximum number. The success rate is calculated as the number of successful runs divided by the total number of runs. Table 2 lists the success performance (SP) and average computational time of each algorithm required to reach such a predefined threshold, where FEs represents the number of function evaluations. In addition, the success performance is defined as SP = mean (FEs for successful runs)*(# of total runs)/(# of successful runs) [26]. As can be seen, PSO-LVIW and RCGA yield the best success rate of 98.0%, but the standard PSO achieves the best SP. Although PSO-TVAC has the worst success rate, it also performs an acceptable reliability of 92.0%. Besides, the standard PSO performs significantly faster than PSO-LVIW and RCGA on the average computational time. To sum up, with almost the same best fitness value, the standard PSO with $w = 1$ and $c_1 = c_2 = 2$ outperforms the other three approaches on the efficiency and reliability for the considered NCS with uncertainty.

Table 2. Comparisons on success performance and convergence speed.

	Rate (%)	FEs	SP	Run time(sec.)
Std. PSO	96.00	95.78	**99.7708**	**0.7609**
PSO-LVIW	**98.00**	97.58	99.5714	0.9004
PSO-TVAC	92.00	**91.66**	99.6304	0.8098
RCGA	**98.00**	97.42	99.4082	1.0051

5 Conclusions

Rather than using complex Lyapunov-Krasovskii stability criterion and solving the difficult LMI problem, this study applies the PSO algorithm to search the optimal control gain matrix for NCS with uncertainty, where the feedback controller is used to compensate the latency and packet dropout. The simulation results on an inverted pendulum system show that the PSO based NCS could perform good stability and inhibit effectively the effect of latency and data packet dropout. In addition, the optimal control gain obtained by the standard PSO performs the best success performance and the least computational time on the considered problem.

References

1. G.C. Walsh, H. Ye, and L.G. Bushnell, IEEE Trans. Contr. Syst. Technol., **10** (3), 438-446 (2002)
2. P. Seiler and R. Sengupta, IEEE Trans. Autom. Control, **50** (3), 356-364 (2005)
3. D. Huang and S.K. Nguang, IEEE Tran. Autom. Control, **53** (3), 829-834 (2008)
4. C. Peng and T.C. Yang, IEEE Trans. Fuzzy System, **18** (2), 326-335 (2010)
5. H. Wu, L. Lou, C.C. Chen, S. Hirche, and K. Kühnlenz, IEEE Trans. Ind. Electron., **60** (2), 554-566 (2013)
6. Y. Shi, J. Huang, and B. Yu, IEEE Trans. Ind. Electron., **60** (12), 5864-5874 (2013)
7. R. Yang, G.P. Liu, P. Shi, C. Thomas, and M.V. Basin, IEEE Trans. Ind. Electron., **61** (1), 512-520 (2014)
8. T.H. Chen, Applied Mathematics & Information Sciences, **9** (1L), 133-139 (2015)
9. H. Li, G. Wu, P. Shi, and Y. Gao, IEEE Tran. Cybern., **45** (11), 2378-2389 (2015)
10. J. Qiu, H. Gao, and S.X. Ding, IEEE Trans. Ind. Electron., **63** (2), 1207-1217 (2016)
11. W. Stallings, *Data and Computer Communications Prentice Hall*, **9(e)** (Prentice Hall, N.J., 2010)
12. S. Boyd, L.E. Ghaoui, E. Feron, and V. Balakrishnan, *Linear Matrix Inequalities in System and Control Theory* (PA: SIAM, Philadelphia, 1994)
13. J. Kennedy and R. Eberhart, IEEE Int. Conf. Neural Netw., 1942-1948, (1995)
14. F. Afshinmanesh, A. Marandi, and M. Shahabadi, IEEE Trans. Antennas Propagat., **56** (7), 1845-1852, (2008)
15. R.R. Pitre, X.R. Li, and R. Delbalzo, IEEE Trans. Aerosp. Electron. Syst., **48** (3), 2551-2565, (2012)
16. N.V. George and G. Panda, IEEE Trans. Instrum. Meas., **61** (12), 3378-3386, (2012)
17. T. Zhou and W. Sun, IEEE Trans. Sustainable Energy, **5** (2), 408-415, (2014)
18. L.X. Zhang, H.J., Gao, and K.N. Okyay, IEEE Trans. Ind. Inform., **9** (1), 403-416 (2013)
19. Q.C. Zhong, D. Rees, and J. Dyn. Syst. Meas. Control, **126**, 905-910 (2004)
20. Z.H. Pang, G.P. Liu, D. Zhou, and M. Chen, IEEE Trans. Ind. Electron., **61** (9), 4867-4877 (2014)
21. K. Tanaka and H.O. Wang, *Fuzzy Control Systems Design and Analysis: A Linear Matrix Inequality Approach* (Wiley & Sons, Inc., N.J. 2001)
22. Y. Shi and R.C. Eberhart, IEEE World Congr. Comput. Intell., 69-73 (1998)
23. Y. Shi and R.C. Eberhart, *Proc. IEEE Int. Congr. Evol. Comput.*, Washington, DC, **3**, 101-106 (1999)
24. A. Ratnaweera, S. Halgamuge, and H. Watson, IEEE Trans. Evol. Comput., **8** (3), 240-255 (2004)
25. Z. Michalewicz, *Genetic Algorithms + Data Structures = Evolution Programs* (Springer-Verlag, New York, 1992)
26. A. K. Qin, V.L. Huang, and P.N. Suganthan, IEEE Trans. Evol. Comput., **13** (2), 398-417 (2009)

Grain bulk density measurement based on wireless network

Fangming Wu, Bingfang Wu[a] and Leidong Yang

Key Laboratory of Digital Earth Sciences, Institute of Remote Sensing and Digital Earth, Chinese Academy of Sciences, 100101, Beijing, China

Abstract. To know the accurate quantity of stored grain, grain density sensors must be used to measure the grain's bulk density. However, multi-sensors should be inserted into the storage facility, to quickly collect data during the inventory checking of stored grain. In this study, the ZigBee and Wi-Fi coexistence network's ability to transmit data collected by density sensors was investigated. A system consisting of six sensor nodes, six router nodes, one gateway and one Android Pad was assembled to measure the grain's bulk density and calculate its quantity. The CC2530 chip with ZigBee technology was considered as the core of the information processing, and wireless nodes detection in sensor, and router nodes. ZigBee worked in difference signal channel with Wi-Fi to avoid interferences and connected with Wi-Fi module by UART serial communications interfaces in gateway. The Android Pad received the measured data through the gateway and processed this data to calculate quantity. The system enabled multi-point and real-time parameter detection inside the grain storage. Results show that the system has characteristics of good expansibility, networking flexibility and convenience.

1 Introduction

Inventory checking of grain storages expends a great deal of manpower and material resources in China, but the accuracy of measurement results is still not satisfying, and the actual density of the grain storages are difficult to determine. Sensors offer a potential solution to this underlying problem. The grain bulk density sensor was designed according to the relationship between permittivity and bulk density of grain [1], and the electromagnetic wave transmission method was used to estimate grain permittivity, and then calculated bulk density.

A wireless network is any type of communication network that uses wireless data connections for connecting network nodes thus, obviates the costly process of introducing cables into a building [2]. ZigBee is a low-cost, low-power, wireless mesh network standard targeted at the wide development of long battery life devices in wireless control and monitoring applications [3]. ZigBee devices can transmit data over long distances by passing data through a mesh network of intermediate devices to reach more distant ones, whereas Wi-Fi is a wireless local area network based on IEEE 802.11 standards [4]. Devices which can use Wi-Fi technology includes personal computers, smartphones, tablet computers and digital cameras.

Decision-makers need accurate information on amount and quality of stored grain to prevent storage losses, which can be a major hindrance to food security. Grain bulk density sensors can be

[a] Corresponding author : wubf@radi.ac.cn

used to measure grain bulk density, which is a good indicator of stored grain conditions. However, multi-sensors should be inserted into the grain storage to quickly collect measured data during the inventory checking of stored grain. This can be costly and sometimes give erroneous results. Therefore, in this study, the ZigBee and Wi-Fi coexistence network's ability to transmit data collected by density sensors was investigated.

2 System design

2.1 System architecture

A wireless network system was developed and it consisted of six sensor nodes, six router nodes, one gateway and one Android Pad, to measure grain bulk density and calculate quantity as shown in Figure 1. ZigBee network was used for communication among sensor nodes, router nodes and gateway. The Android Pad connected with the gateway through Wi-Fi to send commands and receive data.

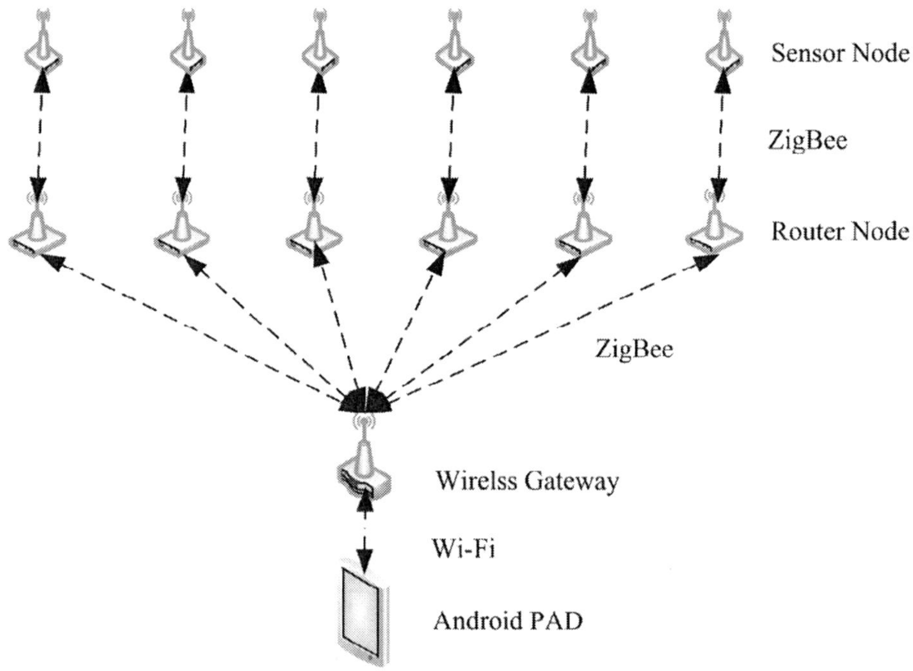

Figure 1. Grain bulk density measurement system architecture.

The sensor nodes received data from the environment and sent it to the router. As well as running an application function, a router node could act as an intermediate router, thereby passing on data from other devices. The gateway had to be set in advance to be able to receive data from sensor or router nodes.

To communicate and transmit data wirelessly, each sensor or router node required a ZigBee module. In addition, to be able to use the entire module is used, such devices require additional power from the battery.

2.2 System components

The sensor node consisted of a CC2530 ZigBee module, AD8302 phase detector, AD8532 operation amplifier and TPS60211 power manager. The CC2530 is a true system-on-chip (SOC) solution for

ZigBee applications. The CC2530 combined the excellent performance of a leading RF transceiver with an industry-standard enhanced 8051 MCU [5].

The router node consisted of a CC2530 ZigBee module, AD9851 Direct Digital Synthesizer (DDS) and MAX2650 amplifier. The shortwave signal was generated by an AD9851, and controlled by a CC2530 ZigBee module through the SPI interface.

The wireless gateway consisted of a CC2530 ZigBee module and a USR-WIFI232 Wi-Fi module. ZigBee worked in different signal channels with Wi-Fi to avoid interferences and was connected to the Wi-Fi module through the UART serial communications interfaces in gateway.

The hardware specifications of the system used in this study are shown in Table 1;

Table 1. System hardware specifications.

SN.	Hardware	Specifications
1	sensor node	CC2530 ZigBee module AD8302 phase detector AD8532 operation amplifier TPS60211 power manager
2	router node	CC2530 ZigBee module, AD9851 DDS MAX2650 amplifier
3	wireless gateway	CC2530 ZigBee module USR-WIFI232 Wi-Fi module
4	Android Pad	CPU: Marvell PXA986 GPU : PowerVR SGX540 RAM :1GB Flash ROM 8GB OS :Android v4.1.2 (Jelly Bean)

2.3 Wireless network configuration

The Gateway had a user interface for system configuration called Manager System, which is similar to configuration application on routers in general. By default, Manager System could be accessed using a web browser via WiFi (using 10.10.100.254). The initial view of Manager System appeared shortly after logging in with correct credentials.

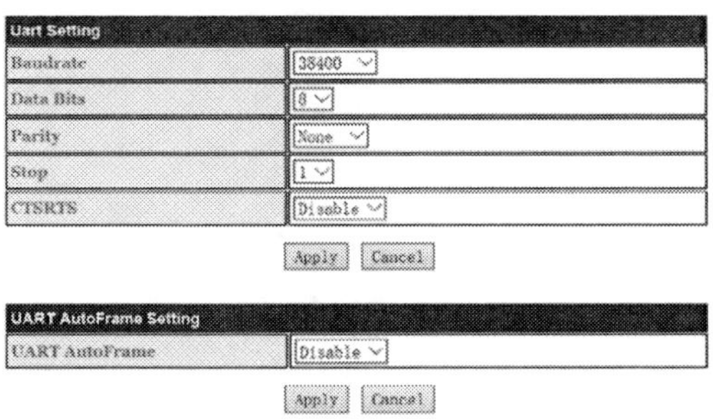

Figure 2. Gateway UART Settings.

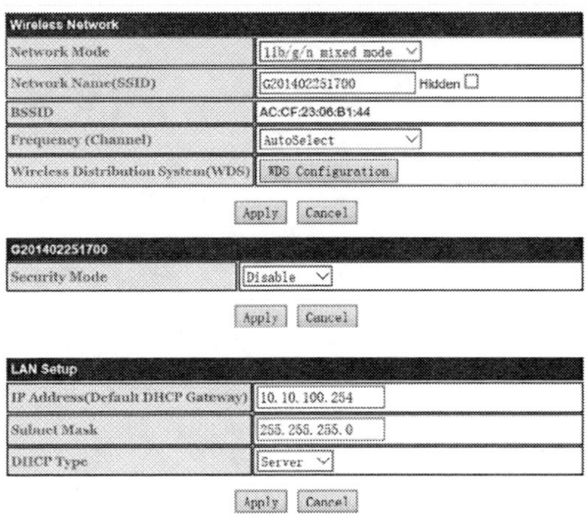

Figure 3. Gateway Network Configuration.

DOI: 10.1051/matecconf/201711901052

The pre-requisite UART settings included the Bandrate, Data Bits and Stop Bit (Fig. 2). Furthermore, network configuration such as Network Name (SSID) and Channel were performed in the interfaces tab in the Manager System, as shown in Fig. 3.

3 Experiment result

In the study, four sensor nodes and four router nodes connected to a wireless gateway were used. The Gateway was then connected to an Android Pad via a Wi-Fi network. The resultant network topology is shown in Figure 4.

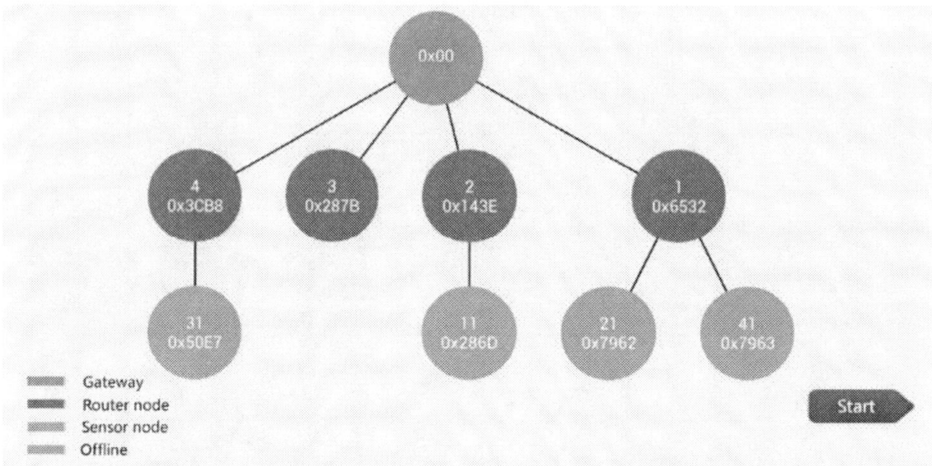

Figure 4. Network topology in this study.

The short address of gateway was 0x00, and each sensor node or router node had a different short address. Pressing the "Start" button was done to receive measured data after all devices were connected to the gateway.

The Android Pad received measured data from each sensor node through the gateway and processed this data to calculate quantity. The result of the measurement is given in Figure 5.

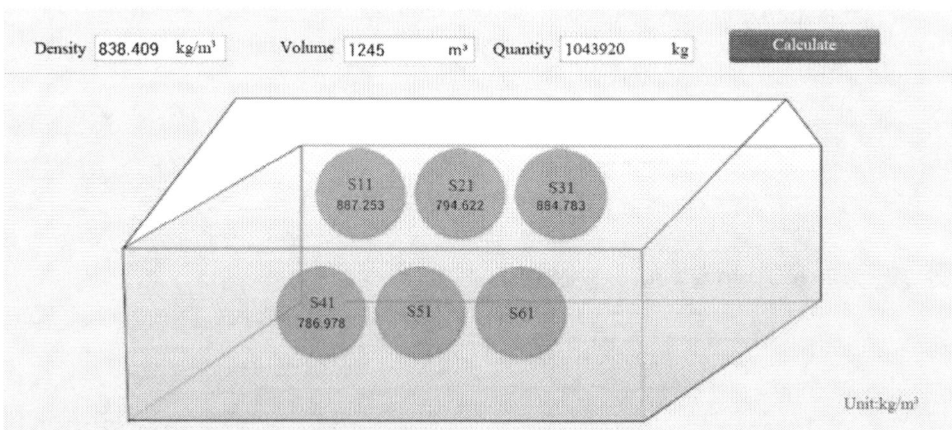

Figure 5. Result of the measurement.

The Android Pad had a MySQL local database to store the values of the measurement. A local database can be very vulnerable to damage and memory limitations of storage, so it is advisable to back-up the data to an external storage.

4 Conclusion

A grain bulk density measurement system based on wireless network was developed in this study. It reliably proves that Wi-Fi network and ZigBee network can be used together. By joining networks, an Android Pad can be employed to send commands, receive data and manage measured values. The system performed multi-point and real-time detection inside the grain storage. These results show that this system has the characteristics of good expansibility, networking flexibility and convenience. It is recommended, however, to explore the potential for using 4G/3G/GPRS to remotely monitor the grain bulk density in future.

Acknowledgements

This research was supported by the Special Fund for Grain-Scientific Research in the Public Interest under Project 201413003-6.

References

1. K. Sacilik and A. Colak, Powder Technology, **203**, 365-370 (2010)
2. G. Miao, J. Zander, K.W. Sung, and B. Slimane, Fundation of Mobile Data Networks (2016)
3. What is ZigBee? The Zigbee Alliance, http://www.zigbee.org/what-is-zigbee/
4. What is Wi-Fi(IEEE 802.11x)? Webopedia Definition, http://www.webopedia.com/TERM/W/Wi_Fi.html

MATEC Web of Conferences 119, 01053 (2017) DOI: 10.1051/matecconf/201711901053
IMETI 2016

A study on static structural of non-rotor unmanned aerial vehicle

Yu-Lan Chen[1,a], Jiunn Fang[2], Yau-Ren Shiau[3] and Wei-Chun Hung[1]

[1]*Ph.D. Program of Mechanical and Aeronautical Engineering, Feng Chia University, Taichung 407, Taiwan, ROC*
[2]*Department of Aerospace and Systems Engineering, Feng Chia University, Taichung 407, Taiwan, ROC*
[3]*Department of Industrial Engineering and Systems Management, Feng Chia University, Taichung 407, Taiwan, ROC*

Abstract. Multi-rotor unmanned aerial vehicles (UAVs) have become common in recent years, but non-rotor UAVs are still relatively rare. This study used an aerodynamic model to replace the multi-rotor design concept and examine load bearing capabilities, specifically for delivery of medical goods by non-rotor UAV. We use static structural simulations for static analysis and fatigue analysis to investigate the capabilities of different structural materials. The simulation results combined fatigue strength verifications, structural safety factor calculations, and finite element analyses to calculate parameters like total deformation, stress ratios, and strain magnitudes to analyze structural fatigue life. The analyses indicated that the factors of safety and fatigue lifetimes of two structural materials far exceeded standard values. They can be used to optimize designs that prevent plastic deformation and breakage. Time and money are saved in the research and development process by carrying out these calculations and static analyses of physical properties before prototypes are actually built.

1 Introduction

In recent years, multi-rotor UAVs controlled with the help of GPS, inertial navigation, and remote control systems have been widely used in aerial photography and videography for environmental monitoring, weather observations, etc. They are more convenient and cheaper than full-size helicopters or reconnaissance aircraft.

There is a plethora of commercially available, single or multi-rotor UAVs with increasingly sophisticated automatic control and monitoring systems that cover a vast array of potential mission capabilities. However, whether UAVs can be used to deliver medical supplies when natural disasters such as earthquakes, landslides, and floods knock out ground transportation, has become a key question in rescue operations.

Most UAVs today have multiple rotors, and in the presence of a cargo carrying function, their cargoes are generally suspended below the aircraft. In this study, aerodynamic models of non-rotor plate load UAVs are used to replace the more common multi-rotor design. In order to reduce research and development time and improve product quality and reliability, static and fatigue analyses are carried out using ANSYS structural simulations.

[a] Corresponding author : wearefamily999@gmail.com

In the Industry 4.0 era, new and improved observation and reconnaissance systems are being constantly developed and brought out on to the market. UAVs are playing an increasingly important role in the prevention and control of environmental disasters and severe weather events triggered by global climate change.

Meteorological sensors aboard UAVs were able to obtain which data of air temperature, pressure, relative humidity, wind speed, etc. during missions at the outskirts of Typhoon Sinlaku [1]. Archaeological surveys carried out with LiDAR data and UAV-assisted photographic marking were faster and cheaper than traditional methods. UAVs also enabled the construction of 3D models of mining sites and hydraulic systems [2]. Moreover, UAVs can provide perspectives from various altitudes and angles to aid in the detection and tracking of vehicles. Being able to verify three significant characteristics of images: edges; optical flow; and local feature points, they assist in improving accuracy and enable significant improvements in vehicle identification and tracking capabilities [3].

Before a product is launched in the market, a prototype is made. However, in addition to the money and time required to make the prototype, tests and analyses must be performed to guarantee its physical capabilities and design effectiveness. Nowadays, computer aided engineering (CAE) plays an important role in that process. In particular, CAE is used for: 1) looking at stress with static and dynamic finite element analyses (FEA); 2) thermal and fluid dynamics analyses and calculations (computational fluid dynamics or CFD); 3) motion and power analyses in multibody mechanics; 4) finite element analysis and boundary element methods in acoustic analyses; 5) simulation of mechanical activities in manufacturing execution systems (MES), etc. Software for these applications included ANSYS, CATIA, NASTRAN, SolidWorks and PTC Creo (Pro/Engineer).

Fishing ships that spend long periods of time at sea need excellent freezers with low power consumption. At the beginning of the design phase of the refrigeration deck, SolidWorks was used to set up a model of a refrigerator. Then, a mesh was created for the model using ANSYS Workbench, so as to calculate the steady state heat transfer rate through the walls of the refrigerator, find an optimal insulation configuration, and determine its effectiveness [4]. In order to determine the dynamic stability of the compressor and reduce wear and tear, a CAE model with constraints was set up to calculate parameters, estimate overall inertia, moments of inertia, contact forces between the rotor and the casing, and dynamic stability. Every parameter of interest could be tested and analyzed using such a simulation. This way, it was possible to see how a dual plane dynamically balanced design improved on the original design [5]. The flux patterns, flux density, and field intensity in magnetic levitation systems used for mass transit were determined using finite element methods. They revealed the working forces on U type actuators and I type rails under various working conditions [6]. An important aspect in loudspeaker design is magnetic field patterns. Models were constructed in SolidWorks to demonstrate the actual magnetic field distribution in loudspeakers. Then, by executing an ANSYS Workbench simulation, it was possible to solve for electrical current density, overall magnetic flux density, and direction of magnetic field. The use of ANSYS Workbench in the verification and optimization process allowed for a quick analysis of the magnetic system, and helped in arriving at detailed conclusions about the comparative magnetic induction strength of sound coils in different positions [7]. In the design of treadmills, in order to reduce the possibility of injury due to poor jogging habits, finite element analysis (FEA) is used in static and dynamic models to compare the effectiveness of the equipment's weight sensors at measuring reaction forces on the user's feet in various positions. Highly integrated embedded systems give treadmills the required responsiveness to enable a new level of interaction between the treadmill and the runner, based on reaction force monitoring [8]. By using ANSYS Parametric Design Language (APDL) to analyze usable structures, we discovered fatigue fractures in UAV landing gear, contour effectiveness, and the stress and fracture growth rates associated with fatigue fractures. By performing static analyses, we located critical regions [9].

2 Methods

This study used PTC Creo to build a geometric model and set simulation conditions like materials, loads, constraints, mesh, etc. in the non-rotor aerodynamic model. With a focus on load characteristics, the model was imported into ANSYS for structural statics and fatigue analyses [10–14].

2.1 Creation of the geometric model

The Bernoulli Principle, which states that when the velocity of a fluid increases, its pressure decreases, is widely used in aeronautical engineering. Flying machines push air downwards to create lift. When the force of lift is greater than the machine's weight, the machine ascends. Optimal design seeks to maximize or minimize the value of an objective function within the range of the set constraints. These constraints are determined by practical demands on performance. For mechanical engineers, design optimization means that geometric parameters like lengths, thicknesses, radii, and nodes are variables. These parameters along with principles followed in mechanics, such as applied forces, are combined to optimize stress distributions and deformations [15, 16].

This study constructed a 3D aerodynamic model of the non-rotor UAV. As shown in Fig. 1, it comprises a 380 mm diameter and 102.87 mm height plate for carrying loads, an engine compartment, and 4 jet intakes and exhaust valves. Using this model, the static structural analysis was carried out.

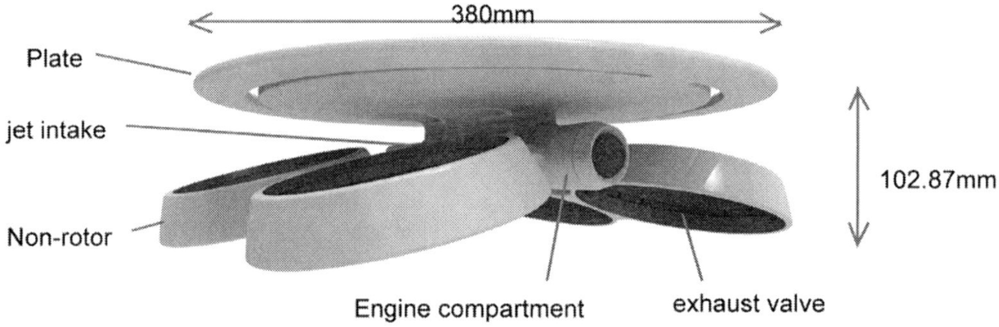

Figure 1. Non-rotor UAV model.

2.2 Setting simulation conditions

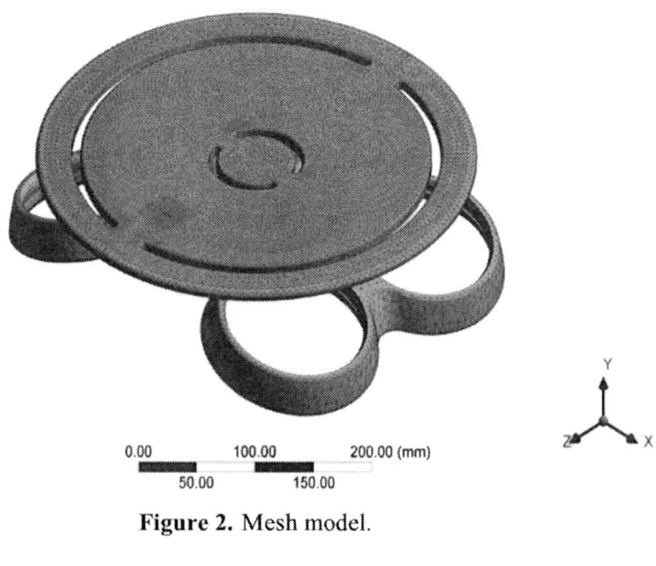

Figure 2. Mesh model.

The materials selected include: Aluminum Alloy, which is widely used in transportation, household electronics, construction, and mechanical hardware industries for its rigidity; and titanium alloy, which is used in the aviation, automobile, and biomedical equipment industries. Aluminum has a high strength to weight ratio and is commonly used in structural applications. Titanium is strong, light, highly resistant to fatigue and corrosion, and non-magnetic. It can also tolerate temperature fluctuations because of its low coefficient of thermal expansion (Table 1). Aluminum alloy and titanium alloy are selected based on the above characteristics and advantages as the structural materials for UAV.

In the mesh division and calculation settings, physical preferences were set to mechanical. The solver target used APDL, with relevance being set to 100. In the relevance center, the parameter was set to fine, while smoothing was set to medium and mesh transition to fast. Accurate analyses and modeling were of primary concern (Figure 2.).

Table 1. Parameters and properties of materials.

Properties	Aluminum Alloy	Titanium Alloy
Density	2.77e-006 kg mm^-3	4.62e-006 kg mm^-3
Coefficient of Thermal Expansion	2.3e-005 C^-1	9.4e-006 C^-1
Specific Heat	8.75e+005 mJ kg^-1 C^-1	5.22e+005 mJ kg^-1 C^-1
Thermal Conductivity	-	2.19e-002 W mm^-1 C^-1
Resistivity	-	1.7e-003 ohm mm
Reference Temperature	22 ℃	22 ℃
Young's Modulus	71000 MPa	96000 MPa
Poisson's Ratio	0.33	0.36
Bulk Modulus	69608 MPa	1.1429e+005 MPa
Shear Modulus	26692 MPa	35294 MPa
Temperature	Yes	Yes
Shear Angle	No	No
Degradation Factor	No	No
Tensile Yield Strength	280 MPa	930 MPa
Compressive Yield Strength	280 MPa	930 MPa
Tensile Ultimate Strength	310 MPa	1070 MPa
Compressive Ultimate Strength	0	0

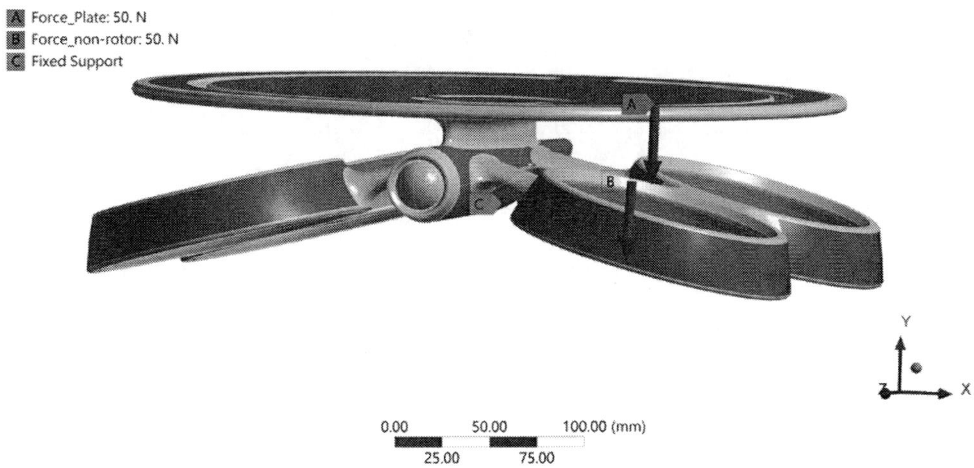

Figure 3. Loads and constraints.

For constraints, we selected fixed support with the engine compartments (Fig. 3C) fixed to constrain their Ux, Uy, and Uz degrees of freedom. The cargo platform (Fig. 3A) load was set to 50 N, with the force direction being downwards. The loads applied to the four airflow vents (Fig. 3B) were 50 N. Even here, the force direction was downwards.

3 Results and discussion

The material analysis data derived by importing the parameters above is shown in Table 2, they are Nonlinear Effects and Thermal Strain Effects, Bounding Box Length X 386.86 mm, Length Y 102.87 mm, Length Z 380.66 mm, Volume 1.6567e+006 mm³, Mass 4.5889 kg and 7.6538 kg, Centroid X -5.3181e-002 mm, Centroid Y 15.743 mm, Centroid Z 2.9233 mm, Nodes 318211, Elements 186514, and None Mesh Metric etc.. Static structural analysis revealed the deformation occurring upon application of stresses. This led to the structural statics and fatigue analysis simulation results, in Table 3, as described in 3.1 Static Analysis.

Table 2. Material analysis.

Assignment / Item		Aluminum Alloy	Titanium Alloy
Material	Nonlinear Effects	Yes	Yes
	Thermal Strain Effects	Yes	Yes
Bounding Box	Length X	386.82 mm	386.82 mm
	Length Y	102.87 mm	102.87 mm
	Length Z	380.66 mm	380.66 mm
Properties	Volume	1.6567e+006 mm³	1.6567e+006 mm³
	Mass	4.5889 kg	7.6538 kg
	Centroid X	-5.3181e-002 mm	-5.3181e-002 mm
	Centroid Y	15.743 mm	15.743 mm
	Centroid Z	2.9233 mm	2.9233 mm
	Moment of Inertia Ip1	37896 kg·mm²	63205 kg·mm²
	Moment of Inertia Ip2	78201 kg·mm²	1.3043e+005 kg·mm²
	Moment of Inertia Ip3	48333 kg·mm²	80613 kg·mm²
Statistics	Nodes	318211	318211
	Elements	186514	186514
	Mesh Metric	None	None

Table 3. Structural statics and fatigue analyses.

Item		Aluminum Alloy	Titanium Alloy
Total Deformation	Min.	0 mm	0 mm
	Max.	5.078e-002 mm	3.7671e-002 mm
Equivalent Elastic Strain	Min.	1.3043e-011 mm/mm	4.3847e-010 mm/mm
	Max.	3.243e-005 mm/mm	3.85e-005 mm/mm
Equivalent (von-Mises) Stress	Min.	7.9588e-007 MPa	4.1547e-005 MPa
	Max.	2.2943 MPa	3.6813 MPa
Safety Factor	Min.	> 15	> 15
Life	Min.	1.e+008 cycles	Not applicable

 DOI: 10.1051/matecconf/201711901053

3.1 Static Analysis

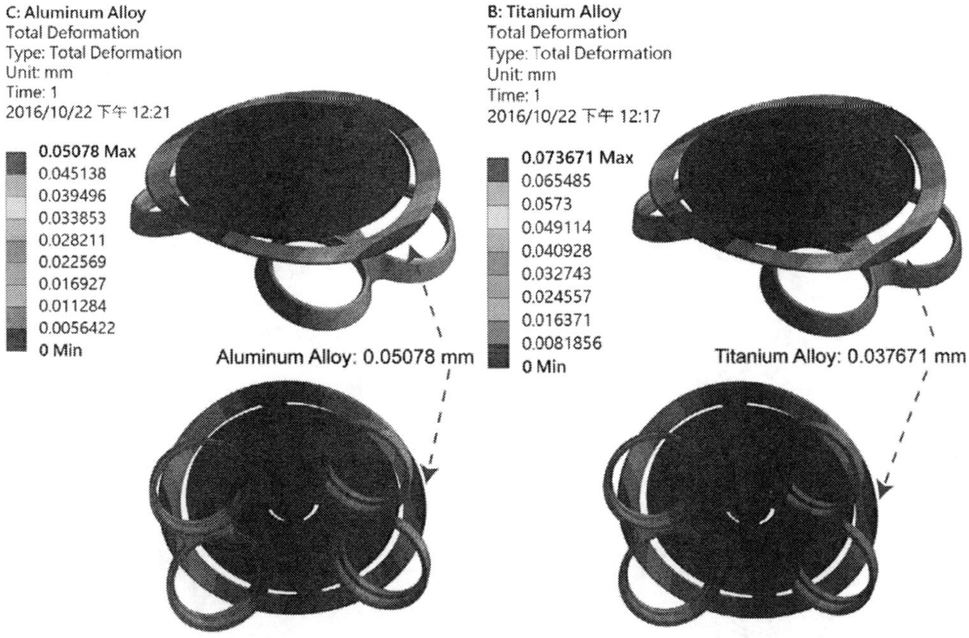

Figure 4. Total deformation (Top view and bottom view)

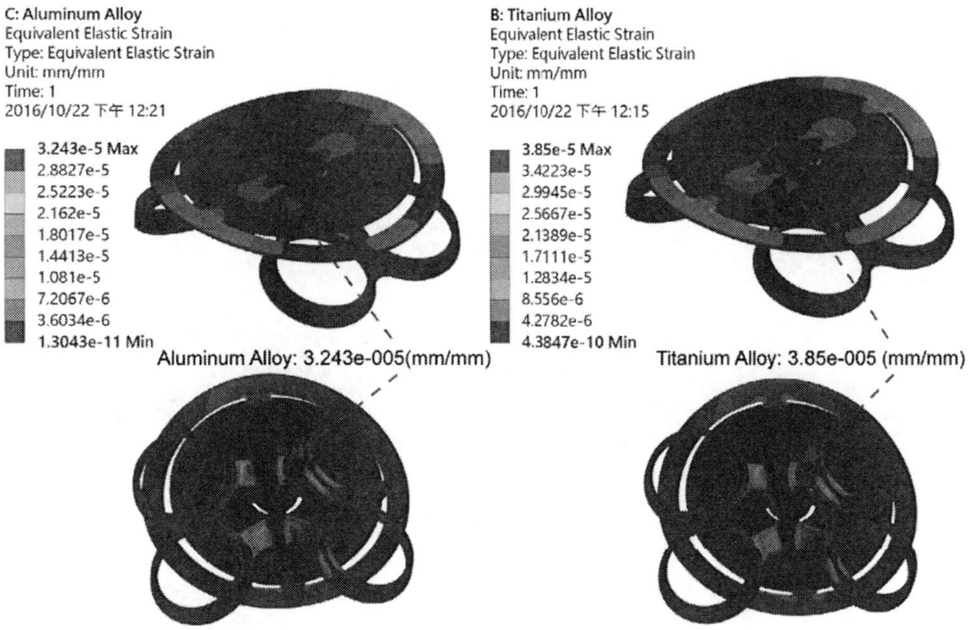

Figure 5. Equivalent elastic strain (Top view and bottom view)

Static analysis can confirm the most likely location of material fatigue. The 50 N force acting on the cargo platform and the 50 N combined forces acting through the airflow ports are shown in Figure 3. The structure itself is composed of 4.5889 kg of aluminum alloy or 7.6538 kg of titanium alloy. Total deformation under load and maximum stresses are in the curved areas. Elastic deformation in the aluminum alloy was 5.078e-002 mm and in the titanium alloy, it was 3.7671e-002 mm. From Fig. 4, we can see that the location of maximum deformation is at the free edges of the cargo platform and not at the point where the weight of the cargo acts. This met the design goals. The maximum equivalent elastic strain was 3.243e-005 in the aluminum model and 3.85e-005 in the titanium model (Fig. 5). Maximum equivalent stress was 2.2943 MPa in the aluminum model and 3.6813 MPa in the titanium model (Fig. 6). If the maximum elastic stress values are lower than the materials' yield stresses with safety factors of 1.25, plastic deformation and breakage can be avoided.

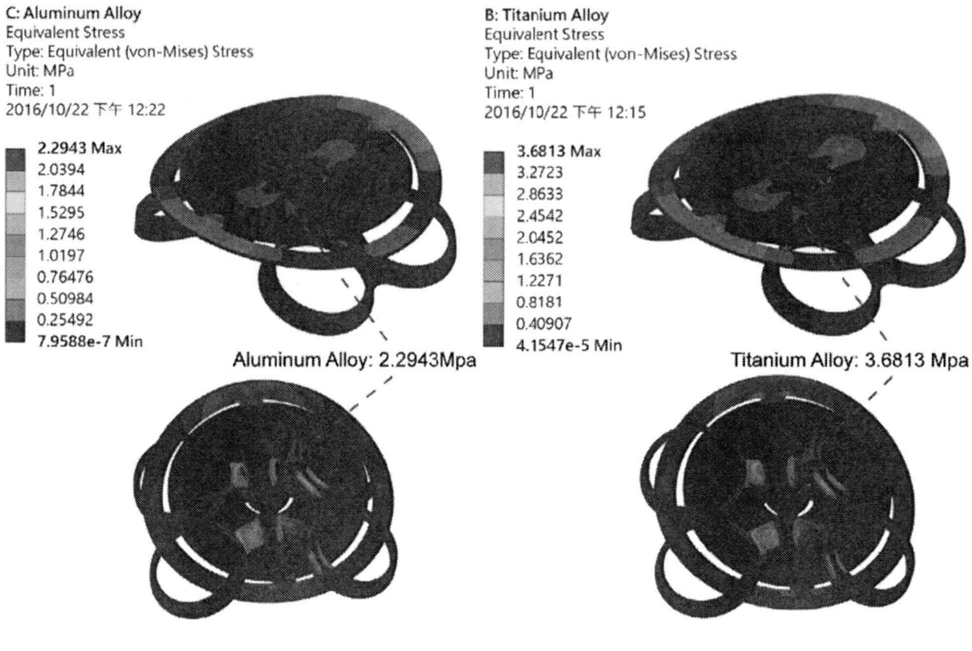

Figure 6. Equivalent Stress (Top view and bottom view)

3.2 Fatigue Analysis

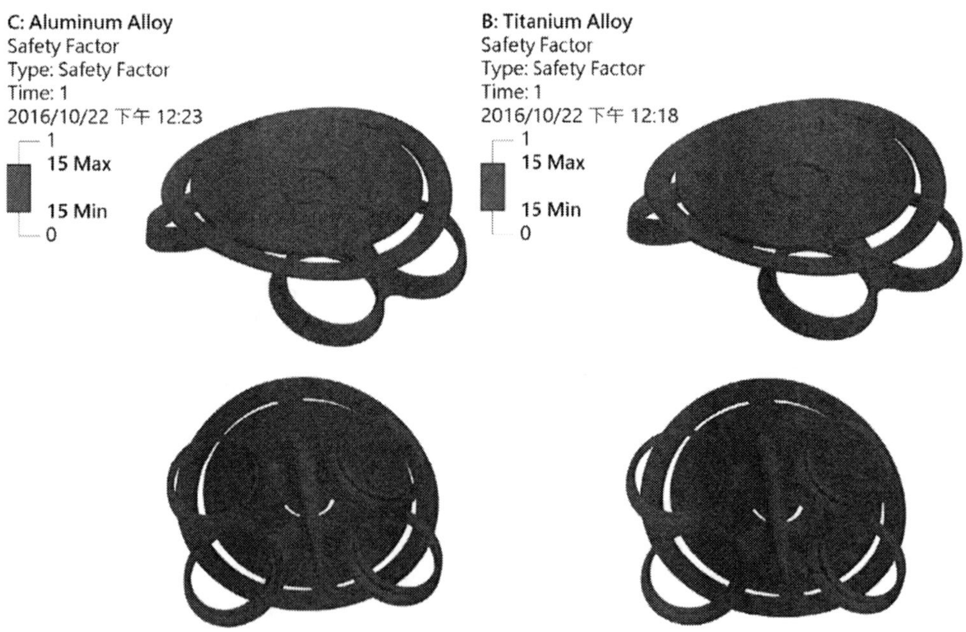

Figure 7. Safety factor (Top view and bottom view)

In general, minimum safety factors are in the range of 1.25 - 3.0. In the aerospace industry, safety factors must be greater than 1.25 and must be accompanied by strict quality control measures to guarantee the operational safety of the components. According to the results of this study, both materials have factors of safety exceeding 15 (Fig. 7). Fatigue life analysis results as shown in Fig. 8 were obtained using stress intensity factor and mesh distribution. Aluminum alloy has a fatigue lifetime of 100,000,000 cycles. When the repeated loading and unloading of titanium alloy does not cause strain above the fatigue minimum, its fatigue lifetime remains unaffected [17].

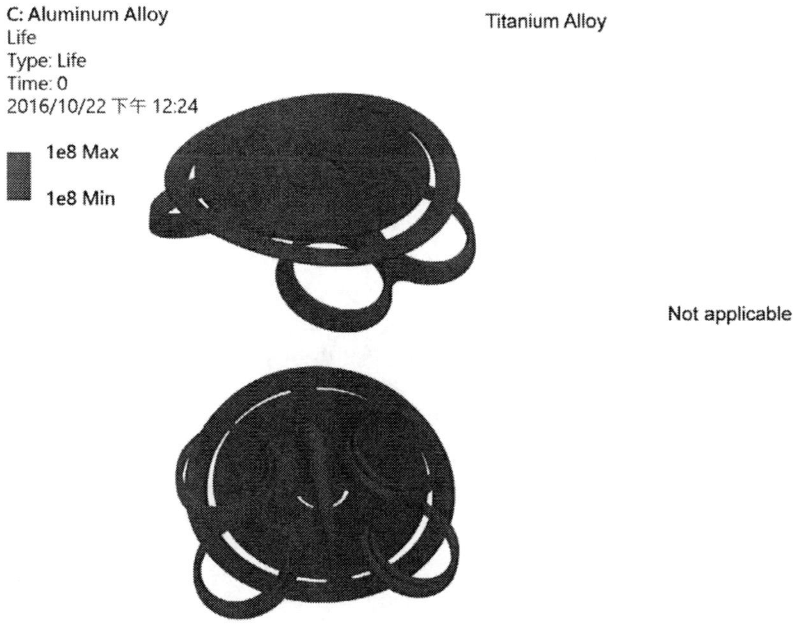

Figure 8. Lifetime of Material Fatigue (Top view and bottom view)

4 Conclusions

The static structural analysis simulation in this research focused on discussing factors of safety and fatigue lifetime of non-rotor UAVs. In the comparative analysis of aluminum alloy and titanium alloy structures, the design load was 50 N with constraints and mesh designs used being identical. An aerodynamic model was used in place of a multi-rotor UAV in the simulation. With load bearing being the vehicle's objective, fatigue intensity verification, structural factor of safety calculations, and finite element analyses were carried out to obtain the total deformation for the aluminum alloy this amounted to 5.078e-002 mm and in case of titanium alloy 3.7671e-002 mm. Equivalent stresses for aluminum and titanium alloys were 2.2943 MPa and 3.6813 MPa, resp. Equivalent strain magnitudes were 1.3043e-011 - 3.243e-005 for aluminum and 4.3847e-010 - 3.85e-005 for titanium, analysis data could be show that can avoid plastic deformation and breakage. These parameters were used to analyze fatigue lifetimes, analyses show that both materials have safety factors and fatigue lifetimes that by far exceed current standards.

From the above analysis and data that suitable structural and materials via application of static structure analysis can be known in advance and avoid plastic deformation and breakage and can be used for optimal design. By calculating and testing the physical properties of the vehicle before actually making a prototype, it is possible to save time and money that would otherwise be wasted on structural redesign. This makes the research and development process more efficient.

In future projects, it is recommended that the engine compartment be designed in more detail. For example, blades, wind, and noise designs could be added, as well as dynamic and modal analyses for completeness.

References

1. Y. Li, S. Ma, and Z. Sun, Analysis of essential meteorological elements surrounding Typhoon Sinlaku by unmanned aerial vehicle, Atmos. Clim. Sci., **6** (1), 29 (2015)
2. J. Fernández-Lozano and G. Gutiérrez-Alonso, Improving archaeological prospection using localized UAVs assisted photogrammetry: An example from the Roman Gold District of the Eria River Valley (NW Spain), J Archaeol Sci: Rep, **5**, 509-520 (2016)
3. L.Wang, F. Chen, and H. Yin, Detecting and tracking vehicles in traffic by unmanned aerial vehicles, Autom. Constr., **72**, 294-308 (2016).
4. Y. Cai, C. Huang, and S. Yang, An investigation of mopso and Ansys into optimizing frozen tank of fishing vessels, SNAME (Taiwan), **33** (1), 43-50 (2014)
5. Z. Gu, Y. Wu, and Y. Zhan, Dynamic balance design of CAE verification for Scroll Compressors, CSSV (Taiwan), 28-33 (2014)
6. K.B Pabitra and B. Subrata, Design and ANSYS software based simulation of U-I Type actuator and rail esed in electromagnetic levitation system, Int. J. Appl. Sci. Eng, **12** (3), 225-239 (2014)
7. L. Zhang, Y. Niu, and J. Zhou, Optimization analysis of the magnetic circuit in micro loudspeaker based on ANSYS Workbench, Electronic Design Eng., **3**, 6 (2015)
8. K.H. Chen, J. Fang, and S.W. Yeh, Design of an unobtrusive reaction force measurement and its application on treadmills, IEEE Trans. Instr. Meas., **61** (7), 1955-1965 (2012)
9. A.S. Vignesh and A.Vivek, A study on fatigue fracture failure of UAV landing gear, i-Manag J. Mech Eng., **6** (1), 9 (2015)
10. S. Nallusamy, N.M. Prabu, K. Balakannan, and G. Majumdar, Analysis of static stress in an alloy wheel of the passenger car, Int. J. Eng. Res. Afr., **16**, 17-25 (2015)
11. M. Ozsoy, K.Pehlivan, M. Firat,; N. Ozsoy, and V. Ucar, Structural strength and fatigue life calculation of Y32 bogie frame by finite element method, Acta Phys. Polon. A, **128** (2B), B-327-B-329, (2015)
12. J. Shah and J.M. Tartaglia, A case study of finite element analysis and designing with data from the AFS strain-life fatigue database, Int. J. Met., **9** (4), 7-18 (2015)
13. D. Croccolo, M. De Agostinis, and N. Vincenzi, Experimental analysis of static and fatigue strength properties in press-fitted and adhesively bonded steel-aluminium components, J Adhes Sci Technol, **25** (18), 2521-2538 (2011)
14. M. Barcaro, G. Meneghetti, and N. Bianchi, Structural analysis of the interior PM rotor considering both static and fatigue loading, IEEE Trans. Ind. Appl., **50** (1), 253-260 (2014)
15. K. Rokosz, T. Hryniewicz, W. Malorny, J. Valiček, and M. Harničárová, ANSYS analysis of stress and strain after cones plastic deformation, Tehnički vjesnik, **22** (2), 503-508 (2015)
16. Z.S. Lan, W.H. Lai, The design and application of UAV for disaster prevention, JoCICHE, Ch Civ Eng J., **27** (3), 191-199 (2015)
17. The S-N Curve. High-Cycle Fatigue. Available online: http://www.efunda.com/formulae/solid_mechanics/fatigue/fatigue_highcycle.cfm (accessed on 25 Judy 2016)

MATEC Web of Conferences **119**, 01054 (2017) DOI: 10.1051/matecconf/201711901054
IMETI 2016

Analysis of postural stability for Chinese crews working on moving ocean platforms

Zhaofang Liu[a], Hong Lin

Food Science and Engineering College, Ocean University of China, 266100 Qingdao, China

Abstract. Moving ocean platforms, such as ship, fishing vessels and offshore platforms etc, suffer seriously from sea loads as wind, wave, currents and other environmental factors. Platforms' motions induced from various sea loads have significant effects on crews aboard. Severe motions of the platform can make crew lose balance or cause motion induced interruptions (MIIs). This paper studies postural stability of crews on moving platforms based on Chinese anthropometric standards. The relationships between height of center of mass (COM), moment of inertia and human's height, mass of human body are analysed. The postural stability indicator, tipping coefficient is obtained of human body on moving platform by applying Chinese anthropometric standards. Then the stability domains of human bodies with translational is derived considering the physical parameters of Chinese adults. The work can be helpful for estimation of MII probability and safety of crews on moving ocean platform.

1 Introduction

Floating platforms are widely used in marine transportation, fishery, exploitation of ocean resources and oceanographic survey. Due to the loads of wind, wave and current of ocean environment, motions of ocean platforms in six degree of freedom (sway, surge, heave, pitch, roll and yaw) occur frequently. These motions have significant effects on the health and safety of crews aboard. Motions of ocean platforms have various negative effects, such as Motion Induced Interruptions (MIIs), Motion Induced Fatigue (MIF) and Motion Sickness (MS) to human(Wertheim,1998) [1]. MIIs were defined as the occasions when sailors on ships with strong motion are easily to become unstable or lose balance and have to stop their work in order to avoid further damage(Graham,1990) [2]. In China, there are huge number of workers as sailors, fishermen, and offshore crews working on moving floating ocean platforms. All the people mentioned above are facing serious health and safe hazards from these motions.

There is plenty of research work on human motion on moving platforms. Graham (1990) [2] made the definition of MII and pointed out it had obvious effects on comfort and performance of human aboard. He developed the model for MII with simplified physical parameters of human body. It provided a useful way to quantify human performance on ships and predicted the number of motion-induced interruptions. Crossland used the model to compare the number of MIIs between experiment and estimation, and found they agreed fairly well(Crossland, 1998)[3].

[a] Corresponding author : lzf@ouc.edu.cn

Haward, Lewis and Griffin(2009)[4] studied the whole human vibration under the motion of platforms and analysed the relationship between response and factors such as exciting frequency, amplitude, direction and so on. Nawayseh and Griffin((2006)[5] studied the effect of frequency, magnitude and direction of oscillation on the postural stability of standing people. Complex models as articulated inverse pendulum model were also adopted to study the control of human body(Wedge and Langlois,2003)[6]. Yao et al studied the human response on ocean platform based on the multi-body dynamics model [7]. An review of the state of art on postural of human body on moving platform was given (Erica Nocerino,2010)[8].Based on existing research some guide rules were given in ABS for habitability of crew on moving platform or offshore installation (ABS, 2010)[9].

Most of the existing MII models neglect the actual physical parameters of human body, especially the variation of people from different countries. In order to evaluate human performance accurately, this paper studies the stability domains of human on moving ocean platform. Physical parameters of Chinese crews are considered based on national standard of Chinese adult anthropometry. Then, stability domains of human could be obtained by solving equations of the MII model.

2 Human parameters based on Chinese standard

The three-dimension human body model and anatomical planes are shown in Figure 1(a). In this paper only the motion in coronal plane is studied and the coordinate frames of human body O-XY is shown in Figure 2(b) and the origin point locates on the gravity centre of human body.

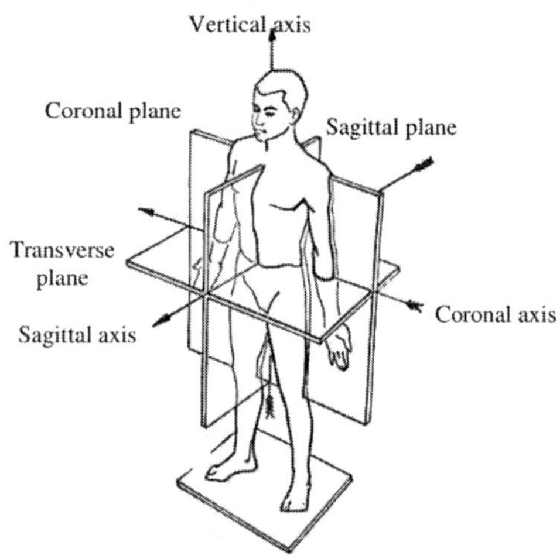

(a) Basic planes and axises in human body model

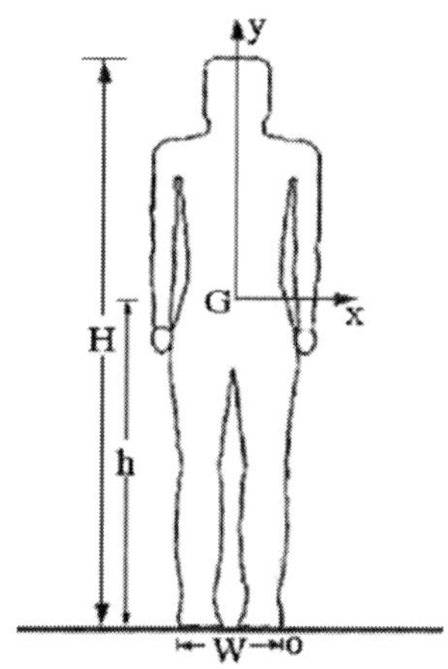

(b) Coordinate system of human body

Figure 1. Scheme of human body.

Basic planes and axes in human body model are shown in Fig 1(a). In this paper only the motion in coronal plane is studied (Fig 1(a)); and the coordinate frames of human body O-XY is shown in Fig 2(b), and the origin point locates on the gravity centre of human body.

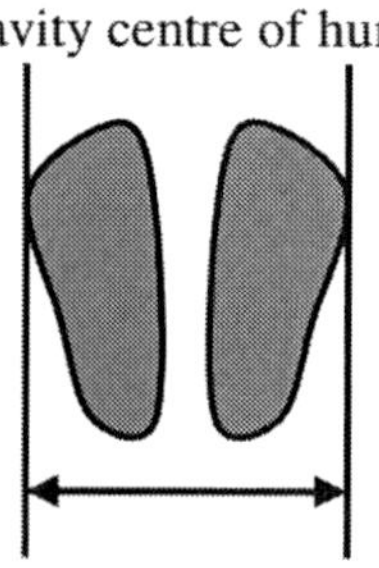

Figure 2. Feet width with natural standing.

Dimension parameters and physical parameters of human body have an important role on postural stability such as height of COM, weight and inertia of moment of human body et al. Different postures can also have a significant effect on stability. These parameters are closely related to the factors as nation, age, gender and occupations.

Width of feet, dimension and area of feet occupying et al can also affect the postural stability obviously. In this paper only natural standing posture is considered, which is standing up straight with feet about shoulder width apart. According to convention of ergonomics, the width of feet or shoulder width W (shown in Figure 2) is a function of the height of human body H as

$$W=0.222H \tag{1}$$

As to Chinese, several national standards on human factors have been carried out such as GB 10000-1988[16], GB/T 12985-1991[17] and GB/T 17245-2004[13]. In the latter standard the values of average height of COM, moment of inertia along with standard abbreviation are given. Moreover, the standard also provided function reflecting the relationship between height of COM, inertia of moment with mass, height and so on based on Chinese anthropometry data. The relationship between height of mass centre and body mass, height is:

$$h = H - (B_0 + B_1 m + B_2 H) \tag{2}$$

Where m is the mass of human body, H is the height of human body, B_0、B_1 and B_2 are regressive coefficients. The coefficients for male and female are shown in Table 1.

Table 1. Regressive coefficients of height of COM of Chinese adults.

Gender	Constant term B_0	Coefficient of mass B_1	Coefficient of height B_2	Multiple correlation coefficient R
Male	-32.2975	-0.4443	0.478	0.833
Female	-95.1467	-0.5457	0.5311	0.921
Remark: Unit of mass is kg, unit of height is mm.				

According to equation (2), values of the height of COM calculated by Graham's model[2] and the present model are shown in Figure 3. Figure 3(a) shows that height of COM increases with the height of human body when the mass is constant value (60Kg). Figure 3(b) shows that the height of COM increases with the mass when the height is constant value as 1700mm. In this figure, the height of COM increases slightly with the mass in Chinese adult's model while it maintains constant in Graham's model. From these two figures, as a whole, it is clear that the height of mass centre is larger in Chinese adult's model than the model of Graham's model.

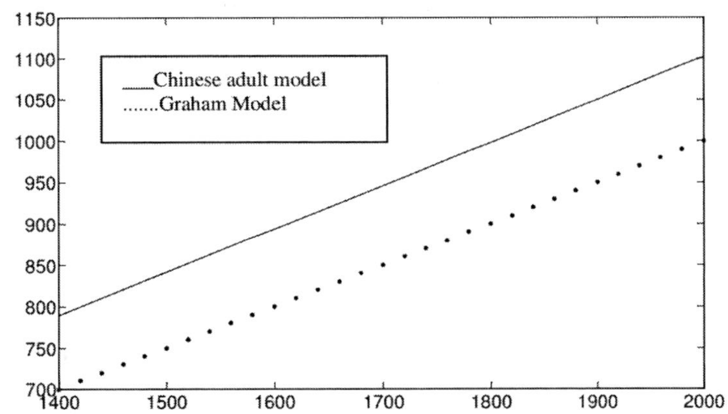

(a)

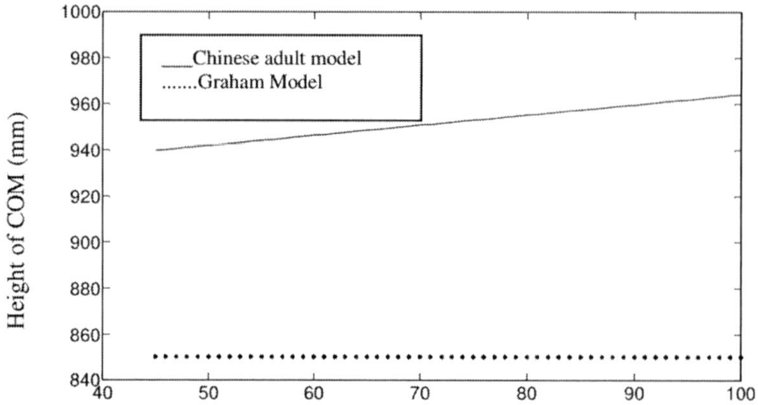

(b)

Figure 3. Relationship between Height of mass center and mass, height (a)human body with mass of 60kg, (b) human body with height of 1700mm.

According to Chinese adult's anthropometry standard (GB/T 17245-2004), the moment of inertia is related to mass and height as below:

$$I_i = C_0 + C_1 m + C_2 H \quad (i = x, y, z) \tag{3}$$

Where C_0, C_1 and C_2 are the regressive coefficients. The coefficients of equation (3) can be found in Chinese adult's anthropometry standard (GB/T 17245-2004) [13]. For the moment of inertia with respect to COM on coronal plane the coefficients are shown in Table 2.

Table 2. The regressive coefficients of moment of inertia of Chinese adults[13].

Gender	**Moment of inertia**	**Constant term** C_0	**Coefficient of mass** C_1	**Coefficient of height** C_2	**Complex coefficient** R
Male	I_G	-25397472.8	130503.9	16396.1	0.971
Female	I_G	-17309877.5	129946.20	11468.6	0.964
Remark: Unit of mass is kg, unit of height is mm, unit of inertia is $kg \cdot mm^2$. I_G is moment of inertia about sagittal axis through the COM.					

Because motions in coronal plane are studied, moment of inertia about sagittal axis through foot edge point *O* or *O'* is:

$$I_O = I_G + m(w^2 + h^2) \tag{4}$$

Where *w* is half of a person's feet width *W* with the natural standing posture, *h* is the height of COM.

Based on equation (4), we can obtain the value of the moment of inertia about supporting foot edge point O in standing human body. It illustrates the relationship between moment of inertia I_O and the mass/ height of human body in this model and Graham's model. Figure 4(a) shows that I_O increases with the height of human body when the mass is constant value (60Kg). Figure 4 (b) shows that the height of COM varies as the mass increases when the height is constant value as 1700mm. From these two figures, it is clear that the moment of inertia is also larger in Chinese adult's model than the model of Graham's model. It should be noticed that the height and moment of inertia are closely related to the standing posture, especially the width of feet separated in the Chinese adult model in this paper.

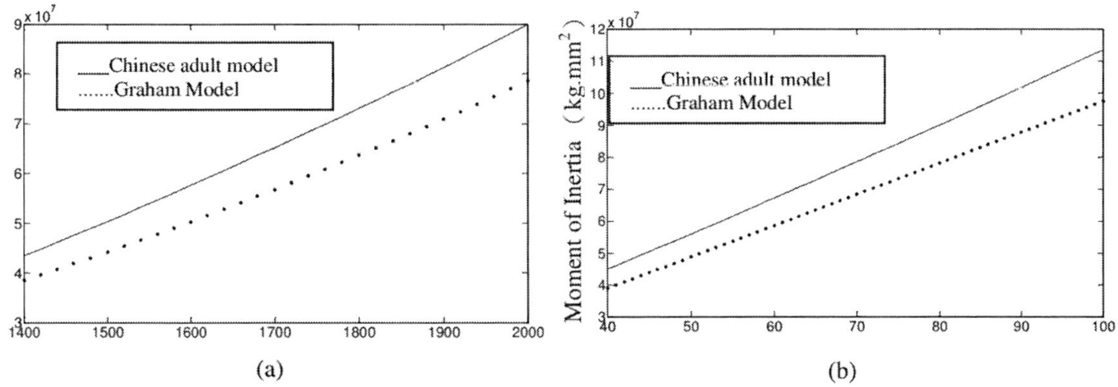

Figure 4. Relationship between Moment of inertia and height, mass of human body (a) human body with the mass of 60kg, (b) human body w ith the height of 1700mm.

3 Human body model for MIIs estimation

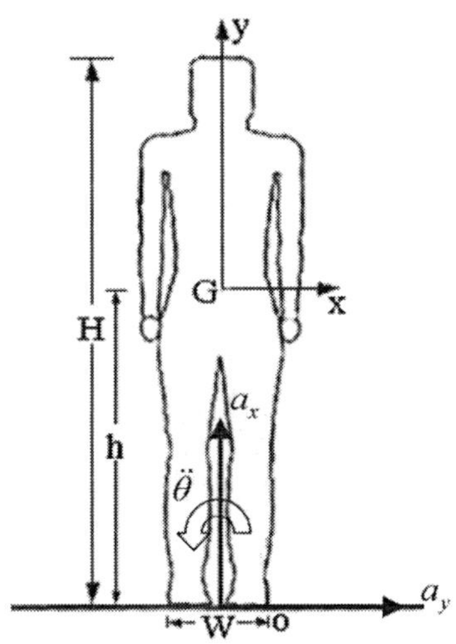

Figure 5. Standing human model.

The model for the analysis of stability of standing human with the translation and rotation excitation is shown in Figure 4. Given the large friction coefficient between boots bottom of crews and deck, the possibility of slip motion is neglected.

According to human on ocean platform suffering from the motions in translation about x, y axis and rotation about z, the motion equation of human body rotating about O can be described as:

$$I_O\ddot{\theta} - F_x h - F_y w = 0 \tag{5}$$

Where θ and $\ddot{\theta}$ are the angle and angular acceleration of human body,

I_O is the moment of inertia about one foot,

F_x is the total force on horizontal direction,

F_y is the total force on vertical direction,

h is the height of a person' COM from the deck,
w is the half of a person's standing width.

When θ is small enough, F_x 、 F_y can be simplified into

$$F_x = m(a_x + g sin(\theta)) \approx m(a_x + g\theta) \tag{6}$$

$$F_y = m(a_y + g\cos(\theta)) \approx m(a_y + g) \tag{7}$$

The motion equation of human body can be obtained as:

$$\frac{I_O}{mh}\ddot{\theta} - (a_x + g\theta) - \frac{w}{h}a_y > \frac{w}{h}g \tag{8}$$

In this situation, the human body will tip to the left direction.

Similarly, when human body tips about point O', the motion equation is:

$$-\frac{I_{O'}}{mh}\ddot{\theta} + (a_x + g\theta) - \frac{w}{h}a_y > \frac{w}{h}g \tag{9}$$

Human parameters in these equations have been calculated and can be applied in the MII prediction. If ocean platforms only have translational motion, then equation (8) could be amended as:

$$-a_x - \frac{w}{h}a_y > \frac{w}{h}g \tag{10}$$

Similarly, equation (9) could be amended as:

$$a_x - \frac{w}{h}a_y > \frac{w}{h}g \tag{11}$$

In equation (10) and (11), we can find that the ratio $\frac{w}{h}$ is an essential factor to affect the stability of human, which is defined as tipping coefficient *e*. It is obvious that probability of crews suffering from MII on ocean platform would increase with tipping coefficient. Based on equation (2), (12) and (13) the tipping coefficient is:

$$e = \frac{w}{h} = \frac{0.111H}{H - (B_0 + B_1 m + B_2 H)} \tag{12}$$

The relationship between tipping coefficient, height and mass is shown in Figure 6.

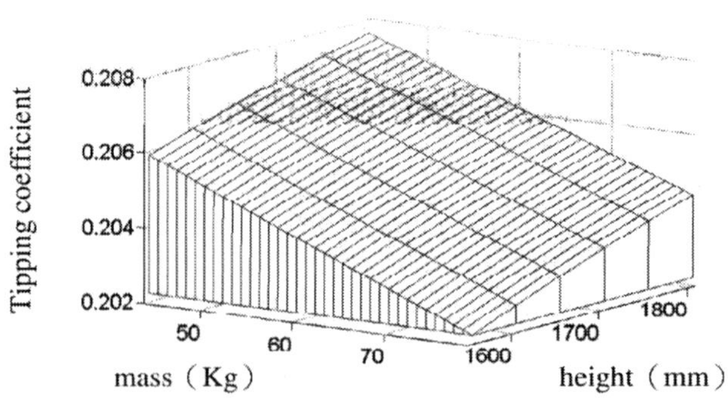

Figure 6. Relationships between tipping coefficient and height, mass.

4 Stability domains

Assuming the coefficient of friction is sufficiently large and there is no slipping between human and deck, crews will move together with ocean platform. So the human body would have the same motion parameters with ocean platform. These motion parameters as a_x, a_y, and θ and $\ddot{\theta}$ can be regarded as the excitation that crews suffer from motions of ocean platform. In this paper, the model of human body considered has the height of 1700mm and the mass of 60kg.

4.1 Under the excitation of translational acceleration

When the rotational motion is neglected, only translational motions with horizontal acceleration a_x and vertical acceleration a_y are considered. According to equation (10) and (11), two oblique straight lines could be obtained, as line 1 and line 2 in Figure 6. The value of intersection points C_1, C_2 are $\pm\frac{w}{h}g$ on x axis. This is to say the stability domain is closely related to the tipping coefficient. If the tipping coefficient increases, the domains would be larger, as is shown by the dash lines. So large tipping coefficient is helpful for stability of human body.

When the vertical acceleration equals -g the human body will be in the condition of weight loss and the contact loss between feet and floor as point *D* in fig 6. It is unstable whatever the horizon acceleration.

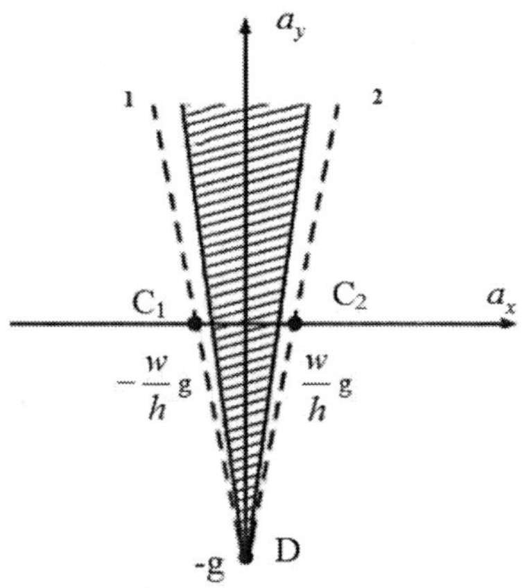

Figure 7. Stability domains under the excitation of translational acceleration.

5 Conclusions

In this paper, the postural stability for Chinese crew working on moving ocean platform is studied. Some human parameters as the height of COM and moment of inertia are calculated based on the standard of Chinese adults' anthropometric parameters. Effects on tipping coefficients of these parameters are investigated. Stability domain was obtained with excitation of translating accelerations. The work can be used as a guide to estimation the possibility of MII and to build an index to describe the stability degree of crews on moving platform.

References

1. A.H. Wertheim, Working in a moving environment, Ergonomics, **41** (12), 1845-1858 (1998)

 DOI: 10.1051/matecconf/201711901054

2. R. Graham, Motion-induced interruptions as ship operability criteria, Naval Engineers Journal **102** (2), 65-71 (1990)
3. P. Crossland and K. Rich, Validating a model of the effects of ship motion on postural stability, *Proceedings of International Conference on Environmental Ergonomics*, SanDiego (1998)
4. B.M. Haward, C.H. Lewis, and M.J. Griffin, Motions and crew responses on an offshore oil production and storage vessel, Applied Ergonomics, **40**, 904-914 (2009)
5. N. Nawayseh and M.J. Griffin, Effect of frequency, magnitude and direction of translational and rotational oscillation on the postural stability of standing people, Journal of Sound and Vibration, **298** (6), 725-754 (2006)
6. J. Wedge and R.G. Langlois, Simulated the effects of ship motion on postural stability using articulated dynamic models, *The Society for Modeling and Simulation, International Summer Computer Simulation Conference, SCSC Liophant & SCS*, Montreal, 20-23 July (2003)
7. B. Yao, Z. Chang, et al., Dynamic response of human body on offshore platform, Mechanical Design, **29** (3), 432-439 (2012) (In Chinese)
8. E. Nocerino, Human postural stability onboard ship as seakeeping criterion, Stance control model and procedure for validating it: a proposal[D], Università degli studi di Napoli Federico II (2010)
9. American Bureau of Shipping, Guide for Crew Habitability on offshore installations, Huston (2012)
10. National Standard Organization of China, GB/T 10000-1988, Human Dimensions for Chinese adults (1988)
11. National Standard Organization of China, GB/T 12985-1991, General rules of using percentiles of the body dimensions for product design (1991)
12. National Standard Organization of China, GB/T 17245-2004, Inertial parameters of adult human (2004)

Research of Chengdu modern industrial architectural heritage

Li-Yang Lu and Yun-Zhang Li[a]

College of Architecture and Environment, Sichuan University, Chengdu 610065, China

Abstract. It is industry, fruit of civilization, symbolizes the progress of the society of human beings. The industrial architectural heritage is an important part of the culture and signals the advance of the science and technology in the industry. The developing process of modern industry in Chengdu highlights the manufacturing progress in Sichuan even the inland China. With the development of the cities, the upgrade, the relocation and reconstruction of the enterprises, protecting the industrial architectural heritage has become a major case on the studying of the progressing procedure of manufacturing. A good conservation and reuse on the heritage will promote the economy and culture in the region. The article systematically elaborates the history of modern industrial development in Chengdu, summarizes the existing types of the architectural heritage in the industry; analyzes the layout, spatial form, styles; presents two representative industrial architectural heritage and a reformation area; points out current situation the heritage sits in and the protection measures on it, as well as its limitations and some related suggestion.

1 Introduction

In 1840, the Opium War broke out, which made China step into the Modern History stage. Under the promotion of Self-Strengthening Movement, the Qing government established a number of modern military and civil industries in 1861[1], which opened the gate of Chinese modern industry.

But Chengdu, located in the hinterland of Western Sichuan Plain, was under the economic limit for geographical factors. All this delayed the development speed of modern industry in Chengdu.

The opening of Chongqing port in 1890 enhanced the formation and development of the commodity market in Chengdu. Due to years of battle between feudal warlords, the modern industry of Chengdu, even Sichuan, lagged behind the port cities hugely.

Many port cities fall to Japan's control since China's Anti-Japanese War which began on July 7, 1937. In order to ensure material demand, 155 factories were selected to move into Sichuan.

Chengdu's modern industry went into a surging period, but the development speed of its heavy industry was remained at a slow speed. Speaking of the shipping conditions, Chengdu was next to close-range Chongqing. After then-government relocated its capital to Chongqing, the city became the only comprehensive industrial base of China. Its hilly terrain helps conceal factories, facilities and munitions during wartime[2]. After the war, most of the factories returned to the cities where they used to locate in.

[a] Corresponding author : 307941240@qq.com

At the time, Chengdu's economy suffered from inflation and repression from then-government-run industries, coupled with the proliferation of foreign goods. The industrial output of Chengdu was only 108 million yuan before 1949, in which the handicraft counted 14,000 households, making up 98.8% of the sum. Few contributions were made by modern industry to Chengdu's economy[3].

In 1953, Chengdu was included as one of the eight national key cities. Nine industrial infrastructural projects were settled in Chengdu, with 4 electronics industries centralized in Eastern Suburbs of Chengdu. In 1955, the State Council formally approved the East Suburb of Chengdu as one of the three national electronic industrial base.

In 1964, the main task was to re-planning and rebuilding modern industry, known as the Third Front Movement plan. As the commanding center of Third Front Movement in Southwest China, Chengdu completed a thorough transformation from consumer-oriented city to an industrial-oriented city[4].

2 Characteristic of industrial architectural heritage in Chengdu

2.1 Type

Before the liberation of the country, Chengdu's economy was dominated by the specialization of workshops and manufacturing rather than steam engine-powered modern industry. The production methods and operating models of the specialization of workshops and manufacturing didn't bring the modern industrial buildings.

2.2 Layout

The modern industry in the West was reformed from traditional handicraft and manufacture. The modern industry of Chengdu, which was roughly copied from the West at the beginning, undermined the natural transition from local manufacturing. It resulted in a shaky foundation and misdistribution for the industry[5]. But as a result of unified planning and construction after 1949, the layout of Chengdu's industrial plants was concentrated.

Chengdu's plain terrain is conducive to productive activities. It offers open space which is beneficial to achieve structure and production activities. The buildings of plant arranged horizontally or longitudinally. The plan of one single block was also neat and simple.

2.3 Space form

Not only a large number of production buildings, factories and warehouses, but also production-assist buildings, such as office buildings and dormitories exist in industrial plants for the multifunction involving production, dwelling and management.

The high green coverage rate could cut down the pollution of production and living to a minimum and prevent the damage to the workers' health. Furthermore, the folks are accustomed to housing pattern in courtyard. High green rate reaps huge fruits in improving micro-climate, cleaning air and reducing noise, ameliorating working and dwelling conditions.

Production buildings take steel structure or reinforced concrete structure. In order to ensure the access of the equipment for lifting and transporting, an open interior is needed. Diverse functions, connected with each other, can be found in one block, sharing the flexible designing on the floor height. Side-windows and skylights, designed for demanding illumination condition, constitute a rich form of profiles.

Production-assist buildings use brick and concrete structure. The sense of efficient space using calls for internal corridors. And the increasing thought of electric light, fire code, demanding mechanical smoke extraction, forms the prototype of the modern public buildings.

2.4 Features

In the 1930s, the trend of modern Western architecture began to affect the trends of Chinese architecture. In the 1950s, the Soviet Union aided China 156 key projects, which brought in the Soviet socialist principle.

The principle urges buildings with sound functionality, reasonable budget and ideological consideration, "reflecting the greatness and beauty of socialism," Si-Cheng Liang noted "in the form of national "as the interpretation of Soviet socialism[6].

Architects were attempting to combine the traditional style with western style. Under the prevailing economic background and the assimilation of western building thoughts, the image of Chengdu industrial buildings suggests three primary characteristics:

(1) Significantly modern features, no decorations, no ornaments, were built to ensure production activities;
(2) Layout, facade designing, materials take the form of Russian (European) style, with some traditional Chinese elements in certain details;
(3) Traditional Chinese architectural elements were used (especially for the roofs), a modern structure, regardless of traditional scale, combined with the European symbols.

3 Case analysis

3.1 Chengdu measuring & cutting tool factory building

Chengdu Measuring & Cutting Tool Factory, built in 1956, was a key enterprise of in measuring and cutting instruments for New China.

The building was used as offices and workshops. Its structure as well as its outer wall was in imitation of the Harbin Measuring & Cutting Tool Factory building. The building faces northwest, sits straight along Fuqing Road. Brick and concrete structure with total width, depth and gross floor area reach 20.31m, 19.45m and 11748.96m2. Exterior walls consist of red bricks, while beams and columns are made of reinforced concrete and precast reinforced concrete panels floors. The outer wall was painted beige at first, then changed to red in 1990 as shown in Figure 1.

Figure 1. Chengdu Measuring & Cutting Tool Factory building and the right side elevation.

Followed the classical composition form of European buildings, the landscape and portrait of the building are three-patterned, substantially symmetrical layout, the middle six-storey (partial eight-storey) with a vase top and arched windows on each side.

The southern part of the block consists of three-storey offices, mechanism red tile roof with fences on the cornices. The northern part of the block features three-storey workshops with higher floor height, flat roof with parapets on its upper side. The farthest part is work cabins and lounge. Eight asymmetrical stairs were arranged according to their different functions.

 DOI: 10.1051/matecconf/201711901055

The designer balanced aesthetics with practicability to ensure the efficiency of project funding. It achieves a good visual effect by using traditional elements on the roofs, eaves and railings. The Figure 2 shows the ground floor plan and Positive elevation of Chengdu Measuring & Cutting Tool Factory building.

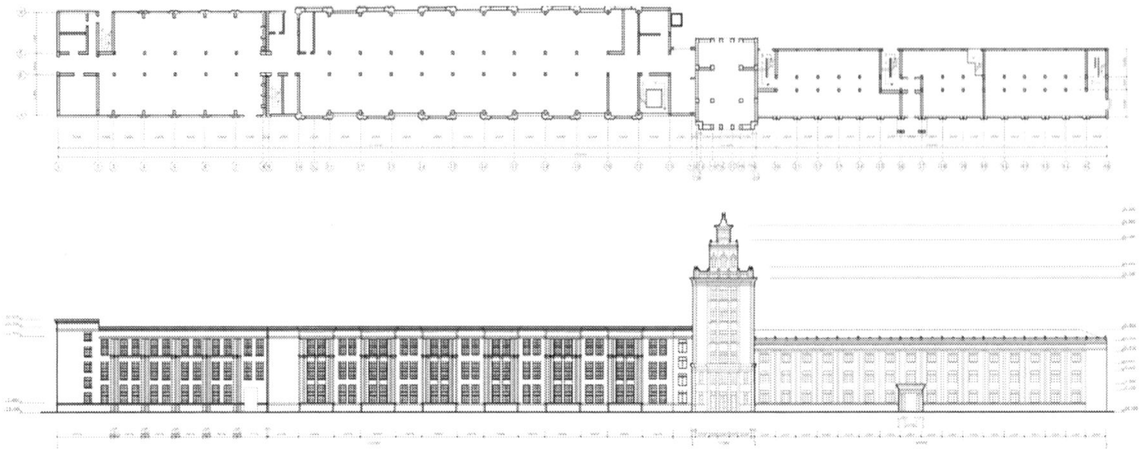

Figure 2. Ground floor plan and Positive elevation of Chengdu Measuring & Cutting Tool Factory building.

3.2 Chengdu vehicle factory building

Founded in 1952, Chengdu Vehicle Factory was a large state-owned enterprise, which is Top 500 of China's transportation equipment and railway vehicle company.

Chengdu Vehicle Factory building is a two-storey Soviet-style building, sits from northeast to southwest, covered an area of 693 m^2. It is made of brick and wood, shapes like a reversed letter T. Its length, width and height are 42m, 14m and 14.6m, with the single-storey bulge 9.5m in length and 11m in width.

The base of black brick walls is surfaced with wipe terrazzo. Walls support beams and hip roof covers small tiles. The northeast part of the building forming a square balcony up there with decorative sculptures. A wooden door with frame carved patterns is at the end of the central corridor. Middle part of the building lays a cement staircase to the second floor. Both southeast and northwest sides of the building have two small balconies. The Figure 3 shows the ground floor plan, as well as the status photos of Chengdu Vehicle Factory building.

Figure 3. Ground floor plan, the main entrance and side entrance of Chengdu Vehicle Factory building.

Chengdu Vehicle Factory building is used as the vehicle-checking room of the Ministry of Railways in Chengdu.

DOI: 10.1051/matecconf/201711901055

3.3 Chengdu memory of eastern suburbs

Chengdu Memory of Eastern Suburbs transformed from Red-light Factory which was one of 156 major projects aided by the Soviet Union. It was put into operation in November 1958, known for the base of electric device and birthplace of domestic television tube. Chengdu Memory of Eastern Suburbs consists of tube-assemble workshop, fitter workshop, infrastructure division office, warehouse, cylinder workshop, gas-production workshop, material-assemble workshop, indicator-assemble workshop, processing workshop and factory building, which are shown in the Figure 4.

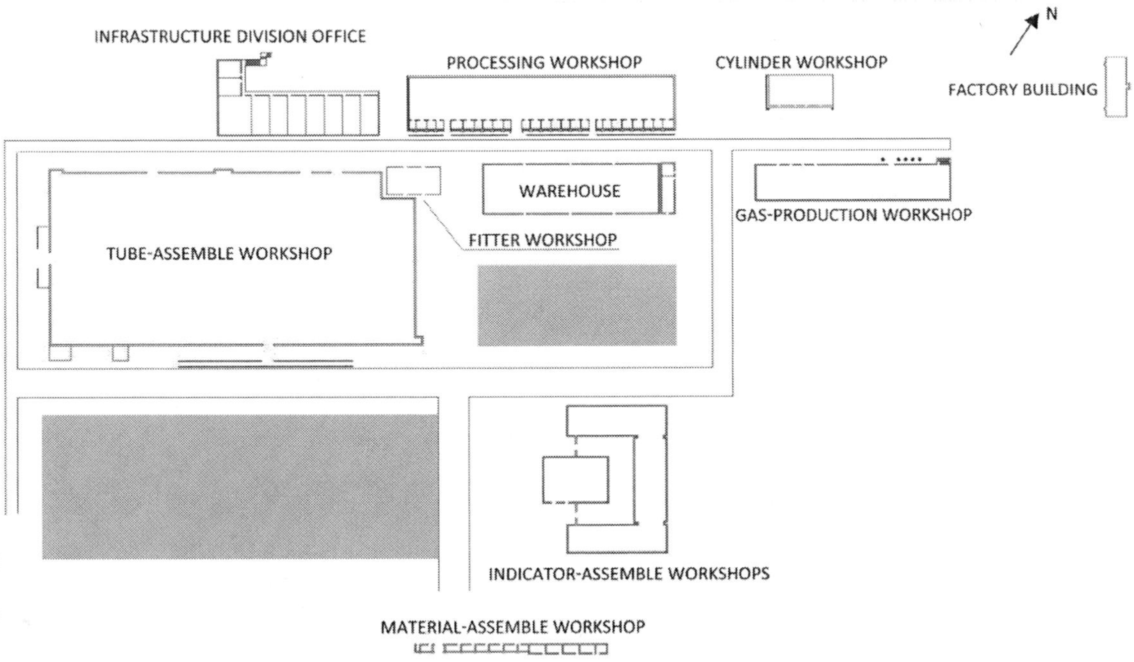

Figure 4. Ground floor plan of Red-light Factory.

Tube-assemble workshop faces southwest, single-layer, flat-topped, brick wall covering with lime, 187m in length, 100m in width, 9m in height, covers an area of 18700 m^2.

The two indicator-assemble workshops cover an area of 6072 m^2.The layout of the two is 凸-shaped. The first one is a four-storey building with flat roof, symmetrically 凹-shaped, reinforced concrete structure, 92m in length, 34m in width, 21m in height. Another one is located inside the former one, with a rectangular layout, 66m in length, 28m in width, 22.5m in height.

The factory building is on a three-storey reinforced concrete structure with a flat top. The whole building is symmetrically arranged by the middle main entrance, 100m in length, 16m in width, 14m in height.

4 Protection of industrial architectural heritage in Chengdu

4.1 Current situation

Since 1990, a few cities stepped into the post-industrial period gradually. Under the guidance of macroeconomic policy obeying socialist market economy, the traditional industrial structure of Chengdu started to change.

The 1998 edition city plan considered Chengdu as a modern city for losing its historical and cultural characteristics, which resulted in ignoring the protection of modern architecture which out of the conventional category of cultural relics[7].

In June 1, 2007, modern industrial architectural heritage, for the first time, was listed in the Seventh Batch of Provincial-level Protective Cultural Relics announced by Sichuan Provincial Government.

In July 16, 2012, the Eighth Batch of Provincial-level Protective Cultural Relics indicated a significant increase in the proportion. It was a meaningful expansion and extension of relic category.

4.2 Protection legislation

In 2004, Chengdu revised the city plan, pinned the city as a historical and cultural one, putting forward the requirements of heritage protection explicitly. In July 28, 2005, Planning Administration of Chengdu leaded to compile the protective outline. According to the guidance on Strengthening the Protection of Excellent Modern Architecture published by the Ministry of Construction, Protection Planning of Chengdu Excellent Modern Architecture took effect by the end of 2005, including quite a few industrial architectural heritage.

4.3 Protection method

4.3.1 Reconstruction into museum

The reconstruction of the heritage into a museum is a crucial protective way to preserve the modern industrial architecture.

Eastern Suburbs of Chengdu Industrial Civilization Museum, reconstructed by repairing workshops of Hongming Machine Factory, is the first classic case of reconstructing industrial plants to a theme museum, which highlights the industrial process of Eastern Suburbs of Chengdu from the 1950s to the 1980s.

4.3.2 Renovation into public service

Designed to provide or assist in production, each building can be renovated into public service in social life after the closure of industrial activities.

Located at the intersection of North Road and Second Ring Road, Yudong Shikong Badminton Arena is reconstructed based on producing workshops of Electric Welder Machine Factory. On account of lacking in sports areas and facilities nearby, the badminton arena provides convenience for folks around.

4.3.3 Updating into cultural industrial zone

Cultural industrial zone is a vivid revelation of the past and retain the urban history to some extent, as well as being beneficial to enhance the updating of obsolete urban district to stimulate regional vitality and creativity.

After the industry adjustment of Eastern Suburbs of Chengdu, Chengdu Memory of Eastern Suburbs grew from the Red-light Factory. Located on a 218 acre land parcels, the park has become the largest industrial updating project in southwest with the gross floor area accounting for 189,000 square meters. The original of about 14 million square meters of plants are remained, which give full play to regional integration and aggregation efficiency, becoming the first digital music-themed park of China. The Figure 5 shows the state of the Chengdu Memory of Eastern Suburbs.

 DOI: 10.1051/matecconf/201711901055

Figure 5. Chengdu Memory of Eastern Suburbs.

5 Conclusion

Eastern Suburbs of Chengdu is the centre for modern industrial architectural heritage with abundant facilities full of strong historic culture. Besides a minority of industrial buildings scattered or far away from the urban area are on the fringe of the effective protection. Due to the lax of protection awareness when comparing with the urban planning and construction, the industrial architectural heritage on the fringes, isolated by the tremendous changed urban framework, is at risk. Moreover, the reconstruction and the renovation are also limited by modern environment.

Located on the 31st, Sanguantang Street, the watchtower is the heritage of Self-Strengthening Movement. Being one of the four watchtowers built in 1878 and owned by Sichuan Machinery Bureau, the surviving one has been beleaguered by advertisement and garbage one year after being cataloged in the First Batch of Excellent Modern Architecture in Chengdu. The deficiency of relevant data and information, limited by the geographical factors and uncertainty, caused by the complicated historical background, industrial architectural heritage on the fringes is often ignored when having cultural relics survey. After the closure of production activities, some of these were wiped out[8].

The Nanbao forge, built in 1958, is a historical relic of Great Leap Forward movement, provides visible data of iron smelting technology. Only one of the original more-than-10 survives.The protection of industrial architectural heritages requires industrial diversity, the echelon of the dates and the variety of its types. Only when fulfilled all these can we complete the history of modern industrial development.

While domestic protection focuses on the preservation of buildings and the facilities, production process, interaction between residents and environment, local memory, technological innovation and major events, all mentioned above has not been particularly presented. The spatiality is the basic assets of industrial heritage, while the meaning and memory it represents is the potential cultural assets indeed. The profound emotion and intimate connection between buildings and people consists of unique history each block on behalf of. The emotional awareness could evolve into protection consciousness. Only after pouring attention into cultural assets to stimulate the sense of involvement and historical identity can we have the industrial architectural heritage being well-preserved and the history of Chengdu modern industry and the process of modern architecture development can pass down for long generations.

References

1. H.M. Chen, Geographical distribution and influence of Chinese modern industry (1927-1861), Economic Research Guide, **15**, 63-65 (2013)
2. D.F. Xu, The chongqing industrial heritage conservation reuse and urban revitalization, PhD Thesis, Chongqing University, China (2012)
3. W. Zhong, From "three and a half chimney" to leading the economic development of western industrial -- Chengdu industrial development, Chengdu Daily, **T63** (2011)

4. X.Z. Li, The 60 years of chengdu: from third front movement to urban-rural integration, China Business News, **23** (2009)
5. H.Y. Liu and Z.F. Wang, The unique way to the formation and development of China's modern industry, Journal of Jiangsu Normal University(Philosophy and Social Sciences Edition), **1**, 21-24 (1996)
6. G.H. Ji, Architecture of New China and the Soviet Union- import of Soviet architectural theory and impact on China in 1950s, Cities and Towns Construction in Guangxi, **2**, 70-74 (2013)
7. B.Y. Liu, A review of industrial architecture heritage research in China, New Architecture (2012)
8. X.Y. Chen, Reuse of industrial buildings, Sichuan Building Science, **39**, 271-274 (2013)

MATEC Web of Conferences **119**, 01056 (2017) DOI: 10.1051/matecconf/201711901056
IMETI 2016

Effect of highly reflective roofing sheet on building thermal loads for a school in Osaka

Jihui Yuan[1,2,a], Kazuo Emura[1] and Craig Farnham[1]

[1]*Dept. of Housing and Environmental Design, Graduate School of Human Life Science, Osaka City University, Osaka 558-8585, Japan*
[2]*School of Environmental Science and Engineering, Huazhong University of Science and Technology, Wuhan, Hubei 430074, China*

Abstract. Currently, urban heat island (UHI) phenomenon and building energy consumptions are becoming serious. Strategies to mitigate UHI and reduce building energy consumptions are implemented worldwide. In Japan, as an effective means of mitigating UHI and saving energy of buildings, highly reflective (HR) and green roofs are increasingly used. In order to evaluate the effect of roofs with high reflection and thermal insulation on the energy conservation of buildings, we investigated the roof solar reflectivity of the subject school in Osaka, in which the HR roofing sheet was installed on the roof from 2010. Thermal loads, including cooling and heating loads of the top floor of school, were calculated using the thermal load calculation software, New HASP/ACLD-β. Comparing the thermal loads after HR roofing sheet installation to previous, the annual thermal load decreased about 25 MJ/m^2-year and the cooling load decreased about 112 MJ/m^2-year. However, the heating load increased about 87 MJ/m^2-year in winter. To minimize the annual thermal load, thermal insulation of the roof was also considered be used together with HR roofing sheet in this study. The results showed that the combination of HR roofing sheet and high thermal insulation is more effective to reduce the annual thermal load.

1 Introduction

Urban heat island (UHI) is a well-documented climatic change phenomenon [1]. UHI intensity in hot climates may rise up 10°C [2,3], resulting in increased discomfort, higher pollution levels while it has a serious impact on the cooling energy consumption of buildings [4]. Important research has been carried out to document its influence on the urban climate [5,6]. As the same UHI phenomenon has become serious, the energy consumption in buildings is also increasing rapidly. Among strategies of mitigating UHI and reducing energy consumption of buildings, highly reflective (HR) materials [7,8] applied to building roofs or retro-reflective (RR) materials [9-11] used for building facades have been researched globally. HR or RR building envelope materials can reduce the cooling load in the summer due to their high solar reflectivity. However, the HR and RR envelopes can also increase the heating load in the winter [12].

In three major cities (Tokyo, Nagoya, Osaka) of Japan, under global warming in progress at the rapid rate, the temperature of Osaka rises 2.1°C in 100 years which is higher than the national average

[a] Corresponding author : yuanjihui@hotmail.co.jp

temperature 1.0°C, this difference 1.1°C is thought to be due to heat island effect [13]. The global contribution from buildings towards energy consumption, both residential and commercial, has steadily increased to between 20% and 40% in developed countries, and has exceeded the other major sectors: industrial and transportation [14]. Buildings in the urban area are a major contributor to the UHI phenomenon. Focusing on the anthropogenic heat emissions, a study showed that exhaust heat from buildings in Japan accounted for approximately half of the anthropogenic heat in the city [15]. Thus, to reduce energy consumption in buildings plays an important role in countering the UHI effect. Many studies to subject the UHI effect have also been implemented widely in Japan [16-19].

In order to mitigate UHI effect and reduce energy consumption of buildings, HR roofing sheets were installed on the roofs of 70 schools in Osaka, Japan from 2010. The solar reflectivity of HR roofing sheets installed on roofs was investigated [20]. In order to evaluate the effect of HR roofing sheet on the energy conservation of buildings, this study aims to calculate and compare the thermal load of building before and after HR roofing sheet installation on the subject school roof. Additionally, to minimize the thermal load of buildings, it is necessary to consider applying appropriate thermal insulation together with the HR roofing sheet to the roof in this study.

2 Subject school and HR roofing sheet

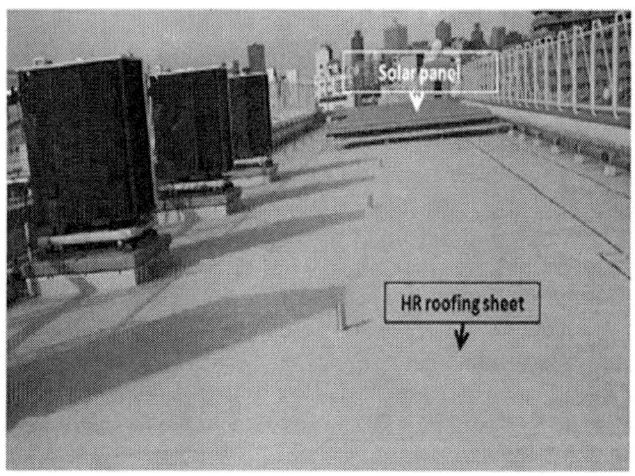

Figure 1. The roof appearance of subject school coated with HR roofing sheet.

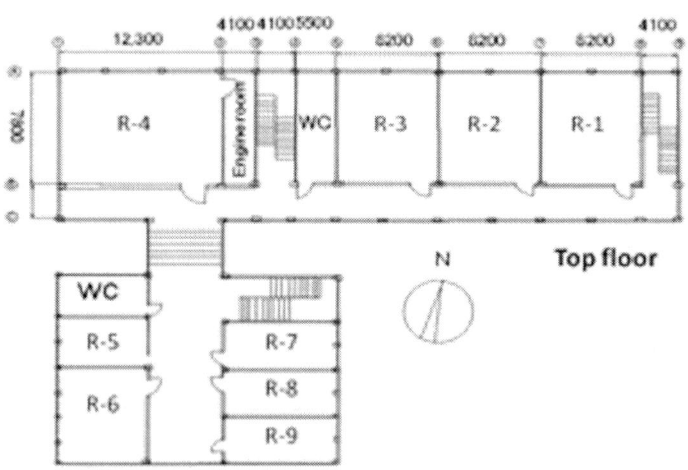

Figure 2. Ground plan of the top floor of subject school with a floor area of about 683 m^2.

The roof appearance of the subject school covered with HR roofing sheet is shown in Figure 1. The ground plan of the top floor (the 4th floor) of the school building with a floor area of about 683 m^2 is shown in Figure 2. Structures of the roof, floor, exterior wall and interior wall are shown in Table 1.

The HR roofing sheet has high solar reflectivity in the near-infrared (NIR) range regardless of the solar reflectivity in the visible (VIS) range.

According to the previous research [20], the solar reflectivity of HR roofing sheet in visible range (400-780 nm) was about 0.5, about 0.58 in near-infrared range (780-1700 nm) and about 0.54 in all wavelengths (400-1700 nm) by the investigation of previous study.

Table 1. Structures of roof, floor, exterior wall and interior wall of school buildings.

	Case A: before construction of HR roofing sheet	**Case B: after construction of HR roofing sheet**
Roof	In order from indoor to outdoor: Rock wool (12 mm) + Gypsum board (9 mm) + Hollow layer + Ordinary concrete (150 mm) + mortar (15 mm) + Asphalt (10 mm) + Light concrete (60 mm)	In order from indoor to outdoor: A + HR roofing sheet (3 mm)
Floor	In order from top to bottom: Plastic tile (3 mm) + Mortar (15 mm) + Ordinary concrete (150 mm) + Hollow layer + Gypsum board (9 mm) + Rock wool (12 mm)	
Exterior wall	In order from indoor to outdoor: Mortar (20 mm) + Ordinary concrete (150 mm) + Mortar (20 mm) + Tile (8 mm)	
Interior wall	Mortar (20 mm) + Ordinary concrete (120 mm) + Mortar (20 mm)	

3 Thermal load calculations

Table 2. Conditions of the thermal load calculations.

External condition	Standard weather data: Expanded AMeDAS Weather Data of Japan			
	Location: Osaka, Japan			
Orientation and area of window	Orientation of window: Northwest 15 degrees Window area: 3.2 m ×1.8 m = 5.76 m^2			
Amount of fresh air	Infiltration		Amount of fresh air into room ($25m^3$/hour-person; Weekday 8:00~17:00)	
	0.5 times/hour		1.4 times/hour	
Internal heat generation	Body (heat generation: 108 W/person)		Lighting (efficiency: 0.9)	OA equipment (Sensible heat)
	Number of person: 36 person		10 W/m^2	5.8 W/m^2
Air conditioning service	Weekday 8:00~17:00			
Temperature and Humidity settings		Heating season (Dec~Mar)	Cooling season (June~Sept)	Middle season (April~May; Oct~Nov)
	Temperature (°C)	22	28	26
	Humidity (%)	40	60	50

In this research, the software, New HASP/ACLD-β, is used to calculate the thermal loads of buildings. As EnergyPlus is used in U.S., New HASP/ACLD-β is a simulation program often used to calculate the indoor temperature, humidity and thermal loads of buildings, which has been developed for the purpose of evaluating the energy consumption of air-conditioning in buildings [21]. Parameters in New HASP/ACLD-β calculation program, such as the solar reflectivity of exterior walls, structure of building walls, operating condition of air conditioning, could be rewritten easily. To calculate the thermal loads of buildings by New HASP/ACLD-β, the following input data are needed for the calculation period.

- Hourly meteorological data: air temperature, absolute humidity, direct solar radiation, diffuse solar radiation, cloud cover, wind direction and wind speed, etc.
- Building data: the longitude, latitude of the building, wall dimensions and directions, roof and overhang dimensions, ground height, solar reflectivity of ground, solar radiation absorption and long-wave emissivity of outer walls, the structure and materials of outer and inner walls, ceiling height, distance to and height of neighboring buildings, and schedule of indoor activities and occupancy, etc.

For this study, monthly and annual thermal loads (annual cooling and heating loads) of subject school before and after construction of HR roofing sheet were calculated using New HASP/ACLD-β. Conditions of the thermal load calculations are shown in Table 2.

4 Results and discussion

Monthly and annual thermal loads (annual cooling and heating loads) before and after construction of the HR roofing sheet in the subject school were calculated and shown in Figure 3 and Figure 4, respectively.

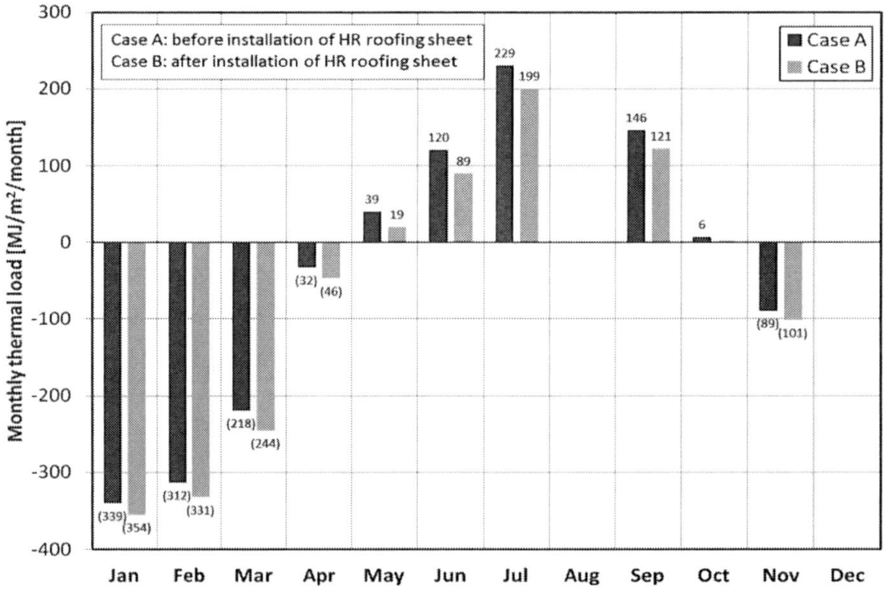

Figure 3. Monthly thermal load of subject school (Case A: before construction of HR roofing sheet; Case B: after construction of HR roofing sheet).

It showed that: i) the thermal load from May to October (here noted that September is summer vacation of school) decreased when the roof was covered with HR roofing sheet (Case B). Among these months, July has the largest reduction of monthly cooling load, about 30 MJ/m^2-month; ii) compared to the case before the roof was retrofitted with HR roofing sheet (Case A), annual total thermal loads decreased about 25 MJ/m^2-year and the cooling load decreased about 112 MJ/m^2-year after the roof was constructed with the HR roofing sheet (Case B). However, annual heating load increased about 87 MJ/m^2-year after construction of HR roofing sheet (Case B).

The result indicated that the HR roofing sheet is more effective for cooling load in the summer period. However, it is disadvantage for heating load in the winter period.

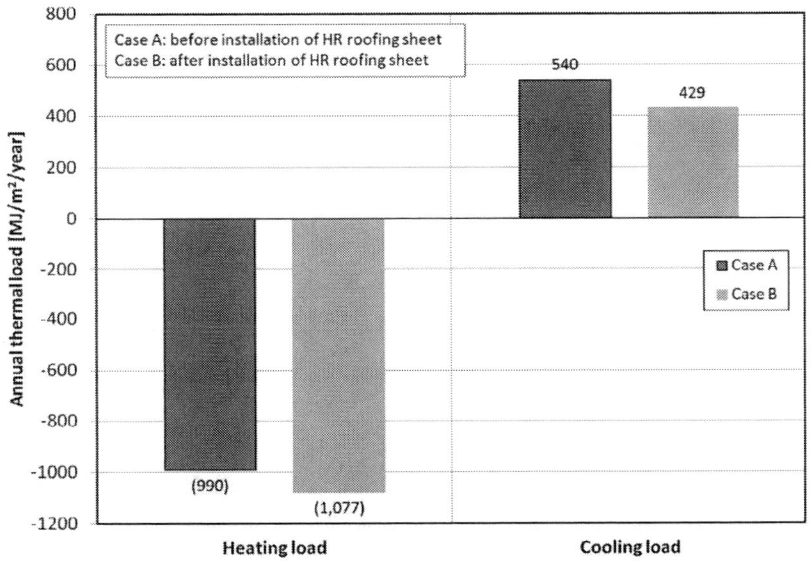

Figure 4. Annual cooling and heating loads of subject school (Case A: before construction of HR roofing sheet; Case B: after construction of HR roofing sheet).

5 Refurbishment for reducing thermal loads

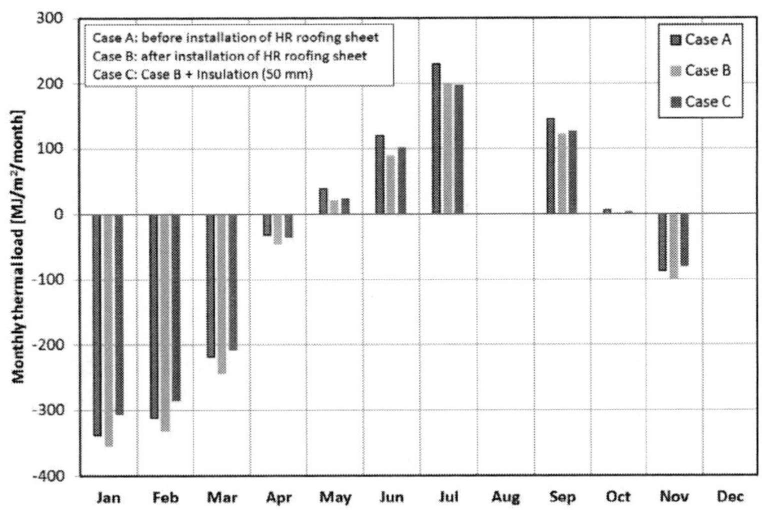

Figure 5. Monthly thermal load of subject school (Case A: before construction of HR roofing sheet; Case B: after construction of HR roofing sheet; Case C: Case B + Insulation thickness of 50 mm).

In order to minimize the total thermal load and heating load, according to our previous study [22], this study considers adding thermal insulation with a thickness of 50 mm to the roof of the subject school combined with the HR roofing sheet (Case C).

Monthly and annual thermal loads (annual cooling and annual heating loads) under three cases (Case A; Case B; Case C) were calculated and shown in Figure 5 and Figure 6.

It showed that: i) the total thermal load decreased about 168 MJ/m^2-year (cooling load decreased about 91 MJ/m^2-year and heating load decreased about 77 MJ/m^2-year) for Case C, compared to Case

A; ii) the total thermal load decreased about 144 MJ/m^2-year (cooling load increased about 20 MJ/m^2-year and heating load decreased about 164 MJ/m^2/-ear) for Case C, compared to Case B.

The result indicated that the appropriate combination of HR roofing sheet and thermal insulation of roof is more effective to reduce the annual thermal load, which is directly related to energy conservation of buildings, compared to HR roofing sheets alone or an added thermal insulation of the roof alone.

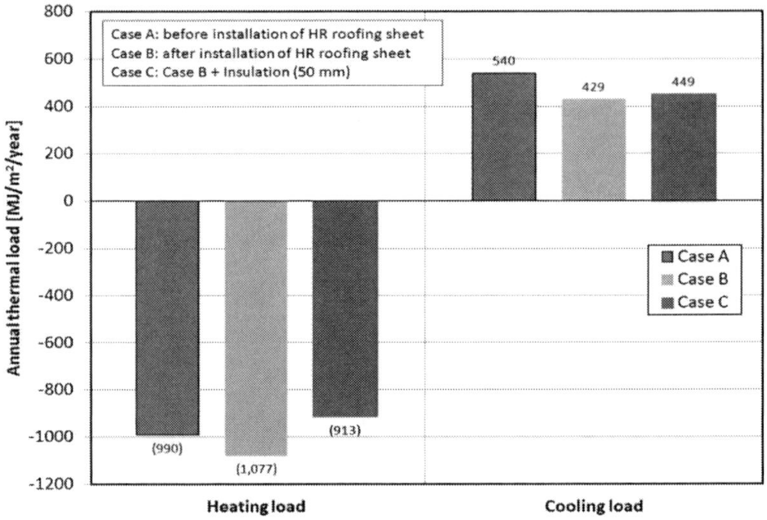

Figure 6. Annual thermal load of subject school (Case A: before construction of HR roofing sheet; Case B: after construction of HR roofing sheet; Case C: Case B + Insulation thickness of 50 mm).

6 Conclusions and future research

The thermal load of subject school before and after construction of the HR roofing sheet was calculated using the software, New HASP/ACLD-β in this study. For retrofitting the roof of the subject school to minimize the thermal load, thermal insulation of the roof with a thickness of 50 mm is considered to be added to the roof combined with the HR roofing sheet.

The cooling and the total thermal loads of subject school decreased after the construction of HR roofing sheet on the roof of subject school. However, the heating load increased after the construction of the HR roofing sheet. Thus, it is considered that the HR materials are effective in the summer and opposite in the winter.

For the location of Osaka (34.7°N), if the thermal insulation could be considered be used together with the HR roofing sheet, it is sufficient to combine the insulation thickness of approximately 50 mm with the HR roof sheet (solar reflectivity was investigated to be about 0.50) to reduce the thermal load directly related to energy consumptions for the subject school.

In order to apply the HR materials more widely, due to the high-cost of current HR materials applied to roofs, the development of low-cost HR materials should be an important issue for the future research. In addition, if the HR material could be applied to building facades, the further work should focus on preventing the "light pollution" to the neighboring buildings, and the reflection directional characteristics of these HR materials should be studied.

Acknowledgment

The authors are sincerely grateful to the Foundation of Construction Materials Industrial Promotion Tostem Corporation, the Architecture Research Foundation of Takenaka Scholarship Foundation of Japan, and the Urban Studies of Osaka City University for their support. We are also grateful to the

Department of Urban Development Department and Education Committee of Osaka for their support during measuring.

References

1. M. Santamouris, A. Synnefa, and T. Karlessi, Sol. Energy, **85** (12), 3085-3102 (2011)
2. I. Livada, M. Santamouris, K. Niachou, N. Papanikolaou, and G. Mihalakakou, J. Theor. Appl. Climatol., **71**, 219-230 (2002)
3. P. Mihalakakou, H. Flokas, M. Santamouris, and C. Helmis, J. Appl. Meteorol,, **41**, 519-527 (2002)
4. S. Hassid, M. Santamouris, N. Papanikolaou, A. Linardi, N. Klitsikas, C. Georgakis, and D. Assimakopoulos, Energy Build., **32**, 131-141 (2000)
5. M . Santamouris, K. Paraponiaris, and G. Mihalakakou, Clim. Change, **80**, 265-276 (2007)
6. H. Akbari, S. Konopacki, and M. Pomerantz, Energy, **24**, 391-407 (1999)
7. A. L. Pisello, F. Cotana, and L. Brinchi, Energy Procedia, **45**, 453-462 (2014)
8. M. Santamouris, Energy and climate in the Urban Built Environment (2001)
9. F. Rossi, B. Castellani, A. Presciutti, E. Morini, M. Filipponi, A. Nicolini, and M. Santamouris, Appl. Energy, **145**, 8-20 (2015)
10. J. Yuan, K. Emura, and C. Farnham, J. Build. Phys., **38** (6), 500-516 (2014)
11. Y. Han, J. E. Taylor, and A. L. Pisello, Energy Build., **102**, 380-389 (2015)
12. M. Dabaieh, O. Wanas, M. A. Hegazy, and E. Johansson, Energy Build., **89**, 142–152 (2015)
13. Osaka District Meteorological Observatory, Ministry of the Environment Report: Research on the environmental impact of business considering the heat island phenomenon in 2002 (2003). Available at: http://www.jma.go.jp/jma/index.html
14. P. L. Luis, O. José, and P. Christine, Energy Build., **40** (3), 394-398 (2008)
15. Ministry of Land, Infrastructure and Transport and the Environment, The measurement report of heat island by anthropogenic heat suppression in 2012 (2013) (in Japanese)
16. T. Inoue, Build. Technol., **684**, 142-145 (2007) (in Japanese)
17. M. Nishioka, S. Inoue, and K. Sakai, J. Environ. Eng., **73**, 1249-1254 (2008) (in Japanese)
18. H. Sakai, K. Emura, N. Igawa, and H. Iyota, *Proceedings of the Second International Conference on Countermeasures to Urban Heat Islands (SICCUHI)*, held September 21-23, in Berkeley, California, CD-ROM 211450, 1-6 (2009)
19. J. Yuan, C. Farnham, and K. Emura, Energy Build., **103**, 107-117 (2015)
20. J. Yuan, K. Emura, and H. Sakai, J. Build. Phys., **37** (2), 170-184 (2012)
21. The Society of Heating, Air-Conditioning and Sanitary Engineers of Japan (SHASEJ), Operation Manual of New HASP / ACLD- β (2004) (in Japanese)
22. J. Yuan, C. Farnham, K. Emura, and M. A Alam, Build. Environ., **103**, 228-237 (2016)

An explorative study of the ecological design of residence architecture in Ming and Qing dynasties in Yangzhou

Man Lu[1,2,a], Gang Wang[3] and Jian Zhou[1]

[1]*School of Civil and Transportation, Hohai University, Nanjing, Jiangsu Province, 210098, China*
[2]*School of Architecture, Southeast University, Nanjing, Jiangsu Province, 210096, China*
[3]*Nanjing Gulou Audit office, Nanjing, Jiangsu Province, 210008, China*

Abstract. Presently ecologicalization has become a key aspect of architecture designs. As studies deepen, the bountiful ecological ideas and design methods represented in Chinese traditional architectural culture are increasingly valued. There remain as of today quite a few residence mansions built in Ming and Qing dynasties in Yangzhou- a city famous for its history and culture. The distinctive regional culture and architectural style that those architectures embody has received worldwide attention in the academic field. By weighing on the shaping of architectural forms, choice in connection genres and structuring measures as well as sorting through and studying on the ecologicalized architectural design practices such as reemploying local materials, deeper understanding of traditional architecture will be achieved for us to better inherit and develop traditional architectural culture. Hence in-depth studies on contemporary ecological architecture and sustainable development will be of greater practical significance. Furthermore, it is beneficial to expand ideas on ecological design, truly integrate architectures into the setting to give expression to the ecological and culture features of them and provide theoretic support for creating new architecture with diversified culture implications and ecological features in light of actual conditions in an orderly manner.

1 Ecologization and its practical significance in traditional Chinese architecture

So called ecologic architecture is theories and methodologies on architectural design and planning based on ecology that studies the symbiotic relationship between creatures in nature and the environment; in other words, it is a branch of ecology that explores the balanced development of life activities and functions on earth into architecture, reflecting the value orientation of modern architecture community [1]. Chinese ancient philosophy has early on realized that humans and nature are an inseparable whole, emphasized that the relationship between humans and nature is defined by the connections between them. The Chinese philosophic concept of "nature and humans being oneness" is thus a view of plain ecology. Chinese ancient philosopher Lao zi said "Humans follows earth, earth follows heaven, heaven follows Dao, Dao follows naturalness". The saying indicates that humans and nature all have their own order and ultimately follow the ever natural laws. While nature

[a] Corresponding author : mmmm0526@sina.com

is all inclusive and acts as the object, humans coexist with nature and acts as the subject. "Nature and humans being oneness" asserts that the subject integrates with the object into a whole. Nature is seen as the greater entity and the human as a smaller entity. Human life continues as it is in the constant exchanges of energy and information with nature. Heaven, earth and humans are an integrated ecologic system featuring organic circulation and metabolism. The ecologic view represented in traditional Chinese architecture is a reflection of traditional Chinese views, aesthetic concept and technology on architecture.

The traditional Chinese residence architectures are the emnodiment of ecological residence. Residents of all ethnic groups in all areas have taken factors into full consideration such as local natural environment, climate, geology, land form, architecture materials, technology, customs and living habits. Ancient Chinese architecture plays a positive role in the harmony of ancient ecosystem, from which we could find many traces of primeval ecological architectures including: the organic close combination between ancient Chinese architectures and the surroundings; the courtyard layout of architectures is consistent with family structure (the consistency on the level of materials and spirit); the architecture structure is sustainable with individual components fixable and replaceable when broken; most of materials for constructing houses are local, which saves energy and reduces pollution. All of the above are of great practical significance for and shed light on the current ecologic architectural technique studies.

2 Ecologic design strategy in the planning of residential architecture of Yangzhou in Ming and Qing dynasty

2.1 Location selection and layout

Figure 1. Outer facade of Lumen Wang Mansion.

Figure 2. The entrance gate and shrine of Wangshi Garden.

Figure 3. Fire preventing alley.

Houses built in Ming and Qing dynasties in Yangzhou are found in both the old and new towns just as Qinglong- one emperor in Qing dynasty described "most salt merchants live in the new town while the old habitants live in the old town" [2]. However, these mansions were often chosen to be built close to the marketplace or along the river to save commute cost. The residential architecture built in the busy marketplace use high walls to block the noise from the outside and protect the privacy by installing no window on the outer wall (Fig 1). Jicheng (a famous landscape architect in Ming Dynasty) wrote in his book Yuanye "though living in the neighborhood of bustle, no noise can be heard behind the closed door" [3]. Furthermore, the entrance gate is not on the same longitudinal axis with the etiquette gate with the former usually located towards the southeast corner of the mansion. Upon entering the mansion, one will be faced with a solid wall with a shrine embedded in it (Fig 2). The sound from the streets are blocked or reflected by the wall, which ensures the quietness within the house. As to the

 DOI: 10.1051/matecconf/201711901057

layout, the front and middle parts of the mansion are the space for meeting guests while the rear and bilateral parts are for living functions. With the locations of the mansions soundly planned according to the local situation, the mansions coexist with each other without losing their independence. Under the condition of not exceeding land quota, all residence architectures are built in parallel with fire-preventing alleys built which not only helps prevent fire but also contributes to ventilation and taking the heat in summer. (Fig 3)

2.2 Facing direction and spacing

Figure 4. General layout of Wangshi Garden.

Figure 5. General layout of Dingshi and Mashi's Former Residence.

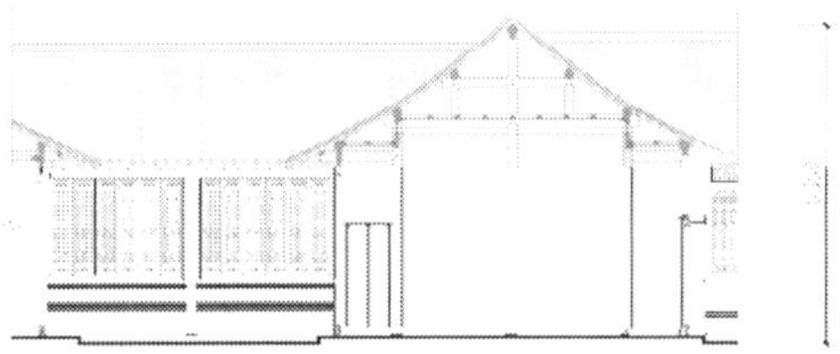

Figure 6. Fire preventing alley.

Figure 7. Yiyu Hall.

Figure 8. Fuhai Hall.

The ideal facing direction of architecture in Yangzhou area is between SE 15° and SW 15°. Research shows that houses built in Ming and Qing Dynasty all met such a standard, typically represented by Ge Garden , Wangshi Garden (Fig 4), Dingshi and Mashi's Former Residence (Fig 5). Even for those mansions that are subject to restrictions of land size, the facing direction of gate or the layout of the house is adjusted to ensure that the architecture has a sound facing direction.

Although there were many Huizhou businessmen living there in Yangzhou in Ming and Qing dynasty and Huizhou style architecture are common in the residential architectures in Yangzhou, quite a few differences are found to exist between them in the field research. For example, the length of houses in Yangzhou area is at a ratio of 1:1 to the height of the house (Fig 6). The width of the architect is larger than its length, which makes the architect a flat rectangle to ensure ample sunlight and ventilation. On the other hand, Huizhou style residence features narrow spacing between courtyards while Yangzhou residences are more spacious.

3 Ecological design strategies in construction of residential architectures of Yangzhou in Ming and Qing dynasty

3.1 Carpentry system in frame structure

Carpentry system in the residential architectures of Yangzhou in Ming and Qing dynasty can be divided into three types: post- and- lintel, cross-and-tie and the combination of both which is most common. Post-and-lintel construction can provide broader indoor space but costs more wood; cross-and-tie construction features sound integrity and economic wood usages but requires dense arrangement of pillars; the combination combines the advantages of both, which makes it ideal for constructing sound, cost effective and spacious indoor space.

Carpentry system gets rid of walls and enhances the connections between the architect and environment. For example, four-sided halls for scenery viewing purpose such as Yiyu Hall of Ge Garden (Fig 7) and Fuhai Hall of He Garden (Fig 8) have windows on four sides. When sitting inside the halls, one is overwhelmed by the views of four seasons from four directions. It also makes it possible to partition the indoor space more freely. For example, a single space can be divided into four or five when it appears to have only three; or use varied kinds of screen panels to create personalized space.

Furthermore, wood structure is easy to maintain. Despite its short longevity caused by years of rotting, mortise and tenon joint are easy to uninstall thus making maintenance easier; also due to certain level of mobility of mortise and tenon joint, wood structure might tilt under foreign force but not fall easily, therefore it has strong anti damage ability. Almost all the components in the carpentry system can be replaced individually, which reduces maintenance cost significantly and extends the longevity of the architecture.

3.2 Walls and flooring featuring thermal insinuation

"He He Wall" of Yangzhou has distinct local characteristics. Referred to as "clear water good" by the locals, the walls, as thick as 360 mm, are constructed with bricks and mortar without whitewash, which requires advanced craftsmanship. Such walls are better at thermal insulation than average brick walls. The lower section of the wall is built with flat laid bricks and mortar while the above section built with rowlock laid bricks. This way of laying bricks symbolizes "harmony" in the cultural sense and saves materials. When building such a wall, crosspiece and "mu pai shan" (a wooden consolidating structure)placed within the wall, along with "tie ba ju"(an iron consolidating structure) shaped like the Chinese character "丁" and nailed on the outside, together consolidate the wall body, equivalent to ring beam consolidating in modern concrete construction[4].

Due to the humid climate in summer in Yangzhou area, the indoor floorings would easily get moist, thus residential houses built in Ming and Qing dynasties in Yangzhou all adopted moist preventive measures. Usually the halls use quadrel flooring with metal bowl at the bottom layer. The four corners of the quadrel are placed on the inverted bowel, which smoothes the edges of the quadrel and prevents the moist. The bedrooms have better absorbing wooden floorings with metal bowls also placed underneath in addition to vent holes installed along the outer wall, which helps expel the moist and slows the process of rotting away wood floor. In some cases, wooden floorings are added to the quadrel layer so as to make the room warmer in winter and removed in summer to make it cooler.

3.3 Flexible arrangement of door and windows

As frame structures in carpentry system needs no wall to bear the weight, the arrangement of windows can be quite flexible. Generally the middle room in the south of the house has 6 or 8 lattice doors [5]and the room next to it often uses lattice windows. As they cover large area, it helps to enhances the lighting indoors and capture warmth from sunlight in winter. In summer, the wooden barrier under the lattice window could be removed leaving only hollowed paneled railings behind. Together with lattice door, screen panel and wicket, they contribute to natural cross ventilation thus cool down the indoors. The northern part of the house has no windows installed for privacy considerations. For architectures mainly for scenery viewing purpose, they often have openings on the north and south sides while the four sided hall even has windows on all four sides with the French window installed on the south side, which ensures sound indoor lighting and ventilation. The way of constructing houses in Ming and Qing dynasty in Yangzhou did not remain the same as time passed. As an immigration city, the inclusiveness of Yangzhou is also represented in its architecture. For example, the blind window and blind door (Fig 9) in Yuxiu Tower of He Garden boldly adopted western elements which not only better adjusts temperature and lighting indoors (Fig 11) but also protects the privacy of family members.

4 Ecological design strategies in construction of residential architectures of Yangzhou in Ming and Qing dynasty

4.1 Arrangement of local plants

After long time of evolution, local plants, a representation of local features, have gained strong life force and are the fittest for local natural conditions. The local plants grown in the houses of Yangzhou in Ming and Qing dynasties include: Weeping willow, ginkgo, maple, magnolia, Koelreuteria, hackberry, Sophora japonica, elm, privet, Viburnum, osmanthus, loquat, crape myrtle, wintersweet, Chinese rose, peony, rochebruni, water lilies and varied kinds of bamboo, most of which are in sound growth. The strategy to grow local plants is not only economic and beneficial for their survival, but also the significant one in the ecological design of contemporary architectural landscape.

The vegetation system composed of trees, shrubs and grass constitutes a multi-layer ecologic community with the help of well-arranged architecture, waters, rockery and paths. The examples of vertical greening strongly promoted in contemporary ecological designs are present in houses of Yangzhou built in Ming and Qing dynasties. For example, the book loft of Ge Garden takes advantage of the climbing and adhesive properties of Parthenocissus tricuspidata by using it to ornament the outer wall with its dense green leaves. In summer, the flourishing leaves of Parthenocissus tricuspidata can shade the wall to cool down indoors (Fig 10 Book loft of Ge Garden). In winter, the concentrated branches and twigs of Parthenocissus tricuspidata after leaves falling could also form an insinuating layer to help the architecture keep the warmth inside.

4.2 Flexible drainage system as situation dictates

Traditional architecture in Yangzhou relies on natural drainage system. As rain fall from the roof to the ground paved with slates, bricks and pebbles, it seeps into the soil layer on its own, only contributing to the replenishment of underground water but also reducing the heat island effect and adjusts microclimate.

The surface of passage space such as fire preventive alleys is paved with bricks whose direction of arrangement is unparallel to walking direction, which somewhat prevent people from slipping. Moreover as the middle section of the path is higher than either part, it is conductive to rapid discharge of rainwater. For thunderstorms with large precipitation, natural seeping is not quick enough to discharge rainwater falling on the surface of the yard. In this situation, outfalls can be installed (Fig

DOI: 10.1051/matecconf/201711901057

11 Outfall). All those mentioned above can be applied to the construction of modern ecological residential area.

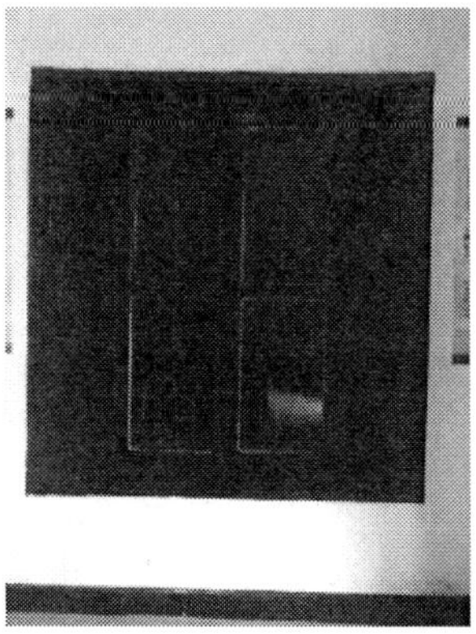

Figure 9. The blind window and blind door

Figure 10. Book loft of Ge Garden

Figure 11. Outfall

5 Conclusion

Residential architecture built in Ming and Qing Dynasties is a vital part of traditional culture of Yangzhou. The unique style grants it the important status in Chinese regional architecture. Under the guidance of philosophical view "nature and human being oneness", Yangzhou residence architecture creates refined living environment compliant with local climate, economic and technological conditions by learning from nature and smart arrangement of architecture elements. In terms of planning, construction and landscape design, it reasonably uses the resources available to create sound micro ecologic community.

In the modern residential environmental design, the method of separating the static and dynamic sections should be widely used, leading a series of space experience —— from pleasant shared space previously to groups before reaching the living area.Green shared space in residential areas can not only regulate the micro-climate, but also get dwellers relieved physically. Additionally, it creates a positive space for self-monitoring of the neighborhood resulting in ecological settlements.Modern constructing techniques and materials make it possible to prefabricate houses, whose structural system is parallel to the one of Yangzhou traditional architecture. In fact, the life-span will be extended as each component can be selected or replaced. Moreover, the cost of enclosure will be reduced leading to residential ecologicalization.It is an inevitable trend for ecological building to choose refined local materials, which can save the expenditure of transportation and maintenance and reflect the characteristics in context to arouse a sense of belonging. Such ecology is both in physical universe and on psychological level.

To salute and analysis the traditional architecture, the essence of ecological and livable concepts ought to be selected and merge with advanced technology to create a suitable living space for modern life, rather than adding symbolic elements to new buildings awkwardly.The summary of the experiences in all the above respects provides reference for contemporary ecological design and also helps promote residential architectural culture of Yangzhou.

References

1. B. Edwards, *Sustainable Architecture* (*Trans v*), 3-324 (Chinese architecture publishing house, Beijing, 2004)
2. C.H. Zhang, *Studies on the Development Path of Residence Architecture in Yangzhou Area*, **50** (Southeast University Publishing House, 2011)

3. J. Cheng and Z. Chen, *Garden Construction*, **60** (Chinese architecture Industry Publishing House, Beijing, 1988)
4. J. Yang and H. Du, *The Charm of Ge Garden*, **59** (Yangzhou Guangling Publishing House, 2006)
5. Lattice Door: *The use of lattice door dates back to the late Tang Dynasty. Its lower part is a solid water-proof board (apron) while the upper one is hollow for forming a light transmission part. And it replaced board door for its nice lighting performance.*

Designing anti-sliding durable type of asphalt mixture based on the maximum peak of angular distribution rate

Duan yi Wang, Yong tan Yu, Gang Wang[a], Chi Xu and Shen shen Shao

School of Civil Engineering and Transportation, South China University of Technology, Guangzhou 510641, China

Abstract. In order to design a anti-sliding and durable asphalt mixture, we independently develop two sets of laboratory equipment: the Laser Contour Detector of Asphalt Pavement and the Accelerated Rub Tester, with which we collect the plate surface contour every two hours for 5 times in the first 8 hours in the rub test, and then we make contour curves and analyze the angular distribution statistics of surface contour and thus we propose characterizing the skid resistance performance and durability of the asphalt pavement based on the maximum peak of angular distribution rate, which has been proved to be reasonable. By designing different gradations of SMA-13 and GAC-13 asphalt mixture and using the maximum peak of angular distribution rate to evaluate the skid resistance performance of different degrees of kneading, the asphalt mixture with better skid resistance has been made. That provides a reference for engineering application. The laboratory research develops a method for designing a anti-sliding and durable asphalt mixture based on the maximum peak of angular distribution rate, which makes up for the defects of the current stage where there is no ways to test the skid resistance performance and durability while designing the asphalt mixture.

1 Introduction

Most Chinese highways use asphalt pavement whose skid resistance performance is directly related to the safety of road traffic, and the skid resistance performance is an important index of road function, therefore the study on the skid resistance performance of asphalt pavement is very important. The problem of the skid resistance performance of asphalt pavement is serious, especially the rapid attenuation of the skid resistance durability. There are many factors that can influence the skid resistance durability, among them the asphalt mixture gradation design and pavement construction technology have the bigger impacts [1~5]. The average texture depth of the road would decay rapidly from 1.0mm to 0.6mm and the cornering ratio would also decay rapidly from 60~70 to 40~50 after carrying traffic for two years[6~7]. All those are not accord with the requirements of trunk road traffic safety and disobey the industry requirements. In order to solve those problems, this thesis developed two sets of experimental equipment Laser Contour Detector and Accelerated Rub Tester. Focusing on the factors like different degrees of roughness and different kneading time, we use the road profile peak angle to study the skid resistance and durability of the asphalt pavement[8~11], aiming to establish related quantitative evaluation way to evaluate the skid resistance performance and durability of asphalt mixture and also attempting to put forward new indexes to design the asphalt pavement with good skid resistance and durability.

[a] Corresponding author : 1873620714@qq.com

In order to adapt to the current unceasing skid resistance development situation of the big structure pavement, this thesis used statistics of the macro profile peak angle to study the macroscopic outline of asphalt pavement. Focusing on the roughness of different road surface and different knead time and combining with the asphalt pavement laser detector and acceleration load knead machine, it also established relevant quantitative evaluation method and attempted to put forward new indexes to evaluate skid resistance durability of the asphalt pavement.

2 Objective and scope of this study

(1) Two sets of independent research and development laboratory equipment: First: Asphalt Pavement Laser contour detector, Second: Accelerated rub tester.
(2) Proposed maximum peak angular distribution rate to characterize the asphalt pavement It skids durability. using the angular distribution of the maximum peak rate evaluation of skid resistance rub their different stages.
(3) Rub in the room combined with accelerated loading test machine, laser contour detector, forming a peak with maximum angular distribution rate Durable skid design of asphalt mixture method, make up this stage of asphalt mixture design phase does not validate its durability against sliding of defects.

3 Laboratory tests and data collection

3.1 Working principle introduction of the laser contour detector

The self-developed laser contour detector in our research is made up of three parts: measurement system, control system and data acquisition system. It could control measurement points within the measurement range of the asphalt pavement surface and store measuring data into computer in real time. It is as shown in figure 1~3.

Figure 1. Laser contour detector.

Figure 2. Laser contour detection plate specimen measured rut.

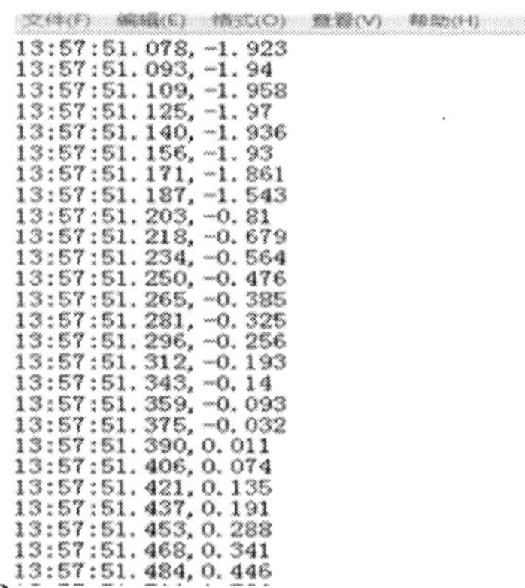

13:57:51.078, -1.923
13:57:51.093, -1.94
13:57:51.109, -1.958
13:57:51.125, -1.97
13:57:51.140, -1.936
13:57:51.156, -1.93
13:57:51.171, -1.861
13:57:51.187, -1.543
13:57:51.203, -0.81
13:57:51.218, -0.679
13:57:51.234, -0.564
13:57:51.250, -0.476
13:57:51.265, -0.385
13:57:51.281, -0.325
13:57:51.296, -0.256
13:57:51.312, -0.193
13:57:51.343, -0.14
13:57:51.359, -0.093
13:57:51.375, -0.032
13:57:51.390, 0.011
13:57:51.406, 0.074
13:57:51.421, 0.135
13:57:51.437, 0.191
13:57:51.453, 0.288
13:57:51.468, 0.341
13:57:51.484, 0.446

Figure 3. Measurement data.

3.2 Working principle introduction of the acceleration load kneading machine

When the tester is working, the driving screw steel bar will make the specimen platform in horizontal movement repeatedly with the effect of the new motor while the old motor will go on driving the traction wheels in vertical movement, so the tire could make back and forth movement on the test rut board. Besides, the thermostatic water bath equipment is added and its adjustable temperature range is from indoor temperature to 70° C. The lateral movement of the apparatus is 10cm/min; The wheels travel 42±1 times per minute. The wheel weighs 42-100kg(adjustable). At the same time, considering the real tire rolling on the road has the thread, we make thread on the smooth tire in order to imitate the actual driving wheel. It is shown as figure 4-a. This tester can be operated simply and fast and the simulation environment is effective.

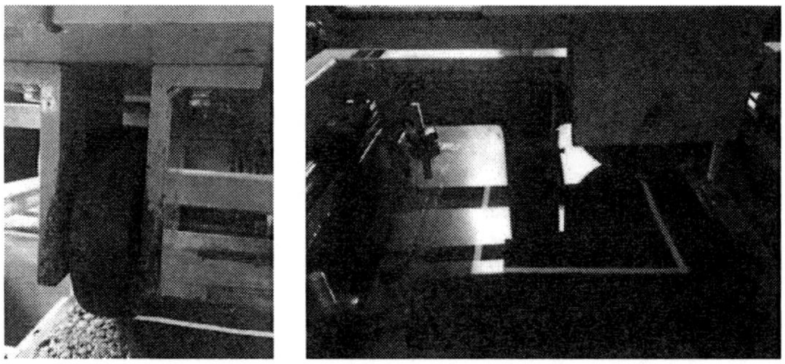

(a) (b)

Figure 4. Kneading machine and working principle.

In this research process, we use the self-developed accelerated rub tester as well the laser contour detector, and improve the gradation composition of AC-13 and its design method based on CAVF method and broken material gradation design theory. So we get the asphalt mixture whose skid resistance surface gradation is of discontinuous 4.75 mm sieve holes. For distinguishing from the traditional standard grading, we call the AC-13 of discontinuous gradation (4.75 mm) the skeleton dense type gradation GAC-13(Gap-graded Asphalt Concrete) whose grading curve is as shown in table 1. We compare the dense-suspended type of AC-13 with this kind of GAC-13, using two kinds of rut boards made by these two kinds of asphalt mixture. The size of these rut boards is 300×300×50mm which has 100 kg wheel load. After kneading the rut boards of GAC-13 and AC-13 under 60°C water environment in 0, 2, 4, 6, 8 hours, we dry out the rut boards and test them with the laser contour detector.

Table 1. Gradation of two asphalt samples（%）

Mixture	Sieve（Square Hole Sieve *mm*）									
	16	13.2	9.5	4.75	2.36	1.18	0.6	0.3	0.15	0.075
AC-13	100	95	78.5	51.5	35	25.5	18	13.5	10	6
GAC-13	100	95.2	62.5	26.9	22.1	16.4	12.1	9.2	7.3	6.3

To show their difference of roughness, we use the self-developed 3-D laser texture measuring equipment to simulate 3-D diagram of the pavement of the AC-13 and GAC-13, which is shown as figure 6. We also figure out the MTD of the rut board of the AC-13 and GAC-13 is 1.36mm and 0.85mm respectively. We use indoor sanding method to measure the initial skid resistance performance of these two rut boards molded indoor. The results show that texture depth of GAC-13 rut board is 1.2mm, while that of AC-13 is 0.92mm. Measured by tilting friction pendulum tester, the initial BPN of GAC-13 rut board is 78, while that of AC-13 is 70. It is quite clear that the initial skid resistance performance of GAC-13 rut board is better than AC-13 rut board. Therefore, we choose these two kinds of rut boards which have significant differences on initial texture depth and BPN as the simulation road in the experiment.

3.3 Principle of the profile peak angle calculation

The profile curve of the rut board surface could be got after denoising and filtering process of the date collected by the laser contour detector. Then, statistical analysis is carried out on peak angle value of the contour curve. The calculation of peak angle value α is as shown in figure 5 and formula (1):

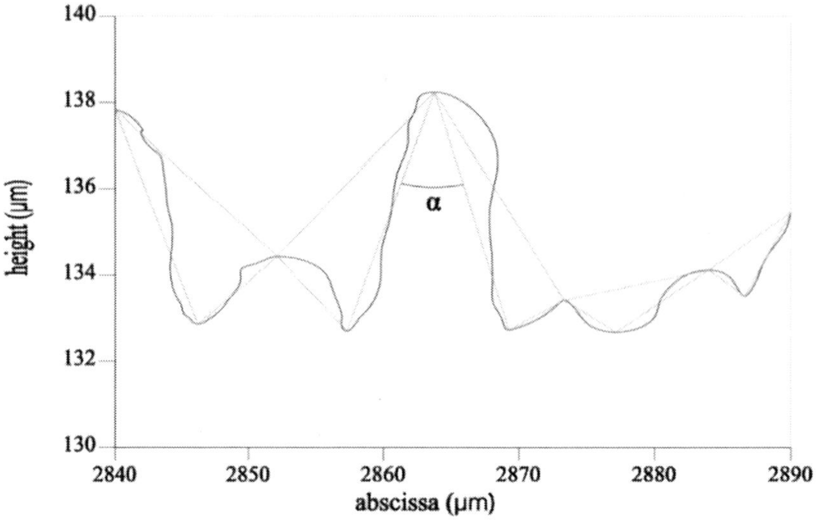

Figure 5. Schematic of outline peak angle.

$$\alpha = \tan^{-1}\left|\frac{x_i - x_{i-1}}{y_i - y_{i-1}}\right| + \tan^{-1}\left|\frac{x_{i+1} - x_i}{y_{i+1} - y_i}\right|, i = 1,2,3 \ldots n \qquad (1)$$

In this formula： x_i
is the value of the No.*i* extreme point on x axis. When the No.*i* point reaches to the summit, the No.*i-1* point and the No.*i+1* point are at the bottom; y_i is the value of No. *i* extreme point on y axis.
Each wave crest of the contour curve could be considered as an aggregate particle on the pavement, while the profile peak angle value represent the form of prominent aggregate particles. The smaller the angle value is, the narrower the aggregate particles are, while the bigger, the wider.

3.4 Data collection

After importing the measured data of laser contour detector into the MATLAB computer programs, we could got a TXT file (figure 7) by the statistical analysis. The specific operation is as shown in figure 6.

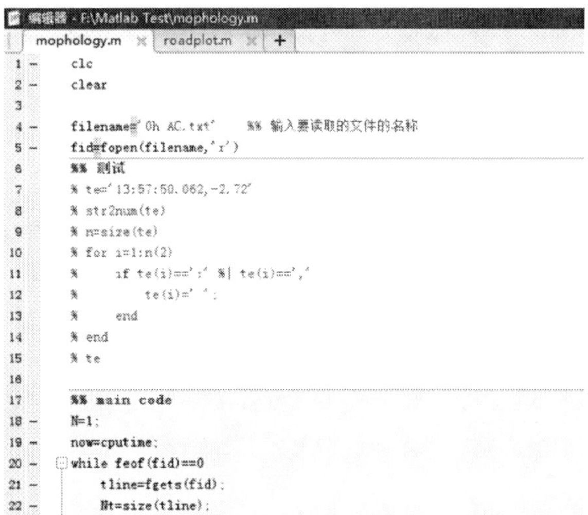

```
编辑器 - F:\Matlab Test\mophology.m
mophology.m  roadplot.m
clc
clear

filename='0h AC.txt'     %% 输入要读取的文件的名称
fid=fopen(filename,'r')
%% 测试
% te='13:57:50.062,-2.72'
% str2num(te)
% n=size(te)
% for i=1:n(2)
%     if te(i)==':' %| te(i)==','
%         te(i)=' ';
%     end
% end
% te

%% main code
N=1;
now=cputime;
while feof(fid)==0
    tline=fgets(fid);
    Nt=size(tline);
```

Figure 6. MATLAB data processing.

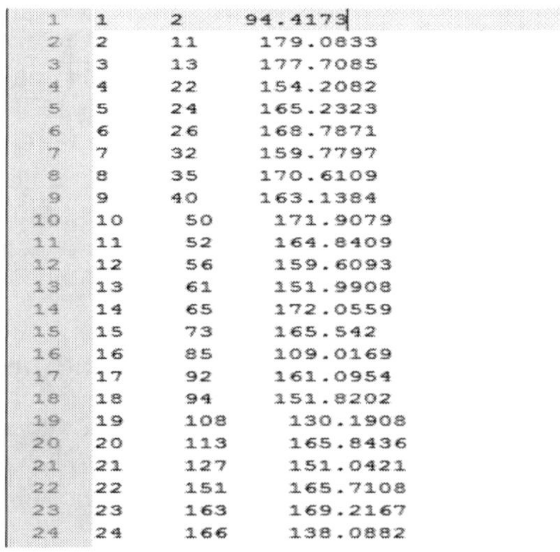

1	2	94.4173
2	11	179.0833
3	13	177.7085
4	22	154.2082
5	24	165.2323
6	26	168.7871
7	32	159.7797
8	35	170.6109
9	40	163.1384
10	50	171.9079
11	52	164.8409
12	56	159.6093
13	61	151.9908
14	65	172.0559
15	73	165.542
16	85	109.0169
17	92	161.0954
18	94	151.8202
19	108	130.1908
20	113	165.8436
21	127	151.0421
22	151	165.7108
23	163	169.2167
24	166	138.0882

Figure 7. Data output format.

According to the measurement scheme and test method introduced above, we measure the profile of the dried rut boards with laser contour detector after knead tests in every two hours, namely, in 0,2,4,6,8 hours and map the profile curve and calculate the angle α. The profile peak angle directly determines the macro profile of rut boards. In every test area, there are 1000 test lines and each of them has thousands of peaks, so in every test of each rut board there are thousands values of α. In this paper, the number of the profile peak angles in every 10 degrees are counted, which means there are 18 sections: 0-10°, 10-20°, ..., 170-180°. Every probability distribution in each area is calculated to illustrate the distributed situation of profile angles. The statistics of the surface profile angles of these two rut boards are as shown in figure 10 and 11. We take the largest peak angle range of 150 ~ 180 degrees as the analysis object of this study, and define it as the maximum peak of angle distribution rate. Table 2 is the rate of maximum peak angular distribution and the final rate of change.

3.5 Asphalt Mixture gradation design

Generally, the anti-sliding performance of SMA-13 and GAC-13 asphalt mixture with skeleton dense structure in the above layer is better than AC-13 asphalt mixture with dense-suspended structure. In this section, we use SMA-13 and GAC-13 gradation as the research object, trying to design the asphalt mixtures with different skeleton structure and evaluate the anti-sliding performance and durability of them with maximum peak of angular distribution rate so as to to select the best asphalt mixture of anti-sliding layer. In the paper written by Liwan Shi[12], through the research of contact force of the backbone chain in SMA-13 mixture, he found that the aggregate under 2.36mm almost didn't participate in the composition of main chain. Therefore, in the test of this section, we choose two kinds of aggregate of SMA-13 mixture which are 9.5mm and 4.75mm to study the best anti-sliding gradation. Considering GAC-13 has the same skeleton dense structure, we design it in the same way, and use AC-13 asphalt mixture with dense-suspended structure as comparison study.

The raw material of asphalt mixture is made by diabase and SBS modified asphalt mentioned in the former test. All the specimen is selected with the nominal maximum particle size of 13.2 mm; the SMA mineral powder dosage is 10.0% while that of GAC-13 and AC mixture is 4% respectively. SMA is made from lignin fiber, with 0.3% dosage of the mixture and mixed within the aggregate. The asphalt dosage of SMA is 6.0%, while that of GAC-13 and AC-13 is 4.8% each. According to the above design points, 3 pieces of rut board with each kind of gradation are made. We conduct knead test in 0, 2, 4, 6, 8 hours, carry out laser test in different kneading phase and write computer program MATLAB to make statistic analysis of maximum peak angular distribution rate. We use the average value of these three parallel tests as the final analysis object. The pass percentage of different gradation in different level is shown in table 2~3.

Table 2. SMA-13 quality pass percentage of different graduation in different level.

Grading type	Sieve (Square Hole Sieve *mm*) (%)									
	16	13.2	9.5	4.75	2.36	1.18	0.6	0.3	0.15	0.075
SMA-1	100	95.0	70.0	35.0	22.5	19.0	16.5	14.0	12.8	10.0
SMA-2	100	95.0	65.0	35.0	22.5	19.0	16.5	14.0	12.8	10.0
SMA-3	100	95.0	60.0	35.0	22.5	19.0	16.5	14.0	12.8	10.0
SMA-4	100	95.0	60.0	30.0	22.5	19.0	16.5	14.0	12.8	10.0
SMA-5	100	95.0	60.0	25.0	22.5	19.0	16.5	14.0	12.8	10.0
SMA-6	100	95.0	55.0	25.0	22.5	19.0	16.5	14.0	12.8	10.0
SMA-7	100	95.0	50.0	25.0	22.5	19.0	16.5	14.0	12.8	10.0
AC	100	95.0	78.5	51.5	35.0	25.5	18.0	13.5	10.0	6.0

Table 3. GAC-13 quality pass percentage of different graduation in different level.

Grading type	Sieve (Square Hole Sieve *mm*) (%)									
	16	13.2	9.5	4.75	2.36	1.18	0.6	0.3	0.15	0.075
GAC-1	100	95.2	70.0	35.0	22.1	16.4	12.1	9.2	7.3	6.3

GAC -2	100	95.2	65.0	35.0	22.1	16.4	12.1	9.2	7.3	6.3
GAC -3	100	95.2	60.0	35.0	22.1	16.4	12.1	9.2	7.3	6.3
GAC -4	100	95.2	60.0	30.0	22.1	16.4	12.1	9.2	7.3	6.3
GAC -5	100	95.2	60.0	25.0	22.1	16.4	12.1	9.2	7.3	6.3
GAC -6	100	95.2	55.0	25.0	22.1	16.4	12.1	9.2	7.3	6.3
GAC -7	100	95.2	50.0	25.0	22.1	16.4	12.1	9.2	7.3	6.3
AC	100	95.0	78.5	51.5	35.0	25.5	18.0	13.5	10.0	6.0

4 Test results and discussion

4.1 Skid resistance durability evaluation of the asphalt pavement based on the macro profile peak angle

The following table 4 is maximum peak of angle distribution rate and the final rate of change after different kneading time.

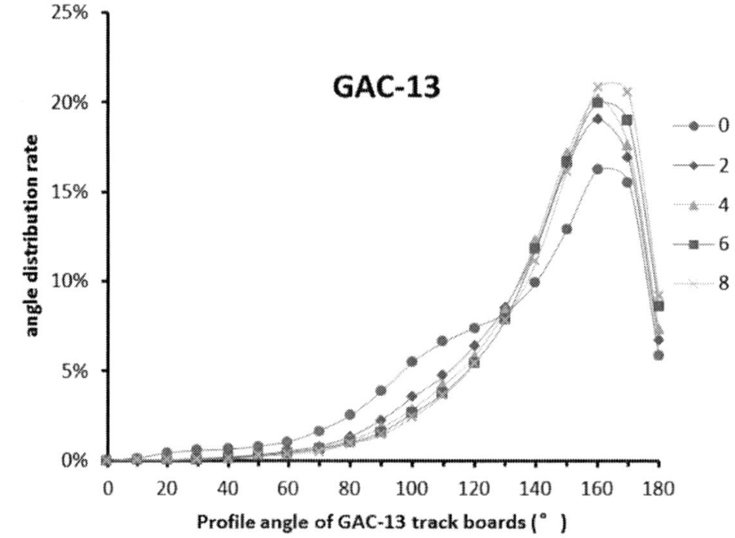

Figure 8. GAC-13 Road surface contour angle distribution chart.

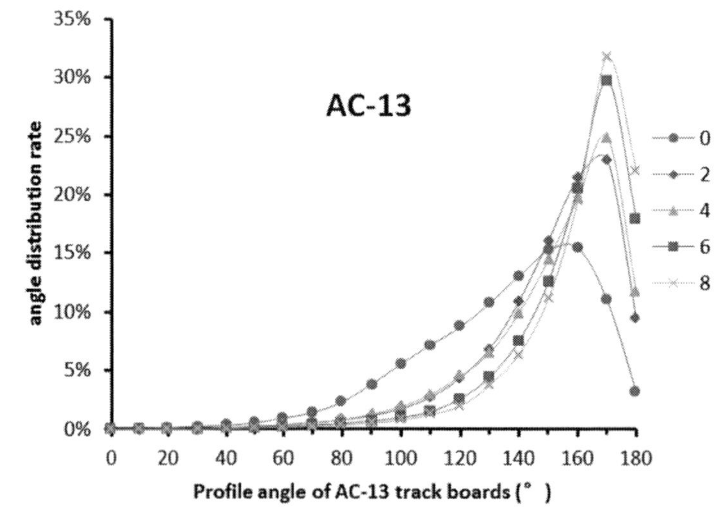

Figure 9. AC-13 Road surface contour angle distribution chart.

Table 4. Rate of maximum peak angle distribution and the final rate of change.

Grading Type	Evaluation Index	0h	2h	4h	6h	8h	Rate of change (%)
AC-13	Angle range	150-180	150-180	150-180	150-180	150-180	
	rate (%)	54.9%	70.0%	70.8%	80.8%	84.4%	34.98%
GAC-13	**Angle range**	150-180	150-180	150-180	150-180	150-180	23.43%
	rate (%)	50.6%	59.2%	61.7%	64.2%	66.0%	

After kneading the rut boards of AC-13 and GAC-13 in 0, 2, 4, 6, 8 hours, we dried them under the normal temperature and used sanding method and tilting friction pendulum tester to measure the MTD and BPN of the two rut boards whose variation trends are as shown in figure 10 to 11, and the correlation analysis is shown in figure 12.

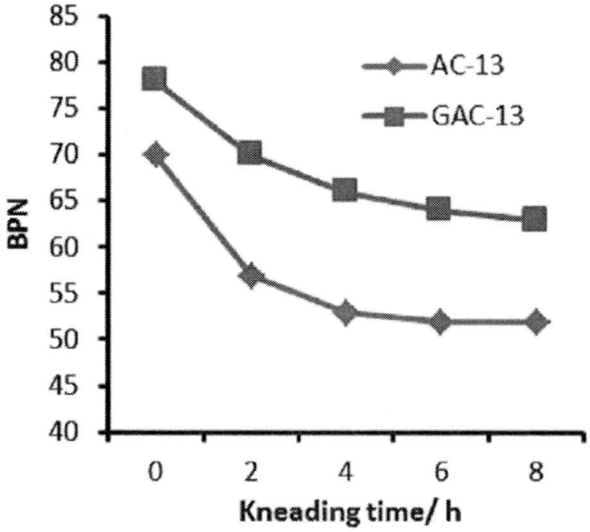

Figure 10. Texture Depth trends.

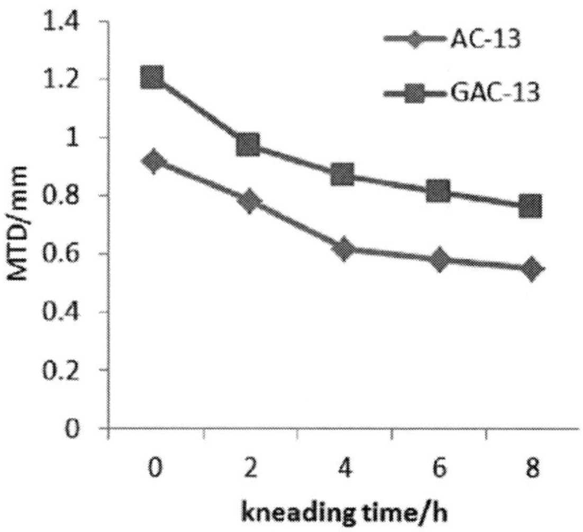

Figure 11. Pendulum value trends.

As we can see from figure 8 and figure 9, the analysis of profile peak angle value α shows good unimodality which is an obvious peak, and the profile peak angle value is basically in line with the gamma distribution. The result shows that the acute angle change is not obvious and the peak angle distribution ratio is small, and the obtuse angle would increase with kneading time going by, which means that the pavement profile angle is getting flat and its skid resistance performance is declining. Therefore, the skid resistance performance attenuation degree of GAC-13 is lower than that of AC-13. Besides, the initial and final skid resistance performance of GAC-13 is better than that of AC-13. In this contrast test, the MTD and BPN of the two rut boards are fading, which is to say the skid resistance performance is declining. The correlation between maximum peak angle distribution rate and BPN as well as MTD of GAC-13 is very good , and R^2 has achieved 0.99. The correlation between maximum peak angle distribution rate and BPN as well as MTD of AC-13 is good , and R^2 has exceeded 0.84. so it is a verification to the proposal that we can use maximum peak of angle distribution rate to represent pavement skid resistance performance. In summary, it's reasonable and accurate to use the maximum peak angle distribution rate to represent the skid resistance performance and durability of the pavement.

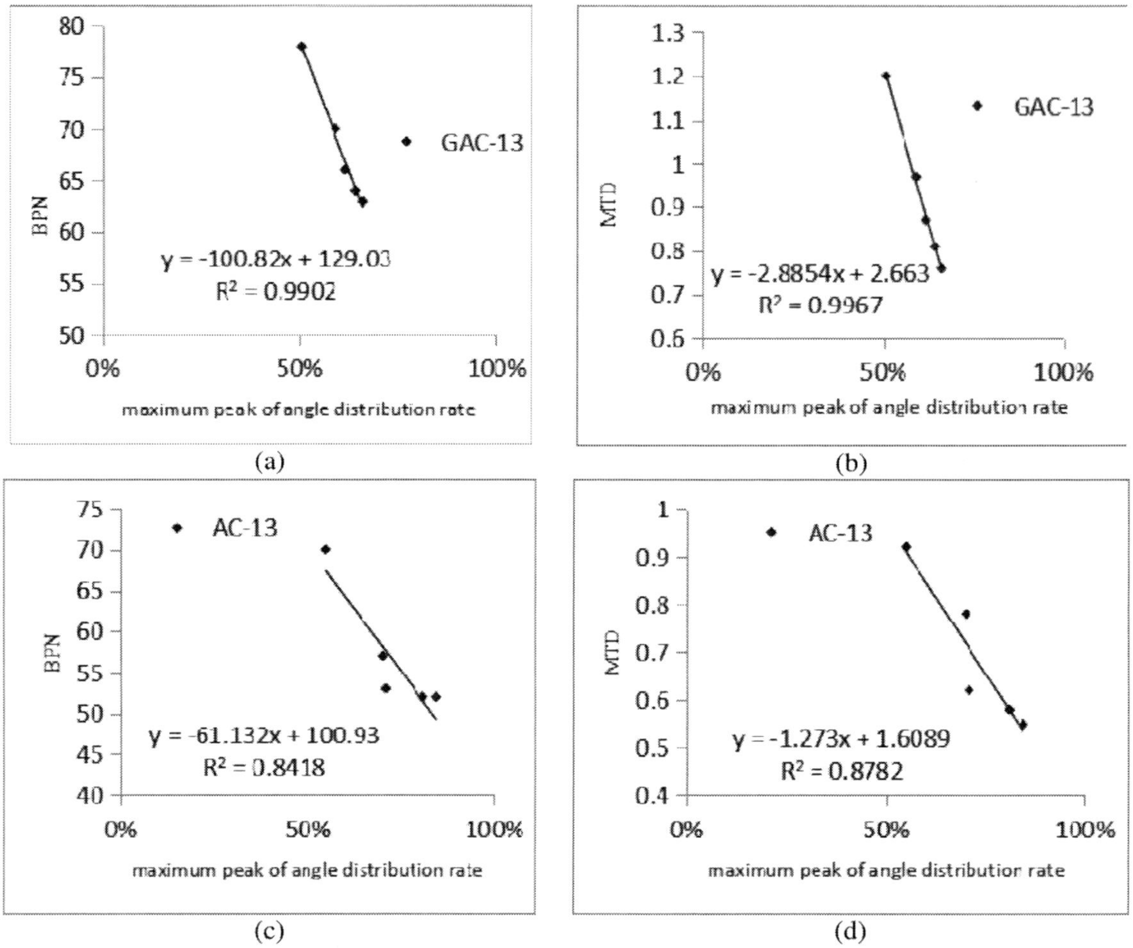

Figure 12. Correlation analysis of GAC-13 and AC-13.

4.2 Designing anti-sliding durable type of asphalt mixture based on the maximum peak of angular distribution rate

The anti-sliding performance and durability of the anti-skid layer of asphalt pavement (layer above) is directly related to the safety of the road traffic. But nowadays, in the asphalt mixture design phase, there has no special anti-sliding design and anti-sliding performance verification. This paper used the self-developed laser detector and acceleration load knead machine, and treated maximum peak of angular distribution rate as the anti-sliding performance evaluation index in the asphalt mixture design phase, so as to make up the absence of anti-sliding design and validation in anti-sliding layer asphalt mixture design phase nowadays.

Maximum peak of angular distribution rate in different stages of different gradations is shown in Table 5~6.

We can see from table 5: (1) The maximum peak of angular distribution rate of SMA is much lower than that of AC, no matter in the beginning or after 8 hours kneading time. It means that SMA has better anti-sliding performance and durability; (2) Comparing the increasing trend of maximum peak of angular distribution rate of SMA rutting plates of 7 different gradations , we can see that before rubbing phase, the maximum peak of angular distribution rate of number 1~3 rutting plate is gradually increasing, while the increase of that of number 5~7 rutting plate is not obvious. The pass percentage of 9.5mm and 4.75mm sieve holes in number 1~3 rutting plate is large, that is to say the aggregate is a bit over-fine. With the increase of coarse aggregate in 1~4 gradation, the initial and final maximum peak angular distribution rate have decreased, and the trend is obvious. Meanwhile, the maximum peak angular distribution rate of 1~4 gradation after 8 hours kneading time is still decreasing. In summary, the maximum peak of angular distribution rate of 1~4 gradation is decreasing with the enlargement of aggregate gradation, and the later phase of kneading test also has the same rule. So we can say the anti-sliding performance and durability of SMA 1~4 has increased with the enlargement of aggregate gradation while through the data statistics of 5~7 plate, we can see that even if the aggregate gradation has the same enlargement, the change of maximum peak of angular distribution rate is not obvious, and comparing with number 4, it has a increase trend which still exists after 8 hours kneading time. Therefore, increasing the mixture ratio of number 5~7 has no substantial contribution to anti-sliding performance and durability of the pavement, which, instead, would damage the anti-sliding performance and durability and make them reduce in the later period; (3) Therefore, we suggest using SMA-4 as the gradation of the anti-sliding layer of the pavement, and making proper adjustments of the pass percentage of 9.5mm and 4.75mm sieve holes for the pavement engineering, under the insurance of the effective anti-sliding performance and durability.

Table 5. Statistical table of maximum peak of angular distribution rate in different stages of SMA-13.

	Grading Type							
	SMA-1	**SMA-2**	**SMA-3**	**SMA-4**	**SMA-5**	**SMA-6**	**SMA-7**	**AC**
0 (h)	40.8%	41.6%	42.8%	41.5%	42.4%	42.9%	42.7%	54.9%
2 (h)	49.5%	48.5%	49.4%	49.1%	50.5%	51.5%	51.2%	70.0%
4 (h)	53.2%	52.9%	52.6%	51.5%	52.2%	53.2%	54.3%	70.8%
6 (h)	55.5%	54.5%	54.9%	53.2%	54.0%	54.5%	55.2%	80.8%
8 (h)	58.1%	57.9%	56.5%	55.8%	56.1%	56.8%	57.2%	84.4%

Table 6. Statistic table of the maximum peak of angular distribution rate of GAC-13 in different stages.

	Grading Type							
	GAC-1	**GAC-2**	**GAC-3**	**GAC-4**	**GAC-5**	**GAC-6**	**GAC-7**	**AC**
0 (h)	48.9%	49.6%	50.8%	50.6%	51.3%	51.5%	51.7%	54.9%
2 (h)	58.3%	59.2%	60.3%	59.2%	58.5%	59.3%	60.1%	70.0%
4 (h)	61.1%	61.9%	62.6%	61.7%	62.5%	61.0%	62.0%	70.8%
6 (h)	65.1%	65.6%	65.4%	64.2%	65.0%	66.4%	65.5%	80.8%
8 (h)	67.9%	67.1%	66.5%	66.0%	66.4%	67.1%	67.2%	84.4%

We can see from Table 6: (1) The maximum peak of angular distribution rate of GAC is much lower than that of AC, no matter in the beginning or after 8 hours kneading time. It means that GAC has better anti-sliding performance and durability as a anti-skid layer; (2) Comparing the increasing trend of the maximum peak of angular distribution rate of 7 different gradations of GAC rutting plates, we can see that before rubbing phase, the maximum peak of angular distribution rate of number 1~3 rutting plate is gradually increasing, while the increase of that of number 4~7 rutting plate is not obvious. The pass percentage of 9.5mm and 4.75mm sieve holes in number 1~3 rutting plate is large, that is to say the aggregate is a bit over-fine. With the increase of coarse aggregate in 1~4 gradation, the initial and final maximum peak of angular distribution rate have decreased, and the trend is obvious. Meanwhile, the maximum peak of angular distribution rate of 1~4 gradation after 8 hours kneading time is still decreasing. In summary, the maximum peak of angular distribution rate of 1~4 gradation is decreasing with the enlargement of aggregate gradation, and this rule also works with the later phase of kneading test. So we can say the anti-sliding performance and durability of GAC 1~4 has increased with the enlargement of aggregate gradation while through the data statistics of number 5~7 mixture, we can see that even if the aggregate gradation has the same enlargement, the change of maximum peak of angular distribution rate is not obvious, and comparing with number 4, it has a increase trend which still exists after 8 hours kneading time. Therefore, increasing the mixture ratio of number 5~7 has no substantial contribution to anti-sliding performance and durability of the pavement, which, instead, would damage the anti-sliding performance and durability and make them reduce in the later period; (3) Therefore, we suggest using GAC-4 as the gradation of the anti-sliding layer of the pavement, and making proper adjustments of the pass percentage of 9.5mm and 4.75mm sieve holes for the pavement engineering, under the insurance of the effective anti-sliding performance and durability; (4) In terms of GAC-13 and SMA-13 pavement, the pass percentage of 9.75 mm sieve holes is changed in number 1~3 mixture; the pass percentage of 4.75 mm sieve holes is changed in number 3~5 mixture and that of 9.75 mm sieve holes is changed in number 5~7 mixture. In this way, we continuously thickening the mixture gradation, and find that anti-sliding performance and durability of the pavement changes significantly when the pass percentage 4.75 mm sieve holes is changed, which means that reasonably controlling the pass percentage of 4.75 mm sieve holes is the key to the anti-sliding performance design of asphalt mixture; (5) To sum up, the anti-sliding performance: SMA>GAC>AC, and SMA-4 and GAC-4 have excellent anti-sliding performance and durability, which is commended as anti-sliding layer of the pavement. SMA-4 is more preferable when the conditions permit.

5 Conclusion

(1) Based on laser contour detection technology and self-developed accelerated rub tester, and a lot of data analysis, this paper propose that the maximum peak angular distribution rate could represent the anti-sliding performance and durability of asphalt pavement. Meantime, the correlation analysis between BPN and texture depth shows that they have a significant relation. It is sufficiently accurate to use maximum peak angular distribution rate to evaluate the anti-sliding performance and durability of asphalt pavement. It remedies the defect that the measure of the BPN and texture depth has being affected by uncontrollable factors like human and external environment.
(2) We use the maximum peak angular distribution rate to represent the skid resistance performance and durability and design different graduations of anti-sliding pavement layer, especially changing the pass rate of two gradations of sieve holes of 9.5mm and 4.75mm. A conclusion can be made that the on the aspect of anti-sliding performance: SMA>GAC>AC, and SMA-4 and GAC-4 have excellent anti-sliding performance and durability, which is recommended as anti-sliding layer of the pavement. SMA-4 is more preferable when the conditions permit. Meanwhile, with proper adjustments, two gradations of sieve holes of 9.5mm and 4.75mm are used in the pavement engineering.
(3) According to the laser contour detection test and method for evaluating anti-sliding performance of the pavement, we form a set of laboratory system, which is able to design and verify the anti-sliding performance and durability of the asphalt mixture in the design stage, which provides guidance for the engineering application.

References

1. J. W. Hall, K. L. Smith, and L. Titus-Glover, *NCHRP*: *guide for pavement friction*, Washington DC: Transportation Research Record (2009)
2. *Guidelines for the management of road surface skid resistance*, Sydney: Austroads Incorporated (2005)
3. E. Farber, M. S. Janoff, S. Cristinzio, J. G. Blubaugh, W. Reisener, and W. Dunning, *Determining pavement skid-resistance requirements at intersections and braking sites*, Washington DC: National Cooperative Highway Research Program, 154 (2004)
4. T. R. Nesbitt and D. J. Barron, Prediction of driving traction performance on snow, SAE Technical Paper (1980)
5. T. R. King and F. E. Matyja, *Tread design effect on winter traction*, PA: Society of Automotive Engineers (1981)

6. P. Cao and X. P. Yan, Theory analyses of effect of topography on skid resistance of asphalt pavements, Tribology, **29** (4), 306-310 (2009)
7. Z. L. Zhao, Z. Q. Zhang, and C. S. Hu, Influence of gradation on anti-skidding performance of asphalt pavement, Journal of Chang'an University, **25** (1), 6-9 (2005)
8. Y. Zhang, Research of gradation optimization on anti-sliding layer of asphalt pavement, Chang Chun: Ji Lin University (2007)
9. X. S. Du, *Study of pavement anti-skid performance based on tire and road interaction*, Harbin: Harbin Institute of Technology (2012)
10. A. Q. Zhang and Z. Y. Ao, Tire contact pressure distribution and test methods, China Rubber Industry, **48** (6), 368-374 (2001)
11. M. D. Beer, L. Kannemeyer, and C. Fisher, Towards improved mechanistic design of thin asphalt layer surfacings base on actual tyre/ pavement contact stress-in motion(SIM) data in south Africa, *Proceedings of 7th Conference on asphalt pavements for Southern Africa* (1999)
12. L. W. Shi, Based on the functional properties of micromechanics asphalt research, Guangzhou: South China University of Technology (2014)

Selective attrition in life satisfaction among elderly people: the harmonisation of longitudinal data

Chih-Ping Li[a]

Department of Health Industry Management, Kainan University, Taoyuan 33857, Taiwan

Abstract. Methodological research of the determinants of attrition makes it possible to identify which predictors of elderly people are most likely to be lost to follow-up in longitudinal studies. The aim of the study is to explore and compare the attrition in longitudinal aging population studies in the UK and Taiwan. Data draw from the *Nottingham Longitudinal Study of Activity and Aging* (NLSAA) and the *Survey of Health and Living Status of the Elderly in Taiwan* (SHLSET). Results from bivariate analyses indicate that no statistically significant association was found from logistic regression analyses between attrition and other factors in the second (1989) and third (1993) wave of the NLSAA survey. However, logistic regression did reveal an association between attrition and having lived with someone else at baseline in the SHLSET. Thus, living with others has effect on attrition and is related to re-interview in follow-up study.

1 Introduction

This research is a cross-national comparison of longitudinal studies, designed to compare the attrition, to identify the physical, mental, social, environmental factors contributing to attrition in old age.

Attrition refers to the loss of respondents from a survey [1]. According to Matthews et al. (2004, p. 2), "Attrition can be defined as the loss of relevant individuals occurring after definition of the population to be included in a study. [2]" There are two major types of subject attrition. One is selective attrition, which is described as the tendency of certain individuals in a study to be more likely to drop out of a study than others, because of some inherent characteristic; e.g., they choose no longer to participate because they do not have the disease of interest, and therefore are less interested in the study; or they move, because they are among the healthy people in the study and are still employed. The other major form of attrition is non-selective attrition, which is when people cannot participate because of some relatively random event, or some event deemed unrelated to the study outcomes of interest, like death in a car accident. In general, selective attrition is non-random and introduces statistical bias because the sample is no longer representative of the population and can detract from the internal and external validity of longitudinal study [3-5]. Non-selective attrition occurs randomly in a sample and in the population. Among the two types of attrition, mortality is considered to be non-selective attrition, and is unavoidable among older participants in longitudinal studies [6].

[a] Corresponding author : luciali@mail.knu.edu.tw

Subject attrition is an important issue in longitudinal studies, because altering the composition of the initial randomly selected sample may introduce biases and adversely affect how representative the sample is of the population of interest. This is particularly a problem in gerontology, as older people may die, or become unable to participate in ongoing surveys due to a variety of age-related diseases and disabilities, e.g., Alzheimer's disease or cardiovascular disease or cerebral vascular disease [7]. Therefore, the longitudinal patterns of non-response were quite complex among older people.

Factors associated with attrition in longitudinal surveys have been investigated in a number of studies [7, 8]. For example, Slymen et al. showed that attrition is associated with older age, lower income, and poorer health. However, much less attention has been paid to examining the impact of attrition on specific research outcomes [2]. It raises the question as to whether the results from responses gathered from participants who are re-interviewed in follow-up surveys, after a certain amount of attrition, can be generalized to the initial population of older people from which the study sample originally was drawn (in this instance, in the UK and Taiwan), and how serious a problem this is. Therefore, the aim of this study examines the effect of selective attrition from baseline to follow-up on the outcomes among the older people' studies, provides a better understanding on how the occurrence and characteristics of attrition may affect life satisfaction assessments in older people, and compared attrition among older people in the UK and Taiwan.

2 Methods

2.1 Participants and ethics

Data were derived from the *Nottingham Longitudinal Study of Activity and Aging* (NLSAA) [9] and the *Survey of Health and Living Status of the Elderly in Taiwan* (SHLSET) [10].

The NLSAA was based on a 1985 survey that had 1,042 participants aged 65 years and over (i.e., born before 30th September, 1920). Follow-up surveys were conducted on those who still were alive and living in Nottingham in 1989 and 1993. In 1989, 690 subjects were re-interviewed (response rate= 88%). In 1993, 410 took part in the interviews, to yield a response rate of 72% [9].

Data were extracted from SHLSET and provided by the Bureau of Health Promotion, Department of Health, Taiwan. This project was conducted jointly by the Taiwan Provincial Institute of Family Planning, in Taichung, and the Population Studies Centre (PSC) and Institute of Gerontology at the University of Michigan, and founded by the US National Institute on Aging (NIA), in 1989, 1993, 1996, and 1999 [10].SHLSET involved a nationally representative cohort of 4,049 older people, initially 60 years old and older (i.e., born before 31^{sth} December 1928) during a baseline survey conducted in 1989. Of this number, 3,155 subjects were re-interviewed in 1993 with a response rate of 91%.

The participants in the NLSAA were born in or before 1920, and the subjects in SHLSET born in or before 1928. Therefore, the sample for comparative study contained all participants born before 1920 who were interviewed in 1989/1993. To identify individuals born before 30 April 1920, an algorithm was used to calculate and create an exact age, and those participants born before 15th April 1920 for the Taiwan-based sample who were interviewed in 1989 an 1993. The 15th April date was used because the exact date of birth was not recorded in SHLSET. The final sample contained 1,438 subjects (35.5%, n=4,049) in 1989 and 1,003 (31.8%, n=3,155) in 1993. An additional 410 participants (10.1%, n=4,049) in 1989 and 299 participants (9.5%, n=3,155) in 1993 were dropped due to missing data relating to date of birth and reported age.

2.2 Instruments

The data collection instrument for the NLSAA was a structured interview questionnaire, composed of 161 questions (including 314 items of data) the addressed demographics, mobility, physical health, psychological health (cognitive status, depression, and morale), social functioning, use of health and personal social services, customary physical activity, and physical measurements [9, 11].

The SHLSET questionnaire was composed of eight sections including marital history and other background characteristics; household schedule, social economic exchanges; health, health care utilization and behaviours; occupational history; activities and general attitudes; residence history; economics/financial well-being; and emotional and instrumental support [10].

The NLSAA and the SHLSET both used shortened versions of the Life Satisfaction Index, a commonly-used measure of well-being and morale in studies of older people [12], adapted for local use, such that the NLSAA used the 13-item LSI-Z scale [13]and the SHLSET used the 10-item LSI scale [14]. For comparative purposes, a common scoring system for each response category and question across the two studies was adopted and the scores for the two scales were standardised to a 100-point scale.

Other variables from the two studies were compared and variables with similar questions and response categories were harmonised using established methods [13, 15]to create new variables covering demographics and socioeconomics, physical and mental health and social activity: further details of which are available elsewhere [16].

2.3 Data analyses

To understand as clearly as possible the differences between participants who continued to accept being interviewed and those who did not, it was imperative to distinguish between the different types of attrition. To assess attrition effects, Goodman and Blum [17]suggested using multiple logistic regression to determine whether attrition influences the sample based upon one or more variables. This is because logistic regression models generally deal with binomial dependent variables, assigning a value of '0' to one outcome (e.g., not re-interviewed) and a value of '1' to the other (e.g., re-interviewed). In this study, Mann-Whitney U tests and chi-square analysis was used as a first step to compare those who were re-interviewed versus those who were not re-interviewed with respect to a variety of continuous variables. Then, as a second step, chi-squared tests were used to compare those who were re-interviewed versus those who were not re-interviewed among categorical variables in the NLSAA and the SHLSET data sets. Finally, logistic regression analysis was performed to assess the relative contributions of, and hence the likelihood of bias caused by, variables deemed different between the two groups (re-interviewed versus not) on selective attrition as it existed at four-year and eight-year follow-up for the NLSAA study (1989 versus 1985; and 1993 versus 1985), and four-year follow-up for the SHLSET study (1993 versus 1989).

3 Results

The mean age of the NLSAA sample was 78 (SD=5.78), ranging from 69 to 98 in 1989 and 81 (SD=5.17), ranging from 73 to 101 in 1993. The majority of participants were female (62.3% in 1989, 66.1% in 1993). Almost half (47.4% in 1989, 50.6% in 1993) were widowed, and 42.7% (1989) and 39.8% (1993) were married. The majority of participants reported in good health (41.5% in 1989, 38.6% in 1993) and average (19.4% in 1989, 21.7% in 1993).

The sample in SHLSET, the mean age was 76 (SD=5.42), ranging from 69 to 98 in 1989 and 79 (SD=4.96), ranging from 73 to 99 in 1993. Over half of the participants were male (53.4%) in 1989 and 52.9% in 1993. The majority of the participants were married (51.3% in 1989, 56.1% in 1993) and about half of sample was widowed (42.6%) in 1989 but 38.9% in 1993. About half of the sample reported in average health (40.7% in 1989, 36% in 1993) and approximately 20% said they were in fair and good health in 1989 and 1993.

3.1 Exploded from Mann-Whitney U tests for continuous variables

Table 1 shows the results of Mann-Whitney U tests to determine whether there were significant differences in 1985 between those who were alive and re-interviewed in 1989 and those who were alive but not re-interviewed in 1989 in the NLSAA study. There was no significant difference in the mean rank for age (Z statistic=-0.511; p=0.609), standardised SAD score (Z statistic=-1.58; p=0.113), standardised LSI score (Z statistic=-0.905; p=0.366), or number of people living in the household (Z statistic=-0.737; p=0.461) in 1985 between survivors interviewed versus survivors not re-interviewed in 1989.

There were significant differences in 1985 between those who were alive and re-interviewed in 1993 and those who were alive but not re-interviewed in 1993 in the NLSAA study.There was an association between whether 1993 survivors were interviewed or not and age (Z=-2.692; p=0.007) and standardised LSI score (Z=-2.488; p=0.013) in 1985. Older survivors were less likely to be re-interviewed, as were those with a lower baseline standardised LSI score (Table 1).

Table 1. Mann-Whitney U tests comparing attrition in continuous variables in the NLSAA.

Variable	Alive not interviewed Mean Rank (N)	Alive interviewed Mean Rank (N)	Z statistic	*p* value
From 1985 to 1989				
Age	364.78 (65)	379.25 (690)	-0.511	0.609
Standardised SAD score	406.01 (62)	363.40 (671)	-1.585	0.113
Standardised LSI score	345.65 (62)	371.15 (675)	-0.905	0.366
Number of people living in household	359.04 (65)	377.61 (686)	-0.737	0.461
From 1985 to 1993				
Age	306.29 (139)	264.39 (410)	-2.692	0.007
Standardised SAD score	278.35 (129)	264.04 (405)	-0.962	0.336
Standardised LSI score	240.04 (130)	278.89 (408)	-2.488	0.013
Number of people living in household	253.96 (137)	280.70 (410)	-1.934	0.053

Table 2 presents the mean rank for the continuous variables and the results of the Mann-Whitney U tests for the SHLSET data in 1989. The only statistically significant difference between 1993 survivors interviewed and those not interviewed related to the number of people living in the household in 1989 (Z=-6.033; p<0.001) in 1989. People in 1989 who had more people living in the household were more likely to be re-interviewed in 1993.

Table 2. Mann-Whitney U tests comparing attrition in continuous variables in SHLSET, 1989-1993.

Variable	Alive not interviewed Mean Rank (N)	Alive interviewed Mean Rank (N)	Z statistic	*p* value
Age	516.23 (99)	553.89 (1001)	-1.125	0.260
Standardised SAD score	538.37 (92)	526.46 (962)	-0.360	0.719
Standardised LSI score	486.02 (93)	536.41 (970)	-1.526	0.127
Number of people living in household	367.72 (99)	568.05 (1000)	-6.033	<0.001

3.2 Exploded from Chi-square tests for categorical variables

Separate Chi-square tests were used to test each null hypothesis that there was no association between participants who were alive and not interviewed and participants who were alive and interviewed from the baseline interviewing date.

Table 3 showed that Chi-square tests compared baseline (1985) categorical variables between people alive and re-interviewed in 1989 versus those alive but not re-interviewed in the NLSAA study. There was an association between whether 1989 survivors were re-interviewed or not and self-rated health (χ^2=4.967; p=0.026) in 1985. Survivors in 1989 who were re-interviewed were more likely than their non-re-interviewed counterparts to have reported good (49.6%, n=335) or excellent (16.7%, n=113) health in 1985, and less likely to have rated their health as poor (11.1%, n=7), fair (12.7%, n=8), or average (22.2%, n=14).

Table 3. Prediction of attrition among elderly people in the NLSAA study.

Study	Variable (reference category)	Category	Odds ratio	95% CI	*p* value
NLSAA	Self-rated health				0.043
1985-1989	(Excellent)	Poor	0.221	0.071–0.687	0.009
		Fair	0.759	0.266–2.168	0.606
		Average	0.460	0.179–1.185	0.108
		Good	0.769	0.326–1.813	0.548
NLSAA	Age		0.972	0.934–1.010	0.150
1985-1993	Marital status				0.941
	(Married)	Single	0.724	0.258–2.037	0.541
		Widowed	0.904	0.413–1.978	0.801
		Separated/Divorced	0.872	0.245–3.106	0.832
	Living status (Live alone)	Live with someone	1.427	0.659–3.090	0.367
	Comparison with peer group health				0.219
	(More healthy)	Less healthy	0.564	0.265–1.202	0.138
		About as healthy	0.796	0.395–1.604	0.523
	Reading newspaper or journals (No)	Yes	1.613	0.723–3.600	0.243
	Attending religious group (No)	Yes	1.630	0.989–2.688	0.055
	Attending clubs (No)	Yes	1.378	0.879–2.162	0.162
	Standardized LSI score		1.008	0.968–1.050	0.685

The results of Chi-square tests comparing baseline categorical variables between people who were alive and re-interviewed in 1993 and people who were alive but not re-interviewed in 1993 in the NLSAA study. There was a statistical association between being re-interviewed among 1993 survivors, versus not, and the following 1985 variables: marital status (χ^2=5.040; p=0.025), living status (χ^2=4.070; p=0.044), comparison with peer group health (χ^2=5.842; p=0.016), read newspapers or journals (χ^2=4.980; p=0.026), attending religious group (χ^2=4.618; p=0.032), and participation in a club or organization (χ^2=3.975; p=0.046) in the NLSAA study, although one or more of these differences may have been the result of Type I error (detecting a difference that does not truly exist) because of the large number of tests done. However, to err on the side of caution all these variables were included in the regression model.

Survivors in 1993 who were re-interviewed were more likely to have been married in 1985 than those not re-interviewed. Those living with someone else in 1985 were more likely to be interviewed in 1993 than those living alone in 1985. Those who reported being about as healthy as their peers in 1985 were more likely to be interviewed in 1993 than people who felt more or less healthy than their peers in 1985. Participants who were reading newspapers and/or journals, and attending a religious group, and/or other club or organization were more likely to be re-interviewed in 1993 than those not re-interviewed.

The results of chi-square tests comparing baseline (1989) categorical variables between people who were alive and re-interviewed in 1993 and those who were alive but not re-interviewed in the SHLSET study. Significant differences were apparent for marital status (χ^2=7.542; p=0.006), socioeconomic class (χ^2=6.884; p=0.009), comparison with peer group health (χ^2=4.791; p=0.029), and reading a newspaper or journal (χ^2=9.796; p=0.002).

Those re-interviewed in 1993 were more likely to have been married, be semiskilled/unskilled /others, and read newspapers or journals than survivors who were not re-interviewed in 1993. People who reported in 1989 that they felt about as healthy as their peers were more likely to participate in the 1993 re-interview than those who felt they were more and less healthy. Those who read a newspaper or journal in 1989 were more likely to be re-interviewed in 1993 than people who were not reading newspaper or journals.

Table 4 shows the results of the logistic regression models for baseline (1989) SHLSET variables to determine if any baseline variables predicted whether survivors were re-interviewed in 1993 or not. In the SHLSET study, the overall model (χ^2=36.518; df=9; p<0.001) with all independent variables entered, was significant. There was a significant association between 1993 re-interview and the number of people living in the household in 1989 (p<0.001), in that those who lived with others in 1989 were more likely to be re-interviewed in 1993 (OR=1.232; 95% CI=1.108–1.370). In brief, living with someone else in 1989-1993 was a significant predictor of a 1993 re-interview.

Table 4. Prediction of attrition among elderly people in the SHLSET study.

Study	Variable (reference category)	Category	Odds ratio	95% CI	*p* value
SHLSET	Marital status				0.066
1989-1993	(Married)	Single	1.165	0.241–5.642	0.849
		Widowed	0.481	0.278–0.834	0.009
		Separated/Divorced	0.798	0.210–3.035	0.740
	No. of people living in household		1.232	1.108–1.370	<0.001
	Socioeconomic class				0.074
	(Professional/intermediate)	Skilled/manual & non-manual	1.575	0.717–3.460	0.258
		Semiskilled/unskilled/others	2.309	1.121–4.757	0.023
	Comparison with peer group health				0.468
	(More healthy)	Less healthy	1.579	0.718–3.473	0.256
		About as healthy	1.013	0.583–1.759	0.963
	Reading newspaper or journals (No)	Yes	0.787	0.426–1.456	0.446

4 Discussion

In any analysis of longitudinally-collected data, it is crucial to look for sources of bias that might result from subject attrition, and it is important to discriminate between mortality and non-mortality-related attrition, such as non-response or moving away. A major challenge in studies of older people is that death is a common source of attrition. In this study, there was substantial mortality-related attrition from 1985 to 1989 and from 1985 to 1993 in the NLSAA sample; and from 1989 to 1993 in the SHLSET sample. However, mortality-related attrition occurred in both samples, as well as in the population from which the samples came, and seems unlikely to have led to significant bias.

Attrition is always a concern in any longitudinal study, as it was in both of the studies included in the present research. That attrition occurred was never a question, it occurs in all large longitudinal studies. Rather, the question was, does attrition significantly affect the comparison between the UK and Taiwanese studies examined here? Mann-Whitney U tests showed that there were no significant differences between those who survived and were re-interviewed and those who survived but were not re-interviewed in the NLSAA study follow-up survey that transpired in 1989. However, age and level of life satisfaction were different between those who were re-interviewed and not re-interviewed in the 1993 NLSAA follow-up survey; and in the SHLSET study, older people who lived with someone in 1989 were more likely to be re-interviewed in 1993 than those who had lived alone.

Chi-square analysis yielded somewhat different results, in that significant differences were evident in the proportion rating their health as good relative to peers between those who were re-interviewed and those who were not in the 1989 NLSAA survey. Even more baseline differences were found in terms of being re-interviewed for NLSAA in 1993, with a statistically significant effect of attrition on marital status, living with others, reporting oneself as healthy relative to peers, reading newspapers or journals, and attending religious groups, clubs or organisations.

No statistically significant association was found from logistic regression analyses between attrition and other factors in the second (1989) and third (1993) wave of the NLSAA survey. However, logistic regression did reveal an association between attrition and having lived with someone else at baseline in the SHLSET. The attrition showed its effect on 'living with some people' among older people in Taiwan. It may reflect the outcome of an individual's living status to affect whether participant accepts an interview or not. Older people live with someone that may be willing to accept interview because someone else could help and answer the questions for them. Older Taiwanese might accept an interview by proxy due to health problems, lower levels of education or other reasons. Thus, the use of a proxy in ageing research is a challenge for future research.

With regard to non-mortality-related attrition, refusal can be problematic, if those who refuse to participate further in a study are systematically different from those who continue to participate; and this is commonly the case. However, in these analyses of the effect of attrition, it seemed to have a relatively minor effect in terms of the overall difference between the British and Taiwanese samples.

The research in this paper had several limitations. First of all, a limitation was that the data were old, collected in 1985, 1989 and 1993, and the situation may have changed in both countries since that time. However, individuals 65 years and over comprise a more diverse group than those in any other age group. The reason for this variability derives from the variety of experiences older people have had over time, and their broad range of health status and/or functional ability. For example, an 85-year old might be frail and housebound because older people themselves as age their health changes, versus a 65-year old who might still being highly active, or even still working. However, these data still have their value, and can help us to understand the complex conditions or status of older people at the time the data were collected.

Another limitation, as with most research of this kind, is that it is likely that certain participant characteristics, related to both dependent and independent variables (so called 'confounders'), were not examined in this research. It must be kept in mind that the presence of such confounding variables - known or unknown - could explain at least part of the relationships detected in this research. For example, it was not possible to assess certain disease states. It is possible some disease states that were not examined (e.g., alcohol abuse) could be related to quality of life and a variety of other factors; and not examining these conditions may limit the extent to which these findings can be interpreted.

Finally, the selected predictors of attribution might not explain the causes of attrition. Given the lack of knowledge regarding which factors of attrition are most important, it is possible that potentially important components were missing in this research; for instance, social relationships, the environment, fear of crime, and so on. This research was designed to examine attrition among older people in the UK and Taiwan. As a result, the questions associated with attrition did not seem complete. Limitations arise because the studies might fail to recognize variations between countries in the importance of attrition. The lack of comparable data (e.g., diseases, social support), often restricts international research as it attempts to explain multifaceted of attrition.

Acknowledgements

I am grateful to Professor Kevin Morgan for access to the *Nottingham Longitudinal Study of Activity and Ageing* in the UK. The NLSAA was initiated with a foundation grant from the Grand Charity. Additional support for the study was provided by Help the Aged, PPP Charitable Trust and Trent Regional Health authority. This research also utilized the data set from the *Survey of Health and Living Status of the Elderly in Taiwan*, provided by the Bureau of Health Promotion, Department of Health Taiwan.

References

1. D.J. H. Deeg, Attrition in longitudinal population studies: does it affect the generalizability of the findings? An introduction to the series, J Clin Epidemiol, **55**, 213-215 (2002)
2. F.E. Matthews, et al., Attrition and bias in the MRC cognitive function and ageing study: an epidemiological investigation, BMC Public Health, **4**, 1-10 (2004)
3. A.E. Barry, How attrition impacts the internal and external validity of longitudinal research, J Sch Health, **75**, 267 (2005)
4. D. Feng, M. Silverstein, R. Giarrusso, J.J. McArdle, and V.L. Bengtson, Attrition of older adults in longitudinal surveys: detection and correction of sample selection bias using multigenerational data, J Gerontol B Psychol Sci Soc Sci, **61**, S323-S328 (2006)
5. A.J.r.d. Wilson and I.L. Webber, Attrition in a longitudinal study of an aged population, Exp Aging Res, **2**, 367-387 (1976)
6. D.J.H. Deeg, T.van Tilburg, J.H. Smit, and E.D. de Leeuw, Attrition in the longitudinal aging study Amsterdam: the effect of differential inclusion in side studies, J Clin Epidemiol, **55**, 319-328 (2002)
7. M.D. Chatfield, C.E. Brayne, and F.E. Matthews, A systematic literature review of attrition between waves in longitudinal studies in the elderly shows a consistent pattern of dropout between differing studies, J Clin Epidemiol, **8**, 13-19 (2005)
8. D.J. Slymen, J.A. Drew, J.P. Elder, and S.J. Williams, Determinants of non-compliance and attrition in the elderly, International Journal of Epidemiology, **25**, 411-419 (1996)
9. K. Morgan, *The Nottingham Longitudinal Study of Activity and Ageing (NLSAA) Survey Interview and Data Catalogue* (University of Sheffield, Britain, 1993)
10. Taiwan Provincial Institute of Family Planning, 1989 Survey of health and living status of the elderly in Taiwan: Questionnaire and survey design, *Comparative Study of the Elderly in Four Asian Countries* (Taiwan Provincial Institute of Family Planning, Population Studies Centre and Institute of Gerontology, University of Michigan, Taichung, 1989)
11. K. Morgan, The nottingham longitudinal dtudy of activity and ageing: a methodological overview, Age Ageing, **27**, 5-11 (1998)
12. K.A. Wallace and A.J. Wheeler, Reliability generalization of the Life Satisfaction Index, Educ Psychol Meas, **62**, 674-684 (2002)
13. N. Minicuci, et al., Cross-national determinants of quality of life from six longitudinal studies on aging: The CLESA project, Aging Clin Exp Res, **15**, 187-202 (2003)
14. C. Chen, Aging and life satisfaction, Soc Indic Res, **54**, 57-79 (2001)
15. P.A. Bath, D. Deeg, and J.L. Poppelaars, The harmonisation of longitudinal data: a case study using data from cohort studies in The Netherlands and the United Kingdom, Age Ageing, **30**, 1419-1437 (2010)
16. C.P. Li, Quality of life among older people in the UK and Taiwan. Thesis, University of Sheffield, Sheffield, UK (2009)
17. J.S. Goodman and T.C. Blum, Assessing the non-random sampling effects of subject attrition in longitudinal research, J Manage, **22**, 627-652 (1996)

Predictors of changes in quality of life among older people: comparison between the UK and Taiwan

Chih-Ping Li[a]

Department of Health Industry Management, Kainan University, Taoyuan 33857, Taiwan

Abstract. The purpose of this study was to compare the factors affecting change in quality of life in old age in the UK and Taiwan. Data were derived from the 1989 and 1993 waves of two nationally representative samples: the *Nottingham Longitudinal Study of Activity and Ageing* (NLSAA) in the UK and the *Survey of Health and Living Status of the Elderlyin Taiwan* (SHLSET). Quality of life was assessed using comparable measures of life satisfaction. Analysis consisted of stepwise multiple regressions, Mann-Whitney U, Chi-square, and multivariate logistic regression models to identify the factors related to changes in quality of life.There are common variables in that participants in both studies selected the factors "better perceived health relative to peers" and "improvement in self-rated health" in the stepwise regression models. Additional variables (e.g., age in the UK and satisfaction with income, walking difficulties, and having a TV or radio in Taiwan) were selected individually at different time periods in the two studies.Related to these various factors and change in quality of life, this research has helped to identify the prospect of clarifying the most important factors affecting the change in quality of life among older people in the two countries.

1 Introduction

Quality of life is a broad, multidimensional concept, which may be affected by such objective conditions and subjective experiences as physical functioning, social factors, perceived health, and psychological well-being. Understanding which factors affect changes in quality of life the most may be important in determining ways to improve life satisfaction as a large percentage of the population reaches their senior years.

The aging population and life expectancy are increasing, but a variety of factors may affect overall quality of life during later life [1]. In essence, older people tend to experience general health problems, difficult financial situations, limited social relationships, disruptive changes in their roles in society, loss of basic functional ability, and anxiety as they age and approach the end of life. Consequently, such problems have become major concerns in the field of gerontology and for health authorities in many countries.

Older people frequently perceive certain factors as being of greater importance to their overall quality of life. A variety of factors (morale, loneliness, physical and mental health, social engagement, etc.) may contribute to, and affect, a change in quality of life, but a key question is this: "Which

[a] Corresponding author : luciali@mail.knu.edu.tw

factors are the most important?" This study seeks to provide a better understanding of what factors affect change in quality of life among older people.

To understand the differences in predictors of change in life satisfaction, the author proceeded to (1) identify the factors that predict a change in life satisfaction among older people in the UK and Taiwan and (2) understand how these factors are different between the two countries.

2 Methods

2.1 Participants and ethics

Data were derived from the Nottingham Longitudinal Study of Activity and Ageing (NLSAA) and the Survey of Health and Living Status of the Elderly in Taiwan (SHLSET). Ethics was approved by the Department of Information Studies, in accordance with the University of Sheffield research ethics policies.

The NLSAA study was set up in 1985, with follow-up interviews in 1989 and 1993 [2, 3]. Of 8,409 older people (aged 65 and over) in three areas of Nottingham, UK, 1,299 were randomly selected for the interview, and 1,042 agreed to participate in the 1985 baseline survey (response rate = 80.2%), of whom 690 were still alive and participated in the follow-up survey in 1989 (response rate = 88.3%) [3, 4].

The SHLSET was jointly conducted by the Taiwan Provincial Institute of Family Planning, the Population Studies Center (PSC), and Institute of Gerontology at the University of Michigan [5]. The baseline survey of SHLSET was conducted in 1989. The sample size was 4,412 people (age 60 years and over), and the response rate was 91.8%, with 4,049 participants interviewed.

To obtain comparable samples from the two studies, individuals born before 30 April 1920 and who were interviewed in 1989 were included in this study. The final samples contained 690 individuals from the NLSAA and 1,438 individuals from SHLSET.

2.2 Survey assessments

The data collection instrument for the NLSAA was a structured interview questionnaire, composed of 161 questions that addressed demographics, mobility, physical health, psychological health (cognitive status, depression, and morale), social functioning, use of health and personal social services, customary physical activity, and physical measurements [2].

The SHLSET questionnaire was composed of eight sections such as marital history and other background characteristics; household schedule, social and economic exchanges; health, health care utilization, and behaviors; occupational history; activities and general attitudes; residence history; economics/financial well-being, andemotional and instrumental support[3, 4].

The NLSAA and the SHLSET both used shortened versions of the Life Satisfaction Index, a commonly used measure of well-being and morale in studies of older people [6]. In the NLSAA, the 13-item Life Satisfaction Index-Z (LSIZ) [7] was used. For the SHLSET, the life satisfaction scale was assessed using a 10-item scale in 1989 and a 4-item scale in 1993 [8]. For comparative purposes, a common scoring system for each response category and question across the two studies was adopted and the scores for both scales were standardized to a 100-point scale.

Other variables from the two studies were compared, and variables with similar questions and response categories were harmonized using established methods [9, 10] to create new variables that cover demographics and socioeconomics, physical and mental health, and social activity [11].

2.3 Data analyses

Change in standardized life satisfaction score was the dependent variable used in the third part of the analyses. The change in life satisfaction was produced from the life satisfaction standardized score in

1989 minus the life satisfaction standardized score in 1993, so that a positive value indicated an increase in life satisfaction from 1989 to 1993 and a negative value indicated a decrease in life satisfaction from 1989 to 1993.

Twenty-five independent variables from the data harmonization process were used: gender, marital status, living status, number of people living in the household, socioeconomic class, satisfaction with income, self-rated health, perceived health relative to peers, smoking, arthritis or rheumatism, heart trouble, stomach problems, dizziness, high blood pressure, urinary incontinence, walking difficulties, use of a walking aid, loneliness, standardized SAD score, friends, reading newspapers or journals, attending a religious group, attending some other club or organization, having a TV or radio, and having a pet. These independent variables were assessed to determine whether they were associated with a change in life satisfaction index between 1989 and 1993. For example, a new response category was created as "no change," "became widowed or separated/divorced," "became married," and "two changes" for the marital status variable according to the difference factors between 1989 and 1993. With regard to diseases, for instance, a new response category of "new problem," "no change," and "no longer a problem in 1993" was generated for the arthritis or rheumatism problem variable. However, the age variable used was the participants' age in 1993, the gender and socioeconomic class variables were those used in 1989, and the standardized SAD score used the difference between 1989 and 1993 (the score in 1989 minus the score in 1993).

Change in life satisfaction was analyzed in two ways: first as a continuous variable, for which stepwise multiple regression with criteria as probability-of-F-to-enter <= 0.050 and probability-of-F-to-remove >= 0.100 was used to identify the factors related to life satisfaction as a continuous variable in the two studies separately.

To identify which factors affect change in life satisfaction, the variable "change in standardized LSI score" was dichotomized into "decrease" and "no change/increase" in life satisfaction and, and using both bivariate and multivariate analysis, assessed for the relationship between this new dichotomized dependent variable and each of the harmonized variables from the 1989 and 1993 surveys.

Bivariate analysis was conducted using the Mann-Whitney U test for continuous independent variables. Pearson chi-square analysis was used for categorical independent variables. In each instance, it was assessed whether a relationship between the variables was significant or not. For multivariate analysis, logistic regression analysis was performed to identify the most important factors affecting change in life satisfaction.

Because of the large number of tests carried out, α was set to 0.01 in order to reduce the likelihood of a Type I error. However, any variables that had a p value of 0.05 or less were included in the logistic regression model to ensure no potentially important factors were omitted. Even though these variables were significantly associated with low/high life satisfaction, they may have an important modifier effect on the other variables.

3 Results

The one-sample Kolmogorov-Smirnov test was used to test whether the distribution for the change in the standardized LSI score was normal. The finding shows that change in the standardized LSI score for the NLSAA study, the SHLSET study, and the combined dataset exhibited a distribution that was not normal. In the SHLSET study, one person had a change of 100 points because he/she had a score of 0 points in 1989. On the other hand, there were two people who had a change of -100 points became they went from 100 points in 1989 to 0 points in 1993. This may well have happened because these two people could not answer the four questions in the LSI in 1993. On comparing the mean score for the change in the standardized LSI, it was found that older people in the NLSAA study had a smaller reduction in the standardized LSI score than the older people in the SHLSET study.

The change in the standardized life satisfaction score in the NLSAA study was more peaked than would be expected for a normal distribution.

3.1 Change in life satisfaction as a continuous variable

Table 1 shows the results of stepwise regression analysis to identify predictors of change in life satisfaction in the NLSAA study between 1989 and 1993. Stepwise regression was conducted and the final combination of variables was perceived health relative to peers, self-rated health, and age in 1993. This combination of variables significantly predicted change in life satisfaction, $F_{(3, 306)}=13.647$, $p<0.001$, with all three variables significantly contributing to the model.

Three variables were included in the final model to predict change in standardized life satisfaction score in the NLSAA study, and this model explained 10.9% of the variance in change in life satisfaction. The change in perceived health relative to peers variable (6.7% of the variance) explained the largest share of the variance, followed by change in self-rated health (3.0%) and age (1.2%). In the regression model 3, the relationship between the age (1993) variable and the change in standardized life satisfaction score was negative (-0.122), indicating that older people had a reduction in life satisfaction.

Table 1. Factors affecting change in life satisfaction in the NLSAA (N=381).

	Model	r	Adjusted R^2	Change in R^2	Beta	95% CI	*p* value
1	Change in perceived health relative to peers	0.265	0.067	-	0.265	4.818–11.438	<0.001
2	Change in self-rated health	0.321	0.097	0.030	0.188	2.028–7.844	0.001
3	Age in 1993	0.344	0.109	0.012	-0.122	-0.958–-0.070	0.023

Note: The excluded variables include gender, marital status, socioeconomic class, number of people living in household, satisfaction with income, smoking, arthritis or rheumatism, heart trouble, stomach problems, giddiness problem, high blood pressure, urinary incontinence, walking problem, walking aid, loneliness, standardized SAD score, having a pet, having friends, reading newspapers or journals/magazines, attending a religious group, and participating in a club or organization.

Table 2. Factors affecting change in life satisfaction in SHLSET (N=488).

	Model	r	Adjusted R^2	Change in R^2	Beta	95% CI	*p* value
1	Change in self-rated health	0.258	0.064	–	0.258	7.517–15.458	<0.001
2	Change in satisfaction with income	0.308	0.091	0.027	0.169	4.499–14.372	<0.001
3	Change in walking problems	0.334	0.106	0.015	-0.135	-15.816–-3.116	0.004
4	Change in perceived health relative to peers	0.352	0.116	0.010	0.128	1.387–10.736	0.011
5	Having a TV or radio	0.368	0.126	0.010	0.107	2.149–20.501	0.016

Note: The excluded variables include age, gender, marital status, socioeconomic class, number of people living in household, smoking, arthritis or rheumatism, heart trouble, stomach problems, giddiness problem, high blood pressure, urinary incontinence, walking aid, loneliness, having a pet, having friends, reading newspapers or journals/magazines, attending a religious service, and participating in a club or organization.

Table 2 presents the results of stepwise regression analysis to identify predictors of change in life satisfaction in the SHLSET study between 1989 and 1993. The variables are included in the final model. The adjusted R squared value was 0.126, indicating that 12.6% of the variance in change in life satisfaction was explained by the five variables in the model. Self-rated health (6.4% of the variance) explained more than half of this variance, followed by satisfaction with income (2.7%), walking problems (1.5%), perceived health relative to peers (1.0%), and having a TV or radio (1.0%). In the regression model 3, the relationship between the walking difficulty variable and the change in standardized life satisfaction score was negative (-0.135), implying that a person with walking problems had a reduced life satisfaction score. This combination of variables significantly predicted change in life satisfaction, $F_{(5, 450)}=14.100$, $p<0.001$, with all five variables contributing significantly to the model.

In the SHLSET study, five variables contributed significantly to changes in life satisfaction: higher self-rated health, satisfaction with income, perceived health better than peers, and having a TV or radio were related to an increase in life satisfaction, whereas walking difficulties predicted a negative change in life satisfaction.

3.2 Change in life satisfaction as a categorical variable

In the NLSAA survey, the results of Mann-Whitney U tests assessed the relationship between the continuous independent variables and change in life satisfaction between 1989 and 1993. Mann-Whitney U tests revealed a statistically significant difference in the mean rank for the change in standardized SAD score (Z statistic = -4.663; $p < 0.001$) between those with no change or an increase in life satisfaction versus those with a decrease. Older people who experienced a decrease in life satisfaction were more depressed than the others.

The results of the Mann-Whitney U tests also assessed the relationship between the continuous independent variables and change in life satisfaction between 1989 and 1993 for the SHLSET survey.The results were similar to those of the NLSAA study. Older people who had higher levels of depression (Z=-3.660; $p<0.001$) were more likely to experience a decrease in life satisfaction.

The association was between change in independent variables (except gender and socioeconomic class) and change in life satisfaction between 1989 and 1993 in the NLSAA study. A significant association existed between changes in life satisfaction and changes in self-rated health ($\chi^2=4.199$; $p=0.040$), perceived health relative to peers (χ =7.607; $p=0.006$), and arthritis or rheumatism($\chi^2=4.033$; $p=0.045$). Older people who rated their health as worsening (60.7%), perceived their health relative to peers as becoming worse (66%), and no change in arthritis or rheumatism (54%) were associated with decreased life satisfaction.

The association was between changes in independent variables and change in life satisfaction between 1989 and 1993 in the SHLSET dataset. There was a significant association between changes in life satisfaction and change in self-rated health ($\chi^2=9.855$; $p=0.002$), perceived health relative to peers ($\chi^2=11.746$; $p=0.001$), having a TV or radio ($\chi^2=4.136$; $p=0.042$), reading a newspaper or journal ($\chi^2=6.013$; $p=0.014$), and participating in a club or organization ($\chi^2=4.064$; $p=$ 0.044).Participants in Taiwan whose level of self-rated health had decreased (63.6%), who reported that their health relative to their peers became worse (65.5%), who no longer had a TV or radio (62.5%), who no longer read a newspaper or journal/magazine (70.7%), and who no longer attended a club or organization (76.7%) were more likely to have a decrease in life satisfaction.

Table 3. Selected variables associated with reduced life satisfaction in the NLSAA study (1989–93).

Variable (reference category)	Category	Odds ratio	95% CI	*p* value
Self-rated health				0.052
(No change)	An improvement	0.740	0.427–1.284	0.284
	A worsening	1.465	0.923–2.324	0.105
Perceived health relative to peers				0.074
(No change)	Better health	1.058	0.615–1.820	0.838
	Worse health	1.797	1.080–2.990	0.024
Arthritis or rheumatism(No change)				0.113
	No longer present	0.682	0.365–1.276	0.231
	New problem	0.555	0.299–1.030	0.062
Change in standardized SAD score		1.031	1.013–1.050	0.001

Logistic regression models were used to identify which selected variables were the best predictors of change in life satisfaction from 1989 to 1993 using data from each of the two studies separately.

The selected independent variables were associated with change in life satisfaction between 1989 and 1993 in the NLSAA study. When all the selected variables were considered together, they were significantly associated with changes in life satisfaction (χ^2=33.497; *df*=7; *p*<0.001). Table 3 shows a significant association between negative or lower changes in life satisfaction and level of depression (*p*=0.001).People who had higher levels of depression (OR=1.031) were more likely to experience decreased life satisfaction. The other variables did not have a statistically significant association.

Table 4 presents the selected variables associated with change in life satisfaction between 1989 and 1993 in the SHLSET dataset.The overall model with all the variables entered was significant (χ^2=55.169; *df*=11; *p*<0.001). As shown in Table 4, there was a significant association between changes in life satisfaction and self-rated health (*p*=0.016) and perceived health relative to peers (*p*=0.009); the same association was true with change in levels of depression (*p*=0.018).Compared to people who experienced no change, people whose self-rated health diminished (OR=1.437) were more likely to have decreased life satisfaction; and people whose health relative to their peers worsened (OR=1.631) were more likely to experience decreased life satisfaction. People who had higher levels of depression (OR=1.009) were more likely to experience decreased life satisfaction.

Table 4. Selected variables associated with reduced life satisfaction in SHLSET (1989-93).

Variable (reference category)	Category	Odds ratio	95% CI	*p* value
Self-rated health				0.016
(No change)	An improvement	0.826	0.579–1.177	0.289
	A worsening	1.437	1.012–2.041	0.043
Perceived health relative to peers				0.009
(No change)	Better health	0.872	0.607–1.252	0.457
	Worse health	1.631	1.140–2.333	0.007
Change in standardized SAD score		1.009	1.002–1.017	0.018
Newspaper or journal				0.133
(No change)	No longer read	0.833	0.425–1.633	0.596
	Started reading	1.874	0.983–3.571	0.056
TV or radio				0.069
(No change)	No longer have	1.900	0.799–4.516	0.146
	Acquired	1.730	0.971–3.083	0.063
Club or organization				0.193
(No change)	No longer attending	0.887	0.396–1.987	0.771
	Became able to attend	2.250	0.922–5.490	0.075

4 Discussion

The purpose of this study is to understand which factors are associated with a change in the quality of life among older people, and understanding the changes in the quality of life and the factors that predict these changes may help to develop ways to improve the quality of life of older people. There are common variables in that in the stepwise regression models, participants selected the factors "better perceived health relative to peers" and "improvement in self-rated health" in both studies. Additional variables (e.g., age in the UK and satisfaction with income, walking difficulties, and having a TV or radio in Taiwan) were selected individually at different time periods in the two studies. These variables tended to be selected in the regression model, and account for a small proportion of the variance, and therefore, are related to changes in life satisfaction. The study location variable was identified as a predictor of change in life satisfaction; however, this result suggests that older people in the UK tended to have decreased life satisfaction as they aged when compared with their counterparts in Taiwan.

In the logistic regression models, an increased standardized SAD score was related to decreased life satisfaction in the two studies, and worsened self-rated health and worsened perceived health relative to peers were associated with decreasing life satisfaction among older people in the SHLSET. These results suggested that mental health was related to a change in life satisfaction among older people in both countries; however, in the case of people in Taiwan, change in perceived health relative to their peers had an effect on life satisfaction.

This research examined change in life satisfaction among older people from 1989 to 1993 and compared the two studies from the UK and Taiwan.Perceived better health relative to peers and improvement in self-rated health were associated with increased life satisfaction in both stepwise regression models. Additional variables were selected individually at different points in the two studies. These variables were selected in the regression models and account for a small proportion of the variances, and therefore, have a relationship with change in life satisfaction. In the logistic regression models, an increased standardized SAD score was related to decreased levels of life satisfaction in the two studies, and worsened self-rated health and worsened perceived health relative to peers were associated with decreased life satisfaction among older people in SHLSET.

The result agreed with previous studies in which it was determined that age, self-rated health, and social activities are associated with changes in life satisfaction [12, 13], but very little is known about the relationship between change in depression level and change in life satisfaction. People who grew older that rated health worse, perceived health worse relative to peers, and had new walking difficulties were more likely to show a decreased life satisfaction level. As discussed earlier, self-rated health and perceived health relative to peers were difficult to assess and might confound other unmeasured variables. Although these variables had only minor variances identified as affecting a change in life satisfaction, small life satisfaction changes may be statistically significant, especially with large studies.

A limitation of this research is that it did not explore how many changes in the standardized LSI scores could be interpreted in terms of small, moderate, or large changes in life satisfaction as reported by older people. Further research is needed to understand what factors affect the different levels of life satisfaction.

Another limitation, as with most research of this kind, is that it is likely that certain participant characteristics, related to both dependent and independent variables (so-called “confounders”), were not examined in the research. It must be kept in mind that the presence of such confounding variables – known or unknown – could explain at least part of the relationships detected in this research. For example, it was not possible to assess certain disease states. It is possible that some disease states that were not examined (e.g., alcohol abuse) could be related to quality of life and a variety of other factors; and not examining these conditions may limit the extent to which these findings can be interpreted.

Future studies could resolve some of the more minor shortcomings of the combined data sets used in the present analyses. In the future, measures incorporating the length, breadth, and profundity of quality of life in different countries would significantly improve the understanding of ageing populations.

5 Conclusions

Lifesatisfaction is a dynamic process that changes according to an individual’s circumstances [14]. Our results support this contention among older people, who appeared to evaluate their overall life satisfaction based upon such circumstances as their financial status or whether they had a TV or radio. Financial strain and low income have been found to be key factors causing low life satisfaction among older people [15, 16]. However, a decreased income might have had a stronger effect in decreasing the levels of life satisfaction in old age.

The present research indicates what factors are associated with life satisfaction among older people. There are numerous determinants of life satisfaction in older people, and most of these hold true in both the United Kingdom and in Taiwan. Aspects of physical health are important, as are one's mobility and one's mental health. Satisfaction with one's personal financial status also appears to play a significant role. In addition, these relationships largely seem to hold true over time. However, the numerous interactions between these various contributors to health are complicated. In other words, what is the horse, what is the cart, and what is the chain linking the two is not entirely clear; further research is warranted. This study has helped tease out general trends and the most important variables.

Aspects of physical and mental health were associated with changes in life satisfaction among people in the two countries. In particular, depression affected change in life satisfaction among people as they aged. Although there are few previous studies regarding change in life satisfaction, there was some evidence in this research that changes in perceived health and mobility might be associated with increased depression because depression might be a result of disease and lessen activities of daily living, thus leading to low life satisfaction.

Related to these various factors and change in quality of life is the prospect of clarifying the most important factors affecting the change in quality of life in old age. Older people in different countries may have specific problems because of local differences in conditions and life experience. A contribution of this research is improving the understanding of the factors that adversely affect people's quality of life in old age. However, such findings underscore the importance of recognizing the need for older people to maintain their quality of life. Such factors as participating in activities with others, maintaining health, developing social relationships from community-based schemes, requiring resources from social support systems are some ways that older people can maintain their quality of life and mitigate some of the factors that indicated a lower or worsening life-satisfaction as they aged.

The next major step with regards to these results is to determine why certain associations happen in these people in the UK and/or Taiwan. For instance, loneliness appears to be more common among older people in the UK. Studies will be required to test alternative explanations systematically for the influence of psychosocial variables on physiological processes and quality of life. Qualitative fieldwork to investigate how older people deal with loneliness could prove useful. In this research, although depression was a major influence on quality of life, the levels of depression did not affect the relationship between quality of life and mortality in the two studies. Even though some investigators have explored the relationship between depression and mortality, few studies have looked at the effect of depression on quality of life and its relationship to mortality: the association between these factors could be investigated in future studies.

Acknowledgements

I am grateful to Professor Kevin Morgan for access to the *Nottingham Longitudinal Study of Activity and Ageing* in the UK. The NLSAA was initiated with a foundation grant from the Grand Charity. Additional support for the study was provided by Help the Aged, PPP Charitable Trust and Trent Regional Health authority. This research also utilized the data set from the *Survey of Health and Living Status of the Elderly in Taiwan*, provided by the Bureau of Health Promotion, Department of Health Taiwan.

References

1. Y. Hellström, G. Persson, and I.R. Hallberg, Quality of life and symptoms among older people living at home, Journal of Advanced Nursing, **48**, 584-593 (2004)
2. K. Morgan, *The Nottingham Longitudinal Study of Activity and Ageing (NLSAA) Survey Interview and Data Catalogue* (University of Sheffield, Britain, 1993)
3. K. Morgan, The nottingham longitudinal study of activity and ageing: a methodological overview, Age and Ageing, **27**, 5-11 (1998)

4. K. Morgan and P.A. Bath, Customary physical activity and psychological wellbeing: a longitudinal study, Age and Ageing, **27**, 35-40 (1998)
5. Taiwan Provincial Institute of Family Planning, "1989 Survey of health and living status of the elderly in Taiwan: Questionnaire and survey design," *Comparative Study of the Elderly in Four Asian Countries* (Taiwan Provincial Institute of Family Planning, Population Studies Centre and Institute of Gerontology, University of Michigan, Taichung, 1989)
6. K.A. Wallace and A.J. Wheeler, Reliability generalization of the Life Satisfaction Index, Educational and Psychological Measurement, **62**, 674-684 (2002)
7. K. Morgan, H.M. Dallosso, and T.H.D. Arie, Mental health and psychological well-being among the old and very old living at home, British Journal of Psychiatry, **150**, 801-807 (1987)
8. C. Chen, Aging and life satisfaction, Social Indicators Research, **54**, 57-79 (2001)
9. P.A. Bath, D. Deeg, and J.L. Poppelaars, The harmonisation of longitudinal data: a case study using data from cohort studies in The Netherlands and the United Kingdom, Ageing and Society, **30**, 1419-1437 (2010)
10. N. Minicuci, et al., Cross-national determinants of quality of life from six longitudinal studies on aging: The CLESA project, Aging Clinical and Experimental Research, **15**, 187-202 (2003)
11. C.P. Li, Thesis, University of Sheffield, Sheffield, UK (2009)
12. A. Bowling, M. Farquhar, and E. Grundy, Associations with changes in life satisfaction among three samples of elderly people living at home, International Journal of Geriatric Psychiatry, **11**, 1077-1087 (1996)
13. D.K. Mroczek and A. Spiro, III, Change in life satisfaction during adulthood: findings from the Veterans Affairs Normative Aging Study, Journal of Personality and Social Psychology, **88**, 189-202 (2005)
14. R. Fernández-Ballesteros, Quality of life: the differential conditions, Psychology in Spain, **2**, 57-65 (1998)
15. E. Aurelia and B. Baldazzi, Unequal perceived quality of life among elderly Italians: different satisfaction levels in selected spheres life, Social Indicators Research, **60**, 309-334 (2002)
16. K.L. Chou and I. Chi, Financial strain and life satisfaction in Hong Kong elderly Chinese: moderating effect of life management strategies including selection, optimization, and compensation, Aging and Mental Health, **6**, 172-177 (2002)

Suppression of the dynamic response of cracked beams with active control

Meng-Ju Lin[1,a], Yu-Cheng Chen[2] and Lin-Tsang Lee[2]

[1]*Department of Applied Mathematics, Tunghai University.No.1727, Sec.4, Taiwan Boulevard, Taichung, Taiwan*
[2]*Department of Applied Mathematics, National Chung-Hsing University, Taichung, Taiwan*

Abstract. Recently, studies on the dynamic responses of cracked beams are quite worthy of being conferred. The theoretical dynamic properties of cracked beams previously obtained by La-DQM method were utilized to evaluate the maximum amplitudes of cracked beams. A comparatively new concept which utilized the controller gain to suppress the transverse vibration response of cracked beams in the perspective of proportional control scheme was proposed. A simulation study which applied an active control strategy to suppress the amplitude of vibration on the cracked beams with MATLAB SIMULINK software has been successfully accomplished. The maximum dynamic response of cracked beams is found to be inversely proportional to controller gain. Increasing the controller gain properly to control the vibration of the cracked beams is suggested in this study.

1 Introduction

Mechanic vibration has been widely applied in various fields of the industry such as: vibration analysis of vehicles, engine vibration and noise analysis, safety control of bridge vibration, stability of the micro flexible mechanical system, etc. It has become an important subject to study the damage to the beam in order to understand the integrity of beam structure that will result in the risk of structural collapse. In recent years there have been a lot of scholars observing dynamic behaviors of beam based on vibration mode, proposing theories for solving the vibration problem of cracked beam by predicting the location and size of cracks based on the natural frequency. In summary, the study on the vibration behavior of cracked beam is a subject worthy of investigation. A mathematical model including interaction between solids and fluids was proposed by Dimarogonas et al. [1]. Besides, a variety of methods were proposed to deal with the crack problems of structures. The general stiffness matrix of Euler-Bernoulli is applied on the local flexibility of beam. The linear spring models have been successfully applied in transverse cracks beams with conditions to simulate cracks by Mohammad [2]. Rizos et al. [3] employed the measured vibration modes to identify the crack locations, depths, and vibration frequencies in order to obtain the dynamic response of cracked beams. Narkis [4] employed the variation of the first two natural frequencies to obtain the locations of the cracks in the simply supported beam. The cracks are simulated as rotational springs and the beam is divided into two sections, while the proposed theoretical method is verified by finite element method. Lin et al. [5] proposed a theoretic model to describe the non-linear, three-dimensional response of a

[a] Corresponding author : vbenz2006@thu.edu.tw

suspended cable supporting an array of discrete masses. It is applied to obtain the eigenvalues of that complicated system, and the transfer matrix of this method can be applied to the free vibration of complex mass system. Lin et al [6] employed the transfer matrix to obtain the natural frequencies of any number of cracks. This is an easy and effective method to obtain the solution of vibration mode, which will result in the numeric results very similar to the experimental results. Later on, Wang et al. [7] proposed simplified sectional flexibility factor to simulate the study on single open crack, where the stress intensity factor is simplified, and the transfer matrix is used to establish the eigenvalue equation. It provided an easy and effective way for analyzing cracked beam. Bellman et al. [8] was the first to propose DQM in 1971 to approximate the integral item to be solved by the "quadrature" approach while approximating the first order differentiation by extending the sum of linear function. Lee et al. [9] applied local adaptive differential quadrature method (La-DQM) to the study of identification of beams with multiple cracks which led to fairly good results.

As an important public facility, bridges provide people the traffic convenience at the cost of the problems of vibration and noise. Experts have developed a variety of methods and theories to avoid the negative effects of vibrations. Among them, the passive control methods and the active control methods are the commonest [10]. The passive control is mainly achieved by changing the property parameters, mass, damping and strength of the system itself. And the active control utilizes the vibration signals of the sensor and counteracts the vibration by the control force generated according to the results of the calculation of the controller [11]. Several scholars utilized discretization of finite difference to conduct the active control of cantilever beam and simulate the dynamic state of structure of the beams through the discretization process. After this, they employed active control to do the control [12-14].

An active control model was proposed by Tavakolpour et al. [15] to suppress the dynamic response of cracked beams. The methodology for the active vibration control of a flexible plate structure employed by Tavakolpour et al. is known as high gain feedback (HGF) regulator. Proportional control scheme is then applied to analyze the final response on the maximum amplitude of cracked beam which can be used to minimize the negative effect of vibration. The amplitude of cracked beam is reduced by increasing the controller gain. This study is to provide a suitable controller gain for proportional control scheme to suppress the dynamic response of cracked beams. The control performance is evaluated by varying the proportional gain until it reaches the optimum value where attenuation percentage of the maximum amplitudes of the cracked beam is obtained.

2 Transverse vibration response of a cracked beam

2.1 Calculation of eigenvalues

Considering an Euler Bernoulli beam with length L with a crack located at X_1 thus separating this beam into L_1, L_2 as shown in Figure 1. In accordance with literature [2], the governing equation for every section of separated beam can be shown as:

$$EI\frac{\partial^4 Y_i(X,T)}{\partial X^4}+\rho A\frac{\partial^2 Y_i(X,T)}{\partial T^2}=0,\ X_{i-1}<X<X_i, i=1,2 \tag{1}$$

where E is the modulus of elasticity , I is the moment of inertia of the Euler Bernoulli beam cross-section, ρ is the density of material and A is the cross-section area.

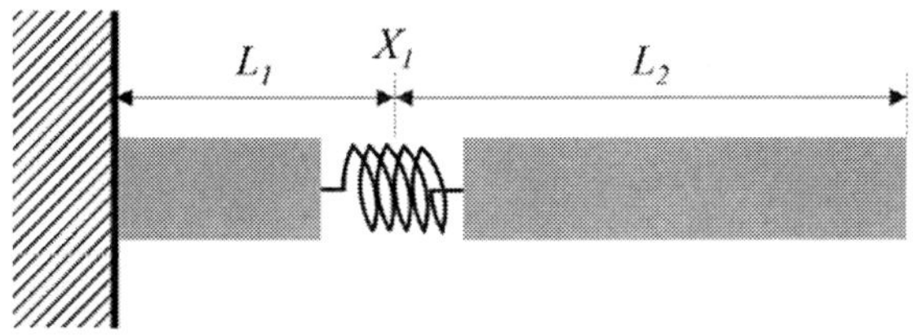

Figure 1. Schematic diagram of a cantilever beam with one crack.

The following quantities are introduced for non-dimensional analysis:

$$y=\frac{Y}{L},\ x=\frac{X}{L},\ t=T,\ x_i=\frac{X_i}{L},\ y_i=\frac{Y_i}{L},\ l_i=\frac{L_i}{L}. \tag{2}$$

The external force $f(x,t)$ is applied to the right free end of the cantilever beam, the governing equation for every section of separated beam can be denoted in the non-dimensional form as shown below:

$$\frac{EI}{L^4}\frac{\partial^4 y_i(x,t)}{\partial x^4}+\rho A\frac{\partial^2 y_i(x,t)}{\partial t^2}=f(x,t),\quad x_{i-1}<x<x_i,\quad i=1,2. \tag{3}$$

By applying separation of variables $y_i(x,t)=w_i(x)e^{j\omega t}$, and letting the external force $f(x,t)=0$ then the governing equation of (3) will lead to

$$\frac{EI}{L^4}w_i^{(IV)}(x)-\rho A\omega^2 w_i(x)=0, \tag{4}$$

and the differential equation for free vibration can be written as

$$w_i^{(IV)}(x)-\lambda^4 w_i(x)=0,\quad x_{i-1}<x<x_i,\ i=1,2 \tag{5}$$

where $\lambda^4=\frac{\rho A\omega^2 L^4}{EI}$ is eigenvalue, and ω is the natural frequency, and the compatibility conditions of the locations of cracks are as shown below:

$$\begin{aligned} &w_1(x_1^-)=w_2(x_1^+),\\ &w_1''(x_1^-)=w_2''(x_1^+),\\ &w_1'''(x_1^-)=w_2'''(x_1^+),\\ &w_2'(x_1^+)-w_1'(x_1^-)=c^* w_2''(x_1^+). \end{aligned} \tag{6}$$

Where c^* is the non-dimensional crack section flexibility coefficient as

$$c^*=\frac{EIc}{L}=\frac{2\pi(1-\upsilon^2)[1-(1-\frac{a}{h})^2]}{9(1-\frac{a}{h})^2}\frac{h}{L} \tag{7}$$

,

and $c = \dfrac{2\pi(1-\upsilon^2)\left[1-(1-\frac{a}{h})^2\right]h}{9EI(1-\frac{a}{h})^2}$ is the section flexibility coefficient , a refers to the depth of the crack.

Every section of beam can be connected by employing transfer matrix method [5] for calculation of eigenvalue of the entire system. The general solution of equation (5), for each segment is as shown below:

$$w_i(x) = A_i \sin[\lambda(x-x_{i-1})] + B_i \cos[\lambda(x-x_{i-1})] + C_i \sinh[\lambda(x-x_{i-1})] + D_i \cosh[\lambda(x-x_{i-1})]$$
$$x_{i-1} < x < x_i, \ i=1,2. \tag{8}$$

Where A_i, B_i, C_i and D_i are constants. These constants in the (i+1)th segment ($A_{i+1}, B_{i+1}, C_{i+1}$ and D_{i+1}) are related to those in the ith segment (A_i, B_i, C_i and D_i) through compatibility conditions in equation (6). The eigenvalue can be obtained in accordance with the transfer matrix of literature [6].

2.2 Generalized coordinates

Based on the reason, $g_r(t) \approx 0$ for $r \geq 6$, so the transverse vibration response of a cracked beam can be shown as:

$$y_i(x,t) = \sum_{r=1}^{5} w_{ri}(x) g_r(t), \ i=1,2. \tag{9}$$

Substituting equation (9) into equation (3), the generalized coordinates function $g_r(t)$ of cracked beam can be expressed via orthogonality and normalization as:

$$\omega_r^2 g_r(t) + \frac{d^2 g_r(t)}{dt^2} = n_r(t), \tag{10}$$

where $n_r(t) = \int_0^{l_1} w_{r1}(x) f(x,t) dx + \int_{l_1}^{1} w_{r2}(x) f(x,t) dx$.

The initial conditions of Euler Bernoulli beam are:

$$g_r(0) = 0, \frac{dg_r(0)}{dt} = 0 \tag{11}$$

Based on the model with the initial conditions (11), and performing Laplace transform to equations (10) and (11), the equation (10) becomes:

$$\omega_r^2 g_r(s) + s^2 g_r(s) = N_r(s). \tag{12}$$

The transfer function of the flexible beam system can be obtained from equation (12) as:

$$G_r(s) = \frac{g_r(s)}{N_r(s)} = \frac{1}{\omega_r^2 + s^2}. \tag{13}$$

3 Proportional control scheme

Experts and scholars have developed numerous methods and theories to avoid the negative impacts of vibration. The proportional control scheme is considered in this study as the scheme for controlling the vibration of flexible beam system. When an external force is applied to the beam, a counter force will be generated by the actuator to compensate the displacement caused by the external force. This dynamical system attempts to reduce the amplitude of cracked beam. The closed-loop active vibration control system is as shown in Figure. 2. In the proportional controller, the control action is proportional to the error, and we can represent the controller as a gain, *K*. The transfer function of the flexible beam system will be plant, *G*. The sensor, G_s, measures the output of the system we are controlling. The error expression tells us how much the output deviates from the input. The device which performs the subtraction to compute the error, *e*, is a comparator. The controller will generally try to reduce the error of displacement between the reference value and its actual value until we get the minimum error. The effectiveness of the controller action depends on the optimum setting of its proportional gain.

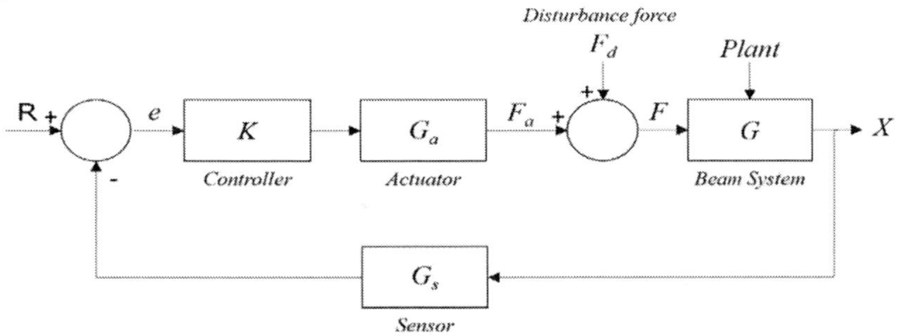

Figure 2 . Proportional Control Scheme.

For the closed-loop system in Fig. 2, the reference input R is set to zero. The input to the actuator G_a is $-G_sKX$, the output is $F_a = -G_aG_sKX$, and F_a , F_d of the flexible beam system are actuating force and disturbing force respectively. Thus, the total external force fed into the flexible beam system is given by:

$$F = F_a + F_d = -G_aG_sKX + F_d \tag{14}$$

Thus the displacement *X* of the flexible beam can be shown as:

$$X = GF_d\left\{\frac{1}{1+G_aG_sKG}\right\} \tag{15}$$

The beam displacement can be reduced by adjusting the gain of controller *K* in accordance with equation (15), and the less displacement indicates less beam vibration. Equation (15) indicates that the *X* value is inversely proportional to controller gain, *K*.

The entire control system can be simplified based on the unity of actuator and sensor. The equation after simplification can be shown as:

$$X = GF_d\left\{\frac{1}{1+KG}\right\} \tag{16}$$

Based on equation (16), the displacement of the flexible beam is reduced by increasing the controller gain, *K*.

4 Numerical analysis and discussions(simulation results)

The proposed methods are used to solve the beam vibration problems with un-cracked and cracked beams. In order to validate the method in this article, the numerical results are compared with the available data. First is the case of the un-cracked cantilever beam shown in Figure 3 with length $L=0.3m$, height $h=0.02m$, width $b=0.02m$, area $A=0.02\times0.02m^2$, Young's modulus $E=2.06\times10^{11}N/m^2$, and mass density $\rho=7800kg/m^3$. The external force $f=10\sin(\omega_f t)$, $\omega_f=2\pi\neq\omega_r$ and $r=1,2,3,\cdots\cdots$, is applied to the right free end of the beam. The natural frequencies of an un-cracked cantilever beam obtained via Theoretical method and La-DQM are as shown in Table 1. With proportion control scheme in this article, the percentage of the maximum vibration amplitudes of un-cracked beam can be reduced by adjusting higher controller gain, *K* value. The percentage reductions of the maximum vibration amplitudes before and after the control over the beam with different proportional gains are obtained as shown in Table 2.

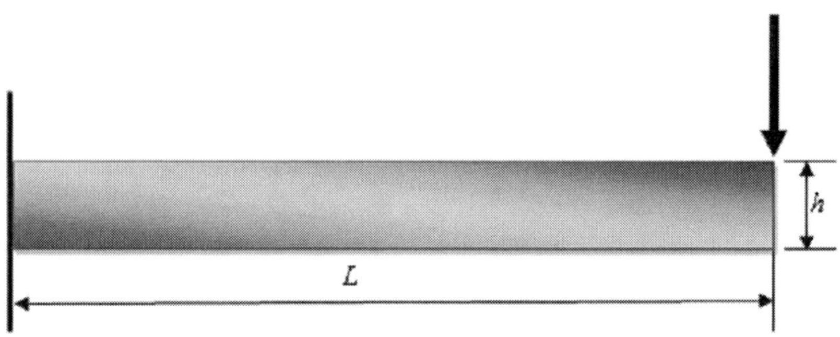

Figure 3. Illustration of an un-cracked cantilever beam.

Table 1. The natural frequencies of an un-cracked cantilever beam

Method	First frequency(Hz)	Second frequency(Hz)	Third frequency(Hz)
Theoretical method by this article	184.45	1156.12	3237.19
La-DQM	184.48	1156.12	3237.08

Table 2. The percentage reductions of maximum vibrations before and after the control over the un-cracked cantilever beam.

K value	Maximum vibration before the control	Maximum vibration after the control	Percentage reduction (%)	rate of reduction per unit gain (%) (Percentage reduction/ *K* value)
10,000	3.9844E-06	3.9587E-06	0.645016	6.45016E-05 (0.645016/10000)
20,000	3.9844E-06	3.9225E-06	1.553559	7.76779E-05
50,000	3.9844E-06	3.8175E-06	4.188836	8.37767E-05
100,000	3.9844E-06	3.6530E-06	8.317438	8.31744E-05
500,000	3.9844E-06	2.8339E-06	28.875113	5.77502E-05
5,000,000	3.9844E-06	7.5218E-07	81.121875	1.62244E-05

Next, the cantilever beam with a single crack is chosen as shown in Figure 4 with length $L=0.3m$, height $h=0.02m$, width $b=0.02m$, area $A=0.02\times0.02m^2$, Young's modulus $E=2.06\times10^{11}N/m^2$, and mass density $\rho=7800kg/m^3$ .The crack is located at X_1=0.14m from the fixed end with the depth $a_1=0.01m$. The crack depth ratio is a/h=0.01/0.02 =0.5 and the

DOI: 10.1051/matecconf/201711901061

crack location X_1/L=0.14/0.3 = 0.4667. The external force $f = 10\sin(\omega_f t)$, $\omega_f = 2\pi \neq \omega_r$ and $r = 1,2,3,\cdots\cdots$, is applied to the right free end of the cantilever beam. The natural frequencies of the cracked beam as shown in Table 3 are obtained via Numerical results, Experiment results, Theoretical results and results of La-DQM. With the proportion control scheme in this article, the percentage reductions of maximum vibration amplitudes before and after the controller with different proportional gains are obtained. The percentage reductions of maximum vibrations before and after the control over the beam based on different *K* values are as shown in Table 4.

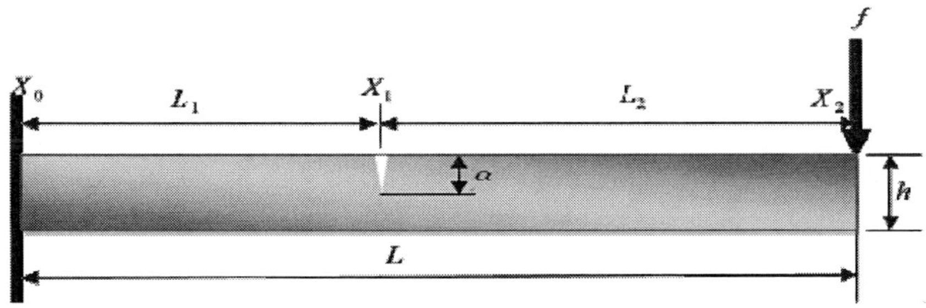

Figure 4. Illustration of a cracked cantilever beam.

Table 3. The natural frequencies of a cracked cantilever beam.

Method	First frequency(Hz)	Second frequency(Hz)	Third frequency(Hz)
Numerical results from[6]	173.88	988.45	3211
Experiment results from[3]	171	987	3034
Theoretical method [5] by this article	178.06	1043.74	3219.98
La-DQM[8]	178.06	1043.74	3219.99

Table 4. The percentage maximum vibrations before and after the control over the cracked cantilever beam.

K value	Maximum vibration before the control	Maximum vibration after the control	Percentage reduction (%)	rate of reduction per unit gain (%) (Percentage reduction/ *K* value)
10000	3.9844E-06	3.9587E-06	0.645016	6.45016E-05 (0.645016/10000)
20000	3.9844E-06	3.9225E-06	1.553559	7.76779E-05
50000	3.9844E-06	3.8175E-06	4.188836	8.37767E-05
100000	3.9844E-06	3.6530E-06	8.317438	8.31744E-05
500000	3.9844E-06	2.8339E-06	28.875113	5.77502E-05
5000000	3.9844E-06	7.5218E-07	81.121875	1.62244E-05

This study is to obtain transverse vibration responses via natural frequencies of un-cracked and cracked beams obtained in accordance with aforementioned conditions of beams, and then the proportion control scheme will be applied to carry out beam control based on different *K* values of controller with *K*=10000 、20000 、50000, 100000, 500000 and 5000000. The percentage reductions of the maximum amplitude before and after the control are obtained for the comparison between the transverse vibration response relationship diagrams of cracked and un-cracked beams before and after the control. The results reveal that proportional control scheme has effectively reduced the vibration amplitudes of un-cracked beam and cracked beam. The higher the proportional gain results into the better attenuation percentage of the maximum vibration amplitudes of un-cracked beam and cracked beam. The comparisons between the transverse vibration response diagrams of un-cracked and cracked beams before and after the control with different proportional gains are as shown in Figure 5

DOI: 10.1051/matecconf/201711901061

to Figure 10. The black solid line and black dotted line in the figure represent the transverse vibration responses of the uncontrolled and controlled un-cracked beams respectively. The red solid line and red dotted line in the figure represent the transverse vibration responses of the uncontrolled and controlled cracked beams respectively.

Proportional control scheme has been proven its capability of reducing the maximum vibration amplitude of un-cracked beam and cracked beam. The performance of the controller in suppressing the maximum vibration amplitudes of beams depends on the proportional gain. That is to say, increasing the proportional gain will effectively reduce the maximum amplitude of the system. It is noted that, although the higher the proportional gain gets the better attenuation percentage of the maximum vibration amplitudes of the cracked beam but it will lead to higher costs and higher actuator force. Based on these reasons, the highest suitable value of the proportional gain obtained is 100000, as shown in Table 2 and Table 4, which will effectively restrain the maximum vibration amplitude of un-cracked beam and cracked beam with the percentages at 6.498656% and 8.317438% respectively. It has the more effective decline the vibration amplitudes and the better rate of decline per unit proportional gain.

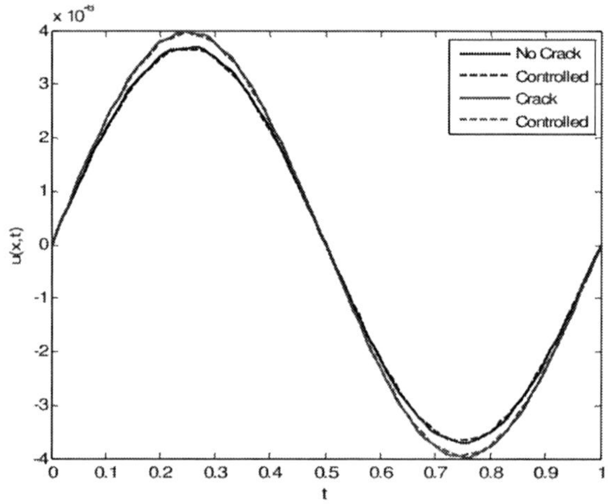

Figure 5. The transverse vibration response diagram of un-cracked and cracked beams before and after control with K=10000.

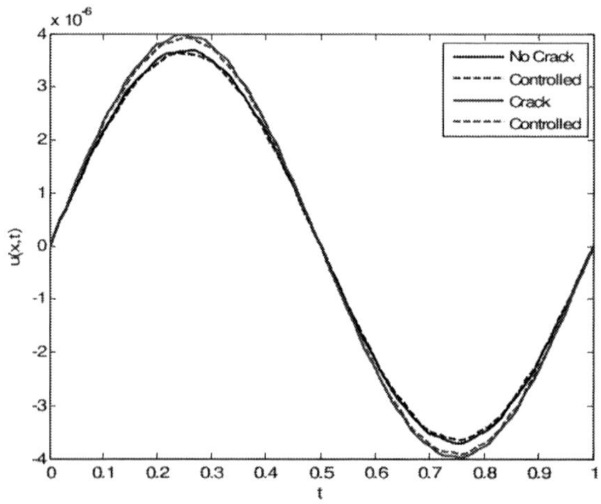

Figure 6. The transverse vibration response diagram of un-cracked and cracked beams before and after control with K=20000.

DOI: 10.1051/matecconf/201711901061

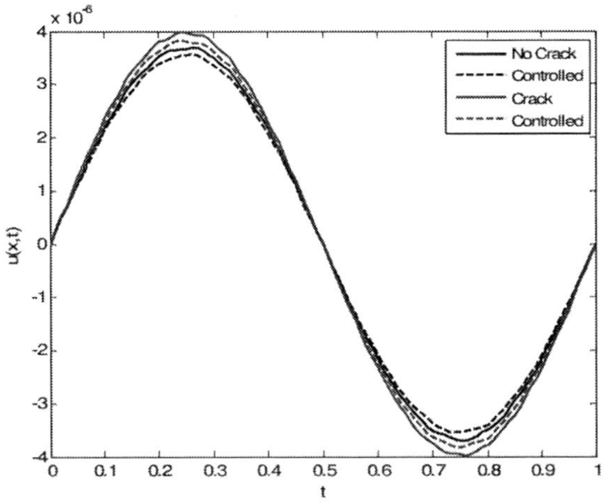

Figure 7. The transverse vibration response diagram of un-cracked and cracked beams before and after control with K=50000.

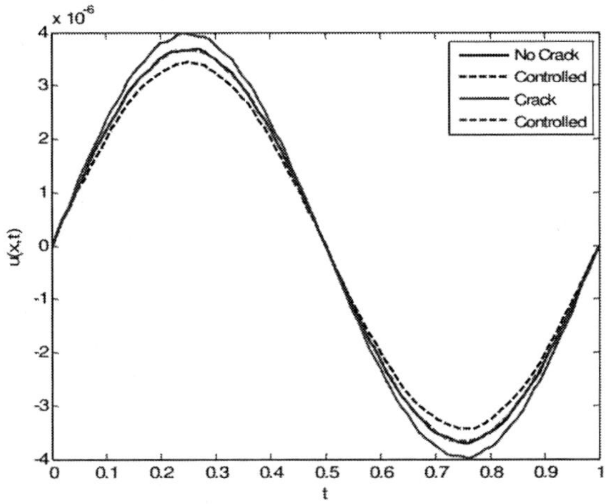

Figure 8. The transverse vibration response diagram of un-cracked and cracked beams before and after control with K=100000.

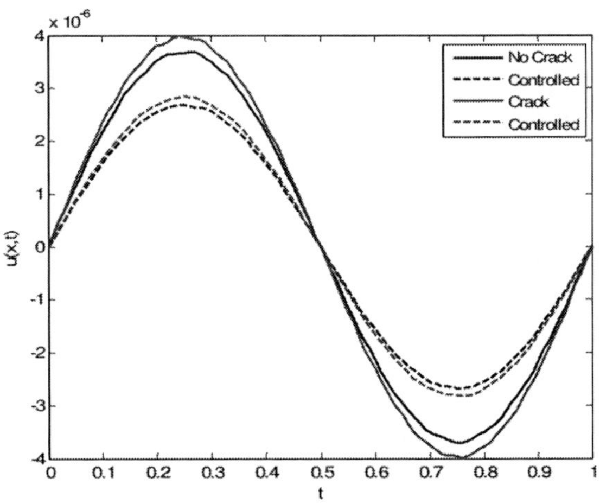

Figure 9. The transverse vibration response diagram of un-cracked and cracked beams before and after control with K=500000.

DOI: 10.1051/matecconf/201711901061

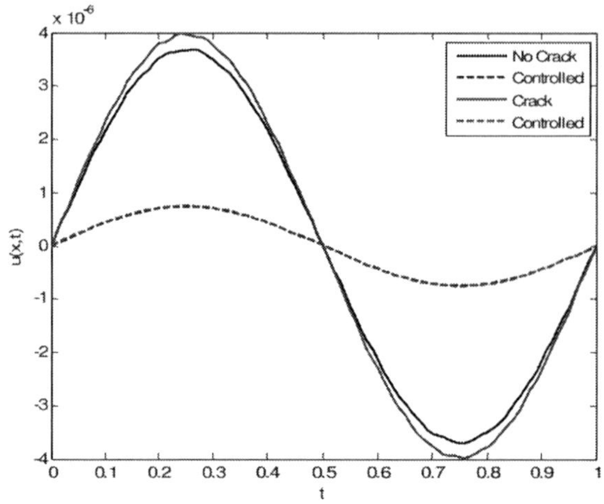

Figure 10. The transverse vibration response diagram of un-cracked and cracked beams before and after control with K=5000000.

5 Conclusions

A simulation study has been successfully accomplished by applying an active control strategy to suppress the amplitude of vibration on the cracked beam with MATLAB SIMULINK software. This simulation model of cracked beams which was represented through the operation of massless rotational spring and the two beams are connected by the compatibility equations. It was demonstrated that the proposed active controller effectively controlled transverse vibration response of the flexible cracked beam system. From the results, it reveals that proportional control scheme has significantly reduced the maximum vibration amplitude of cracked beams. However, the controller gain needs to be selected appropriately because if the proportional gain is increased beyond some values it may result in higher costs and higher actuator force. From Table 4, it is seen that the highest suitable value of the proportional gain obtained is 100000. It can effectively suppress the maximum vibration amplitude percentage of cracked beam with the percentage of 8.317438% while maintaining better rate of reduction for per unit proportional gain value. Simulation results demonstrate that the proposed scheme is applicable effectively to vibration of cracked beam. According to the result of this research, active control system can be applied successfully to the suppression of dynamic response of cracked beam. This study offers a promising approach to realize control system in real time application.

References

1. A. D. Dimarogonas and S. A. Paipetis, *Analytical methods in rotor dynamics* (Applied Science, N. Y., USA, 1983)
2. M. H. Dado, Appl. Acoust., **51**, 381 (1997)
3. P. F. Rizos, N. Aspragathos, and A. D. Dimarogonas, J. Sound. Vib., **138**, 381 (1990)
4. Y. Narkis., J. Sound Vibr., **172**, 549 (1997)
5. H. P. Lin and N. C. Perkins, J. Sound. Vib., **179**, 131 (1995)
6. H. P. Lin, S. C. Chang, and J. D. Wu, J. Sound. Vib., **258**, 987 (2002)
7. C. S. Wang and L. T. Lee, J. Appl. Math., **2012**, 16 (2012)
8. R. B. J. Math., Anal. Appl., **34**, 235 (1971)
9. T. L. Lin and J. Y. Wu, Proc. Inst. Mech. Eng. E J. Process Mech. Eng.,1 (2014)
10. A. Preumont, *Vibration control of active structures: an introduction* (Springer, 2011)
11. B. J. Grobbelaar, G. D. Wood, and C. P. Constancon, J. Julie., 18 (1992)
12. M. S. Saad, H. Jamaluddin, and I. Z. Mat Darus, *Active vibration control of flexible beam system using proportional control scheme in finite difference simulation platform* (IEEE, 2011)

13. M. F. Hassan, M. Mailah, R. Junid, and N. A. Alang, WCE (2010)
14. A. Tavakolpour, M. Mailah, and I. Z. Mat Darus, WSTOCR, **6**, 184 (2011)
15. A. R. Tavakolpour, M. Mailah, and I. Z. Mat Darus, Int. Rev. Mech. Eng., **3**, 579 (2009)

AUTHOR INDEX

AUTHOR INDEX

EDP Sciences – Web of Conferences
17, Avenue du Hoggar
Parc d'Activité de Courtabouf
BP 112
F-91944 Les Ulis Cedex A
France

ISBN 978-1-5108-4639-5

9 781510 846395

CURRAN ASSOCIATES INC.
proceedings.com